T0299262

Pricing in General Insurance

Based on the syllabus of the actuarial profession courses on general insurance pricing – with additional material inspired by the author's own experience as a practitioner and lecturer – ***Pricing in General Insurance, Second Edition*** presents pricing as a formalised process that starts with collecting information about a particular policyholder or risk and ends with a commercially informed rate.

The first edition of the book proved very popular among students and practitioners with its pragmatic approach, informal style, and wide-ranging selection of topics, including:

- Background and context for pricing
- Process of experience rating, ranging from traditional approaches (burning cost analysis) to more modern approaches (stochastic modelling)
- Exposure rating for both property and casualty products
- Specialised techniques for personal lines (e.g., GLMs), reinsurance, and specific products such as credit risk and weather derivatives
- General-purpose techniques such as credibility, multi-line pricing, and insurance optimisation

The second edition is a substantial update on the first edition, including:

- New chapter on pricing models: their structure, development, calibration, and maintenance
- New chapter on rate change calculations and the pricing cycle
- Substantially enhanced treatment of exposure rating, increased limit factors, burning cost analysis
- Expanded treatment of triangle-free techniques for claim count development
- Improved treatment of premium building and capital allocation
- Expanded treatment of machine learning
- Enriched treatment of rating factor selection, and the inclusion of generalised additive models

The book delivers a practical introduction to all aspects of general insurance pricing and is aimed at students of general insurance and actuarial science as well as practitioners in the field. It is complemented by online material, such as spreadsheets which implement the techniques described in the book, solutions to problems, a glossary, and other appendices – increasing the practical value of the book.

Pricing in General Insurance

Second Edition

Pietro Parodi

CRC Press is an imprint of the
Taylor & Francis Group, an informa business

Designed cover image: Lloyd's of London interior, courtesy of Christina Lees

Second edition published 2023
by CRC Press
6000 Broken Sound Parkway NW, Suite 300, Boca Raton, FL 33487–2742

and by CRC Press
4 Park Square, Milton Park, Abingdon, Oxon, OX14 4RN

CRC Press is an imprint of Taylor & Francis Group, LLC

First edition published by CRC Press in 2014

Library of Congress Cataloging-in-Publication Data
Names: Parodi, Pietro, author.
Title: Pricing in general insurance / Pietro Parodi.
Description: Second edition. | Boca Raton : CRC Press, [2023] |
Revised edition of the author's Pricing in general insurance, [2015] |
Includes bibliographical references and index.
Identifiers: LCCN 2022047427 (print) | LCCN 2022047428 (ebook) |
ISBN 9780367769031 (hardback) | ISBN 9780367769048 (paperback) | ISBN 9781003168881 (ebook)
Subjects: LCSH: Insurance–Rates. | Insurance premiums. | Pricing.
Classification: LCC HG8065 .P37 2023 (print) | LCC HG8065 (ebook) |
DDC 368/.011–dc23/eng/20230126
LC record available at https://lccn.loc.gov/2022047427
LC ebook record available at https://lccn.loc.gov/2022047428

ISBN: 9780367769031 (hbk)
ISBN: 9780367769048 (pbk)
ISBN: 9781003168881 (ebk)

DOI: 10.1201/9781003168881

Typeset in Palatino
by Newgen Publishing UK

In memory of my father, Renzo, and of my friend, Paolo Albini.

Contents

Part IV General-Purpose Techniques

Preface to the Second Edition

I'm so darn glad He let me try it again
'Cause my last time on earth, I lived a whole world of sin
I'm so glad that I know more than I knew then
Gonna keep on tryin'
'Til I reach my highest ground

– **Stevie Wonder, Higher Ground**

Why a Second Edition?

I was very pleased with the reception of the first edition of *Pricing in General Insurance*. True, it was never going to make it to the *New York Times* best sellers' list: even relatives and close friends told me that they found the preface so effective that they felt it gave them all the knowledge they needed – and wanted – about the subject, and they read no further.

On the positive side, I have been surprised at how many in our small world of non-life actuarial pricing have read this book or know of this book and have reached out to me with appreciative feedback.

As for myself, since the book's publication I had become increasingly unhappy with what I perceived to be its shortcomings (especially material that I thought was missing) and I soon started posting corrections and additional material on the book's website. For this reason, I was very grateful when the publisher encouraged me to produce a second edition: that gave me the long-awaited opportunity to make the improvements I had in mind.

The most obvious shortcoming of the first edition – not counting the errors and typos that have been found and corrected and will presumably be replaced by new ones – is that, by the time I had completed it, my experience in pricing had only been with insurance and reinsurance brokers (with a couple of years of parallel teaching experience). That had provided me with exposure to a wide variety of insurance buyers, to sophisticated techniques, and to a number of concepts one doesn't normally come across working in an insurance company, such as insurance optimisation. However, my experience was lighter – and my knowledge second-hand – when it came to other activities that pricing actuaries working for an insurance company are involved with, such as rate change calculations and the development and maintenance of pricing models for underwriters: this was reflected on the lesser emphasis these topics were given in the first edition, and on my reliance on other sources.

In the first edition I was also rather timid about discussing machine learning: at the time, it seemed an idiosyncratic interest of mine (Parodi, 2009; 2011; 2012a), and I did not want to make a fuss about it, so I just snuck the topic subtly here and there in Chapter 12 (What Is This Thing Called Modelling?), Chapter 16 (Severity Modelling) and Chapter 26 (Generalised Linear Models). In this new edition, I have tried to explain more explicitly the basic principles around machine learning and why machine learning is helpful in actuarial practice, although not always in the way that people would expect. I still don't want to give an in-depth treatment of the subject in this book and make machine learning the main story, especially because there are now specialised books that do this job brilliantly (Wüthrich and Merz, 2022); I'll simply make the case that machine learning (whose principles are explained more at length in Chapter 12) provides the right framework when thinking about actuarial modelling, and will use it to shed light on techniques such as generalised linear models (GLMs) that are sometimes wrongly considered as outdated precursors to more modern techniques such as neural networks or ensemble boosting.

Finally, many chapters needed modernisation and significant beefing up – or sometimes simplifying. A detailed list of chapter-by-chapter changes is included in the next section.

Naturally, a non-fiction book is ultimately like the cortical homunculus in neuroanatomy, namely a representation of the subject where the proportions in the treatment of the various topics is to some extent distorted by the author's experience and knowledge, and this second edition will be no exception. Hopefully, however, the homunculus has now gained near human-like proportions and will not attract funny looks anymore.

A Detailed List of Changes

The table below shows – at a good level of detail – what has changed from one edition to the other. The significance of the change is summarised with a rating that goes from ◐ to •••• and whose meaning is as follows:

- ◐ = no or very minor changes;
- • = minor changes (terminology, small additions/deletions of material ...);
- •• = some rewriting, with inclusion/deletion of significant material;
- ••• = substantial rewriting and/or addition/deletion of material;
- •••• = extensive rewriting and addition/deletion of material, or new chapter.

Intermediate ratings are captured again by the symbol ◐. For chapters that have a new number I have put the previous number between parentheses, e.g. '21 (20)'.

Chapter		Significance and details of changes	
1	The Pricing Process: A Gentle Start	•••	This chapter has been enriched to include an easy introduction to both exposure and experience rating, and to explain briefly why for the most part we don't use the supply-and-demand definition of pricing. The experience rating portion also includes a slightly less curt treatment of IBNR.
2	Insurance and Reinsurance Products	••	Details of the standard classification systems (PRA, Lloyd's, Solvency II) and of the individual products have been moved to an online appendix.
3	Cover Structure	•	Sharpening up some cover concepts, e.g. the difference between excess and deductible. Removal of some lame jokes that I couldn't bear reading again.
4	Insurance Markets	◐	Minor changes.

Chapter		Significance and details of changes	
5	Pricing in Context	•◐	Inclusion of the risk management function and minor changes.
6	Risk Theory	••	The introduction to the chapter has some rewording to put it into context and there's some coy reference to history. There are some more details on the individual risk model. Extensions to the individual and collective risk models are also discussed.
7–9	Familiarise with Risk Data Requirements Setting Inflation	•	Some rewording and sharpening up.
10	Data Preparation	•	Added a brief section on exposure data transformation.
11	Burning Cost	••••	This chapter has undergone a major rewriting and is now the largest chapter in the book: (a) the methodology has been articulated more precisely; (b) the treatment of large loss calculations has been expanded; (c) a small section has been added to summarise the burning cost method into a single formula; (d) a whole section devoted to portfolio pricing has been included; (e) an expanded treatment of the calculation of earned premium/exposure has been included.
12	What is This Thing Called Modelling? (A gentle introduction to machine learning)	•••◐	The machine learning section was only lightly touched upon in the first edition. Now there is a more meaningful introduction to machine learning techniques – and especially model selection – through an intuitive, non-insurance example (polynomial fitting), with pointers to where machine learning concepts are used throughout the book.
13	Claim Count IBNR	•••	The triangle-free approach to claim count development has been significantly expanded beyond the simple exponential case, and now takes up one half of the chapter. As for triangle-based development, a fifth method for dealing with incomplete diagonals has been included.
14	Frequency Modelling	••◐	Introduced the multivariate Bernoulli distribution. Significant rewriting of the frequency model selection part.
15–16	IBNER Analysis, Severity Modelling	◐	

Chapter			Significance and details of changes
17	Aggregate Loss Modelling	••	Included Monte Carlo results for our case study. Introduced analytical calculations for the coverage modifications. The R code for the numerical calculations (Monte Carlo, Fast Fourier Transforms) has been moved to an online appendix.
18	Measuring Uncertainty	◐	
19	Premium Building	•••	The section on the premium formula and cost-plus methodology has been rewritten. The connection between the profit loading and the return on capital has been sharpened up. The capital allocation section has also been rewritten. Some practical material has been added to the table on risk measures. The game theory allocation principle has been relegated to an online appendix, and has been replaced by the Euler allocation principle.
20	The Pricing Cycle	••••	This is a new chapter devoted to the pricing cycle and especially rate change calculations, which is now dealt with in considerable detail.
21 (20)	Non-Proportional Reinsurance	••	Added analytical formulae for the mean and variance of losses to a layer for the standard distributions (Pareto and GPD). Some added commentary on the log-logistic distribution for the settlement pattern (minor). A few terminological changes.
22 (21)	Exposure Rating (Property)	••••	This chapter has been significantly expanded. Methods for calibrating MBBEFD exposure curves have been included, as well as methods to simulate from such curves. Exposure rating for direct insurance has been expanded and now reflects closely the standard approach to property rating through rates on value, exposure curves and (where needed) deductible impact tables. The issues around scale-independent losses and MPL uncertainty are also addressed.
23 (22)	Exposure Rating (Casualty)	••◐	A whole section has been included on the derivation of ILF curves from severity models using standard from-the-ground-up models such as lognormal, lognormal/GPD, and empirical/GPD.
24 (23)	Specific Lines of Business	•	Added a brief section on construction cover.
25 (24)	Cat Modelling	••◐	A section on pricing using cat rates/hazard maps (rather than cat models) has been introduced. The concept of PLTs/YLTs has been included, and the part describing the simulation of losses for ELTs and PLTs has been rewritten.

Chapter		Significance and details of changes	
26 (25)	Credibility	●●◐	Removed the non-insurance example for the general concept, which was unnecessarily simplistic, and replaced with a simple (and shorter) insurance example. Expanded the treatment of Bayesian credibility, with the inclusion of the Poisson/Gamma model for claim count credibility. Added formulae for multi-source credibility and a discussion of the limitations of the credibility approach. Some terminological changes.
27 (26)	Rating Factor Selection: GLM, GAM and Regularisation	●●●●	This chapter has undergone extensive rewriting. To start with, the initial example explaining what difference rating factors make has been drastically simplified. A section describing the non-actuarial approach to rating factors has been included, in the hope that it sheds some light on the rest of the chapter. The model validation section has been significantly enriched and new sections on model structuring and model structural checks have been included. The two original examples in the first edition – too simplistic and using a terminology at odds with the rest of the chapter – have been replaced with a single, clearer practical example that allows to see the various techniques at work. A treatment of generalised additive models (GAMs) with a practical example has been included. The connection with machine learning has been further clarified.
28 (27)	Multilevel Factors and Smoothing	●	Added a very brief section at the end which explains the connection to unsupervised learning and specifically to clustering.
29 (28)	Multiple Lines of Business	●	Moved R code for copulas to an appendix.
30 (29)	Insurance Optimisation	◐	Very minor changes in wording.
31	Pricing Models	●●●●	This is a new chapter that is in a way the point of convergence of the whole book, as it builds on the material developed in the rest of the book to show how pricing models are structured, and how they are developed, calibrated and maintained. It also provides a taxonomy of models across various dimensions (experience/exposure, type of risk covered, type of client …)

The Book's Double Vocation, and the Meaning of '*'

The main intended readership remains the same as for the first edition: actuarial students[1] and actuarial practitioners. Underwriters, brokers, and risk managers with an interest in the statistical aspects of pricing, especially if they have a quantitative background, might also find some use for it as a reference.

Consequently, this book continues to have a double vocation as a *textbook* and a *reference book*. With respect to the first edition, however, more 'reference' material has been included. Some of it is a bit more technical than the rest and has been prefixed with scare asterisks ('*') to signal that it can be omitted at first reading and can indeed be used for reference as needed; some of it has been moved to the Questions part of the chapter. When questions are prefixed by '*', however, it simply means that these are theoretical questions (easy or not).

The Book's Practical Nature and Style

Despite these changes, this book remains practical in nature. Many textbooks dealing with mathematical topics still show traces of the influence of Nicolas Bourbaki.[2]

> Bourbaki's extremely austere and idiosyncratic presentation of the topics discussed in each of the chapters – from which diagrams and external motivations were expressly excluded – became a hallmark of the group's style and a main manifestation of its thorough influence. [...] Concepts and theories were presented in a thoroughly axiomatic way and systematically discussed always going from the more general to the particular, and never generalizing a particular result.
>
> **Corry**, *2009*

This writing style is very effective if your reader is an AI. Having instead busy human beings working in a business environment in mind, I have tried to turn that on its head and write a book rich with diagrams, signposts, striving to explain the motivation behind every topic. The terms 'axiom', 'lemma' and 'corollary' will not appear (again) in the book, and any mathematical generalisation is always preceded, or accompanied, by special cases, simple examples, and by spreadsheets where appropriate (see *online material*). Where the content in this new edition becomes more technical and the density of formulae increases it is because the book also strives to provide actual recipes for real-world implementation in spreadsheets or code rather than some handwaving in the direction of a solution.

How to Navigate this Book

The chart below shows how the book is structured and how the different chapters are connected between them. Note that the names in each box reflect the topic of each chapter rather than the exact title.

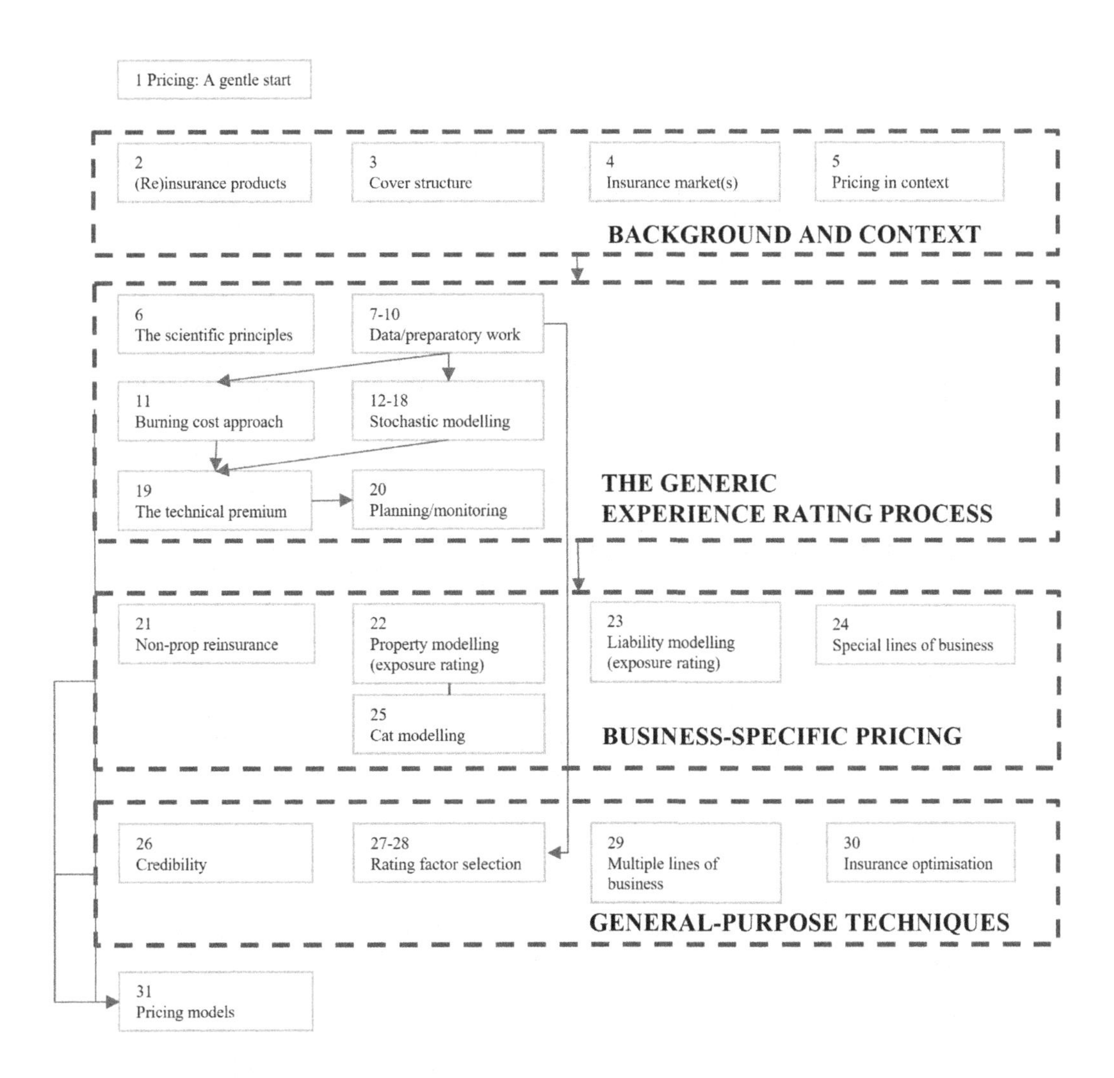

Depending on their specific interests, here are some example routes that readers might want to follow:

- Experience rating route (commercial lines): 6 → 20 [+24] [+29] [+31]
- Property exposure rating (PER) route: 6 → 10+21+25 [+29+31]
- Liability exposure rating (LER) route: 6 → 10+23 [+29+31]
- Hybrid (exposure/experience) rating route: PER + LER + 26
- Personal lines/SME route: 6 → 10 + 27 + 28 + 31
- Applications of machine learning *principles* route: 12 → (14 +) 16 + 27 +28
- *Reader-curious-to-know-what-pricing-is* route: 1 → 5
- *Impatient-reader-curious-to-know-what-pricing-is* route: 1

Additional Online Resources

This book comes with additional online resources, which are accessible to all, whether they have purchased the book or not. The material can be found by accessing this website:[3]

https://pricingingeneralinsurance.net

In this website readers will be able to find:

- An **errata** document, where I will periodically add the errors that readers and I will find in the second edition. I will also leave the errata document for the first edition available separately, for those who own the first edition and are happy to stick with it.
- Various **appendices** with material that either did not make the final cut (among which, material which was included in the first edition but has now been removed) or that is better kept outside of the printed book so that it can be updated as needed, such as the glossary and a description of insurance products. This includes R code for some of the algorithms described in the book.
- **Downloads** for many of the chapters, i.e. spreadsheets with data and – where useful – the calculations discussed in the book. By the way, whenever I feel uncool about sharing spreadsheets rather than (or in addition to) code in more modern programming languages, I tell myself that if a spreadsheet was good enough for Jerry Nelson to design the interlocking mirrors of the Keck telescope in Mauna Kea, Hawaii,[4] it is certainly good enough for me to build and share simple pricing prototypes and examples!
- The **solutions** to (a good number of) the problems at the end of each chapter.
- Other material that I may be inspired to add in the future.

Acknowledgements

First of all, a big thank you to my family for supporting me through a second round of writing that proved to be nearly as time consuming and holiday-disrupting as the first round 9 years ago!

Secondly, I would like to thank:

- The UK Institute and Faculty of Actuaries for their continuous support in this endeavour.
- Jean-Stéphane Bodo, Douglas Lacoss and the management at SCOR, for the encouragement I have consistently received for my research, teaching and writing activity.
- The students and staff at the Bayes (formerly Cass) Business School for providing the continuous testing ground and feedback for the material that I thought necessary to be added and amended for this second edition.

A significant number of individuals have helped me with their advice and feedback for this second edition. Without expecting to be exhaustive, while expecting to be a victim to recency bias, I would like to thank – in chapter order:

- Massimo Vascotto, who has helped me tread the dangerous territory of topics at the border between actuarial and underwriting (Chapters 1–9), providing plenty of much-needed sanity checks, recommending references, and allowing me to sharpen up the terminology;
- Kushan Fozdar for helpful discussion on portfolio pricing (Chapter 11);
- Zhongmei Ji, who has provided crucial help with the lasso experiment in Chapter 12, and Xiaochen Kou for reviewing the chapter;
- Joseph Lees for his review of Chapter 13 and for his advice on how it should be restructured. I am also thankful to Kenneth Wells for suggesting Method 5 for triangle-based development, which I was glad to be able to add to the chapter;
- Tony Neghaiwi for helpful discussion around the Euler principle, and Philipp Arbenz for several conversations around IFRS17, which were helpful when writing on the premium formula (Chapter 19);
- Didier Sedenio for helpful discussion on the business planning section of Chapter 20;
- Francis Cheung, Jean-Stéphane Bodo, Renaud Ambite, Eileen Kee, Samantha Tanner, Douglas Lacoss, Stefan Küttel, and Michael Trachsler for the helpful and at times challenging discussions on the rate change methodology (Chapter 20), and Peter Watson, who also led an implementation of the rate change calculations, helping to clarify delicate points;
- Laura Huang for reviewing Chapter 23; also Ilya Kolmogorov and Renzo Cermenati for helpful discussion on ILFs;
- Matthew Hazzard, Greg McNulty, Jonathan Adams and the other members of SCOR's credibility working party for inspiring discussions (Chapter 25);
- Eileen Kee, Jakir Patel, and Tom Gannon for their advice on Chapter 26 on catastrophe modelling;
- Jan Iwanik for his inputs on the practical aspects of generalised linear models (and for sharing his experience with personal lines insurance) for Chapter 27;
- Julian Glover for his review of Chapter 31 on pricing models and for providing technological sanity checks on model architecture.

Finally, I would like to thank those who have contacted me pointing out errors and typos in the text of the first edition. I can only mention by name the most prolific error-spotters: Michael Fackler, Raymond Poole, Arun Srikrishna Vihar, Andrei Kobelev, and Otto Beyer.

Notes

1 Although the syllabus is modelled around the syllabus of the general insurance pricing exam for the actuarial profession in the UK (without claiming to be a perfect match), based on the experience with the first edition most of the material should be relevant across actuarial bodies.

2 Nicolas Bourbaki is the name of a fictional mathematician behind which lied a 'collective' of mathematicians – among which some all-time greats such as Cartan and Weil – who embarked from the 1940s on an ambitious project of writing a series of maths textbooks. Their influence is the main reason why it is possible today to open certain graduate maths textbooks and read entire chapters without any idea as to what the book is about and why one needs to learn about that topic.

3 The website was originally named 'pricingingeneralinsurance.com'. I won't bother to relate the story of how I lost ownership of that domain name – a story of credit card expiry dates, annoying practices of my previous domain name provider and other minor irritations. It might be that in the future the old domain name will return to me, or that that story will repeat itself with the current domain name, in which case please try to find the website under some new suffixes: I doubt someone will find that *just-in-case-purchasing* all the domains starting with 'pricingingeneralinsurance' is a profitable business idea.
4 www.economist.com/obituary/2017/06/29/obituary-jerry-nelson-died-on-june-10th

Biography

Pietro Parodi has worked in general insurance pricing since 2005. He is currently Head of Methods and Tools for SCOR Specialty Insurance, and since 2012 a part-time lecturer at Bayes (formerly Cass) Business School. He has previously worked at Willis, Aon Benfield and Swiss Re. He is a fellow of the Institute and Faculty of Actuaries in the UK.

In his previous professional incarnations, he was a network administrator/IT project manager and a research scientist in the fields of artificial intelligence and neuroscience. He has a MSc and a PhD in Physics from the University of Genoa, Italy. Prior to university, he studied as an electronics technician at St John Bosco's secondary school in Genoa, where he could have developed some serious practical skills had his attention not been hijacked by the maths around Laplace transforms.

In 2012 he received the Brian Hey Award for his paper *Triangle-free reserving*, which he presented at GIRO.

He lives in East London with his wife and his four children.

Preface to the First Edition

Most people with some interest in the history of science know that in 1903, Albert Einstein, unable to find a job at the university, started working as an 'assistant examiner' in the patent office of Berne, where he worked efficiently enough to be able also to produce four papers, one on Brownian motion (now a standard topic in the actuarial profession exams on derivatives) and three on relativity theory and quantum mechanics (not on the syllabus). According to one biographer (Isaacson, 2007), an opportunity had also arisen for Einstein to work in an insurance office, but he hastily turned that down arguing that that would have meant eight hours a day of 'mindless drudgery'. Details on the exact job description for the position are not available, but since that was in 1903, we can rule out the possibility that Einstein was offered a position as a pricing actuary in general insurance. And apart from the historical impossibility (actuaries were first involved in general insurance around 1909 in the United States, to deal with workers' compensation insurance), the job of the pricing actuary in general insurance is *way* too exciting – contrary perhaps to public perception – to be described as mindless drudgery: pricing risk means *understanding risk* and understanding risk means *understanding (to some extent) how the world works* (which is after all what Einstein was after): to give but a very simple example, pricing a portfolio of properties of a company requires some understanding of what that company does (is it producing gunpowder or hosting data centres?) and what perils (natural and man-made) these properties are exposed to in the territories the company operates in.

This book is based on my experience as both a practitioner and a part-time lecturer at Cass Business School in London. It was written to communicate some of the excitement of working in this profession and to serve the fast-expanding community of actuaries involved in general insurance and especially in pricing. It comes at a time when this relatively new discipline is coming of age and pricing techniques are slowly crystallising into industry standards: the *collective risk model* is widely used to estimate the distribution of future total losses for a risk; *extreme value theory* is increasingly used to model large losses; *generalised linear modelling* is the accepted tool to model large portfolios with a significant number of rating factors – to name but a few of these techniques.

This book was written by a practitioner with actuarial students specialising in general insurance[1] and other practitioners in mind, and its aim is therefore not foundational (such as is, for example, the excellent *Loss Models: from Data to Decisions* by Klugman et al. 2008) but practical. As a matter of fact, I have tried to keep constantly in mind Jerome K. Jerome's wry remark on the lack of practicality of foreign language teaching in British schools during his (Victorian) times: 'No doubt [students] could repeat a goodly number of irregular verbs by heart; only, as a matter of fact, few foreigners care to listen to their own irregular verbs recited by young Englishmen' (Jerome, 1900). This book, therefore, rarely dwells on the more abstract points about pricing, mathematical definitions, and proofs (which might be thought of as the irregular verbs of our profession) beyond the bare minimum needed to develop an intuition of the underlying ideas and enable execution – rather, it is rich in step-by-step methods to deal concretely with specific situations and in worked-out examples and case studies.

Pricing as a Process

One of the main efforts of this book is to present pricing as a process because pricing activity (to an insurer) is never a disconnected bunch of techniques, but a more-or-less formalised process that starts with collecting information about a particular policyholder or risk and ends with a commercially informed rate.

The main strength of this approach is that it imposes a reasonably linear narrative on the material and allows the student or the practitioner to *see pricing as a story* and go back to the big picture at any time, putting things into context. It therefore allows students to enter the pricing profession with a connection between the big and small picture clearer in their minds, and the practitioner who is switching from another area of practice (such as pensions) to hit the ground running when starting to work in general insurance.

At the same time, the fact that there is roughly a chapter for each building block of the detailed risk pricing process should help the practitioner who is already involved in general insurance pricing to *use this textbook as a reference* and explore each technique in depth as needed, without the need of an exhaustive read.

In practice, this is achieved as follows. A few introductory chapters (Section I: Introductory Concepts) set the scene; the most important is Chapter 1, which provides an elementary but fully worked-out example. Then, Section II: The Core Pricing Process introduces a modern generic pricing process and explores its building blocks: data preparation, frequency analysis, severity analysis, Monte Carlo simulation for the calculation of aggregate losses, and so on, with the emphasis always being on how things can be done in practice, and what the issues in the real world are. Alternative, more traditional ways of pricing (such as burning cost analysis) are also explored.

After learning the language of pricing, one can start learning the dialects; Section III: Elements of Specialist Pricing goes beyond the standard process described in Section II and is devoted to pricing methods that have been devised to deal with specific types of insurance or reinsurance product, such as exposure rating for property reinsurance. If the student has started becoming a bit dogmatic about the pricing process at the end of Section II and has started thinking that there is only one correct way of doing things, this part offers redemption: the student learns that tweaking the process is good and actually necessary in most practical cases.

Finally, Section IV: Advanced Topics deals with specific methods that are applied in certain circumstances, such as in personal lines where there is a large amount of data and policyholders can be charged depending on many rating factors, or when one wants to price a product that insures many different risks at the same time and the correlation between these risks is important.

Acknowledgements (First Edition)

First and foremost, I would like to thank my wife, Francesca, for her support and the many creative ways she devised for shielding me from familial *invexendo*, simply making this endeavour possible; and my children, Lisa, David, Nathan and Sara, for those times they managed to breach that shield. Thanks also to the rest of my family for helping in different ways, with special gratitude to my aunt for the crucial time she let me spend in her country house in Trisobbio in the summer of 2013, *retir'd from any mortal's sight* – except of course for the compulsory Moon Pub evening expeditions with my friends.

I am grateful to the staff and the students at Cass Business School for providing me with the motivation for starting this textbook and for the feedback on the lecture notes that became the foundation for this pursuit.

Many thanks to Phillip Ellis, Eamonn McMurrough and all the colleagues at the Willis Global Solution Consulting Group, both in London and Mumbai, for providing so many opportunities of tackling new technical and commercial problems, experimenting with a large variety of different techniques and engaging in critical discussion. I also thank them for their encouragement while writing this book.

My deep gratitude to the Institute and Faculty of Actuaries for its invaluable support in this endeavour. Special thanks to Trevor Watkins for encouraging me and overseeing the process, and to Neil Hilary, who reviewed the material, guided me through the process of ensuring that this book be as relevant as possible to the profession and the students of the General Insurance practice area and encouraged me and advised me in countless other ways.

I would like to thank Jane Weiss and Douglas Wright for their initial review of the material and their encouragement. Special thanks to Julian Lowe for his guidance and for several important structural suggestions on the first draft.

Many people have subsequently helped me with their feedback on various versions of the manuscript. I am especially grateful to Julian Leigh and Sophia Mealy for painstakingly reading the whole manuscript and for their insightful comments on both content and style. I am also thankful to the following people for precious suggestions on specific topics of the book: Paolo Albini (frequency analysis), Cahal Doris (catastrophe modelling), Tomasz Dudek (all the introductory chapters), Junsang Choi (catastrophe modelling), Phil Ellis, LCP (various presentational suggestions), Michael Fackler (all the reinsurance content and frequency/severity modelling), Raheal Gabrasadig (introductory chapters), Chris Gingell (energy products), Torolf Hamm (catastrophe modelling), Anish Jadav (introductory example, products), Joseph Lees (burning cost), Marc Lehmann (catastrophe modelling), Joseph Lo (from costing to pricing, plus all the introductory chapters), Eamonn McMurrough (pricing control cycle), David Menezes (aggregate loss modelling, dependency modelling), Cristina Parodi (claims management), David Stebbing (aviation), Andreas Troxler (dependency modelling) and Claire Wilkinson (weather derivatives).

Note

1 Specifically, the contents of this book are based on the syllabus for the ST8 exam ('General Insurance Pricing') of the Actuarial Profession in the United Kingdom and India, with additional material.

1

The Pricing Process: A Gentle Start

'All beginnings are hard', according to a rabbinic saying cited by Chaim Potok at the outset of one of his novels, *In the Beginning*. Inevitably, this is true for pricing as well. The aim of this chapter is therefore to make the reader's life a tad easier by providing a 'pop' version (that is, a version light on maths and specifics) of the subject of pricing in general insurance through a few pricing examples, while ensuring that the main ingredients of the pricing process are included.

After a brief reminder of what economics theory says about pricing (Section 1.1), we will see basic examples of the main pricing techniques: exposure rating (pricing based on rates applied to exposures, Section 1.2), experience rating (pricing based on past loss experience, Section 1.3), and hybrid pricing (a combination of exposure and experience rating, Section 1.4).

We will complete the chapter in Section 1.5 with a bird's eye view of the pricing process, which will be useful to put the rest of the book's contents into context.

1.1 Pricing in Economic Theory

Economic theory tells us that there is a *law of supply*, stating that there is a direct relationship between price and quantity supplied: if the price of a good increases, the supply of it will also increase; and there is a *law of demand*, stating that as the price of a good increases, the demand for it will decrease. The *equilibrium price* of a good is that at which supply meets demand or, in mathematical terms, the supply curve intersects the demand curve (Landsburg, 2002), as illustrated in Figure 1.1.

Economic theory also tells us that while the optimal price is determined by the balance of supply and demand, a supplier of the good in general needs to ensure that the price it charges is above the (marginal) production cost of the good,[1] so as to achieve a profit (Figure 1.1).

Ultimately, the *actual premium* that an insurer charges for its products (policies) in a free market[2] will be determined by the law of supply and demand; and indeed, the economic

1 Selling above cost is in the interest of the supplier; however, the supplier may be tempted to sell below cost to increase its share of the market. This practice is called 'predatory pricing' and is illegal in many markets.

2 There is an argument that the approximation of the market for insurance as a free market is not actually very good because the demand curve for insurance product is relatively flat (Boor, 2000), but we will not enter into this discussion here.

DOI: 10.1201/9781003168881-1

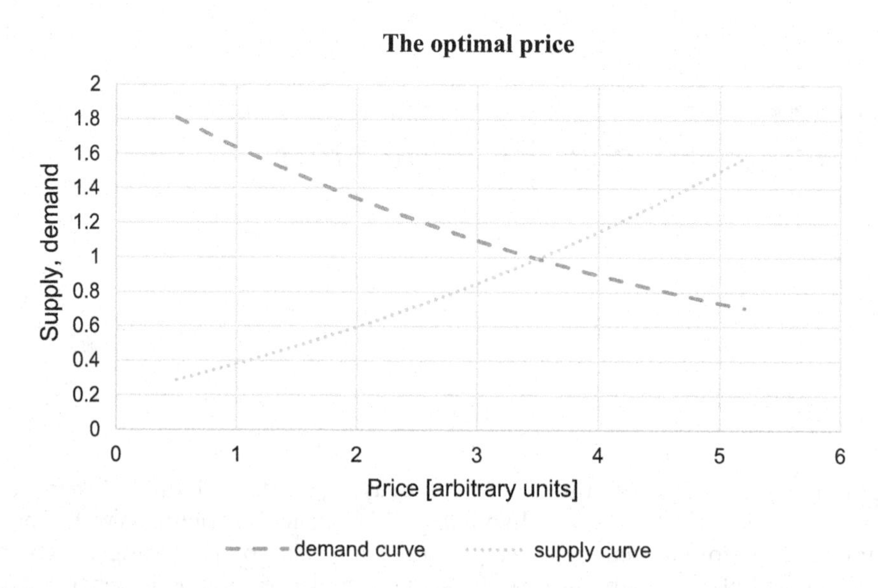

FIGURE 1.1
According to price theory in economics, the optimal price is the price at which the demand curve meets the supply curve.

aspects of premium determination will be reflected in this book in the discussion of the insurance market, the underwriting cycle, and price optimisation (Chapter 19).

The focus of this book, however, will be on the calculation of the production costs of insurance. Insurance business is different from manufacturing business: while for manufacturing business the costs of producing a good are by and large known before the product is sold, in the case of insurance they are to a significant extent unknown at the moment of sale, as they require the estimation of the expected value of future claims, which are the result of what is, for all intents and purposes, a random process (Werner and Modlin, 2016). Most of the content of the book concerns, therefore, the process of *risk costing* – i.e. the calculation of what we will call the *technical premium*, that is the premium that based on actuarial calculations should be charged for a policy contract in order to cover the costs (claims and other expenses) and achieve the desired level of profit.

1.2 Exposure Rating

Traditionally, pricing is done by using rating tables that specify *rates* to apply to *risk exposures* to determine the premium to be charged. 'Exposure' is a metric that measures the amount of risk taken on, and the metric used varies depending on the line of business: it could be the sum insured for building and contents insurance, or the number of vehicles in a fleet for commercial motor insurance, or total wage roll for employers' liability. On top of that, the rating tables will specify *premium adjustments* to account for the characteristics of the policyholder and for the specifics of the cover (limit of indemnity, excess/deductible, and so on).

Let's take a simple, drastically simplified example. Assume that an underwriter wants to price an employers' liability annual policy incepting on 1 January 2025 for a manufacturing company with an estimated 2025 wage roll of £165M, with a 20% of its staff being clerical and the remaining 80% manual within the metallurgical sector. The limit of indemnity (i.e. the maximum amount that the insurer will pay for an employers' liability claim) is set at £10M. The underwriting guidelines might contain the following instructions on how to calculate the technical premium. The instructions will typically be incorporated in a pricing tool.

RATING GUIDELINES (EXAMPLE)

Base rates. The *base rate* (i.e. premium per unit of exposure for the base level case) for employers' liability is $BR_c = 0.1\%$ of the total wage roll for clerical workers and $BR_m = 0.8\%$ of the total wage roll for manual workers.

Risk adjustment factors. Depending on the *type of industry* (heavy, average, light), the base rate for manual workers is multiplied by a factor f_R whose value is 1.5, 1, 0.5 respectively.

Increased Limit Factor (ILF). The minimum limit of indemnity for employers' liability is $L_0 = £5M$. If the actual limit L is larger than the minimum value, the rate should be multiplied by:

$$\text{ILF}_L = \left(1 + \log_2\left(\frac{L}{L_0}\right)\right)^{0.2}$$

Based on these instructions, the premium for the policy mentioned above would be:

$$\begin{aligned}\text{Premium} &= \text{ILF}_L \times \left(\text{BR}_c \times \text{Wage Roll}(\text{clerical}) + f_R \times \text{BR}_m \times \text{Wage Roll}(\text{manual})\right)\\ &= 1.149 \times (0.1\% \times 20\% \times £165\text{M} + 1.5 \times 0.8\% \times 80\% \times £165\text{M})\\ &= £1.86\text{M}\end{aligned}$$

Note that in practice the pricing tool will be more complex: it will incorporate more adjustments for risk and coverage, and will include an allowance for commissions/brokerage fees. It may also separate the pure premium (i.e. the expected claims) from the other expenses and the desired profit. However, *our extremely simplified example gives the essence of what exposure rating looks like*. More sophisticated types of exposure rating will be explored in Chapters 21 (property business) and 22 (casualty business).

The fact that traditional pricing is done in this way leaves many questions unanswered, the most important of which is:

Where do the rates and factors used in exposure rating come from?

This is not an easy question to answer. Ultimately, these rates are an attempt to ensure that the technical premium takes into account the production costs (expected claims and expenses) and the desired level of profit. For this attempt to be successful, the rates (along with the correction factor) need to provide a good estimate of the future loss experience. As for any estimate of a random variable, an obvious place to look for is historical loss experience. Actuarial students are told *ad nauseam* that 'the past is not always a good guide to the future': however, it is often the only sensible guide, and one that in any case will be disregarded at one's peril.

So the rates used in exposure pricing will need to be to some extent consistent with the historical past experience of the relevant portfolio of clients, whether internal or external. Exposure rating is often, therefore, rooted in experience rating (the derivation of rates based on past loss experience of a client or a portfolio).

1.3 Experience Rating

Let's turn now to experience rating – the pricing process driven (or at least informed) – by the historical loss experience of the client; specifically, we will illustrate it through a very basic version of *burning cost analysis*, itself the most basic experience rating technique. We'll use again the example of Section 1.2 – the annual employers' liability policy incepting on 1 January 2025 for a manufacturing company with a £165M wage roll, with a limit of indemnity set at £10M.

This time, however, the prospective insured has provided as part of the submission (the documentation relevant to the risk) the table in Figure 1.2, which gives the total losses incurred by year, from the ground up.

Based on this historical loss experience, what price should the underwriter charge for cover in 2025 (assume that the policy will incept on 1 January 2025 and will last one year) assuming that the insurer has a *target loss ratio* (the ratio of expected losses and premium) for employers' liability business of 75%?

A naïve method that comes to mind to reply to the question above is simply to take the average of the total losses over the period 2015 to 2024, which is approximately £1.26M. To this we have to apply the target loss ratio, yielding £1.26M/75%~£1.68M.

Why is this not satisfactory? The reason is that we have ignored many elements that are essential for the assessment of the risk. Even a cursory look at the actual losses in the last few years would make an underwriter cringe: out of the last 5 years, 4 years show total losses greater than the average (£1.26M), and 2 years show total losses greater than the premium we plan to charge. Something is obviously not right. So what have we forgotten?

Policy year	Losses (£)
2015	890,882
2016	1,055,998
2017	936,637
2018	1,153,948
2019	1,290,733
2020	1,548,879
2021	1,973,818
2022	1,448,684
2023	1,831,386
2024	442,957
Average	1,257,392

FIGURE 1.2
Losses incurred for each policy year, as at 30 June 2024. Note that these are the total losses (i.e. in aggregate) for each year. The average of losses over the period 2015 to 2024 yields £1,257,392.

There is an extensive list of elements that we need to consider. The most obvious ones are set down below. We have subdivided the various steps of the pricing process into four main categories: *risk costing (gross of deductibles, limits, etc.)*, which is the estimation of the future losses regardless of who pays for them; *risk costing (ceded – net of deductibles, limits, etc.)*, which is the estimation of that portion of the losses that is covered by the insurer; *determining the technical premium*, which leads to the premium suggested by the actuary, based on the losses and other relevant costs; and finally, *commercial considerations and the actual premium*, which explains why the premium suggested by the actuary may be overridden.

1.3.1 Risk Costing (Gross)

1.3.1.1 Adjusting for 'Incurred but Not Reported' Claims

As a bare minimum, we will need to take into consideration that data for the year 2024 is not complete: only losses reported until 30 June 2024 have been recorded. Because we only have half a year of experience, we will need to multiply the 2024 figure by *at least* two so that it covers the whole year and not only the first 6 months. The actual correction should be more than that because it takes a while until a claim is reported, and therefore the final amount for year 2024 will be larger than what we see at the end of the year.

For that matter, even the 2023 figure may not be fully complete, and because in some cases liability claims take several years to be reported (asbestos-related claims – which were reported up to 30 years after the period when the worker was working with asbestos – are a case in point), there might be missing claims in *all* the policy years.

These missing claims are part of what are called incurred but not reported (IBNR) claims. This includes both the missing claims (incurred but not *yet* reported, IBNYR) and a correction for claims that have already been reported but which have not been settled yet and therefore their amount is an estimate rather than a firm figure. This correction is called incurred but not enough reserved[3] (IBNER) and – despite the 'enough' in the name – it may be either positive or negative, depending on whether the reserves tend to be underestimated or overestimated.

We need some method to incorporate IBNR claims into our analysis. This is typically done by introducing appropriate multiplication factors (these are often called *loss development factors*) that specify by what number the total losses for each year should be multiplied to estimate the ultimate total claim amount.

These factors may be estimated based on the experience of the client or – when the data is not enough to give a stable estimate, as is often the case – on the analysis of the insurer's portfolio for this type of product, perhaps with some smoothing. This type of analysis may be provided by the reserving function. For our case, we will assume that monthly loss development factors (LDFs) are available for the employers' liability portfolio. If m is maturity of each underwriting number – i.e. the number of months from the beginning of the underwriting year to the as-at date – then LDF(m) is the factor by which we have to multiply the total claim amount for that year to project that amount to its (estimated) ultimate value. The results are shown in Figure 1.3.

If LDFs are not available, one quick and dirty trick is to exclude the most recent year(s) of experience and calculate the average based on the rest. In our case as a first approximation, we could exclude the losses of 2024 (as this year is too immature) and take the

3 IBNER is also sometimes spelled out as 'incurred but not enough *reported*'. This is possibly to contrast it with IBNYR.

Policy year	Losses (£)	Maturity (months)	Loss Development Factor	IBNR-Adjusted Losses (£)
2015	890,882	114	1.001	891,773
2016	1,055,998	102	1.001	1,057,054
2017	936,637	90	1.003	939,447
2018	1,153,948	78	1.005	1,159,718
2019	1,290,733	66	1.009	1,302,350
2020	1,548,879	54	1.017	1,575,210
2021	1,973,818	42	1.031	2,035,006
2022	1,448,684	30	1.057	1,531,259
2023	1,831,386	18	1.197	2,192,169
2024	442,957	6	4.288	1,987,991
Average	1,257,392			1,467,198

FIGURE 1.3
Using loss development factors (LDFs) calculated at portfolio level we project the total claim amount for each year to its projected ultimate value, incorporating an allowance for incurred but not reported claims.

average over the other years. This would lead to the following estimate: Average (2015–2023) = £1,347,885 ~ £1.35M.

1.3.1.2 Adjusting for Claims Inflation

Another simple correction we should make is to account for the time value of money. Obviously, a claim that occurred in 2015 would have, if it occurred again under exactly the same circumstances in 2025, a much larger payout, because of what is called *claims inflation*. Claims inflation is in general *not* the same as the consumer price index inflation (CPI) – the index that attempts to track variations in the cost of living.[4] This is because claims inflation depends on factors that the CPI is not necessarily capturing. For example, property claims will depend on cost of repair materials, cost of repair labour and the like. Liability claims will depend on wage inflation, cost of care inflation and court inflation, which is the tendency on the part of the courts to award compensations that are increasingly favourable to claimants.

Let us assume that claims inflation for the table above is 5%. We are not interested in discussing here whether this is a good assumption or not.

So how are claims revalued in practice? The procedure goes as follows:

1. Set the revaluation date (the date at which claims have to be revalued) to the average day at which claims will occur in the policy under consideration.

   This will normally be the midpoint of the policy (for instance, if a policy starts on 1 October 2025 and is an annual policy, the revaluation date will be 1 April 2026). However, if we know for a fact that claims have strong seasonality (for instance, motor claims occur more frequently in winter) the average point will be shifted. However, again, it is important to remember that claims inflation is never known for sure, and all these calculations are mired in uncertainty, and it is therefore easy to be too clever.
2. Assume that all claims in each policy year equally occur at the midpoint of each policy year (or whatever point the seasonality suggests).

4 CPI (the name commonly given to the general inflation index in English-speaking countries) is of course known under different names in different countries, and may be calculated differently/accompanied by other indices.

Policy year	Losses (£)	IBNR-Adjusted Losses (£)	Revaluation factor	IBNR-Adjusted, Revalued losses (£)
2015	890,882	891,773	1.629	1,452,698
2016	1,055,998	1,057,054	1.551	1,639,491
2017	936,637	939,447	1.477	1,387,563
2018	1,153,948	1,159,718	1.407	1,631,723
2019	1,290,733	1,302,350	1.340	1,745,148
2020	1,548,879	1,575,210	1.276	2,009,968
2021	1,973,818	2,035,006	1.216	2,474,568
2022	1,448,684	1,531,259	1.158	1,773,198
2023	1,831,386	2,192,169	1.103	2,417,962
2024	442,957	1,987,991	1.050	2,087,391
Average	1,257,392	1,467,198		1,861,971

FIGURE 1.4
Calculating the average after revaluing the losses to current terms.

We will see later that this assumption can be dropped if we have a complete listing of the claims rather than the aggregate losses for each year.

3. Revalue the total claim $X(t)$ for each year t to the amount $X^{rev@t^*}(t) = X(t) \times (1+r)^{t^*-t}$ where t^* is the year of the policy we are pricing, t is the policy year whose historical claims we wish to revalue and r is the claims inflation.

   Things get only slightly more complicated if, instead of a roughly constant inflation, we want to use an inflation index that is not constant, I(t) [I(t) could be, for example, the CPI or a wage inflation index]. In this case, the formula above becomes $X^{rev@t^*}(t) = X(t) \times I(t^*)/I(t)$, where $I(t^*)$ is the estimated value of the index at time t*.

In Figure 1.4 we see how this changes the numbers for our simple example. We have introduced a column 'Revalued losses' whose value is equal to that in the column 'IBNR-Adjusted Losses' multiplied by the revaluation factor (a power of 1.05). As a result, the average has gone up to approximately £1.86M.

1.3.1.3 Adjusting for Exposure Changes

Something is still not right with the revalued losses above: there seems to be an upward trend in the number of aggregate losses per year, as shown clearly by the chart in Figure 1.5. For all we know, this might actually be a purely random effect: the upward trend is not huge and there is a lot of volatility around the best-fit line. However, there might be a more obvious reason: the profile of the company may have changed significantly from 2015 to 2024, and the difference in the total losses may reflect this change.

It is not easy to quantify the risk profile of a company, but there are certain measures that are thought to be roughly proportional to the overall risk, which in turn translates into the total expected losses. These measures are called *exposure measures*. As an example, if your company insures a fleet of cars, you expect that the overall risk will be proportional to the number of vehicles in the fleet or, better yet, to the vehicle-years (by which, for example, a vehicle which joins the fleet for only half a year counts as 0.5 vehicle-years). This measure seems quite good, but it is not perfectly proportional to the risk: for example, certain cars are considered riskier than others, if only because they are more expensive and therefore when they are damaged it costs more to repair them, and certain drivers are more dangerous than others. One might therefore devise fancier ways to find a general measure of

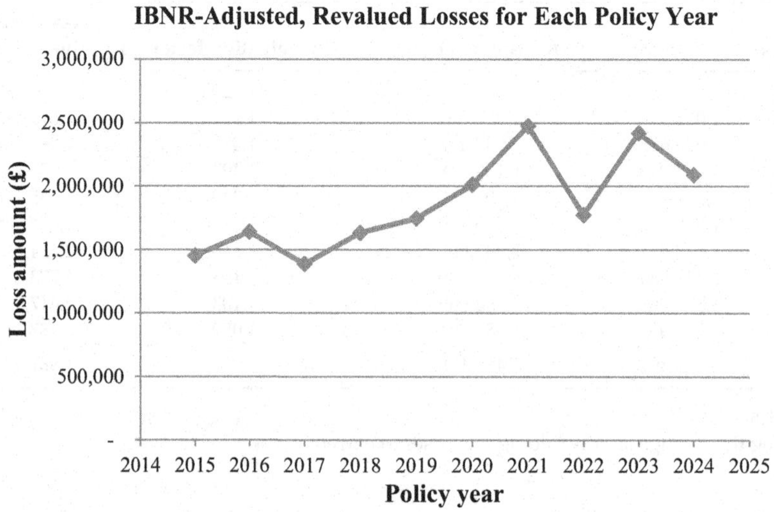

FIGURE 1.5
The total losses show an increasing trend, even after adjustment for IBNR and revaluation – an exposure effect?

exposure. However, simple is often better, and there is often not enough evidence in the data to justify doing something too clever.

In the case of employers' liability, an exposure measure that is often used for burning cost analysis is wage roll (the total amount of salary paid to the company's employees). Another good one is the number of employees (or, better still, employee-years, which – analogously to vehicle-years – gives appropriate weights to part-timers and employees joining/leaving during the year).

To go back to our example, we will assume that we have exposure in the form of wage roll because it gives us a chance to deal with another aspect of the exposure adjustment process – exposure on-levelling.

Because wage roll is a monetary amount, the issue arises of how it should be corrected to be brought to the value that it would have during the policy period. What one normally does is to on-level[5] past exposures with an inflation factor (either a single inflation or an index) much in the same way as we did for the losses. Notice, however, that *the inflation factor we apply to exposures does not need to be the same as that we apply to losses*: in the case of wage roll, for example, it makes sense (unless we have company-specific information on their wage policy) to on-level it based on the wage inflation for that country. If we assume that wage inflation is 3.9%, we can produce an on-levelled exposure as in Figure 1.6.

The last column in Figure 1.6 was obtained by adjusting the total claims by the exposure – that is, imagining what the losses in a given year t would have been if the exposure had been that estimated for the policy renewal period t^*. Assuming again that $t^* = 2025$, we can write

$$X^{\text{IBNR-adj,reval,exposure-adj}}(t) = X^{\text{IBNR-adj,reval}}(t) \times \frac{RE(t^*)}{RE(t)} = X^{\text{IBNR-adj, reval}}(t) \times \frac{E(t^*)}{E(t)} \times \frac{I_w(t)}{I_w(t^*)} \quad (1.1)$$

5 On-levelling is the same as revaluing, but this term is often used for exposures while revaluing is used for claims, to reduce (not always successfully) confusion.

Policy year	Losses (£)	IBNR-Adjusted Losses (£)	+ Revalued losses (£)	Exposure (wageroll, x £100k)	On-levelled exposure (wageroll, x £100k	+ Exposure-adjusted losses(£)
2015	890,882	891,773	1,452,698	676	991	2,418,561
2016	1,055,998	1,057,054	1,639,491	773	1,091	2,480,125
2017	936,637	939,447	1,387,563	913	1,240	1,846,467
2018	1,153,948	1,159,718	1,631,723	988	1,291	2,084,801
2019	1,290,733	1,302,350	1,745,148	1,146	1,442	1,997,278
2020	1,548,879	1,575,210	2,009,968	1,233	1,493	2,221,428
2021	1,973,818	2,035,006	2,474,568	1,453	1,693	2,411,324
2022	1,448,684	1,531,259	1,773,198	1,556	1,745	1,676,428
2023	1,831,386	2,192,169	2,417,962	1,618	1,747	2,284,146
2024	842,957	1,987,991	2,087,391	1,702	1,768	1,947,657
2025 (est.)				1,650	1,650	
Average	1,257,392	1,467,198	1,861,971			2,136,821

FIGURE 1.6
A simple look at this table shows the reason behind the year-on-year increase in total losses: the exposure has steadily gone up, even after on-levelling to consider wage inflation has been performed. Note that the £165M figure in the bottom row is next year's estimated exposure as provided by the client. Our recalculated average over once the change in exposure is taken into account is now £2.14M.

where $RE(t)$ is the revalued exposure at time t, which can be obtained from $E(t)$ (the exposure at time t) by $RE(t) = E(t) \times (1 + r')^{t^*-t}$, r' being the wage inflation, or more in general by $RE(t) = E(t) \times I_w(t^*)/I_w(t)$, where $I_w(t)$ is the wage inflation index. Note that by definition, $RE(t^*) = E(t^*)$.

Based on this exposure correction, our new forecast for next year's total losses is given by the average losses corrected by exposure and claims inflation:[6] £2.14M.

1.3.1.4 Adjusting for Other Risk-Profile Changes

As we mentioned above, even if the exposure remains the same, the risk may change because, for example, the proportion of employees in manual work changes or because, for example, new technologies/risk-control mechanisms are introduced that make working safer. Ideally, one would like to quantify all these changes by creating a 'risk index' $R(t)$ that can be associated with every year of experience, so that the total losses can then be modified for each year by multiplying by $R(t^*)/R(t)$, much in the same way as we did for claims inflation and exposure adjustments.

Underwriters will be looking for such indices (more or less informally), but their exact quantification will be difficult: in the case of commercial insurance and reinsurance, they are part of the negotiations between brokers and underwriters.

In the case of personal lines insurance, the sheer amount of data makes it more feasible to assess the risk posed by each policyholder, and this takes us into the realm of rating factor analysis, which will be addressed in a subsequent unit. Going back to our initial example, let us assume for simplicity that there have been no significant changes in the risk profile.

6 Even better, one should consider the *exposure-weighted average*, which is generally more stable as it gives more weight to years with larger exposures (see Chapter 11 for the definition). In this case, it wouldn't make much difference, as the result would be £2.12M.

1.3.1.5 Changes for Cover/Legislation

Adjustments need to be made to past claims experience if significant changes of cover are made, for example, if certain types of claims were previously covered and now are not. However, one should not automatically adjust the historical data to eliminate large and unusual movements if this means excluding 'uncomfortable' past claims.

Also, adjustments may have to be made for changes in legislation. As an example, the Jackson reforms for liability cases prescribe fixed legal fees for claims below a certain amount, in an attempt to reduce litigation costs: if the liability policy you're trying to price includes legal costs, you will need to look back at all past claims and see what the retrospective effect of this reform would have been to predict your liabilities for next year.

1.3.1.6 Other Corrections

Adjustments are also made to correct for unusual experience (say, for large losses that have happened everywhere in the market but were not experienced by the client in the 10-year period), for different weather conditions, for currency effects and for anything else that we think is appropriate. These corrections are common practice among underwriters, but one should always bear in mind that these corrections, although they purport to increase the reliability of our estimates, also add various layers of uncertainty and of arbitrariness to our estimates. Actuaries are obliged to clarify what assumptions their estimates are based on and to communicate the uncertainty with which they make certain adjustments.

Bringing all considerations together and going back to our initial example, we have seen that after taking into account IBNR, claims inflation and exposure adjustment, and disregarding other corrections such as risk profile changes and large losses, we estimate for 2025 gross losses (that is, from the ground up, without taking any retention into account) of

Expected gross losses ~ £2.14M

However, for the premium to be calculated, we are only interested in the portion of the losses that are ceded (transferred) to the insurer – which is why, at this stage, an analysis of the ceded (insured) losses is required. This is normally not an easy task without the appropriate mathematical techniques, but in our illustrative example we will be able to address it without calculations.

1.3.2 Risk Costing (Ceded)

As we have stated at the beginning, we only insure the first £10M on each loss and in aggregate for each year.

In our case, that doesn't affect the expected losses, because the largest total revalued loss (before exposure adjustment) is £2.47M (in 2021), which is way below £10M in aggregate and therefore obviously on an individual loss basis as well. Therefore our estimate for the expected ceded losses is the same as that for the expected gross losses (although of course a full identity will not strictly be true if losses above £10M are at all possible):

Expected ceded losses ~ £2.14M

1.3.3 Determining the Technical Premium

To move from the expected ceded losses to the technical premium, that is, the premium that should be charged for the risk from a purely technical point of view based on the company's objective, we need to consider the costs incurred by the insurer for running the business and specifically underwriting the risk and handling the claims; consider the cost of any reinsurance arrangements; consider the income that derives from being able to invest the premiums while waiting for claims to occur and then being settled; consider the profit that the firm needs to make to guarantee sufficient return on the capital provided by the investors; finally, when the business is sold through an intermediary such as a broker, a percentage of the premium needs to be paid to them.

Chapter 19 will address the problem of building a premium formula which takes into account all these elements. For the purpose of our simple example, however, we will assume that the company embeds all these expenses (except for the broker's fees, which need to be treated separately and that we will ignore here, assuming that the business is sold directly to the client) into a quantity called the target loss ratio.

In simple terms, the **loss ratio** is the ratio between losses and premium. However, the devil is in the details, and there are various definitions of loss ratio depending on what you exactly put into the 'losses' term and what you put into the 'premium' term.

For the purpose of our example, the **expected loss ratio** can be defined as the ratio between the expected ceded losses and the premium charged:

$$\text{Expected loss ratio} = \frac{\text{Expected Ceded Losses}}{\text{Premium Charged}}$$

The target loss ratio can be defined as the expected loss ratio that you obtain when the premium charged equals the technical premium. Therefore, if the target loss ratio is set by the company, we can use it to calculate the technical premium:

$$\text{Technical Premium} = \frac{\text{Expected Ceded Losses}}{\text{Target Loss Ratio}}$$

In our example, the target loss ratio is 75%. Therefore, the technical premium is:

$$\text{Technical premium} = \frac{£2,136,821}{0.75} \sim £2.85\text{M}$$

1.3.4 Commercial Considerations and the Actual Premium

The premium that the underwriter is going to charge might be quite different from the technical premium calculated above because of commercial considerations: it could be less because of the desire to retain old business/acquire new business, or it could be more to signal that the insurer is not happy with a particular risk or with a particular level of retention by the insured.

On the other hand, this flexibility might be limited by the need for the underwriter to achieve a certain return on capital for the portfolio for which they are responsible and other company guidelines.

Market considerations may also lead to decline the underwriting of the risk, if the market conditions (what we will later on call the 'underwriting cycle') are such that it is only possible to underwrite it at a loss or at a profitability level that is unacceptable to the insurer.

1.3.5 Limitations of this Elementary Pricing Example

As mentioned at the beginning, this elementary pricing exercise is only a simplified version of what pricing looks like in reality.

A more sophisticated process will be illustrated later on in the book. For now, it is worth noting what the main limitations of this simplified approach are. We have already hinted at some of them in the previous sections, such as the absence of an allowance for potential large losses. The main limitation, however, is that the calculations are based on aggregate information, that is, the total claims reported in each year, ignoring the information on the individual claims that make up the aggregates.

This contrasts with the modern way of looking at risk, which is to consider separately the frequency and the severity of claims. This has several advantages. For example, if the total losses exhibit an increasing trend over the years, it enables us to identify whether this derives from an increase in the number of claims or in the nature (and therefore the severity) of their claims or as a combination of the two elements. It also gives us a much more solid basis on which to estimate the future volatility of the loss experience.

Also, the absence of individual loss information makes it impossible to properly price insurance policies with non-proportional elements – for example, policies that pay only the portion of the claim that lies above a certain retention threshold, such as an excess or a deductible (note that the presence of an excess is indeed quite common, for example, in comprehensive motor policies). If we have only aggregate claims information, there are certain things we cannot do, such as analysing the effect of changing the retention threshold or estimating the expected losses to a layer of insurance.

Despite these limitations, this example is useful because it shows a simple version of burning cost that is not too different from what underwriters and actuaries often do. Even when more sophisticated techniques are used, a burning cost is a good exercise to go through as it provides a sense check on any subsequent actuarial analysis.

1.4 Mixing Exposure and Experience Rating

We now have two estimates for the technical premium, one based on exposure (£1.86M – Section 1.2) and the other based on experience (£2.85M – Section 1.3.3). Which one are we supposed to use?

It depends on which one we trust the most. The exposure rating estimate may be based on a larger and more solid set of portfolio and market data and tradition, but it will never be fully relevant to the client and the specifics of its risk. On the other hand, the experience rating estimate may be based on limited data and may not take adequately into account changes in the risk profile over the years.

A third option is to select a technical premium that takes into account both estimates, perhaps by taking a weighted average of the two, where the weight given to the two estimates depends on how much we trust them. If Z (a number from 0% to 100%) represents the relative level of credibility that we give to the experience rating estimate, we can write down the selected technical premium as:

$$\text{Tech Prem (selected)} = Z \times \text{Tech Prem (experience)} + (1 - Z) \times \text{Tech Prem (exposure)}$$

For example, if the credibility that we give to our experience rating estimate in Section 1.3.3 were 85% (we will gloss over for the moment how this number can be determined), the selected technical premium would be:

$$\text{Tech Prem (selected)} = 85\% \times £2.85\text{M} + 15\% \times £1.86\text{M} = £2.70\text{M}$$

This is an example of the application of *credibility theory*, a popular technique among actuaries for mixing a client estimate based on its historical loss experience with a portfolio/market estimate (in our case an exposure rating estimate), balancing the relevance of the portfolio/market estimate against the uncertainty of the client estimate. Naturally, this approach is only useful insofar as (1) a linear combination is a sensible way of combining the two estimates of the technical premium (it often isn't) and (2) we have a robust methodology to calculate the *credibility factor* Z (we often don't). Chapter 25 will provide more information on the credibility approach and its limitations.

Credibility theory, however, is not the only way to mix experience and exposure rating. Experience rating is rarely *purely* based on experience. For example, since burning cost is better at capturing the behaviour of small (attritional), typical losses than that of large, unusual losses, it may include mechanisms to limit the burning cost calculations to losses below a certain threshold while using portfolio/market experience (or exposure rating) above that threshold. So in this book the term 'experience rating' will often be used for any type of pricing that is based on the historical loss experience *to a significant extent.*

1.5 High-Level Pricing Process

Although the examples shown in Section 1.2 to 1.4 are very basic, they are sufficient to enable us to outline the main elements of the pricing process. At a very high level, the pricing process (whether through exposure rating, experience rating or a combination of the two) will consist of these steps, which are also illustrated diagrammatically in Figure 1.7:

1. First, familiarise with the risk and (where applicable) with the insured's business. Then,
2. given information on claims, exposure and cover …
3. … estimate the cost of taking on the risk …
4. … take into account the related cost of running the business (underwriting, claims management, payrolling, etc.) and investment income …

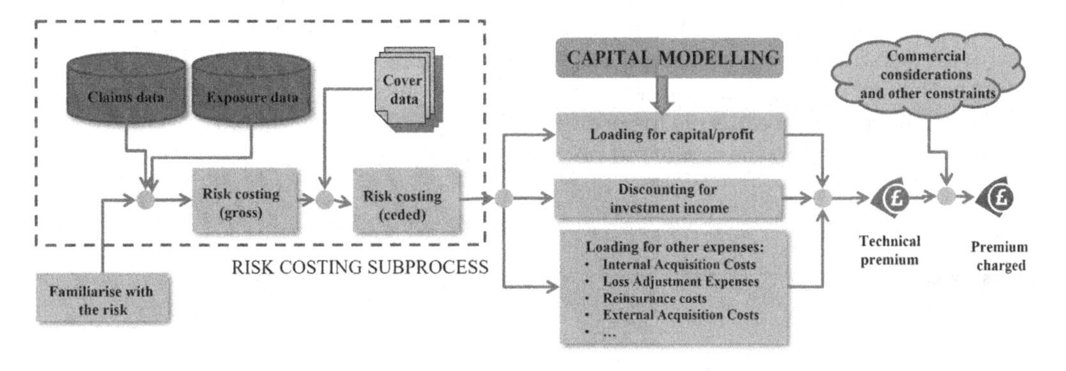

FIGURE 1.7
The high-level pricing process. The pricing exercises discussed in Sections 1.2–1.4 can be seen as examples of this framework, albeit in a simplified form. Not all elements were used for all examples: for example, claims data were not used for exposure rating, and the 'Loadings for capital/profit' step was implicitly incorporated in the target loss ratio.

5. ... add an allowance for profit (which is informed by capital considerations and requires inputs from the capital modelling function and other company functions), thus obtaining the technical premium ...
6. ... and based on the technical premium and commercial considerations, determine the actual premium to be charged.

Figure 1.7 will hopefully help see risk costing – the activity in which actuaries are involved most closely – in the context of the overall pricing process, which is ultimately a business decision as to whether to underwrite a risk, and at what terms.

The main contribution of actuaries and statisticians to pricing has been to transform the risk costing process from a back-of-the-envelope calculation or sometimes even a guessing game based on underwriters' intuition to a technically sound process based on data and modelling – or at least, that is the case when data are sufficient and sufficiently accurate!

It should also be observed that the pricing process described above is itself part of a more general control cycle in which the adequacy of premiums is monitored, and there are strong interactions between the pricing function and the reserving and capital modelling functions. We will look at several of these issues in Chapter 20.

1.5.1 Portfolio Pricing Process

The details of the pricing process outlined in Figure 1.7 will look different depending on the type of insurance contract that is being priced, the technique being used and the type of client. We have already seen how risk costing looks different depending on whether it is based on the exposures only (exposure rating), on the historical loss experience (experience rating), or a mixture of the two (credibility rating, or other ways of mixing experience and exposure rating).

We have also hinted at the fact that pricing can be done for a single risk (as our employers' liability policy) or for a portfolio of risks. We have seen that in the context of exposure rating: in order to be able to calibrate the rates and the adjustment for exposure rating, a portfolio of similar risks must be portfolio-rated. Another example of portfolio pricing can be found in treaty reinsurance: the insurance company may pay another insurance company (the reinsurer) to take on a proportion of all its employers' liability losses (proportional reinsurance) or all its losses above a certain threshold and up to a certain limit (excess-of-loss reinsurance). Personal lines pricing, too, cannot be done on the basis of experience rating alone, and the setting of the rates therefore depends on portfolio-level experience rating with techniques such as generalised linear modelling (Chapters 26 and 27).

Most of the techniques for account pricing for single risks, such as the burning cost analysis of which we saw an example in Section 1.3.1 and stochastic modelling (Chapters 13 to 17) can be transported to the context of portfolio pricing.

1.6 Questions

1. *The 'pub quiz'.* MerryPints is a pub company, owning a large chain of pubs, three breweries and a small building (used as headquarters and for administration purposes) in the country of Freedonia (currency: Freedonian dollars, FRD). MerryPints wants to buy Commercial Property cover for its pubs. The policy year is

due to incept on 1 January 2023, and the fiscal year starts on 1 April 2023. The cover is from the ground up and for the full value of the property.

The historical exposure is as follows:

Fiscal year incepting on:	Number of pubs at start of fiscal year
1/4/2018	453
1/4/2019	470
1/4/2020	498
1/4/2021	502
1/4/2022	510
1/4/2023 (estimate)	Same as previous year

The claims experience as at 30 September 2022 is as follows:

Policy year incepting on:	Total paid (FRDk)	Total O/S (FRDk)	Recoveries (FRDk)
1/1/2018	581.56	0.00	28.37
1/1/2019	888.29	1.46	2.05
1/1/2020	486.59	6.14	0.59
1/1/2021	436.02	172.28	0.88
1/1/2022	75.95	285.19	0.00
Grand total	2,468.41	465.08	31.88

The historical amount of each claim is split into a paid component, an outstanding (O/S) component and recoveries from a third party.

i. Calculate the total incurred claims amount observed for each policy year.
ii. Assuming a 3.5% claims inflation for property claims, an internal expense ratio of 10% of the expected losses, and a profit of 18% as a percentage of the premium, estimate the technical premium for policy year incepting 1 January 2023 using burning cost analysis. State any assumption you make.

You subsequently receive information according to which the claims experience summarised in the table above only includes claims related to pubs, and you receive assurance that no claims were made from 1 January 2018 onwards involving breweries and the headquarters.

iii. On the basis of what you already know about the data and in the light of this additional piece of information, explain the limitations of the data and its consequences on the premium calculations.

Having looked at your burning cost analysis, your underwriters suggest that you compare your results with the results obtained using exposure rating, with the rates recommended in the underwriting guidelines:

Type of building	Premium rate (as a % of the insured value)
Pub	0.28%
Brewery	0.12%
Office building	0.40%

The underwriters also provide you with this breakdown for the exposure as at 1 April 2022.

Insured value	Type of building	Number of buildings
FRD 200k – FRD 399k	Pub	110
FRD 400k – FRD 599k	Pub	170
FRD 600k – FRD 799k	Pub	180
FRD 800k – FRD 1,000k	Pub	50
FRD 1,750k	Office building	1
FRD 25,000k	Brewery	1
FRD 42,000k	Brewery	1
FRD 45,000k	Brewery	1

iv. Making the necessary assumptions, calculate the technical premium using exposure rating.

Part I

Background and Context

2

Insurance and Reinsurance Products

To learn how to do pricing, actuaries need a good working knowledge of the products that they'll have to price. There are several aspects to this: one aspect is the *knowledge of the risk underlying a product*. For example, household insurance is an insurance product, and to price household insurance, one needs to know that it protects against perils such as fire, flood or theft, and needs to know what the factors are that make one house riskier than another.

Another aspect is the *knowledge of the technicalities of the policy* and how this affects the risk. The most obvious elements to be considered are the coverage modifiers, such as the deductible and the limit. Of course, a motor insurance that pays only claims of more than £500 is not the same as a policy that pays the full amount. As a pricing actuary, you need to be able to quantify the effect of increasing or decreasing the deductible (or the limit).

Yet another aspect, which is quite wide, is *the environment your business operates in*. This includes just about everything from the way the insurance market is regulated, to the participants in the marketplace, to the general economy, to your own company.

In this chapter, we will look at the first of these aspects: the products available and their corresponding risks, and we will acquaint ourselves with the classes of business normally underwritten by insurance companies, syndicates and others. We will deal with the other aspects in Chapters 3 through 5.

2.1 Classification of General Insurance Products: General Ideas

There is not a natural or standard taxonomy of insurance products in the same way that, for example, the phylogenetic nomenclature provides a natural taxonomy of animals (as opposed to classifying animals by, e.g. size or territorial distribution or how funny they look).

Insurance classification depends very much on the perspective from which we look at insurance products, and there are therefore as many different classification systems as there are legitimate perspectives. Regardless of which one is used, there are unsatisfactory aspects to it: products that should be together are split and products that should be split are brought together artificially. As an example, should 'aviation liability' be classified in the same broad category of liability along with employers' liability or with 'hull insurance' based on the broad category of aviation?

DOI: 10.1201/9781003168881-3

In the rest of the chapter, we will look at a few different ways to classify insurance; many more are possible. The ones that we have chosen to illustrate are among the most useful and will also help us understand different aspects of today's insurance world.

A more detailed description of the individual insurance products can be found online in the Appendices section of the book's website (Parodi, 2016).

2.2 Products by Category of Cover

According to this classification method, insurance products can be classified into four broad categories: liability, property, financial loss and fixed benefits, depending on the type of losses that are indemnified and the type of compensation itself. This is used, for example, by the UK actuarial profession in their training material for General Insurance Pricing (Institute and Faculty of Actuaries 2010).

2.2.1 Property Insurance

Property cover indemnifies the insured for loss or damage to the insured's property, such as one's house or one's car. Therefore, property insurance is probably the most straightforward to understand: your mobile phone charger, which was left plugged in, overheats and starts a fire in your house; your house burns down; you make a claim on your home insurance policy; your insurer (hopefully) pays you so that you can rebuild your house from scratch.

As the example above shows, although there are always a number of contractual subtleties that need to be taken into account, you do not need to be a lawyer to understand the core of how property insurance works – actually, you do not even need to know much about the legal system of the country where the loss occurs: the principles behind property insurance are roughly the same across all territories.

Regardless of where and when you consider the birth of insurance to have occurred – in the provision made in the Code of Hammurabi in 1790 BC for compensation to merchants robbed by brigands; around 350 BC in Greece with marine contracts in which a loan on the ship and its cargo was waived in the case of a loss; in the first stand-alone contracts of insurance written in Genoa in the fourteenth century – in all known accounts (see, for example, Thoyts, 2010), this birth was in the context of insuring property for the purpose of removing some of the uncertainties of commercial enterprise.

In particular, the insurance of ships with their cargo – for centuries, the most important type of insurance and that around which the London Market was born – can be seen as a form of property insurance cover. It is very difficult to imagine marine commerce having expanded to the extent it did without being adequately protected by insurance – who would dare risk a whole ship with its cargo sending it on a perilous journey without the financial security provided by insurance?

Even today, property insurance is pervasive and makes it possible for all sorts of expensive equipment to be employed in all sorts of undertakings: ships, trains and planes for commerce and travel; satellites for communications; machinery in industrial plants; the industrial plants themselves; large construction projects; large scientific projects and so on. Much of what goes on today and goes under the heading of 'technological progress' would be seriously stifled were it not for insurance in general and for property insurance in particular.

Despite its simplicity, there are – as mentioned above – some legal subtleties that should be kept in mind. These are addressed in the next section.

2.2.1.1 Legal Framework

Principle of Indemnity. This is one of the leading principles of insurance in general and states that the purpose of insurance is to reinstate (at least partially) the insured to the financial position (s)he had before the loss occurred. The principle of indemnity is relatively straightforward when applied to property insurance: compensation must be provided to enable you to replace the lost good or put you back in the financial position you were before losing the good. This implies sometimes that you will not be paid the market price for a house that burned down, but you will get enough money to enable you to rebuild it. The principle of indemnity also implies that you should not be able to make money out of insurance by getting more compensation than what you insured was worth.

Insurable Interest. The principle of insurable interest means that you are not allowed to wager on the possibility of losses for properties that you have no interest in, such as the house of your neighbour down the road, which you might suspect will be damaged in the next big flood.

Exclusions. The main question about property insurance is, what are the risks we are covering against? Originally, many policies covering commercial properties did so against fire risk, which was seen as the main type of risk to be concerned about (and certainly the Great Fire of London in 1666 created a large market for fire insurance in the United Kingdom), but this was gradually extended to cover other types of risk such as explosion, arson, flood and earthquake.

Modern policies have gone a significant step beyond this and are now 'all risks', which means that they focus on the outcome (the property damage) rather than the cause (the peril): anything that causes damage will be reimbursed. These policies, however, contain exclusions against risks that the insurer does not wish to deal with, normally because they have a systemic nature: they would cause damage to too many of its insured for the insurer to be able to cover all of them at the same time. Examples of these are exclusions for war, civil war and nuclear activity.

In addition to damage, a separate cover for *theft* is often provided as a separate section of a commercial policy package.

Underinsurance. Another subtlety regards the amount of compensation that the insured is entitled to and under what circumstances. As an example, if the insured has underestimated the total value of the property insured and files a partial loss claim, her/his claim may be reduced in the same proportion as the sum insured bears to the actual total value. Note that the insurer will also want to avoid *over-insurance* (the overstatement of the value of the property insured) because this leads to moral hazard: it would be better for the insured to make a claim rather than keeping the property.

General Average. This compensates the insured in the case where the insured was forced to dispose of some of the insured goods to salvage the overall enterprise. A commonly cited case of this is that in which the insured jettisons part or all of the cargo to prevent the ship from capsizing.

2.2.1.2 Pricing Characteristics

From the point of view of pricing, property policies are characterised by short-tail claims (losses are usually immediately discovered) and (at least in the case of buildings and plants) a relatively low frequency of significant claims. The severity of the claims is usually capped by the insured value of the property, or more accurately by what the underwriter or the risk engineer estimate to be the maximum possible loss (MPL). The accurate pricing of property often requires the use of catastrophe models and of market severity curves (the so-called 'exposure curves'), as addressed at length in Chapter 21 (property exposure rating) and Chapter 24 (catastrophe modelling).

2.2.1.3 Examples of Property Policies

There are a large number of covers that fall under the heading of property insurance. Here are some examples:

- Home insurance building and contents
- Commercial property
- Marine hull
- Marine cargo
- Aviation hull
- Aviation cargo
- Motor own damage
- Livestock
- Bloodstock
- Crop insurance
- Construction All Risks/Erection All Risks
- Goods in transit

We will meet some of these covers again in subsequent sections.

2.2.2 Liability Insurance

Liability insurance indemnifies the insured for legal liability related to bodily injury, property damage, and consequential financial loss caused by the insured to a third party because of negligence or other types of tort or breach of statutory duty. Legal expenses are normally covered as well, both for the defence of the insured and as claimant's legal expenses, normally considered as part of the indemnity.

2.2.2.1 Legal Framework

Because it is meant to cover legal liabilities, modern liability insurance presupposes a well-developed legal environment and, for this reason, it is a more recent development, dating back to the late nineteenth century, although early examples (see Parsons, 2002) can be found in horse carriage third-party liability insurance in Paris (1825) and employers' liability for steam boiler explosions in the UK (the Steam Boiler Assurance Co. was founded in 1858).

As argued, for instance, in the book by Thoyts (2010), it is also quite possible that the relevant legal framework was able to develop exactly because of the existence of a more sophisticated insurance industry, in what can be described as a virtuous (or vicious, depending on the standpoint) circle.

2.2.2.1.1 *Thou Shalt Not Injure Thy Neighbour*

The legal framework under which liability claims operate is that of *tort law*. Tort comes from the Latin word for 'twisted', the idea being that something has gone wrong and has to be straightened out – or put right. In practice, this means that people must be indemnified for the wrongs of which they were victims.

There are many categories of tort, but the most important for liability insurance is that of *negligence*, that is, the idea that someone had a duty of care towards their neighbour and has somehow failed to exercise it. When this type of tort was introduced, it was to attempt to repackage the evangelical principle 'Thou shalt love thy neighbour as thyself' (Matthew 19:19) into a form that could be used by lawyers and judges (albeit, perhaps, with a different focus and purpose). As clearly stated in the argument by Lord Atkin for the legal case that started it all, *Donoghue v. Stevenson* (House of Lords [1932] UKHL100), indeed

> The rule that you are to love your neighbour becomes in law, you must not injure your neighbour [...]. You must take reasonable care to avoid acts or omissions which you can reasonably foresee would be likely to injure your neighbour.

The circumstances from which the ruling by the House of Lords stemmed could be straight from an episode of *Horrible Histories*.[1] One Glaswegian lady, Mrs. Donoghue, went to a bar in Paisley, seven miles out of Glasgow, and had the misfortune of drinking from a bottle of ginger beer that contained a decomposed snail. She fell sick and demanded compensation from the beer manufacturer. However, compensation was at the time provided in the case of breach of contract or fraudulent acts on the part of the producers. No contract existed between Mrs. Donoghue and the manufacturer, and the bottle manufacturer had not acted fraudulently, and this pretty much seemed to exhaust the lady's possibilities of legal actions. Enter Walter Leechman, a lawyer who had already sued (unsuccessfully) another drink manufacturer because of a dead mouse present in their own version of ginger beer (one would be forgiven for thinking that ginger beers must have been zoos at the time). The case made its way through all the levels of Scottish law and landed in the House of Lords, and the outcome was Lord Atkin's decision mentioned above.

The existence of legal liability due to the law of tort created the need for insurance products that would protect manufacturers against the legal consequences of their negligence, including the costs incurred in legal defence. A case like that described above, for example, would be covered today by a *product liability* policy. The law of tort, however, is much more general than that and applies, as we have mentioned, to many types of relationship, such as that between employers and their employees (which is covered by an *employers' liability* policy in the United Kingdom), drivers and their victims (covered by motor third-party liability) and the like.

Once the general principle was established, the law of tort of negligence could then be used to seek compensation for negligence in a number of widely different contexts, such as in the workplace (employers' liability), in the delivery of professional services (professional indemnity) or on the road (motor third-party liability).

More examples of liability insurance policies are given below. It is worth mentioning that, unlike property insurance, the characteristics of liability insurance depend very much on the jurisdiction where it provides cover because it is inextricably linked to its legislation.

1 A pointer for the non-Brit: http://horrible-histories.co.uk/.

2.2.2.1.2 *What Does Liability Insurance Actually Cover?*

The underlying principle is again that of *indemnity*: reinstating the insured to the position that the insured was in before the loss occurred. This may seem more complicated than in the case of property, as the loss may involve not only damage to tangible property but also bodily injury, pain and suffering, the inability to cater for oneself and the like. A price tag must therefore be attached to such things as good health that are not financial in nature or that cannot be simply bought with money.

However, it should be noted that it is *the insured* and not *the victim* that must be reinstated to her/his initial financial position: the victim receives from the insured compensation for a number of heads of damages (see below), and the insurer indemnifies the insured for the compensation that the insured had to pay to the victim, thereby restating the insured to the financial position that (s)he had before indemnifying the victim, although very often the insurer indemnifies directly the third party and then, if applicable, recovers the deductible from the insured.

Depending on the policy, liability insurance may cover heads of damages such as

- Damage to third-party property
- Bodily injury (cost of care, loss of earnings, pain and suffering)
- Financial loss
- Pain and suffering
- Punitive or exemplary damages that are not compensatory in nature (United States in particular)
- Claimant's legal costs

Note that some types of liability insurance are compulsory, to ensure that the tortfeasor has access to sufficient funds to compensate the victim as necessary, so that the latter has the possibility of seeking redress. Motor third-party is compulsory in most developed countries for this reason.

2.2.2.2 Pricing Characteristics

From the point of view of pricing, these policies are characterised by long-tail claims (discovery may not be immediate and, after discovery, settlement may take a long time), whereas the frequency varies greatly depending on the type of liability.

The indemnity limit under the policy is, in some territories, unlimited (for example, motor third-party liability for bodily injury in the United Kingdom) or in any case very high, without a natural limit as in the case of property.

As mentioned previously, legislation and legal precedents have a strong effect on the compensation awarded and must be taken into account in a pricing exercise.

2.2.2.3 Examples of Liability Insurance

- Public liability
- Product liability
- Employers' liability
- General liability (US term for public liability and product liability)
- Directors' and officers' liability
- Professional indemnity
- Motor third-party liability

- Aviation liability
- Marine liability
- Environmental liability

2.2.3 Financial Loss Insurance

Financial loss insurance indemnifies the insured for financial losses arising from the perils covered by the policy. Therefore, the *principle of indemnity* still holds, but the idea here is that the loss is in itself financial, unlike the case of property, where the loss typically involves a physical object (a building, a ship, a laptop ...) that has a quantifiable value, and unlike liability, where the loss is the amount required to satisfy the duty to compensate a victim for the effects of, for example, bodily injury or property damage, which one can then attempt to quantify.

A typical example of financial loss is the loss of profit resulting from the fact that a fire has damaged a manufacturing plant and has caused an interruption in the production activity. This is normally covered as part of commercial property policies under the heading 'business interruption'.

2.2.3.1 Examples

- Business interruption (sold along property insurance)
- Trade credit
- Mortgage indemnity guarantee
- Fidelity guarantee

2.2.3.2 Pricing Characteristics

From a pricing point of view, this type of cover is often characterised by a strong dependency on systemic factors: for instance, trade credit risk and mortgage indemnity guarantee are likely to be big in both frequency and severity during a recession; another recent example is provided by business interruption claims as a result of the COVID-19 pandemic.

2.2.4 Fixed Benefits Insurance

Fixed benefits insurance provides a fixed amount of cover for a given injury (e.g. loss of limb) or loss of life as a result of an accident. The usual principle of indemnity, therefore, does not (and cannot) apply.

Fixed benefits are more common in life insurance than in general insurance but by legal definition fixed benefit insurance is classified as general insurance; it is found sometimes in isolation but most often as part of another policy, such as private motor or travel insurance.

2.2.4.1 Examples

- Personal accident
- Sickness insurance
- Health cash plans (not, however, part of general insurance – we mention it here because it is one of the categories of insurance used by the insurance regulator in the United Kingdom, the Financial Conduct Authority [FCA])

2.2.4.2 *Pricing Characteristics*

From a pricing point of view, the main characteristic of this risk is that losses are of fixed value given the injury incurred, and that puts defined constraints on the loss models, especially on the way we deal with loss severity.

2.2.5 'Packaged' Products

It is important to note that most insurance policies – especially personal lines policies – will bundle many of these cover categories together. For example,

- A comprehensive motor policy will pay for both liability to third parties *and* own property damage (for example, when you accidentally bump your car into the wall of your house) *and* (possibly) include some fixed benefits if you hurt yourself (Personal Accidents Insurance).
- Travel insurance is even more complicated, as it includes elements of property (such as baggage), liability (such as personal liability, legal expenses), financial loss (such as loss of money) and fixed benefits (personal accident) plus elements of life insurance and cancellation or curtailment of the travel.

2.3 Products by Type of Customer

Insurance can also be classified by type of customer, that is, personal lines (for individuals), commercial lines (for companies) and reinsurance (for insurers).

Personal lines and commercial lines business are together called 'direct' business, in the sense that they are non-reinsurance.

This classification is interesting for actuaries because the type of techniques used in personal lines insurance, commercial lines insurance and reinsurance can be dramatically different. Also, the type of work and the type of work environment can be quite different.

2.3.1 Personal Lines

In personal lines, the customer is a private individual, like you and me. As such, policies are highly standardised and mass-produced (sometimes in millions, if for example you are a large insurer writing a decent share of the motor market in the United Kingdom).

2.3.1.1 *Examples*

- Private motor
- Home insurance (building and contents)
- Extended warranty
- Private marine and aviation
- Travel insurance

2.3.1.2 *Pricing Characteristics*

Personal lines insurers often deal with a large mass of data, allowing the deployment of sophisticated statistical techniques to analyse the customers' experience. Specifically, pricing techniques can be developed that classify customers with a rather fine structure: the selection of appropriate factors according to which differentiating the price offered to different customers (the so-called 'rating factors') indeed becomes the main analytical challenge for personal lines insurance.

A key characteristic of pricing for personal lines insurance is that risks (individual policies) are not rated on the basis of individual experience but by looking at a portfolio of risks and then modifying the rate depending on the rating factors as discussed above.

Another characteristic of pricing is the availability of massive data sets, which by their sheer size make it necessary to use more sophisticated IT systems than those used in commercial lines insurance. As an example, it is not uncommon for an actuary in commercial lines insurance to price a risk on the basis of data stored in a spreadsheet, whereas professional database systems (such as Oracle or SQL Server) are the norm in personal lines.

Another interesting characteristic of personal lines insurance is that owing to the sophistication of statistical work that can be done in personal lines, actuarial teams will have both actuaries and professional statisticians in their make-up.

2.3.2 Commercial Lines

Unlike personal lines, the policyholder of commercial lines insurance is typically not an individual but a company or an organisation (such as a county council).[2] Although an organisation and an individual are quite different entities, an organisation has, like an individual, the need to protect its property and to protect its assets against legal liability and financial losses, also in the interest of its shareholders.

The main difference is that the insurance needs of an organisation are inevitably more complex: for example, as a legal entity, an organisation (and possibly also a sole trader) has a duty of care both towards the public (hence the need for public, product and environmental liability) and towards its own employees (employers' liability).

Also, especially for large corporate organisations, a commercial policy is itself what in personal lines insurance would be considered a portfolio of risks: for example, a commercial motor policy may cover a large fleet of cars, an employer's liability policy may cover the employer for its liability towards a large number of employees, and a commercial property policy may cover a long list of properties spread worldwide. As we will see shortly, this has consequences on the way one goes about pricing commercial policies.

2.3.2.1 Examples

- Commercial motor
- Commercial property
- Employers' liability
- Public liability
- Product liability
- General liability (US term for public/product liability)
- Environmental liability
- Professional indemnity
- Directors and officers
- Business interruption
- Trade credit
- Construction
- Marine
- Aviation

2 The exception is that of a sole trader.

2.3.2.2 Pricing Characteristics

The crucial difference between commercial lines and personal lines pricing is that in commercial lines it is often possible to model each individual policy (for example, a commercial motor policy for a company) based at least partly on the historical loss experience of that specific policy. We're saying 'partly' because the information of one particular policy can be combined with portfolio information (i.e. information derived from similar policies) in various ways, which will be described later on in the book (see, for example, a discussion on credibility in Chapter 25 and on the use of market/portfolio information in Section 16.5 and Chapters 21 and 22).

As for the availability of data, it very much depends on what classes of insurance we are dealing with: motor and employers' liability are normally data-rich, whereas the volume of losses for property or professional indemnity may be quite limited.

Rating factors, the most important tool in the box for personal lines insurance, are also used in commercial lines, but more sparingly, owing to the relatively small number of policyholders. When they are used, the number of factors is itself limited.

Note that commercial insurance also includes sole traders and small businesses for which policies can be very small and may be similar to personal lines.

2.3.3 Reinsurance

In reinsurance, the policyholder is itself an insurer or even a reinsurer in the case of retrocession insurance.

Reinsurance is normally subdivided into treaty reinsurance and facultative reinsurance. In *treaty reinsurance* there is an agreement between insurer and reinsurer to insure, for example, all policies written in a given year/period. In *facultative reinsurance*, each insurance contract is negotiated and the risk transferred on its own merits. The distinction is relevant to us because most actuarial pricing work is done on treaty reinsurance.

Reinsurance can be further subdivided into *proportional reinsurance* (in which the insurer cedes to the reinsurer a predefined proportion of the premium and the claims of each risk) and *non-proportional reinsurance* (in which only insurance losses above a certain threshold are passed on to the reinsurer).

2.3.3.1 Examples

The examples listed below will be defined and explained in Chapter 3.

- Proportional reinsurance
  - Quota share
  - Surplus reinsurance
- Non-proportional reinsurance
  - Excess of loss (XL)
  - Aggregate XL
  - Catastrophe XL
  - Stop loss

2.3.3.2 Pricing Characteristics

Non-proportional reinsurance business has to deal almost by definition with little data, especially when the threshold above which an insurer cedes its losses is high. Special techniques such as exposure rating have been developed to allow estimating expected losses when there is no available experience data.

2.4 Standard Classification Schemes

Besides the generic classification schemes that we have discussed in Sections 2.2 and 2.3 we should mention that a number of classification schemes have been adopted by various organisations for reporting purposes. Examples of these are:

- The Prudential Regulation Authority (PRA) classification (Annex 11.3 to the Interim Prudential Sourcebook for Insurers), used in the transaction of UK business.
- The Lloyd's of London classification (which, following a longstanding tradition, subdivides all non-life insurance into marine and non-marine!).
- The Solvency II classification (the regulatory framework for insurance developed by the European Union).

The PRA classification is described in some detail in the Appendices section of the book's website (see Parodi, 2016), which also includes an informal description of all the insurance products in such classification scheme.

2.5 Questions

In responding to the questions in this chapter, please note that most of the information required is not in the chapter and you might have to do some personal research.

1. Find out how non-life insurance obligations are classified according to
   a. Lloyd's of London
   b. The Solvency II regulatory regime being proposed by the European Union by doing personal research. (*Hint: for Solvency II, you can refer to the Quantitative Impact Study 5 'QIS5' or any updates that might be available at the time you are reading this.*)
2. Can you explain the difference between *employers' liability insurance* and *workers' compensation insurance*? Name three countries where employers' liability is sold and three countries where workers' compensation is sold.
3. Find out who the main specialist suppliers of credit insurance are, then pick one of them and make a list of the products they offer.
4. Find out what the policy 'IT All Risks' is and what it covers. How would you classify it according to the PRA scheme?
5. What is *specie* insurance?
6. What is a *bankers' blanket bond* policy?
   a. Why is it called a 'bond'?
   b. Why is it called 'blanket'?

    c. What does it cover exactly? (Find out specific policy summaries or articulate descriptions from actual insurers.) Try and capture the overall coverage in a matrix like those in Section 2.2.5.
    d. Why do you think there is a need for it?
    e. How would you classify it in terms of cover?
    f. What do you think are the main risks involved?
7. What is *jewellers' block* insurance?
    a. What does it cover exactly? (Find out specific policy summaries or articulate descriptions from actual insurers.)
    b. Why do you think there is a need for it?
    c. How would you classify it in terms of cover?
    d. What do you think are the main risks involved?
8. What do these two aviation insurance policies cover?
    a. Loss of licence
    b. Loss of use
9. Doing some personal research, explain what air traffic control liability is.
10. It is stated in the text that the principle of indemnity cannot be applied to the case of personal accident because loss of life, health or limb cannot be quantified. How is it that it can be applied to the case where these same losses are caused by a third party, in liability insurance?
11. Describe what the legal framework of liability insurance is and how the principle of indemnity applies to liability insurance.
12. Explain briefly what these liability insurance products cover: employers' liability, public liability, aviation liability and professional indemnity, and give two examples of claims for each of these products.
13. Give an example of a personal lines insurance liability product, and explain what it covers.

3

The Cover Structure

> [Definitions] don't help with clarity. They only lead to a pretentious, false precision [...] and not [to] genuine clarity. For this reason, I am against the discussion of terms and definitions, but I'm rather for plain, clear speaking.
>
> **Karl Popper (interview)**

In very simple terms, an insurance policy is a contract between the insurer and the policyholder that defines what is covered, for how much premium and under what conditions. It will specify the policyholder, the insured interest (e.g. a building) is, the list of perils covered (e.g. fire, lightning, explosions, and aircraft collision). It will also mention a list of *exclusions*: for example, car insurance cover may be limited or even withdrawn if the policyholder drinks and drives or deliberately jumps off a bridge; they will typically limit in various ways the amount paid for claims.

Exclusions have an obvious impact on the risk, but it is often hard to assess their impact quantitatively, and very often actuaries find themselves in the position of having to assume that the policies they are pricing have a standard wording and that the wording does not change over the years. However, changes in wording do have an effect and wherever possible they have to be taken into consideration by the pricing actuary.

This chapter is concerned with one specific type of exclusion that has paramount importance for pricing actuaries and whose effect *can* be assessed to a significant extent with actuarial techniques: these are the coverage modifiers, such as excesses and limits. For example, your car policy might require that you bear the cost of the first £250 of each claim and those exceeding £800 for your car radio. All these coverage modifiers have been collated here under the heading of 'cover structure'.

Understanding policy structure is essential for pricing because in almost all cases the cost of risk, which is the most relevant, is almost never the gross amount of a loss but the amount that is paid by an insurer.

Cover structures tend to be quite different depending on who the policyholder is: we have therefore subdivided the treatment of this into three headings: personal lines (in which the policyholders are individuals), commercial lines (in which the policyholders are corporate entities, ranging from multinationals such as Coca-Cola to small companies, charities and sole traders) and reinsurance (in which the policyholders are insurers or even reinsurers).

Let us start with the simplest case – that in which the policyholders are individuals.

DOI: 10.1201/9781003168881-4

3.1 Personal Lines

For personal lines, the most common coverage modifications are rather simple: normally, there is an excess (aka deductible, especially outside the UK[1]) and a policy limit. The only complication is that there may be different excesses and different limits for different sections of the policy/types of risk: for example, a typical motor policy might have the following excesses/limits:

Excess amounts

- Accidental damage £325
- Fire and theft £325
- Windscreen replacement £75
- Windscreen repair £10

Cover limits

- Legal protection limit £100,000
- Car audio/satellite navigation/etc. £1000

From a mathematical point of view, things work as follows:

If a policy has an excess D, and the insured has a loss X, then the actual compensation the insured receives (and therefore the loss that is 'ceded' to the insurer) is $X_{\text{ceded}} = \max(X - D, 0)$, whereas the rest of the loss, $X_{\text{retained}} = \min(X, D)$, is retained by the policyholder.

If there is also a limit L (there almost always is, with the exception of policies for which unlimited liability is provided, such as bodily injury under UK motor insurance), the compensation is capped by L: this leads us to the general (albeit extremely simple) formula for the ceded and retained amount:

$$\begin{aligned} X_{\text{ceded}} &= \min\left(L, \max\left(X - D, 0\right)\right) \\ X_{\text{retained}} &= \min\left(X, D\right) + \max\left(X - D - L, 0\right) \end{aligned} \tag{3.1}$$

3.1.1 Purpose of the Excess Amount

The excess discourages reporting small claims, bringing down the costs of managing the claims and avoiding petty claims. Theoretically, it should also lead to more cautious behaviour as the policyholder pays a price for each claim: however, this is probably achieved just as effectively by the no-claim discount bonus or by the desire to avoid increases in premium (or simply by the desire not to have an accident!).

From the point of view of the insured, the excess has at least one benefit: because it helps to keep both the compensation and the insurer's claims handling costs low, the premium is reduced. The policyholder should therefore choose the excess at a level that exploits the

1 A distinction is normally made in the UK between the *excess*, which is not deducted from the limit and therefore works like in Equation 3.1, and the *deductible*, which works like an excess but is deducted from the limit, so that the amount ceded should be written as $X(\text{ceded}) = \min\left(L - D, \max\left(X - D, 0\right)\right)$. The difference will often be negligible for pricing purposes. In this book we will sometimes speak of 'deductible', 'excess', 'retention' and 'attachment point' interchangeably and clarify how it works depending on the case. Whenever the impact is likely to be material, the actuary should always consult the exact wording of the contract to ensure that the correct formula is applied.

reduction in premium but that does not hurt too much financially (causing instability to one's finances) when the losses happen.

3.1.2 Purpose of the Limit

Individual policyholders are not too concerned about limits, especially if they are very large amounts, as we tend to assume that nothing too extreme will happen to us. On the other hand, insurers are keen to limit their potential liability in case of an accident, and they do not like policies with unlimited liability, which they will issue only if they are required by law. The purpose of a limit is for the insurer to decrease the uncertainty of their results and decrease the capital that needs to be held.

Therefore, the insurer will probably tend to put as much downward pressure on the policy limit as possible – for this reason, this is one of the areas where the legislator and the regulator are most likely to impose constraints.

When shopping around, customers must be careful that the limit is adequate to their needs – that is, that they are not too exposed to very large claims.

3.2 Commercial Lines

Commercial insurance policies can be as simple as personal lines policies but can also have quite complex coverage modifiers. Before delving into the issues related to the exclusion structure (deductibles, limits, etc.), let us observe that a commercial lines policy may be sold on a number of different bases, depending on which claims attach to a policy.

3.2.1 Policy Bases

The policy basis defines which claims attach to a policy, especially as regards to timing. It seems obvious that an annual policy incepting on 1 January 2014 covering all motor claims for the car fleet of a company will cover all motor claims in 2014, but is it all claims that actually *happen* in 2014, or is it all claims that *are reported* in 2014, or something more complicated, such as claims related to vehicles that were already part of the fleet at the inception of the policy?

As you can imagine, there are endless subtle variations as to what the legal details of the insurance contract may provide for. However, in commercial lines, the more widely used bases are the occurrence basis and the claims-made basis.

3.2.1.1 Occurrence Basis

An occurrence basis policy covers all losses that occur during the policy period, regardless of when they are reported (unless the policy includes a sunset clause, which prevents reporting claims after a given date). This is very much like your standard car insurance or home insurance policy and is typically used for property insurance and most types of liability (such as employer's liability, public liability, or motor third-party liability).

Note that, despite the fact that losses are recoverable regardless of reporting date, there are normally limitations as to the time frame within which a known must be reported (sunset clause).

3.2.1.2 Claims-Made Basis

A claims-made policy covers all losses reported to the insured during the policy period, regardless of when they occurred. This is typically used for products such as professional indemnity and directors' and officers' liability (D&O), although it may occasionally be used for more standard types of liability (e.g. pharma liability).

There is a well-defined time frame within which one has to report a claim. For example, a law firm that receives a request for compensation by a disgruntled client needs to report this to its insurer within, say, 1 or 2 weeks for the claim to be valid. The contract may also specify a retroactive date (such that a claim arising from an event occurring before that date will not be covered) and an extended reporting period (defining for how long the insured is allowed to report the claim after the expiry of the policy contract).

3.2.1.3 Other Bases

Other, more complicated, policy bases can be found in the market to cater to the idiosyncrasies of different products. As an example, the 'integrated occurrence' (e.g. the 'Bermuda form' liability policy) basis is a basis sometimes used for product liability to cater to the possibility of a large number of claims attached to the same product. We will explain this in further detail in Chapter 23 (Pricing Considerations for Specific Lines of Business).

3.2.2 Basic Policy Structure

A quite common policy structure has three elements: an each-and-every-loss (EEL) deductible, an annual aggregate deductible (AAD) and a limit (L). Furthermore, there could be an annual aggregate limit (AL).

3.2.2.1 Each-and-Every-Loss Deductible and Self-Insured Retention

Let us first assume that the only feature our policy has is an EEL deductible. The EEL works like the excess/deductible described above for personal lines policies: the insured retains the first EEL amount of each loss and cedes the rest to the insurer. In formulae, the retained amount for each loss is $X_{retained} = \min(X, \text{EEL})$ and the ceded amount is $X_{ceded} = \max(X - \text{EEL}, 0)$. Obviously, the EEL deductible may be at a far higher level than for personal lines customers: sometimes by several million pounds.

A closely related concept to the EEL deductible is the Self-Insured Retention (SIR). Numerically, this will work as the EEL deductible (and therefore we will not distinguish them further in this book), but the two concepts differ mainly in terms of process: in the case of the SIR, the insured is responsible for paying any amount up to the SIR, and the insurer pays the rest (if any); in the case of the EEL deductible, the insurer pays the whole indemnity, and then recovers from the insured an amount up to the deductible.

3.2.2.1.1 What is the Purpose of EEL, and How Should It Be Chosen?

The EEL is designed so that the insured retains most (small) losses in-house, ceding only the large ones to the insurer. This allows the insured to pay smaller premiums, and it allows insurers to reduce the costs of managing the policy because the number of claims to manage will be much smaller.

What 'most' means is left deliberately vague, as different companies might have different appetite for retention. The general idea is to avoid money swapping, i.e. the pointless transfer of expected (rather than unexpected) losses to an insurer, which will cover them adding costs and profit (see the online Glossary for more detail).

A more scientific approach is one that minimises the 'total cost of risk' – we will get back to this in Chapter 29.

3.2.2.2 Annual Aggregate Deductible

The other feature that many commercial lines insurance policies have is the AAD, which provides a cap on the overall amount retained in a given year by the policyholder. This feature is important and the setting of the AAD is one of the main reasons why actuaries are involved in commercial lines insurance in the London Market.

If $X_1, X_2, \ldots X_n$ are the losses incurred in a given year, the amount retained below the deductible becomes (ignoring for now the limit L on each loss):

$$\text{Total retained} = \min\left(\text{AAD}, \sum_j \min\left(X_j, \text{EEL}\right)\right) \tag{3.2}$$

And the total ceded is

$$\text{Total ceded} = \sum_j X_j - \min\left(\text{AAD}, \sum_j \min\left(X_j, \text{EEL}\right)\right) \tag{3.3}$$

It should be noted that although the AAD is also used in reinsurance, it does not work in the same way in that context. In reinsurance the AAD stops the reinsured from claiming the amount recoverable under the deductible, until an aggregate amount equal to the AAD has been reached. The AAD as normally used in commercial lines insurance has a different effect: all amounts below the EEL are retained and when the accumulated amount reaches the AAD the insured is entitled to recover the ground-up amount of each claim it incurs. This difference may sometimes cause some terminological confusion.

3.2.2.2.1 *What is the Purpose of the AAD, and How Should It Be Chosen?*

The AAD provides further stability and security to the insured company by imposing a cap on how much money it retains. To be effective, *from the point of view of the insured*, this AAD should be something that is occasionally exhausted (otherwise, the AAD has no effect) but it is not hit too often (otherwise the benefits of having an EEL are unclear – the same effect could be obtained by simply capping the retention at the amount AAD – and all the volatility is basically ceded to the insurer, at a cost which is likely to be high). The insured will normally be happy if the AAD is hit 5% to 20% of the time.

From the point of view of the insurer, the tendency is instead to position the AAD at rather high levels, so that it is hit as rarely as possible (e.g. $\leq$5% of the time): this is because the insurer is exposed to the additional volatility arising from frequency effects, such as a systemic increase of the number of claims. An insurer will often have guidelines for its underwriters and an AAD breach may be frowned upon.

3.2.2.3 Each-and-Every-Loss Limit

Typically, policies will also have a limit for the compensation that the insured may receive for a single loss and possibly on the aggregate.

If there is an individual limit L but no AAD, the total loss amount is split among retained below the deductible, retained after the deductible and ceded as per the following formulae:

$$\begin{aligned}&\text{Total retained below the deductible} = \sum_j \min\left(X_j, \text{EEL}\right)\\&\text{Total retained above the limit} = \sum_j \max\left(X_j - L, 0\right)\\&\text{Total ceded} = \sum_j \min\left(L - \text{EEL}, \max\left(X_j - EEL, 0\right)\right)\end{aligned} \tag{3.4}$$

Note that in a commercial lines context the EEL is *normally* subtracted by the limit so that the maximum amount ceded is *L*-EEL – however, the contract wording should be consulted to ensure that one is using the correct calculation method.

If there is an AAD as well as an EEL deductible and a limit *L*, the exact workings depend crucially (although this may have limited quantitative impact) on the exact wording of the contract. Specifically, it depends on whether the limit works on a 'drop-down' or on a 'stretch-down' basis.

If the policy works on a drop-down basis, once the AAD is reached the insured still has a cover of size *L*-EEL: that is, instead of being covered between EEL and *L* as was the case with the first claims before the *AAD* was reached, it is now covered between 0 and *L*-EEL. The 'layer' of insurance has 'dropped' down to zero but still remains of the same size it had originally (*L*-EEL).

If the policy works on a stretch-down basis, once the AAD is reached the insured has a cover of size *L*: instead of being covered between EEL and *L* as in the first few claims, it is

Example of Retention/Cession Calculations in a Policy Year (Drop-Down Basis)

Insurance structure	
EEL	50,000
AAD	250,000
L	500,000
Basis	Drop-down

	Loss 1	Loss 2	Loss 3	Loss 4	Loss 5	Loss 6	Loss 7	Total losses
Gross	**1,150,000**	**61,000**	**19,000**	**73,000**	**1,457,000**	**299,000**	**1,208,000**	**4,267,000**
Retained below EEL (before AAD)	50,000	50,000	19,000	50,000	50,000	50,000	50,000	319,000
Cumulative retained (after AAD)	50,000	100,000	119,000	169,000	219,000	250,000	250,000	250,000
Retained below EEL (after AAD)	**50,000**	**50,000**	**19,000**	**50,000**	**50,000**	**31,000**	**-**	**250,000**
Amount of cover available	450,000	450,000	450,000	450,000	450,000	450,000	450,000	450,000
Ceded	**450,000**	**11,000**	**-**	**23,000**	**450,000**	**268,000**	**450,000**	**1,652,000**
Retained above limit	**650,000**	**-**	**-**	**-**	**957,000**	**-**	**758,000**	**2,365,000**

FIGURE 3.1
In the example above, we assume that there have been seven losses in the year under consideration, whose ground-up values are in the row denoted as 'Gross'. The 'Retained below EEL (before AAD)' row is calculated as min (EEL, Gross). Row 'Cumulative Retained (after AAD)' is obtained by cumulating the elements of the row and capping them at AAD. The 'Retained below EEL (after AAD)' row is calculated simply as the difference between two successive values of the row above and gives the amount retained for a given loss after allowance has been made for the AAD. The row 'Amount of cover available' is, in the drop-down case, simply the limit (£500,000) minus the EEL. The row 'Ceded' is the amount ceded to the insurer, capped at *L* (the limit). The amount retained above the limit is calculated as the gross amount minus the retained below EEL (after AAD) minus the ceded and is shown in the row 'Retained above Limit'.

Example of Retention/Cession Calculations in a Policy Year (Stretch-Down Basis)

Insurance structure		
	EEL	50,000
	AAD	250,000
	L	500,000
	Basis	Stretch-down

	Loss 1	Loss 2	Loss 3	Loss 4	Loss 5	Loss 6	Loss 7	Total losses
Gross	**1,150,000**	**61,000**	**19,000**	**73,000**	**1,457,000**	**299,000**	**1,208,000**	**4,267,000**
Retained below EEL (before AAD)	50,000	50,000	19,000	50,000	50,000	50,000	50,000	319,000
Cumulative retained (after AAD)	50,000	100,000	119,000	169,000	219,000	250,000	250,000	250,000
Retained below EEL (after AAD)	**50,000**	**50,000**	**19,000**	**50,000**	**50,000**	**31,000**	**-**	**250,000**
Amount of cover available	450,000	450,000	450,000	450,000	450,000	469,000	500,000	500,000
Ceded	**450,000**	**11,000**	**-**	**23,000**	**450,000**	**268,000**	**500,000**	**1,702,000**
Retained above limit	**650,000**	**-**	**-**	**-**	**957,000**	**-**	**708,000**	**2,315,000**

FIGURE 3.2
This example is the same as in Figure 3.1, but now the policy structure works on a stretch-down basis. Everything works the same up to the row 'Amount of cover available': in the stretch-down case, however, we see that when the AAD is exhausted, the cover is gradually increased up to the full limit L = £500,000. Specifically, when Loss 6 (£299,000) occurs, the first £31,000 is retained, at which point the annual aggregate cap is exhausted. The difference between the higher point of the cover (£500,000) and £31,000, which is £469,000, is ceded to the insurer, and the rest is retained above the limit. For the subsequent losses (only Loss 7 in this case), no amount is retained below the EEL deductible and the first £500,000 is ceded to the insurer.

now covered between 0 and L. The original layer has 'stretched' in size – from L-EEL it has increased to L.

Formulae for the case where both an AAD and a limit L are present can be written down (see Question 3) but they are hairy and unenlightening, so they are better illustrated by an example. Figure 3.1 shows how these calculations can be performed in practice in the case where the structure is 'drop-down'. The spreadsheet with the calculations is provided in the online Download section of the book's website.

Note that although the calculations in Figure 3.1 depend on the order in which losses occur, the final result (the split between retained below the EEL, ceded, retained above the limit) does not change if the order is changed. Figure 3.2 shows how the same calculations in the 'stretch-down' case work. The difference, in this case, is small.

3.2.2.3.1 What is the Purpose of the Limit L, and How Should it Be Chosen?

As the name itself suggests, the function of the policy limit L is to limit the liability of the insurer to a single claim. Insurers are reluctant to write contracts with unlimited liability because the absence of a clearly defined worst-case scenario makes it difficult to determine, for instance, how much capital should be set aside for large claims and makes buying reinsurance more challenging. Reinsurers, who deal with very large claims as a matter of course, are also averse to unlimited liability for the same reason. The insurance industry has often lobbied against the legal requirement for motor third-party liability to be sold with unlimited liability in such countries as the United Kingdom, France, Belgium, and Finland. Therefore, insurers will set a limit to their liability in the insurance contract whenever they are legally allowed to do so.

How should an organisation choose the policy limit to the insurance it buys? Theoretically, the policy limit could simply be chosen so that the probability of breaching it (the 'occurrence exceedance probability', which we will meet again in Chapter 24 on catastrophe modelling) is below a certain pre-agreed level. However, estimating this probability of having losses above a given (large) amount is not easy and cannot be

The Role of the Aggregate Limit

Insurance structure	
EEL	50,000
AAD	250,000
L	500,000
AL	1,000,000
Basis	Drop-down

	Loss 1	Loss 2	Loss 3	Loss 4	Loss 5	Loss 6	Loss 7	Total losses
Gross	**1,150,000**	**61,000**	**19,000**	**73,000**	**1,457,000**	**299,000**	**1,208,000**	**4,267,000**
Retained below EEL (before AAD)	50,000	50,000	19,000	50,000	50,000	50,000	50,000	319,000
Cumulative retained (after AAD)	50,000	100,000	119,000	169,000	219,000	250,000	250,000	250,000
Retained below EEL (after AAD)	**50,000**	**50,000**	**19,000**	**50,000**	**50,000**	**31,000**	**-**	**250,000**
Amount of cover available	450,000	450,000	450,000	450,000	450,000	450,000	450,000	450,000
Ceded before AL	450,000	11,000	-	23,000	450,000	268,000	450,000	1,652,000
Cumulative ceded after AL	450,000	461,000	461,000	484,000	934,000	1,000,000	1,000,000	1,000,000
Ceded after AL	**450,000**	**11,000**	**-**	**23,000**	**450,000**	**66,000**	**-**	**1,000,000**
Retained above limit	**650,000**	**-**	**-**	**-**	**957,000**	**202,000**	**1,208,000**	**3,017,000**

FIGURE 3.3
This example is the same as in Figure 3.1, but we have added the 'aggregate limit' feature to the policy structure. Everything works the same up to the row 'Ceded before AL': in this case, however, the 'Cumulative Ceded after AL' caps the ceded amount at £1M and, as a consequence, there is no ceded amount after that (see 'Ceded after AL' after Loss 6). The overall consequence is an increase in the amount 'Retained above the Limit'.

done with any accuracy based on the historical loss experience of a single insured/reinsured: it requires the use of relevant market loss curves or the output of a catastrophe model.

When these are not available, the choice of the policy limit may be driven by scenario analysis where probabilities and severities are estimated based on expert judgment and market information rather than through actuarial calculations.Often, however, the limit will simply be chosen by benchmarking the organisation against its peers: after all, a risk manager is less likely to be blamed for choosing a specific limit if this falls within standard market practice.

3.2.2.3.2 Aggregate Limit

Some policies have an AL as well as an individual loss limit. This has the purpose of further limiting the uncertainty of the insurer's results by imposing an overall cap on the insurer's liability over a specific period (the policy period). Also, the AL takes care of both the severity *and the frequency* of exceptionally large losses, whereas the limit only takes care of their severity. Therefore, going back to the example of Figure 3.1, an insurer might impose not only an individual loss limit of £500,000 but also an overall limit of £1M, so that the amount that it loses in one particular year will never exceed the latter figure. Figure 3.3 provides an example of how this works in practice.

3.2.3 Excess Insurance

This is insurance that works much like non-proportional reinsurance. The insurer is compensated for any losses incurred that are greater than a certain amount D called attachment point (but also retention or excess) up to a limit L. The attachment point (excess) is normally a large amount and can be in excess of a primary layer written by another insurer. The amount ceded to the insurer is $X_{ceded} = \min(L, \max(0, X - D))$. As this is very similar to excess-of-loss reinsurance, we will leave a more in-depth discussion of this type of coverage to Section 3.3.

3.2.4 Other Coverage Modifiers

In Section 3.2.1–3.2.3 we have looked at the most common coverage modifiers for a commercial lines policy. In this section, we look at other slightly less common but still important coverage modifiers that you must know if you are going to price commercial lines policies.

3.2.4.1 Non-Ranking Each-and-Every-Loss Deductible

Non-ranking (NR) EEL deductible works as the EEL, but the amount that is retained does not contribute towards the AAD. Of course, this works only in conjunction with a standard (i.e. ranking) EEL, which *does* count towards the AAD (otherwise having an AAD would be pointless because it would never be reached).

Given a loss X, we therefore have $X_{\text{retained}} = \min(X, \text{NR EEL} + \text{EEL})$, $X_{\text{ceded}} = \max(0, X - \text{EEL} - \text{NR EEL})$ exactly as before but at the aggregate level we have

$$\text{Total retained below the deductible} = \sum_j \min\left(\max\left(0, X_j - \text{NR EEL}\right), \text{EEL}\right) \quad (3.5)$$

What is the point of the non-ranking each-and-every-loss (NR EEL)? This is usually a small amount, much smaller than the EEL, and it allows the insurer to ignore all petty amounts both from the point of view of managing them and of building up to the AAD. Most importantly, the insured is incentivised to exercise risk management controls even on small claims as it has to retain a share of them regardless of whether the AAD has been reached or not.

Also, the NR EEL commonly appears when a large company with many distinct business units wants to retain the smallest claims inside each business unit but deal with the larger ones centrally.

3.2.4.2 Residual (or Maintenance) Each-and-Every-Loss Deductible

In some cases, after the AAD has been reached, there might still be a residual (or maintenance) deductible, i.e. a retained amount for each claim, obviously lower than the EEL (otherwise, the AAD would have null effect). This is meant to provide the insured with an incentive to manage risks carefully even after the AAD is reached by making sure it has an ongoing stake in each claim.

3.2.4.3 Deductible as a Percentage of Loss

The deductible may sometimes be specified as a percentage of the loss instead of a fixed amount. The simplest incarnation of this structure is that by which the insured agrees to retain $100q\%$ of each loss and as a result $X_{\text{retamed}} = qX$ and $X_{\text{ceded}} = (1-q)X$.

Often, however, a minimum and a maximum monetary amount are also specified, so that the retained amount never falls below the amount m and above the amount M. In this case the retained and ceded amount can be written as $X_{\text{ceded}} = X - \min(M, \max(m, qX))$.

3.2.4.4 Yet More Coverage Modifiers

There is, of course, no limit to how complex a policy structure can be. We have deliberately omitted, at this stage, to speak about structures such as non-monetary deductibles (such as number of days of interruption for business interruption policies) and round-the-clock reinstatements for professional indemnity policies. We will speak about these structures in Part III, in the context of the specific lines of business where they normally arise.

3.3 Reinsurance

Reinsurance (and especially treaty reinsurance) policies have among the most complex structures around, with such features as indexation clauses, premium reinstatements, and the hours clause that are difficult for those that are not insurance professional to grasp. But then again, reinsurance policies *are* purchased by insurance professionals!

3.3.1 Treaty-Specific Policy Bases

There are several types of contract in treaty reinsurance, depending on which losses are covered by the policy. The most widely used types of contract are 'losses occurring during' (LOD), 'risk attaching during' (RAD) and 'claims made'. We look at each of these in turn, having in mind the example of an annual risk excess of loss (risk XL) reinsurance policy incepting 1 July 2014, just to make things more concrete.

3.3.1.1 Losses Occurring During

All losses (to the relevant layer) occurring during the (reinsurance) policy period will be recovered from the reinsurer, regardless of the inception date of the *original* policy and the time at which the loss is reported. For example, if the reinsurance policy period is 1 July 2014 to 30 June 2015, the policy will cover the reinsured for all losses occurred between 1 July 2014 and 30 June 2015, even though some of these losses may be for policies written, for example, in 2013, and regardless of the time when these losses are reported to the reinsurer or when the loss estimate exceeds a certain threshold.

Note that some contracts will qualify this by including a *sunset clause*, that is, a specified date by which losses need to be reported to the reinsurer for the claim to be valid.

This policy basis is very much the same as the occurrence basis policies you find in direct insurance.

3.3.1.2 Risk Attaching During

All losses (to the relevant layer) originating from policies written during the (reinsurance) policy period will be recovered from the reinsurer, regardless of the occurrence date and the time at which the loss is reported. For example, if the reinsurance policy covers the period 1 July 2014 to 30 June 2015, a loss occurring on 15 February 2016 but originating from a policy incepting on 1 April 2015 will be covered, but a loss occurring on 3 March 2015 but originating from a policy incepting on 1 April 2014 will not.

This type of basis is specific to reinsurance, as the policy period depends on the policy period of the *underlying* direct policies rather than on the timings (occurrence date, reporting date, etc.) of the loss itself.

3.3.1.3 Claims Made

All losses (to the relevant layer) reported during the (reinsurance) policy period will be recovered, regardless of the occurrence date and the policy inception date. As an example, if the reinsurance policy covers the period 1 July 2014 to 30 June 2015, a loss which becomes known to the reinsured on 11 January 2015 and is reported within the permitted time frame, will be covered even if the incident occurred 5 years earlier and the original policy incepted in 2009.

3.3.2 Non-Proportional Reinsurance

Most non-proportional reinsurance arrangements are based on the excess-of-loss structure, by which all losses (whether individual or in aggregate) below a value called *the attachment point* are ignored, and the payout is the difference between the loss and the deductible, capped at a value called *the limit* (although the limit may be infinite). Examples of this structure are risk XL, aggregate excess-of-loss (aggregate XL), stop loss and catastrophe excess-of-loss (cat XL).

3.3.2.1 Risk Excess of Loss

In this type of reinsurance, the insurer is compensated for any losses incurred that are greater than a certain amount D called attachment point (but also retention, excess or deductible), up to a limit L (possibly infinity, in which case the XoL contract is called 'unlimited'). So, for example, in the simplest incarnation of this structure, the amount ceded to the reinsurer would be $X_{\text{ceded}} = \min(L, \max(0, X - D))$. The attachment point in a reinsurance layer therefore works as the excess in personal lines insurance (Equation 3.1).

Sections 3.3.2.1.1 to 3.3.2.1.5 show some of the distinguishing features of Risk XL reinsurance.

3.3.2.1.1 Multiple Layers of Reinsurance

You will often have several layers of reinsurance, on top of each other. As an example, a typical structure for motor reinsurance in the United Kingdom is 1M xs 1M, 3M xs 2M, 5M xs 5M, 15M xs 10M, Unlimited xs 25M (all in pounds), organised in a stack, as in Figure 3.4. These structures are also found in commercial insurance but are probably not as widespread as in reinsurance.

3.3.2.1.2 Annual Aggregate Deductible

The AAD is not often used in Risk XL, and when it is used, it is applied to losses greater than the attachment point and not to the ground-up losses – the ground-up losses are actually unknown to the reinsurer unless they are above the reporting threshold. Thus, for example, in a risk XL policy with D = £0.5M, L = £0.5M and AAD = £1M (normally described as a £0.5M xs £0.5M xs £1M), the reinsured retains the first £0.5M of each loss plus the first £1M that go into the layer, and the rest is ceded to the reinsurer.

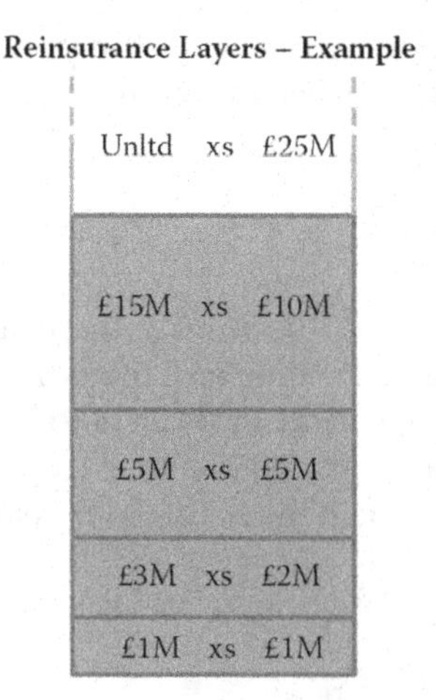

FIGURE 3.4
A typical motor liability reinsurance programme in the United Kingdom.

Reinstatement – Example

Claim amount	Recovery	Additional premium	Cumulative % of layer exhausted
£2.6M	£0.6M	£60k	20%
£9M	£3M	£240k + £30k = £270k	120%
£4M	£2M	£100k	~187%
£11M	£3M	£20k	~287%
£7M	£0.4M	–	300%

FIGURE 3.5
The schedule of recoveries and reinstatements for the example described in the text. Overall, the recoveries amount to 3 × £3M = £9M, whereas the premium paid (initial + reinstatements) is £750k.

3.3.2.1.3 Reinstatements

The ceded amount (not the retained) can be limited, in certain excess-of-loss contracts, to a certain aggregate amount. This is usually defined by the number of reinstatements. For example, if your layer of £3M xs £2M has one reinstatement, that means that once a full £3M has been paid (on one or more claims), another £3M in the aggregate is available. Ultimately, what matters is that the reinsurer makes available £6M for that layer. Once that is also exhausted, no more reinsurance is available. With two reinstatements, the total amount available will be £9M, and with k reinstatements it will be $(k + 1) \times$ £3M. Thus far, this seems simply a quaint way of saying that our layer has an AL of $(k + 1) \times$ £3M, and also unwieldy, as it limits the AL to a multiple of £3M. However, this way of expressing things is used in the industry, most likely because the reinstatements are often not free but may require a reinstatement premium: whenever the layer is exhausted, a reinstatement premium (expressed as a percentage of the original premium pro rata as per the period of cover and the amount of the reinstated limit) is paid, and the limit is thus reinstated. The reinstatement premium will usually be paid in proportion to the amount that needs to be reinstated and in proportion to the period of cover remaining, as in the example below.

Assume for example that we have bought a £3M xs £2M layer paying a premium of £300k with two reinstatements, the first with a 100% reinstatement premium and the second with a 50% reinstatement premium. Assume also we have losses of £2.6M, £9M, £4M, £11M, and £7M. The first loss pays £0.6M and triggers a reinstatement payment of 100% of £0.6M/£3M × £300k = £60k; the second pays £3M (of which the first £2.4M exhausts the layer, and the remaining £0.6M is paid but against payment of a reinstatement premium of 100% of £2.4M/£3M × £300k = £240k for the first part, 50% of £0.6M/£3M × £300k = £30k for the second part, with a total of £270k) and so on, as in Figure 3.5, until all reinstatements are exhausted.

Note that free reinstatements are more common on lower layers than higher layers and it is normal to have more reinstatements available on lower than higher layers. Also, reinstatement, and therefore payment for it, is normally automatic under the treaty.

Also note that in many UK contracts, such as motor reinsurance, it is customary to have unlimited free reinstatements, except for specific types of claims (such as acts of terrorism).

3.3.2.1.4 Indexation Clause (London Market)

Large bodily injury claims tend to be settled over a long period: 2 to 15 years is not uncommon. A claim which occurs today might therefore be paid and settled at a much higher value than the amount at which similar claims are settling now. As a result, reinsurance layers may end up being used much more frequently than intended at the time of designing them. Reinsurers writing excess-of-loss liability products, therefore, seek to protect themselves from this consequence of claims inflation by the indexation clause, which links both the excess and the limit of the layer to an index, normally a wage index (such as the average weekly earnings index for the UK), calculated at the time at which the claim is settled. It is a bit like clay pigeon shooting: you have to point your gun to where your target is going rather than to where it is now if you want to hit it. Note that this type of clause typically applies only to treaty (not facultative) reinsurance.

From the point of view of the insurer that purchases excess-of-loss reinsurance, the ultimate effect of the indexation clause (when applied to a full reinsurance programme and not necessarily to an isolated layer) is to make reinsurance a bit cheaper and to reduce the amount of cover it gets.

There are different types of clauses that can achieve this. In the United Kingdom, the most common is the London Market index clause (LMIC). In simplified terms, this works as follows: if the index is, for example, 150 at the start date of the policy (this is called the *base index*), and the index is 180 when the claim is settled, the excess and the limit will be multiplied by 180/150 = 1.2, so that, for example, a layer £3M xs £2M becomes a £3.6M xs £2.4M (see Figure 3.6) and a loss that is settled at, say, £3.5M, gets a compensation of only £3.5M to £2.4M = £1.1M and not of £3.5M to £2M = £1.5M as it would if the index clause were not applied. Of course, if the loss were £7M, and therefore exhausting the layer, the compensation would be £3.6M and not £3M, therefore actually higher; however, there may be a higher layer (purchased with the same reinsurer or a different one), in which case the overall amount paid by the reinsurer(s) will be lower.

The exact workings of the LMIC are a bit trickier than this. First of all, provisional awards and periodic payments are considered as separate settlements, and therefore different indices must be used for each of them. Second, the property damage component of the claim does not fall into the index clause, but the multiplication factor for the excess and the limit is calculated based on all the payments. These are normally second-order effects (apart from periodic payments), in the sense that the bodily injury component tends to be the largest part of a large claim for Casualty treaties, but it makes some corrections to the simplified formula above necessary.

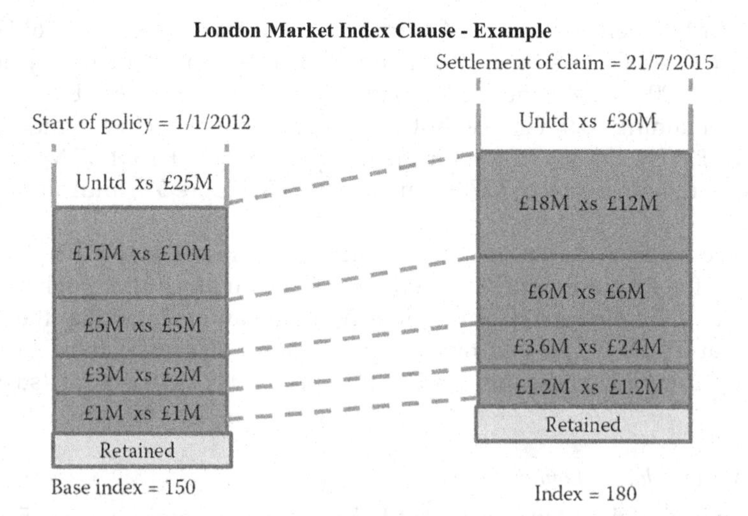

FIGURE 3.6
The indexation of layers according to the LMIC (a rather simplified view). If a claim occurs during the period of the policy and is settled on 21 July 2015 as £3.5M, the recovery to the £3M xs £2M layer is not £1.5M but £3.5M to £2.4M = £1.1M. If all the layers above are reinsured with the same indexation clause, it does not matter to the reinsured whether the reimbursement comes from the first layer or the second. The overall recovery is £3.5M to £1.2M = £2.3M, which is to be compared with £3.5M to £1M = £2.5M if the indexation clause were not there.

The index clause also comes in a slightly more complicated version, the London Market *severe inflation clause,* which works as above but with the provision that nothing happens until the index is more than a certain percentage above the base index. For example, with a severe inflation clause of 10%, and with the base index of 150, the layer is not modified until the index reaches 165. If the index becomes, as in the example above, 180, the layer's endpoints are multiplied by 180/165 and not by 180/150. However, there exists a variation of this clause, the *franchise inflation clause,* in which once the specified value of the index (165 in our example) is breached, the layer's endpoints are indexed by the full amount (i.e. by multiplying them by 180/150 in our example).

The purpose of both the severe inflation clause and the franchise inflation clause is to limit the effect of the indexation clause to cases wherein inflation is really high.

3.3.2.1.5 Other Types of Indexation Clauses

As the word suggests, the LMIC is the indexation clause typically in use in the London Market. Another type of indexation clause is the so-called *European indexation clause* (EIC), which is common across the rest of Europe. The main difference between the EIC and the LMIC is that whereas the attachment point is calculated based on the value of the index at the settlement time in the LMIC, the attachment point is calculated based on payment pattern in the EIC. More specifically, the index by which the layer's attachment point and limit are adjusted is the weighted average of the index over the payment period, the weight being the incremental paid percentage (as in Figure 3.7).

Another difference between the London Market and the European index clause is that the EIC applies to both the bodily injury and the property damage part of the claim rather than bodily injury only compared with the LMIC. Both the EIC and the LMIC can have a severe or franchise inflation clause.

Interestingly, there is normally no indexation clause for US casualty despite the large claims inflation in the US legal environment, the reason being that its introduction was

Index Calculation in the European Indexation Clause

Year	Paid (incremental)	Index
1	0.6%	1.020
2	0.9%	1.061
3	6.3%	1.103
4	28.4%	1.147
5	18.0%	1.193
6	26.2%	1.241
7	8.3%	1.290
8	6.7%	1.342
9	4.6%	1.396
10	0.0%	1.451
11	0.0%	1.510
12	0.0%	1.837
Applied index (IEC)		1.212

FIGURE 3.7
In the EIC, the attachment point is indexed by the average of the index over the relevant period, weighted by the incremental paid percentage: index (EIC) = 0.6% × 1.020 + 0.9% × 1.061 + … + 4.6% × 1.396 = 1.212. For comparison, if we assume that the claim was settled in year 9, the LMIC value of the index would be 1.396.

robustly resisted by American insurers. Instead, the reinsurance market has often adopted other contract mechanisms such as the sunset clause and the claims-made policy basis to address (among other things) the large inflation of liability losses.

A good summary (although not fully up-to-date) of the different types of indexation clauses and more details on the historical reasons behind the introduction of one type or the other of indexation clause can be found in Lewin (2008).

3.3.2.2 Aggregate Excess of Loss

The term 'Aggregate excess of loss (Agg XL)' means different things to different practitioners and organisations. For some, Agg XL is simply a Risk XL cover where an annual aggregate deductible is retained by the reinsured (see Section 3.3.2.1.2).

For others, it is a type of XL reinsurance similar to Risk XL, except that it is triggered when the aggregate (rather than individual) losses consequential to a given event exceed a certain amount. In this section we will focus on this latter meaning of Agg XL, which also blurs into stop loss cover (Section 3.3.2.4).

The purpose of Agg XL is to protect the insurer from the accumulation of a number of risks which individually may be relatively small (for example a number of cars being destroyed by a flood, each of which costing between £3,000 and £60,000) but whose total payout may be very large.

The standard structure looks like risk XL: for example, £1M (aggregate) xs £1M (aggregate). However, the structure may include an interior excess as well, in the sense that the aggregates may apply only to losses exceeding a certain amount. This of course simplifies the management of this cover as small losses do not need to be recorded.

3.3.2.3 Catastrophe Excess of Loss

This is similar to aggregate XL but the level of cover is much higher as it is designed to protect against catastrophic loss occurrences/events such as floods, windstorm, earthquakes, and hurricanes. This type of contract typically applies to property, although casualty catastrophe losses are possible.

The definition of 'loss occurrence' is unique to this type of reinsurance: it covers all accumulated losses from an event of a type covered by the treaty within a period that is usually a multiple of 24 hours, for example, 72 hours for a windstorm. The clause of the contract that defines the duration of this period and the way that losses can be aggregated is called the *hours clause*.

The reinsured normally has the option to choose at what date and time the number of allowed consecutive hours start and can therefore maximise the reinsurance recovery, as illustrated in Figure 3.8. Subject to the limit and possible reinstatements of such a limit, the reinsured has also the option to declare another loss occurrence as long as this is wholly outside the period of the first loss.

Here is an example of an hours clause from an actual contract (with minor modifications to make it simpler). Note how in this particular contract loss occurrence is defined as an individual loss but with an option to aggregate small losses into one aggregate loss, at least for certain perils. Note also that the hours clause also specifies the perils to which the cover applies.

Loss occurrence is to be understood to mean all individual losses arising out of one catastrophe. The duration of any loss occurrence will however be limited to

- 72 consecutive hours for hurricane, typhoon, cyclone, windstorm, rainstorm, hailstorm, and/or tornado
- 72 consecutive hours for earthquake
- 72 consecutive hours for riots and civil commotions
- 168 consecutive hours for flood regardless of cause

[...]

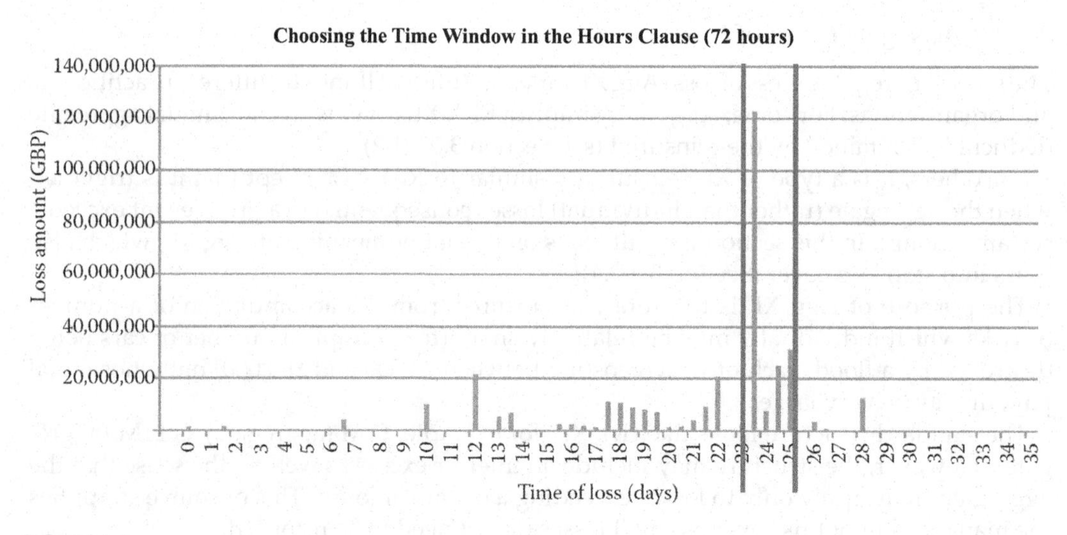

FIGURE 3.8

In cat XL reinsurance, the reinsured normally has the option to define the moment at which to start the clock when calculating the period covered by the hours clause – in practice, this means that the insured can optimise the time window to cover the largest total amount of claims possible. In the figure above, the individual claims are indicated as spikes whose height is proportional to the loss amount: the graph shows the optimal choice of the 72 hours clause.

In all cases, the reinsured may choose the date and time when any such period of consecutive hours commences and if any catastrophe is of greater duration than those periods, the reinsured may divide that catastrophe into two or more loss occurrences, provided no two periods overlap and provided no period commences earlier than the date and time of the happening of the first recorded individual loss to the reinsured in that catastrophe.

The structure of the example contract is

Layer 1:
£10,000,000 each and every loss occurrence
excess of
£7,500,000 each and every loss occurrence
Layer 2:
£22,500,000 each and every loss occurrence
excess of
£17,500,000 each and every loss occurrence
Layer 3:
£40,000,000 each and every loss occurrence
excess of
£40,000,000 each and every loss occurrence

In cat XL reinsurance, it is uncommon to have unlimited free reinstatements of the various layers of reinsurance. In our example contract, the provision was for one full reinstatement under payment of a reinstatement premium. Again from the contract:

Reinstatement provisions:

One full reinstatement per layer calculated at pro rata of 100% additional premium hereunder in accordance with the provisions set out below.

Total amounts recoverable:
Layer 1: £0,000,000
Layer 2: £45,000,000
Layer 3: £80,000,000

Reinstatement provisions applicable to all layers:

Should *any portion* of the limit be exhausted by loss, the amount so exhausted shall be automatically reinstated from the loss occurrence date to the expiry of this agreement, conditional on the payment of the reinstatement premium by the reinsured to the reinsurers when such loss settlement is made.

The term 'pro rata' shall mean pro rata only as to the fraction of the limit reinstated, regardless of the date of the loss occurrence [...]

The index clause is normally not present in property cat XL contracts, as the core costs of catastrophes will be paid shortly after the events.

3.3.2.4 Stop Loss Reinsurance

The purpose of this type of reinsurance is to limit the losses of the reinsured in a given year, usually across different types of causes, or different events, or even different classes of insurance. It is normally formulated in terms of a loss ratio: typically, it will compensate the reinsured whenever the loss ratio for one or more classes of business exceeds a given percentage (e.g. 110%), up to a given limit percentage (e.g. 130%). Alternatively, the cover might be based on a monetary amount, in which case it is sometimes referred to as Agg XL (see Section 3.3.2.2).

Given its all-encompassing nature, this type of cover tends to be expensive and difficult to obtain, although this will of course depend on the current situation of the market and its appetite for premium. One area, however, in which it is rather common is agricultural insurance, especially crop reinsurance.

3.3.3 Proportional Reinsurance

The basic idea of proportional reinsurance is that the reinsured will cede a proportion of each risk to the reinsurer and that it will pay a premium that is a proportion of the overall premium. This allows the reinsured to write a larger number of risks, diversify its portfolio, and reduce its capital requirements.

The proportion of each risk ceded to the reinsurer may be the same for each risk (quota share) or be different for each risk, subject to certain rules (surplus).

3.3.3.1 Quota Share

This is a very simple type of reinsurance in which both claims and premiums are shared proportionally between insurer and reinsurer, and the proportion is constant across all risks. As an example, a quota share reinsurance arrangement may specify that the premium be split in the proportion 30%/70% between the insurer and the reinsurer. As a consequence, all claims will also be split in that proportion.

The main advantage of quota share reinsurance is that it allows the insurer to diversify its risk and to improve its solvency ratio. However, one disadvantage is that it shares with the reinsurer even small risks, which could well be retained by the insurer, whereas retaining too much of the very large risks. Surplus reinsurance refines the concept of quota share to deal with this problem.

It should be noted that pure quota share arrangements are rarely found in the market. Proportional treaties often include features such as sliding scale commissions, profit commissions, loss corridors (see, for example, Clarks [1996] for a definition of these terms) that reduce the 'proportional' nature of quota share contracts.

3.3.3.2 Surplus Reinsurance

We saw that one disadvantage of quota share reinsurance was that the same percentage of the risk was shared between insurer and reinsurer regardless of the risk size – this was very simple from an administrative point of view but has the unhappy result that very small risks that may sit comfortably in the insurer's portfolio are shared with the reinsurer, whereas risks that are way too large for the insurer may still be too large after they are shared.

Surplus reinsurance addresses this by making it possible for the proportion of risk retained by the insurer to vary within certain parameters, so that the proportion ceded can be higher for larger risks and smaller/zero for smaller risks.

The typical parameters of a surplus treaty are:

Maximum retention, R_{max} or simply R	The maximum amount of risk that can be retained by the reinsured.
Minimum retention, R_{min}	The minimum amount of risk that can be retained by the reinsured. By setting this to a suitable value, this has the effect of making sure that the reinsured retains all smaller risks (those below R_{min}) and retains a significant interest in *all* risks.
Number of lines of cover, L	This provides a (complicated) way of defining the maximum amount of risk that can be ceded to the reinsurer. For example, if the insurer retains an amount $r \in [R_{min}, R_{max}]$ for a risk, the maximum amount that it can cede to the reinsurer is $L \times r$, and the maximum risk that can be covered overall is $(1 + L) \times r$, which includes the retained bit. The maximum risk that can be covered in general is therefore Max risk covered = $(1 + L) \times R_{max}$.

How do we calculate the premium split and the claim split? Usually, we start from the expected maximum loss (EML), which is the largest loss that can be expected from a given risk.[2] If EML $< (1 + L) \times R_{max}$, the risk can be insured in the treaty. Assume we choose to retain a portion r of that risk. Therefore, we need to buy EML $- r$ cover, which can be expressed as a number $k = (\text{EML} - r)/r$ of lines. ($k \leq L$ by construction.) The premium P and each claim X are then split between the insurer and the reinsurer in this way: $P_{retained} = P/(1 + k)$, $P_{ceded} = P \times k/(1 + k)$, $X_{retained} = X/(1 + k)$, $X_{ceded} = X \times k/(1+k)$.

As an example, if R = £10M, R_{min} = £500k, $L = 9$, we are able to cede to the treaty any risk below $(1 + L) \times R$ = £100M. If we have a risk with EML = £30M, this falls within the limits of the treaty. Also assume that the premium for taking on this risk is £300k. The insurer has to decide how much it wants to retain. Let us assume it decides to retain r = £5M. This means that it has to buy $k = (\text{EML} - r)/r = 5$ lines of reinsurance,[3] which is less than 9 and therefore within the treaty limits. The premium is then split as follows: the insurer retains £300k/6 = £50k, whereas £250k is ceded to the reinsurer. If a loss of, say, £15M occurs, the loss is split in the same fashion: £15M/6 = £2.5M is paid by the insurer, and the remaining £12.5M is paid by the reinsurer (or to be more precise, £15M is paid by the insurer of which £12.5M is recovered from the reinsurer).

Good faith is necessary for this kind of policy: reinsurers will want to prevent insurers from reinsuring a larger proportion of those risks they consider 'bad' and keeping a higher proportion of the 'good' risks.

2 The definition of EML may depend on the (re)insurer. The Comité Européen des Assurances defines the EML as the maximum loss that can be expected when the normal loss prevention mechanisms work as expected (unlike the MPL which gives the loss under the most adverse circumstances). See the online glossary.

3 Note that k does not have to be an integer.

3.4 Questions

1. A medium-sized UK-based insurance company underwrites mainly commercial and industrial property and motor and liability insurance. Outline, with reasons, the types of reinsurance it is likely to buy for each line of business and (if suitable) for the combined portfolio of risks.
2. Three common bases for policies in treaty reinsurance are losses occurring during (LOD), risk attaching during (RAD), and claims-made.

   i. Explain how these bases work.

   An annual excess of loss reinsurance policy with inception date = 1 October 2012 is underwritten for the layer £1.5M xs £0.5M. The table below shows some of the claims of the reinsured.

   ii. Calculate the total amount that can be claimed by the reinsured for the claims in the table, under the assumption that the reinsurance policy is: (a) on an RAD basis; (b) on an LOD basis and (c) on a claims-made basis. Explain the calculations.

   You can assume that there is no indexation clause.

Claim ID	Inception date of original policy	Loss date	Reporting date	Amount (£)
1	01/07/2012	24/03/2013	08/12/2013	647,935
2	01/10/2012	21/02/2013	18/08/2013	204,973
3	01/10/2012	03/07/2012	03/07/2012	179,985
4	01/01/2012	19/12/2012	17/03/2013	812,868
5	01/01/2012	11/02/2013	09/10/2013	37,036
6	01/07/2012	28/01/2012	03/02/2012	2,058,075
7	01/10/2012	14/01/2013	13/08/2013	219,190
8	01/01/2012	20/02/2013	06/03/2013	665,041
9	01/04/2012	11/08/2013	07/02/2014	1,152,750

   iii. Assume now that this is excess of loss liability cover and that there is a London Market standard indexation clause. The clause is based on the average weekly earnings. What is the amount that is going to be ceded to the £1.5M xs £0.5M layer for Claim no. 9, assuming that
   - Payments of £500,000, £200,000 and £452,750 will be made on 13 May 2014, 14 September 2015 and 10 February 2016, respectively, at which point the claim is settled.
   - Index at 1 October 2012 = 143.
   - Index at 11 August 2013 = 147.
   - Index at 7 February 2014 = 149.
   - Index at 13 May 2014 = 150.
   - Index at 14 September 2015 = 155.
   - Index at 10 February 2016 = 157.

Explain your calculations. You can assume that the claim is only bodily injury and that therefore the index clause applies to the whole claim.

3. (*) Re-write Equations 3.4 for the case where the policy structure includes an EEL deductible, a limit L, an annual aggregate deductible AAD, and an annual aggregate limit AL.

4

The Insurance Markets

This is a short chapter in which we describe the major participants in the insurance market. This is important for many reasons: first, because products are not entities in a vacuum but are offered by a limited number of participants (as an example, most trade credit policies are sold by a handful of large insurance companies); second, when you design a product and you price it, you need to know the business context you are operating in (e.g. the competition and the organisations you need to deal with) and the clients that are going to buy it, whether they are private individuals, non-insurance companies (possibly through their risk management department), public organisations or other insurance/reinsurance companies.

Last but not least, you are probably working (or are going to work) for one of these participants!

4.1 Major Participants in the Insurance Market

The insurance market, and specifically the UK insurance market, is made up of three groups of participants: the buyers, the intermediaries and the insurers (Thoyts, 2010).

4.1.1 Buyers of Insurance

As we have seen in Chapter 3, the buyers of insurance can be private individuals, corporate organisations (companies, public bodies, charities, etc.) or other insurers.

Private individuals will buy personal lines insurance products such as (to mention the most common) motor insurance, travel insurance, household insurance, and extended warranty. Depending on their lifestyles, they may also be interested in other products such as payment protection insurance, pet insurance, or private aviation insurance.

Most corporate companies and public bodies in the United Kingdom will buy these four basic commercial lines insurance products, which may well be bought in a single package by sole traders and small companies:

- Employers' liability (compulsory)
- Public and product liability
- Commercial property (possibly with business interruption)
- Commercial motor

DOI: 10.1201/9781003168881-5

Depending on their specific trade, territories, size, and anything else affecting their risk profile, these organisations will also be interested in other, more specific products.

Figure 4.1 gives a list of industrial sectors (according to the Industry Classification Benchmark [ICB] introduced by Dow Jones & Co. and FTSE Group – two providers of stock market indices – in 2005) and illustrates a few insurance products that are either specifically for that sector (such as the operator extra expenses insurance for oil and gas

Industry sector	**Specific products (examples)**
Oil and gas Exploration and production, equipment and service, pipelines, renewable energy...	Onshore/offshore property Control of well (OEE) Removal of wreckage Loss of production income (LOPI)
Basic materials Chemicals, forestry, paper, metals, mining	Business interruption cover is essential Environmental liability Political risks insurance (trade-related cover, assets' protection) Specie insurance Fidelity and crime guarantee (especially for precious materials, such as gold or pearls)
Industrials Construction, aerospace, defence, electronic equipment, industrial transportation (rail, marine, trucks...)...	Construction all risks Business interruption Satellite cover (aerospace)
Consumer goods Cars and parts, food and beverage, personal and household goods	Product liability has of course special emphasis
Health care Health care providers, medical equipment, pharmaceuticals...	Medical liability (a type of PI) for health care providers Product liability is especially important for pharmaceutical companies and is sometimes rebranded as 'pharma liability' and amended to reflect the special nature of the risk
Consumer services Retailers (including food), media, travel and services (airlines, hotels, restaurants, gambling...)	Aviation (airlines) Weather derivatives (hotels)
Financials Banks, insurers, real estate...	Professional indemnity, D&O, Financial Institutions (a package of PI and D&O) Bankers' blanket bond (banks) Cyber risks insurance Reinsurance (insurers)
Information technology Software and services (e.g. software developers, internet providers), technology hardware and equipment...	IT All Risks, e.g. if a technology company offers data warehousing capabilities
Telecommunication services Fixed line TC, mobile TC	Satellite cover
Utilities Electricity, gas, water...	Environmental liability Section 209 liability for water companies (this refers to a section of the Water Industry Act 1991, dealing with losses caused to third parties by escapes of water from pipes)

FIGURE 4.1
A list of industrial sectors according to the ICB introduced by the Dow Jones and FTSE in 2005, with a few examples (not meant to be exhaustive) of relevant insurance products.

companies) or are particularly important and may need to be tailored for that sector (such as product liability for pharmaceutical companies).

As an actuary, you might find yourself working for a buyer of insurance: not in personal lines obviously (there is no such thing as the 'family actuary') but perhaps in the risk management department of a large company, having to decide on the insurance purchase for the whole group. Also, as an actuary in an insurance company, you might find yourself in the *reinsurance function*, involved in the purchase of reinsurance.

4.1.2 Insurers

Let us start by taking a snapshot view (Figure 4.2) of the major providers of insurance in the world. Although they are not the only participants in the insurance world, they are the hinge around which everything else revolves.

4.1.2.1 Insurance and Reinsurance Companies

The first suppliers of insurance that come to mind are, of course, the insurance companies. These are companies that provide insurance to individuals and also to large companies or small companies that have to buy, for example, employers' liability or have a fleet of cars, properties, and so on, that they need to insure.

Most of the larger insurers have a presence in the London Market (see Section 4.2.1). Companies that deal with personal lines, however, have their main location in less expensive areas, such as in the outskirts of London or York, Norwich, and elsewhere.

Major Providers of Insurance

Insurance companies	Provide insurance to both companies and individuals Examples: Axa, AIG, Allianz, Ace, Aviva, Generali, RSA
Reinsurance companies	Typically international companies Examples: Swiss Re, Munich Re, Berkshire Hathaway, Hannover Re, Scor
Lloyd's syndicates	Lloyd's of London is not a company but a market. The participants are called 'names', which can be individuals or corporate, and they are organised in structures called 'syndicates'. Originally, the names operated on an unlimited liability basis, staking their own fortunes on the insurance they underwrote Examples: Catlin, Kiln, Liberty Mutual
Pools	Participants share premiums and losses in agreed proportions. The insured's liability is not limited Examples: OIL, Pool Re, Flood Re
Captives	Captives are insurance providers wholly owned by ('captive to') another company, whether a non-insurer or an insurer to manage and retain part of their risks Examples: Saturn Insurance Inc. (BP), Aviva Re Ltd (Aviva)
Protection and indemnity (P&I) clubs	Mutual associations of shipowners that provide Marine cover. Currently have 90% of the Marine Liability business. Also provide technical assistance to shipowners Examples: UK P&I Club, London P&I Club

FIGURE 4.2
The major providers of insurance by category, with some name-dropping.

Reinsurance companies are typically international, very large companies that sell reinsurance. All the main reinsurers have a significant presence in the London Market.

It is a very useful exercise to acquaint oneself with the main participants in the insurance and reinsurance markets and get a fair idea of their size. Rather than including too much information here that will become rapidly obsolete, Question 3 at the end of this chapter asks you to find information on the web on the current top 10 insurance/reinsurance groups. As you will see, the order of magnitude of the premiums written by a top insurer or reinsurer company is more than \$10B: for comparison purposes, consider that one third of the world's countries have a gross domestic product which is below that figure!

4.1.2.2 Lloyd's of London

Another important market – arguably the most important of all, at least historically – is Lloyd's of London. Much about the way that Lloyd's of London is set up seems to be peculiar and can be appreciated only by considering the circumstances in which it came about. We can include only a few historical strokes here – a more thorough account can be found, for example, in Bernstein (1996), Thoyts (2010), and Duguid (2014).

The name 'Lloyd's of London' derives from Edward Lloyd (1648–1713), the owner from 1687 onwards of an increasingly successful coffee house where insurance of ships was arranged with the blessing and even the assistance of the owner. Lloyd not only provided coffee and tea but also reliable shipping news, general maritime intelligence (which eventually developed in a bulletin, *Lloyd's List*, which is still published today) and even paper and ink to record business transactions (Bernstein, 1996).

A shipowner who wanted to insure their commercial endeavour would liaise with a broker who would in turn engage with a wealthy individual who was willing to take on the risk and put their signature at the bottom of the insurance contract: hence the name 'underwriter'.

About one hundred years later, in 1771, a number of these underwriters joined forces by setting up the Society of Lloyd's, a self-regulated society whose members put £100 each to transact business. With time, the members of Lloyd's came to be called Names.

Yet another hundred years later, Lloyd's of London was incorporated as a society and corporation (the Corporation of Lloyd's) by the government through the 1871 Lloyd's Act. As a consequence, the society gained the legal powers to enforce its regulations upon its members.

4.1.2.2.1 Fast Forward to the Present Time

Today as it was then, Lloyd's of London is an insurance market (rather than a company, as some may think), a meeting place where insurance business is transacted. The goal of the Corporation of Lloyd's is indeed not to transact insurance but to provide the infrastructure and the services for this insurance market to work (the paper and ink in Lloyd's coffee house) and to regulate the market. The corporation also manages (through the Council of Lloyd's) the Central Fund, which is a capital of last resort that can be used when Lloyd's members are unable to meet their insurance liabilities.

Members of Lloyd's are still called 'Names' and can be *individual names*, that is, wealthy individuals investing their money in insurance, typically on an unlimited liability basis (i.e. the member cannot separate their own wealth and the wealth invested in the business),[1]

1 Traditionally, all individual names joined the market on an unlimited liability basis. New members cannot join on an unlimited liability basis but it is possible to join *as an individual* on a limited liability partnership basis. Old members can also convert from unlimited liability status to limited liability status.

or *incorporated names*, that is, corporate companies that invest their company on a limited liability basis. Note that incorporated names have been admitted as members of Lloyd's only since underwriting year 1994.

Unlike in the eighteenth century, Lloyd's names do not carry the risk by themselves but are organised in *syndicates*, which are the business units of Lloyd's of London. Each syndicate's member is responsible only for his or her share of the risk underwritten and not for the other members' share. Combining wealth into syndicates obviously makes it easier to insure large risks. At the time of writing, there were 96 active Lloyd's syndicates.

A number of professionals other than the Lloyd's members are involved in Lloyd's activities:

- As in the coffee house's times, all risks (with some exceptions) must be placed through Lloyd's brokers (a type of insurance broker specific to Lloyd's) rather than directly.
- Each individual name at Lloyd's is represented and advised by a member's agent, which is in practice a company. At the time of writing, there were only four member's agents: Alpha Insurance Ltd., Argenta Private Capital Ltd., Hampden Agencies Ltd. and ICP General Partner Ltd., Corporate members do not need member's agents but normally refer to a Lloyd's adviser.
- A key role at Lloyd's is played by the managing agents, which are companies that create and manage syndicates and oversee their activities. Their remit includes appointing underwriters and making sure that the syndicate abides by regulations, pays claims, and purchases reinsurance as necessary. A managing agent can manage more than one syndicate. At the time of writing, there were 57 managing agents at Lloyd's.

The Lloyd's market functions according to the slip system, which is described in Section 4.2.1.1. Much else could be said about how the Lloyd's market works and is regulated: how the syndicates are created and closed after 3 years and their liabilities are transferred through a reinsurance-to-close (RITC) arrangement; and how the financial security of Lloyd's is guaranteed through a three-tier system. The interested reader can find a good exposition of all this in Thoyts (2010) and in even greater detail in Lloyds' training material.

4.1.2.3 Captives

Captives are insurance providers that are completely owned by another company, which can either be a non-insurer (for instance, BP owns Saturn Insurance Inc., domiciled in Vermont, US) but sometimes by an insurer (for instance, Aviva owns Aviva Re Ltd., domiciled in Bermuda).

Some of the reasons why captives are used are listed below.

Risk management and lower insurance spend. A captive allows a group with wide-ranging operation to manage their risks together and buy insurance externally only when it is efficient and economic to do so for the group, for instance, when cover is too expensive.[2] This allows a lower insurance spend for the group but allows it to provide protection to individual units that could not retain as much risk.

Access to reinsurance. It allows a non-insurance group direct access to the reinsurance market.

2 See Chapter 30 for a more detailed explanation of what is meant by an efficient insurance spend.

Losses above EEL = £50,000	Ceded to the insurance markets
Losses below EEL = £50,000	Retained by the captive
Losses below £1000	Retained by the business unit (e.g. these could be branches in different territories)

FIGURE 4.3
A typical example of how risk is shared among the marketplace, the captive, and the balance sheet of a business unit.

Unavailability of cover. Captives provide cover when adequate insurance or reinsurance cover cannot be found in the insurance market.

Risk control. Captives can help a group manage its risks in-house by putting in place risk control mechanisms, such as incentives for people to drive more safely or driving lessons to reduce motor risks.

Tax reduction? Historically, captives have also been used for tax reduction purposes. However, current international tax regulations and public opinion are making this less of a factor.

Lighter regulatory burden. Captives are normally domiciled in locations with a lighter-touch regulatory environment, which translates into lower costs and potentially lower capital requirements.

4.1.2.3.1 How Are Captives Normally Used?

When we say that a risk is retained by a company, what we mean is often it is retained at least partly by the captive owned by the group of which that company is part, rather than all in the balance sheet.

Here is an example of how things work. Suppose you have a large company X, which owns a captive C and suppose that all property losses except the smallest ones are managed by the captive. The captive, however, is not comfortable retaining all the losses because it does not have the same ability to diversify that an insurer has (see Section 4.1.7).

So it will, for example, retain all the losses lower than £50,000. The losses above that are going to be ceded to the insurance market. So the risk is shared as illustrated in Figure 4.3.

Another common arrangement is a fronting arrangement. This happens, for example, when a company wishes to retain some of its risks but is bound by law (or prefers) to purchase the whole lot from the insurance markets. This is the case, for example, for employers' liability, which in the United Kingdom an employer needs to buy from an authorised insurer from the ground up to a limit of at least £5M. To retain some of the risk, for example, for all losses lower than £100,000, what the company may do is to arrange for the captive to *reinsure* all the losses[3] of the insurer lower than £100,000. The result is therefore that the insurer is 'fronting' the insurance, but it is the captive that ultimately pays for the claims lower than £100,000. This is not simply a trick to circumvent the law – the insurer is indeed ultimately responsible for paying the claims of X's employees, and it does that directly – and what then happens between the captive and the insurer is much less relevant to the employee who has made the claim. Among other things, this means that if

3 In this case, we speak of a 'reinsurance captive'.

company X went bankrupt, it would be the insurer, and not the victim, who was left with the bill to pay – or in more technical terms, it is the insurer, and not the victim, that has to bear the credit risk – which is the purpose of the legislation that requires employers' liability from the ground up. The law allows for a reimbursement agreement between insured and insurer, so the insurer pays the claim to indemnify the injured employee and then recover the agreed amount from the insured. The insurer requires a letter of credit from the insured to avoid the credit risk.

In calculating the premium in case of reinsurance of a UK EL policy, the insurer will need to factor in the Mess Levy which it has to contribute too by law. This is a non-trivial exercise as it is calculated by the MIB retrospectively based on the claims from the mesothelioma fund and the market share of the UK EL market of the insurer.

4.1.2.4 Pools and Self-Retention Groups (Excluding Captives)

A pool is an arrangement by which a number of individuals or companies decide to share premiums/losses in an agreed proportion. This was quite popular with shipowners for example. A self-retention group is an insurer that has been set up with the purpose of insuring the risks of its members. A captive is an example of such a self-retention group in which the only member is the company itself: we are considering here the case in which the company is set up for the benefit of a number of independently trading companies.

The difference between pools and a self-retention group is not clear-cut and we could perhaps say that a self-retention group is a pool arrangement with a company structure. The following example can legitimately be seen as a model of both.

A specific example: OIL. The best way to illustrate how this works is perhaps with a concrete example, that of Oil Insurance Limited (OIL), a Bermuda-based entity formed in 1972 by 16 energy companies. The formation of OIL was triggered by two industry accidents – an oil spill in Santa Barbara, California, and a refinery explosion in Lake Charles, Louisiana – which made insurance coverage hard to come by in the markets. By 2012, OIL had increased its membership to more than 50 members (www.oil.bm/about-oil/at-a-glance).

The OIL policy covers, among other things, physical damage to property, construction, control of well, pollution liability, and windstorm risks. The members of OIL pay a premium based on their size and on the overall claims experience, and they can then use OIL's capacity to pay for their claims.

Other examples. Other examples of pools are Pool Re, a pool set up by the insurance industry in cooperation with the UK government to enable insurers to keep providing property cover against acts of terrorism, and Flood Re, a not-for-profit flood fund (which was being set up by the insurance industry and the UK government at the time this book was being written), which would ensure that flood insurance remains available and affordable for owners of properties at high risk of flood.

4.1.2.5 P&I Clubs

These are associations of shipowners that provide marine cover, especially marine liability. They were set up at a time when marine liability cover was hard to come by in the market and they decided to provide it to themselves by mutual associations. Despite the fact that this is no longer the case, P&I clubs still have the lion's share of marine liability insurance because they have, in the meantime, become technically very skilled at assessing this cover. Besides providing insurance, they also provide technical assistance to shipowners.

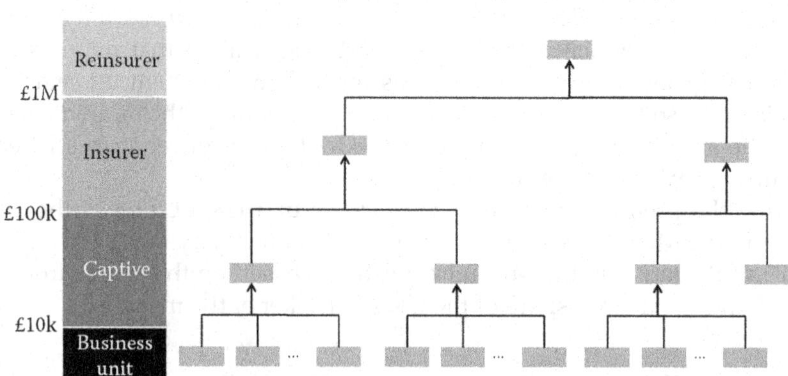

FIGURE 4.4
How volatility is transferred from one layer of protection to the next in a typical commercial insurance situation. The smallest losses are borne by the individual business units of a company, then they are passed to the captive (an in-house insurer owned by the group), and the largest losses are ceded to the insurance market, and the very largest are ceded to the reinsurers. Each layer in the hierarchy ideally holds as much volatility as it is comfortable with (the 'risk appetite').

4.1.2.6 Diversification: How Risk Is Spread in the Insurance Market

What makes insurance useful is that risk is transferred from those to whom a loss might cause an unwanted and possibly impairing financial volatility to those financial institutions that can deal with it without experiencing excessive volatility. This is achieved through diversification and distribution of the risk.

This is perhaps best understood with a couple of examples. The first one is from personal lines insurance:

- You and I do not want to drive around with the possibility that we may face paying a claim of £100,000 because of damaging someone else's Porsche: a swing of £100,000 in our finances may make us bankrupt. However, for an insurer that has many policyholders and strong finances, a swing of £100,000 in their fortune might be no more than a tiny percentage of their overall results.
- However, if I cause a loss of £10M by inflicting multiple deaths or injuries, this might be an unacceptable dip in the profits for an insurer as well. However, if the insurer in turn passes on the largest losses to a reinsurer, or even a panel of reinsurers, there is sufficient diversification in any reinsurer's book of business that a loss of (its share of) £10M is a small percentage of its underwriting results.
- And the reinsurer might in turn reinsure part of its losses to other reinsurers, until each very large loss is spread over virtually the whole insurance market. This is the way in which even losses of many billion dollars arising from huge catastrophes such as hurricane Katrina or the Twin Towers attacks of 11 November 2001 can be absorbed by the market without bankrupting particular insurers.

Another example, this time from commercial lines insurance, is illustrated in Figure 4.4.

4.1.3 Intermediaries

Traditionally, insurance has been purchased in the United Kingdom through brokers, independent financial advisors, and tied agents.

This is still very much the case today for commercial lines and reinsurance business, although a number of organisations will negotiate their insurance needs directly with the (re)insurance carriers. For personal lines insurance, the last decades have seen a distinct move towards direct insurance purchase, first through the use of insurance sold directly on the telephone (such as Direct Line) and then through the Internet, in a phenomenon sometimes referred to as *disintermediation*.

The last decade has seen the growth in importance of so-called *aggregators*, which are websites that offer customers the possibility of comparing the policies of many different providers and purchasing insurance through them. Question 5 at the end of the chapter invites you to do some personal research into which of the insurance aggregators in the United Kingdom are currently the most important.

4.1.3.1 Brokers

Brokers are intermediaries whose job is to advise clients on their insurance purchase, make sure they get the best deal in the market, in terms of both cover and price and actually execute the transaction. They are from the legal point of view the client's agents, in the sense that they have the authority to carry out negotiations and (after consulting with the client) purchase the insurance on their behalf, without the client contacting the insurer directly. Brokers are required to act in the client's best interest, not the insurer's, or their own.

Although in the case of personal lines business, the broker's job is mainly that of matching existing insurance policies to customers' needs and finding the best deal in economic terms, in the case of corporate clients, it is more complex and may include advising on levels of retention and limits, tailoring the features of the policy to the client's need (by making sure, for example, that a specific peril is not excluded) and negotiating the price of a policy with the insurance carrier(s).

Lloyd's brokers are insurance brokers who are authorised to transact business with Lloyd's. (Only an authorised Lloyd's broker can do business with Lloyd's.) Many brokers are both company brokers and Lloyd's brokers.

Insurance companies also use brokers, who are called in this case *'reinsurance brokers'*. One complication with reinsurance broking is that many reinsurance policies are co-insured (that is to say, several parties agree to split premiums and claims of the policy in pre-agreed percentages) and therefore the broker must negotiate the participation of several parties to the deal and discuss the economic terms with all of them. (This will be expanded upon in Section 4.2.1.1. when we explain how the slip system works.) This is not an exclusive feature of reinsurance, but it is in reinsurance where this becomes more common.

If you are an actuary working for a professional broker, your job will certainly involve helping the brokers in their advisory role, for example, helping them determine the correct policy structure, assess whether the policy the clients are currently buying is good value for money, help negotiate the price with carriers and manage clients' expectation as to what they can reasonably expect in terms of premiums.

4.1.3.2 Tied Agents

Unlike brokers, who have the capacity and even the duty to look for the best deal for the client using theoretically the whole insurance market (subject to some constraints, such as the credit rating of the insurance carriers), tied agents are, as the name suggests, tied to a specific carrier and are authorised to transact business only on its behalf. Therefore, when a customer goes to a tied agent, that customer has already decided which insurer to deal with. Actually, as far as customers are concerned, they *are* dealing with the insurer because the insurer holds full responsibility of the actions of the tied agent. Note that the tied agent works for the insurer, as opposed to the broker who works for the insured.

Tied agents are normally found in personal lines insurance.

4.1.3.3 Independent Financial Advisors and Consultancy Firms

The job of independent financial advisors (IFA) and consultancy firms is to give the client (typically a private customer in the case of IFAs and an organisation in the case of consultancy firms) advice on a range of financial decisions, ranging from mortgage to investments to audits to insurance purchase. The distinction between these and professional brokers is sometimes blurred:

- In the case of consultancy firms, the difference between the consultant and the broker is often that the consultant will advise the client under payment of a fee but will not transact the business, whereas the broker will also transact the business and the advice may be offered as part of the commission (a percentage of the insurance premium paid[4]).
- In the case of IFAs, the authority to execute transactions may vary and may depend on the advisor's qualifications and remit.

4.2 Major Insurance Markets

The largest insurance markets are – not surprisingly – located in the largest economies: the United States, Germany, Japan, the United Kingdom, France, and others. A list of the top 10 countries by premium volume as of 2011 can be found, for example, in Swiss Re's sigma publication no. 2/2011 or in the *Reactions* report (available at: www.reactionsnet.com/pdf/2012Directory.pdf).

There are also some specific markets that are important. One of the most important is the London Market.

4.2.1 London Market

The London Market is not the overall UK market (or even the overall insurance market in London) but is a specific market based in the City of London that specialises in reinsurance and large commercial and international risks, such as energy or aviation, large corporate companies, and business that nobody else would insure. For instance, satellite risk was initially insured only in the London Market. To a large extent, the London Market specialises in complex business that needs to be negotiated face-to-face.

Who belongs to the London Market? Examples include all of Lloyd's syndicates, the London Market operations of larger insurance groups and the reinsurance divisions of

4 However, some brokers today are paid a fixed fee instead of a brokerage as a percentage of the premium, also to avoid a potential conflict of interest in a hardening market.

larger groups. A significant number have their operations in the same building as the International Underwriting Association in Minster Court, the London Underwriting Center (LUC). Much of the London Market is transacted through the so-called slip system, which is described next.

4.2.1.1 Slip System

The slip system works as follows:

1. The broker prepares a slip with the main features of the risk. The name 'slip' seems to suggest something rather flimsy but these days it normally includes a substantial amount of information.
2. The broker shows the slip to one or more underwriters, who quote a premium.
3. The insured (with the broker's advice) selects a *lead underwriter* and a 'firm order' price.
4. The lead underwriter accepts a share of the risk (e.g. 20%) and stamps and signs the slip.
5. The broker approaches other underwriters to offer the risk.
6. The follow underwriters accept a share of the risk and 'write under' the lead's line.
7. This goes on until 100% of the order is reached or at the inception date, at which point
   a. if the business underwritten is more than 100% (oversubscribed), all the written lines may (subject to the insured agreeing to this) be scaled down by the same proportion, or
   b. if the business underwritten is less than 100% (undersubscribed) it may be necessary to place the *shortfall cover* at different terms.

Figure 4.5 shows an example of the part of a slip on which the underwriters indicate the share of the risk they are willing to accept.

The first column indicates the underwriter, represented as a syndicate number. The first underwriter is the lead underwriter and is usually someone who is known in the

Underwriter (syndicate number/company)	Type	Percentage written	Revised percentage signed
855	Lead	20%	18.18%
707	Follow	15%	13.64%
XXX Ltd	Follow	20%	18.18%
288	Follow	15%	13.64%
YYY Ltd	Follow	15%	13.64%
872	Follow	10%	9.09%
1091	Follow	15%	13.64%
Total		110%	100%

FIGURE 4.5
The typical appearance of the part of the slip that shows the underwriters' shares.

market as a specialist of the type of risk concerned and whose judgment is trusted by other underwriters.

As you can see, at the end of the broker's tour, 110% of the risk has been placed, at which point the risk of the total accepted percentage is normalised to 100% by reducing each percentage by the same proportion.

4.2.2 Bermuda and Similar Markets

This has become the domicile of many important insurance companies. It is the most important domicile for captives and reinsurance captives. The main reasons are taxation, because the tax treatment in Bermuda is quite favourable to corporate companies, and its lighter-touch regulation.

Other markets with similar characteristics are the Cayman Islands, Barbados, Dublin, Guernsey, Vermont, the British Virgin Islands, and Luxembourg, although none has achieved the size and importance of the Bermuda market.

4.3 How Is Insurance Sold?

4.3.1 Personal Lines

Personal lines insurance products are nowadays often purchased online using an insurance aggregator (see Question 4). The way these work is that insurers agree to have their quotations appear on the aggregator website and pay a fee to the website owner when a sale is made.

Traditionally, personal lines insurance was placed through agents. You went to an agent of a large insurer, and you explained your needs, and insurance was arranged for you. More recently, this has happened on the telephone and more recently still through the Internet (not necessarily the aggregators, where different companies compete).

There are brokers for personal lines insurance as well. They discuss the individual's needs, and they take a percentage of the premium from the insurance companies.

As we said, agents and brokers have a smaller role today because of the Internet, but this way of selling has not been completely abandoned. For instance, if you buy a house and you arrange for a mortgage through the estate agent that has sold you the house, you are likely to be offered to buy household insurance through them – so they are acting as the insurer's tied agent.

4.3.2 Commercial Lines and Reinsurance

As we have mentioned previously, brokers are the most likely route through which one buys commercial lines and reinsurance products. To deal with Lloyd's, clients actually *have* to go through a broker. However, for more standardised products, it is not uncommon for clients with a mature risk management function to deal directly with the insurer.

Going through brokers has the obvious advantage that they can become effective advocates of a client's risks as they deal with many similar risks in the market and therefore they wield more negotiating power. Even if you are an insurer, you do not necessarily have a clear idea of what the reinsurance market is. Also, a broker can analyse (actuarially as well) your current insurance arrangement not only from a cost point of view but also from an effectiveness point of view – is the insurance structure that you are buying adequate to your needs?

The obvious disadvantage is that brokers will charge a 'brokerage fee', which can be a substantial portion of the overall premium.

4.4 Underwriting Cycle

Competition in the insurance market has had some unintended consequences. If you look at the performance of insurers over a number of years measured by the incurred loss ratio (total incurred claims divided by earned premiums over a given year), you will notice that the performance does not show an annual random variation around a central value as one would expect if the volatility in performance were simply due to the volatility of loss experienced: rather, it exhibits a cyclical behaviour with peaks and troughs that are not quite periodic (the time lag between two peaks changes; the vertical distance between a peak and the following trough also changes). This is exemplified in Figure 4.6, where we have looked at the incurred loss ratio (= incurred claims/earned premium) of commercial motor insurance of the whole UK insurance industry over the period 1985 to 2009.

This phenomenon is ubiquitous across most classes of insurance and is called the 'underwriting cycle'. As a consequence of the existence of this cycle, insurance market participants speak about a 'soft market' when insurance premiums are low and therefore we are in the peak section of the cycle (the loss ratio is inversely proportional to the premiums charged) and a 'hard market' when insurance premiums are high.

What is the reason for the cycle? One commonly offered explanation is the following, which is based on the balance of supply and demand of insurance, and which is applied here to a single line of business, such as commercial motor liability. A more in-depth exploration is offered in Fitzpatrick (2004).

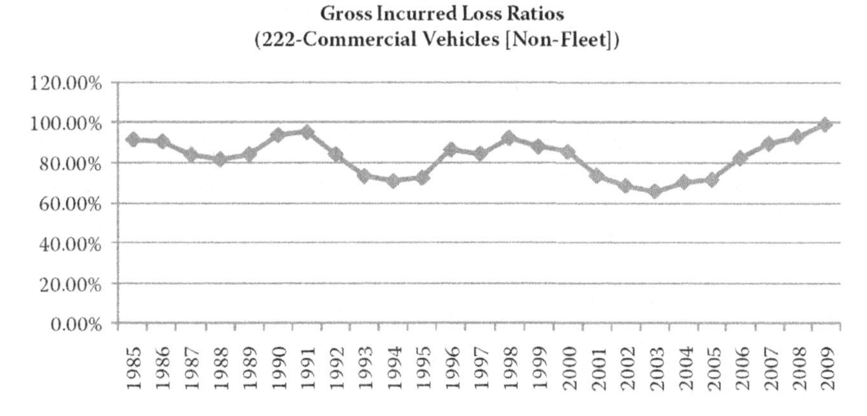

FIGURE 4.6
Commercial motor loss ratio for the whole UK market over the period 1985 to 2009, as derived from information from the former FSA returns. A cyclical (although not periodic) pattern can be easily spotted.

i. When the market is hard for a particular class of business (insurance prices are high), there is an incentive for insurers to enter the market or to move into that class of business or underwrite more of it (with the same capital the insurer can write more premium, as the premium per unit of risk is higher). We also say that the market has plenty of *capacity*.
ii. The increased supply of insurance progressively brings down the price.
iii. Eventually, the market becomes soft and insurance prices become too low to provide an appropriate return on capital and may, indeed, lead to losses being made.
iv. Therefore, insurers stop trading, or stop writing that class of business, or simply reduce the amount they underwrite in that class of business, and hence the amount of capacity in the market is reduced.
v. The decreased supply of insurance pushes up the prices, until eventually the prices become attractive again and the market is hard once more.

A pictorial version of this description is shown in Figure 4.7.

Despite the reasonableness of this explanation, there is no firm consensus on the mechanisms behind the underwriting cycle and many different factors quite different from those outlined above may contribute to the onset of cycles. A list of possible drivers for the underwriting cycle is given below:

Low barriers to entry. If there were no new entrants in the market, changes in capacity would be more limited and there would be less downward pressure on prices.

Delay between writing business and determining profitability. It takes time to realise your price is too low or too high, so it takes time for a company to correct for mistakes. There are obvious issues of corporate governance here, as the underwriter may be the one who sets the prices but there will be others in the company who have the power to enforce a change of policy, and it may take time to achieve a correction.

Premium-based capital regulations. In the current formula-based regulatory regime in the United Kingdom, the capital that the insurer has to put aside is proportional to the premium written. As a consequence, if the premiums are too low, this may lead to insufficient levels of capital and a higher probability of default.

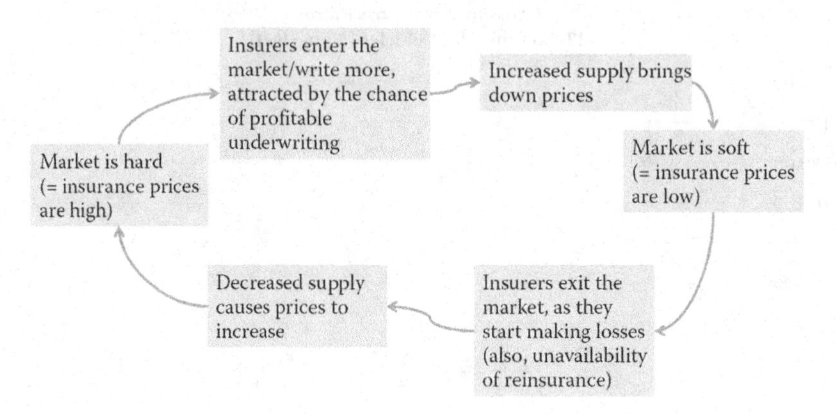

FIGURE 4.7
A possible explanation of the underwriting cycle.

Catastrophes. Catastrophes may suddenly reverse a soft market (especially in property insurance) into a hard one by reducing capital and therefore capacity; hence, allowing premium increases. This is quite a common self-correcting mechanism in reinsurance in which, for example, large hurricanes (such as Hurricane Andrew in 1992 and Hurricane Katrina in 1995) lead to reinsurance premium increases. A similar surge in reinsurance rates was noticed in the aftermath of the 9/11 terrorist attack of the World Trade Center. Paradoxically, it may be more difficult for reinsurers to maintain profitability if no catastrophes happen because, in the absence of catastrophes, there is a constant downward pressure on reinsurance prices and the market may slide into loss.

Unwillingness of insurance carriers to lose market share. In a falling market, market share can sometimes be maintained only by decreasing premiums' levels and financially weaker insurers will not be able to sustain the lower prices for as long as those with stronger capital positions (in this context, too, Keynes' dictum holds: *the market can stay irrational longer than you can stay solvent!*). Those who remain have basically pushed away some of the competition, reduced the capacity of the market and hence achieved a higher equilibrium in the price for their product.

Macroeconomic factors. For some lines of business, such as professional indemnity or mortgage indemnity guarantee, there will be systemic economic factors at play that override any internal mechanism behind the underwriting cycle, much in the same way as a catastrophe may override the property insurance underwriting cycle. As an example, an economic recession will lead to an increase of Professional Indemnity claims and therefore and may lead a number of participants withdrawing from the market until things have calmed down, leading to a hardening of the market, disrupting the normal cycle.

Ultimately, one would do well to note that *the underwriting cycle is not really a cycle* and can be simply attributed to the market going through periods of overcorrection of past errors or voluntary sacrifices of participants to maintain the market share and the need to recoup losses made during soft market periods.

4.4.1 How Does the Underwriting Cycle Affect Pricing?

First of all, the very existence of the underwriting cycle is a clear confirmation that ultimately the premium charged for a policy is a commercial and not a technical decision: if the premium were based on a formula involving the expected losses, the volatility, expenses and profits such as those that you often find in actuarial papers (and that you will find in this textbook as well) would be roughly constant. However, that is obviously not the case.[5]

Indeed, not all classes of business are equally affected by the underwriting cycle. Personal lines business is subject to less violent swings. This is because given the large amount of data, it is possible to price personal lines business more scientifically and accurately than commercial lines and reinsurance business and it will not take long to determine whether the business is loss-making or not. For the London Market, which is based on far scarcer

5 The situation is slightly more complex than this because a formula-based premium would only be guaranteed to flatten the underwriting cycle if the assumptions (e.g. frequency and severity parameters) were perfectly known – otherwise, there would still be oscillations as the market grapples with herd behaviour errors in pricing. However, in this case, the underwriting cycle would eventually converge into a flat line.

data, pricing is far more uncertain, and as argued, for example, in a forthcoming article (see Zhou et al., 2012 for an initial presentation of results), there is more downward pressure on prices. Given the delay between the time when business is written and the time when profitability is determined, there is scope for making expensive mistakes.

4.5 Questions

1. A company X (whose main offices are in London) has three business units: one in Gloucestershire, one in Essex, and one in Yorkshire. The company buys property insurance through a captive. Each business unit has a deductible of £10,000 and the captive has a deductible of £100,000. All the rest is ceded to the insurance market.

   Describe who pays how much in the case of these (accepted) claims:

   a. A claim of £1,500 in Gloucestershire
   b. A claim of £25,000 in Essex
   c. A claim of £1,300,000 in Yorkshire

2. The same company X described in Question 1 buys employers' liability insurance from the ground up. Through a fronting arrangement, the captive reinsures all losses of less than £100,000. All the rest is ceded to the insurance market.
   Describe who pays how much in the case of these (accepted) claims:

   a. A claim of £25,000 in Essex
   b. A claim of £1,300,000 in Yorkshire
   c. A claim of £2,200,000 in Gloucestershire, after which (and for reasons unrelated to the claim) the company goes into liquidation

3. Find out who the 10 top insurance companies in the world are by net premium written (or other relevant measures) and the top 10 reinsurance companies (by the same measure).
4. Find out who the 10 top reinsurance brokers in the world are by total revenues. How does that tie up with the list of the 10 top insurance brokers? What are the main differences?
5. Find out who the top 4 insurance aggregators in the UK are, and for each of them, find out (i) their total revenue; (ii) what types of coverage they offer and (iii) what their relative position is as sellers of motor insurance, home insurance, and travel insurance.
6. Go on one of the major aggregators' websites and get a quote for a pet insurance or a travel insurance policy. What questions are you asked? How are the policies different apart from their prices?
7. Find out the name of the so-called 'big four' consultancy firms and find out the scope of their advisory activity: is insurance advice the main source of their business? What else are they advising on? Pick one of these large consultancy firms and try and find out how many actuaries work for it.

8. A large UK-based pharmaceutical company is reviewing its risk management strategy.
   a. Write down the insurance products that it is likely to consider purchasing to limit the volatility of its results, briefly stating why they are relevant to the company's risk management structure.
   b. Explain why the company might not buy some of these products after all.
   c. Explain what a captive is and why the company might decide to use one to as a way of self-insuring some of the risks.

5

Pricing in Context

The job of pricing actuaries would be easier if they could remain oblivious to everything except the portfolio of risks that they have to price. However, it so happens that the price of insurance – and even the pure cost of risk – is inextricably linked with a number of factors that are out of the control of the individual or the company that wishes to transfer the risk and to the company that takes this risk on.

If you want to price a risk sensibly, there is no other solution than getting 'streetwise' – at least in the limited sense of developing an understanding of what factors out there influence the pure cost of risk and on the premium actually charged. After all, the famous adage goes 'the past is not always a good guide to the future', and the reason why it is not, is that circumstances constantly change, and there is really no option other than keeping abreast of all these changes and being aware of the effect that they might have on your professional activity.

5.1 Regulatory Environment

Of all the things that can affect pricing, regulation is perhaps the most obvious, as it imposes constraints on what can be insured, how much capital must be held by insurers and possibly how products should be priced.

In the United Kingdom, for example, insurance is regulated by the Prudential Regulation Authority and the Financial Conduct Authority, while in the European Union the regulatory framework is Solvency II.

Here is a short list of possible ways in which the regulatory framework affects insurance in general. Note that some of the following requirements (such as premium restrictions and information restrictions) will affect pricing directly; others (such as capital requirements) will affect pricing indirectly; others yet (such as deposit requirements) are unlikely to affect pricing in any significant way.

5.1.1 Product Restrictions

Regulation restricts the type of business that can be underwritten/the products that can be sold. For example, the suretyship product (see online appendix on insurance products) has been forbidden in some European territories until recently.

DOI: 10.1201/9781003168881-6

5.1.2 Premium Restrictions

Regulation may limit or control the premium rates that can be charged. Motor insurance prices in Turkey, for instance, were liberalised only in 2008: prior to that, the premiums were tariff based, that is, the insurance supervisory authority decided what premiums should be charged. Even after 2008, certain limitations exist on premiums that may be charged, and the premiums are reviewed by the authority every three months and must be frozen in between successive reviews (Gönülal, 2009).

5.1.3 Information Restriction

Regulation may restrict the information that the insurer is allowed to use as a basis for pricing. As an example, the European Court of Justice has recently forbidden insurers to use sex as a rating factor in insurance purchases, which has a significant effect for certain policies such as motor insurance; and as another, related example, only a few decades ago, race was an accepted rating factor for motor insurance!

5.1.4 Capital Requirements

Regulation may prescribe a minimum amount of capital that an insurer has to hold for the risks it underwrites. It may also prescribe the types of assets the insurer must hold and may determine that high-risk assets cannot be counted towards the regulatory capital.

In the United Kingdom, the regulator currently has prescriptions for the amount of capital to be held by insurers. A new European-wide regulatory regime, Solvency II, has been set to start for a long time now and will have onerous requirements not only on the capital amount to be held by insurers but also on all the firm's processes, which will need to be aligned with the insurer's capital model. Specifically for pricing, the firm will need to be able to demonstrate that all pricing decisions and assumptions are documented, monitored, and linked to the firm's capital model.

5.2 Legal Environment

Claims processes and amounts are affected by legislation and court decisions. This could be ignored almost completely by actuaries looking at past statistics to model the claims process if the legal environment did not change. However, the legal environment constantly changes and, therefore, a loss that occurred 10 years ago would warrant a very different payout were it to occur today, even regardless of inflation.

In Chapter 2 (Insurance products) we have already come across one clear case in which a ruling by the House of Lords (*Donoghue v Stevenson*, [1932] UKHL100) created the legal environment for liability insurance and created an altogether new type of insurance claims.

We have not attempted a separation of legislation and courts effects. Separating the two is quite difficult, especially in countries such as the United Kingdom where court rulings create precedents and become basis for future decisions.

Here are a few other ways by which the legislation environment can affect the claims production process.

5.2.1 Changes in the Legal Environment May Increase or Decrease the Number and Severity of Claims

Changes in legislation and court decisions may *increase* the frequency and severity of claims, by extending the scope of compensation or by introducing new counts of liability. Examples:

- In 1999, legislation was passed that ruled that the discount rate applied to lump sum compensations accorded for liability claims was to be the risk-free discount rate, as it could not be assumed that the claimant or the claimant's guardians would be able to achieve investment returns above that. A series of actuarial tables, nicknamed the 'Ogden tables' after Sir Michael Ogden, QC, was used to calculate the lump sum, with the exact discount rate set by the Lord Chancellor. As a consequence, all liability claims became instantly more expensive from that time onwards, as a larger payment was necessary to achieve the same present value of the compensation (a severity effect). It should be noted that such rules applied not only to future claims but also to every claim that was still open. The legislation around this has been updated multiple times since then.
- Allowing for the possibility of no-win, no-fee claims in the late 1990s opened the floodgates to a large number of claims that wouldn't otherwise have been made (a frequency effect).

Changes in legislation and court decisions may also *decrease* the frequency or the severity of claims, by the compulsory introduction of health-and-safety regulations or by restricting the types of cases that can be brought to court. Examples:

- The frequency of motor accidents has steadily decreased in frequency for a number of years owing to the introduction of drink-driving rules, speed limits and controls.[1]
- The introduction of compulsory seat belts in 1983 in the United Kingdom has considerably reduced the number of fatalities, although again this may have been compensated by an increase of other types of motor claims – such as pedestrian injuries – caused by drivers feeling 'safer' and driving faster.
- The Jackson reform for liability claims, which was introduced in 2013 as a partial remedy to the problem of the no-win, no-fee claims mentioned above, should help bring down the number of claims, for example, by banning referral fees in case of personal injury claims.

5.2.2 Court Rulings Filter Down to Out-of-Court Claims

Also, one should be aware of the possible misconception that court rulings will only affect claims that are brought to court. It is true that most liability claims are settled in out-of-court negotiations, but the court decisions are used as a benchmark for these negotiations.

5.2.3 Lost in Translation

Another way in which the legal environment affects the number and the severity of claims, has to do with the jurisdiction where the claim is made and accepted. A lot of the features

1 It should be noted, however, that this reduction in the number of accidents hasn't translated into a reduction of the number of claims, which have actually increased, arguably due to the impact of the 'claims management companies', actively encouraging and soliciting claims.

of the amounts and treatments of types of claim vary significantly from jurisdiction to jurisdiction.

As an example that we have already mentioned, the severity of liability claims can be much greater in the United States than in other countries owing to the inclusion of punitive damages (rarely used in the United Kingdom) and to the fact that juries are also involved in deciding the level of compensation. The lack of strong social security system and the dependence of privately funded health care providers also contributes to the increase in indemnity costs to liability insurers. As a result of this, claimants engage in *jurisdiction shopping* where there is scope for doing that, for example, by going out of one's way to make aviation claims in a US court.

5.2.4 Court Inflation

The legal environment also affects the severity of claims by one feature called 'court inflation', which is a tendency for courts to award more generous compensations to claimants (especially for such things as pain and suffering) as time goes by, over and above the general level of inflation that applies to the goods relevant to the claims (such as medical care costs).

5.2.5 General Trends in Behaviour and Awareness

The legal environment is not a product of legislation and court awards only: it captures general trends in behaviour and awareness of compensation. Examples:

a. The last few decades have seen an increased propensity to claim, that is, an increased tendency to seek compensation when compensation is likely (this is sometimes referred to as the 'compensation culture').
b. The propensity to make fraudulent claims (including the overstatement of the claimed amount, for example, in a house contents claim) is often a function of the social acceptability of making these claims – whether the public attitude is of condemnation or of sympathy (for example, because the insurance companies may be perceived to try to get away from legitimate claims and therefore it is just playing tit-for-tat).
c. The way in which advertisement (an example being no-win, no-fee deals) increases awareness of the legal rights of claimants and makes them more likely to seek compensation for claims.

5.3 Larger Economy

As hinted previously in Section 4.4.2, when listing the possible causes of an underwriting cycle, the performance of some classes of business is to some extent linked to general economic conditions. Examples:

- *Creditor insurance* products (payment protection, various income protection insurances) and *credit insurance* products (trade credit, mortgage indemnity guarantee) are obviously going to be affected by an economic recession as many people are made redundant or their purchasing power decreases otherwise and they are unable to repay their debts.

- An economic recession may lead to a flurry of *professional indemnity* claims from people who have been given wrong investment advice, or D&O claims from disgruntled shareholders, and even if these claims are not upheld, defence costs can be significant. Also, during a recession, employers are more likely to lay off employees, who in turn are more likely to bring an employers' liability claim because they have no concern about straining their relationship with the employer and need to replace the lost salary.
- Indirectly, most insurance products are affected by economic conditions, as under economic duress there is an increased tendency to file *fraudulent claims* during a recession (such as setting fire to a shop that is losing money), or simply a *greater incentive to file legitimate claims* that would otherwise not be pursued.

5.3.1 How Does This Affect Pricing?

The pricing actuary cannot be oblivious to general economic conditions, especially when writing specific classes of business that depend on the economy. Specific methodologies must be developed for those classes of business that have such systemic dependence, for example, credit risk. Experience rating, which is based on creating a model based on the analysis of past loss experience, will need to be replaced by different techniques or at least complemented with a consideration for the possibility of economic downturns and with informed projection of what the economic climate will be in the coming policy year. We will see examples of this in Chapter 24, Pricing Considerations for Specific Lines of Business.

5.4 Investment Conditions

Investment is very relevant to the setting of premiums. The reason is that there is always a delay between premium receipt and claim payment – during that time, the money received as premiums can be invested and yield a return. This increases the underwriting profits that the insurer makes and allows the insurer more leeway in setting the premiums.

Also, the capital held by insurers for regulatory purposes can be invested (albeit in safe enough assets) and that also has an effect on the overall insurance profit. This has led some to argue that underwriting risk is better described as borrowing money at an interest rate related to the operating ratio and investing it at a higher rate (see, for example, Briys and De Varenne 2001). This is described in more length in Section 19.1.d.

These observations about the relevance of investment apply, to some extent, to all standard products from all classes of business, but they are especially important in long-tail business such as liability, where the delay between the occurrence of a claim and its payment (and therefore between premium receipt and claim payout) can be considerable.

5.4.1 How Does This Affect Pricing?

As it will be shown in Section 19.1.d, investment returns should be taken into account explicitly when setting insurance premiums, especially for long-tail business and in periods where investment returns are high – hence the need to carefully consider investment conditions.

5.5 Currency Movements

Currency movements may affect the performance of insurers writing policies in different territories. As an example, a multi-territory motor fleet policy may be underwritten by a UK insurer and the policy premium set in British pounds. However, claims will arise in the local currency of different territories depending on where these claims occur.

To show how this causes a problem, assume that

- The overall premium can be somehow allocated to different territories and that the premium for Germany is £800k.
- The premium is received at time $t = 0$ and the claims are all paid exactly 1 year later, at $t = 1$ y.
- The total amount of claims is €950k, and this was roughly the amount the insurer had expected.
- The British pound/Euro exchange rate is 1 £ = 1.25 € at time $t = 0$ and 1 £ = 1.15 € at time $t = 1$.

Under these circumstances, the expected loss ratio would have been

$$\text{LR}(\text{constant exchange rate}) = \frac{\frac{€950\text{k}}{1.25}}{\frac{€1\text{M}}{1.25}} = 95\%$$

Whereas the effective loss ratio, after accounting for the change of currency, is

$$\text{LR}(\text{variable exchange rate}) = \frac{\frac{€950\text{k}}{1.15}}{\frac{€1\text{M}}{1.25}} = 103\%$$

Therefore, a business that should have been profitable has turned into a loss-making one, not because something has gone wrong with the loss experience but because of currency movements.

5.5.1 How Does This Affect Pricing?

The example above shows that – although pricing actuaries are in no better position than anybody else to predict what the exchange rate will be in the future – they may need to consider the risk of adverse changes in the exchange rate calculations when modelling different profitability scenarios. It should be noted that the proper response to the risk of adverse currency exchange rate changes can also be offset by holding assets in the same currency as the liabilities.

5.6 Natural Environment

Ever since the birth of insurance with its focus on sea journeys, the natural environment has played a major role in insurance, be it through normal weather events and seasonality or through major catastrophes such as extreme weather events and geophysical events.

5.6.1 Weather

Actuaries cannot ignore the weather – not even when they stay indoors, happily crunching numbers. Weather is the source of many insurance claims, either directly (a hailstorm ruining a harvest and causing a crop insurance claim) or indirectly (many more car accidents occur in winter than in summer, due to poor visibility and poor road conditions; more subsidence claims happen in the summertime, when the terrain is drier).

Even excluding catastrophic weather events such as hurricanes or floods, basically all types of weather events, such as rainfall, hailstorm, wind, snow, drought and cold/hot temperatures, or sustained periods of those can cause losses to crops, tourism, and buildings. It is not necessary for the event to be extreme to be potentially damaging: the disruption of expected weather patterns may suffice.

Weather is a systemic risk, as it affects many locations and many people at the same time. There are also some systemic elements of a more global nature such as global warming, which might have an effect on long-term tendency for claims.

5.6.2 Space Weather

There is another type of weather that we normally do not give too much thought to and that, to be fair, does not normally cause us so many difficulties as the one for which we have daily forecasts from the media, but which can occasionally be quite disruptive: space weather.

The sun is responsible for such things as solar winds (low-density emissions of plasma, easily rebuffed by the Earth's atmosphere), solar flares and coronal mass ejections (intense bursts of particles that normally take place in concurrence with an increase in sunspot activity, which in turn follows a cycle of about 11 years, as illustrated in Figure 5.1). These phenomena normally cause nothing more harmful than the spectacular northern lights, but when the *magnetic storms* and *radiation storms* they cause are particularly severe, they may result in severe interference with electronic devices and constitute a health hazard.

When dealing with weather risk, however, it is not even enough to confine yourself to our solar system! Radiation from the galaxy – high-energy particles that are produced as a result of the explosion of supernovae – can also be dangerous for aircraft and satellites. Interestingly, this radiation is shielded by the solar wind during periods of high solar activity and reaches the solar system only during the quiet periods: it is therefore countercyclical.

Here are just a few examples of the disruptive effects of space weather:

- Power grids: for instance, in 1989 in Quebec, a major magnetic storm caused the power grid to shut down abruptly and it could not be brought back to operation for 9 hours.

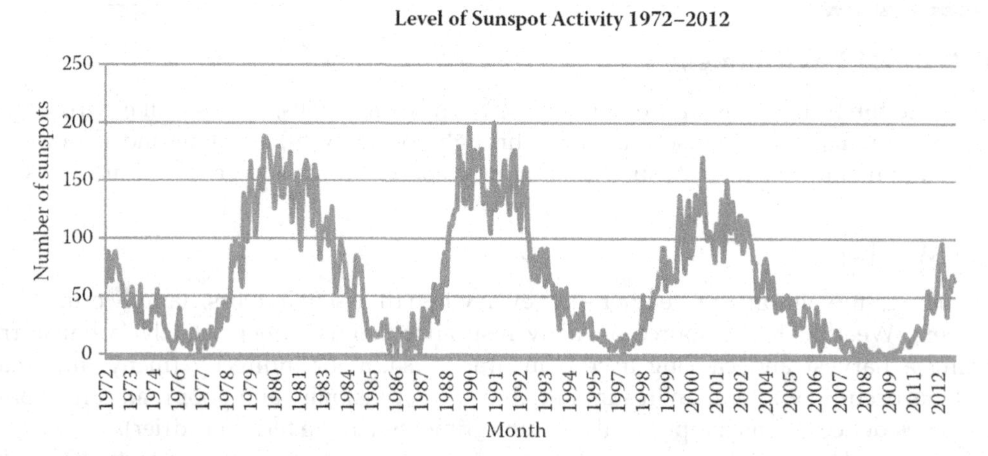

FIGURE 5.1
Number of sunspots as a function of time. (Based on publicly available data from NASA: http://solarscience.msfc.nasa.gov/greenwch/spot_num.txt.)

- Aviation: for example, solar and galactic radiation interferes with GPS navigation systems and may expose crew and frequent flyers to high levels of radiation. (Airlines regularly monitor radiation levels in the cabin during transpolar flights.)
- Communications: satellites that are used for communications and GPS purposes are occasionally damaged by solar and galactic radiation. To give you an idea of the numbers involved, according to the Space Weather Prediction Center (a US government organisation) in the period 1994 to 1999, insurers had to pay out $500M in compensation for insurance claims related to space weather (NOAA 2013).

On the positive side, space weather has a much higher degree of predictability than terrestrial weather because its manifestations are closely correlated with the solar cycle, which is of 11 years.[2] A peak in sunspot activity is expected in the summer of 2013, and hence we expect more losses related to magnetic storms (and less, say, to galactic cosmic rays) in this period.

If you want to know more, an entertaining read on space weather is the joint Lloyd's/RAL Space report *Space Weather*, written by Mike Hapgood (2010), to raise awareness of space weather risk among businesses.

5.6.3 Natural Catastrophes

Natural catastrophes (affectionately, or curtly, denoted as 'nat cats' by practitioners) include such phenomena as earthquakes, floods, wind, volcanic eruptions, tsunamis, heat waves, landslides, and winter storms (see, for example, the online repository of natural catastrophe information available at http://site.ebrary.com/lib/disaster/home.action); therefore, a mixed bag of geophysical phenomena and things that could go under the heading of 'severe weather'.

2 Technically the solar cycle is actually of 22 years, because the magnetic field changes direction every 11 years – that, however, doesn't seem to have any bearing on space weather.

Natural catastrophes are one of the major concerns for insurers, especially for property insurance as well as for business interruption and liability, for their potential accumulation of risks. Indeed, one of the key things that a property insurer will need to check when considering a portfolio of properties for cover is their location and the consequent exposure to natural catastrophes. It is important for insurers to make sure that their overall portfolio of properties does not have an excessive accumulation risk in one particular catastrophe-prone area (or in any area, for that matter). Question 2 encourages you to think about what areas are susceptible to what types of catastrophes.

Natural catastrophes are also a concern (and a business opportunity) for reinsurers who sell Cat XL protection specifically to protect insurers from the effect of catastrophes.

Note that there has been a long-standing increase in the cost of major catastrophes, as more development has been made in catastrophe-prone areas, and as the proportion of insured houses has risen.

Also, the geophysical landscape itself is not stationary – for example, there is a noticeable increase in the number and severity of North Atlantic hurricanes, and global warming may be upsetting the climate and make long-term predictions of trends difficult.

5.6.4 How Does This Affect Pricing?

The natural environment is directly relevant to pricing when pricing weather derivatives (Section 23.2) or catastrophe covers such as Cat XL: the environment itself is exactly what we are trying to model. Catastrophes also need to be priced in when pricing property products. In this case, catastrophes are only one of the perils (others are, for example, fire and explosion).

In all cases in which catastrophes are involved, catastrophe modelling techniques need to be used. These are different in principle from standard actuarial techniques because they are not based on the policyholder's past losses but on scenarios produced by scientific models that combine the knowledge of past catastrophic events with the knowledge of the geophysical mechanisms that produce the catastrophe (Chapter 25). Actuaries, therefore, need to work alongside catastrophe modellers ('cat modellers') when pricing property risk.

Apart from catastrophes, the seasonality of weather may need to be taken into account whilst doing pricing of any class of business, for example, when using the number of motor losses incurred during the springtime and summertime of a given year to estimate the number of losses that will be incurred over the whole year in the United Kingdom: if you do not take into account the fact that normally you get more motor losses in winter, you will probably underestimate the total number of losses.

5.7 Pricing in the Corporate Context

Pricing is not an isolated activity in the company but is related to what other actuaries and other professional figures in the company do. In this section, we look at some of the interactions between pricing and other corporate functions – which ultimately means, between pricing actuaries and other professionals in the company. We will do that by listing what the main functions of an insurer are and comment on the interaction between these functions and the pricing function.

5.7.1 Planning Function

The planning function sets operational objectives in the context of the values, purpose and broad strategic objectives of the business (such as the composition of the portfolio), which lead to guidelines and parameters for pricing decisions. The underwriting cycle enters the debate at this stage because it must be reflected in the financial objectives and targets.

5.7.2 Underwriting Function

Underwriters decide (within the limits of their authority and in collaboration with the other functions) which risks the company should accept and how they should be priced. Underwriters have an increasingly tighter relationship with pricing actuaries in a market such as the London Market.

Actuaries will normally provide underwriters with an estimated cost for the policy, and possibly with a technical premium, and the underwriter will then make the commercial decision within certain parameters. On the other hand, underwriters provide actuaries with the context and with assumptions, such as on what is exactly included in the policies (in terms of cover, territories, exclusions, etc.) and what losses should therefore be included in the historical loss analysis.

Also, actuaries and underwriters will work together in product development and the pricing of non-standard products, such as multi-class, multi-year products that modify risk profiles and cash flows.

5.7.3 Claims Function

The claims management function produces individual loss estimates for all claims of the company according to the claims manual, which outlines the processes on how to deal with reported claims and defines authorisation levels, that is, who needs to assess the amount of a claim depending on its characteristics and the likelihood that the claim exceeds a certain threshold, say £25,000.

5.7.3.1 Interaction with the Pricing Function

The claims management function provides, either directly or through an IT system, all the historical loss information that actuaries use. Also, claims people are in the best position to explain the company's practice on how claims are reserved, reported, and so on, which will help pricing actuaries in the interpretation of data.

The pricing actuary needs to understand the effect of any changes in claims handling that may distort loss data, such as new factor estimates or changes in legislation that affect future claims cost. Pricing/reserving actuaries might then provide statistics on claims amounts by type, on claims inflation trends, and so on, which can be used as part of the control cycle to set the initial estimate for the claim.

Also, actuaries (especially reserving actuaries) will be able to feed back to the claims people their findings on the presence of systematic overreserving or underreserving trends.

5.7.4 Actuarial: Reserving

The job of reserving actuaries is that of estimating the likely amount of outstanding liabilities across a portfolio of risks and advising the management on this amount. This is done at the portfolio level.

5.7.4.1 Interaction with the Pricing Function

The output of the reserving process is important for pricing purposes. As an example, in the traditional burning cost analysis (a pricing technique discussed in Chapter 11), the projection of every year's reported claims to ultimate is normally done based on portfolio-level loss development factors provided by the reserving function.

Also, reserving is part of the control cycle for pricing because it is only by following the development of the total claims for each year that we find whether the assumptions made in pricing were borne out on average. The pricing function can, on the other hand, help the reserving function in setting some of the assumptions, such as the expected severities of large losses that might be reported with long delays, such as asbestos-related claims.

5.7.5 Actuarial: Capital Modelling

The capital modelling function advises on how much regulatory and economic capital[3] the company needs to hold to abide by regulations and remain solvent and how this capital should be serviced by the different lines of business. Capital modellers, like reserving actuaries, need to look at the risks, not contract by contract, but at the portfolio level, focusing on the large losses and their correlations.

5.7.5.1 Interaction with the Pricing Function

Capital modelling affects pricing especially in the determination of the premium (see Chapter 19), which will normally need to cover the expected losses plus a loading for expenses and profits. We will speak about expenses later, but as far as profits are concerned, one way to look at the profit loading is as a return on the capital that the investors have put into the company, and which is necessary to keep the company solvent. Capital loads are therefore determined for different lines of business, and the pricing actuaries and underwriters will need to make sure that this loading is incorporated equitably across all contracts.

On the other hand, pricing actuaries might need to provide loss models similar to those used in the pricing of contracts but based on the overall portfolio and focusing on the large losses only. Pricing actuaries will need to price the risks of a portfolio much in the same way as a reinsurance pricing actuary would price an excess-of-loss reinsurance policy (see Chapter 21).

5.7.6 Finance

The finance function is in charge (among other things) of raising the capital necessary to run the company and choosing the projects that make the best use of the company's

3 The regulatory capital is the capital that needs to be held to satisfy the regulator. The economic capital is the capital that the company actually decides to hold to remain solvent and achieve a good credit rating. Obviously, the economic capital must be larger than the regulatory capital.

resources based on risk and return. It is in charge, among other things, of financial reporting/accounting, treasury management, relationship with the regulator and relationship with the rating agencies.

5.7.6.1 Interaction with the Pricing Function

Most of the relationships between the finance function and the pricing function are indirect, with some exceptions, one of which is expense allocation. The premium charged for a policy contract will normally include an allowance for expenses, which include costs related to the claims handling process, the policy management process, and a contribution to the infrastructure expenses of the organisation. Much in the same way as capital, expenses are usually allocated from the group down to portfolios, often by the accounting function. The underwriter and the pricing actuaries can then work out how to split them at the level of the individual contract.

5.7.7 Management Information

The management information function analyses the performance of the company in terms of loss ratios for different categories, renewal rates, conversion rates, portfolio composition and the like and reports to the management of the company, enabling the management to make informed strategic decisions.

5.7.7.1 Interaction with the Pricing Function

The management information function receives from pricing actuaries, up-to-date information on price rates and price rate changes. It feeds back to the pricing actuaries the key factors listed above, which allows the pricing actuary to revise their assumptions (such as on frequency, severity or on rating factors) as necessary.

It is important for the pricing actuary to understand the sources of management information and details of calculations. Even for a simple concept such as loss ratio, there are numerous variations, such as incurred loss ratio, paid loss ratio, and ultimate loss ratio on an underwriting year or accident year basis.

5.7.8 Investment Function

The investment function decides what investments are the most adequate to hedge the company's liabilities, purchase the assets and track their performance. Investment actuaries may be part of this function, advising on the investment mix that is suitable to hedge the company's current and future liabilities.

5.7.8.1 Interaction with the Pricing Function

Pricing actuaries (along with reserving actuaries) may advise the investment function on the expected future liabilities for the portfolio mix. The investment function may provide pricing actuaries with investment information that can be used for discounting future liabilities (as explained, for example, in Chapter 19).

5.7.9 Reinsurance Function

The reinsurance function is in charge of inwards and outwards reinsurance. The outwards reinsurance function decides on how much reinsurance to buy and negotiates prices with the reinsurance underwriters. The inwards reinsurance function is in charge of selling reinsurance, if the company is in that business. Note that *reinsurers* have inwards and outwards reinsurance functions as well.

5.7.9.1 Interaction with the Pricing Function

First of all, pricing actuaries may be actually pricing reinsurance as well, whether to negotiate the purchase of outwards reinsurance or to price inwards reinsurance. Furthermore, the reinsurance function will advise the pricing function as to what allowance one should make in the technical premium for the cost of reinsurance (see Chapter 19 for details). Finally, reinsurance pricing actuaries might provide (unless Chinese walls need to be in place) direct insurance pricing actuaries with information on tail losses, and direct insurance actuaries might provide reinsurance pricing actuaries with information on ground-up losses for the construction of increased limit factor curves (see Chapter 23).

5.7.10 Risk Management Function

According to a popular framework used by companies (see, for example, IIA (2013)), there are three lines of defence against risk in insurance companies: the first line of defence owns the risk; the second line of defence oversees the risk; and the third line of defence provides an independent assurance.

In the context of an insurance company, the underwriters and the management provide the first line of defence, while the internal audit function provides the third line of defence. The second line of defence is provided by the risk management[4] function, whose job is to '[facilitate] and [monitor] the implementation of adequate risk management practices by operational management and assists risk owners in defining the target risk exposure and reporting adequate risk-related information throughout the [organisation]' (IIA, 2013).

Here are a few examples of activities that the risk management function of an insurance undertaking leads or participates in:

- the definition of the risk appetite and tolerance of the company;
- portfolio steering, for example avoiding the excessive concentration of catastrophe risk for a given peril/territory combination;
- the identification and measurement of *emerging* risks;
- the sign-off process of large transactions;
- the assessment of underwriting quality and pricing adequacy;
- the validation of important risk models, including pricing models.

4 Risk management, in case a reminder is needed, deals with the identification of, quantification of, and implementation of controls around the risks faced by a corporation. These include, among others, business risks (the core risks related to conducting the business and dealing with the competition), credit risks (the risk that debtors won't pay), market risks (the risks arising from investment performance, interest rate changes, etc.) and operational risks (the risks internal to the organisation arising from people, systems, and processes).

5.7.10.1 Interaction with the Pricing Function

The risk management function has mostly an indirect relationship with pricing, through the control it exercises on the underwriting function (see list above). However, it also has a more direct relationship with actuarial pricing on issues such as pricing adequacy and the validation of pricing models.

5.7.11 Other Functions

The functions indicated above are only the most obvious functions that the pricing actuary needs to work with. Other functions include

- The *marketing function*, which identifies successful insurance products and promotes them in the marketplace
- The *sales function*, which coordinates the sales operations of the company
- The *infrastructure functions*, such as IT, HR, Legal and Compliance, which are necessary for the functioning of any company

Pricing actuaries will mostly have an indirect relationship with these functions. However, the work of these offices is fully relevant to the pricing function. For example, the use of certain rating factors is not lawful, and policy wording may have consequences on the expected losses for a policy contract.

5.8 Other Things You Need to Keep Up With

Political risk. The political situation all over the world is relevant to many insurance products such as trade credit, marine, property, and personal accident. The political situation is also immediately relevant to terrorism cover, riot cover, and others, which is part of many property policies.

Latent claims. Asbestos was not considered to be a problem at all when employers' liability policies were written, but emerged as a huge issue in the 1990s, almost crippling the insurance industry in the United Kingdom.

Medical risks. One such risk is that of pandemics (COVID-19 is of course a recent example); also, there is the possibility of new risks, such as genetically modified organisms, mobile phone radiation, or nanotechnology: will one of them turn into another asbestos?

Scientific and technological advancements. Every new advancement brings with it new risks and new risk solutions. Think of aeroplanes – because of them, we now have to think of the effect of high levels of radiations from space or volcano emissions: something we could have ignored a century ago. Advances in scientific knowledge can also establish a previously missing causal link between, say, the use of a product and subsequent disease, leading to new types of claims.

5.9 Questions

1. It is now illegal in the United Kingdom (and the rest of Europe) to use sex as a rating factor, for example, in motor insurance. Still, there are some insurance providers, which were traditionally aimed at the women's markets, that are still around. Do some personal research to check what happened to them after the change in legislation.
2. Identify, for each of these territories, the two or three most likely natural catastrophe events and discuss if the risk is homogeneously spread over the territory or localised in a specific area:
   a. United Kingdom
   b. United States
   c. Japan
   d. Italy
   e. Germany
   f. A territory of your choice
3. Is space weather more or less predictable than terrestrial weather? Why?
4. Doing personal research, find the list of the top 10 natural catastrophes by death toll and by insured losses in the last 30 years. Comment on how similar these two lists are.

Part II

The Generic Experience Rating Process

6

The Scientific Basis for Pricing: Risk Theory

In Chapter 1 we looked at an extremely simplified, and old-fashioned, way of doing pricing, by considering the total losses incurred in a policy over the years. By a series of enhancements, this method can actually be developed to the point where it actually provides a reasonable first-cut indication of what the correct price may be, provided the loss experience is sufficient: this is the so-called burning cost analysis, which will be illustrated in Chapter 11. However, to put pricing on a sound scientific basis, we need to understand and model the process by which losses are generated. This is what **risk theory** – the stochastic modelling of insurance losses – is about.

In this chapter, we introduce risk theory as a useful method to capture theoretically the loss-generation process. Section 6.1 (Aggregate Loss Models) is devoted to studying the properties of the two main models of risk theory – the collective risk model and the individual risk models – and shows examples of situations in which they can be applied. Risk costing can then be reformulated (Section 6.2) as the problem of finding the correct distribution for the number of claims (the frequency distribution) and of the individual loss amounts (the severity distribution) and the value of the necessary parameters. Section 6.3 illustrates the revised risk costing process based on the frequency/severity analysis.

6.1 Aggregate Loss Models

The road through which risk theory became the main framework for loss modelling in general insurance has been rather bumpy. The story is interesting but is outside the scope of this book; it has however been told succinctly and effectively in Turnbull (2017), to which the interested reader is referred.[1]

The main idea of risk theory is to model the aggregate losses over a period as a compound stochastic process where both the number of losses over a given period and the size of each individual loss are random variables.

1 Earlier attempts at providing such historical background myself have quickly proven that I'm no historian and have therefore been abandoned.

DOI: 10.1201/9781003168881-8

The main aggregate loss models used in general insurance are the individual risk model (Section 6.1.1) and Lundberg's collective risk model (Section 6.1.2). Good in-depth references for this material in this chapter are the book *Loss Models: From Data to Decisions* by Klugman et al. (2008) and the classic *Risk Theory* by Beard et al. (1984).

6.1.1 Individual Risk Model

We have n risks on the book, and each of them has a probability $p_1, \dots p_n$ of having a loss during the policy period. If a loss occurs, the amount could either be a random amount or a fixed amount. Only one loss is possible for each risk. The total losses for each year will then be

$$S = X_1 + \dots + X_n$$

Where X_j is the loss for risk j, which will be = 0 if risk j has no loss. The X_js are often written as the product of two random variates:

$$X_j = I_j Y_j$$

where $I_j = 1$ if risk j has a loss, 0 otherwise, and Y_j is the loss amount if a loss occurs (therefore, $Y_j > 0$ always). Y_j is called 'loss given default' in the context of credit insurance – we'll shortly see why.

We make the following two additional assumptions: that the loss events are independent and that the severities are independent (but not necessarily identically distributed). We can summarise all the assumptions as follows:

Individual Risk Model: Assumptions

IRM1 – There is a finite number n of risks.

IRM2 – Only one loss is possible for each risk j, and this happens with given probability $p_j = \Pr(I_j)$ $(j = 1, \dots n)$.

IRM3 – The loss events are independent – that is, the fact that a given risk has a loss does not have an effect on whether or not another risk has a loss.

IRM4 – The severities Y_j are independent but they are not necessarily identically distributed.

The expected losses over a period and the corresponding variance (remembering that the loss events and the severities are independent) are given by

$$\mathbb{E}(S) = \sum_{j=1}^{n} p_j \mathbb{E}(Y_j)$$

$$\mathbb{V}(S) = \sum_{j=1}^{n} p_j \mathbb{V}(Y_j) + p_j(1 - p_j)\mathbb{E}(Y_j)^2 \tag{6.1}$$

Although the assumption of a finite number of risks in the book is sensible (there is always a finite number of policyholders, even though the number of risk units may increase/decrease during the year, for instance in the case of fleets of cars or employees in a company), the assumption that each risk has at most one loss in the year is often too strict. Consider fire insurance for properties: the same property may have more than one loss in a given period. However, there are cases in which this is a perfectly sensible model because the loss is an event that by definition only happens once: life insurance is perhaps the most obvious example. We are not concerned with life insurance in this book, but there are some life-related examples that do fall within the remit of general insurance: one is accidental death cover. The other, which we look at in Section 6.1.1.1, is cover against the death of animals. Yet another example (Section 6.1.1.2) concerns credit defaults, since bankrupt is to companies what death is to living beings.

6.1.1.1 Example: Horse Insurance

Horse establishments that breed horses for racing or other purposes will normally have insurance against the death of their animals. This is a type of property insurance. The loss amount is normally fixed, and this is agreed at the outset with the insurer, and reflects the value of the horse, whether it be a racehorse, a stallion, or a broodmare.

> *Question*: Which assumption of the individual risk model may not be appropriate and in what circumstances? How would you deal with that?
>
> *Answer*: The main issue here is in the fact that loss events are independent. Although normally horses will die independently, it is quite easy to imagine ways in which their deaths may be correlated: for example, if they are poisoned by some of the food they are given or if they fall victim to an epidemic. The effect here is to increase the probability of dying for all (or some) of the horses dying simultaneously. There are many ways to deal with this: a very easy one is to assume that occasionally there may be common shocks that increase the probability of death for all horses at the same time – and try to quantify the occurrence of those and the increase in probability.

6.1.1.2 Another Example: Credit Insurance

Death is a type of loss that can only happen once to any individual person or animal, but it is not the only one. Another one is default – which happens when one officially ceases to be able to pay one's debt. When a company defaults, from the point of view of its creditors, it ceases to do business and all that happens is that you may get a share of the proceeds of bankruptcy, so that the debt is partially repaid. The type of insurance that deals with that is called credit insurance. This may be more or less interesting than insuring racehorses, but it is certainly a more common example, and we present it as the second example and not the first only because it is more complex.

Credit insurance is a type of financial loss insurance, by which the insured gets compensation if the third party (typically a company) with which it does business is unable to pay its debts and therefore declares bankruptcy. The proceeds of the liquidation of the company are then distributed among the creditors, who may then be able to recover some of the losses.

The loss amount is the loss given default (LGD), which will be equal to the amount of the exposed debt minus any recoveries. Credit insurance is more complicated than the insurance of horses because the LGD is variable and not fixed as the amount payable on the death of a horse is.

Question: Which assumption of the individual risk model may not be appropriate and under what circumstances? How would you deal with that?

Answer: Again, loss events may not be independent – for example, lots of bankruptcies may derive from a common cause, such as a recession (in 2008, trade credit events certainly increased in number). The severities also may be related – if something really big happens, it may be that not only several companies will go bankrupt but also that their bankruptcies may be more severe, leaving a different residual value and therefore generating a different LGD. Again, the correlation in frequency may be dealt with by modelling a common shock to the system that increases all probabilities at the same time. As to the LGD, one may have (speaking simplistically) an LGD distribution for 'the good times' and one for 'the bad times'.

6.1.2 Collective Risk Model

This is the main model used in non–life insurance. As in the case of the individual risk model, there are going to be a number of risks in the book, but the idea of the collective risk model is to shift the focus from the risks to the losses. We ignore the question of how many risks there are and what the individual probability of loss and severity of loss is for each risk, and instead we simply assume that there are going to be a number N of losses in each particular year, with N being a discrete random variable: $N = 0, 1,\ldots,$ according to some distribution F_N of claim counts: $N \sim F_N(\lambda_1, \lambda_2 \ldots)$ where $\lambda_1, \lambda_2 \ldots$ are the parameters of the claim count distribution. Examples of commonly used frequency distributions are Poisson (with just one parameter: the rate) and the negative binomial distribution (with two parameters: the rate and the variance/mean ratio).

Each of the losses will have a severity X_j, according to some severity distribution F_X, which is the same for all losses (in other words, the losses are independent and identically distributed random variables): $X_j \sim F_X(\mu_1, \mu_2 \ldots)$ where $\mu_1, \mu_2 \ldots$ are the parameters of the severity distribution. Examples of commonly used distributions are the lognormal distribution, the Gamma distribution, and the Pareto distribution. We also assume that the loss number N and the severity X of the losses are independent (i.e. the severity of a given loss does not depend on how many losses there were in a given period). We summarise all the assumptions in the following list:

Collective Risk Model: Assumptions

CRM1 – The number of losses N can be described as a discrete random variable with distribution F_N.

CRM2 – The loss amounts X are independent, identically distributed random variables with common distribution F_X.

CRM3 – The loss number N and the severity X of the losses are independent (i.e. the severity of a given loss does not depend on how many losses there were in a given period).

According to the collective risk model, the total losses for each year (or any other period, for that matter) will then be

$$S = X_1 + \ldots + X_N \qquad (6.2)$$

where N is the number of losses in a given year, and X_j is the amount of a given loss. Equation 6.2 looks very much like the expression for the individual loss model, the main difference being that here N is a random variable.

The problem of finding the distribution of the aggregate losses is not a trivial one and will be addressed in Section 17.1. However, there are two important questions about the aggregate loss distribution that we can answer simply. The first question is: what is the total amount of losses that we expect in a given year? In other words, if we were to repeat this experiment for variable tens of thousands of years and the loss distribution remained the same, what would the long-term average of the total losses be?

This question can be answered simply by calculating the mean $\mathbb{E}(S)$ of the random variable S, which can be easily found to be equal to the mean $\mathbb{E}(N)$ of the number of losses times the mean $\mathbb{E}(X)$ of the loss amount:[2]

$$\mathbb{E}(S) = \mathbb{E}(N)\mathbb{E}(X) \tag{6.3}$$

The other question that we can easily answer is what the year-on-year volatility of S around the mean $\mathbb{E}(S)$ is going to be. This is given by the standard deviation of S or, equivalently, by the variance $\mathbb{V}(S)$. Again, it is easy to see[3] that the variance $\mathbb{V}(S)$ is given by

$$\mathbb{V}(S) = \mathbb{V}(X)\mathbb{E}(N) + \mathbb{E}(X)^2\,\mathbb{V}(N) \tag{6.4}$$

Finally, the moment generating function of S, defined as $M_S(t) = \mathbb{E}(e^{tS})$, is given by[4]

$$\mathrm{M_s}(t) = \mathrm{M_N}\left[\log \mathrm{M_X}(t)\right] \tag{6.5}$$

The moment generating function is interesting for at least two reasons:

First, as the name suggests, it allows one to calculate all moments of the aggregate loss distribution by calculating the derivative of $M_S(t)$ at $t = 0$ as many times as necessary, for example, $M_S'(0) = \mathbb{E}(S)$, and in general $M_S^{(k)}(0) = \mathbb{E}(S^k)$. By using Equation 6.5, you can then express $\mathbb{E}(S^k)$ as a function of the moments of X and N and have alternative derivations, for example, of Equations 6.3 and 6.4 (see Question 1).

Second, the moment generating function is formally similar to the Laplace transform and to the Fourier transform of the underlying distribution. This opens up specific possibilities on how to calculate numerically the aggregate loss distribution, as we will see in Chapter 17: specifically, this makes it possible to use Fast Fourier Transform (FFT) techniques, which are very efficient.

Because the collective risk model is the main risk model used in non–life insurance, it is not difficult to find actual examples where the collective risk model is used. Most lines of business are priced with the collective risk model in mind. Examples are employer's liability, public and product liability, motor (private/commercial), property (residential/commercial), aviation, marine, many reinsurance lines of business ... just about everything, really. Even situations which could be modelled more rigorously with the individual risk model, such as the horse insurance illustrated in Section 1.1, can be approximated with the collective risk model. Below, we have chosen motor fleet losses as an illustration.

2 Using an elementary property of the conditional mean (see for example Klugman et al., 2008) we can write $\mathbb{E}(S) = \mathbb{E}[\mathbb{E}(S \mid N)] = \mathbb{E}[N\,\mathbb{E}(X)] = \mathbb{E}(N)\,\mathbb{E}(X)$.

3 Using an elementary property of the conditional variance (see, for example, *Loss Models* by Klugman et al., 2008) we can write $\mathbb{V}(S) = \mathbb{E}(\mathbb{V}(S \mid N)) + \mathbb{V}(\mathbb{E}(S \mid N)) = \mathbb{E}(N\mathbb{V}(X)) + \mathbb{V}(N\,\mathbb{E}(X)) = \mathbb{E}(N)\,\mathbb{V}(X) + \mathbb{E}(X)^2\mathbb{V}(N)$.

4 The proof goes as follows: $M_s(t) = \mathbb{E}(e^{tS}) = \mathbb{E}[\mathbb{E}(e^{tS} \mid N)] = \mathbb{E}[\mathbb{E}(e^{tX_1} e^{tX_2} \ldots e^{tX_N} \mid N)]$ = (because all vairiables involved are independent) $\mathbb{E}([\mathbb{E}(e^{tX})]^N) = \mathbb{E}[\mathbb{E}[M_X(t)^N] = \mathbb{E}(e^{N\log M_X(t)}) = M_N[\log M_X(t)]$.

6.1.2.1 Example: Modelling Motor Fleet Losses

We have a fleet of K cars in a company, and the whole fleet is insured against third-party liability (own damage is retained). Every year, we have a number N of losses, which is a discrete random variable. Every loss will be of a certain amount X_j, another random variable, with an underlying severity distribution.

> *Question*: Are the assumptions valid? When do you expect them to break down? How would you go about correcting for that effect?
>
> *Answer*: None of the assumptions will ever be completely valid. However, the thorniest one is probably the independence of N with the loss amounts X_j. It is quite easy to picture situations in which these are not independent. For example, we know for a fact that bad weather is a factor that affects the number of motor losses. To the extent that the severity of losses might depend on whether, for example, because of low visibility the accidents *may* tend to be more serious, then there is an obvious correlation between N and X_j, driven by their dependency on a common factor. By the same reasoning, if in a given year there is a particular cause (say snow) that causes an unusual number of claims, and this cause tends to produce claims with a different severity distribution $X_{j,\text{snow}}$ from the general one, this will cause some degree of dependency among losses in a given year. As you will have noticed, in both examples, we have a common element that drives the dependency.
>
> Even when common factors are absent, the assumption of independent severities won't be perfect. For example, if the most expensive car in the fleet is written off mid-year, the largest possible property damage claim will now be the value of the second most expensive car in the fleet: therefore, the shape of the severity distribution (at least as far as the own damage part of the claim is concerned) will have changed.

Another question relates to how the size of the fleet affects the model. Suppose, for simplicity's sake, that we have modelled the number of losses of our fleet as a Poisson variable with rate λ and the severity losses as a lognormal distribution with parameters μ and σ.

> *Question*: How does the size K of the fleet enter the risk model above?
>
> *Answer*: It does not – K will not appear explicitly anywhere in the collective risk model describing third-party liability losses for our fleet. That does not mean, of course, that the fleet size is irrelevant to the future loss experience: the fleet size will be taken into account during the creation of the loss model, when we have to determine the parameters λ, μ and σ. Specifically, the Poisson rate λ is likely to be proportional to the current fleet size.

When one looks more closely at the details of the application of the collective risk model to motor fleet losses, some additional difficulties emerge.

First, each vehicle can have more than one loss, but if it crashes too badly, it will have to be written off and will therefore not be able to have further losses – in other words, the underlying fleet has changed and so the overall risk changes and the initial model will not, strictly, be valid anymore (although it will probably be replaced). Actually, we can generalise this a bit further, by considering that vehicles may be removed from or added to the fleet for a number of different reasons. Although in our model we have assumed that the fleet has K

vehicles, the number K is never constant – vehicles are taken off the fleet and others are added.[5] In other words, the overall risk changes continually and so should the model. In practice, this is only a problem if the changes are significant, and the usual (approximate) approach is to price the policy to a certain fleet size and then adjust the final premium at the end of the policy to the actual average over the year.

Also, a motor loss normally has several components that (depending on the use we want to make of a model) might need to be modelled separately – such as own damage (excluded in our example), third-party property damage, and third-party bodily injury.

6.1.3 Extensions of the Collective and Individual Risk Models

The collective risk models and the individual risk model are quite powerful, and their use is ubiquitous; however, as we have seen, their underlying assumptions are seldom fully realised. Extensions to these models are possible – for example, Bolancé &Vernic (2020) extend the collective risk model to the case where there is a Sarmanov-type dependence[6] between the number of claims and the severities.

In the rest of this section we'll look at different ways in which the basic version of the CRM and IRM can be extended.

6.1.3.1 Common-Shock Model

In most cases, the breaches to the assumptions of the collective and individual risk models can be overcome during calculation by the introduction of a simple workaround, such as using the already-cited common-shock model (Meyers, 2007). This assumes that the claim count and severity variables of the individual/collective risk model are dependent on one another not through a direct correlation but because they are affected by the same underlying variable (the 'shock' variable). Two examples for the individual risk model (IRM) and the collective risk model (CRM) are described below.

6.1.3.1.1 Example for the IRM: Credit Insurance

We saw in Section 6.1.1.2 that the LGDs in a credit risk model appear to be correlated, and that they appear to be correlated to the probability of default (when there is a large number of defaults, creditors recover less of their money). Instead of trying to model this correlation explicitly, we can assume that the number of claims and the severities are affected by a common variable that describes the general health of the economy. In its

5 To take this into account, insurers measure the size of motor fleets not by the number of vehicles but by the number of vehicle-years, according to which each vehicle contributes to the count only for the fraction of the year when it was on the book.

6 Following almost verbatim Bolancé &Vernic (2020), a Sarmanov-type dependence between each severity X_i and N is one where $(X_i, N)_{i\geq 1}$ are identically distributed with the following Sarmanov probability density function:

$$f_{X,N}(x,n) = \begin{cases} p(0) & n = x = 0 \\ p(n) f(x)\left(1 + \omega\psi(n)\phi(x)\right) & n \geq 1, x > 0 \end{cases}$$

With ψ, ϕ bounded non-constant kernel functions, ω a real number and $f(x)$ a probability density function. For the expression above to define a valid pdf, the constraints $\sum_{n\geq 1}\psi(n)p(n) = 0$, $\int_{-\infty}^{\infty}\phi(x)f(x)dx = 0$, and $1 + \omega\psi(n)\phi(x) > 0$ must also be satisfied.

simplest incarnation, this might be a simple binary variable R that is equal to zero when the economy is normal and equal to 1 when the economy is in a recession. The parameters of the claim count and severity distribution will then depend on R, and that creates a correlation among the severities, and between the severities and the number of claims.

6.1.3.1.2 Example for the CRM: Motor Fleet Insurance

We saw in Section 6.1.2.1 that bad weather may cause an increase in both the number of motor claims and their severity. We may assume that the parameters of both the claim count distribution and the severity distribution are affected by a common variable W related to the weather. This could for example be the number of days in the year in which the temperature has dropped below zero, or some measure of precipitation.

The ideas behind the common-shock model will make their appearance again in the rest of the book, and specifically in Sections 14.4.3, 23.3 and 30.2.4.3.

6.1.3.2 Hierarchical Risk Models

Another possible extension of the risk models is the additional of other random variates (besides the common shock). While the basic collective risk model has the number of losses N and their severity X as the fundamental variates, there are many situations in which it makes sense to consider N and X (either or both) as compound variables.

As an example, to model business interruption losses one should consider the number N of losses, the time T (in days) for which business is interrupted, and the amount Y lost per day during the interruption, all as random variates. The severity X would then arise from the convolution of T and Y.

Notice however that examples like this are not real extensions of the collective risk model – they simply show that it is sometimes helpful to look at its deeper structure for modelling purposes, defining a hierarchy of variables. Also note that while theoretically valid, in practice a three-variate approach to business interruption is hardly ever used as it is difficult to calibrate.

6.1.3.3 Collective Risk Model with Time-Dependent Parameters

Another, subtler way – specific to the CRM – in which the assumptions fail is perhaps more obvious in the case of the modelling of property losses, and we hinted at it when we looked at motor fleet losses.

As we will see at length in Chapter 21, in the case of property insurance we have a finite number of locations (buildings, plants, etc.), each with its own severity curve. A Poisson model (see Chapter 14) is normally adequate to model the number of losses N (note that each property can in general have more than one loss during the policy period), while the overall severity X for the 'portfolio' of locations depends on the severity distribution of the individual locations.

This is normally a good approximation. However, if one thinks a bit more deeply about the situation at hand, it becomes clear that once – for example – a location is hit by a total loss that location needs to be written off for the rest of the policy period and the overall severity distribution changes. This is perhaps most obvious when we imagine that the most valuable property be struck: if that property is written off, the maximum possible loss for the rest of the policy period is the value of the second-most-valuable location.

How does one deal with this? Theoretically, one needs an extension of the CRM in which the parameters of both the claim count distribution and the severity distribution depend on the state of the portfolio at a given time. When one property is struck (especially with a large or total loss), we need to update both the number of properties (and therefore the Poisson rate) and the severity distribution.[7] This is not impossible to do (and can be achieved with numerical methods such as Monte Carlo simulation) but it is complex and requires in-depth knowledge of how a specific loss impacts the possibility of having another loss at a given location. Even the calculation of the expected aggregate loss cannot be calculated simply by using Equation 6.3.

Given the difficulty of incorporating the right assumptions into the model, the (potential!) gain in accuracy is unlikely to be worth the increase in complexity, and this time-dependence is typically disregarded.

6.2 Applications to Risk Costing

The heart of pricing lies in quantifying the risk. An underwriter who has to decide how much to charge for a policy needs to assess what the likely losses will be in a given year, how much volatility there is around that number and how often the end of the year will result in a profit. We call the activity of quantifying the risk 'risk costing' in that it helps determine the cost of the risk taken on by the underwriter. The cost of risk is related to the price of a policy in the same way as the cost of producing a dishwasher is related to the price at which the dishwasher is sold – the price is ultimately a commercial decision and does not necessarily bear an obvious relationship to the cost, but not knowing the cost of producing a dishwasher almost inevitably leads to poor commercial decisions.

It might seem that we have already solved the problem of risk costing – after all, we have given methods to derive the aggregate losses from a frequency and a severity model. However, of course, we have not addressed the most important questions: (i) where do the parameters of the frequency and severity models come from and (ii) how should the outputs of the aggregate loss model be used?

What we have achieved in this chapter is simply a justification of the widespread use of the frequency/severity model in risk costing, and we have seen how the frequency and the severity model can actually be used to produce an aggregate loss model that yields the expected cost of risk and its volatility. This now allows us to review the simplistic risk costing process of Chapter 1 and produce a new process that includes frequency/severity modelling.

6.3 Risk Costing Process for the Frequency/Severity Model

Now that we have taken a look at the scientific basis for risk costing, we can incorporate the frequency/severity approach in the risk costing (sub)process. The new risk costing process is shown in Figure 6.1. Note that some of the content of this flowchart will be

7 Note that when every loss is a total loss then the situation is best modelled using an IRM. For this reason, a collective risk model with time-dependent parameters can be considered as a hybrid model between a pure IRM and a pure CRM.

obscure at this point; however, things will become clear in the next few chapters, and the general flow should be easy to grasp.

Although the process in Figure 6.1 does not capture all the possible variations of the risk costing process (and we will see several variations of this process in Part III of the course, which deals with the pricing of specific types of businesses such as non-proportional reinsurance), it is a rather standard process that one should be therefore become accustomed to.

Note that the risk costing subprocess described in Figure 6.1 refers to the case of experience rating. When exposure rating is used, the process simplifies considerably as the parameters of the frequency and severity models are calculated based on the exposures only and are therefore almost immediately available (see Chapters 21 and 22 for details).

The rest of Part II of this book will mostly be devoted to looking at the various stages of the risk pricing process for experience rating illustrated in Chapter 1, with the risk costing subprocess exploded as in Figure 6.1.

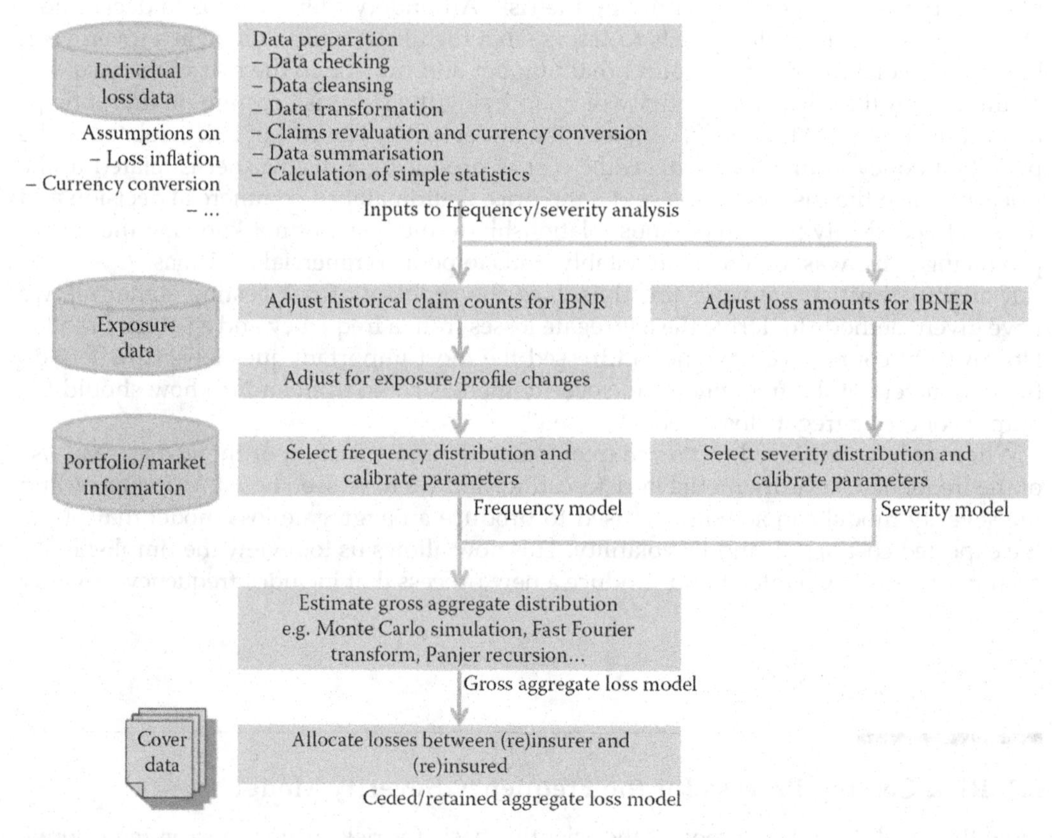

FIGURE 6.1
The risk costing subprocess according to the frequency/severity approach. The main difference between this and the simple example is that frequency and severity are analysed separately and are then combined with, for example, a Monte Carlo simulation. Note that in practice there will be many variations to this simple process, but it is fair to say that this is the 'vanilla' version of the frequency/severity approach.

6.4 Questions

1. (*) Prove that for the collective risk model $\mathbb{E}(S) = \mathbb{E}(X)\,\mathbb{E}(N)$ and that $\mathbb{V}(S) = \mathbb{E}(N)\,\mathbb{V}(X) + \mathbb{V}(N)\,\mathbb{E}(X)^2$ *using the moment generating function* $M_S(t)$ *of* S. In the formulae above, N, X and S are the random variables for the number of claims, the severity, and the aggregate losses, respectively.
2. One of the most popular risk models for pricing is the collective risk model for total losses, S.
   a. Explain what the collective risk model is and what its underlying assumptions are.
   b. Discuss the validity of the collective risk model's assumptions when used to model employers' liability losses.
   c. One method by which it is possible to approximately calculate the distribution of S when (as is normally the case) an exact solution is not available is Monte Carlo simulation. Outline the workings of this method, including an explanation of how the 95th percentile of the aggregate loss distribution can be derived. *Note*: This will be addressed at length in Chapter 17.
   d. State two other methods that you could use for approximately calculating the distribution of S. *Note*: This will be addressed at length in Chapter 17.
3. Describe briefly all the steps of the standard risk costing process based on the collective risk model for the experience rating of a portfolio of risks (e.g. a company which takes a motor insurance policy), including inputs (types of data) and outputs. Feel free to use a flowchart or a list as you prefer, but make it clear what the inputs and outputs of the different parts of the process are. You can ignore the role of portfolio/market information.

7

Familiarise Yourself with the Risk

It is difficult to overemphasise the importance of familiarising oneself with the risk one is trying to price before delving into the technicalities of pricing (Figure 7.1), and still it is difficult to write something sufficiently general about it. This is the phase of pricing that is least susceptible to being automated, as it requires getting to know your client[1] (whether you are the underwriter, a broker or a consultant) with a mind that is as open as possible. There is no real prescription but what you want, in a nutshell, is to make sure that whilst you eagerly apply what the underwriters and the brokers may call with some complacency 'your actuarial magic' to the data you are provided with, you do not ignore the context in which your client operates and the potential risks it faces.

It should also be emphasised that knowing your client well is not useful only for pricing a particular risk, but it helps in identifying other insurance needs of the client and in general (if you have an advisory role and not only a selling role) to advise on the overall risk management strategy.

The questions that you may ask of a prospective client are different if you are pricing a personal lines policy rather than a commercial lines/reinsurance policy and, in the case of mainstream policies such as car insurance or home insurance, it may not seem necessary. However, even in that case, you must ensure that you have a preliminary idea of what the portfolio of individual risks looks like.

In the following we'll briefly discuss some simple steps that one should take to familiarise oneself with the risk. The reader interested in a wider discussion on this topic is referred to the short book by Serena (2018).

7.1 Things to Look Out for (Commercial Lines/Reinsurance)

In commercial lines and reinsurance, it is customary to look at each client individually, although smaller clients might be treated in a streamlined fashion. Here are a number of things you might want to consider before or whilst you are looking at the data related to your client.

1 In this book, we will use the term 'client', 'policyholder', and 'insured' (or 'reinsured') quite interchangeably to indicate the organisation, or the individual, who purchases an insurance (or reinsurance) product or inquires about purchasing one.

DOI: 10.1201/9781003168881-9

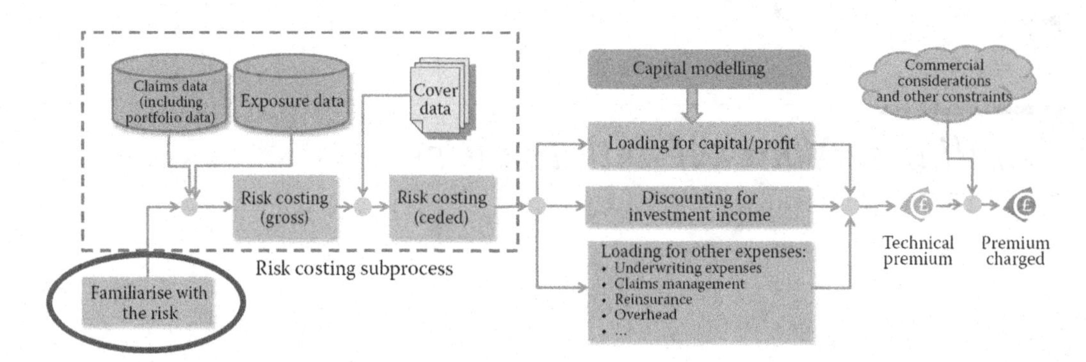

FIGURE 7.1
The pricing process (high level).
The first thing to do in a pricing exercise is to familiarise with the risk; that is, getting to know your client and its industry (this comes from Figure 1.7).

7.1.1 What Does the Client Do, That Is, What Is Its Business?

For an underwriter, it is important to familiarise with the client's business and industry – the client's risks will – to a varying degree depending on what is covered – depend on the type of business they operate. For a broker, or a consultant, it will be important both for risk considerations and to win their clients' confidence on their ability to give them informed advice.

7.1.2 What Are the Client's Main Location(s) and Jurisdiction(s)?

This is obviously related to the client's exposure to risk. The locations where the client operate are very important, for example, for property risk, as it has an impact on whether the client will be affected by geo-dependent perils such as natural catastrophes and terrorism. Information on locations can often be found via map applications on the Web.

The jurisdiction(s) in which the client operates are very important for liability risks as the way the amount of compensation for a claim will be determined depends critically on the jurisdiction. For example, several European countries determine the compensation for bodily injury claims on the basis of tables that relate the compensation to the degree of disability suffered by the claimant, which is unlike the Anglo-Saxon world where compensation is based on negotiation between the parties and possibly by the courts.

Location and jurisdiction are of course correlated but not identical: for example, an airliner operating flights between the US and Germany but with headquarters in Switzerland may be sued for compensation in any of these jurisdictions.

7.1.3 What Are the Client's Main 'Numbers' (Exposures and Financials)?

The 'numbers' mentioned in the title provide a measure of exposure to the risk and may be related to the complexity of insurance needs and amount of coverage needed. Some of these measures of exposure can be found in the client's financial statements, which in turn can normally be found in their website under the 'Investors' section.

The financial statements also give plenty of important information about the company's financial situation which in turn will affect the company's ability to control its risks. Based on the annual financial returns it is possible to calculate a number of useful financial ratios that will provide an indication of the client's liquidity, financial structure, profitability and financial leverage (Serena, 2018).

7.1.4 Have There Been Any Notable Changes in the Risk Profile of the Client over Time?

We have spoken informally of the risk profile in Chapter 1. The risk profile is anything about a particular company or individual that is relevant to its level of risk. For an individual who buys car insurance, factors such as age and type of car affect the individual's risk profile; for a company buying employers' liability, the risk profile is determined by such things as the type of work its employees do and their salaries; for an insurer buying property reinsurance, the risk profile is the type of property portfolio, such as veered towards residential, commercial, or industrial.

This will affect the degree to which past claims experience is a good guide to future claims experience for the risk under consideration.

Specifically, a dramatic change in the risk profile may take place if there have been mergers/acquisitions or divestitures in the company over the years. This is because the merging/parting companies may differ in activity, jurisdiction, loss history, etc. and we need to ensure that the underwriting and pricing approach takes into consideration these factors.

7.1.5 Industry-Related Questions

Some relevant questions will relate not to your clients directly but to *its industry as a whole*: What market share does your client have? What were the top losses for a given line of business in that industry?

For instance, if your client is a pharmaceutical company, it helps to know that the largest amount paid for a mass tort claim to date was, for example, $4.5B, even if your client was not the one that had to pay it. It turns out that to do pricing properly for a given client, you often have to become acquainted not only with your client's business but also with its industry.

(In reinsurance, this is partly different because there is basically *one* industry you have to know rather well: the insurance industry itself and the specific companies that you are trading with. However, some knowledge of the underlying book of the reinsured is still necessary.)

7.1.6 Sources of Knowledge on the Risk

Below are some suggestions on where to find information relevant to a given risk:

- *The Web.* A lot of useful information can be researched on your own simply by looking up the institutional website of the client, or information on the client that can be gathered through news aggregators, consumer reviews, Wikipedia, etc. (Serena, 2018). Note that the availability of such information on the Web does not discharge the client's duty of fair representation of its risks!
- *Client's accounts.* If the client is a public company, it will need to make its financial accounts (balance sheet, profit and loss account) public, for example, on its website.

- *Insurance submission*. The client will normally prepare a submission for the insurer, including claims and exposure information, and useful background information.
- *Conversations with the client/their broker*. There are more things that you will need to discover by talking to your client and by asking questions, many of which will originate during the analysis.
- *Industry publications*. These are not related directly to the client but are helpful for understanding the client's business. They include trade publications (such as the *Oil & Glass Journal* for the petroleum industry) and industry guides (e.g. 'Best Practice Guidelines for the Production of Chilled Foods' for the food industry).
- *Scientific publications*. There are often scientific papers, of an actuarial nature or not, on the risk that you are interested in. Examples are articles on the effect of frost on beet harvests, actuarial papers on extended warranty risk and so on.

7.2 Things to Look Out for (Personal Lines)

In this case, rather than a specific client with a portfolio of risks (such as properties, employees and cars) you have a portfolio of clients, each of whom makes a purchasing decision independently. However, it is still possible to form an idea of what this portfolio of risks is by analysing

- The make-up (profile) of the portfolio, in terms of the distribution risk factors that are expected to be relevant for the risk underwritten, such as age, car model, location, and many others for a portfolio of car policies
- Changes in the profile, because of customers entering/exiting the portfolio
- Any particular large losses relevant to that portfolio

Ultimately, however, this is useful especially for small personal lines portfolios, whereas for a large portfolio, understanding its features and the drivers of the risk is very much part of the core analysis itself.

7.3 Questions

1. On the Internet, look up the list of the top 5 pharmaceutical companies and pick one of them. Let us call that company X. Now imagine that X has come to your underwriter for a quote for product liability. Find out as much as you can about the exposure of X to this risk and answer as many of the questions listed in this chapter as you can. What have been the largest product liability payouts in the pharmaceutical industry?
2. Choose a large high-street retailer Y and find as much as you can about it by looking at public information on the Web, including their financial statements. What is their territorial distribution? Are they present abroad? What have their revenues been in the last 4 to 5 years? How many employees do they have? What would you say their main risks are? Have they ever had a very big loss?

3. Choose a medium-sized UK insurer and try to reply to the following questions: What lines of business are they writing? What are their total revenues? How many employees do they have? What was their gross written premium in 2012? What was their operating ratio? What was their expense ratio?
4. Find out the payouts and circumstances of the five most expensive car insurance claims in the United Kingdom's history.

8

Data Requirements for Pricing

The actual analysis mostly starts with data, and you will need to have appropriate data to undertake your risk costing exercise. As Figure 8.1 shows, there are several sources of data that you will be interested in: some of it will be related to the client[1] (or prospective client), and some will be external to the client (such as portfolio or market information).

One reason why it is important to understand what data you need for a pricing exercise is that you often have to request this data explicitly and make sure it meets certain requirements. Alternatively, if you work for an underwriter, you may be in a position to help define how a claims/policies management system is to be structured.

8.1 Policy and Cover Data

To start with, we need to collect some basic information on the policies that we are supposed to price. In the case of commercial lines, there may be a single policy, such as the employers' liability policy purchased by an organisation. In personal lines, pricing is carried out on a portfolio of policies rather than on each policy in isolation. In reinsurance treaties, we normally price a single reinsurance policy with an underlying portfolio of policies underwritten by the reinsured.

Furthermore, policy data is paramount in other exercises apart from pricing, such as the analysis of the performance of a portfolio.

Regardless of the number of policies we have in the portfolio, this is typically what you should have for each policy:

- Policy ID
- Policy inception date
- Policy expiry date
- Policy basis (occurrence, claims made …)
- Policy limits and deductibles (individual and aggregate), both current and historical

1 In practice, client-related data may be kept by the client itself (for example, by its risk management function), or it might be in the hands of the insurers (both the current insurer and the ones from which the client has purchased insurance in the past).

DOI: 10.1201/9781003168881-10

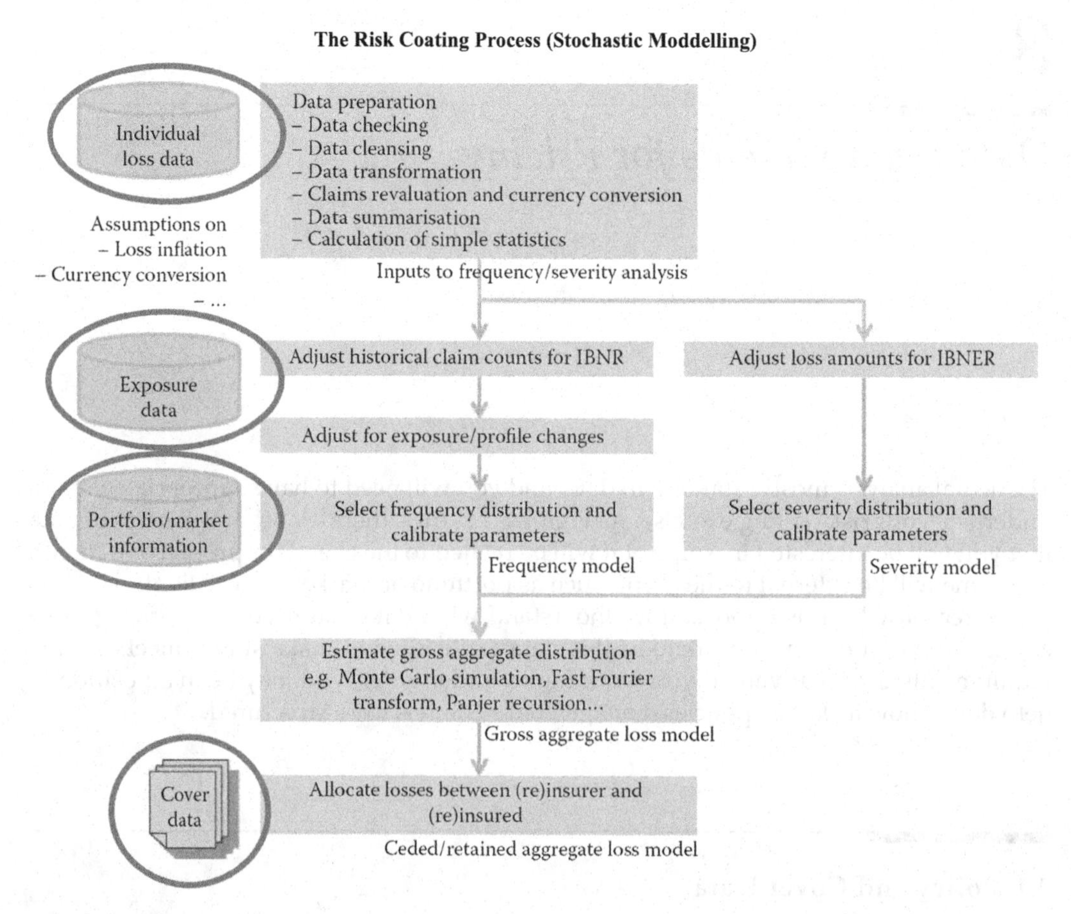

FIGURE 8.1
Data on different aspects of the risk (losses, exposure, cover and portfolio/market information) enters the process at several points. Note that exposure data and portfolio information are used for several of the steps of the process in frequency and severity modelling. For example, portfolio information can be used for tail modelling, frequency benchmarking and incurred-but-not-reported (IBNR) and incurred-but-not-enough-reserved (IBNER) adjustment.

- Participation percentage (if the insurer does not underwrite 100% of the risk, but co-insures the risk with others)
- Relevant exposure
- Rating factors (if applicable)
- Gross premium
- Brokerage fee
- Type of coverage
- Territories
- Exclusions

In the case of reinsurance, we also need other information such as the index clause, the hours clause, and the reinstatements. This will be addressed in more detail in Chapter 21 (Experience Rating for Non-Proportional Reinsurance).

Example of policy data					
Policy ID	**Policy_7022**	**Policy_16141**	**Policy_20434**	**Policy_7870**	**Policy_19714**
Inception Date	01/01/2008	27/07/2008	29/07/2008	01/01/2009	01/01/2009
Expiry Date	01/01/2009	27/01/2010	28/07/2015	01/01/2010	01/01/2010
Territory	Spain	Australia	United Kingdom	Russian Federation	United States
Client	Client_2	Client_7	Client_15	Client_17	Client_23
Class of Business	PI	D&O	D&O	D&O	PI
Insured's Industry Sector	Project management	Manufacturing	Energy	Technology, Media, Telecomms	Legal Services - Solicitor
Currency	EUR	AUD	GBP	USD	USD
Limit Amount	3,000,000	10,000,000	20,000,000	15,000,000	20,000,000
Excess Amount	0	10,000,000	0	15,000,000	3,000,000
Layer Type	Primary	Excess	Primary	Excess	Primary
Gross Premium	8,468	15,281	208,750	82,500	1,547,100
Brokerage Deductions	1,111	2,717	46,969	11,813	274,180
Net Premium	7,356	12,564	161,781	70,687	1,272,920

FIGURE 8.2
Example of policy data (multi-policies case).

Figure 8.2 shows an example of policy data that has actually been used for the analysis of the performance of a portfolio of financial institutions policies (mostly professional indemnity [PI], directors' and officers' liability [D&O], financial institutions' insurance [FI]).

Obviously, not all the items in the list are included every time. For example, it is more common for rating factors to be included in personal lines policies than in commercial lines policies.

8.2 Claims Data

Normally, claims data will include the following. Note that practitioners sometimes refer to 'loss' and 'claim' to mean two different things, the first referring to the insured's loss and the second referring to the amount the insured claims based on the policy the insured has purchased. However, we will often use the two terms interchangeably here, as is often done.

Field	Notes
Claim ID	Essential to identify the claims correctly, avoid double counting, etc.
Relevant policy ID	Essential to connect to information on rating factors, deductible, limit, rating factor details, etc. Note that in many cases the deductible, limit, etc., of the policy may appear directly as fields in their own right

Field	Notes
Loss date	The date when the loss actually occurred. Essential to assign the loss to the proper policy year (if policy is *losses occurring*) In the case of *claims-made* policies, this may not be available, or may be replaced with a 'trigger date' (or otherwise named) where the action/advice for which the policyholder is now pursued by the claimant was taken/given. In this latter case, the trigger date is used to estimate the reporting lag between the trigger date and the reporting date
Reporting date	The date when the loss was reported to the insurer. This is essential to assign the loss to the proper policy year (if policy is claims made) This may be replaced, or complemented, by the date in which the loss was uploaded onto the system, and can be used to calculate the provision for IBNR losses (Chapter 13)
Quantification date	There is no unique name for this, but what is meant is a field that gives the point in time at which the claim was first assigned an estimated amount This is not necessary if the full history of the estimates is provided anyway (as will normally be the case in a good claims management system), but if not, this field is essential to estimate the reporting delay of non-zero claims Although in many classes of business an estimate is entered for the loss as soon as the loss is put into the system, there are classes of business (such as PI and D&O) for which the amount is entered only when there is enough certainty about the amount itself. The reason is legal: if the claimant's legal team has knowledge of the estimated loss amount, that becomes almost by default the very minimum that the claimant will be asking as compensation
Settlement date (if applicable)	Used to calculate settlement pattern
Claim status (open, closed, reopened.)	Used to determine whose claim amounts are almost certain (apart from reopening) and which are still subject to variations
Paid loss amount to date (or individual payments with dates and type of payment – indemnity, lawyers' fees, adjustors' fees)	This allows one to determine the 'certain' part of the claim (apart from recoveries) and to calculate the payment pattern (if dates of payments are given). It contributes towards the total incurred amount Ideally, all the payments with their respective dates should be provided
Outstanding loss amount to date (or individual outstanding estimates with dates)	Knowing how much of the claim is outstanding provides information on how uncertain the loss amount is and provides input for the analysis of IBNER adjustment Because the outstanding loss amount will be revised periodically (especially in the case of the larger claims), one should ideally receive not only the latest estimate of the outstanding loss but also *the full history of different estimates*

Field	Notes
Incurred loss amount to date (or individual incurred estimates with dates)	Note that Incurred = Paid + Outstanding This may be omitted if Paid and Outstanding are present. In any case, either this or Paid and Outstanding is essential for pricing analysis
Fees amounts	Unless the incurred amount includes not only the indemnity but also the fees (such as legal fees), this field is essential to determine the overall cost of the claim. Ideally this should be split between paid and outstanding as well.
Recoveries (e.g. salvage, recoveries from third parties and others)	This needs to be subtracted from the overall amount to calculate the effect of the claim on the insurer
Currency of payments and outstanding amounts	Essential when more than one currency is involved
Type of claim (e.g. own damage, third party damage or third party bodily injury)	Essential if one wants to determine the effect of adding/removing one part of the policy, and for IBNR calculation purposes. Also, it may be necessary if different deductibles/limits apply to different types of claims (e.g. no deductible on motor third party, £300 deductible on own damage)
Type of peril (such as fire)	Used for costing modifications to the policy

Claims Data Set – Example for Direct Insurance

Policy ID	BZW3143854UF	BZW3143854UF	BZW3143854UF	BZW3143854UF	BZW3143854UF
Claim ID	203306	334247	469379	22150	152899
Claimant name	Name_1	Name_2	Name_4	Name_7	Name_8
Loss date	08-Aug-2005	24-Oct-2007	16-Aug-2007	11-May-2007	13-Mar-2007
Reporting date	13-Sep-2005	19-Dec-2008	28-Jul-2008	16-Dec-2009	15-May-2008
Circumstance	Manual handling	Others	Falling or flying object	Manual handling whiplash	Falling or flying object
Injury description	Sprain	Burn/Scald	Bruise/oedema/swelling		Bruise/oedema/swelling
Territory	United Kingdom	United Kingdom	United Kingdom	United Kingdom	United Kingdom
Settled date	28-Sep-2009	26-Feb-2009	23-Aug-2010		
Loss status	Closed	Closed	Closed	Open	Open
Original currency	GBP	GBP	GBP	GBP	GBP
Total OS	0	0	0	10,484	9000
Total paid	33,166	0	20,216	9044	0
Total incurred	33,166	0	20,216	19,528	9000
Fees	5472	0	3356	2910	0
Recoveries	0	0	1240	510	0
Overall amount	38,638	0	22,331	21,928	9000

FIGURE 8.3
A typical claims data set for direct insurance (also called a 'loss run'). Note that the claims are listed here by column rather than by row (which is more common) – this is just for the table to appear clearer on paper.

Claims Data Set – Example for Reinsurance

Claim ID	Loss date	Loss year	Type	2004	2005	2006	2007	2008	2009	2010	2011
210245	03/07/2004	2004	Paid	–	–	300,300	300,300	300,300	720,261	720,261	720,261
210245	03/07/2004	2004	O/S	4,050,000	2,500,000	469,700	413,578	419,961	–	–	–
210245	03/07/2004	2004	Incurred	4,050,000	2,500,000	770,000	713,878	720,261	720,261	720,261	720,261
211136	04/07/2004	2004	Paid	–	–	–	–	–	–	–	–
211136	04/07/2004	2004	O/S	190,000	140,000	140,000	–	–	–	–	–
211136	04/07/2004	2004	Incurred	190,000	140,000	140,000	–	–	–	–	–
213330	10/07/2004	2004	Paid		–	–	–	–	–	–	–
213330	10/07/2004	2004	O/S		50,000	50,000	–	–	–	–	–
213330	10/07/2004	2004	Incurred		50,000	50,000	–	–	–	–	–
216085	15/07/2004	2004	Paid	–	–	21,773	50,942	50,942	50,942	50,942	50,942
216085	15/07/2004	2004	O/S	50,000	85,000	43,227	–	–	–	–	–
216085	15/07/2004	2004	Incurred	50,000	85,000	65,000	50,942	50,942	50,942	50,942	50,942
217372	12/07/2004	2004	Paid	–	–	–	–	11,146	11,146	11,146	11,146
217372	12/07/2004	2004	O/S	265,000	300,000	300,000	11,146	–	–	–	–
217372	12/07/2004	2004	Incurred	265,000	300,000	300,000	11,146	11,146	11,146	11,146	11,146
217513	02/07/2004	2004	Paid	2172	2172	2172	2172	2172	2172	2172	2172
217513	02/07/2004	2004	O/S	–	–	–	–	–	–	–	–
217513	02/07/2004	2004	Incurred	2172	2172	2172	2172	2172	2172	2172	2172
220236	23/07/2004	2004	Paid	–	–	1,659,825	1,659,825	1,659,825	1,659,825	1,659,825	1,659,825
220236	23/07/2004	2004	O/S	1,500,000	1,500,000	–	–	–	–	–	–
220236	23/07/2004	2004	Incurred	1,500,000	1,500,000	1,659,825	1,659,825	1,659,825	1,659,825	1,659,825	1,659,825

FIGURE 8.4
Typical claims data set for reinsurance (main fields only). As an example to show how this table should be read: claim 210245 is reported for the first time in 2004 and a provisional reserve of £4.05M is established. The reserve is then revised to £2.5M in 2005. In 2006, the overall reserve is yet again revised downward to £770k, and a first payment of £300.3k is made, leaving £469.7k outstanding. Fast forward to 2009, where the claim is settled for £720.3k. For more information on the claims data in reinsurance, see Chapter 21.

In the ideal situation, then, one obtains the full list of movements for all claims. Often, however, one is asked to price a policy based simply on a snapshot of the claims situation for all years under examination at one point in time (i.e. the paid/outstanding amounts are the most recent ones), as in Figure 8.3, rather than having the full history for each claim.

This will hamper our ability to estimate possible IBNER (adverse development of claims), and we will either need to assume that the estimates will not show any systematic deviation from the ultimate amount, or complement this snapshot with other information, such as a number of different snapshots at different years, or aggregate triangulations. We will look at this into more detail when discussing IBNER (Chapter 15). Figure 8.3 shows an excerpt from an actual claims snapshot for commercial lines (data was changed to make it unrecognisable).

For pricing purposes, we would ideally have claims over a period of approximately 10 years – there is no strict rule about this period, but we need to strike a balance between having too short a time window (and therefore little credibility) and too long a time window (the years far back will have little relevance). Having 5 to 10 years of experience is normal in commercial insurance and reinsurance. Less may be needed in personal lines insurance given the large amount of data collected, but that depends on the purpose of the particular investigation; for example, no-claim discount bonuses may need data collected over a large period.

Figure 8.4 shows another excerpt from a standard data set – this time for reinsurance data. The main difference in this case (apart from the fact that we have reduced the number of fields to avoid cluttering the table with too much information) is that for each individual loss, the year-end position of payments and reserves is presented so that one can determine if there is any systematic overestimation or underestimation of reserves.

Of course, there is no reason why one should have data presented in this format in direct insurance as well as reinsurance – normally, however, such a level of detail is probably more common where large losses are involved and where reserves are actually checked and reworked with regularity.

8.3 Exposure Data

8.3.1 What Is Exposure?

We have already discussed exposure measures, but it is time to go through this again in more detail. An exposure measure is a quantity that is roughly proportional to the risk of a policyholder (or a group of policyholders). Examples of exposure measures are the number of vehicle years for motor, the wage roll or number of employees for employer's liability, the turnover for products liability, the total sum insured for property and the like. The idea is that if the exposure (e.g. the number of vehicles in a fleet) doubles whilst everything else (e.g. the 'value profile' of the cars of the fleet, the type of drivers and others) remains the same, the risk also doubles.

Because exposure is only a rough measure of risk, we may want to improve it by taking into account different factors related to the risk. For example, if we are pricing an employer's liability, it is customary for the underwriter to take into consideration not only the total wage roll but also to split that into the number of clerical and manual workers, because they have a different risk profile. If we know, for example, that manual workers have an average cost to the insurer per pound of wage roll, which is roughly four times that of clerical workers, a change in the mix of workforce definitely corresponds to a change in the total risk, and therefore the variable

$$E = 4 \times \text{total wage roll of manual workers} + \text{total wage roll of clerical workers}$$

is a better measure of exposure than the total wage roll.[2] With enough experience, one can introduce more factors in the calculation of exposure, taking into account what exactly these workers are doing, how much they are paid, what type of industry they are in and the like.[3] There is no limit to how deep one can drill, and finding a good measure of exposure may well end up in a full-blown multivariate analysis with generalised linear modelling!

This latest observation perhaps clarifies the true nature of exposure: *exposure is a prior estimate of the risk* – that is, prior to considering the actual experience of the client.

8.3.2 Exposure Data for Experience Rating

When carrying out a pricing exercise, we ideally need

- The historical exposure data over the same period for which we have occurrences or acts committed giving rise to a claim.
- The (estimated) exposure for the renewal year

The ultimate reason why we need historical exposure for experience rating is that we need to be able to compare the loss experience of different years on a like-for-like basis and adjust it to current exposure levels: for example, £130,000 of claims (revalued) in 2008 for a fleet of 1800 cars are roughly equivalent to £195,000 of claims if next year's fleet is estimated to be 2700 cars.

Note that *the measure of exposure we need to use also depends on what loss variable we are trying to project*. Specifically, it may be different depending on whether we are projecting the aggregate losses or the number of claims.

Consider, for example, employers' liability cover.

- In *burning cost analysis*, of which we saw a rudimentary example in Section 1.1, and which will be analysed in more depth in Chapter 11, we estimate next period's aggregate losses based on an exposure-adjusted average of the aggregate losses of previous years. Part of the compensation for a loss is related to loss of earnings and therefore ultimately to salary, wage roll is often chosen as a measure of exposure because it increases if either the number of employees increases or their salary increases. (As we

2 As we saw in Chapter 1, underwriters will indeed have rough rating rules such as Premium = 0.1% × Clerical wage roll + 0.8% × Manual wage roll (the actual numbers will vary depending, for example, on what type of manual work is normally performed in the company).

3 Other common factors that can influence UK EL risk assessment and pricing are: (1) Whether the company have an accredited OHSAS18001 Health & Safety Management System in Place. (2) Whether the company have a formal documented system whereby employees are provided with relevant Health & Safety Training. (3) The percentage of activities that have a written & current (<= 2 years old) risk assessment and corresponding safe systems of work. (4) Whether the company have a dedicated and competent Health & Safety Officer/Advisor. (5) Whether the company have access to an Occupational Health Service and embrace a return-to-work strategy. (6) Whether the company have been served with any improvement or prohibition notices within the last 2 years.

Fiscal year	Turnover (£ × 1000)
2002	10,500
2003	10,856
2004	11,246
2005	11,630
2006	12,047
2007	12,450
2008	12,857
2009	13,341
2010	13,798
2011	14,334
2012 (estim)	14,783

FIGURE 8.5
Example of exposure data for public liability.

have seen, some refinement may be necessary to take into account at least the difference between blue-collar and white-collar employees.)

- In *frequency analysis*, which will be discussed in Chapters 13 and 14, we estimate next period's number of claims based on an exposure-adjusted average of previous years' claim counts. In this case, the number of employees is a better measure of exposure than wage roll because (at least as a first approximation) an increase in wage roll with the same number of employees (and the same proportion of blue-collar to white-collar employees) will not increase the number of claims.

Another example is aviation hull, where average fleet value (AFV) is often used as a measure of exposure. This makes sense for burning cost analysis, but the number of claims for a specific airline is more likely to be proportional to the number of departures (or, all things being equal, number of planes), whereas the average value of an aircraft (AFV/ number of planes) is more likely to affect the severity of claims.

8.3.2.1 Criteria for a Good Measure of Exposure

Traditionally, the Casualty Actuarial Society has identified three criteria to compare the suitability of different measures of exposure. These criteria are an excellent starting point but should not be used blindly, as explained below.

i. The measure is proportional to the expected losses.
This criterion is the most important of the three but should always be used with an eye to the purpose for which exposure is used. The criterion can be used in this form for burning cost analysis but 'expected losses' should be replaced with 'expected claim count' in the case of frequency analysis.
ii. The measure is objective and easy to obtain.
This is important because it is pointless to use a measure of exposure that better captures the risk if there is no guarantee that it can be evaluated reliably and with little effort.
iii. There are historical precedents for its use.
This criterion should also be taken with a pinch of salt. Admittedly, using non-traditional measures of exposure may cause problems of acceptance and change the way that premiums are determined. However, that should not be an excuse for persisting on using inadequate exposure measures – otherwise, the same argument

could be used for discouraging the introduction of new rating factors because they lead to changes in the premium structures.

Whatever measure of exposure is used, it is important that the period over which the exposure is calculated is specified, so that it is subsequently matched to the policy years to which claims are allocated. An example of exposure data for a public liability policy is shown in Figure 8.5.

8.3.3 Exposure Data for Exposure Rating

For certain types of analysis, such as exposure rating (that is, rating based not on the client's historical loss experience but on a reasonably detailed knowledge of the risk), we need exposure information at a more detailed level.

Examples:

- Detailed property schedule (list of properties with sum insured/maximum probable loss, type of building, location, etc.) for property policies
- ... or at least properties by band of value
- Detailed personnel schedule (such as salary, age, etc.) for employer's liability or personal accident group policies ...
- ... or again aggregated by bands

We will look into this again in more detail when we consider exposure rating of property (Chapter 22) and liability rating based on increased-limit-factors (ILF) curves (Chapter 23).

8.4 Portfolio and Market Data

Portfolio (the wider pool of risks written by the insurer) and market information may help us in pricing where the client experience is scant. Both portfolio data and market data refer to non–client data, but we normally speak of *portfolio data* for data owned by the insurer (or an intermediary) and of *market data* for data in the public domain, accessible either freely or by subscription, such as exposure curves published by reinsurers or data available through the Insurance Services Office in the United States.

Here are some examples of how portfolio/market information can be used in practice.

a. If individual claims data sets for the same risk (such as motor or employers' liability) and for different clients (ideally, the whole market) are available, then a portfolio severity curve can be derived, which can be used to complement a client's information, or possibly replace it altogether, when the client's information is scant or non-existent. One common situation in which a client's information is scant almost by definition is the tail of severity distribution – the large losses region – therefore, it may make sense to use client's data for all losses below a certain threshold and then use a 'portfolio tail' above that threshold (see Chapter 17).

b. The property losses of a client can be modelled using an exposure curve (Chapter 22) and large liability losses can be modelled using an 'increased limit factor' market curve (Chapter 23).
c. In credibility theory, portfolio data is combined with the client's data to produce a more reliable estimate of the expected losses and hence a better premium (see Chapter 25).
d. The frequency per unit of exposure of the policyholder can be compared (and perhaps combined, using credibility theory) with that of the portfolio/market. This might apply to either the ground-up frequency or to the frequency for large losses, where the policyholder's experience is necessarily more limited (Chapter 14).
e. When reporting dates are not available for the client, the distribution of reporting delays for the portfolio can be used in its stead (Chapter 13).
f. The payment pattern and settlement pattern for claims can be rather unstable if analysed for a single client – bringing many clients together may help (Chapter 21).
g. Historical loss ratios for different lines of business may provide a guide for what loadings on the expected losses can be achieved (Chapter 19).
h. Portfolio data can be used to estimate claims inflation for a particular risk, which is difficult to do accurately unless the data set is massive (Chapter 9).

8.5 Questions

1. With a foray in the necessary chapters on reinsurance and rating factor selection, describe in what ways the claims data used for pricing personal lines insurance, commercial lines insurance, and excess-of-loss reinsurance are likely to differ.
2. To enable yourself to price aviation risk, you decide to build a database of large aviation losses. (a) What information would you seek to include in the database? (b) Where would you look for that information?
3. What information would you request to price (a) commercial motor risk and (b) professional indemnity risk?

9

Setting the Loss Inflation Assumptions

Experience rating – the type of pricing based on the analysis of past loss experience and which will be the basis of much of this book – is loosely based on the idea that the past is a good guide to the future and that you can therefore use past losses to estimate future losses. This is what we have done, in a very simple way, in the elementary pricing example of Chapter 1.

Even the staunchest supporter of experience rating, however, knows that loss experience will not simply repeat itself identically and that *experience must be made relevant to current conditions.* One thing that one must definitely do to be able to make past losses relevant to today's environment is to bring them up to the value they would have if they happened today (or more accurately, the value they would have if they happened during the policy period). This is similar to what we do when we compare prices from the past to prices today, and we therefore call this type of inflation 'loss (or claim) inflation'. Loss inflation is affected by the behaviour of many other inflation indices but is not normally identical to any of them.

To deal with loss inflation appropriately, one needs to be aware of what, in the wider environment, drives increases (or reductions) in claim payouts as a function of time and needs to be able to tap into the correct information using publicly available inflation indices.

Examples of relevant indices are the consumer price index (CPI), the retail price index (RPI), the average weekly earnings (AWE) index and a number of specialised indices for different classes of risk.

The idea behind loss inflation is to estimate the value that past losses would have if they happened during the policy period. Let us remind ourselves of what we discussed in Chapter 2:

> Obviously, a loss which occurred in 2005 would have, if it occurred again under exactly the same conditions in 2015, a much larger payout,[1] because of a thing called loss inflation. Loss inflation is not the same as the CPI inflation that we have in many countries, as it depends on other factors that the CPI is not concerned about. For example, property losses will depend on a number of factors including cost of repair materials, cost of repair labour, and others. Liability losses will depend on wage inflation, cost of care inflation, and court inflation, that is, the tendency on the part of the courts to award compensations that are increasingly favourable for claimants.

1 This is at least the typical case. There are, however, exceptions: for example, addressing some environmental damage has become cheaper over the years because technological advances make it cheaper to carry out some repairs.

DOI: 10.1201/9781003168881-11

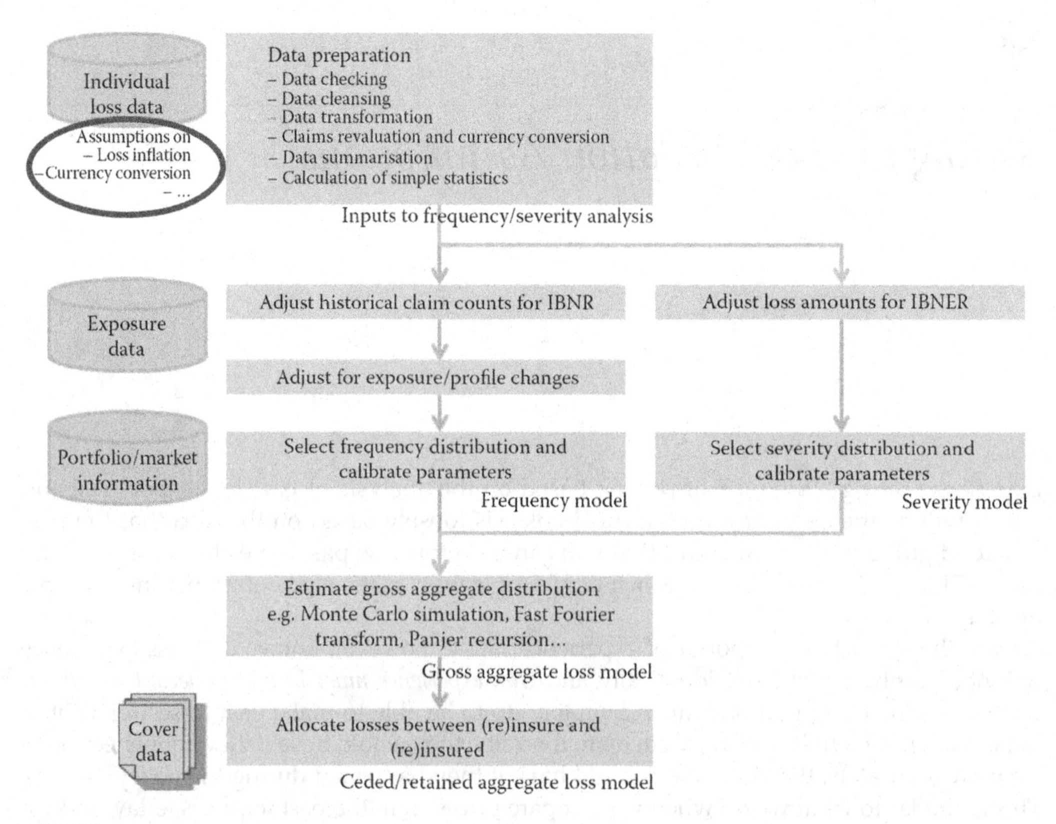

FIGURE 9.1
We must set suitable inflation assumptions before we prepare the inputs for the frequency/severity analysis.

Let us now look at how information can be extracted in practice. There are a number of sources for inflation figures; furthermore, we are sometimes able to extract loss inflation information from the data itself.

Figure 9.1 shows where setting the loss inflation assumption fits into the overall process.

9.1 Sources of Inflation Information

Figure 9.2 shows a number of indices that can be used as proxies for loss inflation for different lines of business in the United Kingdom.

However, each policy must be considered on its own merits, and very specific indices are sometimes used. For example, if you are looking to insure losses that derive from oil spills from ships, the portion of the loss that relates to the lost oil itself might be usefully linked to the *oil price*.

Note that

- To take things such as court inflation into account, sometimes you add a margin to the chosen index. For example, if wage inflation is 4%, a loss inflation of 5% for

Sources of Claims Inflation for Different Classes of Business

Type of insurance	Relevant (but not *fully* relevant) inflation indices
Household contents	Consumer Price Inflation (CPI), Retail Price Inflation (RPI)[a]
Property	Cost of building materials Labour costs → Average weekly earnings (AWE)
Property (Energy)	CEPCI (Chemical Engineering Plants Cost Index)
Medical expenses	Specialist inflation indices for medical care
Motor property damage claims	Cost of parts Motor repair labour → AWE Large claims: cost of whole vehicles
Fixed benefits	No inflation
Liability	For property damage: see Property For bodily injury • Loss of income → AWE[b] • Medical care → Medical care inflation • Pain and suffering → 'court inflation'/'social inflation'
Marine Cargo	Various commodity indices, depending on the materials transported, such as oil
Loading for expenses	Mostly driven by wages → AWE

[a] The main differences between the CPI and RPI relate to

- Population base – the RPI excludes very high and low income households.
- Commodity coverage – the CPI excludes owner occupiers' housing costs.
- The CPI uses a combination of geometric means and arithmetic means, whereas the RPI uses only arithmetic means.

[b] Note that there are 24 AWE indices, depending on the industry.

FIGURE 9.2
Possible sources of loss inflation information for different lines of business.

liability might be found to be adequate. The difference between the two is called 'superimposed inflation'.

- We can either use a raw index or fit a best-fit function (such as an exponential) through the index to produce a smoothed index with possibly a single number for inflation per annum.
- Large-loss inflation may be different from normal loss inflation because different factors may be at play. For example, court inflation will have a disproportionate effect on large losses. This is why inflation assumptions in excess of loss insurance and excess of loss reinsurance are normally harsher than in direct insurance: inflation assumptions of 8% to 10% for large liability losses are not uncommon. This imbalance, if persistent, has some disturbing long-term consequences because an ever-larger proportion of the total loss amount is bound to be allocated to the largest losses.

9.2 Data-Driven Estimates of Loss Inflation

One might think that the obvious way to estimate loss inflation is using the data itself. This makes sense and should be attempted whenever possible. However, in most cases, it is *not* possible unless one possesses a large portfolio of data from different policyholders. It is hardly ever possible for a single policyholder. Sections 9.2.1 through 9.2.3 outline three possible approaches to data-driven estimation of loss inflation.

9.2.1 Statistical Estimation of Loss Inflation Using Robust Statistics

One obvious method to estimate loss inflation is to track the average loss amount for a cohort of losses and design an index based on that (similar to, e.g. the wage inflation index), or calculate an overall trend (e.g. 6% per annum).

This method works as follows:

1. Calculate the empirical average[2] loss amount $\overline{X}(t)$ for historical losses over a number of months/years.
2. Build an empirical index based on this average loss amount: $I(t)=\overline{X}(t)$...
3. ... or, alternatively, assume that the average loss amount $\overline{X}(t)$ for year y grows by an approximately constant inflation rate, $\overline{X}(t)\approx A(1+r)^t$, and determine r by least squares regression techniques. This can be done easily by calculating the slope of the best-fit line through all the available pairs $\left(t_j,\ln\left[\overline{X}(t_j)\right]\right)$. The slope is related to r by the formula $r=\exp(\text{slope})-1$. The fit can then be validated with standard model validation techniques.

Despite the fact that this is conceptually reasonable and computationally straightforward, there are some difficulties:

- If we use the mean as the average $\overline{X}(t)$, a single large loss might distort the estimate in a given year: typically, the mean will jump up and down with no apparent regularity.
- If the data is scant, even more robust statistics such as the median will exhibit numerical instabilities and lead to poor lossinflation estimates.
- If we see only losses above a certain level of deductible, this will completely throw off our estimate because as loss inflation pushes up losses, losses that were previously below the deductible will enter the data set, with the result that we are not comparing the losses on a like-for-like basis. This is especially true in reinsurance, where deductibles may be very large, such as £1M. To have an idea of how much this affects inflation calculations, note that if the underlying severity distribution of more than £1M is a single-parameter distribution [$F(x)=1-(\theta/x)^\alpha$ with $\theta=$ £1M], it can be proved that the distribution remains exactly the same after loss inflation is applied to it – in other terms, there is, in this case, no way of inferring the inflation from statistics based purely on the empirical severity distribution above a fixed threshold.
- Similarly, if only the portion of losses below a certain limit counts, this might skew inflation calculations.

2 We are deliberately not committing to a specific type of average (mean, median, mode, trimmed mean.) at this stage – we will later discuss which type of average is more adequate.

- If the cover changes and the losses are not of the same type, the loss inflation estimate will be irrelevant. This is a problem that the Office of National Statistics in the United Kingdom faces when calculating official CPI and RPI figures. This is why they must be so careful in selecting a proper basket of goods and following them through.

As a partial response to these difficulties, here are some precautions that we should take in calculating loss inflation from data:

a. One should use portfolio/market information such as large data sets of losses from many different policyholders rather than data from a single policyholder, to reduce numerical instabilities as much as possible.
b. Because the empirical index approach (Step 2) tends to give more erratic results from one year to the other, whereas the constant inflation rate approach (Step 3) tends to give more stable results (see, for example, Question 2), the empirical index approach should be used only if the data set is really massive.
c. One should use robust statistics for the average loss amounts $\bar{X}(t)$ that are not affected too heavily by large losses. One example of such a statistic is the median, but other possibilities (such as trimmed means) are also worth considering. This, of course, is not without undesired side effects: for example, the median will by definition focus on the middle part of the distribution and ignore inflationary effects that are specific to large losses.
d. One should use wherever possible ground-up, uncapped losses to prevent deductibles and limits from skewing the loss inflation calculations.

These precautions go some way towards making it feasible to use the statistical method for loss inflation, but one should be aware of a couple of remaining issues:

- Differences in the average loss amount over time may be due to different types of losses driving the average in different periods. For this reason, it is necessary to find large data sets of relatively homogeneous losses, which may sometimes be challenging.
- Another difficulty is that this method is not suitable for finding the loss inflation for losses above a threshold (e.g. £500,000), as is often the case in reinsurance.

These two points are addressed in Sections 9.2.2 and 9.2.3 in turn.

9.2.2 Basket Method, Based on Selected Samples

This is similar to what the Office for National Statistics in the United Kingdom would do: choose a basket of similar 'goods', such as losses that result in similar effects (such as the same type of disability), and see how differently they are priced from one year to the other. Because of the need for selecting very similar losses, the selected sample will necessarily be quite small with respect to the data sets used in statistical methods (Section 9.2.1).

The different components of the losses can be broken down and one can look at what drives inflation for each component. The problem with this method is that it is difficult to find losses that are really similar. Also, even losses that are almost identical in terms of circumstances may result in different compensation levels based on different compensation requests, the insurer's attitude, the negotiations, the judge, and possibly other factors.

Despite these difficulties, this method leads to credible results, especially if the basket sample is large enough for the second-order differences between losses to be evened out. This method (among other methods) was used, for example, in producing estimates of motor bodily injury loss inflation for very large losses in the IUA/ABI Fourth UK Bodily Injury Awards Study (IUA/ABI 2007).

9.2.3 Statistical Approach to Large-Loss Inflation

In Section 9.1, we mentioned that the inflation for large losses is not necessarily the same as that for smaller losses. Because in many circumstances (for example, when pricing excess of loss reinsurance), we are specifically interested in the behaviour of large losses and we do not even have visibility on losses below a certain threshold, we need to be able to estimate large loss inflation from a cohort of losses above that threshold. In what follows, 'large loss' means 'loss above a given threshold', in which the exact value of the threshold will depend on the circumstances, specifically on the line of business and the (re)insurance structure we need to price. We will only give a few highlights here. The reader interested in a more in-depth analysis of inflation for large losses is referred to Fackler (2011).

For large losses, the normal statistical approach for calculating inflation (Section 9.2.1), based on calculating the average loss amount for that cohort at different points in time, is not valid because one of the effects of inflation is to push losses above the selected threshold.

To show why this is a problem, suppose that we want to calculate the inflation of losses of more than £1M. Assume that in year n we observe 10 losses as in Figure 9.3 and that in

Loss ID	Year n	Year $n + 1$	Year-on-year increase
1	830,652	913,718	10.0%
2	924,209	1,016,630	10.0%
3	933,172	1,026,489	10.0%
4	1,013,290	1,114,619	10.0%
5	1,777,693	1,955,462	10.0%
6	1,800,240	1,980,264	10.0%
7	2,030,804	2,233,885	10.0%
8	2,264,346	2,490,780	10.0%
9	2,624,934	2,887,428	10.0%
10	6,675,509	7,343,060	10.0%
Mean (all losses)	2,087,485	2,296,233	10.0%
Mean (losses > £1M)	2,598,117	2,449,846	–5.7%
Median (all losses)	1,788,966	1,967,863	10.0%
Median (losses > £1M)	2,030,804	1,980,264	–2.5%

FIGURE 9.3
The mean and median above a given threshold cannot be used to calculate inflation because the definition of the cohort of losses over which inflation is calculated itself depends on the inflation we are trying to measure. In this specific example, the mean and the median of losses worth more than £1M are affected by the fact that the cohort of losses worth more than £1M in year $n + 1$ includes losses (Loss 1 and Loss 2) that would not have belonged to the equivalent set in year n.

year $n + 1$ we have (stretching the imagination) another 10 losses *of identical characteristics* but whose amount is in all cases 10% higher because of loss inflation. If we calculate the mean and median across all losses, we find that these are 10% higher in year $n + 1$, as expected. However, if we calculate the mean and median of the losses of more than £1M, we find that these have not increased by the same amount; actually (in this specific example) they are lower in year $n + 1$! The reason is that to calculate inflation properly, you need to compare the statistics for two equivalent cohorts in year n and $n + 1$; however, the definition of the cohort over which inflation is calculated includes a monetary amount (the threshold of £1M and higher, losses that are considered large), which is itself subject to the inflation that you are trying to measure.

To overcome this problem, we therefore need to resort to other statistics that are not defined based on a monetary threshold.

An example of such a statistic is the mean of the largest, for example, 100 losses in a year. Other possibilities are the median and the smallest of this loss data set. These statistics will not be affected by threshold-crossing and therefore any increase in them is genuinely related to loss inflation.

However, any of these statistics is affected by other problems:

- Numerical instability if the statistic is based on a small sample, e.g. 10.
- The average or smallest of the losses in the sample will be affected by the ground-up frequency: for example, if one year has an unusually large number of ground-up losses, the top 100 losses for that year are likely to be larger because they correspond to higher percentiles of the overall distribution.

The first of these two effects can be mitigated by using a sample that is as large as possible.

The second effect can be, to some extent, mitigated by taking into consideration market information on the ground-up frequency, where available. For example, if the ground-up frequency has gone from 50,000 to 60,000 (an increase of a factor 1.2), then the 100th largest loss for year n should be compared with the 120th largest loss for year $n + 1$, as they both refer to the same percentile of the loss distribution: (50,000 − 100)/50,000 = 99.8%.

Question 2 shows an example of this approach and encourages you to try this approach on your own simulated data.

9.3 Questions

1. Explain what index (or what flat rate) you would use to revalue the following types of cover, doing personal research where necessary.
   i. Marine Cargo cover for ships transporting (i) oil or (ii) cigarettes
   ii. Contents insurance (UK)
   iii. Personal accident insurance with fixed benefits
   iv. Workers' compensation
   v. Motor own damage, motor third-party property damage and motor third-party liability (UK)
   vi. Property cover for oil refineries
   vii. Employers' liability and public liability reinsurance

2. The following table shows a few statistics related to the loss amount distribution for a risk over the period 2014 to 2023. The loss amounts from which these statistics are derived were generated from a lognormal distribution with parameters 10.3 and 1.5. For each year, a *fixed* number of 500 losses were generated, and a 5% loss inflation was superimposed on each successive year, so that, for example, the loss amounts for 2017 were derived by sampling losses from a lognormal distribution with the parameters above and then multiplying each amount by $1.05^{2007-2004} = 1.16$.

	2014	2015	2016	2017	2018	2019	2020	2021	2022	2023
Mean	93,396	107,148	93,523	93,754	108,782	88,764	105,208	119,736	118,352	139,733
Median	27,008	35,398	33,892	31,944	37,930	34,966	42,755	44,157	41,747	43,156
10th largest	680,164	591,770	812,676	976,665	783,431	977,358	763,855	937,060	1,088,360	1,050,619
Trimmed mean (90%)	26,950	35,205	33,510	31,948	38,325	35,545	42,388	44,618	41,913	43,776

i. Make a chart of the mean, median, and trimmed mean against the year and calculate the year-on-year inflation for each statistic, commenting on the stability of the statistic.

ii. Calculate the loss inflation based on the four statistics provided and compare it with the true inflation.

iii. Run your own experiment with the same parameters as specified above and see if you get similar results. Also, run the same experiment with 100 losses only and with a variable number of losses, based on a Poisson distribution with a Poisson rate equal to 100.

10

Data Preparation

It's a dirty job (but someone's gotta do it).

Iain Archibald Band

Modelling is rarely an automated process by which one transforms a raw set of data into a set of results. Before any modelling can be attempted, a significant amount of exploratory work often needs to be carried out on the raw data. For lack of a standard term, we call this exploratory work 'data preparation' (Figure 10.1). This is often an open-ended exercise but will normally involve the following activities:

- Checking data to determine whether there are errors or anomalies
- Summarising data to provide a bird's-eye view of the risk and understand obvious trends/anomalies
- Preparing a standardised input for the rest of the risk costing process, in particular for the frequency and the severity analysis

To keep things as concrete as possible, we will illustrate data preparation using real-world data sets, in which amounts and names have been disguised beyond recognition for data privacy purposes. The policyholder (or prospective policyholder) discussed here is in facilities management, and the policy is public liability.

An excerpt from the data can be seen in Figure 10.2. The complete version of the data set (with more columns and many more rows) contains 1718 losses, some of which are losses with zero amounts. In the following, we will keep this example in mind to illustrate the various steps of data preparation. We assume that we need to price an annual public liability policy that incepts on 1 April 2015. We assume loss inflation to be 5%. The losses data are given in pounds, but the policy needs to be priced in euros. The exchange rate used here was 1£ = 1.2€.

A typical process will look something like a three-step approach involving data cleansing (Section 10.1), data transformation (Sections 10.2 and 10.3) and data summarisation (Section 10.4).

DOI: 10.1201/9781003168881-12

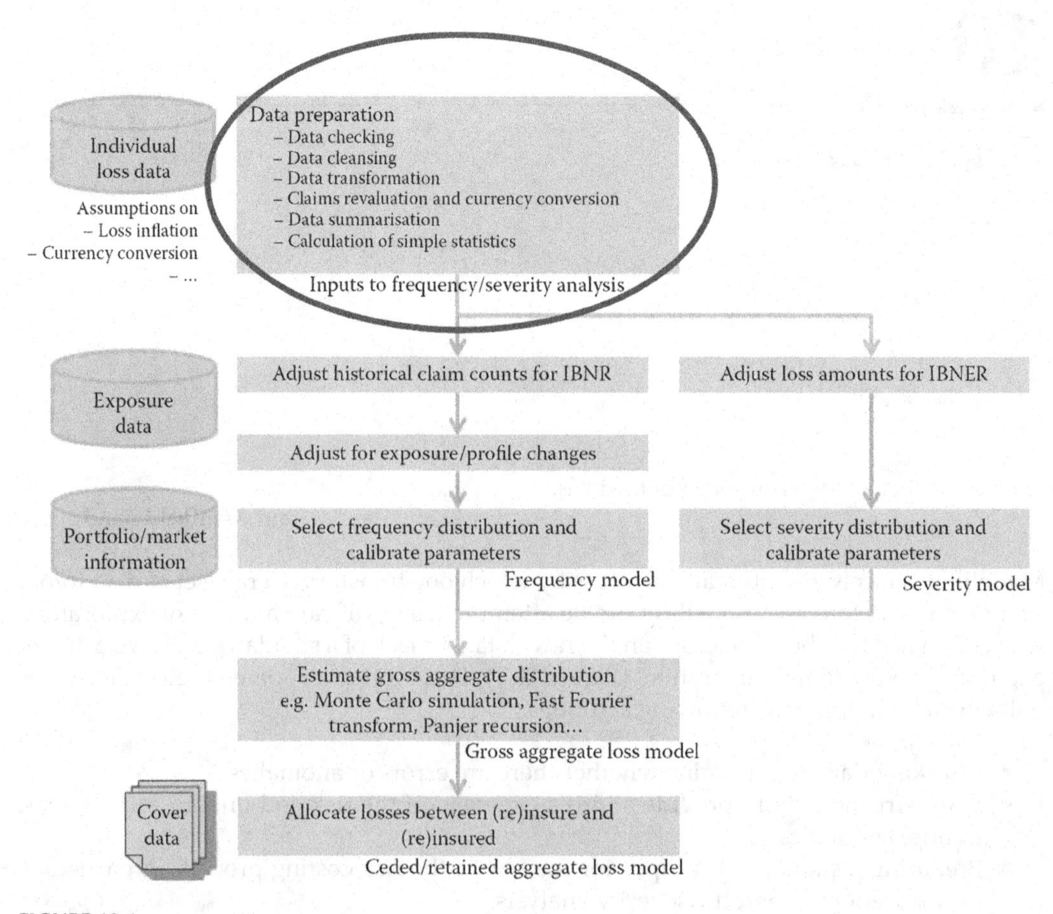

FIGURE 10.1
This is how exploratory analysis fits in the risk costing process.

10.1 Data Cleansing

Loss data are 'cleansed' and obvious mistakes are corrected. In a large company, this may be done by teams that specialise in data cleansing and have systematic ways for checking for anomalies such as data in the wrong format, figures that do not add up (e.g. paid + outstanding not being equal to incurred) and so on. Anything dubious will be referred back to the analyst. If some fields are missing, suitable assumptions must be made about them.

10.2 Loss Data Transformation

Loss data are transformed so that they are ready to use in the rest of the risk costing process. Some examples of data transformations, which should not be seen as exhaustive in all cases, are illustrated in the following sections.

Claim ID	Claimant name	Loss date	Reporting date	Currency	Paid	Outstanding	Total incurred	Fees	Recoveries	Overall amount	Claim status
Claim_1	Name_1	6/13/2009	9/3/2009	GBP	818	17,516	18,333	0	0	18,333	Open
Claim_2	Name_2	8/9/2009	9/3/2009	GBP	0	0	0	250	0	250	Closed
Claim_3	Name_3	12/17/2008	1/26/2009	GBP	161	0	161	0	0	161	Closed
Claim_4	Name_4	11/1/2010	7/29/2013	GBP	372	0	372	0	0	372	Closed
Claim_5	Name_5	2/7/2011	2/14/2011	GBP	0	0	0	0	0	0	Closed
Claim_6	Name_6	1/17/2007	1/18/2008	GBP	15,645	0	15,645	0	0	15,645	Closed
Claim_7	Name_7	3/3/2004	4/23/2004	GBP	28,812	0	28,812	0	0	28,812	Closed
Claim_8	Name_8	7/5/2004	6/23/2005	GBP	2316	0	2316	0	0	2316	Closed
Claim_9	Name_9	6/6/2005	1/23/2007	GBP	13,414	0	13,414	0	1180	12,234	Closed
Claim_10	Name_10	6/15/2007	6/19/2007	GBP	785	0	785	0	0	785	Closed
Claim_11	Name_11	7/21/2004	8/31/2005	GBP	897	0	897	0	0	897	Closed
Claim_12	Name_12	8/30/2006	10/25/2006	GBP	387	0	387	0	0	387	Closed
Claim_13	Name_13	8/4/2004	2/9/2005	GBP	39,309	0	39,309	0	0	39,309	Closed
Claim_14	Name_14	5/4/2006	5/26/2006	GBP	640	0	640	0	0	640	Closed
Claim_15	Name_15	11/7/2007	12/5/2007	GBP	3151	0	3151	0	0	3151	Closed
Claim_16	Name_16	8/31/2007	1/14/2008	GBP	6128	0	6128	0	0	6128	Closed
Claim_17	Name_17	10/13/2004	1/4/2005	GBP	3988	0	3988	0	0	3988	Closed
Claim_18	Name_18	1/7/2004	4/20/2004	GBP	514	0	514	0	8	506	Closed
Claim_19	Name_19	7/1/2008	7/1/2009	GBP	85	0	85	125	13	196	Closed
Claim_20	Name_20	6/25/2007	9/4/2007	GBP	1112	0	1112	0	0	1112	Closed
Claim_21	Name_21	5/25/2009	6/4/2009	GBP	483	28,683	29,167	0	2314	26,853	Open

FIGURE 10.2
An excerpt from PL data for a facilities management company; the data have been changed beyond recognition (the amounts have been changed as well).

10.2.1 Loss Revaluation

The full incurred amount of all losses is revalued to bring them to the average point of occurrence t^* of the renewal policy by using the appropriate inflation index $I(t)$ as discussed in Chapter 9. More specifically, a standard way of doing this is by

a. Calculating the average point of occurrence t^* of the renewal policy. This is *normally the midpoint of the policy*: e.g. for an annual policy incepting on 1 April 2015, the midpoint will be 1 October 2015. However, if the losses exhibit seasonality (e.g. if there are more losses during winter), then t^* might be different from the midpoint. In general, for a policy incepting at $t = 0$ and expiring at $t = T, t^* = \int_0^T t\,v(t)\,dt$, where $v(t)$ is the proportion of losses expected between t and $t + dt$, with the constraint $\int_0^T v(t)\,dt = 1$. Note that if $v(t) = 1/T$ (i.e. no seasonality), then $t^* = T/2$ (the midpoint of the policy).
b. Revaluing the incurred claim amount $X(t)$ to its estimated value at time t^* as follows:

$$X^{\text{rev@}t^*}(t) = X(t) \times \frac{I(t^*)}{I(t)} \tag{10.1}$$

where $I(t)$ is the (known) value of the index at time t and $I(t^*)$ is the *estimated*[1] value of the index at time t^*. Assuming that the latest date for which the index is known is today t^{latest}, and that $t < t^{\text{latest}} < t^*$, we can calculate $I(t^*)$ as

$$I(t^*) = I(t^{\text{latest}}) \times (1+r)^{t^* - t^{\text{latest}}} \tag{10.2}$$

where r is the assumed flat inflation rate going forward, based perhaps on an extrapolation of average inflation rate of the index in the recent period, or the risk-free discount rate, or another educated guess, depending on the nature of the index.

If we can assume a flat inflation rate r, across the whole period, Equation 10.1 simplifies to

$$X^{\text{rev@}t^*}(t) = X(t) \times (1+r)^{t^* - t} \tag{10.3}$$

In our public liability example, with loss inflation assumed to be $r = 5\%$ per annum and the policy incepting on 1 April 2015, a claim that occurred on 21 March 2010 would be revalued by $1.05^{2020/365.25} \sim 1.31$, where 2020 is the number of days from 21 March 2010 to the midpoint of the policy, 1 October 2015. Note that we have used the exact dates for the revaluation: many practitioners, however, will simply assume that all losses in a given policy (or accident) year happen at midpoint of that policy year. In our example, that would lead to a revaluation factor of $1.05^{2015-2019} \sim 1.34$. This approximation will normally be good enough on average although, when inflation is severe and losses are

1 The reason why $I(t^*)$ is only an estimate is that t^* is generally in the future, unless you are pricing a policy that has already expired or is anyway past its average point of claim occurrence.

large, this may occasionally cause a significant difference in the burning cost and the severity model.

10.2.1.1 Which Losses Do We Need to Revalue, What Amount Do We Have to Revalue Exactly?

It is important to note that *all losses* need to be revalued to current terms, whether they are settled or not, and that the revaluation applies to *the full incurred amount*, not the outstanding amount only. This needs to be clarified because this is one area where you are likely to meet professionals from other areas who will challenge this approach and you will need to justify why you do things in a certain way.

The guiding principle is that you want to make an informed guess as to what the value of a claim that occurred at time t would be if it occurred at time t^*. It therefore makes sense to *assume that all the characteristics of the claim apart from the claim amount – for example, the payment pattern – remain unchanged but are translated by $t^* - t$ years (or days)*. In mathematical terms, this is equivalent to assuming that

i. The claim payment and reserving process is *stationary* except for the monetary amounts.
ii. The monetary amounts (such as the past payments and reserves) are all scaled by the same factor, $I(t^*)/I(t)$.

Consider, for example, a claim X occurred in 2005, was reported in 2006 and reserved for £1M, then the reserve changed to £3.5M in 2008, with 50% having been paid in 2010, and another 20% in 2012, and is still open today, with the incurred amount still at £3.5M. Assume that you are pricing an annual reinsurance policy written on a losses occurring basis incepting on 1 January 2014, and that loss inflation is approximately 10%. You therefore want to revalue claim X to 1 July 2014. By the assumption of stationarity, you should consider that this claim will occur in 2014, will be reported in 2015, and is reserved for £1M $\times$ 1.1^9 = £2.36M, then the reserve will change to £3.5M $\times$ 1.1^9 = £8.25M, of which 50% will be paid in 2019 and the other 20% in 2021, at which point it will still be open. It therefore makes sense to revalue the current incurred amount (£3.5M) to £8.25M.

As all assumptions, of course, this can be challenged, and there may be circumstances in which it can be argued that the claim process has drastically changed or that a specific claim, perhaps a very large claim, has gone through an idiosyncratic history of payment and development and needs to be revalued in a different way. A changing inflation environment[2] might also lead to reconsider all estimates previously made for outstanding losses and may make it more difficult than usual to project the price index to the future.

10.2.1.2 Other Considerations

Note that different components of a claim (e.g. property and liability) might need to be revalued at different rates, although one always has to be mindful of the spurious accuracy arising from an analysis that is too detailed. Also note that in claims-made policies, t and t^* can refer to the reporting dates – all other considerations remain the same.

2 An obvious example is the transition in 2021–22 to a high-inflation environment as a consequence – among other factors – of post-pandemic supply chain disruption, the war in Ukraine and increased money supply (see, e.g. *Bulletin de la Banque de France* 239/2 – January–February 2022).

10.2.2 Currency Conversion

All loss amounts are converted into the currency that we intend to use. For pricing purposes, the exchange rate that we use should *normally* be the same for all losses, regardless of when they occurred.[3] However, there are exceptions to this rule,[4] and each situation needs to be assessed on its own merits.

For other types of exercises (such as performance analysis), it may make sense to use the exchange rate at the date when the loss was paid.

10.2.3 Policy Year Allocation

All losses are allocated to the appropriate policy year. For pricing, this is not necessarily the original policy year, but the policy year according to the renewal policy period (normally these are the same): if, for instance, the policy runs from 1 April 2015 to 31 March 2016, we assign a claim occurred on 16 May 2008 to 2008, even if the original policy started from 1 July 2008 and the policy year of that claim would then have been 2007.

The outputs of data preparation may be, for example, the detailed list of revalued losses, the detailed list of delays between loss occurrence and loss reporting and the number of losses for each policy year.

10.2.4 Individual Loss Adjustments for Exposure Changes

Apart from loss inflation, individual severity might need to be adjusted for other factors. We consider here specifically exposure as such factor. In some circumstances, historical information on *individual* loss amounts might need to be adjusted for exposure changes. This is because certain measures of exposure relate to the total risk, others relate to the frequency, and others yet relate to the severity component. In practice, this is often ignored, and the same measure of exposure is used to adjust historical total losses (in burning cost analysis) and historical claim counts (in frequency analysis).

However, in some cases, this is not entirely appropriate. For example, as discussed in Section 8.3.2, in aviation hull, the average value of an aircraft (AFV/number of planes) is likely to affect the severity of losses, and one may want to adjust the value of past individual losses to reflect the changes in the average aircraft value.

This may cause complications because adjusting individual losses with a blanket exposure factor may cause paradoxical results, such as a claim being revalued to an amount larger than the most expensive aircraft in the portfolio. Ultimately, these problems arise

3 To understand why this is the case, consider that the purpose for which you are doing claims revaluation and currency conversion is to determine what past claims would cost if they were to happen today. The currency exchange rate at the time of occurrence, however, has no bearing on the amount that would need to be paid today for a past claim: the *claim amount in current terms in the local currency* only depends on the original value and the claims inflation in the territory of interest, whereas the *claim amount in current terms in the destination currency* only depends on the claim amount in current terms in the local currency and on today's exchange rate.

4 Here is an example from energy business where the cost of claims is typically in dollars because the international energy economy is priced in dollars. However, if claims are paid in local currency, their value will change with the exchange rates. For example, if one Ruritanian Crown was worth $1 3 years ago, 80 cents 2 years ago and $1.05 a year ago and an oil widget costs $1 (ignore inflation) then a claim for one will cost these amounts but be constant when valued at the exchange rates of the time. Note that local costs will still be in local currency, and these may be a significant part of claims, so neither approach is perfect, but this example shows that using constant exchange rates is not always automatically right. (I am indebted to Julian Leigh for this example.)

Policy inception month	4
Revalue at	01-Oct-15
Loss inflation	5%
Destination currency	EUR
Exchange rate	1.20

Class of Business	Loss date	Reporting date	Policy Year	Original currency	Overall amount	Non-zero? (1=Yes)	Overall (EUR)	Overall revalued (EUR)
Public Liability	13/06/2009	03/09/2009	2009	GBP	18,333	1	22,000	29,917
Public Liability	09/08/2009	03/09/2009	2009	GBP	250	1	300	405
Public Liability	17/12/2008	26/01/2009	2008	GBP	161	1	193	269
Public Liability	01/11/2010	29/07/2013	2010	GBP	372	1	447	567
Public Liability	07/02/2011	14/02/2011	2010	GBP	0	0	0	0

FIGURE 10.3
Examples of calculations done during data preparation (columns 4, 7, 8, and 9). It is worth going through the calculations and see if you obtain the same results.

because if the severity of the losses is truly proportional to an exposure measure, maybe a different approach should be considered, such as using exposure rating for property, which bypasses the problem by focusing on the damage ratio, that is, the amount of loss as a percentage of the total value/sum insured/maximum probable loss. We will discuss exposure rating in Chapter 21.

In our example, individual loss severities do not need to be adjusted for exposure changes. Overall, the output of the data transformation will look like Figure 10.3 (only a few fields of the data set have been chosen, for clarity).

10.3 Exposure Data Transformation

As we have seen already in Chapter 1, for loss experience to be useful it must be related to the exposure to which it corresponds. When assessing the historical loss experience of a client, what counts the most are the number of losses or the total loss amount in a given period but the same quantities *per unit of exposure*. Therefore, as part of data preparation, it is important to match losses and exposure.

10.3.1 Matching Exposure to Losses

In the simplest case, matching exposure to losses simply means that the exposure period needs to be aligned to the policy period. This is not an issue for our data set because both the financial year and the policy year start on 1 April.

Suppose, however, that the turnover for a policy is given for the financial year starting 1 April and that the policy year incept 1 July. In this case, assuming that the risk is distributed uniformly over time the exposure aligned to policy year n would be calculated as:

$$\text{Exposure}(\text{PY}=n)=\frac{9}{12}\times\text{Exposure}(\text{FY}=n)+\frac{3}{12}\times\text{Exposure}(\text{FY}=n+1)$$

When the risk is distributed in more complex ways (seasonality and other time dependency) the formula gets more complicated, although non-contrived examples of where this is necessary are limited. We will look at examples of how time dependency can be incorporated in exposure calculations when looking at treaty pricing (Chapter 20) and portfolio pricing (Chapter 11).

10.3.2 Exposure On-Levelling

When exposure is a monetary amount, it needs to be revalued so that it is relevant to the period of the policy that we are pricing, exactly we did it for losses. To avoid confusion between losses and exposure revaluation we will typically refer to exposure revaluation as 'on-levelling'.

The index with which we on-level exposure is not necessarily the same as the one with which we revalue the losses. For example, in the case of employers' liability we might use wage roll as exposure and on-level it using wage inflation, but we may revalue the losses with an index equal to the wage inflation plus a (say) 1.5% superimposed inflation to account for the presence of other sources of social inflation (e.g., medical cost inflation, court inflation).

Clearly, if the exposure is *not* a monetary amount, no on-levelling is necessary.

10.4 Data Summarisation

Simple statistics such as the number of losses (non-zero and zero), total losses (paid/ outstanding/incurred), average amount, maximum amount, standard deviation, median and others, are calculated for each policy year. Also, a list of top losses with their descriptions

Policy year	No. of claims	No. of non-zero claims	Total losses (unrevalued, GBP)	Total losses in EUR (unrev)	Total losses in EUR (reval)	Average amount in EUR (reval)	Max amount in EUR (reval)	StdDev in EUR (reval)
2005	71	71	1,790,069	2,148,083	3,552,162	50,030	2,580,670	277,947
2006	77	77	1,864,182	2,237,018	3,518,246	45,692	2,457,626	256,537
2007	102	101	1,193,531	1,432,237	2,133,735	21,126	657,892	68,807
2008	111	109	992,194	1,190,632	1,679,933	15,412	142,974	21,303
2009	152	148	1,302,797	1,563,357	2,096,518	14,166	239,730	23,145
2010	265	259	2,560,578	3,072,693	3,935,751	15,196	1,171,627	74,683
2011	250	239	1,711,892	2,054,270	2,503,979	10,477	99,067	15,837
2012	255	185	1,704,616	2,045,539	2,363,409	12,775	89,110	12,064
2013	211	180	1,891,232	2,269,479	2,504,243	13,912	167,722	17,940
2014	76	70	639,988	767,986	816,430	11,663	111,938	14,833
Grand total	1570	1439	15,651,077	18,781,293	25,104,406	17,446	2,580,670	93,570

FIGURE 10.4
Simple statistics for the data of Figure 10.2. The number of losses can actually be used in the rest of the costing process. The rest provides a sense-check and shows some obvious trends. The revaluation was done at 5% per annum, and the revaluation date has been chosen to be 1 October 2015. The exchange rate used was 1£ = 1.2€.

Policy year	Paid losses (unreval, GBP)	O/S losses (unreval, GBP)	Incurred losses (unreval, GBP)	Fees (unreval, GBP)	Recoveries (unreval, GBP)	Overall amount (unreval, GBP)
2005	1,795,217	–	1,795,217	125	5,273	1,790,069
2006	1,876,546	–	1,876,546	–	12,364	1,864,182
2007	1,199,009	–	1,199,009	250	5728	1,193,531
2008	995,026	–	995,027	250	3084	992,194
2009	1,187,585	118,868	1,306,453	1875	5530	1,302,797
2010	1,485,836	1,105,759	2,591,595	4250	35,267	2,560,578
2011	1,162,220	547,127	1,709,347	6750	4205	1,711,892
2012	628,707	1,073,697	1,702,404	12,750	10,539	1,704,616
2013	229,205	1,661,905	1,891,110	13,000	12,878	1,891,232
2014	1483	637,173	638,656	4000	2668	639,988
Grand total	10,560,834	5,144,529	15,705,363	43,250	97,536	15,651,077

FIGURE 10.5
Other simple statistics for the data of Figure 10.2, focusing here on the different components of the losses (paid, outstanding, fees, recoveries, etc.).

will help in understanding the risk. The following figures show an example of both data preparation and data summarisation (Figures 10.4 and 10.5).

Note that a look at the summarised losses already allows us to make a number of observations, some of which are listed below. Readers are encouraged to come up with their own list of observations and of questions about the data before reading on.

- The number of losses has increased quite a bit from 2005 to the most recent years – from 71 to around 250 – is this an exposure effect, or is something more complex at play?
- The total losses have a more complicated behaviour – for example, in 2005, the total loss amount was high, but the number of losses was low. Why?
- The average amount varies quite erratically, even on a revalued basis, it goes from €9k to €30k. Why?
- The column 'Max amount' perhaps explains the two preceding points: the maximum value of a claim in a given year varies even more wildly, and hugely affects both the average amount and the total.

This type of observations proves why you need to do data summarisation: producing simple statistics will often throw up issues with the data, show trends, and generally lead to a better understanding of the risk you are analysing.

10.5 Questions

1. You decide to revalue three contents insurance losses to the midpoint of an annual policy that will incept on 1 June 2013 by using the consumer price index (CPI). The losses occurred in the United Kingdom on September 2010, March 2011 and

February 2012 and the amounts are £1500, £3400, and £2200, respectively. On the website of the Office of National Statistics, you find the following values of the CPI.

2. Calculate the revalued amount of the three losses, stating all additional assumptions you have made.

Year	Month	Index value	Year	Month	Index value
2010	Sep	114.9	2012	Jan	121.1
	Oct	115.2		Feb	121.8
	Nov	115.6		Mar	122.2
	Dec	116.8		Apr	122.9
2011	Jan	116.9		May	122.8
	Feb	117.8		Jun	122.3
	Mar	118.1		Jul	122.5
	Apr	119.3		Aug	123.1
	May	119.5		Sep	123.5
	Jun	119.4		Oct	124.2
	Jul	119.4		Nov	124.4
	Aug	120.1		Dec	125.0
	Sep	120.9	2013	Jan	124.4
	Oct	121.0		Feb	125.2
	Nov	121.2		Mar	125.6
	Dec	121.7		Apr	125.9

3. How would your answer to Question 1 have changed if you also had available statistics on the distribution of contents insurance losses over the year, and you knew that
   - 22% of the losses occur between January and March.
   - 25% between April and June.
   - 30% between July and September.
   - 23% between October and December.
4. A UK company writes a commercial motor fleet policy for a company Y in Freedonia, which has suffered from high inflation in the last few years. You have been given the list of the five losses notified to Y in the last few years.

   As part of your pricing exercise, you have to prepare appropriate input for your severity analysis, and you revalue all losses to the relevant date and convert all claim amounts from Freedonian dollars to pounds. Some information (not all of it useful) on loss inflation and exchange rates is provided in the table below.

Claim ID	Loss date	Original currency	Claim amount (FRD)	Loss inflation index at loss date	Exchange rate at loss date (value of 1 FRD in GBP)
Claim_1	11/1/2008	FRD	99,799	102	0.0891
Claim_2	3/3/2009	FRD	2,665,226	117	0.0724
Claim_3	3/10/2009	FRD	510,466	133	0.0581
Claim_4	15/6/2011	FRD	432,408	180	0.0222

Claim ID	Loss date	Original currency	Claim amount (FRD)	Loss inflation index at loss date	Exchange rate at loss date (value of 1 FRD in GBP)
Claim_5	21/7/2011	FRD	921,064	186	0.0191
	Today			201	0.0164
	Mid-term of policy		(projected)	213	
	(Today = 6/3/2013; inception date of policy = 1/7/2013; annual policy)				

- Explain what loss inflation is and why loss inflation for motor losses differs from consumer price inflation.
- Explain (very briefly) what inflation indices are relevant for motor bodily injury losses, and why a combination of these indices may not suffice to account for the observed loss inflation.
- Revalue all losses above and convert to pounds as necessary to price a motor policy that incepts on 1 July 2013.
- Explain why you converted the claim amounts to pounds in the way you did, given that this is part of a pricing exercise.

11

The Burning Cost Approach

Despite its simplicity and its many limitations, the burning cost approach is one of the most widespread pricing techniques, especially in such contexts as the London Market. It provides an estimate of the expected losses to a policy based on the average of the claims experience in past years, possibly with corrections/adjustments to make this experience relevant to the policy period, but without attempting any explicit modelling. In a way, it builds up on the data summarisation exercise that we described in Section 10.3.

We have already come across a rudimentary example of burning cost analysis in Section 1.3 ('Experience Rating'): the input was a table with aggregate employers' liability losses incurred over a number of years. We then adjusted the total claims for each year by adding an allowance for incurred but not reported (IBNR) claims, updated the value of the claims to take claims inflation and exposure changes into account. Finally, we took an average of the total losses over all years. The result was the *burning cost* estimate of the expected losses for the policy period.

In this chapter, we will take a closer look at this process and we will remove some of the simplifications that we made in Chapter 1: the most important improvement is that we are going to make full use of the individual claims information, which will allow us to fully consider the impact of the policy coverage.

Burning cost analysis is perhaps as far as one gets towards costing a risk and its volatility without resorting to stochastic modelling: and it is surprising how much one can achieve by that. Burning cost analysis can be used as a pricing technique in its own right but can also be used to provide a different estimate of the cost of risk alongside frequency/severity modelling analysis. This will have several advantages:

- It will provide a sense-check on the results of frequency/severity modelling.
- The results of burning cost analysis can be used as an intermediate step between the initial summary statistics and the results of frequency/severity modelling to communicate the results of frequency/severity modelling and have the client's or the underwriter's buy-in.

There is not a lot of theory behind the burning cost approach and this technique is best learnt through examples. We will therefore explain it step by step with the aid of a case study. However, in Section 11.1.12 we will collate all the calculations in a few compact formulae as a reference.

The data set for this case study can be found on the book's website.

DOI: 10.1201/9781003168881-13

11.1 Burning Cost Methodology

Our case study involves a UK company for which we have 11 years of public liability loss experience (a mixture of property damage and bodily injury). We have used real-world data, which has been anonymised and rescaled as necessary to make it unrecognisable.

11.1.1 Data and Assumptions

In this case study, we have used the following information on the policy and on the data set:

- The (current) policy inception date is 1 April (the fact that in the past the inception date may have been different is of no relevance to a burning cost exercise, as long as individual loss information is available).
- The aim of the exercise is to price the policy for the policy year 2015 (1 April 2015 to 31 March 2016). This is called the *reference policy year*.
- Claims data are available for the period 1 April 2005 to 30 November 2014 (the latter is called the 'as-at' date of the claims data set).
- The policy is on an *occurrence basis*, i.e. it covers all losses occurred (and not, for example, reported) during the policy period.
- The policy has an each-and-every-loss (EEL) deductible of £100,000, an annual aggregate deductible (AAD) of £3,200,000 and a limit (applied to each individual losses) of £5M.

We have also made the following *assumptions*:

- Claims inflation is 5% p.a.
- The historical turnover of the company has been brought to current terms by revaluing it yearly by 4%.

11.1.2 Input to the Burning Cost Analysis

The inputs to the burning cost analysis are

i. The individual claims listing.
ii. The policy structure and information.
iii. The historical exposure information. Depending on the case, we may also have market/benchmark information on large losses, IBNR, etc.

11.1.3 Step A: Revalue Losses and Apply Policy Modifiers to All Claims

Calculate the retained and ceded amount for each revalued loss, using the claims data set.

Each loss X_j, which has occurred at time t_j, will be revalued to the mid-policy period t^* (in our case, t^* = 1 October 2015) using a claims inflation index $I(t)$, as per Equations 10.1 and 10.2 which we combine in one equation below.

$$X_j^{\text{rev}} = X_j \times \frac{I\left(t^{\text{latest}}\right) \times (1+r)^{t^* - t^{\text{latest}}}}{I\left(t_j\right)} \tag{11.1}$$

As in Chapter 10, t^{latest} is the latest date for which the index is known and r is the assumed inflation after that. In our case, we have assumed a constant inflation throughout at $r = 5\%$ so the equation can be simplified to $X_j^{\text{rev}} = X_j \times (1+r)^{t^*-t_j}$.

For each revalued loss X_j^{rev} we can then calculate the amount retained below the deductible, $\min\left(X_j^{\text{rev}}, EEL\right)$, that ceded to the insurer, $\min\left(L - EEL, \max\left(X_j^{\text{rev}} - EEL, 0\right)\right)$, and that retained by the insured *above* the limit L, $\max\left(X_j^{\text{rev}} - L, 0\right)$. Examples of this calculations for our case study are shown in Figure 11.1.

11.1.4 Step B: Aggregate Loss by Policy Year

The gross, retained (below the limit) and ceded amounts for each year y for a policy like that in our case study (and before any other adjustment and the application of the annual aggregate deductible) can then be calculated as:

$$\begin{aligned} S^{\text{gross}}(y) &= \text{Total gross for year } y = \sum_j X_j^{\text{rev}} \\ S^{\text{ret}}(y) &= \text{Total retained below the limit for year } y = \sum_j \min\left(X_j^{\text{rev}}, EEL\right) \\ S^{\text{ced}}(y) &= \text{Total ceded for year } y = \sum_j \min\left(L - EEL, \max\left(X_j^{\text{rev}} - EEL, 0\right)\right) \end{aligned} \tag{11.2}$$

We can also calculate the total amount retained by the insured above the limit as $S^{\text{retained above L}}(y) = \sum_j \max\left(X_j^{\text{rev}} - L, 0\right)$. However, this component will have no bearing on the subsequent calculations so we will henceforth ignore it.

These calculations can be easily performed using spreadsheet functionalities[1] – see the download section of the book's website for details of the calculations (Parodi, 2016). The results of the calculations for our case study are shown in Figure 11.2, which also shows the number of claims per year.

Note that these formulae – which are suitable for the coverage modifiers assumed in Section 11.1.2 –will need to be adapted if the details of the coverage are different.

11.1.5 Step C: Adjust for Exposure Changes

11.1.5.1 C1: Alignment and On-Levelling of Exposure

Before we adjust our loss figures for changes in exposure, we need first of all to align the exposure period to the policy period. For our case study, this is not a problem, as the period for which turnover is reported (the fiscal year) and the policy period are the same. When that is not the case, the alignment calculations are trivial: the exposure for each policy period will be a linear combination of the exposure of adjacent fiscal periods.

Furthermore, when the exposure is a monetary amount (as in our case study, which uses turnover), we need to rebase the historical exposure figures applying an adequate revaluation factor or index (Figure 11.3), as we do for claims. This process is typically called 'on-levelling' to distinguish it from claim revaluation.

1 In MS Excel, standard solutions for aggregating by year are pivot tables, conditional sum functions (e.g. SUMIFS), and macros.

Data as-at date	30-Nov-14
Policy inception month	4
Revalue at	1-Oct-15
Claims inflation	5%
EEL (£)	100,000
Limit (£)	5,000,000
AAD (£)	2,500,000

Claim ID	Claimant name	Class of business	Loss date	Reporting date	Policy year	Currency	Overall amount	Non-zero? (1 = Yes)	Overall amount (revalued)	Retained	Ceded	Delay (days)
Claim_1	Name_1	Public Liability Bodily Injury	1/17/2007	1/18/2008	2006	GBP	115,645	1	176,827	100,000	76,827	366
Claim_2	Name_2	Public Liability Bodily Injury	3/3/2004	4/23/2004	2003	GBP	28,812	1	50,688	50,688	0	51
Claim_3	Name_3	Public Liability Bodily Injury	7/5/2009	6/23/2010	2009	GBP	2316	1	3140	3140	0	353
Claim_4	Name_4	Public Liability Bodily Injury	6/6/2005	1/23/2007	2005	GBP	12,234	1	20,240	20,240	0	596
Claim_5	Name_5	Public Liability Property Damage	6/15/2007	6/19/2007	2007	GBP	785	1	1177	1177	0	4
Claim_6	Name_6	Public Liability Bodily Injury	7/21/2004	8/31/2005	2004	GBP	897	1	1549	1549	0	406
Claim_7	Name_7	Public Liability Property Damage	8/30/2006	10/25/2006	2006	GBP	387	1	602	602	0	56
Claim_8	Name_8	Public Liability Property Damage	8/4/2004	2/9/2005	2004	GBP	39,309	1	67,748	67,748	0	189
Claim_9	Name_9	Public Liability Bodily Injury	5/4/2006	5/26/2006	2006	GBP	640	1	1014	1014	0	22
Claim_10	Name_10	Public Liability Bodily Injury	11/7/2007	12/5/2007	2007	GBP	3151	1	4633	4633	0	28

FIGURE 11.1
Illustration of Step A of the burning cost calculation method, based on our case study. At the top of the table, the basic assumptions on the policy and claims inflation are stated.

Policy year	No. of claims	No. of non-zero claims	Total losses (reval)	Retained losses (reval)	Ceded losses (reval)
2005	86	86	2,577,699	803,768	1,773,932
2006	92	92	2,684,421	841,800	1,842,621
2007	111	110	1,718,684	1,142,872	575,812
2008	147	146	1,428,759	1,407,011	21,748
2009	196	194	1,876,028	1,765,098	110,930
2010	263	256	3,687,767	2,583,974	1,103,794
2011	263	250	2,465,124	2,465,124	–
2012	265	199	2,454,646	2,454,646	–
2013	225	201	2,722,839	2,562,839	160,000
2014	70	68	921,583	895,583	26,000
Grand Total	1718	1602	22,537,551	16,922,715	5,614,837

FIGURE 11.2
The total number and amount of individual losses (gross, ceded, and retained), aggregated by policy year. The allocation to each policy year is made according to the current (not historical) policy definition: that is, if the policy inception date had been 1 January 2006 and is now 1 April, a claim occurring on 13 February 2006 would be classified as belonging to policy year 2005: this is the most useful criterion to identify trends.

Alignment and on-levelling of exposure

Fiscal year	Turnover (x 1€m)	Policy year	Turnover (x 1€m) - aligned to policy year	Turnover (x 1€m) - revalued @ 4% p.a.
2005	311.19	2005	311.19	409.50
2006	381.98	2006	381.98	483.32
2007	450.23	2007	450.23	547.77
2008	514.58	2008	514.58	601.99
2009	675.84	2009	675.84	760.23
2010	773.96	2010	773.96	837.12
2011	837.05	2011	837.05	870.53
2012	1,057.54	2012	1,057.54	1,057.54
2013	1,134.21	2013	1,134.21	1,090.59
2014	1,164.35	2014	1,164.35	1,076.51
2015	1,265.00	2015	1,265.00	1,124.58

FIGURE 11.3
Exposure data for the public liability policy for the same period as the claims data. Exposure data, which are given for each fiscal year (starting 1 April), must first be aligned to the policy year (here, 1 April as well) in the usual way. Then they are on-levelled at 4% (this number may be company-dependent, for example because of the way the prices of their products/services change). Note that the exposure figure for the renewal year (2015) is an estimate.

For each year the on-levelled exposure for policy year y (denoted $\text{Exposure}^*(y)$) will therefore be given by:

$$\text{Exposure}^*(y) = \text{Exposure}(y) \times \frac{I_E(y^*)}{I_E(y)} \tag{11.3}$$

where $\text{Exposure}(y)$ is the exposure for year y (after alignment) and $I_E(y)$ is a suitable exposure index calculate for policy year y. This exposure index need not be equal to the claim inflation index (5% in our case) nor to the standard price indices. It may even be client-dependent because it may reflect company-specific variables such as product price changes or salary updates for the workforce; as these are normally not shared, generic indices are used: e.g., the average weekly earning index for the relevant industry section may be used to on-level wage roll and GDP growth may be used to on-level turnover. For our case study, we have on-levelled past exposure by 4% per annum, so that the on-levelled exposure will be simply given by $\text{Exposure}^*(y) = \text{Exposure}(y) \times (1 + r_E)^{y^* - y}$ with $r_E = 4\%$ and $y^* = 2015$.

11.1.5.2 C2: Adjusting Losses for Changes of Exposure

We can now adjust the total losses for each year to bring them in line with the exposure relevant to the **reference policy year** y^* (i.e. the policy year of the new (or renewed) policy – $y^* = 2015$ for our case study). The formula we use for the exposure-adjusted total losses for each year are simply the following:

$$\text{Exposure-Adjusted Total Losses}(y) = \frac{\text{Exposure}^*(y^*)}{\text{Exposure}^*(y)} \times \text{Total Revalued Losses}(y) \quad (11.4)$$

(note the double asterisk at the numerator). This calculation applies to the gross, retained, and ceded losses. Figure 11.4 shows the results of these calculations for all years for our case study. As an example, the adjusted gross losses for year 2005 can be calculated as

$$\begin{aligned}\text{Adjusted gross}(2005) &= \frac{\text{Exposure}^*(2015)}{\text{Exposure}^*(2005)} \times \text{Gross}(2005) \\ &= \frac{1265}{461} \times 2{,}577{,}699 = £7{,}078{,}867\end{aligned}$$

11.1.6 Step D: Make Adjustments for IBNR

Based on either client information or (more frequently) benchmark data, we adjust the total losses for each year to take into account the possibility of incurred but not reported (IBNR) claims – either new claims being reported (IBNYR) or of further reserve development of old claims (IBNER)

An example of how this could be done using standard chain ladder techniques is given in Section 11.2. Here, we simply assume that we have benchmark factors to ultimate, which depend on the age of development (maturity) of each policy year. In a pricing context such age of development is almost never an integer number of years: e.g., for our case study the policy year 2014 is the most recent year available and has 8 months of development (from the inception of the policy year to the as-at date); 2013 has 20 months of development; and so on. Therefore we normally need the loss development factors (or LDFs, which are the factors by which we need to multiply the total losses for each year of development to obtain the total losses *projected to ultimate*) for each number of months. These LDFs may be obtained by portfolio studies, possibly carried out by the reserving function. The IBNR-adjusted figures for each year is given by the following formula, which should be applied to the gross, retained, and ceded losses:

$$\text{IBNR-Adjusted Total Losses}(y) = \text{LDF}(y) \times \text{Exposure-Adjusted Total Losses}(y) \quad (11.5)$$

			Original				Exposure-adjusted		
Policy year	No. of claims	No. of non-zero claims	Total losses (reval)	Retained losses (reval)	Ceded losses (reval)	Exposure	Total losses (reval)	Retained losses (reval)	Ceded losses (reval)
2005	86	86	2,577,699	803,768	1,773,932	461	7,078,867	2,207,303	4,871,564
2006	92	92	2,684,421	841,800	1,842,621	544	6,246,061	1,958,685	4,287,376
2007	111	110	1,718,684	1,142,872	575,812	616	3,528,463	2,346,319	1,182,143
2008	147	146	1,428,759	1,407,011	21,748	677	2,669,090	2,628,462	40,628
2009	196	194	1,876,028	1,765,098	110,930	855	2,775,146	2,611,051	164,095
2010	263	256	3,687,767	2,583,974	1,103,794	942	4,954,146	3,471,310	1,482,836
2011	263	250	2,465,124	2,465,124	–	979	3,184,543	3,184,543	–
2012	265	199	2,454,646	2,454,646	–	1190	2,610,253	2,610,253	–
2013	225	201	2,722,839	2,562,839	160,000	1227	2,807,710	2,642,723	164,987
2014	70	68	921,583	895,583	26,000	1211	962,738	935,577	27,161

FIGURE 11.4
The gross, retained, and ceded losses adjusted for exposure changes, bringing everything in line with the renewal exposure (£1,265M). All monetary amounts are in pounds, except for turnover which is in million pounds.

For our case study, the loss development factors (which we assume were given to us by the reserving actuaries and their study of the general liability portfolio) are as below:

Policy year	2005	2006	2007	2008	2009	2010	2011	2012	2013	2014
Maturity (m)	116	104	92	80	68	56	44	32	20	8
LDF	1.001	1.002	1.002	1.002	1.001	1.002	1.007	1.019	1.082	2.529

These LDFs also appear in the column 'Factor to ultimate' in Figure 11.5, and the projected total losses using such LDFs are shown in the last three columns under the header 'IBNR-adjusted'.

11.1.7 Step E: Make Adjustments for Large Losses

If we believe that large losses are either overrepresented or underrepresented in our data set, we can correct for that by using some assumptions on the distribution of large losses.

11.1.7.1 Adding/Removing Large Losses

One simple heuristic for doing that is with the addition or removal of large losses:

i. If, for example, there is a large loss of £10M in our 10-year data set (therefore having an effect – at constant exposure – of £1M on the burning cost), which for some reason (such as portfolio/market information), we believe is a 1 in 100 event, we might want to remove that loss and add back an amount $\frac{1}{100} \times £10\text{M} = £100\text{k}$ to the burning cost. This can be generalised to more than one large loss: in case large losses $LL_1, LL_2 \ldots LL_m$ of estimated return periods $RP_1, RP_2 \ldots RP_m$ (all larger than the observation period) are already present in the data set and we want to scale down their effect, the adjustment might take the form:

$$LLA = -(LL_1 + LL_2 + \ldots + LL_m) + \frac{LL_1}{RP_1} + \frac{LL_2}{RP_2} + \ldots + \frac{LL_m}{RP_m} \quad (11.6)$$

ii. We can do exactly the opposite with a large loss that *has not* happened but is thought to occur in the market with a certain return period.

This adjustment will normally take the form $LLA = \text{LL}/RP$ where LL is the expected size of a large loss of a given return period and RP is the return period.[2] This can also be generalised to a number of large losses $LL_1, LL_2 \ldots LL_m$ that are expected to occur with return periods $RP_1, RP_2 \ldots RP_m$. In this case the large loss adjustment is going to be simply:

$$LLA = \frac{LL_1}{RP_1} + \frac{LL_2}{RP_2} + \ldots + \frac{LL_m}{RP_m} \quad (11.7)$$

2 Note that the size of the correction is tailored to the reference exposure – hence this correction must be made after the exposure correction.

				Exposure-adjusted				IBNR/IBNER-adjusted		
Policy year	No. of claims	No. of non-zero claims	Exposure	Total losses (reval)	Retained losses (reval)	Ceded losses (reval)	Factor to ultimate	Total losses (reval)	Retained losses (reval)	Ceded losses (reval)
2005	86	86	461	7,078,867	2,207,303	4,871,564	1.001	7,087,468	2,209,985	4,877,483
2006	92	92	544	6,246,061	1,958,685	4,287,376	1.002	6,257,498	1,962,272	4,295,226
2007	111	110	616	3,528,463	2,346,319	1,182,143	1.002	3,533,949	2,349,968	1,183,981
2008	147	146	677	2,669,090	2,628,462	40,628	1.002	2,673,283	2,632,591	40,692
2009	196	194	855	2,775,146	2,611,051	164,095	1.001	2,778,950	2,614,630	164,320
2010	263	256	942	4,954,146	3,471,310	1,482,836	1.002	4,964,220	3,478,369	1,485,851
2011	263	250	979	3,184,543	3,184,543	–	1.007	3,207,837	3,207,837	–
2012	265	199	1190	2,610,253	2,610,253	–	1.019	2,659,341	2,659,341	–
2013	225	201	1227	2,807,710	2,642,723	164,987	1.082	3,036,735	2,858,290	178,445
2014	70	68	1211	962,738	935,577	27,161	2.529	2,435,030	2,366,332	68,698

FIGURE 11.5
Correction factors are applied to all immature years to take IBNR into account. For example, the total losses for 2013 (£2,807,710) increased by 8.2% to yield £2,807,710 × 1.082 = £3,036,735.

This heuristic[3] is very simple and there is of course considerable subjective judgment in:

- the choice of the threshold above which a loss is considered 'large';
- the size of the large loss(es) to add when large losses are added;
- the estimation of the return period of the large losses.

*11.1.7.2 Using Increased Limit Factors (ILFs)

A slightly more sophisticated approach – which can be used for liability classes – is by using ILFs. ILFs were mentioned in Chapter 1 and will be addressed at length in Chapter 23; the reader might want to ensure that they are familiar with the content of Section 23.1 before reading on.

This approach requires that an ILF curve (or rule) is available for the type of business that the company is writing. The ILF curve provides an indication of how much the expected losses increases [decrease] if the policy limit is increased [decreased]. Specifically, the expected total losses capped at limit L, $\mathbb{E}(S(L))$, are related to the expected total losses capped at limit L_0, $\mathbb{E}(S(L_0))$, by the simple relationship:

$$\mathbb{E}(S(L)) = \frac{ILF_b(L)}{ILF_b(L_0)} \times \mathbb{E}(S(L_0)) \tag{11.8}$$

Where $ILF_b(x)b$ is the increased limit factor calculated with respect to a reference limit b.[4]

This suggests this heuristic way of dealing with large losses:

(a) Pick a large-loss threshold LLT. There is considerable leeway as to what value LLT should be set at, but the choice should be guided by the data set itself: one criterion is to choose a value below the value of any loss in the data set which could be fairly considered as an outlier ... but not much lower, to avoid getting rid of data that provide solid information historical loss experience. In our case, we have chosen $LLT = £1M$.

(b) Before calculating the burning cost for each policy year, we need to go back to Step A cap each individual (revalued) loss at LLT, $\min(X_{\text{REV},i}, LLT)$. Carry out all the calculations of the losses ceded and retained below the each-and-every-loss deductible based on the capped losses.

(c) The aggregate gross, retained, and ceded losses *for each policy year y* and before the application of any aggregate deductible/limit can be calculated in the usual way (Steps C and D):

$$S_{\text{CAPPED}}^{\text{gross}}(y) = \frac{\text{Exposure}^*(y^*)}{\text{Exposure}^*(y)} \times \text{LDF}(y) \times \sum_i \min(X_{\text{REV},i}, LLT)$$

3 See solution to Question 8 in the book's website (Parodi, 2016) for a possible justification of Equation 11.7 from first principles.

4 This is because by definition $ILF_b(x) = \mathbb{E}(\min(X,x)) / \mathbb{E}(\min(X,b))$. Therefore, the relationship between the limited expected value of individual losses can be written as $\mathbb{E}(\min(X,L)) = \frac{ILF_b(L)}{ILF_b(LLT)} \times \mathbb{E}(\min(X,LLT))$.

Finally, the total expected losses from the ground up can be also written as $\mathbb{E}(S(x)) = \mathbb{E}(N) \times \mathbb{E}(\min(X,x))$, where $\mathbb{E}(N)$ is the expected claim count from the ground up, yielding Equation 11.8.

$$S_{\text{CAPPED}}^{\text{retained}}(y) = S^{\text{retained}}(y)$$

$$S_{\text{CAPPED}}^{\text{ceded}}(y) = S_{\text{CAPPED}}^{\text{gross}}(y) - S^{\text{retained}}(y) \tag{11.9}$$

where y^* is the year of the prospective policy, $\text{Exposure}^*(y)$ is the (on-levelled) exposure for policy year y, $\text{LDF}(y)$ is the loss development factor to adjust year y for IBNR. Note that (unless *LLT* should be inexplicably chosen to be lower than the *EEL* deductible), capping at LLT has no effect on the retained amount. The index i runs over all losses for year y.

(d) Multiply the aggregate gross losses by the ratio of the ILF factors calculated at *LLT* and at L (the policy limit), to have an approximate idea of what the effect of the losses above *LLT* might be:

$$S_{\text{LLA}}^{\text{gross}}(y) = S_{\text{CAPPED}}^{\text{gross}} \times \frac{ILF_b(L)}{ILF_b(LLT)} \tag{11.10}$$

where b is the reference amount for the ILF curve. Note that the policy limit L is used even when grossing up the gross losses because ultimately what we need is an estimate of the effect on the ceded losses. Again, the retained losses require no adjustment.

$$S_{\text{LLA}}^{\text{retained}}(y) = S_{\text{CAPPED}}^{\text{retained}}(y) = S^{\text{retained}}(y)$$

$$S_{\text{LLA}}^{\text{ceded}}(y) = \frac{ILF_b(L)}{ILF_b(LLT)} \times S_{\text{CAPPED}}^{\text{ceded}}(y) + \left(\frac{ILF_b(L)}{ILF_b(LLT)} - 1\right) \times S^{\text{retained}}(y) \tag{11.11}$$

(e) Calculate the average of $S_{\text{LLA}}^{\text{gross}}(y)$, $S^{\text{retained}}(y)$ and $S_{\text{LLA}}^{\text{ceded}}(y)$ to obtain an estimate of the expected gross losses, the expected retained losses, and the expected ceded losses, respectively. Note that all these expected values assume that losses are capped at the limit L. An estimate of the uncapped losses (if it is of interest) could be obtained by using the ratio $\frac{\lim_{L\to\infty} ILF_b(L)}{ILF_b(LLT)}$, but depending on the shape of the ILF curve the limit at the numerator might diverge to infinity: this is the case, for example, when using Riebesell curves.

To reiterate, the approach outlined above only works for liability classes. For property classes, it can be replaced with a criterion using exposure curves, but the process is more complex (see comment in Section 22.8.3).

For our case study we have used (purely for illustration purposes) a Riebesell ILF curve with $\rho = 0.1$. The results are shown in Figure 11.6.

11.1.8 Step F: Make Other Adjustments

Make other adjustments (e.g. for the changing risk profile) as necessary. Adjustments such as for changing the risk profile can be captured by an index, similar to what we do with claims inflation.

For our case study no other adjustments were considered necessary.

	Original					Exposure-, IBNR- and large-loss adjusted			
Policy year	Total losses (reval)	Retained losses (reval)	Ceded losses (reval)	Exposure	Factor to ultimate	Total losses (capped, reval, IBNR, expos)	Total losses (+ ILF Adjusted)	Retained losses (+ILF adjusted)	Ceded losses (+ ILF Adjusted)
2005	2,577,699	803,768	1,773,932	461	1.001	4,715,082	5,883,017	2,209,985	3,673,033
2006	2,684,421	841,800	1,842,621	544	1.002	4,246,620	5,298,516	1,962,272	3,336,244
2007	1,718,684	1,142,872	575,812	616	1.002	3,533,949	4,409,315	2,349,968	2,059,347
2008	1,428,759	1,407,011	21,748	677	1.002	2,673,283	3,335,460	2,632,591	702,869
2009	1,876,028	1,765,098	110,930	855	1.001	2,778,950	3,467,301	2,614,630	852,672
2010	3,687,767	2,583,974	1,103,794	942	1.002	4,850,009	6,051,365	3,478,369	2,572,997
2011	2,465,124	2,465,124	0	979	1.007	3,207,837	4,002,425	3,207,837	794,587
2012	2,454,646	2,454,646	0	1,190	1.019	2,659,341	3,318,065	2,659,341	658,724
2013	2,722,839	2,562,839	160,000	1,227	1.082	3,036,735	3,788,939	2,858,289	930,650
2014	921,583	895,583	26,000	1,211	2.529	2,435,030	3,038,192	2,366,332	671,859

FIGURE 11.6
Increased limit factors (ILFs) are applied to all years to consider adequately of large losses, using Equations 11.10 and 11.11. A Riebesell curve with $\rho = 0.1$ was used for this. *Note*: As mentioned in the text, this choice is arbitrary and was done only for illustration purposes; because of this, we will not use the output of this large-loss correction for the rest of the exercise.

Let us then denote $S^{\text{gross}}_{\text{pre-AGG}}(y), S^{\text{retained}}_{\text{pre-AGG}}(y)$ and $S^{\text{ceded}}_{\text{pre-AGG}}(y)$ the total gross, retained (below the deductible) and ceded losses after all adjustments in Steps A to F have been taken into account, but before any aggregate deductibles/limits have been imposed. These will be considered in Step G.

11.1.9 Step G: Impose Aggregate Deductibles/Limits as Necessary

As a last step before calculating the burning cost, and *after all other adjustments have been made*, we apply the aggregate modifiers of the policies, such as the AAD and the aggregate limit (Figure 11.7). As an example, if there is an AAD, the modified total retained and ceded losses for each year will look like this:

$$S^{\text{retained}}_{\text{post-AGG}}(y) = \min\left(AAD, S^{\text{retained}}_{\text{pre-AGG}}(y)\right)$$

$$S^{\text{ceded}}_{\text{post-AGG}}(y) = S^{\text{ceded}}_{\text{pre-AGG}}(y) + \max\left(0, S^{\text{retained}}_{\text{pre-AGG}}(y) - AAD\right) \tag{11.12}$$

11.1.10 Step H: Calculate the Burning Cost Estimate of the Expected Losses

The burning cost per unit of exposure (p.u.e.) for each year (we are mainly interested in the insured losses, but of course it is possible to define the gross and retained burning cost in the same way) is given by:

$$BC(y) = \frac{S^{\text{ceded}}_{\text{post-AGG}}(y)}{\text{Exposure}^*(y^*)} \tag{11.13}$$

Note that we have divided each year's adjusted losses by $\text{Exposure}^*(y^*)$ and not by $\text{Exposure}^*(y)$. This is because $S^{\text{ceded}}_{\text{post-AGG}}(y)$ already need to incorporate the adjustment for the change in the exposure before applying the aggregate deductible/limit. It is easy to see that if no aggregate feature is present, Equation 11.13 is equivalent to the total losses for year y adjusted for everything except exposure divided by the exposure for year y.

Policy year	No. of claims	No. of non-zero claims	Exposure	IBNR/IBNER-adjusted Total losses (reval)	IBNR/IBNER-adjusted Retained losses (reval)	IBNR/IBNER-adjusted Ceded losses (reval)	AAD/Aggregate limits applied Total losses (reval)	AAD/Aggregate limits applied Retained losses (reval)	AAD/Aggregate limits applied Ceded losses (reval)
2005	86	86	461	7,087,468	2,209,985	4,877,483	7,087,468	2,209,985	4,877,483
2006	92	92	544	6,257,498	1,962,272	4,295,226	6,257,498	1,962,272	4,295,226
2007	111	110	616	3,533,949	2,349,968	1,183,981	3,533,949	2,349,968	1,183,981
2008	147	146	677	2,673,283	2,632,591	40,692	2,673,283	2,632,591	40,692
2009	196	194	855	2,778,950	2,614,630	164,320	2,778,950	2,614,630	164,320
2010	263	256	942	4,964,220	3,478,369	1,485,851	4,964,220	3,200,000	1,764,220
2011	263	250	979	3,207,837	3,207,837	–	3,207,837	3,200,000	7837
2012	265	199	1190	2,659,341	2,659,341	–	2,659,341	2,659,341	–
2013	225	201	1227	3,036,735	2,858,290	178,445	3,036,735	2,858,290	178,445
2014	70	68	1211	2,435,030	2,366,332	68,698	2,435,030	2,366,332	68,698

FIGURE 11.7
The gross, retained, and ceded losses are modified by imposing the aggregate deductible of £3.2M on each policy year. The only years for which the aggregate deductible has an effect are 2011 and 2012. Notice how the ceded amount (Column 10) increases as a consequence of the retained amount (Column 9) being capped.
Note that the AAD has been applied to the figures without the large-loss correction (see comment in Figure 11.6).

The burning cost estimate p.u.e. *for the reference policy year* y^* is then given by the average (exposure-weighted or unweighted) of the burning cost over the past years:

$$BC_{\text{unweighted}}\left(y^*\right)=\frac{\sum_{\text{past } y} BC(y)}{n}$$

$$BC_{\text{weighted}}\left(y^*\right)=\frac{\sum_{\text{past } y} \text{Exposure}^*(y)\times BC(y)}{\sum_{\text{past } y} \text{Exposure}^*(y)} \tag{11.14}$$

where n is the number of past policy years used in the calculations, and 'past y' is the actual set of years used in the calculations ($n=\sum_{past\ y} 1$). The exposure-weighted average will normally be a better estimate of the burning cost, as it gives more weight to those years for which the experience is statistically more significant.

Using the exposure-weighted burning cost p.u.e., the estimated expected losses for the new policy are then obtained by multiplying it by the (expected) exposure for the reference policy year y^*:

$$\text{Expected Ceded Losses}\left(y^*\right)=BC_{\text{weighted}}\left(y^*\right)\times \text{Exposure}^*\left(y^*\right) \tag{11.15}$$

and similarly for the calculations based on the unweighted burning cost.

The calculations for our case study are illustrated in Figure 11.8. Note that:

- the calculations shown are for the case where no large-loss adjustment has been made. The calculations for the case illustrated in Section 11.1.7.2 are formally identical and can be found in the download section of the book's website (Parodi, 2016);
- only years 2005 to 2013 were included. This is because despite the fact that we have introduced corrections for the years that are not fully developed (all years, in this case), we concluded that the 2014 year projection was still too uncertain to be used as part of the calculations. This is of course a matter of case-by-case judgment and not an absolute rule. If the risk profile has changed significantly after a certain date, the policy years before that date may also sometimes be excluded.

For our case study, therefore, the exposure-weighted burning cost estimate of the expected ceded losses is:

Expected Ceded Losses (2015) ~ £984k

Note that if instead we use the unweighted burning cost estimate, the expected losses become:

Expected Ceded Losses (2015) ~ £1.390M (unweighted average)

Such a large difference is unusual, and the reason why this happens here is that the exposure changes dramatically over the years.

Apart from the expected losses, we can also calculate some summary statistics such as the standard deviation and (if the number of years of experience is sufficient) a few percentiles. These additional statistics give an approximate measure of the year-on-year volatility of the total losses, although they fall well short of providing a full picture of the aggregate loss distribution; for example, the full information on how likely it is for total losses to exceed a specific amount. A very rough way of providing this full picture is

Policy year	No. of claims	No. of non-zero claims	Exposure	IBNR/IBNER-adjusted Total losses (reval)	IBNR/IBNER-adjusted Retained losses (reval)	IBNR/IBNER-adjusted Ceded losses (reval)	AAD/Aggregate limits applied Retained losses (reval)	AAD/Aggregate limits applied Ceded losses (reval)
2005	86	86	461	7,087,468	2,209,985	4,877,483	2,209,985	4,877,483
2006	92	92	544	6,257,498	1,962,272	4,295,226	1,962,272	4,295,226
2007	111	110	616	3,533,949	2,349,968	1,183,981	2,349,968	1,183,981
2008	147	146	677	2,673,283	2,632,591	40,692	2,632,591	40,692
2009	196	194	855	2,778,950	2,614,630	164,320	2,614,630	164,320
2010	263	256	942	4,964,220	3,478,369	1,485,851	3,200,000	1,764,220
2011	263	250	979	3,207,837	3,207,837	–	3,200,000	7837
2012	265	199	1190	2,659,341	2,659,341	–	2,659,341	–
2013	225	201	1227	3,036,735	2,858,290	178,445	2,858,290	178,445
2014	70	68	1211	2,435,030	2,366,332	68,698	2,366,332	68,698
Straight average (05-13)				*4,022,142*	*2,663,698*	*1,358,444*	*2,631,897*	*1,390,245*
Standard deviation (05-13)				*1,672,393*	*473,988*	*1,913,783*	*418,879*	*1,917,636*
25th percentile (05-13)				*2,778,950*	*2,349,968*	*40,692*	*2,349,968*	*40,692*
75th percentile (05-13)				*4,964,220*	*2,858,290*	*1,485,851*	*2,858,290*	*1,764,220*
Weighted average (05-13)				*3,703,005*	*2,755,396*	*947,609*	*2,719,375*	*983,630*
Weighted standard deviation (05-13)				*1,491,568*	*459,251*	*1,678,724*	*401,233*	*1,692,280*
25th percentile (weighted, 05-13)				*2,720,773*	*2,551,472*	*13,013*	*2,551,472*	*18,344*
75th percentile (weighted, 05-13)				*4,234,542*	*3,068,002*	*1,331,847*	*3,063,300*	*1,468,201*

FIGURE 11.8
The burning cost is calculated based on an average of years 2005 to 2013, both on a weighted and an unweighted basis. Some other summary statistics are also calculated.

BOX 11.1 WEIGHTED STATISTICS

Weighted Mean

The weighted mean of quantities $x_1, x_2 \ldots x_n$ with weights $w_1, w_2 \ldots w_n$ is defined as

$$\bar{x} = (w_1 x_1 + w_2 x_2 + \ldots + w_n x_n)/(w_1 + w_2 + \ldots + w_n) \tag{11.16}$$

Weighted Standard Deviation

The weighted (sample) variance is defined as

$$s^2 = \frac{\sum_{i=1}^{n} w_i}{\left(\sum_{i=1}^{n} w_i\right)^2 - \sum_{i=1}^{n} w_i^2} \sum_{i=1}^{n} w_i (x_i - \bar{x})^2 \tag{11.17}$$

Weighted Percentiles

The formula for the percentiles depends on the definition of percentile that we use – unfortunately, there is no unique definition although all of them converge to the same value for large data sets.

In the definition we illustrate here, the percentiles are calculated as follows:

i. First, sort out the n values $x_{(1)}, x_{(2)} \ldots x_{(n)}$ and calculate the partial sums of the corresponding weights $w_{(i)}$ (we have used [i] rather than i to indicate that the values have been sorted):

$$W_k = \sum_{i=1}^{k} w_{(i)} \tag{11.18}$$

ii. Then calculate the (weighted) percentiles p_k (a number between 0 and 1) corresponding to the n sorted values:

$$p_k = \frac{1}{W_n}\left(W_k - \frac{w_{(k)}}{2}\right) \tag{11.19}$$

iii. Finally, to find out the value $\widehat{F^{-1}}(p)$ corresponding to an intermediate percentile p (the symbol $\widehat{F^{-1}}(p)$ has been chosen as a reminder that this is an estimate of the inverse of the cumulative density function $F(x)$), interpolate between the values $x_{(k)}$ and $x_{(k+1)}$ in proportion to

$$\widehat{F^{-1}}(p) - x_{(k)} + \frac{p - p_k}{p_{k+1} - p_k}\left[x_{(k+1)} - x_{(k)}\right] \tag{11.20}$$

Note that by definition this method does not allow us to calculate percentiles that are outside the range $[x_{(0)}, x_{(n)}]$.

Figure 11.9 illustrates how these calculations are performed in practice for one column (total revalued losses) of our case study.[5] In applying the formulae above to

5 As at the time of writing, Excel does not have functions for the weighted statistics, so you will need to create the calculations yourself. As for the functions it has for the *unweighted* percentile, as noted above, there is no unique definition for the percentiles and the formulae used in Excel in the functions PERCENTILE.EXC and PERCENTILE.INC produce slightly different results from the formulae above.

Policy year	Exposure	Losses	Losses, exposure-adjusted	Rank, k	Sorted losses, x(k)	Weights, w(k)	Cumulative weights, W(k)	Weighted percentiles, p(k)
2005	461	2,577,699	7,078,867	1	2,610,253	1190	1190	0.08
2006	544	2,684,421	6,246,061	2	2,669,090	677	1867	0.20
2007	616	1,718,684	3,528,463	3	2,775,146	855	2722	0.31
2008	677	1,428,759	2,669,090	4	2,807,710	1227	3949	0.45
2009	855	1,876,028	2,775,146	5	3,184,543	979	4928	0.59
2010	942	3,687,767	4,954,146	6	3,528,463	616	5544	0.70
2011	979	2,465,124	3,184,543	7	4,954,146	942	6486	0.80
2012	1190	2,454,646	2,610,253	8	6,246,061	544	7030	0.90
2013	1227	2,722,839	2,807,710	9	7,078,867	461	7491	0.97
		Straight average	*3,983,809*				*Weighted average*	*3,650,762*
		Standard deviation	*1,693,521*				*Weighted standard deviation*	*1,520,087*
		25th percentile	*2,775,146*				*25th percentile (weighted)*	*2,716,755*
		75th percentile	*4,954,146*				*75th percentile (weighted)*	*4,226,809*

FIGURE 11.9
An illustration of the calculation of the weighted statistics for losses over 9 years. Note how the total losses for each year are first sorted in ascending order, then the cumulative weights are calculated according to this same order, and based on this, the weighted percentiles (and all the other statistics) are calculated. Also notice how the difference between two successive percentiles is not constant, as it would be the case for the standard percentiles.

the estimation of loss statistics, the x_k's can either be interpreted as the burning cost per unit of exposure for each year or as the *exposure-adjusted* projected losses (i.e. historical losses corrected for inflation, IBNR and brought to the renewal exposure): this is indeed equivalent to using the losses per unit of exposure in the calculations and then multiplying the result by the renewal exposure.

addressed in Section 11.3 (Producing an Aggregate Loss Distribution); for a more refined approach, we will need to wait until we develop appropriate stochastic loss modelling skills in Chapters 12 through 17.

Percentiles based on a burning cost exercise also provide a sense-check on the result of more sophisticated stochastic models: for instance, if the more sophisticated model predicts that gross losses of £6M will happen on average every 50 years, we can immediately see from Figure 11.8 that the stochastic model may be missing something because we have losses of *more* than £6M twice in the period 2005 to 2013.

The calculation of the weighted statistics does not involve advanced maths but does require some care. Box 11.1 provides some guidance on how to produce these statistics.

11.1.11 Step I: Premium Calculation Based on the Burning Cost

We can now use the results of the burning cost analysis to determine the premium that the insurer should charge for a given risk. In our case study, for example, the insurer becomes liable for the ceded losses (Column 8 in Figure 11.8) and charges a premium for them. If we assume for simplicity that the insurer's pricing model is simply to achieve a target loss ratio of 80%, the premium can be calculated (using the more stable weighted average method) as

$$\text{Premium} = \frac{\text{BC estimate of the expected losses}}{\text{Target loss ratio}} = \frac{£984\text{k}}{80\%} = £1.23\text{M}$$

*11.1.12 Summary: Burning Cost Analysis in Formulae

The burning cost process as we described it through Sections 11.1.3 to 11.1.11 can be usefully summarised for the mathematically inclined into a few compact formulae rather than in a step-by-step approach as illustrated through Sections 11.1.3 to 11.1.11. While the formulae below are not absolutely general,they are true for the case of a policy with:

- an each-and-every-loss deductible EEL,
- an individual loss limit L, and
- an annual aggregate deductible AAD,

similar formulae can be produced for different cases. Adapting these formulae to specific insurance structures is left to the reader as an exercise (Question 7).

Note that in order to make the formulae more compact we will use a more compact notation for the same objects than in the previous sections: e.g., we will use $\mathcal{E}$ for 'Exposure', $a \wedge b$ for $\min(a,b)$, a_+ for $\max(0,a)$.

The main quantity we want to estimate is the expected ceded losses for the reference policy year y^*. This is given by Equation 11.15, which we rewrite using our shorthand as:

$$\text{Expected Ceded Loss}\,(y^*) = \text{BC}(y^*) \times \mathcal{E}^*(y^*) \tag{11.21}$$

where $\text{BC}(y^*)$ is the *estimated* burning cost per unit of exposure for the reference policy year y^* and $\mathcal{E}^*(y^*)$ is the expected exposure for that year.

The estimated burning cost is obtained as the exposure-weighted average of the burning cost over all (or a sub-set of the) past years y available:

$$\text{BC}(y^*) = \frac{\sum_y \mathcal{E}^*(y) \times BC(y)}{\sum_y \mathcal{E}^*(y)} \tag{11.22}$$

where $\mathcal{E}^*(y)$ is the aligned and on-levelled exposure and $BC(y)$ is the burning cost p.u.e. for policy year y.

$$BC(y) = \frac{\left(\frac{\mathcal{E}^*(y^*)}{\mathcal{E}^*(y)}\text{LDF(y)}\sum_i X_i^{\text{ced},y} + LLA\right) + \left(\frac{\mathcal{E}^*(y^*)}{\mathcal{E}^*(y)}\text{LDF(y)}\sum_i X_i^{\text{ret},y} - AAD\right)_+}{\mathcal{E}^*(y^*)} \tag{11.23}$$

And where in turn:

- $X_i^{\text{ced},y} = (L - EEL)\Lambda\left(X_i^{\text{rev},y} - EEL\right)_+$ is the amount of the i-th loss ceded to the insurer in policy year y, and where in turn $X_i^{\text{rev},y}$ is the value of the i-th loss after revaluation;
- $X_i^{\text{ret},y} = X_i^{\text{rev},y}\Lambda\, E\pi L$ is the amount of the i-th loss retained by the insured in policy year y, before any aggregate deductible;
- LDF(y) is the loss development factor for policy year y, which takes into account IBNR development;
- $\mathcal{E}^*(y)$ is the exposure for policy year y after on-levelling to the reference policy year and alignment between exposure period and policy period;
- $\mathcal{E}^*(y^*)$ is the exposure for the **reference policy year** y^*;
- *LLA* is a large-loss adjustment (negative or positive) applied uniformly to all policy years.

We have ignored adjustments for changes in the risk profile, terms and conditions, etc. to avoid cluttering the formula. These adjustments can be treated by using an index, similarly to inflation.

For liability business, and in case an ILF is used for the purpose of large-loss correction, Equation 11.23 can be re-written as follows:

$$BC(y) = \frac{\frac{\mathcal{E}^*(y^*)}{\mathcal{E}^*(y)}\text{LDF}(y)\left(\frac{ILF_b(L)}{ILF_b(LLT)}\sum_i \widehat{X_i^{ced,y}} + \left(\frac{ILF_b(L)}{ILF_b(LLT)} - 1\right)\sum_i X_i^{ret,y}\right) + \left(\frac{\mathcal{E}^*(y^*)}{\mathcal{E}^*(y)}\text{LDF}(y)\sum_i X_i^{ret,y} - AAD\right)_+}{\mathcal{E}^*(y^*)} \tag{11.24}$$

where:

- LLT is the large-loss threshold;
- $ILF_b(x)$ is the increased limit factor with base b (see Chapter 22);
- $\widehat{X_i^{\text{ced},y}} = (L - EEL)\Lambda\left(\widehat{X_i^{\text{rev},y}} - EEL\right)_+ = \left(\widehat{X_i^{\text{rev},y}} - EEL\right)_+$ is the ceded amount for the capped

Accident year	Development year 0	1	2	3	4	5	6	7	8	9
2003	758,859	6,712,563	7,295,862	8,481,698	8,581,273	8,929,061	9,406,673	9,421,491	9,425,375	9,547,636
2004	588,009	1,786,021	2,187,149	2,365,737	2,474,465	2,842,739	2,842,739	2,882,701	3,398,944	
2005	514,089	1,532,487	2,331,175	8,377,877	8,954,659	9,117,566	9,138,301	9,147,275		
2006	419,422	2,882,030	4,009,785	4,413,923	4,468,089	4,616,335	4,823,964			
2007	261,482	2,089,735	3,050,709	3,684,369	4,130,221	5,036,548				
2008	893,053	2,121,944	4,368,448	4,546,849	6,942,262					
2009	481,366	954,766	2,026,609	2,481,851						
2010	696,678	1,505,950	2,283,808							
2011	4,336,497	5,355,547								
2012	433,625									

FIGURE 11.10
The input to a reserving exercise is, in many cases, a claims development triangle like this, where each figure inside the triangle represents the cumulative claim amount incurred in a specific accident year at different years of development. For example, if the table above represents incurred claims, a total of £3,684,369 in claims that occurred during accident year 2007 had been reported in the first 3 years (AY = 2007, DY = 3). If, instead, the table above represents paid claims, £3,684,369 is the amount of 2007 claims that has been paid in the first 3 years. An alternative representation of the same information is that which shows incremental claims rather than cumulative claims: in that case, for example, the cell with AY = 2007 and DY = 3 would be £3,684,369 – £3,050,709 = £633,660.

revalued loss $\widehat{X_i^{rev,y}} = \sum_i \min\left(X_i^{rev,y}, LLT\right)$;

- $X_i^{ret,y} = X_i^{rev,y} \wedge EEL$ as above.

11.2 Adjusting for IBNR: Development Triangles

We have seen in Section 1.5 that if we have appropriate correction factors, we may adjust the total losses for each policy year to include an allowance for IBNR claims. This section explains how these correction factors may be derived. One way in which it is possible to adjust for IBNR is by using total claims triangulations for past years (or periods of other duration), which the insurer should have for its reserving exercises anyway. An alternative to using development triangles is using triangle-free (or granular) reserving, which we will encounter again in Chapter 13 for projecting claim counts and that the reader can find in its full form for example in Parodi (2014).

The input to the reserving exercise will often be a claims development triangle as shown in Figure 11.10. This triangle can be used to estimate the ultimate total claims for each year by projection techniques such as the chain ladder, Bornhuetter–Fergusson, generalised linear models, or others. A reminder of how the chain ladder works is shown in Box 11.2.

11.3 Producing an Aggregate Loss Distribution

The burning cost is almost by definition a deterministic method, based purely on experience. However, we have seen how several modifications can be made to the method so that

BOX 11.2 THE CHAIN LADDER

The chain ladder is probably the most famous projection technique for estimating the ultimate claims amount in a reserving exercise. Actuaries would never use its results blindly, but the chain ladder methodology is also the basis for several more sophisticated methods, such as the Bornhuetter–Ferguson method and the Cape Cod method.

The chain ladder is based on the idea that the more recent years will develop roughly in the same way as the previous years: specifically, if on average the amount in development year (DY) 1 is 2.787 times the amount as in DY 0 based on years 2003 to 2011 (as is the case in Figure 11.11), then we can assume that the amount for DY = 1 of 2012 will also be 2.787 times the amount for DY = 0 of 2012, that is, it will be £433,625 × 2.787 = £1,208,461. The factor '2.787' in the calculation above is called a '(year-on-year) development factor' as it allows to develop the total claim amount from one year to the next.

This can be applied iteratively to all development years, possibly adding a tail factor to account for further development beyond 9 years of development. In the case of Figure 11.11, we have assumed a tail factor of 1.005. This allows us to produce the so-called 'factors to ultimate', that is, cumulative development factors that allow us to project each year to its ultimate expected value. For example, the factor to ultimate for DY = 0 is

$$F2U(0) = 2.787 \times 1.407 \times \ldots \times 1.005 = 6.942$$

The ultimate loss amount predicted for 2012 is therefore going to be

$$\text{Ultimate}(2012) = F2U(0) \times \text{Reported}(2012) = 6.492 \times £433,625 = £3,010,173$$

Analogously, the ultimate for 2008 will be

$$\text{Ultimate}(2008) = F2U(4) \times \text{Reported}(2008) = 1.168 \times £6,942,262 = £8,105,848$$

We still haven't specified how to calculate the average development factors. This can be done in several different ways (a straight average, a weighted average, a trimmed average) but the standard way is by a weighted average. For instance, the average development ratio between DY = 0 and DY = 1 can be written as

$$DR(0->1) = \frac{6,712,563 + 1,786,021 + \ldots + 5,355,547}{758,859 + 558,009 + \ldots + 4,336,497} = 2.787$$

Note that the sum extends only to those (nine) years for which there is an entry for both DY = 0 and DY = 1 (i.e. 2003 to 2011 – excluding 2012). Analogously, for DR $(k->k+1)$ the sum will extend to those 9–k years for which there is an entry for both DY = k and DY = k + 1.

The output of such an exercise is the ultimate total claims for each year, as shown in Figure 11.11. This output automatically takes IBNR into account and can be used as the input to the burning cost exercise in several different ways.

Accident year	Development year 0	1	2	3	4	5	6	7	8	9	Projected
2003	758,859	6,712,563	7,295,862	8,481,698	8,581,273	8,929,061	9,406,673	9,421,491	9,425,375	9,547,636	9,595,374
2004	588,009	1,786,021	2,187,149	2,365,737	2,474,465	2,842,739	2,842,739	2,882,701	3,398,944	3,443,033	3,460,248
2005	514,089	1,532,487	2,331,175	8,377,877	8,954,659	9,117,566	9,138,301	9,147,275	9,533,951	9,657,620	9,705,908
2006	419,422	2,882,030	4,009,785	4,413,923	4,468,089	4,616,335	4,823,964	4,838,344	5,042,872	5,108,285	5,133,827
2007	261,482	2,089,735	3,050,709	3,684,369	4,130,221	5,036,548	5,175,956	5,191,385	5,410,837	5,481,023	5,508,428
2008	893,053	2,121,944	4,368,448	4,546,849	6,942,262	7,411,461	7,616,604	7,639,308	7,962,239	8,065,521	8,105,848
2009	481,366	954,766	2,026,609	2,481,851	2,768,464	2,955,573	3,037,381	3,046,435	3,175,215	3,216,402	3,232,484
2010	696,678	1,505,950	2,283,808	3,104,665	3,463,203	3,697,266	3,799,603	3,810,929	3,972,026	4,023,549	4,043,667
2011	4,336,497	5,355,547	7,534,365	10,242,402	11,425,232	12,197,416	12,535,030	12,572,396	13,103,860	13,273,836	13,340,205
2012	433,625	1,208,461	1,700,104	2,311,164	2,578,066	2,752,306	2,828,488	2,836,919	2,956,842	2,995,197	3,010,173
										Tail	
Development factor	2.787	1.407	1.359	1.115	1.068	1.028	1.003	1.042	1.013	1.005	
Factor to ultimate	6.942	2.491	1.771	1.302	1.168	1.094	1.064	1.061	1.018	1.005	

FIGURE 11.11
Illustration of the chain ladder method on the claims development triangle of Figure 11.10. By the same reasoning, the average development factor from DY = 1 to DY = 2 as calculated over the years 2003 through 2010 can be applied to AY = 2011 to yield £5,355,547 × 1.407 = £7,534,365. Also, because we have already projected AY = 2012 to DY = 1 (yielding £1,208,461), we can apply the 0 → 1 development factor to that as well, yielding £1,208,461 × 1.407 = £1,700,104.

1. The most straightforward way is to use the last column of Figure 11.11 as the new estimate of the ultimate cost for each year, which then needs to be revalued for claims inflation and adjusted for relevant factors other than IBNR. (This is only possible, however, if we have been working on a triangle of only the claims from the policy we are trying to price.)
2. Another simple way, which is useful where the client's data is insufficient to produce a reliable estimate of the factors to ultimate, is to use benchmark factors to ultimate from a portfolio of similar risks/clients (possibly the whole market, if these statistics are available, as is the case in the United States). Thus, if $F2U^P(d)$ are the benchmark factors to ultimate for a portfolio P, and y is the most recent year that is at least partially earned, and k is the year that we aim to develop to ultimate, the adjusted claim amount for year k with information up to year y can be defined as

$$\text{Adjustment claim amount}\,(k,y) = F2U^P\,(y-k)\times\text{Reported}\,(k) \qquad (11.25)$$

where $F2U$ is the relevant factor to ultimate for year k.

In the case where the factors to ultimate are derived from the client's claims development triangle, Equation 11.25 simply yields the estimated ultimate cost for year k described in Point 1.

We should check that the reported data have similar maturities as the benchmark data. For example, if the data is compiled 9 months through the policy year, then it will be as at 9 months for the latest year, 21 months for the previous year, and so on. However, the triangles used for reserving may be as at 12, 24 and 36 months of development, and so forth. The relation between the exposure period of an individual policy and the exposure period of the reserving data also needs to be considered.

These adjustment methods can theoretically be applied to the gross, retained, or ceded losses separately. However, it is often the case that there is not enough information on the ceded losses to be able to produce stable triangulations. A quick alternative is therefore to use the factors to ultimate for the gross amount and apply them indiscriminately to the gross, retained, and ceded losses.

In our case study, for example, we have assumed that we have been given benchmark factors to ultimate (see column 'Factor to ultimate' in Figure 11.5) derived from some prior portfolio analysis that apply equally well to the gross, retained, and ceded amount, thus obtaining the figures in the last three columns of Figure 11.5.

it takes into account elements that go beyond the experience: for example, adjustments for large losses and for IBNR.

We have also seen how the burning cost can be extended to include (rough) calculations not only of the expected losses but also of their volatility and of the percentiles of the total loss distribution. Obviously, there must be enough years of relevant experience for these statistics to have even a rough meaning, and in any case the percentiles can predict numbers only within the past range of total losses.

Some practitioners extend this even further to obtain a (very rough) indication of the aggregate loss distribution at *all* percentiles, including those that are outside the loss experience. One popular (if very crude) method to do this is by modelling the annual

	Average total losses (£M)	Standard deviation of total losses (£M)	Lognormal distribution parameters	
Gross	3.703	1.492	15.05	0.388
Retained	2.755	0.459	14.82	0.166
Ceded	0.948	1.679	13.52	1.192

FIGURE 11.12
Calculation of the parameters of the lognormal for the total, retained, and ceded losses (after adjustment for exposure and IBNR, but before the imposition of an AAD and aggregate limit) for the example of Figure 11.8. Note that the weighted statistics have been used as an input to the lognormal model.

total losses as a lognormal distribution whose mean is equal to the (weighted) average total losses over all years, and the standard deviation is equal to the (weighted) standard deviation of the total losses over all years.[6] Note that this method can be applied to gross, retained, and ceded losses separately.

To use this method, we need the formulae that relate the mean and standard deviation of the lognormal distribution to its parameters, and vice versa:

$$\mathbb{E}(S)=\exp\left(\mu+\frac{1}{2}\sigma^2\right),\ \mathrm{SD}(S)=\mathbb{E}(S)\sqrt{\exp(\sigma^2)-1}$$
$$\sigma=\sqrt{\ln\left(1+\left(\frac{\mathrm{SD}(S)}{\mathbb{E}(S)}\right)^2\right)},\ \mu=\ln(\mathbb{E}(S))-\frac{1}{2}\sigma^2 \tag{11.26}$$

where S is the random variable representing total losses and SD(S) is the standard deviation of the total losses. In practice, the parameters of the lognormal distribution are estimated using Equation 11.26, but using the empirical estimates of $\mathbb{E}(S)$ and SD(S) derived from the data.

Let us now see how this works in our case, using the results after IBNR adjustments (but before the imposition of an annual cap) of Figure 11.8. One might wonder why we are not modelling the case that is the most relevant to us, that is, the one with an AAD. The main reason is that, in this case, the retained losses are by definition limited to £3.2M, whereas simulating the retained losses from a lognormal distribution might lead to occasional breaches of this limit – it is therefore better to simulate the retained losses before the application of the AAD (Figure 11.1) and to impose a cap afterwards, as shown in the example below. The percentiles of the aggregate loss distributions can then be calculated in the usual way, by using for example spreadsheet functions, simulations, or statistical packages, as shown in Figure 11.13.

Figure 11.12 shows that this method provides for our case study an approximation of some of the percentiles for the gross and retained distribution, which is at least consistent with the empirical percentiles. However, for the ceded distribution, the results are obviously at odds with the empirical percentiles (e.g. the 25th percentile is £208k for the lognormal model and £13k for the empirical burning cost).

6 Alternatively, one can first take the logarithm of the total losses of each year, and calculate the parameters μ and σ of the lognormal in that way. However, this method only works if the total losses are different from zero for all years.

Burning-Cost-Based Lognormal Model

	Gross	Retained	Ceded	Retained (with AAD)	Ceded (with AAD)
Mean					
Std Dev					
10%	2,089,733	2,198,382	101,138	2,198,382	112,365
20%	2,478,423	2,364,448	170,849	2,364,448	196,430
25%	2,644,359	2,430,776	208,505	2,430,776	240,967
30%	2,802,823	2,491,924	249,344	2,491,924	288,337
40%	3,113,445	2,606,278	344,421	2,606,278	394,375
50%	3,434,828	2,717,903	465,818	2,717,903	518,111
60%	3,789,386	2,834,308	630,002	2,834,308	683,845
70%	4,209,344	2,964,375	870,227	2,964,375	954,481
75%	4,461,589	3,038,946	1,040,675	3,038,946	1,109,234
80%	4,760,303	3,124,195	1,270,043	3,124,195	1,341,872
90%	5,645,719	3,360,197	2,145,436	3,200,000	2,155,638
95%	6,499,793	3,568,474	3,307,895	3,200,000	3,325,959
98%	7,616,541	3,818,370	5,385,002	3,200,000	5,494,978
99%	8,465,715	3,994,617	7,452,061	3,200,000	7,356,337
99.5%	9,325,594	4,163,038	10,032,331	3,200,000	9,907,062
			... and Empirical Statistics		
Weighted average (05–13)	3,703,005	2,755,396	947,609	2,719,375	983,630
Weighted standard deviation (05–13)	1,491,568	459,251	1,678,724	401,233	1,692,280
25th percentile (weighted, 05–13)	2,720,773	2,551,472	13,013	2,551,472	18,344
75th percentile (weighted, 05–13)	4,234,542	3,068,002	1,331,847	3,063,300	1,468,201

FIGURE 11.13
The percentiles for gross, retained, and ceded (columns 2, 3, and 4, respectively) are obtained by using a spreadsheet function to calculate the percentile of a lognormal distribution with the parameters in Figure 11.1. The percentile for the retained amount with an AAD of £3.2M (column 5) is derived by capping the retained amount of column 3 ('Retained') at £3.2M, and the percentile for the ceded amount with AAD (column 6) is obtained by simulation: the retained and ceded amounts are sampled from the lognormal distributions with the parameters in Figure 11.1, and for each simulation, the ceded with AAD is calculated as Ceded (with AAD) = Ceded (without AAD) + Retained (without AAD) – Retained (with AAD). Note that, in this instance, the 10th percentile of the retained amount is larger than the gross – a paradoxical consequence of the fact that the ceded, retained, and gross amounts are modelled independently.

The main weakness of this method is that there is actually no particular reason (and certainly, normally, not enough evidence) to support the view that the aggregate loss distribution should be a lognormal distribution. The issue of model uncertainty, therefore, looms large.

On top of this, there is the usual issue of parameter uncertainty because the parameters are calculated over a very limited number of points (normally 5 to 10). Despite the fact that this method is very crude, it is sometimes used amongst practitioners as it is simple and provides an easy answer (although not necessarily the right one) to a range of questions, such as how the AAD should be set.

The other problem with this approach is that if we produce separate models for gross, retained, and ceded losses, we miss out on the obvious relationship between these three components, which are linked by the simple relationship gross = retained + ceded (where retained can be further broken up as retained below and above the deductible).

11.4 Burning Cost for Portfolio Pricing

The burning cost methodology outlined in Section 11.1 was illustrated with a single risk in mind – specifically, a public liability policy for a corporation. The same methodology can also be used to price entire portfolios. This can be done, for example, when calculating the loss ratio of a portfolio for the purpose of estimating its profitability, or when pricing treaty reinsurance (Chapter 20).

The additional difficulty in portfolio pricing is that the process of matching exposure and claims is more complex than for single-risk pricing, where (at least for occurrence-based policies) all that was required was aligning the exposure period to the policy period. In portfolio pricing, however, exposures come from different policies and more care is needed to ensure that we only include the exposure that is relevant to the portfolio.

In this section we will therefore introduce the concept of 'earned exposure' and we will see how that can be used to match exposures and claims at portfolio level. To be clear, the concept of 'earned exposure' makes perfect sense also at single-risk level, but its calculation is normally trivial.

11.4.1 The Concept of Earned Amount

'Earned amount' is a common accounting concept that is related to a possible method by which revenues can be recognised for accounting purposes (see Blanchard, 2005) and is not exclusive to insurance.

In non-life insurance we speak of earned premium and, more generally, of earned exposure. The concept of earned premium emerges because premium is normally received in one lump sum at the start of the policy, while claims covered by the policy occur over a period of time. For accounting purposes it makes sense, therefore, to recognise insurance premiums over time, as the risk of the policy runs off, according to the deferral-matching approach (Blanchard, 2005).

11.4.1.1 Earned Premium

Before defining earned premium, we need to define written premium. The **written premium** for a portfolio $\mathbb{P}$ of policies during a reference period $[t_0, t_1)$ is the premium charged for that portfolio during the reference period.

Note that:

- for premium to be allocated to $[t_0, t_1)$ the inception date must also be within $[t_0, t_1)$;
- for an individual policy, the written premium for the period $[t_0, t_1)$ is simply equal to the premium for that policy as long as the inception date is within $[t_0, t_1)$;
- the written premium can be provided gross or net of fees/commission (i.e., before or after external expenses are subtracted from the premium charged to the customer) and gross or net of reinsurance arrangements.

The **premium earned over a reference period** $[t_0, t_1)$ for a policy or a portfolio $\mathbb{P}$, $EP(\mathbb{P}, [t_0, t_1))$, is the amount of premium written *at any time before t* that should be allocated to period $[t_0, t_1)$ to reflect the proportion of the total risk that falls into $[t_0, t_1)$.

There is some vagueness in this definition as it doesn't specify how the proportion of total risk is determined. We can tighten this definition by defining 'risk' in terms of the expected losses (or a proxy for them) over a certain time frame. This will become clearer once we look at the actual earned premium calculations.

An alternative definition is this: the earned premium in year n is equal to the premium written in year n plus the **unearned premium reserve** (UPR) in year $n-1$ minus the UPR for year n: $EP(n) = WP(n) + UPR(n-1) - UPR(n)$. This is more familiar in an accounting and reserving context and is accurate because 'written premium' and 'unearned premium reserves' are well-defined in that context but in conceptual terms it doesn't add much because you can't really understand what the unearned premium is without understanding what the *earned* premium is!

Finally, note that the earned premium can be defined – as was the case for written premium – as gross or net of any adjustment for external expenses and gross or net of reinsurance.

11.4.1.2 Earned Exposure

The definition of **earned exposure** is very much the same as the definition of earned premium, except that the premium is replaced with a measure of exposure, e.g., number of vehicle years. 'Written exposure', in this case, simply means the exposure associated to the premium written.

The concept of earned premium and earned exposure is not unique to portfolio pricing. It can be used for pricing of a single risk: in this case, the definition and the calculation is the same as for a portfolio (except that the portfolio $\mathbb{P}$ is made of exactly one policy!). Note that for an individual policy written on an occurrence basis (i.e., one paying all claims occurring during the policy period), the earned premium is the same as the written premium. In that case it might still be interesting to determine the earning pattern, but it is obviously less important than for a portfolio.

11.4.2 Earned Amount Calculations

Having discussed how the concept of earned (premium or exposure) amount helps us in matching loss experience and exposure, let us look at how we can actually *calculate* the earned amount from information on the written amount and on our knowledge as to how risk depends on time.

The earned amount calculations depend on the way in which the earning pattern (also referred to here as 'risk intensity') depends on time. Although hybrid (and more complex) scenarios are possible, the main options are that the earning pattern depends on the time from the inception of the policy (e.g. as in an extended warranty policy) or on the calendar time (e.g. as in a catastrophe excess-of-loss policy for which a disproportionate number of losses occur in the hurricane season).

Various reasons of why the earning pattern may depend on time are illustrated in Blanchard (2005). In this section, the earning pattern is assumed to be known. Section 2.2.6 will address the problem of determining earning pattern in practice.

Let us first introduce some terminology.

- The **reference period** (i.e. the period over which the earned premium/exposure needs to be calculated) is $I^{\text{ref}} = [t_0, t_1)$
- The **cover period** for policy i is $I_i^P = [SD_i, ED_i)$
- The **written amount** (premium or exposure) for policy i is P_i (depending on the

purpose of the exercise, P_i should be the **on-levelled written amount**. On-levelling of premium amounts may also include a correction for rate changes.

- The **earned** (premium or exposure) **amount** for policy i over the period I^{ref} is $EA\left(I_i^P, I^{\text{ref}}\right)$
- The earned (premium or exposure) amount for the portfolio $\mathbb{P}$ over the reference period I^{ref} is denoted as $EA\left(\mathbb{P}, I^{\text{ref}}\right)$.
- $\rho(\tau)$ is the **earning pattern** for a policy in which risk depends on the time from inception, τ. (Typically, $\rho(\tau)d\tau$ will be attempting to capture the proportion of the total losses that is expected in the interval $[\tau, \tau + d\tau]$.) Without loss of generality we can assume that the risk intensity over the cover period is normalised: $\int_{SD_i}^{ED_i} \rho(t - SD_i)\,dt = 1.$
- $r(t)$ is the earning pattern for a policy in which risk depends on the calendar date, t. (Analogously to $\rho(\tau)$, $r(t)dt$ will typically attempt to capture the proportion of the total losses that is expected in the interval $[t, t + dt]$.) Without loss of generality, we can assume that the risk intensity over the cover period is normalised: $\int_{SD_i}^{ED_i} r(t)\,dt = 1.$

11.4.2.1 Earned Amount Calculations for a Single Policy

Figures 11.13 to 11.15 illustrate how the earned premium can be calculated for a single policy with different assumptions on the earning pattern. To make this illustration more concrete, we will assume that these facts are true across the three examples, and only the earning pattern will change:

- The start date of the policy is $SD = 1/9/2024$, and the end date is $ED = 31/8/2025$.
- The reference period is the whole of 2025 ($t_0 = 1/1/2025$, $t_1 = 31/12/2025$)
- The written premium of the policy is P = £1.2M.

The most common case is that of a uniform earning pattern – this is a good approximation for most insurance products. In this case, the earned amount as a percentage of the written amount is simply given by:

$$\%\text{Earned} = \frac{\min(ED, t_1) - \max(SD, t_0)}{ED - SD} \tag{11.27}$$

and the earned amount is the percentage earned (%Earned) times the written amount. Figure 11.14 shows a calculation example for this case.

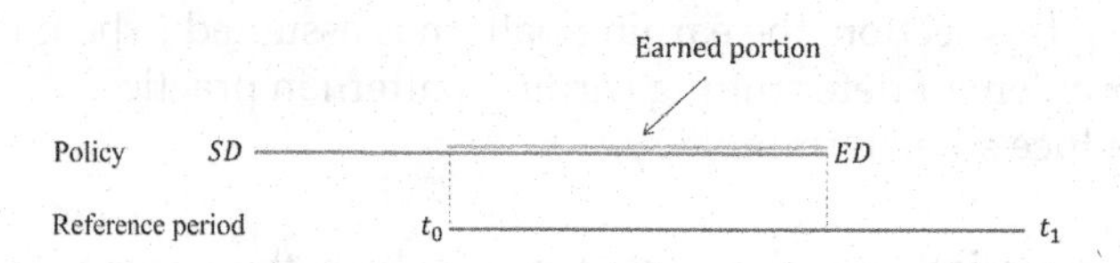

FIGURE 11.14
Earned premium calculation for a single policy with uniform earning pattern. In this case, the earned percentage is simply given by the percentage of the cover period that overlaps with the reference period. Based on the assumptions above, the premium earned by this policy during the reference period is then (31/8/2025–1/1/2025)/(31/8/2025–1/9/2024) × £1.2M = 8/12 × £1.2M=£0.8M.

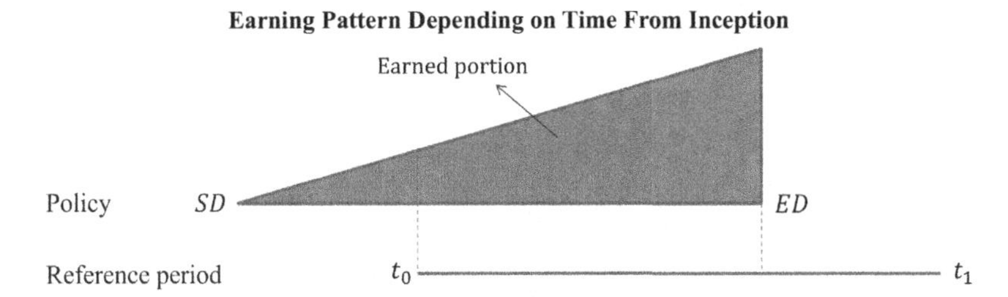

FIGURE 11.15
Earned premium calculation for a single policy with an earning pattern that depends linearly (and starting from zero) on time from inception: $\rho(t) = \alpha t$. The earned premium is given by the integral in the figure, which in this simple case is the premium of the policy multiplied by the percentage of the area of the triangle that corresponds to the reference period: $\textbf{Earned Premium} = \frac{\alpha/2 - \alpha/18}{\alpha/2} \times P = \frac{8}{9} \times P \approx \pounds 1.07\text{M}$. Note how the earned premium does not depend on α– all we need to know is that the relationship is linear!

A more general case is when the earning pattern depends on the time from inception. Examples where this type of dependence occurs are extended warranty policies (where risk has the familiar bathtub behaviour) and construction policies (where risk is very low at the beginning of the project – the 'hole in the ground' stage – and increases as the construction takes shape).

In this case the percentage amount earned can be written down as:

$$\%\text{Earned} = \int_{\max(SD,t_0)}^{\min(ED,t_1)} \rho(t - SD)\,dt \tag{11.28}$$

and the earned amount is the percentage earned times the written amount. Figure 11.15 shows a calculation example.

Another generalisation is the one where the earning pattern depends on calendar time. This is the case, for example, of motor insurance (where the risk is higher in winter) or some cat risk (where, e.g. the risk of hurricane or flood varies significantly depending on the season).

In this case, we have

$$\%\text{Earned} = \int_{\max(SD,t_0)}^{\min(ED,t_1)} r(t)\,dt \tag{11.29}$$

and as usual the earned amount is the percentage earned times the written amount. Figure 11.16 shows a calculation example.

Of course, hybrid situations (where the earning pattern for a policy depends on both calendar time and time from inception) are also theoretically possible.

11.4.2.2 Earned Amount Calculations for a Portfolio

The earned amount (premium or exposure) calculations for a portfolio are a simple generalisation of the calculations for a single policy. Specifically, the earned amount for a portfolio over a given reference period is simply the sum of the earned amounts of the individual policies, each with their own starting and end date:

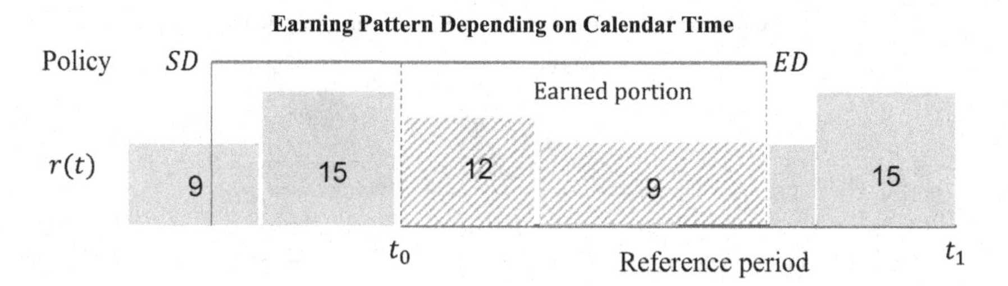

FIGURE 11.16
If the earning pattern depends on the calendar date, the earned premium is the premium P of the policy multiplied by the percentage of the area below the earning pattern that corresponds to the reference period – unlike Figure 11.15, however, the earning pattern does not depend on the policy and has therefore (for clarity) been associated with the reference period. In this example, the relative risk is 12 for the first three months, 9 for the following 6 months, and 15 for the last three months of the reference period. The earned premium is therefore equal to Earned Amount $= \frac{\text{Area (dashed)}}{\text{Area (full policy period)}} \times P = \frac{12\times 3+9\times 5}{12\times 3+9\times 6+15\times 3} \times £1.2\text{M} \approx £0.72\text{M}.$

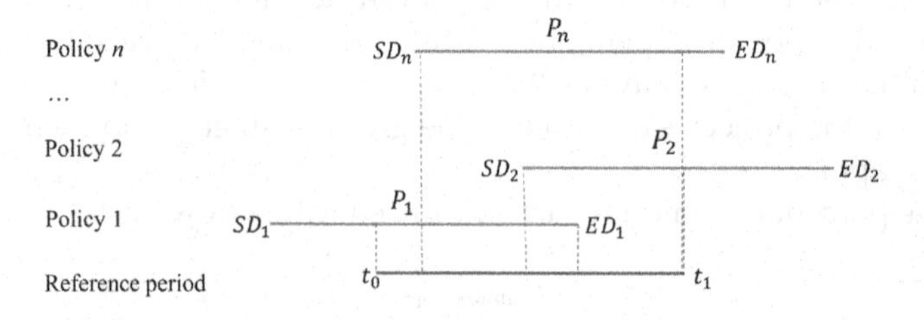

FIGURE 11.17
A graphical illustration of how the earned amount is calculated for a portfolio of policies, for the case where the earning pattern is uniform: for each policy, the earned premium is calculated as in Equation 11.27, and the contributions are summed as in Equation 11.30.

$$EA(\mathbb{P}, I^{\text{ref}}) = \sum_{i=1}^{n} EA(I_i^P, I^{\text{ref}}) = \sum_{i=1}^{n} \%\text{Earned}_i \times P_i \tag{11.30}$$

Equation 11.30 is quite general and covers all types of earning pattern ($\%\text{Earned}_i$ may be calculated by using Equation 11.27, 11.28 or 11.29 in most cases). Figure 11.17 illustrates the process of calculating the earned amount for a group of n policies.

11.4.2.3 Earned Premium Calculations: Continuous Version and Approximations

Although Equation 11.30 is sufficiently general to cover all practical cases, it can be further generalised to the case where premium is written continuously over time. This is more than a pointless mathematical generalisation – it turns out to be useful when the inception date of each policy is not known but we have a good idea of when the premium is more likely to be written. It is also useful for proving the formulae for well-known methods such as 24ths and the 365ths methods.

Let us therefore assume that the premium is written continuously with premium density $p(\tau)$: in other words, the premium written between τ and $\tau + d\tau$ is given by $p(\tau)d\tau$. To avoid unnecessary complications, we assume that the reference period is $I^{ref} = [n, n+1)$ where n is the beginning of year n (i.e. 0:00am of $1/1/n$) and that all policies are annual, as is often the case in practice. Hence, no policy with earliest inception date $\leq n-1$ can have an effect on the reference period. (The generalisation to a generic policy duration is conceptually easy.)

In the case where risk depends on the time from inception Equation 11.30 becomes:

$$EA(\mathbb{P},[n,n+1)) = \int_{n-1}^{n+1} p(\tau)\left(\int_{\max(\tau,n)}^{\min(\tau+1,n+1)} \rho(t-\tau)dt\right)d\tau \tag{11.31}$$

whilst for the case where risk depends on calendar time Equation 11.29 becomes:

$$EA(\mathbb{P},[n,n+1)) = \int_{n-1}^{n+1} p(\tau)\left(\int_{\max(\tau,n)}^{\min(\tau+1,n+1)} r(t)dt\right)d\tau \tag{11.32}$$

In the special case where the earning pattern is uniform, both Equations 11.31 and 11.32 reduce to:

$$\begin{aligned} EA(\mathbb{P},I^{\text{ref}}) &= \int_{n-1}^{n+1} p(\tau)(\min(\tau+1,n+1)-\max(\tau,n))d\tau \\ &= \int_{n-1}^{n} p(\tau)(\tau+1-n)d\tau + \int_{n}^{n+1} p(\tau)(n+1-\tau)d\tau \end{aligned} \tag{11.33}$$

We are now looking at various approximations of this formula that are used in practice.

Uniformly written premium

In the simple case where:

(a) the earning pattern is constant over time;
(b) the premium is written uniformly over the interval $[n-1,n)$ and the total written premium for that year is $WP(n-1)$ (assume it's already been on-levelled as necessary);
(c) the premium is written uniformly over the interval $[n,n+1)$ and the total premium for that year is $WP(n)$;

then we can use Equation 11.33 to obtain:

$$\begin{aligned} EA(\mathbb{P},[n,n+1)) &= WP(n-1)\int_{n-1}^{n}(\tau+1-n)d\tau + WP(n)\int_{n}^{n+1}(n+1-\tau)d\tau \\ &= WP(n-1)\left[\tau^2/2+(1-n)\tau\right]_{n-1}^{n} + WP(n)\left[(n+1)\tau-\tau^2/2\right]_{n}^{n+1} \\ &= \frac{WP(n-1)+WP(n)}{2} \end{aligned} \tag{11.34}$$

This formula (which we will find again in Section 20.3 when discussing treaty reinsurance, is exact if the premium is indeed written uniformly in years $n-1$ and n, but is often a viable approximation when $WP(n-1)$ (resp. $WP(n)$) is the overall premium written in $n-1$ (resp. n), uniformly or otherwise.

The 24ths and the 365ths methods

The **24ths method** is a way of calculating the premium earned during a reference period by assuming that:

(a) all policies are annual
(b) policies are written uniformly over each month, and hence can be assumed to be all written in mid-month
(c) risk is uniform over the year

For example, a policy written in April 2024 will be assumed to contribute 17/24th of its premium to calendar year 2024 (i.e. 8 months and a half, from mid-April to end of December), and 7/24th to calendar year 2025.

The 24ths method can be seen as a special case of Equation 11.33 for which $p(\tau)$ is approximated by a piecewise-constant function:

$$p(\tau) \approx \begin{cases} 12 \times p_1 & n-1 \le \tau < n - \frac{11}{12} \\ 12 \times p_2 & n - \frac{11}{12} \le \tau < n - \frac{10}{12} \\ \dots & \dots \\ 12 \times p_{24} & n + \frac{11}{12} \le \tau < n+1 \end{cases}$$

where $p_1, \dots p_{24}$ are the premium amounts for each of the 24 time intervals of equal length in the time interval $[n-1, n+1)$. Note that the premiums are then annualised (multiplied by 12) for consistency with Equation 11.33, in which $p(\tau)$ is a premium density (premium per year).

This yields the following formula for the 24ths approximation:

$$EA(\mathbb{P}, [n, n+1)) \approx \sum_{k=1}^{12} p_k \frac{2k-1}{24} + \sum_{k=1}^{12} p_{k+12} \left(1 - \frac{2k-1}{24}\right) \quad (11.35)$$

which can be further approximated by replacing the premiums p_k for the intervals of equal length in $[n-1, n)$ with the premiums written in the actual months of different durations (28 to 31 days): $\tilde{p}_{\text{Jan}(n-1)}, \tilde{p}_{\text{Feb}(n-1)}, \dots \tilde{p}_{\text{Dec}(n)}$.

One issue with this method is that it is not adequate in a context where a large portion of premium can be attributed to policies that incept on the first day of the month, which is not uncommon. For example, a 1 January renewal would earn 1/24th of the premium in the following calendar year – an undesired artefact of the method.

The **365ths method** helps overcome this issue by increasing the granularity to the 'day' unit: e.g. a policy incepting on 10/2/2025 will contribute with 325/365 of its premium to the earned premium in calendar year 2025. The assumptions are roughly the same as for the 24ths method, the only difference being that instead of (b) it is assumed that if a policy incepts on day x then it is assumed to be written at the *beginning* of the day. This is very much the same as Equation 11.30 when all policies are annual, and time is discretised at the day level, with P_i representing the total amount of premium written at day i. The difference between leap and non-leap years is ignored.

11.4.2.4 Determining the Earning Pattern in Practice

In the previous section we have introduced formulae for the calculation of the earned premium assuming that the earning pattern was known. In practice, we need to make informed assumptions on the earning pattern. Possible ways of determining the earning pattern are as follows:

- Use past claims experience, either internal or external. For example, to determine the earning pattern as a function of calendar date we may look at past statistics on the number (or total amount) of claims that have occurred in each day of the year, smoothing the statistics as necessary; and to determine the earning pattern as a function of time from inception, we may look at a cohort of policies and calculate the percentage of claims that have occurred at different times from inception.
- Use a loss model – this is especially useful when the non-uniform risk intensity is a consequence not of the uneven spread of the underlying risk but of policy modifiers such as aggregate deductibles/limits.
- Use a reasonable prior assumption – e.g. that the risk is larger by 20% in winter, or that the risk increases linearly by 10% every year from inception – depending on what expert judgment suggests.

In all cases one should remember that the earning pattern is, much like exposure, always a prior estimation of the real (unknown) earning pattern, and that as experience accumulates the earning pattern, too, could be updated.

11.4.2.5 Distortions to the Earning Pattern Caused by Coverage Modifiers

Coverage modifiers such as aggregate deductibles and aggregate limits will distort the earning pattern (see Blanchard, 2005). For example, the presence of an annual aggregate deductible (AAD) in a commercial lines insurance policy will cause (on average) an increase in loss activity towards the end of the policy year, when the AAD is more likely to kick in. Another example is a stop loss reinsurance policy, where cover is triggered by the loss ratio (or the total monetary losses) exceeding a certain amount over the policy period: the threshold is more likely to be triggered towards the end of the policy and that is where the reinsurance premium is earned.

The general way for working out the earning pattern in such situations is to produce a large number of possible scenarios with Monte Carlo simulation (or other suitable numerical techniques), recording for each simulated scenario not only the number of claims and their amounts but also the timing of the claims and the order in which they occur. We can then calculate how the expected losses depend on time by calculating the ceded and retained losses within time t from inception for each simulation and averaging over all simulations.

The solution to Question 6 shows an example of how this can be done.

11.4.3 Burning Cost Analysis (Occurrence-Based Policies)

Once the earned amounts are calculated, burning cost calculations can proceed in the usual way: the burning cost for each policy year is given by the incurred losses (after the usual adjustments for IBNR, inflation etc.) divided by the earned exposure (or original premium, in the case of reinsurance):

$$BC_j = \frac{\text{Incurred losses (adjusted for IBNR, inflation, cover, large losses, etc.) in year } j}{\text{Earned On-Levelled Exposure (or Premium) Amount in year } j} \tag{11.36}$$

The expected loss is then given by the weighted average of the burning costs over the years, multiplied by the estimated exposure for the policy period.

To reiterate what is mentioned in the title of this section, this approach is useful for occurrence-based policies (including loss occurring during (LOD) policies in treaty reinsurance). The case of risk attaching during (RAD) policies in treaty reinsurance doesn't require the calculation of earned amounts as the exposure is given by the written premium or exposure. As for claims made policies, the matching of exposure and claims is slightly more complex and will be addressed later in the book (Chapter 23).

11.5 The Limitations of Burning Cost Analysis

We have shown how such a simple method as burning cost, a common tool used by underwriters from time immemorial, can be and has been adapted to provide a rough basis for pricing, taking into account claims inflation, large losses, IBNR losses, and much more. We have also shown that the results of burning cost give information not only on the average cost of a risk but also on the year-on-year volatility and – with some leap of faith – on the percentiles of the aggregate loss distribution.

So what is still missing in this picture that motivates us to move on to stochastic modelling?

As we have seen, the burning cost method has some severe limitations:

Minimal modelling. The burning cost method relies mostly on past information, with minimal attempt to model prospective risk through various adjustments (inflation, IBNR, and ILFs in its most sophisticated incarnation). This can sometimes be an advantage – fewer assumptions are needed that may distort the picture or make it more controversial – but it ultimately impairs our ability to incorporate our views on future risk. Since burning cost is not model-based, it does not, by definition, attempt to find regularities in the data and predict beyond its range of experience: it simply regurgitates what we put into it. The burning cost–based lognormal model of Section 11.3 is an exception to this, although as we have seen there is no reason to take its results seriously.

No reliable estimation of the aggregate loss distribution. As we have seen, data is sometimes sufficient to estimate a few percentiles of the aggregate loss distribution, but it is not possible to estimate the full aggregate loss distribution reliably based on a burning cost analysis. The failure of the lognormal model to provide a reliable aggregate loss distribution (Section 11.3) is an illustration of this, but the ultimate reason is not that the lognormal model is arbitrary and we should replace it with something closer to reality – the ultimate reason why this approach is inadequate has to do with the fact that no matter how rich the initial claims data set is, the burning cost approach aggregates this information into a small set of numbers (normally between 5 and 10) representing the total losses for each year, and any model based on such a small number of points is doomed to fail to adequately predict the richness of possible outcomes.

Failure to discern frequency and severity trends. It is sometimes the case that there are trends in data over the years. Sometimes these trends are in frequency (the number of road accidents per mile driven may decrease over the years because of speed cameras and other risk control systems) and sometimes they are in severity (the type of injuries may change over the years because of the introduction of new safety system, such as airbags). Sometimes these two effects go in different directions and therefore compensate each other to some extent. Because the burning cost is performed on the aggregate losses, it is ill-suited to disentangling these two effects.[7]

The effect of large losses. Large losses strongly affect the final outcome of a burning cost analysis:

- The inclusion of a large loss may skew not only the burning cost itself but also affect the overall distribution, giving it a much fatter tail.
- The omission of a large loss may underestimate the burning cost and the tail of the distribution.
- Removing a large loss from the data set and spreading it over the years may lead to underestimating the year-on-year volatility; furthermore, it is unclear what probability you should assign to a large loss.

All these difficulties emerge from the fact that, without a reliable severity model, it is impossible to know what the likelihood of a large loss is. The LLA method (Section 11.1.7.1) and the ILF method (Section 11.1.7.2) are an attempt to partly make up for this limitation, at the price of introducing new assumptions with the inevitable uncertainty around them.

11.6 Questions

1. In Box 11.1, we have introduced weighted statistics for the empirical mean, standard deviations, and the percentiles of a data set. Show that by setting all weights to 1, the formulae for the mean, the standard deviation, and the percentiles for the unweighted case are obtained.
2. In Box 11.2, we have seen that in the standard chain ladder method, the weighted average is normally used to calculate the average development ratios between two successive development years.
   a. Produce a definition of the average development ratios as a simple average.

7 It might be argued that the burning cost method does indeed provide separate information on frequency and severity (see e.g. the number of claims column in Figure 11.6) and one can easily conceive of a burning cost method that looks at claim counts and average severities separately, chooses a 'burning cost frequency' and a 'burning cost severity' based on a set of relevant years (possibly a different set for frequency and severity), and then multiplies the two together to obtain the overall burning cost. However, this is not as easy as it sounds, as the average severity is often a rather shaky variable that is sharply affected by the presence or absence of large claims. One ends up looking at other statistics such as the median and the trimmed mean to get a more refined idea of whether there is a genuine difference in severities over the years, and quite quickly finds oneself needing the full severity distribution; at that point the concept of burning cost is completely lost.

b. Calculate the average development ratios for the claims development triangle of Figure 11.10 according to this definition.
c. What do you think is the main disadvantage of the simple average method?
d. Produce an example in which this disadvantage would have great effect on the final projection results.
e. How could the simple average definition be amended to overcome this disadvantage?

3. In Box 11.2, we say that the standard way of calculating the average development ratios in the chain ladder is a weighted average. What weights are used in the definition of $D(0\to1)$ (and, for that matter, for all development ratios)?
4. A medium-sized UK-based company (A) is currently insuring its *public liability* risks with an insurance company (B). You have been asked (as a consulting actuary) to advise company A on whether or not it is paying too much for its insurance.

   The loss data (which is based on information provided on 30 September 2012) is in the following table. The year is the accident year. The loss data has been revalued with a claims inflation of 5% per annum to 1 July 2013. All losses are from the ground up and are expressed in British pounds.

	2005	2006	2007	2008	2009	2010
	64,995	56,284	7115	2237	4801	97,938
	119,630	101,993	14,952	2060	159,535	85,959
	1010	27,036	8576	25,106	55,957	
	259,414	2559	745,954		71,563	
	33,277		346,021		75,680	
	30,319		6582		2634	
	3297		59,881			
	104,119		536			
			81,287			
Total	616,060	187,872	1,270,905	29,403	370,170	183,897

- Claims are typically reported within a year of occurrence.
- Some of the figures in the most recent years are reserved claims (estimates).
- The company turnover has changed little in the last 5 years and no changes are expected in the near future.
- The loss ratio for public liability in the UK market is between 70% and 90%.
- The current insurance policy is as follows:
  i. EEL deductible = £100k.
  ii. AAD = £2M.
  iii. Individual/aggregate limit = £5M.
- The premium proposed by company B for 2013 is £300,000.
- The policy is annual and incepts on 1 January. It covers losses occurring during the policy year.
- Claims inflation is at 5% (already incorporated in the table above).

  i. Describe how a commercial insurance policy with an EEL, AAD and individual/aggregate limit works and explain what it is trying to achieve. Build a simple numerical example with three losses that shows how EEL and AAD work. Make sure that the example is non-trivial (i.e. both the retained and the ceded amount are non-zero, and the AAD is hit).

ii. Using burning cost analysis, determine the expected losses ceded to the insurance programme for next year. State all the assumptions you have made.
iii. Based on the analysis in (ii) and on any additional considerations, estimate the technical premium and comment on whether the insurance is priced in line with market practice.
iv. Explain whether you think each of the values of EEL, AAD and individual/aggregate limit are adequate. Suggest an alternative for any value(s) you consider inadequate and explain your reasoning. Explain qualitatively (or quantitatively where possible) how the changes you propose will affect the price of the policy.

	Premium written					
	01-Jan	01-Apr	01-Jul	01-Oct	Total	Rate index
2006	706	228	432	173	1,540	100.0
2007	893	252	508	183	1,835	100.4
2008	927	218	489	173	1,807	103.3
2009	828	242	447	190	1,707	105.5
2010	777	264	487	181	1,709	105.9
2011	837	233	456	183	1,709	112.7
2012	912	235	544	190	1,882	115.3
2013	1,032	261	552	194	2,039	117.2
2014	1,000	284	491	190	1,965	116.5
2015	1,061	277	581	183	2,102	120.3
2016						*121.0 (estim)*

All monetary amounts in GBPm

5. As part of a pricing exercise for a Commercial Property Risk XL programme incepting on 1 January 2016 and written on a *losses occurring* basis, you have to calculate the earned premium for a number of policy years and compare it to the claims experience. The underlying policies of the reinsured *are all annual* and incept on one of these four dates: 1 January, 1 April, 1 July or 1 October. The amount written for each of these dates is specified in the table below. The table also shows the year-on-year rate changes. You can assume that across-the-portfolio rate changes are only made at 1/1 renewals.
   a. Calculate the earned premium for the 2012 accident year assuming a uniform risk intensity (earning pattern) for all policies
   b. Carry out the same calculation assuming that risk is distributed roughly as follows: 19% from January to March; 20% from April to June; 33% from July to September; 28% from October to December.
6. Consider a simple loss process where the number of claims follows a Poisson distribution with rate $\lambda = 5$ and the individual claim amount follows a lognormal distribution with $\mu = 11$ and $\sigma = 2$. Assume that the insurer offers a simple stop loss protection by which the insurer pays all losses in excess of an aggregate yearly amount of AAD = £2.5M. (In mathematical terms, if S is the annual loss amount, the amount paid by the insurer is $S_{\text{ced}} = \max(0, S - AAD)$.

   Using Monte Carlo simulation, estimate the earning pattern resulting from this situation, and explain why it differs from a uniform distribution.

7. Show how the burning cost formulae shown in Section 11.1.12:
   a. simplify in the case where there is no annual aggregate deductible ($AAD = \infty$);
   b. should be amended if an annual aggregate limit AAL is present.
8. (*) Show how Equation 11.7 can be justified as an approximation of the formula for the total expected losses $\mathbb{E}(S) = \lambda \times \int_0^\infty x f(x) dx$ where λ is the expected number of claims from the ground up and $\int_0^\infty x f(x) dx$ is the expected severity. (Hint: split $\int_0^\infty x f(x) dx$ into two parts, one up to the large-loss threshold LLT and the other above LLT, and approximate the second part as a Riemann sum using losses $LL_1, \ldots LL_m$ as the endpoints of the approximating rectangles.)

12

What Is This Thing Called Modelling?: A Gentle Introduction to Machine Learning

I'm no model lady. A model's just an imitation of the real thing.

Mae West

Before delving into the details of the pricing process and showing how one can produce a frequency and severity model of losses, it makes sense to pause and ask ourselves what a model actually is.[1] In attempting to answer that, it is not easy to go much further, in terms of philosophical depth, than Mae West: 'A model's just an imitation of the real thing'. Others have viewed models as 'representations' (Frigg and Hartmann, 2012), citing Bohr's model of the atom as one such representation, and yet others have referred to them as 'metaphors' (Derman, 2010). More precise mathematical definitions of 'model' are also available: for example, a statistical model can be defined as a mathematical description of the relationship between random variables or more formally as a collection of probability distributions (see Wikipedia, 'Statistical model', http://en.wikipedia.org/wiki/Statistical_model).

My favourite definition of model, however, is this:

A model is a simplified description of reality.

This definition captures the 'imitation' concept and the nature of the relationship between model and reality but also reminds us that models are approximations and as such, we should always be aware of their limitations: something which is not obvious from the more formal definitions of models.

In this chapter, we look at these ideas in more depth, and we will explain what the modelling approach entails in practice and why we need to use modelling at all. We will also provide a quick foray into the problem of selecting a good model through a conceptually simple non-insurance example (polynomial fitting).

At the end of this brief chapter, we will have hopefully planted the seeds of some healthy distrust for the predictive power of complex, clever models and will have clearer ideas on what modelling is actually about. These 'seeds of distrust' will then be developed in following chapters on frequency (Chapter 14) and severity models (Chapter 16), and on measuring uncertainty (Chapter 18). Some of the techniques described here will be taken up again in Chapter 27 in the context of rating factor selection.

1 I am well aware that Karl Popper (one of the loudest voices in my head) hated – for good reasons – questions such as 'what is a model?' and 'what do you mean by model?', but sometimes I just can't help myself.

DOI: 10.1201/9781003168881-14

12.1 Why Do We Need Modelling at All?

We have explained above what a model is, but we have yet to mention why modelling is needed at all. We will try to explain that with a simple example in Section 12.1.1, and in Section 12.1.2 we will mention other uses of models.

12.1.1 Models Allow Us to Make Predictions That Generalise Loss Experience

As we have already mentioned in Section 12.1, what a model does is to take reality (the data) in all its muddiness and complexity and identify the regularities in the data that allow a representation of that data to be worked out which is much simpler but captures its essential features. Perhaps the simplest example is that of a linear model: in many interesting cases, the dependent and independent variable of a phenomenon (say, the average value of a property loss and the value of a property; or the logarithm of the average size of a loss and the logarithm of the claims inflation) will be related by an approximately linear relationship.

Consider, for example, the empirical data points in Figure 12.1. Based on these empirical data points, without any further calculations, how are we going to answer the following questions?

1. *Interpolation*. What value of y do we expect when $x = 3.5$?
2. *Extrapolation*. What value of y do we expect when $x = 16$?
3. *Compact representation*. How can we represent the set of 13 data points more succinctly than by simply listing all the 13 pairs? (This question is a bit theoretical, but it will help illustrate one of the fundamental functions of models.)

If one does not have a model of how the variable y is related to x, there is simply no way of answering any of the questions. Perhaps for Question 1 we may be tempted simply to

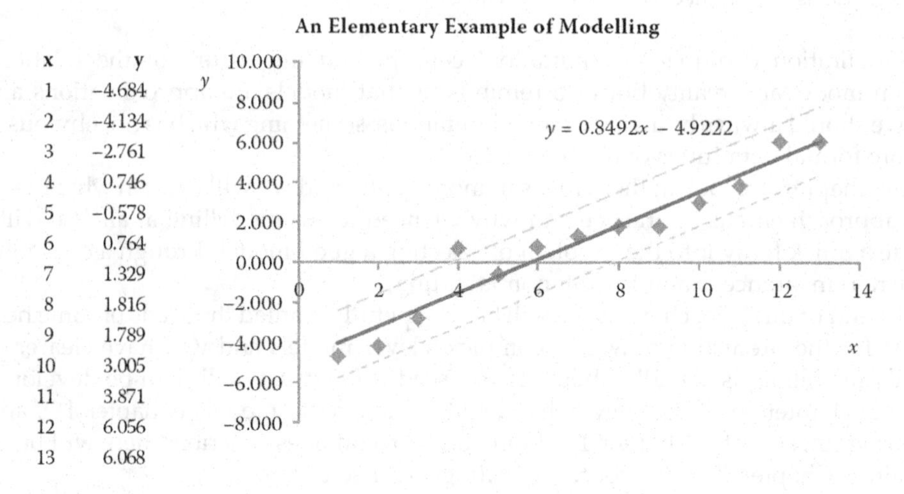

x	y
1	-4.684
2	-4.134
3	-2.761
4	0.746
5	-0.578
6	0.764
7	1.329
8	1.816
9	1.789
10	3.005
11	3.871
12	6.056
13	6.068

FIGURE 12.1
A chart of the 13 data points on the left shows that these points are roughly located around a straight line (solid line), whose equation can be determined by least squares regression. The two dashed lines are the boundary of all points that lie within 1.96 standard deviations of the predicted values.

take the average of the values of y corresponding to $x = 3$ and $x = 4$, which is -1.007, but this is already based on a linear model of the values of y between two values of x and, in any case, we will see that the result wouldn't be very good. As for Question 2, we do not have much to work on: perhaps we can say that the distance between $y(13)$ and $y(16)$ will be equal to the distance between $y(10)$ and $y(13)$, but that also implies linearity. Finally, for Question 3, if there is no rule that relates values of x and values of y, we simply have no way of compacting this information any further than by listing the 13 pairs.

However, if we plot the 13 data points in a chart like that in Figure 12.1, we notice that the 13 pairs of data points are roughly positioned around a straight line. We can then decide to *model* our data set as a straight line with some added Gaussian noise [$\varepsilon \sim N(0,1)$] with zero mean and variance σ^2:

$$y(x) = a + bx + \sigma\varepsilon \tag{12.1}$$

for some values of a, b and σ that we can determine with the method of least squares regression. For the data set of Figure 12.1 (left), the least squares method yields the values: $a = -4.9222$, $b = 0.8492$ and $\sigma = 0.8697$.

This is our (linear) model for the points in Figure 12.1. Once we accept that the linear model is a fair representation of our set of 13 data points, we can attempt to answer Questions 1 through 3:

1. $y(3.5) = 0.8492 \times 3.5 - 4.9222 = -1.950$
2. $y(16) = 0.8492 \times 16 - 4.9222 = 8.665$
3. Equation 12.1 is in itself an answer to Question 3, as it is a compact representation (using just two parameters rather than 13 values of y) of the data set. However, this is only an approximate representation of the data set – it does not reproduce the 13 values of y fully. This is not necessarily a problem because the exact value of y is partly a linear relationship with x and partly Gaussian noise, which we do not need to record exactly.

In some circumstances, however, replying to Question 3 properly implies first negotiating the level of accuracy with which one needs to represent the data set. In this case if, for example, one agrees to accept as a representation of a point (x_i, y_i) the modelled point $(x_i, a + bx_i)$ as long as the point itself is within 1.96 standard deviations of the model: $|a + bx_i - y| < 1.96\,\sigma$, then the data set can be succinctly described as follows: '12 data points with integer values of x from 1 to 13 (excluding $x = 4$) and values of y approximately on the straight line $y(x) = -4.9222 + 0.8492x$, to which Gaussian noise with mean 0 and standard deviation 0.8697 is overlapped; plus the point (4, 0.7460)'.

Through our example, we have seen three possible uses of models: interpolation and extrapolation of our data set, and to represent the data set itself more economically. Although it should be clear to all why interpolation and extrapolation are useful, it might not be clear why a *compact representation* of the data set should be useful. The reason is simply that a compact representation of one data set can be used for application to a different but related data set. As an example, we will see in Chapter 21 how past property losses can be used to produce exposure rating curves, and the compact representation of these exposure curves as two-parameter curves (the so-called MBBEFD curves) can be used to model many different property risks. Therefore, compactness increases the portability of experience.

It should now be clear that *interpolation, extrapolation and compact representation are all about the same thing: using models to make predictions outside experience.*

The example above may also help explain the difference between a model and a theory. A theory is something that goes one step beyond a model, in that it provides an *explanation* of the data and not simply a compact description of it. As an example, Hubble found a linear relationship between redshift and the distance of galaxies – a linear model – but the explanation of this relationship as coming from the expansion of space-time is a theory. It would be good if we had more theories that led us to correct risk costing, but in most cases, in actuarial practice, we simply rely on useful models: nobody has ever produced a theory, for example, of why the widely used lognormal model should be[2] the right description of the severity of losses in some classes of business.

12.1.2 Other Uses of Models

In Section 12.1.1, we have looked at models used as prediction tools, to interpolate data, extrapolate beyond data, or make experience portable to different contexts. When you look at models as prediction tools, it makes sense (as we will see in next section) that a model should be judged on how accurately it predicts reality. If it makes poor predictions, it is a poor and dangerous model.

However, it must be said that models can also be used for different purposes, for which this attitude is too strict. Another oft-mentioned quote on models is by George E.P. Box: 'All models are wrong but some are useful', which combines a healthy distrust for models with the recognition that imperfect models may anyway have other purposes. As noted, for example, by Tsanakas (2012), 'the usefulness of a model is not reducible to the accuracy of its outputs'. Interrogating models help us understand reality, and risk models 'can be used to educate management in aspects of risk, by illustrating concepts, analysing scenario impacts, studying sensitivities and showing the range of possible outcomes' (ibid.). Models may therefore serve many purposes, including educating the users of the model and facilitating the discussion of issues.

12.2 The Modelling Approach

The modelling approach to risk can be contrasted (Klugman et al., 2008) to the so-called empirical approach.

Empirical approach. Assume that the future is going to be exactly like a sample from the past, perhaps after adjustments for trends such as inflation, exposure changes, and others.

Modelling approach. Build a *simplified mathematical description* of the loss process in terms of occurrence, timing and severity based on experience, prior domain knowledge, and data.

The empirical approach is what we have adopted until now, for example, in our elementary pricing example and especially in our treatment of burning cost (Chapter 11).

2 It very likely isn't.

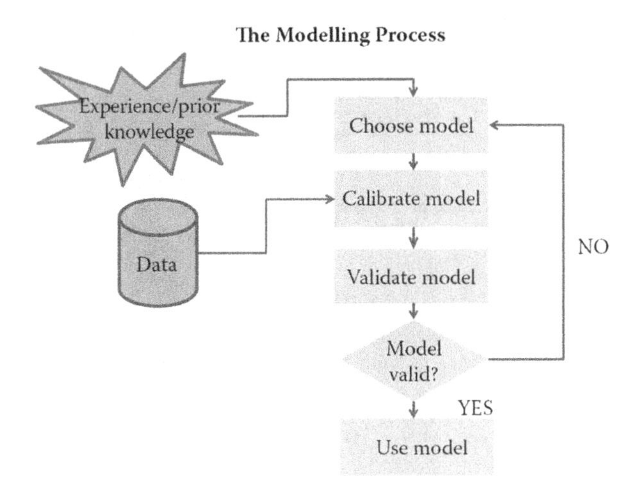

FIGURE 12.2
The modelling process according to Klugman et al. (2008), with some small amendments. Note how the process does not start with data but with experience and prior domain knowledge. (From S.A. Klugman, H.H. Panjer, G.E. Willmot, *Loss Models: From Data to Decisions*, 4. 2008 [Chapter 1, Figure 1.1]. Copyright Wiley-VCH Verlag GmbH & Co. KGaA. Reproduced with permission.)

The modelling approach as defined above – and which we are going to illustrate through the rest of the course – *seems* to suggest that the experience should be used to build a loss model and that this loss model is going to predict the future experience better than the empirical approach. Let us put it another way: if this interpretation of what modelling means is correct, a compact, simplified version of your loss data is going to have more prediction power than the loss data itself! One might be forgiven for thinking that this is nothing short of a magic trick.

Indeed, if you take some data, apply some algorithm that *blindly* takes the data and spits out a model, the result is most definitely not going to help you predict losses any better than burning cost. To be useful, a model must be able to find real regularities in the data, and normally, a good model is borne not out of data alone but out of experience and prior domain knowledge, as in the modelling process outlined in Figure 12.2.

The first thing that jumps to attention when one looks at Figure 12.2 is that the modelling process does not start with data but with the analyst's experience and prior domain knowledge. Thanks to this experience and knowledge, one knows which models are good for modelling insurance losses, either because of some fundamental reason or because they have worked well in the past on many different clients or on market data. The data is then used to calibrate the model – estimating its parameters – and once the model is produced, it must somehow be validated, that is, tested on a data set that has not been used for model selection and calibration (Figure 12.3).

So how does experience and prior domain knowledge 'add value' to the data so that the resulting model has more predictive power than the data itself? There is no unique recipe for this, in the same way in which there is no algorithm for scientific discovery. However, here are a few examples of how this happens in practice.

- Experience might tell us, for example, that a lognormal model is usually quite good to model loss amounts, as it has proven its worth in many circumstances, perhaps with large data sets (note the 'might': I don't actually believe that this is the case!).

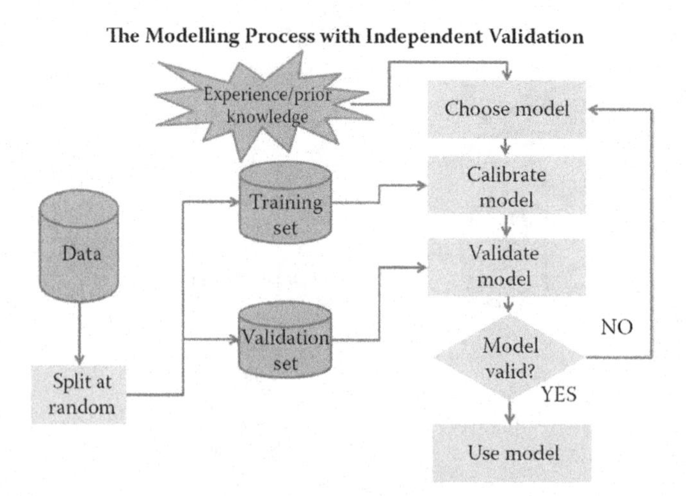

FIGURE 12.3
The modelling process using a more rigorous validation protocol – the main idea is that true validation is validation against a data set, which is different from what has been used to produce a model. Many validation methods (such as the Akaike information criterion, to be discussed in Section 12.3) do not use a validation set but approximate the prediction error in other ways.

- Theoretical results may tell us that, under certain conditions, large losses should be modelled with a generalised Pareto distribution, and that rare independent events should be modelled with a Poisson distribution.
- Knowledge of the risk may tell us that our model should not produce losses above a certain level, for example, because there is a maximum sum insured for a portfolio of properties.
- Even if we have no other reasons to prefer one model to another, we will still have some constraints on the model, for example, that it be as simple as possible: the more compact a well-fitting model is, the more likely it is that it has captured some fundamental regularities in the data and has not simply 'adapted' to the data, and therefore the more explanatory power it is likely to have. This is the idea behind the concept of 'minimum description length (MDL)' and Kolmogorov complexity (Grunwald et al., 2005), and in general a powerful idea informing machine learning.

12.3 How Do You Select a Good Model? A Foray into Machine Learning

The main thing you need to know to judge whether a model is good or not is how strong its predictive power is, or, in more technical terms, how large the prediction error is. This is the distance between the true value of the statistic we wish to measure (such as mean, variance, percentiles …) and the value predicted by the model, *in the range to which the model needs to be applied*. Anything else – the model's simplicity, its beauty, its consistency with models for other situations and whether or not the model is realistic – is ancillary to this overarching criterion of predictive power. Of course, this still leaves plenty of leeway as to the exact formulation of this criterion and the exact definition of distance we should use, what the range of applicability of the model should be, and so on.

Because all models will be affected by prediction error to some extent, the problem that in practice one tries to solve is to select the best model from a set of competing models, and this is the problem we will focus on in this section. Even if the number of models is very large and possibly infinite (but countable), this problem is infinitely simpler than trying to find the best model amongst the (uncountable) set of all possible models.

Model selection is at the core of many activities in actuarial modelling for general insurance: selecting a frequency model, selecting a severity model, selecting the correct rating factors (such as age, sex, profession ...) for a personal insurance policy and much more. For this reason model selection considerations will pop up regularly in chapters ahead (see Section 12.4 for a quick summary).

How does one select the appropriate model? Consider one of the examples listed above, that of selecting a severity model, that is, the appropriate statistical distribution that fits a set of historical losses (we will look at this problem in more depth in Chapter 16). One method we often use for this purpose is to select from a number of statistical distributions (perhaps those provided by a distribution fitting tool) the one which fits the data best. Normally, the fit is very good because some of the distributions at our disposal have three or four parameters, and we can choose between 20 or more distributions, which is like having an extra hidden parameter. Only the weirdest data sets will not find a good match under these circumstances. This is eerily reminiscent of John von Neumann's disparaging dictum: 'With four parameters I can fit an elephant, and with five I can make him wiggle his trunk'.

If model selection were about picking models that look nice when plotted against the historical losses, we should look no further than this approach. However, model selection is actually about finding models that have predictive power when applied to data *you haven't seen* (or used) before: for example, next year's losses or data you have set aside for validation purposes (Figure 12.5).

The problem above is a simple but fully formed example of a machine learning problem. The main goal of machine learning – also known as *statistical learning* – is 'to provide a framework for studying the problem of inference, that is, of gaining knowledge, making predictions, making decisions or constructing models from a set of data' (Bousquet et al., 2004).

Machine learning is a young but vigorous discipline: the 'Stanford school' of statisticians (those who created the bootstrap) have developed a number of novel methods for it; Fields medallists such as Stephen Smale have looked into it and brought mathematical rigour to it. There is plenty of material that we can use almost off the shelf for actuarial analysis; and more than anything else, it brings a totally different attitude to the issue of model selection. It pays to know more about machine learning.

If you were to select a single key fact to keep in mind about machine learning, it should be this: increasing the complexity of your models does not always increase the accuracy of your predictions. Actually, if the model becomes too complex, the opposite happens.

This concept is well illustrated by the classical bias-variance trade-off chart (see Figure 12.4 in Box 12.1). This gives the prediction error as a function of the complexity of a model. The reason why this is called the 'bias-variance trade-off' graph is because models with complexity that is too low have a large bias (i.e. they are too simplistic to capture the patterns in the data) while models that are too complex have a large variance (i.e. they are overfitting because they capture not just the true pattern but also the noise). In both cases, the prediction error is higher than it could be if we chose the model with the right complexity.

Box 12.1 has a more in-depth discussion about this important concept.

***BOX 12.1 THE BIAS-VARIANCE DECOMPOSITION AND TRADE-OFF**

The exposition below follows mostly Hastie et al. (2001), to which the reader is referred for more details. Assume we have

$$Y = f(X) + \varepsilon$$

Where X represents the explanatory factor(s), Y the dependent variable, $f(X)$ the true relationship between X and Y and ε some random noise with 0 expected value and variance σ^2. This is a rather standard situation in actuarial practice – e.g. this is the relationship that is assumed between rating factors and losses in, say, motor insurance. The term ε is explained by the fact that – even though we were able to determine exactly the factors affecting the expected losses, losses are ultimately a random phenomenon – the noise ε captures that randomness.

In practice, however, we do not know the true model $f(X)$, but only an estimated model $\hat{f}_{\mathcal{D}}(X)$ calibrated on a specific training set $\mathcal{D}$. We are interested in calculating the expected error made by approximating Y with $\hat{f}_{\mathcal{D}}(X)$. This depends in general on the point $X = x$ at which we are calculating $f(X)$. Let's call $EPE(x)$ the expected (mean squared) prediction error at $X = x$:

$$EPE(x) = \mathbb{E}_{\mathcal{D},\varepsilon}\left(\left(Y - \hat{f}_{\mathcal{D}}(x)\right)^2 \mid X = x\right)$$

The expectation is taken over the ensemble of the data sets $\mathcal{D}$, each producing their own estimate $\hat{f}_{\mathcal{D}}(x)$, and over the noise ε.

We can rewrite Y calculated at $X = x$ as:

$$Y\big|_{X=x} = f(x) + \epsilon = f(x) + \mathbb{E}_{\mathcal{D}}\left(\hat{f}_{\mathcal{D}}(x)\right) - \mathbb{E}_{\mathcal{D}}\left(\hat{f}_{\mathcal{D}}(x)\right) + \epsilon$$

where $\mathbb{E}_{\mathcal{D}}\left(\hat{f}_{\mathcal{D}}(x)\right)$ is the expected value of $\hat{f}_{\mathcal{D}}(x)$ over the ensemble of data sets. Therefore, the error is given by:

$$Y\big|_{X=x} - \hat{f}_{\mathcal{D}}(x) = f(x) - \mathbb{E}_{\mathcal{D}}\left(\hat{f}_{\mathcal{D}}(x)\right) + \mathbb{E}_{\mathcal{D}}\left(\hat{f}_{\mathcal{D}}(x)\right) - \hat{f}_{\mathcal{D}}(x) + \epsilon$$

The expected squared error is therefore:

$$EPE(x) = \left(f(x) - \mathbb{E}_{\mathcal{D}}\left(\hat{f}_{\mathcal{D}}(x)\right)\right)^2 + \mathbb{E}_{\mathcal{D}}\left(\left(\hat{f}_{\mathcal{D}}(x) - \mathbb{E}_{\mathcal{D}}\left(\hat{f}_{\mathcal{D}}(x)\right)\right)^2\right) + \sigma^2$$

(Note that the first term has no expectation because $f(x)$ is deterministic. Also, notice the absence of the covariance terms, which can be easily shown to be zero.)

We have now written down $EPE(x)$ as the sum of three components:

1. The **(squared) bias** $\left(f(x) - \mathbb{E}_{\mathcal{D}}\left(\hat{f}_{\mathcal{D}}(x)\right)\right)^2$, which is related to the difference between the average model result and the true value. The bias is a consequence of the *inability of the model to capture the patterns seen in the data.*

2. The **variance** $\mathbb{E}_{\mathcal{D}}\left(\left(\hat{f}_{\mathcal{D}}(x)-\mathbb{E}_{\mathcal{D}}\left(\hat{f}_{\mathcal{D}}(x)\right)\right)^2\right)$: this captures the volatility of the model prediction from one sample to the other. It is a consequence of the fact that the model doesn't just capture the intrinsic patterns in the data but also the noise, which by definition will change from one sample to the other.
3. The **(squared) irreducible error** σ^2: this is simply the inherent volatility due to the fact that the process by which the values of Y are realised is a stochastic (random) process. Elsewhere we call this 'process variance'. No matter how good the model is, you will still have volatility around the true value f(X).

We can therefore rewrite the **expected prediction error** in the simpler form:

$$\text{EPE}(x)=(\text{bias})^2+\text{variance}+(\text{irreducible error})^2$$

Note that the calculations above are valid for one point $X = x$. By taking the expected value over the range of values of x we obtain the overall expected prediction error.

THE BIAS-VARIANCE TRADE-OFF

Now that we have explained what the bias-variance decomposition is we are in a position to understand the concept that is central to machine learning – that of the bias-variance trade-off. The reason why we speak of a bias-variance trade-off is

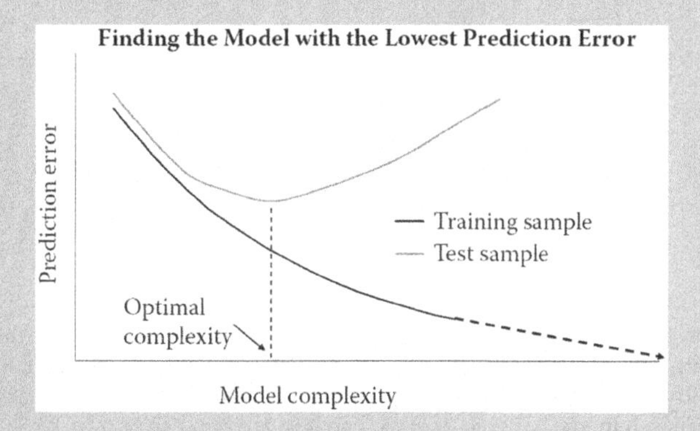

FIGURE 12.4
The bias-variance trade-off graph exposes (among other things) the fallacy implicit in using complex models to fit the existing data. The left-hand side of the graph (lower complexity) is characterised by high bias and low variance; the right-hand side of the graph (higher complexity) is characterised by low bias and high variance. As far as the data is concerned, a distinction is made between the 'training sample' (the data used to select and calibrate the model) and the 'test sample' (the data used to assess the model): it is crucial to keep the distinction in mind. As the complexity of the model grows, the fit with the training sample becomes tighter and tighter: eventually, the model may become so rich that the data is replicated exactly, and the fit is perfect. However, this precision is purely illusory, as becomes obvious as soon as the distance between the model and the data is assessed using data that was not used in calibrating the model ('test sample'). The graph also shows that as a consequence of this trade-off there must be an optimal level of complexity at which the predictive error (the distance between the model and the test sample) is minimal. One of the objectives of model selection is to identify the model with the optimal level of complexity. (With kind permission from Springer Science+Business Media: *The Elements of Statistical Learning*, Chapter 2, 2001, 3, T. Hastie, R. Tibshirani and J. Friedman, Figure 2.11.)

that – given a certain size of the data available – we can typically reduce the bias by using a more complex model which better captures the *true* pattern of the data, but doing so will inevitably cause that model to also pick up some noise; and since noise patterns will change from one sample to the other that will cause the model to perform more poorly on a test set.

On the other hand, we can reduce the variance by forcing it to be less responsive to noise: that can typically be achieved by ensuring that the model is smooth (as opposed to wiggly) and simple. However, that very smoothness and simplicity also makes the model less responsive to genuine patterns in the data.

All this is captured by the bias-variance trade-off graph in Figure 12.4, which shows that models with low complexity/high smoothness have high prediction error because of bias while models with high complexity/low smoothness have high prediction error because of variance – one has to find the happy medium.

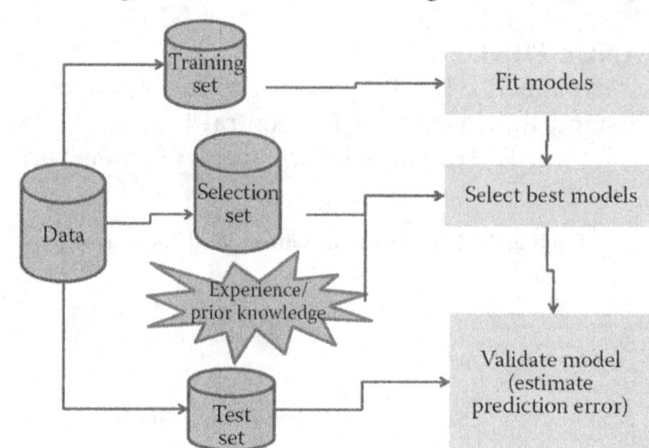

FIGURE 12.5
In a data-rich situation, the discipline of machine learning recommends splitting the data set into three segments: the training set, which is used to fit the models; the validation (or selection) set, which is used to choose the model that performs best amongst those fitted with the training set; and finally a test set, which calculates the prediction error on the chosen model.

Both concepts are perhaps easier to understand with an example: if the data come from a quadratic function with some noise, a straight line will be a poor model of the data, as the curvature will not be captured (of course, it also depends on the scale at which the curvature happens). On the other hand, if we attempted to fit the curve with a 10-degree polynomial will have so much flexibility that it will tend to replicate some of the changes in convexity observed in the data, although these are artefacts of a specific sample. A more elaborated example of this is shown in Section 12.3.1.

The message in the bias-variance trade-off chart is an important one to keep in mind because there is often pressure for actuaries to try and make their models as realistic as possible, and therefore to include as many features of the real world as possible to avoid embarrassing questions of whether they have taken this or that factor into account. However, the truth is that a more 'realistic' model (i.e. one which seems to be closer to reality) may be more helpful as an explanatory tool but is not *necessarily* better at making predictions and ultimately at making decisions, because there may not be enough data to calibrate the parameters of such model reliably.

12.3.1 Model Selection Concepts Through a Simple Example

We will see how the principle illustrated in Figure 12.4 comes to fruition in the selection of severity models (see Chapter 16) and in rating factor selection (Chapter 27). However, before we look at how machine learning can be helpful in selecting actuarial models it is

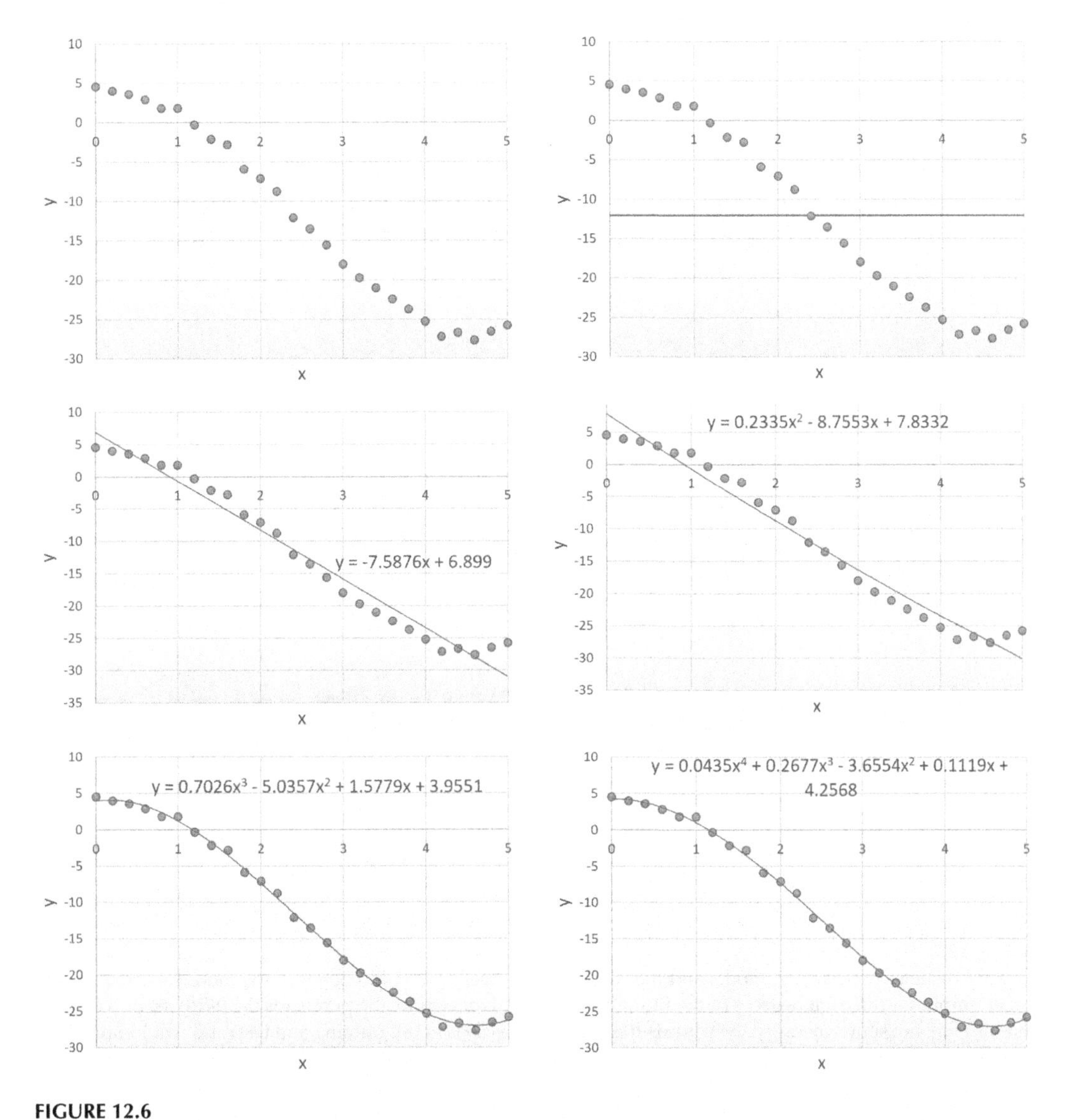

FIGURE 12.6
(Top left) A sequence of data points generated randomly from a cubic polynomial plus some random Gaussian noise. (All the other panels) Polynomials of increasing degree fitted to the sequence of data points, treated as a training set.

useful to look at a simpler example of model selection. Suppose we have a set of points like that in Figure 12.6 (top left) and we want to fit a polynomial through them. What polynomial should we use, and which degree should it have? Once we have decided the degree of the polynomial, the problem of calibrating it is simple – we can just use least squares regression for polynomials (when you plot a chart in Excel, Excel lets you fit a polynomial through the data using this methodology; there is also a function LINEST which does that). However, how do we decide which degree the polynomial should have? Inevitably, the higher the degree of the polynomial, the better the fit to the data will be, as illustrated in Figure 12.6. However, we have already seen that that is not a good way of selecting a model – goodness of fit must be judged against an independent test set.

Therefore, let's do a simple experiment with synthetic data.[3] Let's create a data set made of 26 pairs of data points (x_i, y_i), where the x_i s are the points 0, 0.2, 0.4, … 5.0, and the y_i s are obtained by the equation

$$y_i = 4 + 1.5\,x_i - 5\,x_i^2 + 0.7\,x_i^3 + 0.5\,\epsilon_i$$

where ϵ_i is normalised Gaussian noise (zero mean, unit variance). The result of one particular run of this experiment is the sequence of data points in Figure 12.6 (top left).

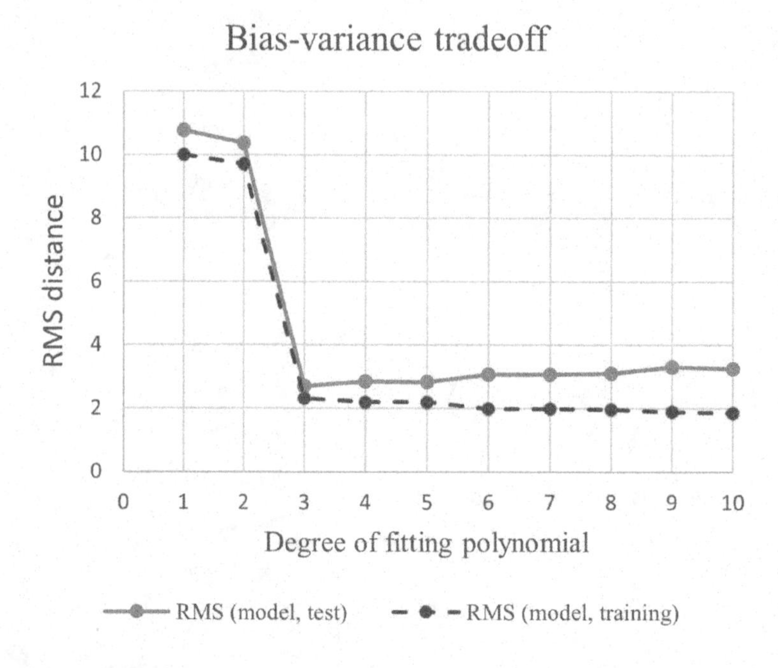

FIGURE 12.7
The RMS distance between the model and the training set (lower graph) and between the model and the *test* set (upper graph) for different degrees of the fitting polynomial. Notice how the points for the 0-degree polynomial (constant) are not shown because the value is too large (RMS ~ 59 for both training and test) and would not allow to see the details of the rest of the graph.

3 Synthetic data are extremely useful to test whether methods are *not* working. They can't show that things will work in real-world situations with real-world data, but if they fail even with controlled, artificial data then we know that those methods are flawed. They are therefore at least a good *falsification* tool.

We then pretend we don't know where this sequence comes from, and we try to fit polynomial functions to it, using Excel's in-built function for trendlines (also available as standalone as LINEST). How do we know where to stop? We find that the higher the degree of the polynomial, the better the fit is. With a polynomial of degree 25 (the number of points minus one) the fit will be perfect! Figure 12.7 shows how the Kolmogorov-Smirnov (KS) distance (a measure of the distance between two distributions, which we will meet again in Chapter 16) reduces progressively as we use polynomials of higher and higher degree. However, the figure also shows that when measured against another sample of 26 points from the same distribution above, the KS distance decreases until the degree is 3 and then increases again, albeit slowly. This is exactly the same message as that in Figure 12.4, but the behaviour is less spectacular: and you will find that in practice these bias-variance graphs normally do not forcefully bounce back beyond the optimal complexity, they just sort of flatten out and you can stop when there is no further meaningful reduction in the error between the model and the test set.

To select your model properly, therefore, you need a mechanism to control complexity. In the following we'll look at some examples of how to deal with complexity. First of all, let's define some concepts and terms.

12.3.1.1 Basic Concepts and Terminology.

The input to the exercise is a data set $\mathcal{D}=\{(x_1,y_1),(x_2,y_2),\ldots(x_n,y_n)\}$ made of observations of the independent variable x and the response y. We want to find the model (function) $f(x)$ that is best at predicting the value of y for a given x.

In order to evaluate how close the model is to the data we introduce a *loss function* $L(f,\mathcal{D})$. The loss function is a measure of the distance between the model f and the data $\mathcal{D}$. Common examples are the mean squared error, the mean absolute error or the negative log likelihood (loglik). Note that in the case of a linear model with Gaussian noise the negative loglik and the mean squared error are actually the same thing. For our polynomial case we will indeed use the mean squared error: therefore, the distance between a polynomial $f_k(x)=a_0+a_1x+\ldots+a_kx^k$ of degree k and the data set is

$$L(f_k,\mathcal{D})=\sum_{i=1}^{n}(y_i-f_k(x_i))^2=\sum_{i=1}^{n}(y_i-a_0-a_1x_i-\ldots-a_kx_i^k)^2$$

12.3.1.2 Empirical Risk Minimisation

One way is to avoid altogether the problem of controlling complexity is by severely restricting the number of possible models and their complexity.

Specifically, empirical risk minimisation works as follows.

1. A small set of models $\{f_1,f_2\ldots f_K\}$ from which the model can be chosen is selected. It is important that this set is made of models with comparable complexity (otherwise the models with more degrees of freedom will have an advantage).
2. For each of these models, evaluate the loss function $L(f_k,\mathcal{D})=\sum_{i=1}^{n}(y_i-f_k(x_i))^2$ *using the training set only* (no validation against an independent data set). Note that f_k will in general have some parameters and these must be chosen so as to minimise $L(f_k,\mathcal{D})$.

3. Select the model $f^* \in \{f_1, f_2 \ldots f_K\}$ such that $L(f^*, \mathcal{D})$ is minimal.

The classic example of empirical risk minimisation is linear modelling in one dimension (best-fit line): in that case, the set of models includes a single model: the straight line $y = ax + b$, and the problem of risk minimisation reduces to finding the parameters a, b of the regression line through the least squares method.

The fact that there is no use of an independent validation set is of course a limitation, but this is partly counterbalanced by the fact that the complexity of the admissible models is restricted a priori – so there is no risk that arbitrarily complex models will be chosen.

In our case, we can for example demand that only polynomials up to, say, degree four should be used. This would not lead to the correct result, because the true model that has generated the synthetic data is a polynomial of degree three; however, the fitted coefficient of x^4 is likely to be small.

This approach is particularly useful in those cases – common to actuarial practice, especially outside personal lines insurance – where data are sparse and it is not practical to remove some of the data for independent set validation. Some analysts, for example, will model *every* loss data set with a lognormal distribution. This strategy – which is an extreme version of empirical risk minimisation – may have its limitations, but it is not without merit: it certainly has more predictive power than picking the 'best' amongst 20 distributions with up to four parameters.

Empirical risk minimisation is also used, for example, in frequency modelling (Chapter 15), where many practitioners use only three models for claim counts: binomial, Poisson, and negative binomial.

12.3.1.3 Structural Risk Minimisation

Another method to control complexity is that of punishing complexity explicitly through a penalty. Instead of choosing the distribution f that has the smallest 'distance' from the data $\mathcal{D}$, as captured by the loss function $L(f, \mathcal{D})$, we can choose the distribution that minimises the modified loss function $L_{\text{mod}}(f, \mathcal{D})$, given by:

$$L_{\text{mod}}(f, \mathcal{D}) = L(f, \mathcal{D}) + p(d),$$

where $p(d)$ is a penalty that depends on the number of parameters d of the model f. Note that here, as well as for the case of empirical risk minimisation, the introduction of a penalty does away with the need to consider separate test/validation sets.

The Akaike Information Criterion (AIC) is an example of this type of approach: for this criterion the loss function is $L(f, \mathcal{D}) = -2\text{loglik}$ (loglik is the maximised log likelihood of the data $\mathcal{D}$ given the model f) and the penalty is $p(d) = 2d$. Since in our example the noise is Gaussian with uniform dispersion σ, this can be written as:

$$\text{AIC}(f_k, \mathcal{D}) = 1/\text{s}^2 \sum_{i=1}^{n} (y_i - f_k(x_i))^2 + 2d = 1/\text{s}^2 \sum_{i=1}^{n} (y_i - a_0 - a_1 x_i - \ldots - a_k x_i^k)^2 + 2d$$

Where the parameters $a_0, a_1 \ldots a_k$ are assumed to be the ones that correspond to the max likelihood for the polynomial of degree k, and s is the estimated value of the dispersion parameter σ.

Using the R package glm (which is designed for generalised linear modelling – see Chapter 27 – but can be used for this exercise by considering the polynomial as a linear combination of the monomials $1, x, x^2 \ldots$), we can calculate the AIC for polynomials of different degrees and pick the polynomial with the lowest AIC.

There are different ways of doing this. The most obvious is by *exhaustive search*. Assuming we only look at polynomials of degree 10 or less (to avoid having to consider an infinite number of polynomials), note that the most complicated polynomial is

$$f_{10}(x) = a_0 + a_1 x + a_2 x^2 + \ldots + a_{10} x^{10}$$

And all other polynomials are special cases of $f_{10}(x)$ where some of the coefficients are set to zero. Excluding the case where *all* coefficients are zero, that leaves us with $2^{11} - 1 = 2047$ polynomials to check. These are a lot of polynomials, and therefore analysts prefer to use some imperfect but useful shortcuts.

One of these shortcuts is called *forward selection*. This is an example of the so-called *greedy approach*. We start from the simple model

$$f_0(x) = a_0$$

and we calculate its AIC. Then we add another coefficient. We have 10 different ways of doing this: $f_0(x) + a_1 x$, $f_0(x) + a_2 x^2$, ... $f_0(x) + a_{10} x^{10}$. For each of these models we can calculate the AIC, and we can pick the model with the smallest AIC. Then we can add another term in the same way, and so on. We stop when none of the models with the added term has an AIC which improves on the polynomial with fewer terms. This is straightforward enough but can get us trapped into local minima.

This is what happens for example when we use this methodology for our simple example. Before we show the results, a notational point: in the R package glm a linear model $v = a_0 + ax + by + cz$ (this is just an example) is denoted using the shorthand

```
v ~ x + y + z
```

which makes the representation leaner. Also note that (again for notational convenience) we have denoted the monomials $x, x^2, \ldots x^{10}$ as `x1, x2` ... `x10`. Finally, note that in the following we will consider a data set of 500 data points (available through the Downloads section of the book's website (Parodi, 2016)), while the example shown in Figure 12.4 only used 26 data points for clarity of illustration).

The results are as follows:

```
Start:  AIC=205.01
y ~ 1

        Df Deviance    AIC
+ x1    1      99.9 114.78
+ x2    1     394.1 150.47
+ x3    1     830.4 169.84
+ x4    1    1215.0 179.74
+ x5    1    1526.9 185.68
+ x6    1    1776.4 189.62
+ x7    1    1976.8 192.39
+ x8    1    2139.5 194.45
+ x9    1    2273.3 196.03
+ x10   1    2384.5 197.27
```

```
<none>        3467.8 205.01

Step:  AIC=114.78
y ~ x1

        Df Deviance     AIC
+ x10    1   48.661  98.081
+ x9     1   50.314  98.950
+ x8     1   52.684 100.146
+ x7     1   55.990 101.729
+ x6     1   60.511 103.748
+ x5     1   66.568 106.228
+ x4     1   74.439 109.134
+ x3     1   84.063 112.295
<none>       99.872 114.775
+ x2     1   94.154 115.243

Step:  AIC=98.08
y ~ x1 + x10

        Df Deviance    AIC
+ x2     1   21.079 78.329
+ x3     1   28.645 86.304
+ x4     1   33.885 90.672
+ x5     1   37.430 93.258
+ x6     1   39.868 94.899
+ x7     1   41.589 95.998
+ x8     1   42.840 96.769
+ x9     1   43.774 97.329
<none>       48.661 98.081

Step:  AIC=78.33
y ~ x1 + x10 + x2

        Df Deviance    AIC
+ x3     1   4.7794 41.747
+ x4     1   4.9736 42.782
+ x5     1   5.6501 46.098
+ x6     1   6.4977 49.732
+ x7     1   7.3730 53.018
+ x8     1   8.2109 55.816
+ x9     1   8.9847 58.158
<none>      21.0786 78.329

Step:  AIC=41.75
y ~ x1 + x10 + x2 + x3

        Df Deviance    AIC
<none>       4.7794 41.747
+ x9     1   4.7356 43.507
+ x8     1   4.7472 43.571
```

```
+ x7     1   4.7601 43.641
+ x6     1   4.7722 43.707
+ x4     1   4.7739 43.717
+ x5     1   4.7792 43.745

Call:  glm(formula = y ~ x1 + x10 + x2 + x3, family = gaussian(), data =
input_data)

Coefficients:
(Intercept)           x1          x10           x2           x3
  4.160e+00    7.691e-01    2.892e-07   -4.468e+00    6.007e-01

Degrees of Freedom: 25 Total (i.e. Null);  21 Residual
Null Deviance:          3468
Residual Deviance: 4.779     AIC: 41.75
```

In a nutshell, we have started from the model $f(x) = a_0$ (AIC=205.01), and we have progressed by adding in turn x, x^{10}(oddly), x^2 and x^3 before the AIC stopped improving. So we have indeed been trapped in a local minimum, because if we then try what we know is the correct model, $f(x) = a_0 + a_1 x + a_2 x^2 + a_3 x^3$ (y ~ x1 + x2 + x3), we find that the AIC is actually lower than the AIC of the model selected via forward selection.

The reason why we get this strange result, by the way, is probably to ascribe to the fact that the variables $x, x^2, \ldots x^{10}$ are obviously strongly dependent. We'll see later on how this causes even bigger problems to other methods such as the lasso regression.

A slight modification of the forward selection method, which is domain-specific because it takes into consideration the fact that we are dealing with polynomials, is simply starting from the polynomial $f_0(x) = a_0$ and then add each time a monomial of higher degree: $f_1(x) = a_0 + a_1 x, f_2(x) = a_0 + a_1 x + a_2 x^2$, etc. until the AIC stops decreasing. This is quicker and seems reasonable as it is reminiscent of the Taylor approximation. However, it doesn't spare us from local minima because if, for example, the correct model were $f(x) = 2 + 4x^3$, we would end up with a full polynomial of degree 3 with spurious coefficients for x and x^2. In our case, however, it would lead to the correct response.

An alternative to forward selection is *backward selection*: that starts from the largest model that we are willing to consider, let's say $f_{10}(x) = a_0 + a_1 x + \ldots + a_{10} x^{10}$, and remove each time the term that causes the remaining polynomial to have the smallest AIC, until the AIC stops decreasing.

The AIC is only one possible penalty-based criterion. A similar approach is the Bayesian Information Criterion (BIC) approach, which uses a harsher penalty $p(d) = \log(n)d$ where n is the number of data points. Both the AIC and the BIC can ultimately be justified as approximations of the prediction error (Hastie et al., 2001).

It can be proven that the BIC gives an asymptotically consistent result, meaning that when the number of data points tend to infinity (and given a set of models which includes the true model) the probability that the correct model is chosen approaches 1. AIC, on the other hand, tends to choose models that are too complex when the number of data points is large. For small samples, however, the BIC penalty tends to choose models that are too simple.

It can be proven that the BIC gives an asymptotically accurate estimation of the expected prediction error, which is ultimately what we need to minimise. Also, the BIC is related to the minimum description length (MDL). According to MDL theory – which can be seen as an application of information theory – one should look for the model that is able to describe the data with the minimum description length, where the description length is a measure of the length of the code needed to specify the data within a certain error. We saw the core of this idea at work in our simple example in Section 12.1.1, where we considered the problem of communicating a list of 13 data points. This 'communication' can be seen as a piece of code, and the time it takes to do this communication an informal proxy for the length of this code.

12.3.1.4 Regularisation

One of the most interesting methods for model selection is certainly *regularisation*: this involves replacing the problem of minimising the loss function $L(f,\mathcal{D})$ with the problem of minimising the regularised loss function:

$$L_{\text{reg}}(f,\mathcal{D})=L(f,\mathcal{D})+\lambda g(\beta)$$

where $g(\beta)$ is a function of the parameters β of the model. The effect of the term $\lambda g(^2)$ is in a way similar to that of a penalty term like in structural risk minimisation, but it is achieved by imposing a certain amount of 'smoothness' to the solution – the extent to which this smoothness is achieved depending on the size of the parameter λ. What this exactly means depends on the specific shape of the function $g(\beta)$.

In our case, the regularised loss function will look something like this:

$$L_{\text{reg}}(f,\mathcal{D})=\sum_{i=1}^{n}\left(y_i-a_0-a_1x_i-\ldots-a_{10}x_i^{10}\right)^2+\lambda g(a_0,a_1\ldots a_{10})$$

Typical functional forms for $g(a_0,a_1\ldots a_{10})$ that one would use for a problem like ours are the following:

$$g_{\text{ridge}}(a_0,a_1\ldots a_{10})=\|\vec{a}\|_2=\sqrt{a_0^2+a_1^2+\ldots+a_{10}^2}\ \text{(ridge regression)}$$

$$g_{\text{lasso}}(a_0,a_1\ldots a_{10})=\|\vec{a}\|_1=|a_0|+|a_1|+\ldots+|a_{10}|\ \text{(lasso regression)}$$

$$g_{\text{elast}}(a_0,a_1\ldots a_{10})=\alpha\|\vec{a}\|_1+(1-\alpha)\|\vec{a}\|_2\ \text{(elastic net regression)}$$

where $\vec{a}=(a_0,a_1\ldots a_{10})$ is the vector of the coefficients and α is between 0 and 1.

Both the ridge regularisation and the lasso regularisation add a term which is simply the norm implied by a metric (l_2 and l_1, respectively) in the space of the coefficients $(a_0,a_1\ldots)$. The elastic net is a convex combination of the two norms.[4]

The ridge regression tries to achieve smoothness (and ultimately simplicity) by penalising coefficients that are too large. In this way, only a few of the dependent variable will have a significant impact – so variables that are irrelevant will not be removed altogether but their impact will be greatly reduced.

4 Other norms could in principle be used, e.g. $\|\vec{a}\|_p=\left(a_0^p+a_1^p+\ldots+a_{10}^p\right)^{\frac{1}{p}}$ (based on the l_p metric) or $\|\vec{a}\|_\infty=\max(a_0,a_1,\ldots a_{10})$ (based on the l_∞ metric).

The lasso regression has a more drastic effect. Not only it puts pressure on the size of the coefficients, but when the regularisation coefficient λ is sufficiently large some of the coefficients will become zero (we speak of *sparsity-inducing regularisation)*. So the method removes fully the variables that do not have sufficient explanatory power, simplifying the model in the most obvious way.

The elastic net does a combination of the two things and has been shown to be effective when some of the variables are strongly correlated.

The disadvantage of the lasso and the elastic net, however, is that (unlike ridge regression) they have no simple analytical solution (basically a matrix inversion) – instead, they need numerical methods to work out an optimal solution.

In the case of lasso and elastic net, there is the added advantage (with respect to structural risk minimisation using AIC) that the selection of the relevant variables (or monomials in this case) is a byproduct of the numerical optimisation.

The question is, of course: how do we select the best value of the regularisation coefficient λ? The problem is as usual to determine the bias-variance trade-off that produce the most predictive model. A standard methodology to do that is through *k*-fold cross-validation. This is a trick to calculate the expected prediction error when one doesn't have the luxury of being able to split the data neatly into a training set and a test set, or even into three sets (training, selection, and validation). The treatment of cross-validation here is mostly lifted from Parodi (2013a).

(k-fold) cross-validation

Cross-validation (and specifically, *k*-fold cross-validation) estimates the prediction error by dividing *at random* the data set $\mathcal{D}$ into K different subsets $\mathcal{D}_1,\ldots\mathcal{D}_K$ (for simplicity, we assume that n is a multiple of K; the modifications where that is not the case are trivial). Each subset $\mathcal{D}_k$ $(k=1,\ldots K)$ is in turn removed from the data set and the model is fitted to the remaining set $\mathcal{D}[-k]=\mathcal{D}-\mathcal{D}_k$. The k-th subset $\mathcal{D}_k$ is used as a test set, and $f^{\mathcal{D}[-k]}(x)$ is the function fitted on $\mathcal{D}[-k]$. The loss function calculated on subset $\mathcal{D}_k$ therefore becomes:

$$L(f,\mathcal{D}_k)=\frac{K}{n}\sum_{\text{all } \frac{n}{K} \text{ points in subset } \mathcal{D}_k} L\left(y_i, f^{\mathcal{D}[-k]}(x_i)\right)$$

The process is repeated for all subsets and the cross-validation estimate of the prediction error is given by the average over the K possible subsets.

$$CV=\frac{1}{K}\sum_{k=1}^{K}L(f,\mathcal{D}_k)=\frac{1}{n}\sum_{\text{all } n \text{ points across all } k \text{ subsets}} L\left(y_i, f^{\mathcal{D}[-k]}(x_i)\right)$$

Note that for each point x_i there is only one subset $\mathcal{D}_k$ for which $f^{\mathcal{D}[-k]}(x_i)$ needs to be calculated, and therefore the expression $f^{\mathcal{D}[-k]}(x_i)$ is unambiguous.

In the regularised case, we do not have a single model to cross-validate but a set of different models $f(x;\lambda)$ indexed by λ. For this set of models, we have a different value of $\text{CV}(\lambda)$ for every λ, and the function $\text{CV}(\lambda)$ is effectively an estimate of the test error curve depicted in Figure 12.4. The optimal model is then $f(x;\lambda^*)$ where λ^* is the value that minimises $\text{CV}(\lambda)$.

One key issue with cross-validation is of course the choice of K. Without delving too deeply into this, let's just say that anything from K=2 to K=n is possible, but typical values are K=5 and K=10.

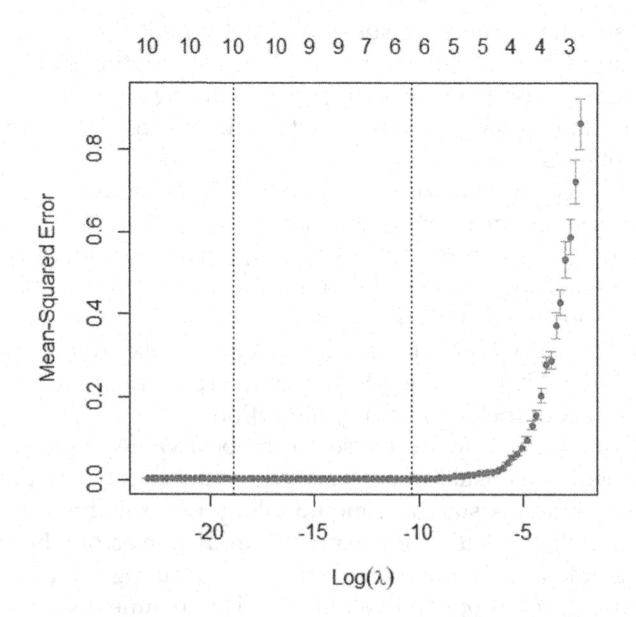

FIGURE 12.8
The bias-variance trade-off chart as built through cross-validation by glmnet. Note that unlike the chart in Figure 12.4, complexity decreases from right to left, since larger values of λ enforce higher sparsity. This is evidenced in the top x-axis, which shows the number of non-zero coefficients corresponding to each value of λ, which is shown in the bottom x-axis (in *natural* log scale). Two values of λ are highlighted in the chart: $\lambda_{min} \sim 6.31 \times 10^{-9}$ (left dotted line), which is the value at which the estimated prediction error is smaller, and $\lambda_{1se} \sim 3.16 \times 10^{-5}$ (right dotted line), which is the value at which the estimated prediction error is first within one standard deviation of the prediction error calculated at λ_{min}. The latter is normally a more robust estimate of the optimal value of the regularisation parameter.

Note that cross-validation is a more general technique than AIC and although it also overestimates the prediction error, it does so to a much lesser extent than the AIC and similar criteria (Hastie et al., 2001).

The calculations can be performed with the R package glmnet, which we will come across again in Chapter 27.

```
# define a sequence of lambdas to use to build the CV plot
> lambda_seq <- 10^ seq(-10,-1,0.1)

# create a cross-validation plot
cvfit <- cv.glmnet(x,y,family='gaussian',alpha=1, lambda=lambda_seq,
                   thresh=1e-14, maxit=5e7)
plot(cvfit)
```

Note that the value alpha = 1 in the R command above corresponds to lasso regression (alpha = 0 corresponds to ridge regression, and 0 < alpha < 1 corresponds to elastic net).

As shown in Figure 12.8, the optimal value of λ (i.e. the value for which the estimated prediction error is within the experimental error) is around 3.16×10^{-5}. The coefficients of the model corresponding to the optimal λ can be obtained as a byproduct of cv.glmnet:

```
# show the coefficients of the optimal model
coef(cvfit, s='lambda.1se')
```

The results are as follows:

```
(Intercept)     4.047200e+00
x1              1.280092e+00
x2             -4.739990e+00
x3              5.830660e-01
x4              .
x5              1.891572e-02
x6              .
x7              .
x8             -4.897238e-05
x9              .
x10             1.183512e-07
```

As the results above show, lasso regularisation (in its basic form) does remove some of the variables (x4, x6, x7, x9) but still retains – like structural risk minimisation with a linear model and AIC penalty did – extraneous variables (x5, x8, x10). Also, the coefficients are not always very close to the true values (which we know because we are using synthetic data). However, the results are a reasonable approximation, and the coefficients of x8 and x10 are very small and this suggests that these could be set to zero (although with some caution, given that x8 and x10 become quite large at the end of the range, $x = 5$).

12.3.1.4.1 Limitations of Polynomial Fitting

The relative lack of precision mentioned above is only partly due to the fact that the data set is quite small (500 data points). Another reason, however, is that while polynomial fitting is a very simple problem to explain and therefore has good didactic value, it is a hard problem to crack, because of numerical instabilities (e.g. Runge's oscillations). Also, the monomials $x, x^2, x^3 \ldots$ are linearly uncorrelated but for obvious reasons strongly rank-correlated, and the lasso is known to perform badly under this circumstance. Indeed, the various parameters involved in the fitting (like maxit – the maximum number of iterations – and thresh – a parameter of the optimisation algorithm) need to be chosen with care in order to obtain sensible results, as readers can verify by themselves using the default fitting parameters.

For these reasons, the problem of approximating functions, which as we have learned in calculus can be theoretically solved by Taylor's expansion, is in practice solved by different techniques such as regression splines, which gives very satisfactory visual results and is not subject to the issues mentioned above. We will come back to regression splines in Chapter 27 where we discuss generalised additive models, another important item in the machine learning's toolkit.

12.4 Actuarial Applications of Machine Learning Concepts/Techniques

Now that we have gone through the basic concepts of machine learning using a toy problem it might be helpful to mention how these can be applied to actuarial practice.

As was argued at the beginning of this chapter and more extensively in Parodi (2012a), machine learning techniques are not simply additional techniques that can be usefully added to the actuary's toolkit: rather, machine learning provides with a solid theory of model building that should inform *all* of actuarial modelling. Specifically, it can be seen that *all risk costing (and specifically pricing) is a form of supervised learning*, i.e. a functional optimisation that aims to discover the functional relationship between explanatory variables and response variables using past observations basically what we have described in Section 12.3.

We will indeed see examples of the techniques described in Section 12.3.1 later on in the book. **Empirical risk minimisation** will be applied:

- to the selection of frequency models (Chapter 14), where we will see that practitioners typically avoid overfitting by demanding that the frequency distribution be in the Panjer class (binomial, Poisson and negative binomial);
- to the selection of severity models (Chapter 16), in the case where practitioners strongly restrict the class of severity distributions to be, for example, composed of a lognormal body and a generalised Pareto tail.

Examples of **structural risk minimisation** will be seen:

- again in the selection of severity models (Chapter 16), where the severity distribution is chosen among a restricted number of types and complexity is penalised by the AIC or similar criteria;
- in rating factor selection (Chapter 27), a crucial exercise in personal lines (and sometimes in small commercial lines), when doing model selection using GLM and GAM with AIC.

Finally, **regularisation** will again be seen in the context of rating factor selection as an alternative to model selection for GLM and GAM via AIC. Furthermore, regression splines are a classic example of objects that emerge as solutions to a regularisation problem (Chapter 27).

All the techniques mentioned above are for supervised learning. The topic of unsupervised learning (which mainly deals with clustering of data points with sufficiently similar properties) is briefly touched upon in Chapter 28.

12.5 Further Readings on the Application of Machine Learning to Actuarial Modelling

We have only touched the surface of how machine learning can be applied to actuarial modelling. The reader interested in this topic is advised to consult the following resources:

- Parodi (2012a, 2012b) provide an introduction to both machine learning and (more generally) AI to general insurance. It is perhaps most useful as a way of understanding why many actuarial problems should be interpreted as machine learning problems, and especially problems of supervised learning problems.

- Richman (2021a, 2021b) also addresses the applications of machine learning to actuarial science (to both life and non-life insurance), with an emphasis on deep learning techniques (which are only mentioned in passing in Chapter 27 of this book).
- Wüthrich and Buser (2021) has an in-depth (and quite technical) treatment of the most popular machine learning techniques used in general insurance: GLMs, GAMs, neural networks, regression trees, and ensemble learning methods.
- Noll et al. (2018) shows the application of a variety of techniques (GLM, neural networks, regression trees, boosting) to a specific case study (French Motor TPL).

13

Frequency Modelling: Adjusting for Claim Count IBNR

We now get into the heart of the frequency/severity model. We first turn our attention to frequency modelling. Because producing a frequency model means calculating the characteristics of the frequency distribution, such as its mean and its variance, we need to know the ultimate claim (loss) count for each year. However, the only information we have is the number of claims *reported* in any particular year. To go from the reported claim count to the *ultimate* claim count, we need to estimate the number of incurred but not (yet) reported (IBNR) claims. The objective of this chapter is to illustrate how this can be done. Figure 13.1 shows how IBNR adjustment fits in the risk costing process – this will help us not to lose track of the overall scheme of things.

13.1 Input to IBNR Adjustment

The data required for the IBNR adjustment are roughly the same as those described for the pricing exercise as a whole, and are spelled out in Chapter 8, except for the information on the exact claim amounts. To recap, we need individual claims information, and for each claim, we need:

- The loss occurrence date.
- The reporting date.
- The *quantification date* (the date at which a non-zero incurred amount was first attached to a claim).
- The *date at which the claim was set back to zero* (if applicable) because at that point the loss will disappear from the count of non-zero losses. This replaces the requirement of the individual transition dates for every claim transition, that is, all times at which the outstanding and paid loss amounts are revised.

In practice, we may not always be able to have all the information above. Frequently, if we have individual-claim information, we may only have a snapshot of the claims information at a point in time, and only the most basic dates (loss date and reporting date) may be available, as in Figure 13.2: in this case, we may need to make some assumptions on the quantification date and on the issue of claims going back to zero.

DOI: 10.1201/9781003168881-15

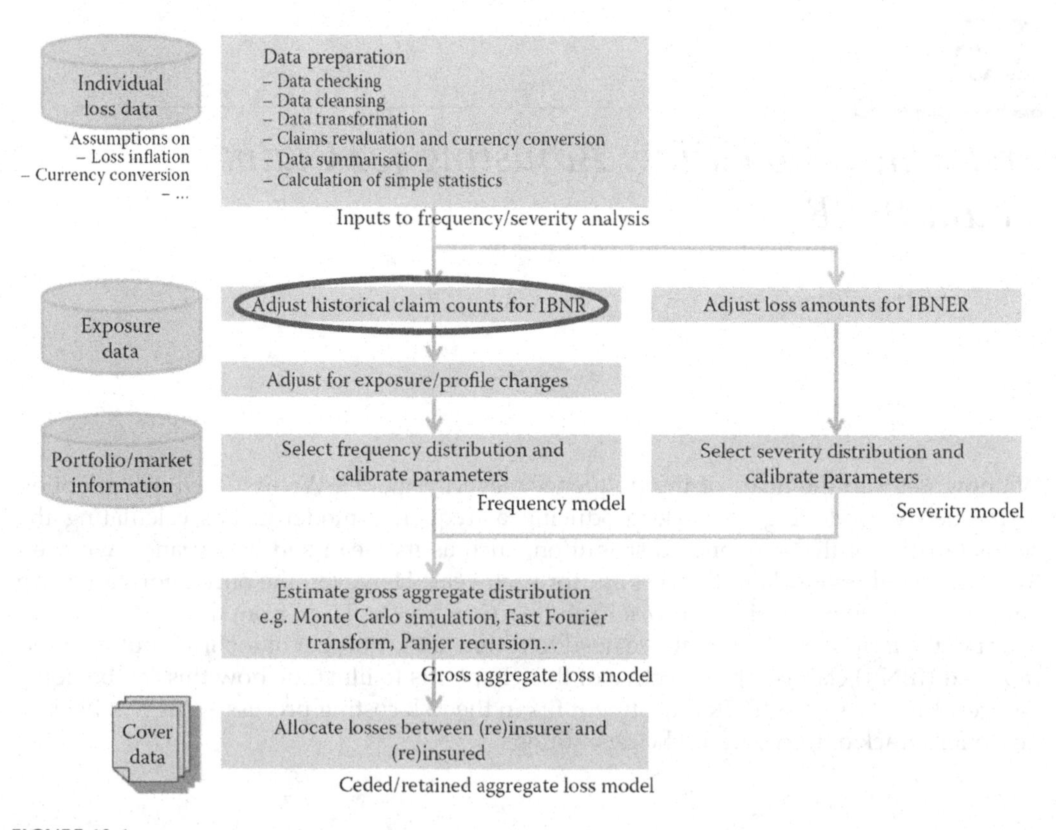

FIGURE 13.1
This is how adjusting for the number of IBNR claims fits in the overall risk costing process. Specifically, it is part of frequency analysis, which eventually leads to a frequency model.

Alternatively, we may be given claim count triangulations such as that in Figure 13.3, in which the number of non-zero claims at different points in time is predigested for us. This is summarised information and therefore not as complete as individual-claim information, but at least it considers the issue of zero claims automatically.

It sometimes happens that we simply do not have enough information to make a judgment on the delays; for example, because we are missing the loss date or the reporting date. In these cases, we will need to use external information to assess IBNR levels.

In the rest of this chapter, we will make use of the case study introduced in Chapter 11, and for which the data set is also available on the book's website.

13.2 Triangle Development Methods for IBNR Estimation

The most common method for the estimation of the number of IBNR claims is based on the development of claim count triangles, using the chain ladder or other similar triangle development techniques for projecting the claim count to ultimate.

(a)

Policy year	Reporting year	Delay (years)
2009	2009	0
2009	2009	0
2008	2008	0
2010	2013	3
2010	2010	0
2006	2007	1
2003	2004	1
2004	2005	1
2005	2006	1
2007	2007	0
2004	2005	1
2006	2006	0
2004	2004	0
2006	2006	0
2007	2007	0
2007	2007	0
2004	2004	0
2003	2004	1
2008	2009	1

(b)

Loss date	Reporting date	Delay (days)
6/13/2009	9/3/2009	82
8/9/2009	9/3/2009	25
12/17/2008	1/26/2009	40
11/1/2010	7/29/2013	1001
2/7/2011	2/14/2011	7
1/17/2007	1/18/2008	366
3/3/2004	4/23/2004	51
7/5/2004	6/23/2005	353
6/6/2005	1/23/2007	596
6/15/2007	6/19/2007	4
7/21/2004	8/31/2005	406
8/30/2006	10/25/2006	56
8/4/2004	2/9/2005	189
5/4/2006	5/26/2006	22
11/7/2007	12/5/2007	28
8/31/2007	1/14/2008	136
10/13/2004	1/4/2005	83
1/7/2004	4/20/2004	104
7/1/2008	7/1/2009	365

FIGURE 13.2
Delay information (in years) for a sample from our case study (a). Delay information (in days) for the same claims (b). Note that the policy in our case study incepts on 1 April; hence, the apparent mismatch between loss date and policy year. In this case, we do not have information on the quantification date and on when previously non-zero claims have been set back to zero. For this particular data set, we will assume that the quantification date is the same as the reporting date and that the issue of claims reverting to zero can be ignored.

	Delay (years)									
	0	1	2	3	4	5	6	7	8	9
2005	50	26	5	3	2	0	0	0	0	0
2006	41	28	13	6	4	0	0	0	0	
2007	65	33	10	2	0	0	0	0		
2008	83	47	8	6	1	0	1			
2009	120	45	16	9	2	2				
2010	149	77	20	7	3					
2011	158	64	19	9						
2012	96	81	22							
2013	102	99								
2014	68									

FIGURE 13.3
In this incremental claim count triangle, each row represents a policy year, and each column represents the delay (in years) from occurrence to reporting. Each number represents the number of claims of a given policy year reported with a given delay. For example, in policy year (PY) 2008 there were 83 claims reported by the end of PY 2008 (no delay), 47 claims reported by PY 2009 (1 year delay), 8 claims reported by PY 2010 (2-year delay), etc. Although there is no such case here, in general, it is also possible to have a negative number of non-zero claims in a given cell, signalling that a claim that was previously non-zero has dropped to zero.

The main assumptions in doing this projection are that

a. All years develop in roughly the same way (so that we can apply past development factors to the currently incomplete years).
b. There is no further development for the earliest year (so that we do not have to worry about developments after a certain number of years).

To apply this method, the first thing to do is to organise the claim counts in a triangle such as that in Figure 13.3. Ideally, such a claim count triangle should be built from *complete* information on the claims, including information on claims dropping to zero, reopened claims, and others. However, in this particular case, these triangles have been built based on our case study, for which a snapshot at a particular point in time was given.

To apply the chain ladder technique, we need to build the cumulative version of the triangle of Figure 13.3, which we show in Figure 13.4.

With the claims count triangle now in a cumulative format, we can apply the chain ladder method to project the number of claims to ultimate. Before doing that, notice that we have a problem: the latest diagonal (the one for which policy year + delay = 2014) is not complete, as the 'as-at' date of the data set is 30 November 2014, and therefore we have only 8 months' worth of data.

There are several ways of dealing with this problem. Here are some:

i. Ignore the last diagonal and apply the chain ladder method to the complete years only.
ii. Gross up the numbers in the last diagonal by 12 (total number of months in a year) divided by 8 (number of months available) = 1.5. (This is really rough and can be further improved if we have more information on the distribution of delays. However, we put it here as an illustration.)
iii. Use periods of 4 months instead of full years and calculate the link ratios for the new triangle. Note that in this case, instead of having 9 complete years and 1 incomplete year, we have 29 complete periods of 4 months. (This method can be adjusted

Cumulative Claim Count Triangle

	Delay (years)									
	0	1	2	3	4	5	6	7	8	9
2005	50	76	81	84	86	86	86	86	86	86
2006	41	69	82	88	92	92	92	92	92	
2007	65	98	108	110	110	110	110	110		
2008	83	130	138	144	145	145	146			
2009	120	165	181	190	192	194				
2010	149	226	246	253	256					
2011	158	222	241	250						
2012	96	177	199							
2013	102	201								
2014	68									

FIGURE 13.4
In this cumulative claims count triangle, each row represents a policy year, and each column represents the delay (in years) from occurrence to reporting. Each number in the triangle represents the total number of non-zero claims that occurred in a given policy year with a delay of up to a given delay. For example, in PY 2008, there were 144 claims that were reported with a delay of up to 3 years, that is, by the end of PY 2011 (see circled figures).

depending on how many complete months we have, for example, if we had 6 full months and 20 days, we might exclude the last 20 days and use periods of 6 months.)

iv. Shift the observation period so that each diagonal, including the most recent one (but apart from the top left diagonal made of one single cell) represents a full year.

v. Shift the development period so that the first development period has 8 months and each diagonal represents a full year.

We'll now look at each of these methods in more detail (Sections 13.2.1–5). A comparison of the results is shown in Section 13.2.6.

13.2.1 Method 1: Ignore the Last Diagonal

Let us start by looking at Method 1. Figure 13.5 shows how this is done and also the full results of the analysis.

For example, the link ratio between delay 3 and delay 4 is calculated as follows:

$$\text{LR}(3 \rightarrow 4) = \frac{86+92+110+145+192}{84+88+110+144+190} = 1.015$$

Although the triangle only goes up to 8 years of delay, we may still get the odd claim reported after 9 years or so. To take this into account, we sometimes introduce a tail factor link ratio (8 → Ultimate), based perhaps on market information or an otherwise educated guess. Because no claim has been reported after development year 4, it is not unreasonable to assume that in this case the tail factor is 1 (no tail):

$$\text{LR}(8 \rightarrow \text{Ultimate}) = 1$$

These link ratios can then be used to project each year to ultimate. To make this easier, it is useful to calculate the cumulative version of the link ratio, the cumulative link ratios (CLR). These are defined recursively as

$$\text{CLR}(n \rightarrow \text{Ultimate}) = \text{CLR}(n+1 \rightarrow \text{Ultimate}) \times \text{LR}(n \rightarrow n+1)$$

Calculation of Development (Link) Ratios with the Chain Ladder Method

	Delay (years)								
	0	1	2	3	4	5	6	7	8
2005	50	76	81	84	86	86	86	86	86
2006	41	69	82	88	92	92	92	92	
2007	65	98	108	110	110	110	110		
2008	83	130	138	144	145	145			
2009	120	165	181	190	192				
2010	149	226	246	253					
2011	158	222	241						
2012	96	177							
2013	102								
									Tail
Link ratios	1.526	1.092	1.039	1.015	1.000	1.000	1.000	1.000	1.000
Cumulative	1.758	1.152	1.055	1.015	1.000	1.000	1.000	1.000	1.000

FIGURE 13.5
Link ratios calculation with the chain ladder methodology: for example, the link ratio between delay 3 and delay 4 is 1.015. The exact calculation is shown in the text.

Projection to Ultimate Removing the Last Diagonal

	Delay (years)									
	0	1	2	3	4	5	6	7	8	Ultimate
2005	50	76	81	84	86	86	86	86	86	86.00
2006	41	69	82	88	92	92	92	92		92.00
2007	65	98	108	110	110	110	110			110.00
2008	83	130	138	144	145	145				145.00
2009	120	165	181	190	192					192.00
2010	149	226	246	253						256.70
2011	158	222	241							254.17
2012	96	177								203.90
2013	102									179.34
Link ratios	1.526	1.092	1.039	1.015	1.000	1.000	1.000	1.000	1.000	
Cumulative	1.758	1.152	1.055	1.015	1.000	1.000	1.000	1.000	1.000	

FIGURE 13.6
Projecting the claims count to ultimate using the chain ladder method. For example, Ultimate (2012) = 177 × 1.152 = 203.90.

with

$$\text{CLR}(n \to n+1) = 1$$

for all n above a given N (in this case, $N = 8$).

For example,

$$\text{CLR}(0 \to \text{Ultimate}) = 1.526 \times 1.092 \times 1.039 \times 1.015 \times 1 \times 1 \times 1 \times 1 \times = 1.758$$

To project each policy year (PY) to ultimate, we now simply have to multiply the number of reported claims in PY by the relevant CLR, that is,

$$\text{Ultimate}(2013) = \text{CLR}(0 \to \text{Ultimate}) \times \text{Reported}(2013)$$

$$\text{Ultimate}(2012) = \text{CLR}(1 \to \text{Ultimate}) \times \text{Reported}(2012)$$

$$\ldots$$

$$\text{Ultimate}(2005) = \text{CLR}(8 \to \text{Ultimate}) \times \text{Reported}(2005)$$

Figure 13.6 shows the results of projecting to ultimate with this method.

13.2.2 Method 2: Gross Up the Last Diagonal

In Method 2, we gross up each element of the last diagonal in the incremental triangle of Figure 13.3 to make up for the missing number of months. In the most simple-minded version of this method, this means that (in our case study, in which we have 8 months of experience in the last year) we multiply by 12/8 = 1.5. A more careful calculation will take into account the actual reporting delay pattern, looking at the average percentage of losses reported in the first 8 months for previous calendar years for the same client or (to have a more stable result) across a portfolio of clients; our rough-and-ready approach, however, is enough to illustrate the point (Figure 13.7). We then proceed to calculate the link ratios in the same manner as with Method 1, yielding the results shown in Figure 13.8.

Grossing Up the Last Diagonal

	Delay (years)									
	0	1	2	3	4	5	6	7	8	9
2005	50	26	5	3	2	0	0	0	0	**0**
2006	41	28	13	6	4	0	0	0	**0**	
2007	65	33	10	2	0	0	0	**0**		
2008	83	47	8	6	1	0	**1.5**			
2009	120	45	16	9	2	**3**				
2010	149	77	20	7	**4.5**					
2011	158	64	19	**13.5**						
2012	96	81	**33**							
2013	102	**148.5**								
2014	**102**									

FIGURE 13.7
The original triangle of Figure 13.3 has been modified so that the elements of the last diagonal are grossed up by a factor of 12/8 = 1.5, to make up for the missing 4 months in policy year 2014: for example, for 2014, Rep(2014) × Gross-up factor = 68 × 1.5 = 102. Of course, this can be done much better by making better use of the pattern by which claims are reported.

	Delay (years)										
	0	1	2	3	4	5	6	7	8	9	**Ultimate**
2005	50	76	81	84	86	86	86	86	86	86	**86.0**
2006	41	69	82	88	92	92	92	92	92		**92.0**
2007	65	98	108	110	110	110	110	110			**110.0**
2008	83	130	138	144	145	145	146.5				**146.5**
2009	120	165	181	190	192	195					**195.7**
2010	149	226	246	253	257.5						**259.6**
2011	158	222	241	254.5							**260.6**
2012	96	177	210								**224.3**
2013	102	250.5									**296.1**
2014	102										**197.2**
										Tail factor	
Link ratios	1.636	1.107	1.043	1.016	1.005	1.003	1.000	1.000	1.000	1.000	
Cumulative	1.934	1.182	1.068	1.024	1.008	1.003	1.000	1.000	1.000	1.000	

FIGURE 13.8
The cumulative triangle and the projection to ultimate for Method 2.

13.2.3 Method 3: Choose an 'Ad Hoc' Observation Period

In Method 3, we reason like this: it is true that the last policy year is incomplete (4 months are missing), but if we choose a different observation period, for example, 4 months instead of a year, we replace an incomplete year with two complete 4-month periods in PY 2014. The results of the calculations are shown in Figure 13.9.

Can we always choose an observation period of adequate length? If, for example, we had 7 months of experience available for the last year rather than 8, because 7 and 12 have no common factors, we would have no choice but to choose periods of 1 month or get rid of one month of experience and choose 6 months as the unit period. There is a trade-off: choosing too small a unit period may lead to instabilities (link ratios would be erratic) and to large triangles, whereas choosing too large a unit period may lead to having to discard some of the experience.

Projection to Ultimate Using 4-Month Periods

Delay (number of 4-month periods)

Period	0	1	2	3	4	5	6	7	8	9	10	11	12	13	14	15	16	17	18	19	Ultimate
2005-0	8	14	18	23	24	25	25	26	27	27	27	27	27	28	28	28	28	28	28	28	28.00
2005-1	10	14	17	18	18	19	19	19	19	20	20	20	20	20	20	20	20	20	20	20	20.00
2005-2	18	26	30	33	34	35	35	37	37	37	38	38	38	38	38	38	38	38	38	38	38.00
2006-0	4	8	11	13	13	14	15	15	16	16	17	17	17	17	17	17	17	17	17	17	17.00
2006-1	13	21	23	26	27	28	30	30	30	33	34	35	35	35	35	35	35	35	35	35	35.00
2006-2	9	19	27	28	32	35	36	36	36	37	40	40	40	40	40	40	40	40	40	40	40.00
2007-0	20	27	32	35	39	40	43	44	45	45	46	46	46	46	46	46	46	46	46	46	46.00
2007-1	18	26	31	33	36	36	39	39	39	39	39	39	39	39	39	39	39	39	39	39	39.00
2007-2	7	15	19	22	24	24	24	24	24	25	25	25	25	25	25	25	25	25	25	25	25.00
2008-0	13	24	29	34	35	37	37	37	37	39	39	39	39	39	39	39	39	39	39	40	40.00
2008-1	29	37	50	54	57	60	60	60	60	61	63	64	64	64	64	64	64	64	64		64.20
2008-2	17	25	32	36	40	40	41	42	42	42	42	42	42	42	42	42	42	42			42.13
2009-0	29	40	45	49	51	54	54	54	54	54	55	55	55	55	55	55	55				55.17
2009-1	33	49	54	61	63	69	70	71	71	73	76	77	77	77	77	77					77.24
2009-2	26	37	46	48	51	51	56	57	58	59	60	60	60	61	62						62.19
2010-0	26	39	48	53	60	63	64	64	66	66	66	66	66	66							66.31
2010-1	34	58	66	73	78	79	81	82	83	84	84	85	86								86.65
2010-2	43	70	79	85	92	96	98	99	99	103	104	104									104.92
2011-0	36	53	69	75	79	81	82	82	84	86	87										88.17
2011-1	37	57	64	69	72	74	76	76	76	77											79.36
2011-2	32	55	64	69	75	79	81	84	86												90.40
2012-0	18	34	43	50	54	56	59	59													62.65
2012-1	18	36	44	51	58	63	65														69.76
2012-2	17	43	54	63	72	75															82.72
2013-0	16	29	50	60	62																71.21
2013-1	12	37	58	71																	87.52
2013-2	15	50	68																		93.62
2014-0	23	49																			83.77
2014-1	19																				55.46
Link ratios	1.707	1.242	1.117	1.073	1.041	1.028	1.011	1.010	1.020	1.017	1.005	1.001	1.003	1.002	1.000	1.000	1.000	1.000	1.003	1.000	
Cumulative link ratios	2.919	1.710	1.377	1.233	1.149	1.103	1.073	1.062	1.051	1.031	1.013	1.009	1.008	1.005	1.003	1.003	1.003	1.003	1.003	1.000	

FIGURE 13.9
The cumulative triangle and the projection to ultimate for Method 3. The projected ultimate is provided here for each period of 4 months. To obtain the projected ultimate for each policy year, simply add up the relevant periods: for example, Ult(2010) = Ult(2010–0) + Ult(2010–1) + Ult(2010–2) = 257.88. Only the first 19 4-month periods were used in the projection because there was no further development in the data after that.

13.2.4 Method 4: Shift the Observation Period

If the full history for all individual claims is available, one can redefine the triangle so that the time elapsed between one diagonal and the other is always 1 year. In our case, this can be done by considering the claim count at 8 months, 20 months, 32 months and so forth (Figure 13.10).

The advantage of this method is that we do not have to worry about the last diagonal being of different length than the others and about discarding recent data.

The disadvantage is, however, that the ultimate claim count does not match the original policy year, and if we wish to estimate the ultimate number of claims for the original policy year, we have to combine approximately the IBNR claim count information of different shifted policy years in some approximate way. In our case, we may, for example, assume that the number of IBNR claims for PY 2012, n_{2012}, is a weighted combination of the number of IBNR claims for shifted year 2011/2012, $n'_{2011/2012} = 13.02$ and shifted year 2012/2013, $n'_{2012/2013} = 33.78$:

$$n_{2012} = \frac{8}{12} \times n'_{2011/2012} + \frac{4}{12} \times n'_{2012/2013} = \frac{8}{12} \times 13.02 + \frac{4}{12} \times 33.78 = 26.86$$

The projected ultimate for 2009 is therefore 177 + 26.86 = 203.86, which is close to the result obtained with Method 1, 203.90. The results for all years are shown in Figure 13.11.

Alternatively, we may abandon the original policy years and continue the frequency analysis using the shifted policy years, making sure we match the exposure to the shifted policy year rather than the original policy year (see Section 14.4.2 for an illustration of the role of exposure in frequency analysis).

Projection to Ultimate Using Shifted Policy Years

Policy Year (shifted)	Delay (years) 0	1	2	3	4	5	6	7	8	9	Ultimate
2004/05	24	42	45	47	48	48	48	48	48	48	48.00
2005/06	51	74	82	88	90	90	90	90	90		90.00
2006/07	72	107	119	125	125	125	125	125			125.00
2007/08	72	113	121	125	128	128	129				129.00
2008/09	105	152	166	170	174	174					174.51
2009/10	119	184	203	210	214						214.62
2010/11	169	244	257	268							273.65
2011/12	116	184	210								223.02
2012/13	95	208									241.78
2013/14	136										250.48
										Tail factor	
Link ratios	1.584	1.095	1.040	1.018	1.000	1.003	1.000	1.000	1.000	1.000	
Cumulative	1.842	1.162	1.062	1.021	1.003	1.003	1.000	1.000	1.000	1.000	

FIGURE 13.10
The cumulative triangle and the projection to ultimate for Method 4. 'Policy year (shifted)' is the 12-month period starting on 1 December of each year and ending on 30 November of the following year. For example, 2009/2010 is the period 1 December 2009 to 30 November 2010. Note that the link ratios have been calculated after removing the claim counts for year 2004/2005, since this year is incomplete: claims data are available from 1 April 2005, whereas the shifted policy year incepts on 1 December 2004.

(a) Shifted-Policy-Year Basis

Shifted PY	Reported	Ultimate	IBNR
2004/05	48	48.00	–
2005/06	90	90.00	–
2006/07	125	125.00	–
2007/08	129	129.00	–
2008/09	174	174.51	0.51
2009/10	214	214.62	0.62
2010/11	268	273.65	5.65
2011/12	210	223.02	13.02
2012/13	208	241.78	33.78
2013/14	136	250.48	114.48

(b) Original-Policy-Year Basis

Original PY	Reported	IBNR (approx.)	Ultimate-method (4)
2005	86.00	–	86.00
2006	92.00	–	92.00
2007	110.00	–	110.00
2008	145.00	0.34	145.34
2009	192.00	0.59	192.59
2010	253.00	3.97	256.97
2011	241.00	10.56	251.56
2012	177.00	26.86	203.86
2013	102.00	87.58	189.58

FIGURE 13.11
Projection to ultimate for Method 4 calculated on a shifted-policy-year basis (a) and realigned to the original-policy-year basis (b).

Projection to Ultimate Using Shifted Development Years
Delay (months)

Policy year	8	20	32	44	56	68	80	92	104	116	Ultimate
2005	54	76	82	85	86	86	86	86	86	86	**86.0**
2006	48	72	81	91	92	92	92	92	92		**92.0**
2007	68	99	107	110	110	110	110	110			**110.0**
2008	86	132	139	144	145	145	146				**146.0**
2009	126	165	182	191	193	194					**194.4**
2010	167	230	245	254	256						**257.0**
2011	165	226	242	250							**253.0**
2012	113	184	199								**210.2**
2013	116	201									**228.9**
2014	68										**113.8**
										Tail factor	
Link ratios	1.469	1.079	1.044	1.008	1.002	1.002	1.000	1.000	1.000	1.000	
Cumulative	1.673	1.139	1.056	1.012	1.004	1.002	1.000	1.000	1.000	1.000	

FIGURE 13.12
The cumulative triangle and the projection to ultimate for Method 5. The development years have been redefined to start at 8, 20, 32 ... months. The way to read this table is as follows: there were 107 losses that occurred in policy year 2007 and that were reported within 32 months.

13.2.5 Method 5: Shift the Development Period

An alternative to shifting the policy year is shifting the development year to start at $d, d+12, d+24$, etc. where d is the number of months (this can also be fractional) available for the latest policy year observed. Figure 13.12 shows the calculations for our case study.

Note that (unlike Method 4) no further adjustment for the shift in development period is needed after projection to ultimate.

		Projected Claim Counts				
Original PY	Reported	Method (1) - ignore last diagonal	Method (2) - gross up last diagonal	Method (3) - choose a shorter period	Method (4) - shift policy years	Method (5) - shift dev't years
2005	86	86.00	86.00	86.00	86.00	86.00
2006	92	92.00	92.00	92.00	92.00	92.00
2007	110	110.00	110.00	110.00	110.00	110.00
2008	146	145.00	146.50	146.32	145.34	146.00
2009	194	192.00	195.68	194.59	192.59	194.45
2010	256	256.70	259.63	257.88	256.97	257.00
2011	250	254.17	260.59	257.93	251.56	252.99
2012	199	203.90	224.31	215.13	203.86	210.16
2013	201	179.34	296.10	252.35	189.58	228.94

FIGURE 13.13
A comparison of the different projection methods for the years over the period 2005 to 2013. The divergence for the last policy years highlights the caution needed in whether to include the last year as part of the calculations at all, at least for the chain ladder.

13.2.6 Comparison of Methods 1 through 5

We can now compare the projection results for the different methods described in Sections 13.2.1 through 13.2.5. Figure 13.13 shows that the different methods yield similar results for the earlier years (as expected), and they start diverging by more than 5% from the 2012 policy year.

13.3 Triangle-Free Methodology for IBNR Estimation

The main issue of the approach outlined in Section 13.2 is that aggregating claim counts in a triangle throws away most of the information we have: by way of illustration, if we have 1000 losses over 10 years, we will compress the information on the reporting delay (which is made of 1000 different numbers) into a small triangle (which is made up of 55 different numbers). Bizarrely, increasing the number of losses (e.g. from 1000 to 10,000) does not increase the amount of information that we can use to project to the final number of claims, although it does increase the stability of the patterns. This information compression is helpful for visualisation purposes and made sense in an age when calculations were performed by hand; it is also resilient enough when the objective is finding an accurately projected number of claims for each year; however, it ties our hands when we are trying to calculate the distribution of possible outcomes, making the uncertainty difficult to assess.

In this section, we explain the workings of an alternative approach. The idea of using granular projection methods has been around for a while, especially among Scandinavian actuaries (Norberg, 1993; 1999). A more extensive bibliography can be found in Antonio and Plat (2012; 2014) and in Parodi (2012c). The specific implementation discussed here is that presented in Parodi (2012c).

The method works as follows:

i. Estimate the reporting delay distribution based on the observed reporting delays (Section 13.3.2)
ii. Use the reporting delay distribution to project the reported claim counts to ultimate (Section 13.3.3)

Some of the theoretical content in the rest of the chapter is a bit more technical than usual. Therefore, in order to help the reader, we will take the following steps:

- before looking in detail at the general case, we will look at a simple special case (Section 13.3.1). Readers who are happy with the assumptions made in this section and are not interested in knowing where the projection formulae come from will find enough information in this section to use the method in a plug-and-play fashion.
- the sections describing the theory around triangle-free projection (Sections 13.3.2–3) are prefixed by asterisks ('*'). This doesn't sound overly helpful but at least it helps signal that the material here is more technical and can be ignored by those who don't need to know the workings under the hood.
- I have included a more practical section, Section 13.3.6, which dispenses with the theoretical mystique and shows how the general method can be applied in practice, in a step-by-step manner. The calculations are also available via a spreadsheet in the Downloads section of the book's website (Parodi, 2016).

13.3.1 A Simple Special Case: Exponential Delays with Uniform Exposure

We will now look at the special case where we can assume that the delays follow an exponential distribution. If you find that the exponential assumption is not adequate – because, for example, the exponential distribution gives too low probability to the largest delays observed or expected, you could consider instead a power-law distribution for the delays, such as the GPD or the Type II Pareto (see Section 13.3.5): the formulae for the projection to ultimate are not more complex, although bias-correction is less straightforward.

We assume that:

i. The underlying distribution of the delays T is $f(t)=\frac{1}{\tau}\exp(-t/\tau)$, and does not change over time.
ii. Losses occur uniformly[1] over the period $[0,a]$

The observed delay distribution is in this case:

$$f_a(t)=\begin{cases} c\left(1-\frac{t}{a}\right)\exp(-t/\tau) & \text{for } t<a \\ 0 & \text{elsewhere} \end{cases} \tag{13.1}$$

where c is a constant that can be easily determined by imposing the condition that the integral of $f_a(t)$ over $[0,a]$ is 1. (Equation 13.1 is a special case of Equation 13.6, which is a simple application of Bayes' theorem.)

1 This assumption is a stronger than that of uniform exposure over the period; however, in most cases it can be considered equivalent.

In practice, we observe an empirical version of $f_a(t), \hat{f}_a(t)$, and we want to derive the true distribution $f(t)$. This can be done, at least in the range $[0,a)$, by inverting Equation 13.1. However, in the case where $f(t)$ is an exponential distribution with mean τ, there is a shortcut: we can estimate the expected observed delay and use it to derive the expected *true* delay. The expected observed delay is given by:

$$\tau_{\text{obs}}(a,\tau) = \int_0^a t f_a(t)\,dt = \tau\left(1 + \frac{e^{-\frac{a}{\tau}} - \frac{\tau}{a}\left(1-e^{-\frac{a}{\tau}}\right)}{1 - \frac{\tau}{a}\left(1-e^{-\frac{a}{\tau}}\right)}\right) \tag{13.2}$$

Although τ cannot be obtained from $\tau_{\text{obs}}(a,\tau)$, analytically, the relationship between τ and $\tau_{\text{obs}}(a,\tau)$ is a continuous and monotone increasing function that can be inverted numerically. We can therefore use this relationship to estimate τ from the empirical mean delay $\hat{\tau}_{obs}(a)$, as illustrated in Figure 13.14.

Having found an estimate for the mean true delay τ, we now have an estimate $\hat{f}(t)$ of the reporting delay distribution for all delays t, not only for the values in the observation window $0 \le t < a$.

It can be shown that under the assumptions of exponential delays and uniform rate density case, the number of claims projected to have occurred during the period $[0,a')$ can be estimated as:

$$\text{projected}_a(a') = \frac{a'}{\min(a',a) - \hat{\tau}\left(\exp\left(\frac{\min(a',a)-a}{\hat{\tau}}\right) - \exp\left(-\frac{a}{\hat{\tau}}\right)\right)} \times \text{reported}_a(a') \tag{13.3}$$

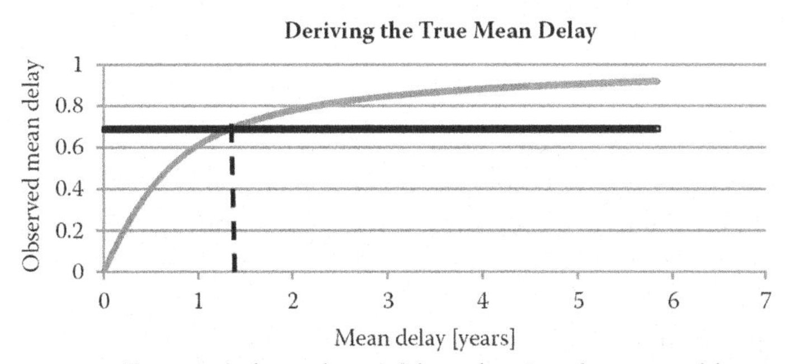

FIGURE 13.14
An illustration of how it is possible to estimate the true mean delay from the observed delay in the case of an exponential distribution. In this specific case, the true delay distribution is known and is an exponential with mean $\tau = 1.5$ years. Claims reporting is observed over a period of $a = 3$ years. The black line shows the function $\tau_{\text{obs}} = \tau_{\text{obs}}(a, \tau)$ with $a = 3$ as for Equation 13.2. The grey line is the observed value of τ_{obs}, $\tau_{\text{obs}} \approx 0.69$. The x value of the intersection between the grey line and the black line gives the estimated value of the true delay τ, $\hat{\tau} \approx 1.35$, to be compared with the true value $\tau = 1.5$. The difference is due to the parameter uncertainty: we have based our estimate on a limited sample of claims.

where:

- $\hat{\tau}$ is the mean delay as estimated from the data;
- $\text{reported}_a\,(a')$ is the number of losses occurred during the period $[0,a')$ *and* observed during $[0,a)$

Notice that a' can be larger or smaller than a. The case $a' > a$ occurs for example when we try to project the most recent year of observation, and this is incomplete – the incomplete-diagonal case that we have contended with at length in Section 13.2.

Note that Equation 13.3 is a special case of Equation 13.16, which is itself a special case of Equation 13.15.

The IBNR claim count is then simply given by the difference between the projected and the reported number of claims:

$$\text{IBNR}_a\,(a') = \text{projected}_a\,(a') - \text{reported}_a\,(a')$$

For example, assume we have a general liability risk with an average reporting delay of $\tau = 0.8$ years. We look at the reported claims as at 31 August 2012, and we want to project to ultimate the experience of the policy year started on 1 January 2011 and finished on 31 December 2011. The number of reported claims for the 2011 policy year is 10 claims. If we assume that the distribution of delays is exponential with an average of 0.8 years and we arbitrarily use the start date of the policy (1 January 2010) as $u = 0$, we can substitute the numbers $a' = 1\ y$, $a = 1.67\ y$, $\tau = 0.8\ y$, $\text{reported}_{167}(1) = 10$ in Equation 13.3, obtaining

$$\text{projected}_{1.67}\,(1) = \frac{1}{1 - 0.8\left(\exp\left(-\frac{0.67}{0.8}\right) - \exp\left(-\frac{1.67}{0.8}\right)\right)} 10 = 13.3$$

and

$$\text{IBNR}_{1.67}\,(1) = 13.3 - 10 = 3.3$$

In practice we often aim at projecting to ultimate each underwriting year separately. In this case, the projected/reported ratio for each year is given by:

$$\frac{\textit{projected}_a\left([k,k+1)\right)}{\textit{reported}_a\left([k,k+1)\right)} = \frac{1}{\min(1,a-k) - \hat{\tau}\left(\exp\left(\frac{\min(k+1-a,0)}{\hat{\tau}}\right) - \exp\left(\frac{k-a}{\hat{\tau}}\right)\right)} \tag{13.4}$$

where $k = 0,\ldots n$. This is a special case of Equation 13.18, which we will see in Section 13.3.3.1. Another interesting special case of Equation 13.18 is Equation 13.24 (Type II Pareto delay distribution).

Let's now take a step back and look at the problem of triangle-free projection in more general terms.

*13.3.2 Estimating the Reporting Delay Distribution (General Case)

The most natural way to go about estimating the reporting delay distribution seems to record the loss date and the reporting date for all the claims reported so far and plotting the empirical cumulative distribution function of these reporting delays: the idea is that this will be an approximation of the true underlying reporting delay distribution.

The problem with this method is that not all reporting delays are observed with the same probability: for example, if reporting delays are observed over a period of 3 years, the probability of observing a delay of more than 3 years is obviously zero. Also, it is very difficult to observe a delay of, for example, 2 years and 11 months, because this requires that the loss happens in the first month of the 3-year period, and that it is observed just before the observation period ends. As a result, *the observed distribution is biased towards small delays.*

This bias can be clearly observed in Figure 13.15, where the true reporting delay distribution for artificially generated data is compared with the empirical cumulative distribution function of the delays observed through a 3-year time frame.

In mathematical terms, if t is the delay between occurrence and reporting and $[0,a]$ is the observation window, we do not observe the true probability density function $f(t)$ of the delays but a sample from the distribution $f_a(t)$, which gives the probability that a delay of length t is observed between $t=0$ and $t=a$. In order to reconstruct the unbiased distribution $f(t)$, we need to be able to determine the relationship between $f_a(t)$ and $f(t)$.

To see how this can be done, let us first introduce some terminology. Let us denote T_0 as the time at which the loss occurs, and T as the random variable representing the delay between occurrence and reporting, so that $T+T_0$ represents the time at which the loss is reported.

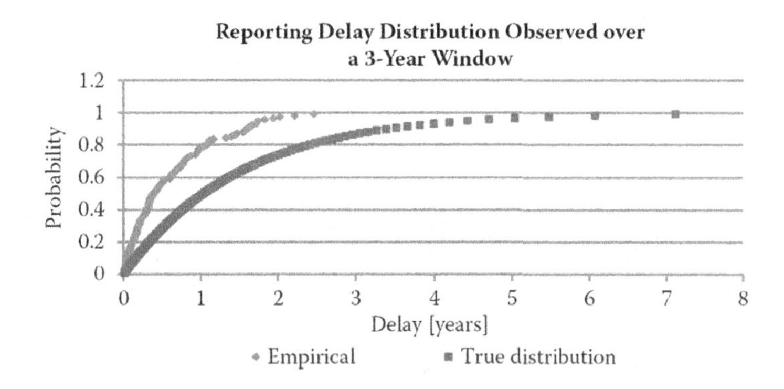

FIGURE 13.15
The result of an experiment with artificially generated delay data. 209 delays are sampled from an exponential distribution with mean delay of 1.5 years (squares). Loss dates are generated uniformly over a period of 3 years, and the losses for which the reporting date falls outside the 3-year time frame are discarded, leaving the 'observed' delays (rhombi). As expected, there are no observed delays of more than 3 years (actually, the largest observed delay is, in this case, around 2.4 years), and the probability of observing a delay of, for example, more than 1 year is only 22%, considerably lower than the actual probability of a delay larger than 1 year occurring $\left(e^{-\frac{1}{1.5}} \approx 51.3\%\right)$.

The function $f_a(t)$ is by definition the probability (density) that the delay T is equal to t given that the time $T+T_0$ at which the loss is reported falls within $[0,a]$. Using Bayes' theorem and assuming $0 \le t \le a$,

$$\begin{aligned} f_a(t) &= \Pr(T=t \mid T+T_0 \le a) = \\ &= \frac{\Pr(T+T_0 \le a \mid T=t)\Pr(T=t)}{\Pr(T+T_0 \le a)} = \\ &= \frac{\Pr(T_0 \le a-t) f(t)}{k(a)} \end{aligned}$$

Where we have noted that $\Pr(T+T_0 \le a \mid T=t)$ is the same as $\Pr(T_0 \le a-t)$, $\Pr(T=t)$ is by definition the probability density $f(t)$, and $\Pr(T+T_0 \le a)$ is simply a normalisation constant $k(a)$ that depends on a(but not on time) and that can be determined (if $f(t)$ is fully known!) by imposing that the integral of $f_a(t)$ between 0 and a be 1.

The probability $\Pr(T_0 \le a-t)$ can be expressed in terms of the **rate density** $\rho(t_0)$, which represents the prior probability density that a loss occurs at time $T_0 = t_0$ within the time window $[0,a]$. In general, such probability will not be uniform because the exposure and the risk profile changes in time.[2] We can therefore write

$$\Pr(T_0 \le a-t) = \int_0^{a-t} \rho(u)\,du \tag{13.5}$$

Where we can assume (without loss of generality) that $\rho(u)$ is normalised so that $\int_0^a \rho(u)\,du = 1$. The observed delay probability density can then be written as:

$$f_a(t) = \begin{cases} \dfrac{\left(\int_0^{a-t} \rho(u)\,du\right) \times f(t)}{k(a)} & 0 \le t \le a \\ 0 & t > a \end{cases} \tag{13.6}$$

Where we have extended the distribution to be zero above a.

The unbiased delay distribution $f(t)$ can be obtained by inverting Equation 13.6:

$$f(t) \propto \begin{cases} \dfrac{f_a(t)}{\int_0^{a-t} \rho(u)\,du} & 0 \le t \le a \\ \text{undefined} & t > a \end{cases} \tag{13.7}$$

Note that $f(t)$ is defined up to an unknown multiplicative constant and it is undefined above a, because any probability density $f^*(t)$ that:

2 Furthermore, we are assuming that any dependency between the timing of different (past) losses can be ignored, which is equivalent to assuming that the underlying process is (in generally non-stationary) Poisson. As will be discussed in Chapter 14, this captures most processes in practice, as most excess volatility can be modelled in terms of systemic shocks to the Poisson rate.

- is equal to $\alpha f(t)$ between 0 and a
- is equal to an arbitrary $g(t)$ above a
- has integral over the whole domain equal to 1, to have a well-defined probability

is going to produce the same 'windowed' distribution $f_a(t)$.

For practical purposes, what we need to project the number of losses to ultimate is an estimate of $F(t)$, the cumulative distribution function of $f(t)$, for all values up to $t = a$ (included). This, too, is only known between 0 and a up to an unknown multiplicative factor. However, this factor can be estimated if we have an estimate of the tail factor φ_{tail}, i.e. the factor by which we have to multiply the number of losses occurred by time $t = a$ to project to the ultimate number of losses. The tail factor is indeed given by:

$$\varphi_{\text{tail}} = \frac{1}{F(a)} \tag{13.8}$$

Therefore if we have an estimate of the tail factor we can write the **cumulative delay distribution function** $\boldsymbol{F(t)}$ as:

$$F(t) = \frac{1}{\varphi_{\text{tail}}} \frac{\displaystyle\int_0^t \frac{f_a(z)}{\int_0^{a-z} \rho(u)du} dz}{\displaystyle\int_0^a \frac{f_a(z)}{\int_0^{a-z} \rho(u)du} dz} \quad (0 \le t \le a) \tag{13.9}$$

A rough estimate of φ_{tail} can be obtained for example by assuming that the delays come from an exponential distribution with mean delay τ, in which case

$$\varphi_{\text{tail}} \sim \frac{1}{1 - \exp(a/\tau)} \tag{13.10}$$

Note that using Equation 13.10 as a first estimate of φ_{tail} does not commit us to use an exponential approximation for the *whole* delay distribution, but simply allows us to add a tail to the observed data. The approximation is good as long as the tail of the delay distribution is not too heavy. In Section 13.3.4 we will see how we can estimate the value of τ from the average of the observed delays under the exponential assumption.

Note that the integral $\int_0^{a-t} \rho(u)du$ (which appears in most of the preceding equations) can normally be approximated assuming that the rate density $\rho(u)$ is piecewise-constant over time (Section 13.3.1.1) or even uniformly constant over time (Section 13.3.1.2).

13.3.2.1 Special Case: Piecewise-Constant Rate Density

A special useful case is the one where $\rho(u)$ can be assumed to be proportional to the exposure and the exposure (on-levelled where necessary) can be considered constant during a given policy year (a common assumption in pricing). Under these assumptions, it is easy to see that the right-hand side of Equation 13.1 can be approximated as:

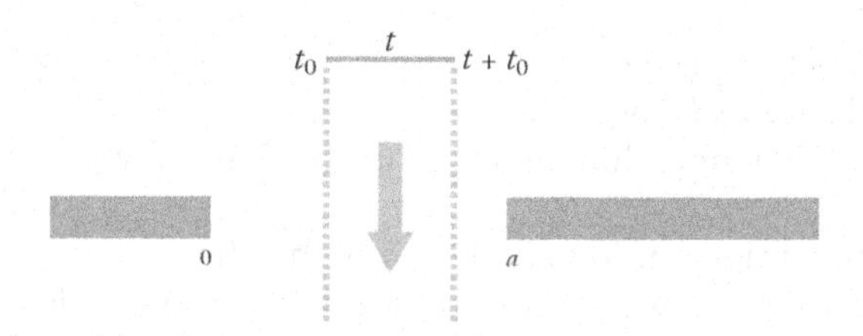

FIGURE 13.16
A geometric illustration of the probability of observing a delay of length $T = t$ in the case of uniform rate density. Imagine you have a machine that generates needles of random length T, with T following a distribution $f(t)$, and drops them so that they remain horizontal. The needle starts at a random point T_0, with an equal probability of starting at any point $T_0 = t_0 \in [0,a)$ inside that interval. The probability of the needle falling inside the slot without touching the borders $[0,a)$, conditional on the delay being $T = t$, is $1 - t/a$ (and 0 if $t \geq a$), and the probability density that a needle of random length $T = t$ falling into the slot is $f_a(t)$. That corresponds to the intuition that if the needle is very short it is likely to fall through as long as its left endpoint is within the slot, whereas if the needle has length close to the size of the slot, a, it will only be able to fall through if the left endpoint is very near the point $T=0$. (Reprinted from Parodi, P., *British Actuarial Journal*, 19, 1, 168–218, 2014. With permission.)

$$\int_0^{a-t} \rho(u)\,du = \frac{\sum_{i=0}^{\lfloor a-t \rfloor -1} \varepsilon_i + \varepsilon_{\lfloor a-t \rfloor} \times \left(a - t - \lfloor a-t \rfloor\right)}{\sum_{i=0}^{\lfloor a \rfloor -1} \varepsilon_i + \varepsilon_{\lfloor a \rfloor} \times \left(a - \lfloor a \rfloor\right)} \tag{13.11}$$

where ε_i is the on-levelled exposure in year i and x is the integer part of x.

13.3.2.2 Special Case: Uniform Rate Density

When the on-levelled exposure doesn't change or changes only moderately over the years we can further approximate Equation 13.1 as:

$$\int_0^{a-t} \rho(u)\,du = \frac{a-t}{a} = 1 - \frac{t}{a} \tag{13.12}$$

Equation 13.12 has a very simple geometric interpretation, which is illustrated in Figure 13.16.

*13.3.3 Projecting Claim Counts to Ultimate (General Case)

To project the reported number of claims to ultimate, we can now use an estimate $\hat{F}(t)$ of the *cumulative delay distribution* $F(t) = \Pr(T \leq t)$, that is, the probability that the delay is less than or equal to t.

First of all, note that if the cumulative delay distribution $F(t)$ and the times $t_1, t_2 \ldots t_N$ at which the claims are reported, we can write

$$E\left(\text{reported}_a\,(a')\right) = \int_0^{\min(a,a')} \sum_{i=1}^{N} \delta(u - t_i) F(a-u)\,du = \sum_{i=1}^{N} F(a - t_i) \tag{13.13}$$

Where the '$\delta(u - t_i)$' in this equation is Dirac's delta (a point mass of probability 1 at $u = t_i$– see the footnote in Section 17.1 for more details). Intuitively, Equation 13.13 can be explained by noting that if a loss occurs at time $u = t_i$, then the likelihood that it is reported by time $u = a$ is $F(a - t_i)$.

If the occurrence times are not known but the total number N of claims occurred during $[0,a')$ are known, the number of losses occurring and expected to be reported during $[0,a)$ is instead given by:

$$E\left(\text{reported}_a\left(a'\right)\right) = N \int_0^{\min(a,a')} \rho(u)F(a-u)du \tag{13.14}$$

Intuitively, Equation 13.14 can be explained by noting that when the timing of occurrence of the losses is unknown, we can assume that on average there will be $N\rho(u)du$ losses within the infinitesimal interval $[u,du)$ and that the expected number of losses reported by $u=a$ will be $N\rho(u)du \times F(a-u)$.

In practice, however, the number of losses occurred during $[0,a')$ is not known, and what we have is the number of losses occurred during that period *and* observed during $[0,a)$, $\text{reported}_a\,(a')$. We can then estimate the number of losses actually occurring during $[0,a')$ by sort-of-inverting Equation 13.10 and replacing the true delay distribution F with the *estimated* distribution $\hat{F}$, yielding:

$$\text{projected}_a\left(a'\right) = \frac{\int_0^{a'} \rho(u)du}{\int_0^{\min(a,a')} \rho(u)\hat{F}(a-u)du} \times \text{reported}_a\left(a'\right) \tag{13.15}$$

Note that the limits of integration for the numerator are different: this covers the case in which the period for which we are trying to estimate the ultimate number of claims extends beyond the observation window, as is the case for example when (as it usually happens) the latest reported year is incomplete.

Also note that Equation 13.15 is indeed only an estimate and that there are many different methods to derive the number of losses based on the reported number of claims.[3]

In the case where the rate density can be considered uniform Equation 13.15 becomes:

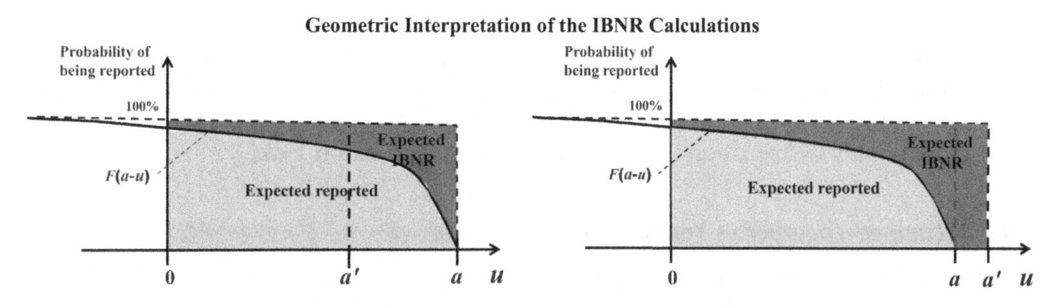

FIGURE 13.17
The denominator of the fraction in Equation 13.16 can be pictured as the light grey area below the CDF ($F(a-u)$) and between the y-axis and the axis $u=a'$ (expected reported) and the numerator as the rectangle with sides a' and 1 (Expected IBNR + Expected reported). The factor to ultimate for the whole period $[0,a']$ is therefore the ratio of the two areas. This covers both the case where $a'<a$ (left) and the case where $a'>a$. (Modified figure from Parodi, P., *British Actuarial Journal*, 19, 1, 168–218, 2014. With permission.)

3 See, for example, Antonio and Plat (2012), in which a similar equation (Equation 14 in their article) is derived through maximum likelihood estimation for the case when the underlying frequency distribution is Poisson, yielding the Poisson rate.

$$\text{projected}_a(a') = \frac{a'}{\int_0^{\min(a,a')} \hat{F}(a-u)du} \times \text{reported}_a(a') \tag{13.16}$$

A geometric interpretation of Equation 13.16 is shown in Figure 13.17.

$$\textbf{Projected} = \frac{\text{Expected Reported+Expected IBNR}}{\text{Expected Reported}} \times \text{Reported}$$

Although Equation 13.15 is general enough that it can be used to project to ultimate the reported number of claims for all cases, it is helpful to have a formula that applies individually to different policy years. Therefore, assume that we start counting policy years at $t=0$ (so the first policy year is $[0,1)$, the second is $[1,2)$ and so on until the latest (generally incomplete) year $[n,a)$, where a is by construction such that $a \le n+1$. We can project the years individually using the following formula for policy year $[k,k+1)$, which holds for k from 0 to n.

$$\text{projected}_a([k,k+1)) = \frac{\int_{k-1}^{k} \rho(u)du}{\int_k^{\min(a,k+1)} \rho(u)\hat{F}(a-u)du} \times \text{reported}_a([k,k+1)) \tag{13.17}$$

which reduces for a uniform rate density to:

$$\text{projected}_a([k,k+1)) = \frac{1}{\int_k^{\min(a,k+1)} \hat{F}(a-u)du} \times \text{reported}_a([k,k+1)) \tag{13.18}$$

Alternatively, we can calculate the projected ultimate for the periods $[0,k)$ and $[0,k+1)$ separately using Equation 13.15 and subtract the two:

$$\text{projected}_a([k,k+1)) = \text{projected}_a([0,k+1) - \text{projected}_a([0,k)) \tag{13.19}$$

Equations 13.18 and 13.19 will in general lead to slightly different results.

In practice, the integrals in the equations above will be replaced by finite sums, as the number of reported claims will be known for discrete values of t (typically, in days) and the estimated delay distribution will also be discrete (see Section 13.3.5).

13.3.3.1 Lumped-Together vs Year-by-Year Projection

One difference between triangle-based and triangle-free methods is that triangle-free methods do not require projecting the number of IBNR claims for each year separately: for example, Figure 13.18(a) shows an example with a single figure for the ultimate number of claims over a period of 10 years. Hence, the typical output of the IBNR analysis will normally look similar to Figure 13.18(b). The advantages and disadvantages of the individual-year presentation of data are explained in the legend of Figure 13.18.

*13.3.4 Uncertainty

The standard error of the projected number of claims can be easily calculated if we can make a reasonable assumption of what the underlying claim count process is: for example, Poisson or negative binomial.

If the underlying claim count process is Poisson with parameter $\lambda(a')$, the number of IBNR claims in the time window $[0,a')$ also follows a Poisson distribution with parameter $\lambda(a') - \text{reported}_a(a')$. Because the variance of a Poisson process is equal to its mean, the variance of the process is also $\lambda(a') - \text{reported}_a(a')$, which can be estimated from the data as $\text{IBNR}_a(a') = \text{projected}_a(a') - \text{reported}_a(a')$.

The square root of this can be used as an estimate of the standard error by which we know the ultimate claim count for the period $[0,a')$:

$$\text{S.E.}\left(\text{IBNR}_a\left(a'\right)\right) \approx \sqrt{\text{projected}_a\left(a'\right) - \text{reported}_a\left(a'\right)} \tag{13.20}$$

(Note that we are not speaking here of the standard error on the estimated Poisson rate, which would be much smaller, but of the standard error of the projected claim count. This is why we do not need to divide by the number of observations.)

In case of an overdispersed Poisson process, such as a negative binomial process, this method can be easily modified by setting the standard error equal approximately to

$$\text{S.E.}\left(\text{IBNR}_a\left(a'\right)\right) \approx \sqrt{(1+r)\left(\text{projected}_a\left(a'\right) - \text{reported}_a\left(a'\right)\right)} \tag{13.21}$$

where $1 + r$ is the variance-to-mean ratio of the process.

An example of the use of these formulae is given in the next chapter, in Section 14.5.

*13.3.5 A Fat-Tailed Model for the Delay Distribution

In Section 13.3.1 we looked at the case where the delay distribution could be modelled as an exponential and the rate density was uniform. This is because the exponential distribution is a simple model with a straightforward mechanism for bias correction and which works well when very long delays are rare.

In this section we hold on to the uniform rate density assumption and we consider another distribution (Type II Pareto) that can be used to model delay distributions, and which allows for a wider range of delays. It is (as the exponential) a 'threshold' distribution, i.e. a distribution such that the PDF is strictly decreasing from $t = 0$. This is normally a good assumption for delay distributions – however, when that is not the case one should consider distributions with a mode at $t \neq 0$, such as the lognormal distribution.

The Type II Pareto is the distribution with cumulative distribution function $F(t) = 1 - \left(\frac{\theta}{t+\theta}\right)^{\alpha}$. and is, like the exponential distribution, a special case of the Generalised Pareto Distribution (see Question 3) with a power-law (rather than exponential) tail. It is slightly more complex than the more widely used single-parameter Pareto ($F(t) = 1 - \left(\frac{\theta}{t}\right)^{\alpha}$) as θ is here a genuine parameter and not a pre-defined threshold but has the advantage of being defined from $t = 0$.

The tail factor can be calculated as:

$$\varphi_{\text{tail}} = \frac{1}{F(a)} = \frac{1}{1 - \left(\frac{\theta}{a+\theta}\right)^{\alpha}} \tag{13.22}$$

The projected/reported ratio for a generic interval $[0,a')$ (using Equation 13.16) becomes:

$$\frac{\text{projected}_a(a')}{\text{reported}_a(a')} = \frac{a'}{\min(a',a) + \frac{\theta}{1-\alpha}\left(\left(1+\frac{\max(0,a'-a)}{\theta}\right)^{1-\alpha} - \left(1+\frac{a}{\theta}\right)^{1-\alpha}\right)} \quad (13.23)$$

The projected/reported ratio for individual years (using Equation 13.18) can be written as:

$$\frac{\text{projected}_a([k,k+1))}{\text{reported}_a([k,k+1))} = \frac{1}{\min(1,a-k) + \frac{\theta}{1-\alpha}\left(\left(1+\frac{a-k}{\theta}\right)^{1-\alpha} - \left(1+\frac{a-\min(k+1,a)}{\theta}\right)^{1-\alpha}\right)} \quad (13.24)$$

Determining the parameters of the Type II Pareto from observed delays

The calibration of the Type II Pareto from the observed delays is less straightforward than for the exponential case because there are two parameters (α and θ) instead of one. However it can be done by the method of moments by calculating the theoretical mean $m = f(\alpha,\theta)$ and standard deviation $s = g(\alpha,\theta)$ of the observed delays, and solving numerically for the values of α and θ that produce the *observed* mean and standard deviation $\hat{m}$ and $\hat{s}$. The use of judgment to ensure that the probability of very long delay makes sense is recommended, even if that requires adding some constraints on the numerical solutions.

13.3.6 Calculations in Practice – An Example

Let us now take a look at how these calculations are carried out in practice, by using our case study. This includes loss data and exposures for the period 1/4/2005 to 30/11/2014.

The first thing to note is that all the integrals in the equations above for the general case need to be discretised. The simplest option is arguably to consider time intervals of one day, as both occurrence dates and reporting date will typically (but not always) be provided in integer date format, so the delays will be multiple of one day.

Let us look at the methodology for the practical calculations.

13.3.6.1 Discretisation of Dates and Delays

We can identify a date d with an integer, counted from an arbitrary calendar day, e.g. the first day of the observation window. So for our case study $d = 0$ would correspond to 1/4/2005.

We will set $d = n_a$ as the *last* day of the observation window, and we will ignore the subtlety that $t = a$ might actually include fractions of the day, as the effect will be minimal. For our case study, $n_a = 3530$ days.

The first day of policy year $[k,k+1)$ will be denoted as n_k, and the last day as $n_{k+1} - 1$. So for example $n_0 = 0$ will denote 1/4/2005, $n_1 = 365$ will denote 1/4/2006, and so on (taking into account the leap years).

We will also discretise delays in the same way, with a delay of $d = k$ meaning that a loss is reported k days after it occurs.

13.3.6.2 Discretisation and Selection of the Rate Density

We will denote as ρ_d the discretised version of the rate density $\rho(u)$, representing the prior probability that a given loss happens on day d. Typically, ρ_d will be set proportional to the exposure for the relevant year, so for example for our case study we would set

$$\rho_0 = \rho_1 = \ldots = \rho_{364} = c \times \text{Exposure}(PY = 2005) = c \times £460.6\text{M}$$

$$\rho_{365} = \rho_1 = \ldots = \rho_{729} = c \times \text{Exposure}(PY = 2006) = c \times £543.7\text{M}$$

$$\ldots$$

Or in compact form:

$$\rho_d = c \times \text{Exposure}(PY = y(d))$$

where $y(d)$ is the policy year containing day d. The constant c could be determined by setting $\sum_{d=0}^{n_a} \rho_d = 1$ so that it represents a probability – however, it is not necessary as ρ_d always appears at both the numerator and denominator of every relevant calculation.

For types of business where systemic effects are important, it is possible to incorporate that information in the definition of the rate density. One way is setting $\rho_d = c \times \text{Exposure}(PY = y(d)) \times \text{SF}(PY = y(d))$, where $\text{SF}(PY = y(d))$ captures the systemic factor: e.g. we could put $\text{SF}(PY = y(d)) = 1$ for 'normal years' and $\text{SF}(PY = y(d)) = m$ with $m > 1$ for 'crisis years'.

13.3.6.3 Estimation of the Observable Delay Distribution

Let $f_{a,d}$ be the discretised version of $f_a(t)$, i.e. the probability that a loss observed within $[0, a)$ has a delay of d days, with d ranging from 0 (loss reported on the day of occurrence) to a (loss occurred at day 0 and reported at time a, measured in days). This can be estimated from the data as:

$$\hat{f}_{a,d} = \frac{\{\#\ of\ losses\ with\ delay = d\ days\}}{\{total\ \#\ of\ losses\} + 1} \tag{13.25}$$

13.3.6.4 Estimation of the Unbiased Delay Distribution

Let f_d be the discretised bias-free distribution $f(t)$, i.e. the probability that an arbitrary loss has a delay of d days. This can be estimated (based on Equation 13.7) as

$$\hat{f}_d = c \times \frac{\hat{f}_{a,d}}{\sum_{i=0}^{n_a - d} \rho_i} \tag{13.26}$$

where c is a constant that can be estimated from the tail factor (see Section 13.3.5.5), and n_a is the last day of the observation window.

13.3.6.5 Estimation of the Tail Factor

An estimate $\hat{\varphi}_{\text{tail}}$ of the tail factor can be obtained through an estimate of $1/F(a)$, e.g. by fitting a distribution such an exponential or a Type II Pareto to the empirical unbiased data and using that fitted distribution to calculate $F(a)$ (without necessarily trusting it across all delays).

In the case of the exponential distribution this can be done simply by calculating the average delay for the observed (biased) distribution and then estimating the average delay of the unbiased distribution by inverting numerically Equation 13.2.

Using the exponential approximation for our case study we find that the average observed delay is

$$\hat{\tau}_{\text{obs}} = 206.3 \text{ days}$$

which corresponds (inverting Equation 13.2) to an estimated bias-free delay of

$$\hat{\tau} = 221.1 \text{ days}$$

which in turn corresponds to a tail factor of

$$\hat{\varphi}_{\text{tail}} = \frac{1}{1-\exp\left(-a/\hat{\tau}\right)} = 1.00000012$$

which is in practice indistinguishable from 1 and we will set to 1 to avoid spurious accuracy. (This result may suggest that the exponential model assumption, although fully consistent with the data, may be too optimistic, and signals that we may have to allow for a fatter tail.)

13.3.6.6 Estimation of the Cumulative Distribution Function

The discretised cumulative distribution function F_d (the probability that the delay is below or equal to d days) can then be estimated (as per Equation 13.9) as:

$$\hat{F}_d = \frac{1}{\hat{\varphi}_{\text{tail}}} \times \frac{\sum_{j=0}^{d} \frac{\hat{f}_{a,j}}{\sum_{i=0}^{n_a-j} \rho_i}}{\sum_{j=0}^{n_a} \frac{\hat{f}_{a,j}}{\sum_{i=0}^{n_a-j} \rho_i}} \tag{13.27}$$

where $\hat{\varphi}_{\text{tail}}$ is an estimate of the tail factor φ_{tail} (see Section 13.3.5.5).

13.3.6.7 Estimating the Projected Ultimate for All Years

We can now use $\hat{F}_d$ to project the ultimate number of claims for the whole period or on an individual year basis. If n_k is the first day of policy year k the projected number of losses for year $[k, k+1)$ is:

$$\text{projected}_a\left([k,k+1)\right) = \frac{\sum_{i=n_k}^{n_{k+1}-1} \rho_i}{\sum_{i=n_k}^{\min(n_a, n_{k+1}-1)} \rho_i \hat{F}_{n_a-i}} \times \text{reported}_a\left([k,k+1)\right) \tag{13.28}$$

(a)

Start of period	End of period	Exposure (on-levelled)	Reported	#days from start to as-at date	#days from end to as-at date	% complete	Factor to ultimate	Projected ultimate
01/04/2005	31/03/2015	8,700.9	1,602	3530	0	90.5%	1.1056	1771.1

(b)

Start of period	End of period	Exposure (on-levelled)	Reported	#days from start to as-at date	#days from end to as-at date	% complete	Factor to ultimate	Projected ultimate
01/04/2005	31/03/2006	460.6	86	3530	3166	100.0%	1.000	86.0
01/04/2006	31/03/2007	543.7	92	3165	2801	100.0%	1.000	92.0
01/04/2007	31/03/2008	616.2	110	2800	2435	100.0%	1.000	110.0
01/04/2008	31/03/2009	677.2	146	2434	2070	99.9%	1.001	146.1
01/04/2009	31/03/2010	855.2	194	2069	1705	99.9%	1.001	194.3
01/04/2010	31/03/2011	941.6	256	1704	1340	99.6%	1.004	257.1
01/04/2011	31/03/2012	979.2	250	1339	974	97.8%	1.023	255.6
01/04/2012	31/03/2013	1,189.6	199	973	609	94.0%	1.064	211.7
01/04/2013	31/03/2014	1,226.8	201	608	244	81.8%	1.222	245.7
01/04/2014	31/03/015	1,210.9	68	243	0	31.5%	3.175	215.9

FIGURE 13.18
An example of calculation for the projected ultimate, both when all years are lumped together (a) and when the calculations are carried out separately for each individual accident year (b). The former method should lead to more accurate overall results in terms of total projected number because the projection of isolated years is affected by significant errors for the more recent years. However, the latter method is useful to identify the IBNR in the claim count for each individual year, and this provides (among other things) some evidence on trends and year-on-year volatility. Note that the total number of projected losses for (b) – in this case, 1814.4 – does not in general match the overall projected number for (a) – in this case, 1771.1. Note that the numbers in grey are the inputs, while the numbers in black are calculated.

The results for the projection across all years are shown in Figure 13.18(b).

The calculations can also be found in detail in the spreadsheet in the Download section of the book's website (Parodi, 2016).

13.3.6.8 Using Exponential Delay with Uniform Rate Density Approximation

Using an exponential distribution approximation with uniform rate density, the calculations simplify drastically. All we really need to do is to (a) calculate the average observed delay; derive the value of the estimated mean delay, $\hat{\tau}$ using Equation 13.2; (c) apply Equation 13.4 (which is repeated below for the reader's convenience) to all years.

$$\frac{\text{projected}_a([k,k+1))}{\text{reported}_a([k,k+1))} = \frac{1}{\min(1,a-k)-\hat{\tau}\left(\exp\left(\frac{\min(k+1-a,0)}{\hat{\tau}}\right)-\exp\left(\frac{k-a}{\hat{\tau}}\right)\right)}$$

Start of period	End of period	Reported	#days from start to as-at date	#days from end to as-at date	% complete	Factor to ultimate	Projected ultimate
01/04/2005	31/03/2006	86	3530	3166	100.0%	1.0000	86.00
01/04/2006	31/03/2007	92	3165	2801	100.0%	1.0000	92.00
01/04/2007	31/03/2008	110	2800	2435	100.0%	1.0000	110.00
01/04/2008	31/03/2009	146	2434	2070	100.0%	1.0000	146.01
01/04/2009	31/03/2010	194	2069	1705	100.0%	1.0002	194.04
01/04/2010	31/03/2011	256	1704	1340	99.9%	1.0011	256.29
01/04/2011	31/03/2012	250	1339	974	99.4%	1.0060	251.50
01/04/2012	31/03/2013	199	973	609	96.9%	1.0321	205.40
01/04/2013	31/03/2014	201	608	244	83.8%	1.1937	239.93
01/04/2014	31/03/2015	68	243	0	26.5%	3.7803	257.06

FIGURE 13.19
Calculations of the projected ultimate by year using the exponential distribution approximation.

Note that no discretisation is needed in this case so all the variables in the formula (including the constant '1'!) are expressed in years.

For our case study, $\hat{\tau} = 0.61$ years, $a = 9.66$ years, and k goes from 0 (policy year incepting 1/4/2005) to 9 (policy year incepting 1/4/2014). The results are shown in Figure 13.19.

13.3.7 Analysis of Zero Claims

The calculations above can be used to derive the ultimate number of claims over a period based on either the number of all claims or the number of non-zero claims, that is, claims that have a zero amount associated with them.

There are several reasons why a claim may be recorded with a zero amount attached to it: the most common reason is that an estimated reserve was initially attached to the claim on a conjectural basis, but the claim has subsequently been dropped and has 'settled as zero'. Another reason is that the claim may be recorded on the system but a figure is not attached to it until there is some certainty about what the value of the claim will be.[4]

13.3.7.1 Approach Based on the Number of All Claims

In this approach, the reported number of all claims is projected as explained in Sections 13.3.1 and 13.3.2 to the ultimate number of claims for each year. When modelling the severity of claims, we need to consider the fact that a proportion of these claims will actually be zero. This proportion can be estimated based on past statistics. The past statistics will have to look both at the closed zero claims that were previously non-zero and to the open zero claims that may eventually become non-zero.

For example, if the expected number of claims (including zero claims) is 50 per year, but past statistics show that only 70% of these claims are eventually settled as non-zero, we can reduce the estimated frequency by 30%, yielding 35 claims per year, and use this for the Monte Carlo simulation.

4 This is often the case in certain lines of business such as professional indemnity or D&O insurance in which there is a risk that the claimant's lawyers 'discover' the amount reserved and use it as a lower bound for their compensation request. We will look at this in more detail in Chapter 24 when discussing Professional Indemnity insurance.

Correction for Claims Dropping to Zero

Policy year	Number of losses	Number of non-zero losses	Proportion of non-zero losses	No. of non-zero losses with paid > 0	Assumed percentage of non-zero claims	Revised number of non-zero losses
2006	713	483	67.7%	482	67.7%	483
2007	626	451	72.0%	450	72.0%	451
2008	548	400	73.0%	397	73.0%	400
2009	542	399	73.6%	380	73.6%	399
2010	597	439	73.5%	382	73.5%	439
2011	548	458	83.6%	252	71.8%	393.3
2012	184	180	97.8%	46	71.8%	132.1
Total 2006–10	3026	2172	71.8%			

FIGURE 13.20
The percentage of non-zero losses is roughly constant over 2006 to 2010 (weighted average, 71.8%) and then increases in 2011 to 2012 – a sign that losses are initially reserved as non-zero but some then drop to zero. As a first approximation, it may be assumed that only approximately 71.8% of the losses reported in 2011 and 2012 will eventually remain non-zero, respectively, 393.3 and 132.1 losses. These figures need to be compared with the number of non-zero losses for which a payment has already been made and are therefore unlikely to drop to zero: in this particular case, the number is higher so we do not need to make corrections.

13.3.7.2 Approach Based on the Number of Non-Zero Claims

In this approach, the reported number of non-zero claims is projected as explained in Sections 13.3.1 and 13.3.2 to the ultimate number of non-zero claims for each year. In this case, we do not have the issue of claims being zero that could become non-zero (because all zero claims have already been excluded), but we still have the issue of non-zero claims that may eventually drop to zero.

One way by which we can address this is again to look at past statistics of how often a claim initially reserved as non-zero is subsequently dropped. Very often, however, we do not have the full history of each claim. A quick workaround is to look at the percentage of zero claims in any particular year. If, for example, we have on average, 30% of zero claims in far-back years and 5% in recent years, this suggests that many of the claims in recent years will eventually drop to zero. We can then adjust the number of claims in recent years to take this into account before projecting the claim count to ultimate.

Note that the number of claims that will eventually drop to zero is normally limited – as a first, good approximation – to those for which no payments have yet been made. All this is illustrated with a numerical example in Figure 13.20.

13.3.8 Comparison between the Triangle-Free Approach and the Triangle-Based Approach to IBNR

Advantages of the triangle-free method

- May be more accurate, especially when the number of claims is low.
- Uses the full delay information, without compressing it into a triangle.
- There is no need for corrections for last diagonal when last diagonal is incomplete.
- Modelling the tail factor is easier.

Disadvantages of the triangle-free method

- It is more complex in the general case, with some non-intuitive maths involved.
- A mechanism for taking into account the number of claims that will eventually drop to zero must be included in the process (this is also true, however, for the triangle-based methodologies if the development triangles are based on a single snapshot).

13.4 Questions

1. You are given the following claim count triangulation, providing the number of claims occurred in each policy year and are reported within a certain number of development years.

Development Year

		0	1	2	3	4	5	6	7	8	9
Policy Year	2004	4	9	9	10	10	10	10	10	10	10
	2005	4	8	9	9	9	9	9	9	9	
	2006	4	6	7	8	9	9	9	9		
	2007	6	14	18	18	18	18	18			
	2008	4	9	10	11	11	11				
	2009	5	8	10	11	11					
	2010	1	3	6	7						
	2011	5	9	11							
	2012	1	6								
	2013	1									

Using the chain ladder method, and assuming 12 months of development are available for 2013:

a. Estimate the number of IBNR claims for each policy year.

b. Without doing any further calculation, comment on the uncertainty of the estimated IBNR for each policy year.

c. Show how your answer would change if only 9 months of development were available for 2013.

State explicitly any other assumption you need to make.

2. The following table records all claims that occurred from 1 April 2010 and were reported by 31 October 2013.

Loss ID	Occurred	Reported	Delay (days)	Delay (years)
R01	29/7/2011	21/12/2012	511	1.399
R02	14/2/2013	10/3/2013	24	0.066
R03	17/7/2011	22/1/2012	189	0.517

Loss ID	Occurred	Reported	Delay (days)	Delay (years)
R04	8/6/2011	5/2/2013	608	1.665
R05	3/8/2011	15/3/2012	225	0.616
R06	13/9/2011	10/10/2011	27	0.074
R07	21/7/2011	10/10/2013	812	2.223
R08	3/11/2011	8/6/2013	583	1.596
R09	15/9/2010	27/7/2011	315	0.862
R10	10/9/2010	17/6/2011	280	0.767
R11	7/11/2011	12/4/2012	157	0.430
R12	10/1/2011	19/3/2011	68	0.186
R13	2/8/2011	19/1/2013	536	1.467
R14	28/7/2010	27/4/2011	273	0.747
R15	11/6/2010	12/7/2011	396	1.084
R16	7/3/2013	20/7/2013	135	0.370
R17	29/11/2010	10/3/2011	101	0.277
R18	13/7/2012	18/9/2013	432	1.183
R19	17/6/2010	17/8/2013	1157	3.168
R20	18/4/2010	16/7/2010	89	0.244
		Average	337	0.923

Assuming that the policy year is from the 1 April to 31 March and that the delays between occurrence and reporting follows an exponential distribution:

a. Estimate the mean delay between occurrence date and reporting date.
b. Estimate the number of IBNR claims for the period 1 April 2010 to 31 October 2013 using the triangle-free method.
c. Build a development triangle based on the loss data with an appropriate observation period and estimate the number of IBNR claims for the period 1 April 2010 to 31 October 2013 using the triangle development method (you can choose the calendar year for 'bucketing').
d. Comment on the relative advantages of the triangle-based and triangle-free IBNR calculation methods in general and as they manifested themselves in the example above.

State explicitly all assumptions that you used throughout the exercise.

3. (*) Show that when the tail distribution is a Generalised Pareto distribution (see Chapter 16) with null threshold: $F(x)=1-\left(1+\xi\frac{x}{\sigma}\right)^{-\frac{1}{\xi}}$,

   a. the tail factor and the projected/ultimate ratio are as follows:

$$\varphi_{\text{tail}}=\frac{1}{F(a)}=\frac{1}{1-\left(1+\frac{\xi a}{\sigma}\right)_{+}^{-\frac{1}{\xi}}}$$

$$\frac{\text{projected}_a(a')}{\text{reported}_a(a')} = \frac{a'}{\min(a',a) - \frac{\sigma}{1-\xi}\left(\left(1+\xi\frac{\max(0,a'-a)}{\sigma}\right)_+^{-\frac{1}{\xi}+1} - \left(1+\frac{\xi a}{\sigma}\right)_+^{-\frac{1}{\xi}+1}\right)}$$

(b) the formulae for the exponential and Type II Pareto can be obtained from the formulae above by setting $\xi \to 0$ (exponential) and $\xi = 1/\alpha$, $\sigma = \theta/\alpha$ (Type II Pareto).

14

Frequency Modelling: Selecting and Calibrating a Frequency Model

The objective of frequency modelling is the production of a model that gives you the probability of a given number of losses occurring during a given period, such as the policy period or a calendar year. The input to frequency modelling will typically be a table like that in Figure 14.1 which, for each past policy year, gives the exposure relevant to that year and the number of losses after adjustment for IBNR and other secular trends (such as known changes in the risk profile). Figure 14.2 shows how frequency modelling fits into the overall experience rating process. Note that as is the case with all models we use for pricing, the frequency model may be based to a varying extent on past losses, but it is a *prospective* model, that is, it tells you what the number of losses will be *in the future*.

Figure 14.3 gives an example of how a frequency model (in this case a Poisson model with an average frequency of 7) might look like: as an example, it tells us that the probability of having exactly four claims is approximately 0.091, and that the probability of having up to four claims is approximately 0.173.

Theoretically, there are infinitely many models that one could use for frequency. In practice, however, because there are only a small number of years/periods of experience that one can use to fit a frequency model (typically 5–10 years, but it can be as small as 1 year) it is not possible to discriminate between many models, and in the spirit of machine learning (see Section 12.3), it is better to rely on a few well-tested models that one can trust and that are analytically tractable and choose from among them rather than having a large dictionary of models that one can choose from on the basis of flimsy statistical evidence.

The three most popular models used in insurance are probably the *binomial distribution*, the *Poisson distribution*, and the *negative binomial distribution*. All three of them are so-called 'counting distributions', that is, discrete distributions with values over the natural numbers only.

The main practical difference between these three models is that they have different variance/mean ratios: this is 1 for the Poisson distribution, less than 1 for the binomial distribution and greater than 1 for the negative binomial distribution. A calculation of the variance/mean ratio of the number of claims over a certain period may therefore be the factor that drives the choice of what model is used, although, as we will soon show, things are more complicated than that.

In practice, the binomial distribution (and its generalisation, the multivariate Bernoulli distribution) is normally used with the individual risk model, whereas the Poisson and negative binomial are used with the collective risk model.

Not only are these three models very simple but they also have some useful properties: all of them are part of the Panjer class, and they are the only distributions belonging to

DOI: 10.1201/9781003168881-16

Policy year	No of losses (adjusted)	Exposure (£m)
2005	86.0	460.6
2006	92.0	543.7
2007	110.0	616.2
2008	146.3	677.2
2009	194.6	855.2
2010	257.9	941.6
2011	257.9	979.2
2012	215.1	1,189.6
2013	252.3	1,226.8
2014	139.2	1,210.9

FIGURE 14.1
A typical input for frequency modelling as part of experience rating.

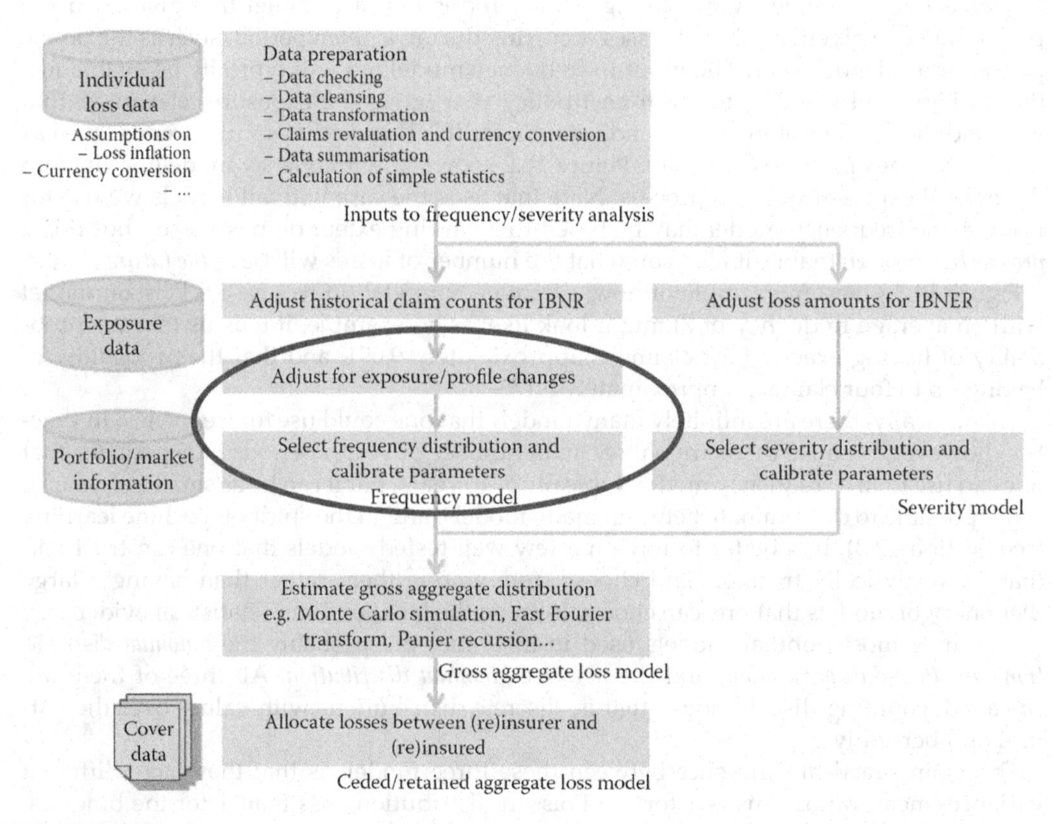

FIGURE 14.2
How selecting and calibrating a frequency model fits in the risk costing subprocess.

this class. The Panjer class (also referred to as the (a,b,0) class) is characterised by the fact that there exist constants a and b such that

$$\frac{p_k}{p_{k-1}} = a + \frac{b}{k}, \; k = 1, 2, 3 \cdots \tag{14.1}$$

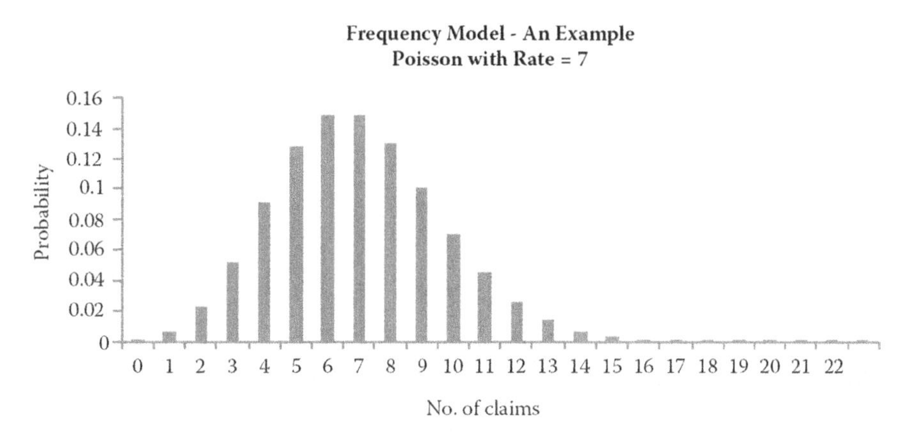

FIGURE 14.3
An example of a frequency distribution – a Poisson distribution with rate = 7. The distribution is obviously asymmetrical – only numbers greater than or equal to 0 have a probability greater than 0, whereas the distribution extends indefinitely to the right.

where p_k is the probability of having exactly k losses, and p_0 is uniquely determined by imposing the condition $\sum_{k=0}^{\infty} p_k = 1$.

As an example, it is trivial to realise that a Poisson distribution with rate λ satisfies Equation 14.1 with $a = 0$, $b = \lambda$. Question 3 asks you to prove this and the analogous results for the binomial and the negative binomial distribution.

This recursion property is used, for example, in the Panjer recursion formula for the calculation of the aggregate loss distribution (Section 17.3.2). Another property that makes these distributions suitable for aggregate loss calculations is that their moment generating function has a simple analytical form and therefore makes them easy to use with the Fourier transform method (Section 7.2.3). More about the properties of the $(a,b,0)$ class and other related classes can be found in Klugman et al. (2008).

Let us now look at each of these distributions in turn.

14.1 Binomial Distribution and the Multivariate Bernoulli Distribution

From a statistical point of view, the binomial distribution $B(n,p)$ is that count distribution that models the probability of having k successes in an experiment repeated n times, with the probability of any given experiment being successful being p. The special-case distribution $B(1,p)$ is called a Bernoulli distribution: or in other terms, the binomial distribution $B(n,p)$ arises from the sum of n Bernoulli variates.

In actuarial practice, the binomial distribution is suitable for describing the situation in which there are n independent and identical risks, each with a probability p of making a claim, with the added constraint that only one claim per risk is possible. Therefore, the maximum number of claims in a given year is n. This makes it suitable for use with the individual risk model (Section 7.1.1) in the case where the probability of a claim is the same across all risks.

Using elementary results from statistical theory (see, for example, Klugman et al. 2008), we can write the probability of having exactly k claims from the n risks as

$$\Pr(N=k)=\binom{n}{k}p^k(1-p)^{n-k} \tag{14.2}$$

The mean and variance of the binomial distribution are

$$\mathbb{E}(N)=np,\ \mathbb{V}(N)=np(1-p)$$

14.1.1 Simulating from a Binomial Distribution

One simple way of producing a random variable from a binomial distribution $B(n,p)$ is as illustrated in the following pseudocode:

```
# Generation of a random B(n,p)-distributed number of claims
Input: n, p
k = 0
For j = 1 to n
r = rand() # produces a random number between 0 and 1
if r < p then k <- k+1
End For
Return k
```

Informally, this pseudocode can be described as follows: go through all the risks and for each risk, assign a loss to it if a generated random number is less than p; then count all the losses.

Most spreadsheets and statistical packages will have shortcuts to produce binomial distributions:

In Excel, one can generate a random binomial variable from distribution B(n,p) using the function BINOM.INV(n,p,RAND()), where RAND() is a uniform random variable between 0 and 1.

In R, it is possible to generate N random variables from B(n,p) using the function rbinom(N,n,p), as in the following example:

```
> no_of_simul = 10
> no_of_risks = 5
> p = 0.3 #                  probability of having a claim
> rbinom(no_of_simul,no_of_risks,p) # no of claims in each simulation
[1] 0 2 3 3 2 0 2 1 1 1
```

14.1.2 Generalisation: The Multivariate Bernoulli Distribution[1]

As mentioned above, the binomial distribution is suitable for an individual risk model for n independent and identical risks, each with a probability p of defaulting. In practice, we

1 Note that while we are describing the multivariate Bernoulli distribution as a generalisation of the binomial distribution, we might equally consider both the binomial distribution and the multivariate Bernoulli distribution generalisations of the Bernoulli distribution.

have seen that the individual risk model also encompasses the case where the n risks have different probabilities of default.

A distribution that caters for this case (which is actually the most common case) is the multivariate Bernoulli distribution (Dai et al., 2013). The multivariate Bernoulli distribution is an n-dimensional distribution with random variable $\vec{Y} = (Y_1, Y_2 \dots Y_n)$, where each component Y_j can take the values 0 and 1.

While in the general case the variables may have a complex dependency structure, we are interested in the case where they are independent, as per the assumptions of the individual risk model (see Section 6.1.1). For this special case, the probability $\Pr(Y_1 = y_1, Y_2 = y_2, \dots Y_n = y_n)$ (which we will write as $p(y_1, y_2, \dots y_n)$ for simplicity) of a specific combination $(y_1, y_2 \dots y_n)$ is given by:

$$p(y_1, y_2, \dots y_n) = \prod_{j=1}^{n} p_j^{y_j} (1 - p_j)^{y_j} \tag{14.3}$$

where p_j is the probability that $Y_j = 1$.

The probability of having k claims from n risks therefore becomes

$$\Pr(N = k) = \sum_{y_1, \dots y_n \text{ such that } y_1 + \dots + y_n = k} p(y_1, y_2, \dots y_n) \tag{14.4}$$

Needless to say, Equation 14.4 reduces to Equation 14.2 when $p_1 = \dots = p_n = p$.

Equation 14.4 is not straightforward to calculate. However, the pseudocode in Section 14.1.1 can be easily extended to the case where each risk 1, … n has a different probability of loss, and therefore to sample multivariate Bernoulli variates, as shown in the following pseudocode.

```
# Generation of a random variate from a multivariate Bernoulli
distribution
  Input: n, p(1), … p(n)
  k = 0
  For j = 1 to n
    r = rand() # produces a random number between 0 and 1
    if r < p(j) then k <- k+1
  End For
  Return k
```

The mean and variance of the multivariate Bernoulli distribution with independent risks are only slightly more complicated than for the binomial model:

$$\mathbb{E}(N) = \sum_{i=1}^{n} p_i, \; G = \mathbb{V}(N) = \sum_{i=1}^{n} p_i (1 - p_i)$$

For the sake of clarity, notice that the multivariate Bernoulli distribution is not the same thing as the multinomial distribution, another generalisation of the binomial distribution where the outcomes are not limited to 0 and 1 (with probability p and 1-p) but more generally to a (finite) number $K > 1$ of outcomes with probabilities $p_1, p_2, \dots p_K$.

14.2 Poisson Distribution

The Poisson distribution is arguably the most important and ubiquitous of all counting distributions. It is used in physics to model the number of events that arise from stationary processes such as cosmic rays counting and shot noise in photodiodes. Unsurprisingly, it is very widespread in risk theory, too: perhaps too widespread, as we will argue later.

For a counting process; that is, a stochastic process $\{N(t), t \geq 0\}$ where $N(t)$ represents the number of events that occur by time t – to be Poisson, it must satisfy the following properties:

1. $N(0) = 0$ (this is simply telling us that we start counting at time $t = 0$).
2. The process is stationary, that is, the distribution of the number of events in an interval $[t_0, t_0 + t]$ of length t should depend on the length t only and not on t_0.
3. The number of events occurring in disjoint time interval are independent.
4. Events are rare, in the specific sense that the probability of having two events in a small time window of length h is negligible: $\Pr(N(h) \geq 2) = o(h)$ and that the probability of having *one* event in that time window is proportional to the length of the window: $\Pr(N(h) = 1) = \lambda h + o(h)$.

The constant λ in Condition 4 is called the *Poisson rate.*

It is quite easy to prove that if Conditions 1 through 4 are satisfied, the probability of having n events in a time interval of length t, $[s, s + t)$, is given by

$$\Pr(N(t+s) - N(s) = n) = \frac{e^{-\lambda t}(\lambda t)^n}{n!} \tag{14.5}$$

On the other hand, if Conditions 1 and 3 are satisfied and Equation 14.5 holds true, Conditions 2 and 4 will also hold and the process is Poisson. A proof of this is required in Question 2 at the end of the chapter.

The mean and the variance of the number of claims in the interval [0,1] are given by

$$\mathbb{E}(N) = \mathbb{V}(N) = \lambda \tag{14.6}$$

The key assumption is that the rate at which losses happen is constant. If the rate is not constant, other interesting distributions, such as the negative binomial, are produced.

14.2.1 Facts about the Poisson Distribution

Here are some interesting facts about the Poisson distribution that are useful for simulation and other purposes:

Fact 1 – The Poisson distribution is the limiting distribution of the binomial distribution (n,p) with $n \to \infty$, $p \to 0$ and $np \to \lambda$.

Fact 2 – When $\lambda \to \infty$, the distribution converges to a normal distribution with mean and variance equal to λ.

Fact 3 – The waiting time is exponential: that is, the time between two successive events of a Poisson process with rate λ follows a distribution

$$T \sim \text{Exp}(\lambda) \tag{14.7}$$

Fact 4 – If the number of events over the interval [0,1] follows a Poisson distribution with rate λ, Poi(λ), then the number of events over the interval [0,*a*] follows a Poisson distribution with rate *a*λ, Poi(*a*λ).

14.2.2 Simulating from a Poisson Distribution

Remember the 'waiting times' property (Fact 3) in Section 14.2.1: the distribution of waiting times follows an exponential distribution with 'time constant' equal to λ: $T \sim \text{Exp}(\lambda)$.

In other terms, the distribution of waiting times has cumulative distribution function (CDF)

$$F(t) = 1 - \exp(-\lambda t) \tag{14.8}$$

Hence, the waiting time corresponding to a given percentile $F(t)$ is

$$t = -\frac{\ln(1 - F(t))}{\lambda} \tag{14.9}$$

This suggests a simple algorithm for generating Poisson variates:

```
# Generating a random number of claims according to the Poisson
distribution
  s = 0; n = 0
  Do While s < 1
  u = rand()
  s = s - ln(u)/lambda
  k = k + 1
  Loop
  Return k
```

The idea of the algorithm above is not to generate the Poisson variate directly but to fit as many randomly generated exponential variates as possible in an interval of unit length.

14.2.2.1 Generating Random Poisson Variates in Excel

Excel does not have a function that allows one to simulate the number of claims coming from a Poisson variate directly as in the case of the binomial distribution, so you need to use a workaround.

As an example, you can use the function POISSON.DIST to calculate the probability that the number of claims is equal to a certain value and the cumulative distribution function, and then you can look up the largest value, which is less than a given random value.

Figure 14.4 illustrates this process. Note that this method is practical only for small values of the Poisson rate – for large values, the lookup table is just too cumbersome. In this case, we can use the property that a Poisson distribution with large values of λ can be well approximated with a normal distribution with mean = λ, variance = λ, and we can generate the random numbers of claims directly from the normal distribution using the function '=ROUND(NORM.INV(RAND(),lambda,sqrt(lambda)),0)' where the function

lambda	3.5				Empirical average	3.530
					Empirical variance	3.461
No. of claims, k	**Prob(no. of claims = k–1)**	**Prob(no. of claims <= k–1)**	**No. of claims, k**		**Random**	**No. of claims**
0	0	0	0		0.867920	6
1	0.03019738	0.030197383	1		0.792287	5
2	0.10569084	0.135888225	2		0.634380	4
3	0.18495897	0.320847199	3		0.340163	3
4	0.21578547	0.536632668	4		0.048506	1
5	0.18881229	0.725444953	5		0.763640	5
6	0.1321686	0.857613553	6		0.948095	7
7	0.07709835	0.934711903	7		0.473987	3
8	0.03854917	0.973261078	8		0.070945	1
9	0.01686526	0.990126342	9		0.403398	3
10	0.00655871	0.996685056	10		0.509458	3
11	0.00229555	0.998980606	11		0.991125	9
12	0.0007304	0.999711008	12		0.403322	3
13	0.00021303	0.999924042	13		0.072211	1
14	5.7355E-05	0.999981397	14		0.519688	3
15	1.4339E-05	0.999995736	15		0.639637	4
16	3.3457E-06	0.999999082	16		...	...

FIGURE 14.4
The generation of random Poisson variates with Excel. The PDF and CDF of the Poisson distribution are obtained by the functions '=POISSON.DIST(k-1,lambda,FALSE)' and '=POISSON.DIST(k-1,lambda,TRUE)' (respectively), where k is the number of claims on the same row and lambda is 3.5. Of course there is a non-zero probability that any number k of claims could occur so the table should be infinite, but it is sufficient to stop at a level for which the probability of mistakenly ignoring the possibility of a larger number of claims is below a certain level (in the table above, for example, this probability is roughly 1 in 1,000,000). The right part of the table is the simulation itself. A random number between 0 and 1 is generated (column 'Random') and a lookup function, which identifies the largest value in the column 'Prob(no. of claims $\leq k-1$)', is smaller than the generated random number and is used to determine the number of claims generated: for example, if the generated random number is 0.63438, the largest value in column 3, which is smaller than 0.63438, is 0.536632668, which corresponds to four claims; hence, the number 4 is generated.

'=ROUND (x, 0)' has been used to round the result to the nearest integer. This method already gives a good approximation for lambda of approximately 30 or higher, as shown, for example, in Figure 14.5.

Alternatively, you can use Equation 14.9 to generate time intervals distributed to an exponential distribution with parameter λ, and count the number that fits into the unit interval (see Question 1).

14.2.2.2 Generating Random Poisson Variates in R

In R, things are as usual more straightforward, and one can generate N numbers drawn from a Poisson distribution with rate lambda by using the function rpois (N,lambda), as in the following example:

```
> N = 10
> lambda = 3.5
> rpois(N,lambda)
[1] 10 2 1 1 7 3 3 7 4 9
```

Random	No. of claims	No. of claims (Gaussian approx)	No. of claims (Gaussian approx, rounded)
0.689583	33	32.70941201	33
0.248927	26	26.28714477	26
0.716111	33	33.12929395	33
0.839607	35	35.43803113	35
0.02414	20	19.18302947	19
0.602516	31	31.42334058	31
0.84691	36	35.60469128	36
0.744793	34	33.60506808	34
0.874179	36	36.27894146	36
0.297228	27	27.08397297	27
0.983151	42	41.63181718	42
0.788423	34	34.38704196	34
0.344678	28	27.81058647	28
0.385797	28	28.41000795	28
0.173317	25	24.84516976	25
0.626825	32	31.77164015	32
...	...	...	...

FIGURE 14.5
A comparison of the number of claims for a Poisson process as simulated by the lookup table method and by the normal approximation for lambda = 30. The latter works by finding the value corresponding to the percentile of the normal distribution defined by the random number x in the first column ('=ROUND (NORM. INV(x, lambda, sqrt (lambda)),0)'), and rounding it to the nearest integer. On average, the difference between the two simulated numbers is less than 0.005 (~0.01% of the value of lambda), and things get better (in terms of percentage) as lambda increases.

14.2.3 How Do You Calculate λ?

Calculating the Poisson rate λ based on the number of claims (including incurred but not reported [IBNR]) projected for the past K years, $U_1, U_2,.., U_K$ ('U' stands for 'ultimate'), is easy when the exposure is constant. The maximum likelihood estimate is the average over all years:

$$\hat{\lambda} = \frac{\text{Total number of claims (including IBNR)}}{K} = \frac{\sum_{j=1}^{K} U_j}{K} \tag{14.10}$$

When the exposure E_j is not constant but depends on the year j, the MLE for the rate per unit of exposure (pue) is given by:

$$\begin{aligned}\hat{\lambda} &= \text{Frequency pue} \times \text{Renewal exposure} \\ &= \frac{\text{Total number of claims (including IBNR)}}{\text{Total exposure}} \times \text{Renewal exposure} \\ &= \frac{\sum_{j=1}^{K} U_j}{\sum_{j=1}^{K} E_j} \times E_{K+1}\end{aligned} \tag{14.11}$$

In practice, things may be more complicated because you may want to give different weights to different years (e.g. years with a lot of IBNR are more uncertain and therefore

Policy year	No of losses (IBNR-adjusted)	Exposure	Exposure-adjusted number of claims
2005	86.0	460.6	236.2
2006	92.0	543.7	214.1
2007	110.0	616.2	225.8
2008	146.3	677.2	270.9
2009	194.6	855.2	284.0
2010	257.9	941.6	344.8
2011	257.9	979.2	328.4
2012	215.1	1,189.6	216.8
2013	252.3	1,226.8	184.9
2014	139.2	1,210.9	
		Estimated exposure (2015)	1,265.0
		Weighted average @ 2015 exposure	256.6

FIGURE 14.6
An example of Poisson rate calculation. The frequency per unit of exposure is calculated as the sum of the ultimate claim count for each year (86 + 92 + ... + 179.3) divided by the sum of the exposures (460.6 + 543.7 + ... + 1226.8), and then multiplied by the current exposure, 1265, giving 256.6. Note that the policy year 2014 has been excluded as these ultimates were obtained with the chain ladder method after removing the last diagonal.

may be weighted less; at the same time, the years with a lot of IBNR are likely to be the most recent, which are also the most relevant). Figure 14.6 shows an example of a calculation for the Poisson rate.

14.3 Negative Binomial Distribution

14.3.1 Technical Generalities

Formally, the *negative binomial distribution* is a discrete probability distribution of the number k of successes in a sequence of Bernoulli trials before $r > 0$ failures occur. If the probability of failure is given by $\frac{1}{1+\beta}(\beta > 0)$, the probability that the number of successes is k is given by

$$\Pr(N=k) = \binom{k+r-1}{k}\left(\frac{1}{1+\beta}\right)^r\left(\frac{\beta}{1+\beta}\right)^k \tag{14.12}$$

This can be seen easily by writing

$$\Pr(N=k) = \Pr(k \text{ successes, } r-1 \text{ failures in } k+r-\text{trials}) \times \Pr(k+r\text{th trial fails}) \tag{14.13}$$

and using basic combinatorial calculus.

The mean and variance of the negative binomial are given by

$$\begin{aligned} \mathbb{E}(N) &= r^2 \\ \mathbb{V}(N) &= r^2\,(1+{}^2) \end{aligned} \tag{14.14}$$

This makes it suitable to model situations where the Poisson distribution does not have enough volatility, in that the variance is *always* larger than the mean.

The negative binomial distribution can be seen as a Poisson distribution whose Poisson rate follows a gamma distribution. More specifically, if N ~ Poi(Λ), Λ ~ gamma (θ,α), then

$$N \sim NB(\beta = \theta,\ r = \alpha)$$

This is very useful for simulation purposes.

14.3.2 Simulating from a Negative Binomial Distribution

The standard way of generating negative binomial variates with parameters β and r is to use the Poisson/gamma method:

1. Sample a value λ of the Poisson rate from a gamma distribution with parameters $\theta = \beta$, $\alpha = r$ (Devroye 1986).
2. Sample a number n from Poi(λ) using the usual algorithm for Poisson variates.

The number n thus generated is distributed according to $N \sim NB(\beta, r)$.

It is not difficult to produce a negative binomial variate in Excel: we can produce a gamma variate by using the function GAMMA.INV(u,r,β) (in Excel 2010) or GAMMAINV(u,r,β) (in Excel 2007), where u = RAND() is a uniform random number between 0 and 1, and then we can generate a Poisson variate with the lookup table method outlined in Section 14.2.2.

However, it is difficult to use this method for the mass production of negative binomial variates for simulation purposes as we would need one lookup table for each negative binomial variate. The obvious way around this is writing VBA code that reproduces repeatedly the functionality of the lookup table or (even better) uses repeatedly the exponential waiting times method outlined in Section 14.2.2 without needing to store a large number of intermediate results.

In R, things are as usual much simpler, and negative binomial variates can be produced straightforwardly using the function `rnegbin(no_of_simul,mu,theta)`, included in the basic library MASS. Here is, for example, the code to produce 10 variates from a negative binomial with mean of 6 and variance/mean ratio of 1.8:

```
> # Generates no_of_simul negative binomial variates
> # with mean = mu and variance/mean ratio = var_mean_ratio
>
> # Load libraries
> library(MASS)
>
> # Input parameters
> mu = 6
> var_mean_ratio = 1.8
> no_of_simul = 10
>
> # Calculations
> theta = mu/(var_mean_ratio-1)
> rnegbin(no_of_simul,mu,theta)
[1] 7 10 8 3 7 5 8 1 7 3
```

14.3.3 Fitting a Negative Binomial (i.e. Calculating r and β) to Past Claim Counts

The parameters of the negative binomial distribution can be estimated from the past claim counts and the relevant exposures using either the method of moments or maximum likelihood estimation.

14.3.3.1 Method of Moments

The mean of a negative binomial distribution, $\mathbb{E}(N) = r\beta$, can be estimated by Equation 14.11 as in the Poisson case. The variance $\mathbb{V}(N)$ can be obtained as the weighted variance of the claim counts of the different years, after rescaling all historical claim counts to reflect the renewal exposure (see Box 11.1).

Once we have estimates $\hat{\lambda}$ and $\hat{\upsilon}$ of $\mathbb{E}(N)$ and $\mathbb{V}(N)$, respectively, we can calculate the estimates of the parameters r, β as follows:

$$\hat{\beta} = \frac{\hat{\upsilon}}{\hat{\lambda}} - 1 \; \hat{r} = \frac{\hat{\lambda}^2}{\hat{\upsilon} - \hat{\lambda}} = \frac{\hat{\lambda}}{\hat{\beta}} \tag{14.15}$$

Note that if $\hat{\upsilon} < \hat{\lambda}$, Equation 14.15 yields values of $\hat{\beta}$ and $\hat{r}$ that are negative and therefore not admissible for the negative binomial distribution.

Also, note that the method of moments gives reasonably good results if the renewal exposure is not too far from the average exposure over the years, but can be too rough otherwise (see the results for our case study in Figure 14.10). If this is the case, the approximation can be improved by calculating the variance/mean ratio using the average exposure as the reference exposure and rescaling later as needed; however, a cleaner solution is to use maximum likelihood estimation (Section 14.3.4.2), for which the variance/mean ratio does not depend on the renewal exposure.

14.3.3.2 Maximum Likelihood Estimation

A more accurate estimation of r and β can be obtained with maximum likelihood as follows. We assume the relationship between exposure and expected frequency to be linear, and described as

$$r = \rho\varepsilon \tag{14.16}$$

where ε is the (known) exposure. Therefore, we want to estimate (ρ,β) given a sample of observed pairs $\{(n_i, \varepsilon_i), I = 1,\ldots m\}$.

Let L_i be the log-likelihood associated to the pair (n_i, ε_i). Straightforward calculations, similar to those found in Klugman et al. (2008), lead to:

$$\frac{\partial L_i}{\partial \beta} = \frac{n_i}{\beta(1+\beta)} - \frac{\rho\varepsilon_i}{1+\beta} \tag{14.17}$$

and

$$\frac{\partial L_i}{\partial \rho} = \varepsilon_i \left(\sum_{k=0}^{n_i - 1} \frac{1}{\rho\varepsilon_i + k} - \ln(1+\beta) \right) \tag{14.18}$$

The parameters can therefore be obtained by imposing the conditions:

$$\sum_{i=1}^{m}\frac{\partial L_i}{\partial \beta}=\sum_{i=1}^{m}\left(\frac{n_i}{\beta(1+\beta)}-\frac{\rho\varepsilon_i}{1+\beta}\right)=0\Rightarrow \rho\beta=\frac{\sum_{i=1}^{m}n_i}{\sum_{i=1}^{m}\varepsilon_i} \quad (14.19)$$

$$\sum_{i=1}^{m}\frac{\partial L_i}{\partial \rho}=\sum_{i=1}^{m}\varepsilon_i\left(\sum_{k=0}^{n_i-1}\frac{1}{\rho\varepsilon_i+k}-\ln(1+\beta)\right)=0 \quad (14.20)$$

Equation 14.20, subject to the constraint imposed by Equation 14.19, can be easily solved numerically to yield $\hat{\rho}$ and $\hat{\beta}$ and from that, $\hat{r}=\hat{\rho}\varepsilon_{m+1}$ where ε_{m+1} is the renewal exposure.

Note that the mean's estimate is the same as that obtained with the method of moments. Also note that if the volatility of historical claim counts is too low, Equations 14.19 and 14.20 together might not have an admissible solution $\hat{r}$ and $\hat{\beta}$ (i.e one for which $\hat{r}>0$ and $\hat{\beta}>0$).

14.4 Selecting a Frequency Model

Having described the most common claim count models and having explained how to calibrate them and how to sample losses from them (something that we will need to be able to do to produce an aggregate loss model), we can now address the problem of selecting the appropriate frequency model based on the client's (or a portfolio's) past loss experience (as summarised, for example, in Figure 14.1).

As we mentioned in the introduction to this chapter, there are theoretically infinitely many models that one could use to model claim counts. However, because there are only a small number of years/periods of experience that one can use to fit a frequency model (typically 5–10 periods) it is not possible to discriminate between many models, and therefore practitioners tend to focus on a few models that are analytically tractable and choose from among them: the *binomial distribution* (or the *multivariate Bernoulli distribution)*, the *Poisson distribution*, and the *negative binomial distribution*.

Our job of selecting a model therefore simplifies to the job of selecting the most adequate among these three models. There are many methods by which we could do that in theory (see, e.g. Question 5). Before proceeding, however, we should disabuse ourselves of the notion that even for this very limited problem we can use any sophisticated data-driven discrimination criterion. Given the scarcity of data, only very basic reasoning and judgment can guide us through this process.

In this section we'll see how even the most basic of methods to select the right model based on data are likely to give ambiguous results (Section 14.4.1), and we will look at the theoretical considerations that should guide the selection process (Section 14.4.2). A practical approach based on these considerations (and some careful use of calibration) is outlined in Section 14.4.3.

14.4.1 The Issues with Using the Variance/Mean Ratio as a Selection Criterion

Because the main thing that distinguishes the three models is the variance/mean (V/M) ratio, this seems to suggest the following decision method:

a. Calculate the variance/mean (V/M) ratio
b. Pick binomial, negative binomial or Poisson depending on whether the V/M ratio is smaller than one, larger than one, or *approximately* 1, respectively.

However, this simple method is not as good as it looks at first sight. First of all, we need to define what we mean by 'approximately 1'. If for example the number of claims is a Poisson process and the claim count is observed each year for 10 years, the observed variance/mean ratio will vary wildly. You can easily verify this by generating 10 Poisson variates with an arbitrary Poisson rate (say, 100) and calculating the variance/mean ratio, like in the following snippet of R code, and then repeating for a good number of times to get a feel for how widely the variance/mean ratio varies from one iteration to the other.

```
> a=rpois(10,100);a; mean(a);var(a)/mean(a)
[1] 108 95 95 98 113 90 97 102 91 99
[1] 98.8
[1] 0.5258659
```

This 'feel' can be confirmed by deriving the actual distribution of the observed variance/mean ratio using a large number of simulations (see Figure 14.7).

The sampling distribution of Figure 14.7 does *not* depend on the Poisson rate and can therefore be used as a statistical test for the null hypothesis H_0 that the distribution is Poisson and, consequently, to clarify in quantitative terms what 'approximately 1' means.

For example, we see that 95% of the time, the variance/mean ratio for 10 samples[2] from the Poisson distribution will fall between 0.31 and 2.14, as shown in Figure 14.7. Therefore,

Distribution of the Variance-to-Mean Ratio

		No. of years						
		2	5	10	20	50	100	200
Percentile	1.0%	0.00	0.08	0.24	0.40	0.59	0.69	0.78
	2.5%	0.00	0.12	0.31	0.47	0.64	0.74	0.81
	5.0%	0.00	0.18	0.37	0.53	0.69	0.78	0.85
	10.0%	0.02	0.26	0.47	0.61	0.75	0.82	0.88
	20.0%	0.05	0.41	0.60	0.72	0.82	0.88	0.92
	30.0%	0.13	0.55	0.72	0.81	0.89	0.92	0.95
	40.0%	0.26	0.69	0.83	0.89	0.94	0.96	0.97
	50.0%	0.45	0.84	0.94	0.96	0.98	0.99	1.00
	60.0%	0.71	1.01	1.06	1.04	1.04	1.03	1.02
	70.0%	1.07	1.21	1.20	1.13	1.09	1.07	1.05
	80.0%	1.65	1.49	1.38	1.25	1.16	1.12	1.08
	90.0%	2.72	1.93	1.64	1.42	1.26	1.18	1.13
	95.0%	3.84	2.36	1.91	1.58	1.35	1.24	1.17
	97.5%	5.08	2.77	2.14	1.72	1.42	1.30	1.21
	99.0%	6.68	3.37	2.46	1.90	1.52	1.36	1.25

FIGURE 14.7
The sampling distribution $F(k; x)$ of the observed variance/mean ratio based on a number k of periods does not depend on the value of the Poisson rate, but only on the number of periods (years) that have been used to calculate the ratio. The table above was produced using 10,000 simulations with a Poisson rate of 100 (simulations performed in R). Note that there is a hidden assumption that the exposure is constant over the periods/years.

2 Because the variance/mean ratio distribution does not depend on the Poisson rate but only on the number of periods, one may be tempted to get around the excessive volatility of the variance/mean ratio by subdividing the observed periods into smaller intervals – for example, looking at the claim count by month instead of by year so as to have more data points. However, things are not that simple in reality, as many loss processes have a seasonality – for example, the number of motor claims is typically higher in winter – therefore using a smaller interval might lead to an increased volatility and this volatility is of a type that we do not necessarily want to capture in our model.

Of course, the very presence of a seasonal volatility means that the Poisson process with constant Poisson rate can only model our loss process roughly, and that a more sophisticated approach (e.g. one where the Poisson rate is allowed to change seasonally) is necessary.

all series of 10 claim counts for which the variance/mean ratio falls within that range can be classified as Poisson at a significance level of 5%. Also:

$$\text{'approximately1'} \overset{@5\%\text{ signif level}}{\Leftrightarrow} 0.31 \le \frac{\text{variance}}{\text{mean}} \le 2.14$$

That's quite a wide range!

This approach would also need to be modified to take account of the fact that the exposure is in general not constant and therefore the variance/mean distribution will in general be different from that of Figure 14.7. This is not a major difficulty, but it adds another layer of complexity – generating an ad-hoc sampling distribution for a particular historical exposure profile.

Ultimately, however, these considerations should lead us to conclude that a suitable frequency model cannot be chosen based on data alone, as the data for this type of analysis is almost always too scarce for a robust conclusion. Theoretical considerations should drive our choice. These considerations are the topic of Section 14.4.1.

14.4.2 Theoretical Arguments that Should Drive Model Selection

In this section, we denote for simplicity a model with variance/mean ratio larger than 1 as 'overdispersed' and one with the ratio smaller than 1 as 'underdispersed'.

14.4.2.1 Underdispersed Models are Inadequate for the Collective Risk Model

First of all, note that – no matter what the variance/mean ratio based on a small sample says – the use of a binomial distribution is justified theoretically only for the individual risk model,[3] where each risk can have a loss only once.

For the collective risk model, when more than one event is possible for a given risk and there is not (unlike in the individual risk model) an overall limit to the number of losses in a given year, it is difficult to justify a process with a variance/mean ratio that is at least equal to that of a Poisson process (i.e. equal to 1).

Why? A Poisson process is characterised by an interarrival time distributed exponentially, which in turn reflects the fact that this process is memoryless: the fact that a loss has occurred does not change the probability that another loss will occur in a given interval after that first loss. The presence of systemic factors, such as the weather or the economic environment, or other clustering effects, can be pictured as breaches of this 'memoryless' property, by which the occurrence of a loss at a given time t increases the likelihood that a loss will also occur in the period immediately after t: this is, intuitively, the cause for the extra variance.

So, how can we picture a stochastic process with a variance/mean ratio less than one? This must be a process for which the memoryless property is again breached but in reverse: the occurrence of a loss makes it less likely that another loss will occur immediately afterwards. This situation is not impossible to envisage: for instance, in property insurance one can imagine that the total loss of a property reduces the pool of properties at least temporarily and therefore makes it more difficult for a loss to happen. However, this is a minor effect (especially where total losses are the exception

3 In general, an individual risk model will have different failure probabilities for the different risks in the portfolio, in which case the frequency model should be a multivariate Bernoulli distribution. This reduces to a binomial distribution in case all probabilities are the same.

and not the norm) in most circumstances, and a Poisson assumption for this case will normally be just fine.

14.4.2.2 The Poisson Model Should Only Be Used in the Absence of Significant Systemic Factors

Is the Poisson process itself adequate? This really depends on the risk. A Poisson process is justifiable only if losses are independent and the Poisson rate is genuinely constant, that is to say, if systemic effects such as bad weather play a negligible role.

For example in commercial property insurance, it may *typically* be safe to assume that (non-catastrophic) fire losses are independent and that we can disregard cross-location systemic effects. In this case a Poisson assumption is reasonable.

However, wherever systemic effects are at play assuming a Poisson distribution leads us to underestimate the volatility of the loss experience.

For example, consider a fleet of cars of a (very) large company having roughly 1000 claims each year. If you modelled the loss process as a Poisson, that would lead to the conclusions that the standard deviation is approximately $\sqrt{1000} \approx 32$, and that the probability of having more than 1100 losses is less than 1 in 1000. However, it takes only a harsh winter to cause a hike in the loss experience because of rain and ice to observe an increase in loss experience of that magnitude.

Most analysts (and especially those who have had some scientific training) will be naturally drawn to the Poisson distribution because it is simple and seems to be the inevitable choice for modelling rare events. However, claims are not elementary particles, and the assumptions of stationary and independent increments are not easy to justify in practice. There are two main problems with using the Poisson distribution:

14.4.3 Recommended Approach to Frequency Modelling

The theoretical considerations in Section 14.4.2 suggest that the choice of the frequency model should not be driven blindly by a statistical test on the variance/mean ratio (or similar criteria) but should be guided by our knowledge of the underlying process.

Building on these considerations, the following two sections outline a recommended approach for both the individual and the collective risk model, which we summarise in Figure 14.9.

14.4.3.1 Frequency Models for the Individual Risk Model

If the relevant model for the risk is the *individual risk model* (such as for credit risk and horse insurance), the claim count model more naturally suited to represent the loss process is the multivariate Bernoulli distribution, which reduces to the binomial distribution in case all default probabilities are the same.

The extra volatility coming from systemic factors such as an economic recession (credit risk) or an epidemic (horse insurance) can be captured, e.g., by using common shocks to the default probabilities (see, e.g. Section 24.2.2.3). Alternatively, and less precisely, the extra volatility may be taken into account by using a Poisson or negative binomial distribution, in which case the simulation will also need to incorporate a mechanism of selecting the risks that give rise to each loss in proportion to the default probability of each risk, so as to take into account the (potentially) different severities.

14.4.3.2 Frequency Models for the Collective Risk Model

If the relevant model for the risk is the *collective risk model* (such as employers' liability, motor, or property), the main question to ask is whether there are systemic factors at play.

14.4.3.2.1 No Systemic Factors at Play – Poisson Distribution

If we can sensibly assume that there are no systemic factors at play, a Poisson model will be sufficient to capture the year-on-year (or period-on-period) volatility. This is true also in the case where the underlying process is a *nonstationary* Poisson process where the Poisson rate changes seasonally *within* the policy period, as the overall number of losses within that period will still follow a Poisson distribution (see Section 5.4.1 in Ross, 2003).

14.4.3.2.2 Significant Systemic Factors at Play – Common Shocks/Negative Binomial Distribution

If, however, the mean frequency per unit of exposure is likely to change from one year to the other in response to systemic factors, then a Poisson model will not be able to capture the extra volatility. A typical example of this is a Poisson process where the Poisson rate is dependent on external factors (economic conditions, weather, pandemics ...). This may also be viewed as a nonstationarity that has an effect *across* different policy periods.

There are different ways of taking this extra volatility into account. The most common are by modelling a common shock (i.e. an increase of the rate affecting all risks simultaneously and thus creating a correlation) and by using a negative binomial – or both.

a. **Common-shock modelling**. Where appropriate (i.e. when this reflects the underlying physical reality of the risk), the extra volatility can be accounted for explicitly in the process of simulating loss scenarios by introducing a common shock to the Poisson rate so that, for example, a year has a certain chance of being a recession year, or a year with a harsh winter. This requires, of course, some knowledge of the likelihood of these common shocks and their impact on the frequency. As an example, the simulation of 100,000 scenarios for the number of motor losses where the effect of a harsh winter is taken into account might look something like in the example of Figure 14.8.

Scenario	Harsh winter?	Poisson rate	No of losses
1	No	114.0	121
2	No	114.0	134
3	**Yes**	**125.4**	**142**
4	No	114.0	110
5	**Yes**	**125.4**	**138**
...	...	...	...
100,000	No	114.0	95

FIGURE 14.8
In this simulation, 100,000 possible scenarios for the number of motor losses are simulated. Each scenario is one with a harsh winter with probability 17%, in which case the Poisson rate is increased by 10% (from 114.0 to 125.4). A random Poisson variate with that rate is then generated in the 'No of losses' column. The relevant spreadsheet can be found online in Parodi (2016; Downloads).

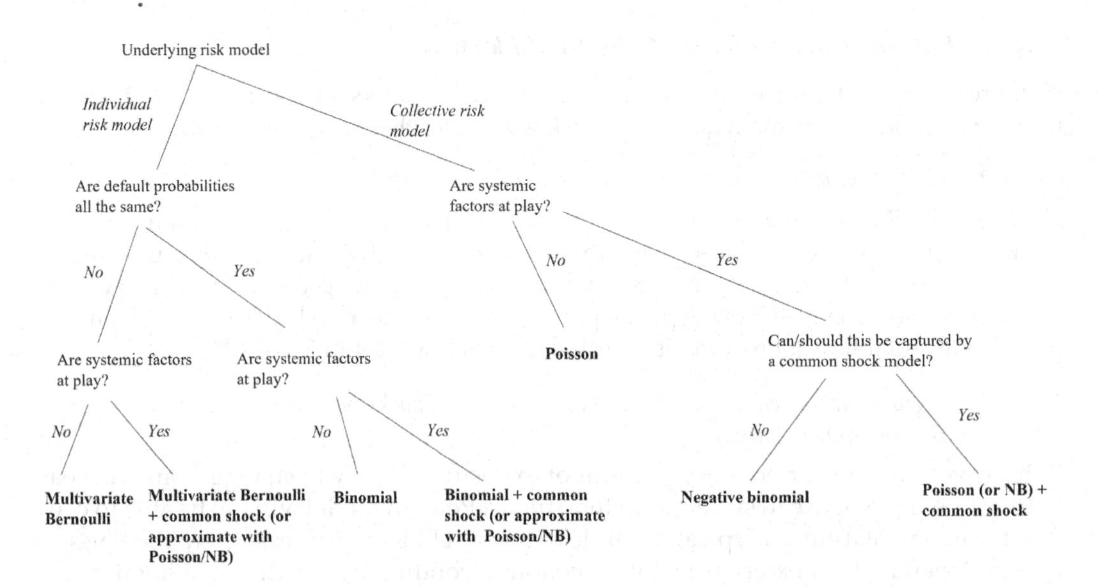

FIGURE 14.9
A succinct representation of the recommended approach for frequency model selection.

b. **Negative binomial distribution**. In all other cases, a negative binomial distribution can be used instead of a Poisson distribution. This implicitly incorporates the nonstationarity of the Poisson rate as the negative binomial can be seen as a Poisson distribution with Poisson rate (at a given level of exposure) sampled from a Gamma distribution. The advantage of this approach is that it doesn't require further exploration of the reason for the extra volatility, and the calibration of the common shock: all we need to do is an estimate of the parameters of the negative binomial distribution.

The suggested calibration process for the negative binomial distribution is as follows:

i. Attempt modelling the claim count as a negative binomial distribution, estimating its parameters $\hat{r}$ and $\hat{\beta}$ from the historical claim counts with the method of moments (Section 14.3.4.1) or maximum likelihood estimation (Section 14.3.4.2).
ii. If this attempt does not yield parameters within the admissible range ($\hat{r} > 0$ and $\hat{\beta} > 0$), use a variance/mean ratio based on judgment or (where available) a benchmark or revert to a Poisson distribution, calculating the Poisson rate λ using Equation 14.11.

The overall recommended approach is summarised in the flowchart in Figure 14.9.

14.5 Results of the Frequency Analysis for Our Case Study

Armed with our toolkit of counting distributions, and with the simple methodology outlined in Section 14.4.3.2, we can now again take up our case study and produce a frequency model. We will model our claim counting process as a negative binomial (as is

Policy year	Reported	Projected ultimate	Exposure (on-levelled)	Exposure-Adjusted Ultimate	Standard error
2005	86	86.0	460.6	236.2	0.0
2006	92	92.0	543.7	214.1	0.0
2007	110	110.0	616.2	225.8	0.0
2008	146	146.1	677.2	273.0	0.9
2009	194	194.3	855.2	287.4	1.5
2010	256	257.1	941.6	345.4	2.7
2011	250	255.6	979.2	330.3	5.9
2012	199	211.7	1,189.6	225.1	8.1
2013	201	245.7	1,226.8	253.3	14.9
2014	68	215.9	1,210.9	225.5	27.2
			2015 exposure	1265.0	
			Weighted average @ 2015 exposure	263.8	
			Variance/mean ratio (method of moments)	8.5	
			Variance/mean ratio (maximum likelihood)	4.8	

FIGURE 14.10
Results of the frequency analysis for our case study, using the general triangle-free methodology. All exposure has been used in the mean and variance calculations – however, in practice the latest year (2014 in this case) could be excluded because it is too immature. Note that the difference between the method of moments and maximum likelihood calculations of the variance/mean ratio is significant: the main contribution to this difference is the fact that the renewal exposure is in this example substantially different from the average exposure over the period.

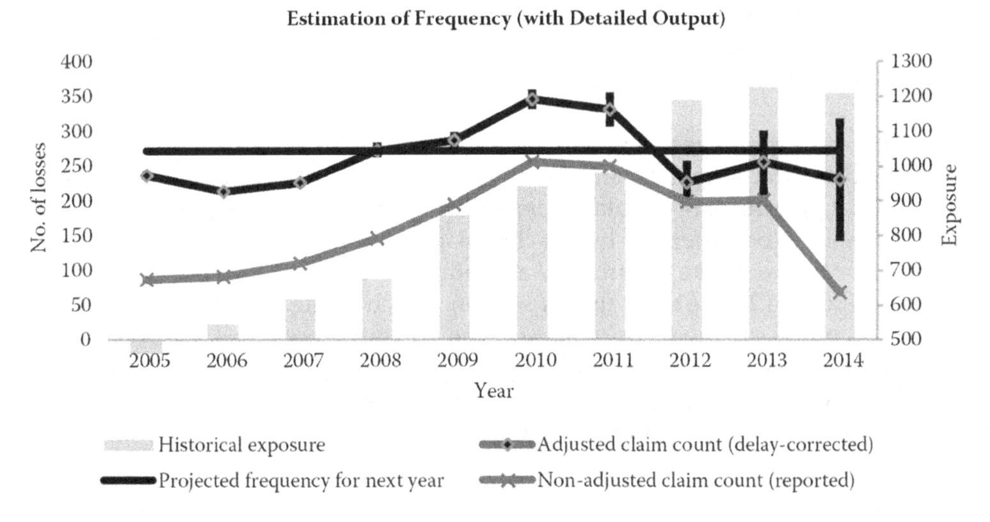

FIGURE 14.11
Results of the frequency analysis for our case study, showing the number of losses before (light grey) and after (dark grey) adjustment, the exposure (shown as grey bars) for each year, and the projected claim count for next year (straight line across). An error bar is shown for those years that are not finalised but IBNR-adjusted, and could therefore turn out to be quite different. The length of the error is set to be twice the standard error on each side.

often the case with liability classes, we can assume that systemic factors are at play, and this is borne out by the data) with the mean and the variance/mean ratio calculated as in Section 14.3.3.

As an input to the calculation, we use the (IBNR-adjusted) number of losses and the corresponding exposure for each year. We do not assume any trend in the data – if a trend

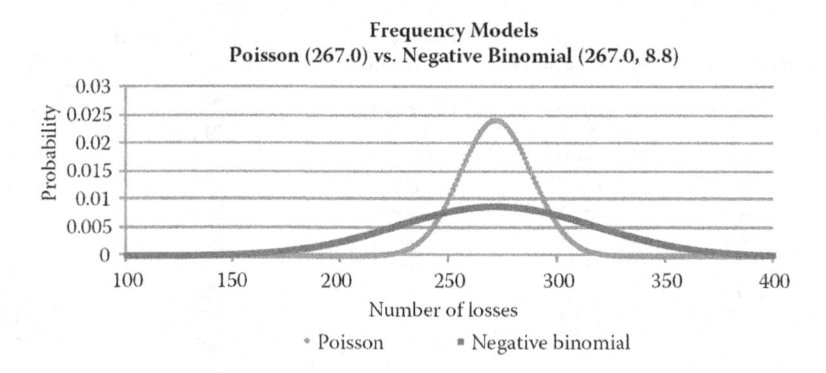

FIGURE 14.12
A comparison of a Poisson distribution and a negative binomial distribution with the same mean, but with different variance. The negative binomial captures the volatility of the experience of our case study much better.

is actually present (for example, if the frequency per unit of exposure is increasing over the years) it is better to dig a bit deeper to find the underlying reason for this trend in terms of the changing risk profile, and detrend the data based on the observed and expected changes in the risk profile (see Step F in Chapter 11) rather than try to assume that any trend observed in the past will simply continue.

The results are shown in Figure 14.10. It is often useful also to show the results in a chart such as in Figure 14.11, which gives an at-a-glance view of the volatility and illustrates how sensible (or not) the choice for next year's frequency (the black horizontal line) is.

Because the more recent years are only projected numbers of claims, it is also good practice to show the uncertainty (the standard error) around these projected ultimates. The way these standard errors are calculated depends on how the ultimates have been derived. If the ultimates have been derived by the chain ladder method, we can use the method in Mack (1993) (or the bootstrap) to produce a standard error estimate.

Finally, in Figure 14.12, we show the consequence of choosing a negative binomial rather than a Poisson process to model our data set. The Poisson process has relatively low volatility, so that it is very unlikely to have years when the number of claims is less than 220 or more than 320. The Poisson process is therefore not an appropriate model for our case study. The negative binomial allows for a much wider volatility, which reflects the actual experience of our case study.

14.6 Calibrating a Frequency Model with No Losses

Sometimes the loss experience for a client is so scant as to be absent. This is especially the case if only losses above a deductible are reported to the captive or to the insurer. It seems therefore that we have no information at all on what our frequency model could be.

However, there is actually some information hidden here, and it is actually quantitative information! Before reading on, try to give some thought to the following question on your own.

Question: A client has had no property losses in the last n years (you have no information on the years prior to that) and the client's exposure has remained constant (e.g. the properties have remained the same and no major changes have been made to them). You want to model the number of losses for next year as a Poisson process. What value of the Poisson rate λ would you choose?

If you used a maximum likelihood estimation to estimate the value of the Poisson rate λ for the question above, you would find the simple answer $\lambda = 0$. However, this answer is not satisfactory, because the likelihood of having a property loss cannot be just nil – your prior knowledge of how these things work in practice tells you that.

Also, two bits of valuable, quantitative information are provided to you:

- Your loss experience spans n years.
- There have been no losses in those n years.

Obviously, this information does not pin down your choice of λ to a single value, but depending on how conservative you are in your estimate, you can use one of the simple heuristic rules below.

14.6.1 Heuristic Rules for Selecting a Value of λ in the Case of No Losses

a. To be cautious, you may assume that in the year immediately before the n years you had exactly one claim. This leads to an estimate of the frequency equal to

$$\hat{\lambda} = \frac{1}{n+1}$$

b. To be a bit less cautious (but perhaps fairer), you might assume that although in the past n years you had no losses, in the previous n years you had exactly 1. This would lead to an estimated Poisson rate of

$$\hat{\lambda} = \frac{1}{2n}$$

c. A slightly more sophisticated, and perhaps less arbitrary, rule is to use λ as the maximum value that makes the probability of having observed no losses in the past n years larger than a given value p, for example, 50%:

$$\hat{\lambda} = \text{argmax}\left\{\Pr\left(\text{no. of losses over } n \text{ years} = 0 \mid \lambda\right) \geq p\right\}$$

For example, if $p = 50\%$, we obtain $\hat{\lambda} = \ln(2)/n \approx 0.69/n$: therefore, we are picking the maximum value of λ for which it is more likely to have observed no losses than having observed at least one loss. A more conservative estimate can be obtained by increasing the value of p.

These heuristic rules, and especially Rule (c), hinge upon the idea that the information available does not allow you to pinpoint a unique value for the Poisson rate but gives you some idea on an upper bound for this Poisson rate consistent with not having observed any losses in n years. There is no unique upper bound either, but it becomes unique once you fix a given level of probability for this to have happened.

14.6.2 The Amending Function Approach

One unsatisfactory aspect of these heuristic rules is that they may not be consistent with the frequency selection we make when we *do* have losses. Consider for example Rule (a): if we have no losses, we pick a Poisson rate of $\hat{\lambda} = \frac{1}{n+1}$, but if we have one loss, we are likely to choose a rate of $\hat{\lambda} = \frac{1}{n}$, simply dividing the number of losses by the number of years – it looks like the difference is too small.

Should we not therefore make an allowance for the possibility that there are unseen losses just outside the observation window in the case where some losses have been observed in the observation window as well? Perhaps we should, and in any case, we should try to be consistent between the two cases. So what we need is actually a recipe that makes us choose sensibly a value of the Poisson rate when the number of losses is zero or small. How to make this choice in a mathematically consistent way has been addressed by Fackler (2009), who argues that the assumed frequency should be in the form

$$\hat{\lambda} = \frac{g(k)}{n}$$

where n is the number of years and k is the number of losses. The function $g(k)$ is called the *amending function* and has the purpose of correcting for the effect of unobserved losses. Desirable properties for $g(k)$ are

i. $\lim_{k\to\infty} g(k) = k$
ii. $g(0) > 0$
iii. $g(k) \geq k$
iv. $g(k+1) > g(k)$

along with the (informal) requirement that the function $g(k+1)/g(k)$ should be reasonably smooth. As an example, our heuristic Rule (b) can be written as $g(0) = \frac{1}{2}$, $g(k) = k \forall k > 0$ and such $g(k)$ trivially satisfies all rules (i) to (iv).

14.7 Questions

1. Using a spreadsheet, simulate a random number of claims from a Poisson process using the property that the interarrival time of claims from a Poisson process is an exponential distribution. *Hint*: Use Equation 14.9 and calculate the cumulative time.
2. (*) Prove that Equation 14.5 holds true if Conditions 1 through 4 in Section 14.2 are satisfied. Also prove that if Conditions 1 and 3 are satisfied and Equation 14.5 is satisfied, then Conditions 2 and 4 must hold true.
3. (*) Prove that the Poisson distribution, the binomial distribution, and the negative binomial distribution all satisfy Equation 14.1 and find the values of a and b in terms of the parameters of these distributions.

4. The table below (Case 1) shows the number of employers' liability losses reported to an organisation over the period 2017 to 2024, and the relevant exposure over the same period. Some IBNR factors, based on portfolio estimates, are also provided.
   a. Based on these figures, advise on a frequency model for the number of losses in 2025, for which you also have been given an estimated exposure, and calibrate the model as needed. State all assumptions and approximations you make.
   b. Explain how your advice would change if the number of claims reported were as in Case 2.

		Case 1		Case 2	
Policy year	**Number of employees**	**Number of claims reported (non-zero)**	**Factor to ultimate**	**Number of claims reported (non-zero)**	**Factor to ultimate**
2017	51,500	174	1.000	183	1.000
2018	55,400	182	1.000	202	1.000
2019	69,800	211	1.001	221	1.001
2020	66,100	226	1.005	226	1.005
2021	64,100	184	1.022	235	1.022
2022	62,800	242	1.083	212	1.083
2023	69,400	209	1.257	205	1.257
2024	66,200	51	3.702	65	3.702
2025 (estimate)	65,400				

5. (*) Discuss (a) how the relationship $\frac{p_k}{p_{k-1}} = a + \frac{b}{k}$, which holds for the Panjer class variates, can theoretically be used to determine whether realisations from a count process should be classified as binomial, Poisson or negative binomial, and (b) the practical feasibility of using this as an empirical criterion.

15

Severity Modelling: Adjusting for IBNER and Other Factors

Having produced a frequency model in Chapters 13 and 14, we can now explore the lower 'branch' of the pricing process (Figure 15.1), which aims at producing a severity model based on the past loss experience of the portfolio under investigations and possibly other information from other relevant portfolios of risks.

15.1 Introduction

The first problem that we need to address is that many of the losses that we may use to estimate the overall distribution of losses (the 'severity distribution') are not final settled values but estimates, as they are composed of a paid and an outstanding component, and the outstanding reserve is an estimate of what the remaining amount will be.

The difference between the final settled amount and the overall estimate (paid + outstanding) is called the 'IBNER' (incurred but not enough reserved – or sometimes, bizarrely, incurred but not enough *reported*) amount.

Naturally, if the policy of the company is to set reserves neutrally and is competent at that, you expect errors to average out and systematic IBNER to be zero. However, if reserves systematically overestimate or underestimate the loss amount, then one needs to correct for IBNER. IBNER will be positive if the reserves are underestimated, and negative if they are overestimated.

How do we know if there is a systematic error? Ideally, we should have the complete history of all reserve estimates and how they change in time. Large losses are typically reviewed at regular intervals and the reserve is updated if any new fact emerges. With small losses, this is less likely, but they will still be given a temporary amount according to some company protocol and they will then be updated when they are settled. In both cases, a good strategy would be to compare the estimated amounts with the final one.

Even if we do not have the reserving history, we may still have for each claim the amount reserved and the information on whether it is closed or paid (either stated explicitly or implicitly by looking at whether a percentage of the amount incurred is still outstanding). In this case, a comparison of the severity distribution of the closed/settled claims with the distribution of the open claims may be interesting.

DOI: 10.1201/9781003168881-17

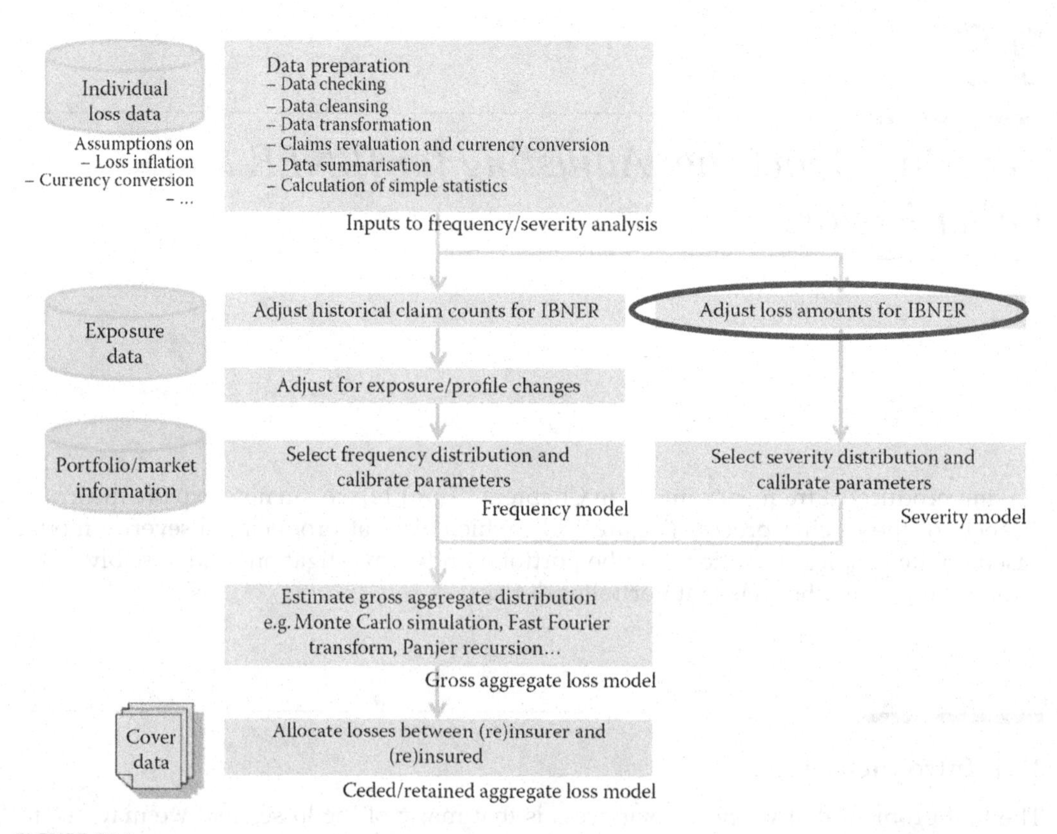

FIGURE 15.1
This figure shows how IBNER analysis fits within the pricing process.

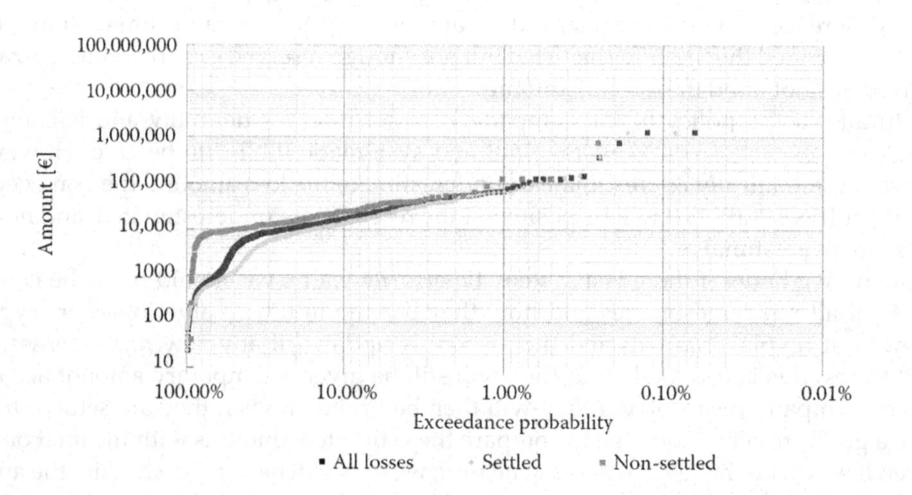

FIGURE 15.2
Each dot in this graph represents a loss. The graph is to be read as follows: for settled (light grey) losses, for example, there are roughly 1% of losses that are above €100k. The graph shows that non-settled (medium grey) losses are usually bigger than the settled losses, especially in the small–medium loss range. This has quite a large effect on the mean: non-settled claims have a mean of €14k compared with €8k for settled losses.

We have done exactly this for our case study and have shown the results in Figure 15.2. In this figure, we are comparing the distribution of settled claims (where the definition of settled was 'more than 95% paid') with the distribution of the non-settled claims. Obviously, the two distributions are, in this case, quite different, and the main difference is that non-settled claims seem much larger than settled claims.

However, by simply looking at the two empirical distributions, it is not clear whether the reason for the difference is that there is a systematic overreserving or that larger claims take longer to settle and therefore tend to be underrepresented in the settled claim data set, or again a combination of the two reasons. This can only be decided with an in-depth statistical analysis of the way that claims are reported, reserved, and settled.

15.2 Methods for Identifying and Dealing with IBNER

There are several ways of identifying IBNER, and to adjust for it. Roughly speaking, IBNER techniques can be grouped under three headings:

15.2.1 Method 1: Ignore IBNER

This is the only method available when you have insufficient information on the development of individual reserves or when you do not know how much has been paid and how much is still outstanding for each claim. By ignoring IBNER, you will naturally introduce an unquantified uncertainty in the analysis, and it should be stated that your analysis does not include a provision for the development of reserves.

15.2.2 Method 2: Just Use Closed Claims

If you have at least information on which claims are settled or almost completely settled, you should consider whether it makes sense to use only the settled claims for your severity analysis. This approach may not be feasible if data is too scanty – you just will not have enough claims to perform a credible analysis.

Also, this approach will almost certainly introduce a bias because larger claims are more likely to be the ones that are open. More complex statistical techniques are therefore needed to remove this bias. As an example, the Insurance Services Office in the United States has developed a methodology to compensate for this bias when building increased limit factor curves, which we are going to describe in Chapter 23. This methodology is described in Palmer (2006) and is based on the Kaplan-Meier product limit estimator for survival probabilities (see, for example, Klugman et al. 2008). We will take a closer look at it in Chapter 23 where we describe the construction of increased-limit-factor curves.

For now, suffice it to say that Method 2 should not be used unless you also have a proper bias removal technique in place.

15.2.3 Method 3: Identify Trends and Adjust for IBNER

The idea here is to identify trends in the estimated reserves for individual claims and use these trends to predict the ultimate value of each claim, much in the same way as that we use to estimate the estimated total incurred claims for reserving purposes.

This method can be used if you have the history of the claim reserves over the years (this is common for very large claims), or at least (for smaller claims) if you have the initial estimate and the settled amount. This will allow you to identify trends and regularities in the way data goes from being fully outstanding to fully paid.

There are many different ways for doing this. Here are two possible methods:

a. Develop individual unrevalued claims triangles with chain ladder–like techniques – the ultimate projected will be the estimated final amount.
b. Use a generalised linear modelling (GLM) style approach to identify the factors that are most relevant to IBNER and predict the final amount.

Let us look at these two methods in more detail.

15.2.3.1 Method 3a: Identify Trends Using Development Triangles

To use this method, which was described in the work of Murphy and McLennan (2006), we need the history of the reserves of a claim over a number of years at regular intervals. Note that if all we have is an initial estimate and a final amount, we can still formally use this method, although another, more straightforward approach, such as looking at the ratio between final cost and initial estimate for a number of claims, is probably more useful. An example of input for this method is given in Figure 15.3.

The first step in performing an IBNER analysis then is to put this data set in another form, by creating a 'triangle' (actually, more like an inverted ladder) like that in Figure 15.4. The differences with the claims listing in Figure 15.3 are that:

1. The columns show the *development year*, namely, the year from the first year in which the claim was reported as non-zero (in the case of reinsurance claims, this often means the first year in which the claim exceeded the reporting threshold and therefore came to the attention of the reinsurer).
2. For every year, there are two columns, one for the estimated incurred and the other for the percentage still outstanding (this is therefore equivalent to giving paid, outstanding, and incurred).
3. For every claim there is a single row.

The meaning of the different fields is probably clearer by direct inspection of Figure 15.4 and the accompanying legends.

Then, we need to:

1. Remove all claims for which the outstanding percentage is 0% (because there is no IBNER for them, by including them, we underestimate the amount of IBNER).
2. Exclude claims that go from non-zero to zero, or vice versa (so as not to mix IBNR and IBNER issues). The idea that the phenomenon of claims going to zero/becoming non-zero is already taken into account by the IBNR analysis of non-zero claims.[1]

1 This is not the only approach that one can take. If the claim count analysis considers all claims, whether zero or non-zero, as one unit, and therefore the number of reported claims only increases, the possibility of a claim becoming nil or non-nil does need to be considered in IBNER analysis (to avoid double counting). Either approach is legitimate, but the important thing is that IBNR analysis and IBNER analysis are consistent in the way they treat zero claims, so that transitions to/from a zero amount are taken into account by either analysis but not both.

Claim ID	Loss date	Loss year	Policy year	Type	2004	2005	2006	2007	2008	2009	2010	2011
210245	7/3/2004	2004	2004	Paid	–	–	300,300	300,300	300,300	720,261	720,261	720,261
210245	7/3/2004	2004	2004	O/S	4,050,000	2,500,000	469,700	413,578	419,961	–	–	–
210245	7/3/2004	2004	2004	Incurred	4,050,000		770,000	713,878	720,261	720,261	720,261	720,261
211136	7/4/2004	2004	2004	Paid	–	–	–	–	–	–	–	–
211136	7/4/2004	2004	2004	O/S	190,000	140,000	140,000	–	–	–	–	–
211136	7/4/2004	2004	2004	Incurred	190,000	140,000	140,000	–	–	–	–	–
213330	7/10/2004	2004	2004	Paid	–	–	–	–	–	–	–	–
213330	7/10/2004	2004	2004	O/S	–	50,000	50,000	–	–	–	–	–
213330	7/10/2004	2004	2004	Incurred	–	50,000	50,000	–	–	–	–	–
216085	7/15/2004	2004	2004	Paid	–	–	21,773	50,942	50,942	50,942	50,942	50,942
216085	7/15/2004	2004	2004	O/S	50,000	85,000	43,227	–	–	–	–	–
216085	7/15/2004	2004	2004	Incurred	50,000	85,000	65,000	50,942	50,942	50,942	50,942	50,942
217042	7/16/2004	2004	2004	Paid	–	–	–	–	–	–	–	–
217042	7/16/2004	2004	2004	O/S	–	100,000	100,000	100,000	100,000	–	–	–
217042	7/16/2004	2004	2004	Incurred	–	100,000	100,000	100,000	100,000	–	–	–
217372	7/12/2004	2004	2004	Paid	–	–	–	–	11,146	11,146	11,146	11,146
217372	7/12/2004	2004	2004	O/S	265,000	300,000	300,000	11,146	–	–	–	–
217372	7/12/2004	2004	2004	Incurred	265,000	300,000	300,000	11,146	11,146	11,146	11,146	11,146

FIGURE 15.3
An example of data set with the history of the reserves for each claim. Paid, outstanding, and incurred are listed separately. Notice that Paid + O/S = Incurred.

Date of loss	Loss year	Year reported as non-zero	Year 1 Estimate	Year 1 O/S %	Year 2 Estimate	Year 2 O/S %	Year 3 Estimate	Year 3 O/S %	Year 4 Estimate	Year 4 O/S %
7/3/2004	2004	2004	4,050,000	100.00%	2,500,000	100.00%	770,000	61.00%	713,878	57.93%
7/4/2004	2004	2004	190,000	100.00%	140,000	100.00%	140,000	100.00%	–	0.00%
7/10/2004	2004	2004	50,000	100.00%	85,000	100.00%	65,000	66.50%	50,942	0.00%
7/15/2004	2004	2004	265,000	100.00%	300,000	100.00%	300,000	100.00%	11,146	100.00%
7/16/2004	2004	2004	2172	0.00%	2172	0.00%	2172	0.00%	2172	0.00%
7/12/2004	2004	2004	1,500,000	100.00%	1,500,000	100.00%	1,659,825	0.00%	1,659,825	0.00%
7/2/2004	2004	2004	1830	0.00%	1830	0.00%	1830	0.00%	1830	0.00%
7/23/2004	2004	2004	111,400	100.00%	80,000	100.00%	80,000	83.31%	80,000	83.31%
7/3/2004	2004	2004	65,000	100.00%	0	0.00%	–	0.00%	–	0.00%
7/28/2004	2004	2004	247,000	100.00%	275,000	100.00%	189,482	0.00%	189,482	0.00%
7/24/2004	2004	2004	280,000	100.00%	320,000	100.00%	58,114	0.00%	58,114	0.00%
7/26/2004	2004	2004	350,000	100.00%	0	0.00%	–	0.00%	–	0.00%
7/26/2004	2004	2004	6790	100.00%	4138	0.00%	4138	0.00%	4138	0.00%

FIGURE 15.4
The claims listing in a format suitable for IBNER analysis. Each column represents a development year. Note that 'Year reported as non-zero' refers to the first year in which the claim exceeded the reporting threshold – a common feature of reinsurance claims.

Year reported as non-zero	Estimate (year 1)	O/S % (year 1)	Estimate (year 2)	IBNER factor 1–2
2006	2000	0.00%	2000	1.00
2006	3000	0.00%	3000	1.00
2006	3,000,000	84.50%	3,750,000	1.25
2006	4500	0.00%	4500	1.00
2006	210,000	70.00%	230,000	1.10
2006	25,000	100.00%	–	–
2006	70,000	100.00%	50,000	0.71
2006	1,200,000	0.00%	1,200,000	1.00
2006	2000	0.00%	2000	1.00

FIGURE 15.5
An example IBNER calculation. The only claims that are relevant for the calculation of the IBNER factor between years 1 and 2 are the third, fifth, sixth, and seventh claim, as they are the only ones with a percentage still outstanding and are non-zero both in year 1 and year 2. The fifth column gives the IBNER factor between year 1 and year 2 for that particular claim: the relevant ones are for us 1.25, 1.10 and 0.71.

Year	IBNER	Cumul IBNER	Period
1–2	1.22	0.82	(1 onwards)
2–3	1.02	0.67	(2 onwards)
3–4	0.84	0.66	(3 onwards)
4–5	0.96	0.79	(4 onwards)
5–6	0.91	0.82	(5 onwards)
6–7	0.89	0.89	(6 onwards)
7 onwards	1.00	1.00	(7 onwards)

FIGURE 15.6
The IBNER factors for our example data set, calculated with the method of the weighted average.

Therefore, for example, if a claim with a non-zero reserve eventually goes to nil, that will decrease the non-zero claim count by one unit.

3. Use, for example, a chain ladder methodology to calculate the link ratios.

An example calculation is shown in Figure 15.5 with a small sample of nine claims. The result will of course depend on which statistic you use for calculating the average IBNER: in the case of Figure 15.5, for example, we have

$$\text{Simple average} = \frac{1.25 + 1.10 + 0.71}{3} = 1.02$$

$$\text{Weighted average} = \frac{3{,}750{,}000 + 230{,}000 + 50{,}000}{3{,}000{,}000 + 210{,}000 + 70{,}000} = 1.17$$

$$\text{Median} = 1.10$$

The results for our example data set (see book's website) are given in Figure 15.6.

Once equipped with the IBNER factors of Figure 15.6, we can then project to ultimate our reserved estimates and calculate the IBNER amount: for example, if X was first reported as non-zero in 2008 and is now (in 2011) estimated as £100,000, its IBNER-corr. value is

$$X(2011) \times \text{Cumul\_IBNER}(4 \rightarrow \text{onwards}) = £100\text{K} \times 0.79 = £79\text{k}$$

Comments

- One may want separately to calculate the IBNER distribution or mean of, say, large losses compared with small losses, or losses with different ranges of outstanding percentage (e.g. <25%, 25%–75%, >75%); however, one must be aware of introducing spurious accuracy – unless the data set is large enough, there will not be enough information to have robust estimates of the IBNER factors.
- If the sample over which IBNER is calculated is too small, for example, with less than 30 data points, one may want to set IBNER = 1, to avoid trusting IBNER factors that are mere statistical fluctuations, unless there are clear tendencies even within small samples (for example, all IBNER factors being larger than 1).
- When there are not enough data points, this method will not give statistically meaningful results. In this case, a solution might be to use industry IBNER factors as a proxy. Yet another alternative is to derive your own market IBNER factors by pooling a number of similar risks together.
- Note that by using roughly the same method, one can also produce a distribution of possible IBNER factors because the intermediate output of this method is a collection of IBNER factors that can be sorted and its percentiles can be calculated. This was done, for instance, by Murphy and McLennan (2006).

The distribution of IBNER factors thus calculated can then be used in the investigation of parameter uncertainty of the severity distribution. We will not get into the details of this here.

15.2.3.2 Method 3b: Identify Trends Using Multivariate Analysis

Method 3a uses a one-size-fits-all approach, in that it puts together all claims and calculates some IBNER averages by reporting lag (in years). However, it is obvious that IBNER will in general depend on many factors, such as

- Development year d (the younger the claim, the more scope there is for it to deviate from the incurred estimate), as already mentioned
- Size of claim x (larger claims may have more uncertainty about them)
- Outstanding ratio r (the larger the outstanding amount, the more conjectural the estimate of the incurred value will be)
- Type of claim t (for example, bodily injury claims will be more difficult than property claims to estimate reliably because the victim's health may change unexpectedly)

One solution, which was mentioned at the end of the section in Method 3a, is to repeat the same chain ladder exercise for different categories of claims, for example, subdividing by claim size, by outstanding ratio, and so on. However, as we also mentioned, if we do this quickly, we could run into problems about the credibility of the sample: there will not normally be enough data to allow for a statistically significant analysis of the difference between the many different cases.

An alternative approach is to use a generalised linear modelling (or other multivariate analysis) approach and write IBNER factors as a linear combination of different variables, possibly further transformed by a function h:

$$\text{IBNER}(d,x,r,t\ldots) = h\left(a_1 f_1(d,x,r,t\ldots) + \ldots + a_n f_n(d,x,r,t\ldots)\right) \tag{15.1}$$

and identify the relevant functional form of $h, f_1, \ldots f_n$ and the parameters $a_1, \ldots, a_n$ that are suitable to describe the IBNER factors. For example, one might consider a very simple model such as

$$\text{IBNER}(d,x,r) = e^{-(a_1 + a_2 d + a_3 x + a_4 r)} \tag{15.2}$$

This is a special case of the more general Equation 15.1, where:

- $f_1(d, x, r, t\ldots) = 1$
- $f_2(d, x, r, t\ldots) = d$
- $f_3(d, x, r, t\ldots) = x$
- $f_4(d, x, r, t\ldots) = r$
- $h(x) = \exp(x)$

Generalised linear modelling will be addressed in a subsequent unit. For now, suffice it to say that this method allows us to consider all variables at the same time with less demanding requirements as to the size of the data set.

15.3 Comparison of Different Methods

We now look briefly at the advantages and disadvantages of the different methods for dealing with IBNER that we have introduced in this chapter.

15.3.1 Method 1: Ignore IBNER

(+) The method is very simple, and only requires a snapshot of the loss run at a given point in time – no need for the full historical development of claims.
(+) It will work fine if there is no systematic overreserving and underreserving.
(+) It is the only option when there are not enough loss data to produce a credible statistical analysis of IBNER, and when no industry IBNER factors are available.
(−) If there is systematic overreserving or underreserving, the results will inevitably be biased.

15.3.2 Method 2: Just Use Closed Claims

(+) Does not use loss estimates and therefore the input to the analysis is more certain.
(−) It does not use all data and therefore 'conceals' evidence.
(−) It should not be used without a bias-correction mechanism to take account of the fact that large claims are the ones that are more likely to take longer to settle.

15.3.3 Method 3a: Identify Trends and Adjust for IBNER by Using Development Triangles

(+) Allows the use of all data, even the estimated amount.
(+) It is relatively simple because development factors can be calculated, for example, by chain ladder techniques.

(−) It requires more data than Method 1.
(−) In its simplest version, it does not take into account the fact that losses of different size and different outstanding percentage may have different developments, and therefore the estimate may be too rough.

15.3.4 Method 3b: Identify Trends and Adjust for IBNER by Using GLM-Like Techniques

(+) Allows the use of all data, even the estimated amount.
(+) Allows the consideration of the effect of different factors such as loss size and outstanding percentage on the way to the loss will develop.
(−) It is a relatively complex method.
(−) The method only works with a significant amount of data.

15.4 Questions

1. You need to produce the input losses for a severity model.
   i. Describe three methods for dealing with IBNER when producing such input.
   ii. Explain the advantages and disadvantages of each method listed in (i).
   iii. Calculate the IBNER factor from year of development, 2 to year of development, and 3 based on the following cohort of claims, using different types of average (weighted average, straight average, median) of the individual development factors, showing your calculations.

Development year 2	Development year 3	Development factor from year 2 to year 3	Outstanding ratio
1419	1419	1.00	0%
1809	1809	1.00	0%
1,041,681	1,332,408	1.28	99%
48,001	48,001	1.00	87%
111,808	111,808	1.00	0%
200,050	200,050	1.00	0%
30,494	2777	0.09	98%
149,398	228,689	1.53	42%
80,387	80,387	1.00	65%
9026	9026	1.00	0%
280	6819	24.36	81%
1419	1419	1.00	0%
978	978	1.00	0%
1355	1355	1.00	58%

   iv. Explain the relative advantages/disadvantages of each of the three types of average mentioned in (iii) *for the purpose of calculating IBNER factors.*

2. As a result of the overall IBNER analysis, you produce the following cumulative IBNER factors.

Development years	**IBNER (cumulative)**
1->onwards	2.44
2->onwards	1.32
3->onwards	1.20
4->onwards	0.95
5->onwards	1.00

Development year = 1 is the first year when the loss was reported.

Apply the relevant factors to the following list of five claims and calculate the amount adjusted for IBNER:

Loss ID	**Loss amount at current position**	**Outstanding percentage**	**Year of loss**	**Year of reporting**	**Year of current position**
1	10,000	50%	2009	2012	2013
2	30,000	15%	2010	2010	2013
3	25,000	0%	2010	2011	2013
4	45,000	10%	2008	2009	2013
5	7000	100%	2011	2013	2013

16

Severity Modelling: Selecting and Calibrating a Severity Model

The objective of severity analysis is the production of a severity model (Figure 16.1) for the loss amounts. A severity model is not only essential to calculate – in combination with a frequency model – the distribution of aggregate losses but also will help us to reply to questions such as

- What is the probability that an individual loss is going to be at least £100,000?
- What level of deductible is adequate for this risk?
- What limit should I purchase for my public liability policy?

All the questions above are not questions about the total losses but about the distribution of individual loss amounts.

Actuaries have of course always been concerned with the severity of individual loss amounts, but in traditional practice, both in reserving and pricing, they have often been happy with considering the average loss amount over a period (see Chapter 12). However, average loss amount analysis, although a useful tool for exploratory analysis and in contexts where the number of claims is very large (such as private motor), is not sufficient to reply to questions like those listed above.

16.1 Input to Severity Modelling

The input to severity analysis is the database of loss amounts:

- After revaluation (Chapter 10)
- After currency conversion (Chapter 10)
- After correction for IBNER (Chapter 15)

An example of such an input in graphical format is shown in Figure 16.2. The losses represented there are those of a real-world data set. We have made no distinction here between open and closed losses.

DOI: 10.1201/9781003168881-18

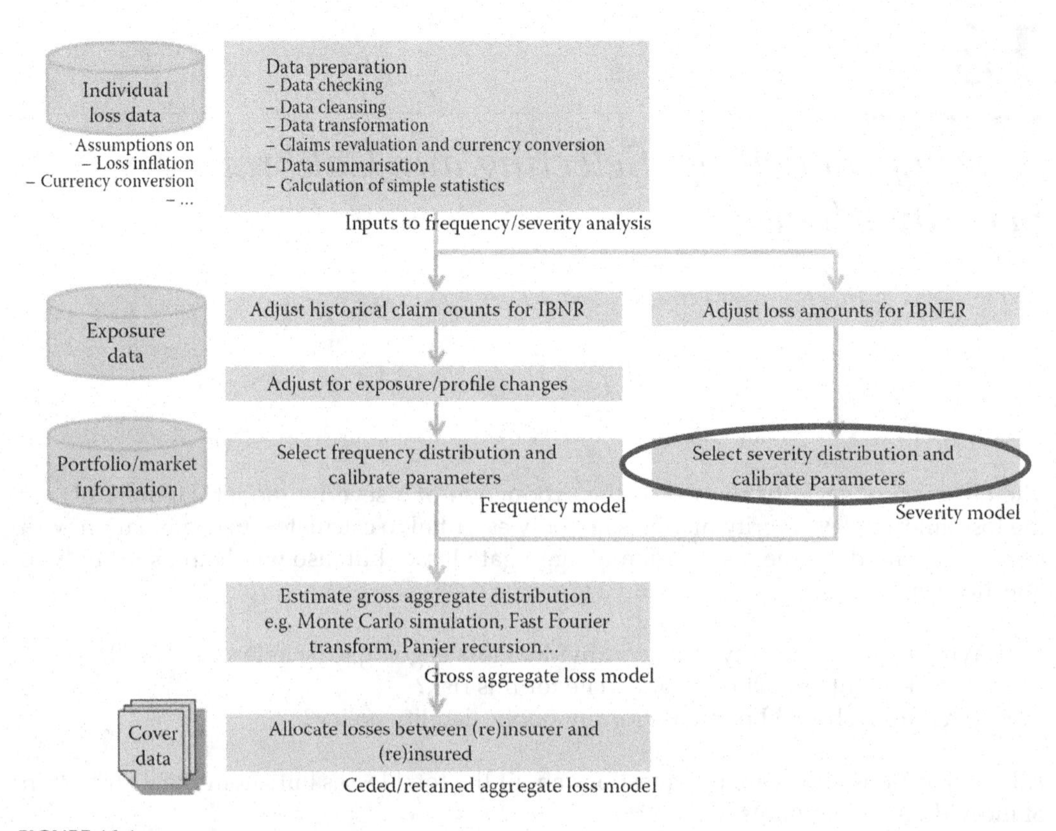

FIGURE 16.1
How the selection of a severity model fits into the overall pricing process.

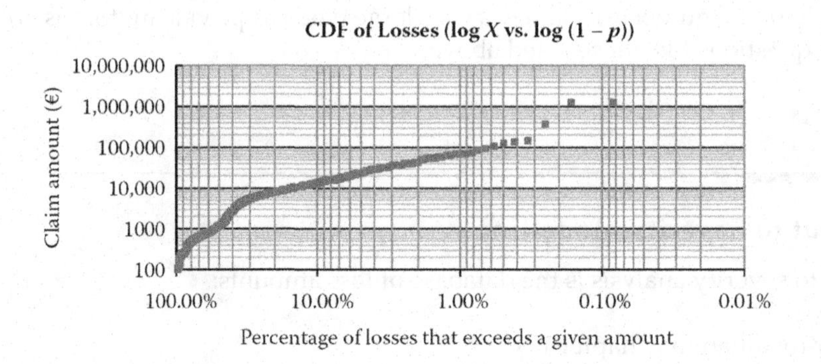

FIGURE 16.2
A graphical representation of the loss data base (or data set). Every dot in this graph represents a loss. For each loss, the value of the x-axis represents the percentage of losses in the database that exceed the value of that loss. For example, roughly 10% of losses are above £10,000. The graph is in log-log scale, in the sense that although the values shown are the original ones, the amount represented is actually log X against log $(1 - p)$, where p is the percentage of losses exceeding a given amount. This log-log representation has many advantages: it allows showing losses over many orders of magnitude, and the distribution is a straight line if and only if the losses come from a single-parameter Pareto distribution, which is useful for analysing the behaviour of the tail.

16.2 Choosing the Right Severity Model: A Subtle Problem

We now need to choose a severity model that adequately represents this data set. How do you do that? An all-too-common approach is to fit the historical losses to a variety of distributions from a distribution-fitting tool and simply choose the one that fits best. As we have already mentioned in Chapter 12, this carpet-bombing approach has several problems:

- It focuses on goodness-of-fit rather than prediction.
- There are too many free parameters that can be optimised to give a spuriously good fit: some distributions have three to four parameters and, normally, one has the freedom to choose amongst 20 to 30 distributions, which is like having an extra (categorical) parameter. This is especially critical for data-sparse situations.
- There is often no reason behind the inclusion of some of the models for our problem at hand – we accept *any* distribution as long as it fits, ignoring what it should be used for. This is not in itself damning – a distribution that is chosen for no good reason might still turn out to be a good one – but becomes so if an adequate validation process is missing.
- There is no consistency amongst different risks – if a Lognormal or a Gamma distribution is a good model for many different clients, surely there is no reason to switch to another distribution for one policyholder just because it happens to fit better in that instance.

To show that this approach does not work in practice, consider the following experiment with an artificial data set (Chhabra and Parodi, 2010), which is assumed to represent liability losses. What this simple experiment does is generate random data from a Lognormal distribution, then fit the data with a distribution-fitting tool: in most cases, the distribution that best fits the data turns out not to be the Lognormal distribution! Because, in this case, the true model ***is*** indeed (by construction) a Lognormal distribution, this shows that there is something wrong with this model selection method.

In more detail, here is how the experiment was devised.

1. Samples of different sizes ($n = 20$, $n = 100$, $n = 1000$) were from a severity distribution, which was known a priori: $X \sim \text{LogN}(9,2)$.
2. A distribution-fitting tool was used to choose the best amongst a set of 22 distributions (without discarding any of them a priori).
3. The parameters of the distributions were calculated using a standard maximum-likelihood estimator.
4. The fit was assessed using a variety of methods: Akaike information criterion (AIC), chi-square, Kolmogorov–Smirnov (KS), Kuiper, and Anderson–Darling (AD), which were all included in the distribution-fitting tool.

The rather surprising results are shown in Figure 16.3, which tells us that even when the sample of losses is quite large ($n = 1000$), the Lognormal is not always chosen as the best distribution that fits … a Lognormal process! Most goodness-of-fit tests will choose a Beta transformed distribution as well, which is a capped distribution, just because it has more parameters and can therefore be more easily adapted to any data set. The AIC chooses the Lognormal because this method imposes a penalty for the number of parameters, thereby

		AIC	Least squares	KS	Kuiper	AD
N = 1000	Lognormal ranks...	#1	#2	#2	#2	#1
	... and the winner is:	Lognormal	Beta transformed	Beta transformed	Beta transformed	Lognormal
	# of parameters:	2	4	4	4	2
N = 100	Lognormal ranks...	#5	#9	#9	#9	#10
	... and the winner is:	Pareto	Burr	Burr	Burr	Loglogistic
	# of parameters:	2	3	3	3	2
N = 20	Lognormal ranks...	#4	#9	#9	#10	#5
	... and the winner is:	Paralogistic inverse	Beta transformed	Beta transformed	Beta transformed	Gamma transformed
	# of parameters:	2	4	4	4	3

FIGURE 16.3
Results obtained with Risk Explorer, with 22 distributions to pick from, using maximum likelihood estimation (MLE), for different sample sizes. The table shows how the Lognormal ranks using different methods (AIC, least squares, and others) to assess the fit, and also what the winning distribution is, and how many parameters it has. It shows this for three different data sets, with a number of data points equal to 1000, 100, and 20, respectively.

	True lognormal Logn(9,2)	Estimated lognormal Logn(8.86,1.78)	Beta Transformed	Burr
Mean	60,945	34,722	16,448	244,294
95%	218,750	131,815	78,649	138,859
99.5%	1,439,375	701,616	78,649	1,891,458
99.9%	3,941,002	1,715,139	78,649	13,883,349

Curve fits based on N = 20

FIGURE 16.4
Comparison of some percentiles as calculated using the original distribution with the true parameters (LogN(9,2)) with other distributions that are frequently picked by the distribution-fitting tool, including the Lognormal itself but with the parameters calculated based on the sample (LogN(8.86,1.78)).

forcing the model to be simpler. The AD method chooses the Lognormal because it focuses on fitting the tail, thereby probably assuming that the Beta transformed distribution does not have a suitable tail because it is a capped distribution.

When the number of data points is smaller (n = 100, n = 20), the Lognormal just does not stand a chance with any method, and it never ranks higher than #4 – which is quite disconcerting. The general rule again is that most methods for assessing goodness-of-fit will tend to pick distributions such as Burr, Beta transformed, and Gamma transformed, which have three or more parameters and are therefore more adaptable. The main exception is AIC, because of this embedded mechanism for punishing model complexity.

One might argue that although this methodology seems inadequate at choosing the right distribution, this may not really matter as long as the predictions made with the chosen method are similar to those made with the true distribution (LogN(9,2)).

Figure 16.4 should duly crush that remaining hope. Based on a sample of n = 20 (a rather extreme case, but not uncommon), we see that especially for high percentiles such as 99.5%, the difference in the prediction is huge. Notice how, for example, a Beta transformed distribution, being capped at around £78.6k (the largest claim in the data set) will produce a 99.5% of around £78.6k, whereas a Burr distribution will predict £1.89M. Of course, predicting the 99.5% percentile based on 20 data points is never a good idea, but the error is less extreme when the correct model is chosen, as the comparison between LogN(9,2) and LogN(8.86,1.78) shows.

16.2.1 General Rules for Model Selection

Here are some of the lessons that can we draw from our simple experiment with Lognormal data:

- Unconstrained distribution fitting to select the best distribution does not work (as illustrated by the fact that the correct distribution – the Lognormal – ranks consistently too low amongst the possible choices of distribution).
- More complex models – that is, models with more parameters – are more likely to be chosen.
- Unsuitable models such as Beta transformed for liability losses (not adequate to represent liability losses because it is limited) are likely to be chosen.
- The only method for assessing goodness-of-fit that sometimes gives sensible results is AIC, because it penalises the complexity of the model.
- The prediction error – especially on the calculation of the highest percentiles – is very large.

This suggests the following strategy for selecting the right model (a complete version of this, which includes Bayesian analysis, can be found in Chhabra and Parodi (2010); however, that is beyond the scope of this textbook).

General Rules for Model Selection	
1. Restrict the number of admissible models	Use experience or theoretical reasons, such as extreme value theory, to look only at interesting models. Exclude distributions with undesirable properties, such as a capped distribution to represent an uncapped phenomenon, or distributions that can produce negative losses. Choose only distributions with some rationale – or which are known to be successful in representing other clients, or the market as a whole.
2. Punish model complexity	According to statistical learning theory, any model more complex than necessary makes poorer predictions. If there is sufficient data, one should partition the data set into a training set (used for parameter estimation) and a test set (used for selection and validation). In this way, models that are too complex will be automatically discarded as they will perform badly against the test data set. If data is insufficient for such a partition, one should consider using a criterion such as the AIC, which punishes complexity by minimising[a] AIC = −2 log(Pr(data)) + 2 × no of parameters.

a See Section 27.3.5.2 for an explanation of why AIC may be a decent substitute for the strategy of partitioning the data set into a training and a test set. In a nutshell, however, it can be shown (see, e.g. Hastie et al., 2001) that the AIC provides a rough approximation of the prediction error.

Distribution	–log Pr(D|M)	AIC	No. of parameters
Logn	215.12	219.12	2
Burr	213.36	219.36	3
Weibull	216.71	220.71	2
Gamma	218.77	222.77	2
Loglogistic	219.01	225.01	3
Frechet	225.05	229.05	2
Paralogistic	229.70	233.70	2
Gumbel	244.54	248.54	2

FIGURE 16.5
Ranking of the admissible models (some models, such as the Normal distribution and the Beta transformed distribution, have been excluded) according to AIC (third column). The Burr distribution would be ranked first … if not for the penalty term of AIC (2× number of parameters), which adds 6 to the AIC for Burr, 4 for Lognormal.

16.2.2 Experiment Revisited

Let us now see how this strategy would have worked if we had used it on our Lognormal data sets. (To be rigorous, we should repeat the experiments with other data to demonstrate that we have not just chosen the method that will work on these particular data sets.)

1. Exclude inverse distributions, threshold distributions (see Section 16.3 for a definition), and distributions with negative values such as Normal.
2. Rank distributions according to the AIC.

The results are shown in Figure 16.5, which demonstrates that if we rank the distributions according to AIC, the Lognormal comes up on top, as it should (third column) – although not by much. This table also shows what the effect of punishing complexity is: the second column shows the goodness-of-fit as assessed by the likelihood of the data only (the first term of the AIC). As you can see, if you used that criterion only, the Burr would come on top, but once you add the penalty for the number of parameters ($3 \times 2 = 6$) the Burr distribution falls into second place, behind the Lognormal.

16.3 Modelling Large ('Tail') Losses: Extreme Value Theory

We have seen in Section 16.1, and especially in Section 16.2.1, how important it is to incorporate experience and theoretical reasons in the selection of the severity model rather than scatter-gunning for a model that fits the data well. However, we haven't given any example of how theory can actually guide us towards the use of a specific severity model.

The modelling of large losses, however, provides an example that shows that this is not just wishful thinking but, on the contrary, the reason for choosing a particular model can be quite compelling.

First of all, to model large losses, one needs a ***threshold distribution***, that is, a distribution with values above a certain positive threshold (as opposed to a distribution over all real numbers, or over all non-negative numbers).

Second, we want a severity distribution whose density decreases with size; for example, if the threshold is £1M, it does not seem sensible that the probability of a loss of £1.5M should be higher than the probability of a loss of £1.2M.

These combined requirements exclude, for example, distributions such as the Normal distribution (whose domain is the whole set of real numbers), and the Lognormal distribution (whose domain is all non-negative real numbers, and whose density increases and decreases, with a mode at $x = \exp(\mu)$). They include distributions such as the shifted exponential distribution, $F(x) = 1 - \exp(-(x - \theta)/\sigma)$, and the single-parameter Pareto, $F(x) = 1 - (\theta/x)^{\alpha}$, which are defined above θ.

These constraints are useful for restricting the number of possible models for tail losses. However, it turns out that we can say much more about tail losses than this: there is a whole branch of statistics – called *extreme value theory* – which deals, amongst other things, with the problem of fitting losses above a (large) threshold.

Extreme value theory was born for a very specific problem in hydrology – predicting how large tides could be, so that it would be possible to build dams that are high enough that water does not spill over except in truly exceptional circumstances – say, once in 1000 years. Not surprisingly, many of the statisticians that came up with valuable results for this problem are from the Netherlands.

16.3.1 Pickand–Balkema–de Haan Theorem

Extreme value theory has many interesting results, but for our present purposes we need only one, the so-called *Pickand–Balkema–de Haan theorem*. The theorem states that no matter what the ground-up distribution looks like, its tail (i.e. the distribution above a given threshold) can always be modelled by the following distribution, which is called the generalised Pareto distribution (GPD):

$$G(x) = \begin{cases} 1 - \left(1 + \xi \dfrac{x - \mu}{\sigma}\right)^{-1/\frac{3}{4}} & \xi \neq 0 \\ 1 - \exp\left(-\dfrac{(x - \mu)}{\sigma}\right) & \xi = 0 \end{cases} \tag{16.1}$$

This is a distribution whose PDF is zero up to a certain value μ, and then declines to zero more or less quickly depending on the value of a certain parameter ξ called the 'shape parameter'.

- If $\xi > 0$, the distribution goes to zero as a power law; that is, quite slowly. Imagine the function $f(x) = 1/x$, which goes so slowly that you cannot even calculate the area under it because it diverges; or $f(x) = \dfrac{1}{x^2}$ (in physics, a law like this describes the strength of interaction between two planets or a planet and the sun, with x being the distance between the two bodies, which goes to zero a bit more quickly, but still slowly enough that very distant objects such as galaxies are kept together by it). $f(x) = \dfrac{1}{x^n}$ seems to go to zero very quickly for large n, but it is still relatively slow when you compare it with an exponential distribution.
- If $\xi = 0$, the distribution goes to zero as an exponential, such as $f(x) = exp(-x)$: that is, really fast. Again, to resort to examples from physics, neutrons and protons in an atomic nucleus interact with a strength that goes to zero with distance with a law similar to this: a very strong force that keeps nuclei together, but which is absolutely

negligible outside the nucleus so that one nucleus does not affect the nucleus of other atoms.

- If $\xi < 0$, the distribution is capped, that is, it is zero above a certain threshold. I have no fundamental force of nature to showcase for this, but no physical analogy is needed to understand the properties of such a distribution!

The other parameter, σ, is just a 'scale parameter', in the sense that it stretches the distribution horizontally without actually changing the way it goes to zero.

Therefore, this method allows you to subdivide distributions according to their tail behaviour: fat-tailed power law distributions (such as Pareto, Pareto II, Burr.), thin-tailed exponential law distributions (such as Lognormal, Normal, Gamma ...) and capped distributions (such as Beta distribution).

This is the Pickand–Balkema–de Haan theorem in a nutshell. A more mathematically accurate version of this theorem is discussed, for example, in Embrechts et al. (1997) and the main differences with the broad-brush outline above is that:

i. The tail of a distribution becomes equivalent to a GPD only asymptotically, that is, when for $\mu \to \infty$.
ii. By 'becomes equivalent', we mean that the tail distribution of $F(x)$, $F_\mu(x)$, converges to a GPD $G_\mu(x)$ in the sense of distributions, that is, the upper bound of the difference of the empirical and modelled distribution goes to zero.
iii. This is true not for all distributions but for all distributions that satisfy certain (rather broad, admittedly) constraints.

In practice, these mathematical clarifications do not significantly dampen the message and if μ is large enough, things will in practice work pretty well. But when is μ large enough?

16.3.2 Identifying the Extreme Value Region

The key difficulty with extreme value theory is indeed understanding when μ is large enough that you can apply it: that is, finding the so-called 'extreme value region'. There are many recipes for finding this region.

The most famous one is probably that of working out the mean excess value function $e(\mu)$ (the mean value of losses exceeding a threshold) and plot it against the value of the threshold μ itself: for a GPD, this should be a linear function of μ:

$$e(\mu) = \mathbb{E}(X-\mu|X\rangle\mu) = \frac{\tilde{A}+\xi\mu}{1-\xi} \tag{16.2}$$

Another, often more effective, approach is that of plotting the value of the shape parameter ξ for various values of the threshold μ, and plotting the (normalised) KS distance at the same time, as in Figure 16.6. The idea behind this second approach is to discard all points of a data set except those higher than a given μ, where μ is chosen equal to each of the data points, and calculate the values of ξ and σ for the remaining points. One then chooses a value of μ around which the estimate of ξ is stable.

Here are a few simple rules on how to choose the threshold: these rules will not determine the threshold uniquely, but should exclude the values that do not make sense.

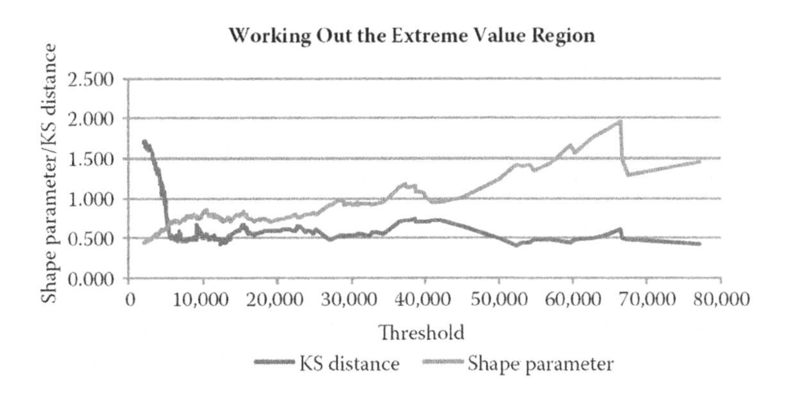

FIGURE 16.6
The value of the shape parameter and of the (normalised) KS distance as a function of the threshold, for different values of the threshold (black line and grey line, respectively). Approximately €10,000 of the value of the shape parameter starts stabilising at around 0.75, and the normalised KS distance decreases to a very low level, around 0.5 – this makes €10,000 a suitable choice for the threshold. On the contrary, note how at more than €50,000 the model does not work: the shape parameter is on the rise and behaves rather erratically, going well above 1 (which is a problem because if the shape parameter is >1 the mean is not defined, making pricing impossible). Note that the shape parameter is calculated by maximum likelihood estimation (as outlined in Section 16.3.4) for different values of the threshold (and excluding all losses below that threshold).

Choose μ such that:

1. The normalised KS distance is low. A rough rule of thumb is that the KS distance should be less than approximately 0.8 for a 5% significance level. The exact value depends on the value of the GPD parameters and is lower than the critical value normally found in statistical tables because the parameters are calculated from the data. See the article by Porter et al. (1992) for an analysis of this modified KS test for the GPD, and more precise quantitative results.
2. The graph of ξ is 'flat', that is, its value is roughly constant around that value of μ.
3. ξ is lower than 1 (if ξ is larger than 1, the mean is not defined and the distribution cannot be used for pricing).
4. There are at least 30 data points larger than μ: you will often see that the graph behaves wildly when there are approximately less than 30 points. This is because of parameter uncertainty, which makes the estimate of ξ very unstable.
5. Choose the smallest value of the threshold for which Properties 1 through 4 apply – this will reduce the parameter uncertainty to a minimum by basing the parameter calculations on a set that is as large as possible.

16.3.3 Creating a Spliced Small/Large Loss Distribution

Now that we have chosen a threshold on which to model the data set with a GPD, we are almost done. Although if we use GPD as part of a ground-up model (or, in any case, as part of a model with a lower threshold), we need to combine it with a small loss distribution. This procedure is called splicing and the combined distribution can be written like this:

$$F(x)=\begin{cases} F_1(x) & 0\le x\le \mu \\ F_1(\mu)+(1-F_1(\mu))G_{\mu,\xi,\sigma}(x) & x>\mu \end{cases} \tag{16.3}$$

where $F_1(x)$ is the distribution for the small losses, such as a Lognormal, or the empirical distribution. This is quite handy when creating an input for a simulation tool, as it allows us to calculate all percentiles of the distribution.

16.3.4 Calculating the Values of the Parameters

Thus far, we have not explained how the parameters of the GPD are actually calculated. This is normally done by maximum likelihood estimation, but there are specialised algorithms that can find the values of the parameters more efficiently. One such algorithm is that by Smith (1987). For more information on this, read *Modelling Extremal Events* by Embrechts et al. (1997) or, for a less formal treatment, the book by Klugman et al. (2008).

16.3.5 Extreme Value Theory: Actual Practice

Is this the way actuaries – especially reinsurance actuaries, who work with large losses as a matter of course – actually do things? Of course, it is not easy to know for sure because companies do not normally advertise the technical details of their pricing methods. However, it is probably fair to say that:

- The GPD model is used by many participants in the London market, in Germany, Switzerland, and in the United States – and its use is on the rise.
- Some large reinsurance companies still use single-parameter Pareto, with some modifications (e.g. changing the parameter continuously, or using a combination of different Paretos) to make it work in practice. In general, practitioners may use the single-parameter Pareto for those cases where there are not enough data points to calibrate a GPD.
- Many practitioners still use a Lognormal model, shifted to start from the analysis level. Although this may give acceptable results in some circumstances, it does not really make sense: the density of the Lognormal distribution has a (skewed) bell shape with a big maximum, and it is not clear what it represents: the probability of a loss above a certain size should decrease with size, not go up and down. And in practice, it never fits a tail very well.

16.3.6 Limitations of Extreme Value Theory

Although extreme value theory drastically reduces the model uncertainty (we know a priori what the right model is), and seems to perform very well in practice, it has some limitations:

Threshold choice. There is some arbitrariness in the choice of the threshold, and because the prescription is only of an asymptotic nature, there is no guarantee that the loss data will not change behaviour as the threshold changes, even if the GPD fit seems to be good.

Parameter uncertainty. Parameter uncertainty causes significant problems in pricing, especially in the case of sparse data – and data are sparse in the tail almost by definition. In reinsurance, for example, the GPD can be used to predict the price of different layers, but the error by which the expected losses to the layer are calculated increases as the layer's excess increases. Eventually, there is a point where the error on the expected losses becomes larger than the expected losses themselves. There is, in other terms, a sort of 'pricing horizon'. For example, when you have 100 losses at more than £500,000, the pricing horizon may be approximately £3 to 5M; however, you still you need to price layers such as £15M xs £10M or unlimited xs £25M.

In particular, the values of the parameter show a strong dependency on a few large data points.

Parameter bias. When the data is sparse, not only is there a large error on the parameter, but there is also a strong bias on the parameters themselves: ξ is underestimated, and σ (which is negatively correlated to ξ) is overestimated.

16.4 A Simple Strategy for Modelling Ground-Up Losses

In Section 16.2.1, we have given some 'general rules for model selection': in this section, we present a specific method for modelling the severity of losses, which is consistent with these rules and is adequate for losses from the ground up.

The method presented here, although somewhat general, should not be construed as a one-size-fits-all method that can be applied indiscriminately to all situations, but rather as an illustration that when it comes to severity modelling, the focus should be not on what distributions we choose, but on the existence of a clear process that guides the choice of the model.

As a practitioner, however, you should have your own process in place that is adequate for your portfolio of risks, for example, a portfolio of motor fleets (or private cars), or a portfolio of property risks.

Note that the existence of a default model is not strictly necessary in the method above – a variation on the approach in Box 16.1 is to *always* use resampling below a certain threshold and model tail risk separately. This alternative approach is, however, only feasible when a sufficient number of points below the threshold is available (see the considerations below). This approach is further formalised in an article by Fackler (2013), in which a class of spliced Lognormal/GPD distribution is introduced for simultaneously modelling small and large losses.

Considerations on the validity of this approach

Now, if you remember the discussion in Chapter 12, a very complex model may provide a good fit but will in general have little predictive power.

BOX 16.1 A SIMPLE APPROACH TO MODELLING LOSS AMOUNTS

1. Decide on a default distribution (e.g. Lognormal, Gamma, etc.), which can be used for modelling losses for any risk in a given portfolio of risks.
2. Model the tail separately as explained in Section 16.3, if necessary.
3. Validate the model using a holdout sample or an approximate method such as the AIC.
4. If the default model fails, use the empirical distribution for losses for the attritional (i.e. small) losses and model the tail separately.

By resorting to using the empirical distribution for the attritional losses in Step 4, are we not exactly falling into that trap? After all, the empirical distribution can be seen as a model with as many parameters as the number of data points. This is a fair point. However, the Glivenko–Cantelli theorem, a very well-known theorem of mathematical statistics (see, for example, Gnedenko, 1998) assures us that the empirical distribution will asymptotically converge to the correct distribution. Therefore, if the number of data points is sufficiently large, choosing the empirical distribution is not an arbitrary choice and it is likely to perform better than, say, the Lognormal model. This is true in general, but it is even truer when the empirical distribution is used only below a given threshold, where there is less chance than in the tail for any model that fits the data points reasonably well (including the empirical distribution itself) to go horribly wrong when tested against a different data sample.

Interestingly, upper bounds involving the KS distance between the empirical distribution $F_n(x)$ and the true underlying distribution $F(x), K(F_n, F) = \sup_{x \in \mathbb{R}} |F_n(x) - F(x)|$ are also available: according to the Dvoretzky–Kiefer–Wolfowitz inequality (see Dvoretzky et al. 1956), the probability that the (non-normalised) KS distance is larger than an arbitrary value $\mu > 0$ *is* $Pr(K(F_n, F) > \mu) \leq 2\exp(-2n\mu^2)$.

Therefore, in the case in which the number of data points is sufficient, we have solid reasons to use resampling from the empirical distribution for the attritional losses and the GPD for the large losses, and we have a non-arbitrary process for selecting the threshold that separates the attritional and large losses.

Question 3 invites you to reflect on why, given that the empirical distribution does such a good job as implied by the Glivenko–Cantelli theorem and the Dvoretzky–Kiefer–Wolfowitz inequality, we do not just use it over the whole range of losses and instead use a GPD for tail losses.

16.4.1 Application to a Real-World Case Study

Let us see how our simple approach works out in practice. The input, as usual, is the loss data base $\{x_1, \ldots x_n\}$.

The first thing we do is to try the Lognormal model and see what happens. To parameterise the Lognormal distribution, we remember that the log of the losses are distributed according to a Normal distribution. Therefore,

1. We take the natural logarithm of all losses: $\{\log(x_1) \ldots \log(x_n)\}$.
2. μ = the average of all $\log(x_j)$.
3. σ = standard deviation of all $\log(x_j)$.

We are going to try the Lognormal model on a real-world case – a data set of public liability claims for a large UK company (see Figure 16.4). We obtain the following values for the Lognormal parameters:

$$\mu = 7.48, \ \sigma = 1.60$$

Figure 16.7 compares the Lognormal model to our data set. Visually, it seems a rather good match, but we need to have a proper statistical test to judge proximity. One such test is the KS test.

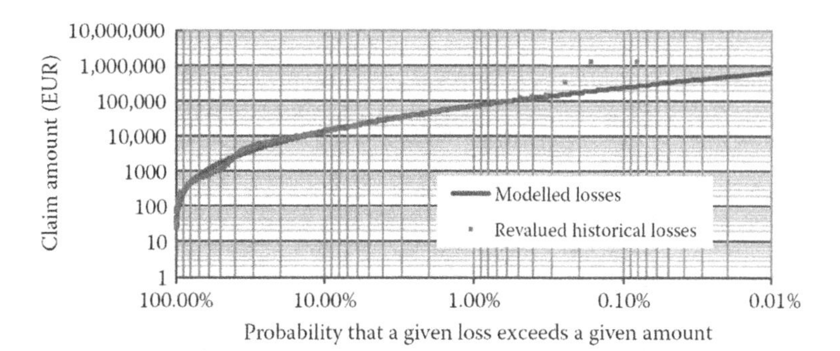

FIGURE 16.7
Visual comparison of our Lognormal model (LogN(7.48,1.60), black line) to our data set of public liability (PL) data (grey dots). Note that the interpretation of the x-axis is slightly different for the loss data (for which the x-axis is the empirical percentage of losses exceeding a certain amount) and for the model (for which it represents the probability that a given loss exceeds a given amount). The fit seems visually rather good, except for some clustering of losses in the €1000 to €4000 range, and two changes in concavity around that area. A proper statistical test reveals indeed that these exceptions are important and that the visual impression of a satisfactory fit is deceptive.

16.4.2 Kolmogorov–Smirnov Test

There are several tests to assess whether a set of data points come from a given distribution or not: the chi-square is probably the most famous. The problem with the chi-square is that, before applying it, you have to put the data into bins and compare the actual and the expected amount for every bin; different binnings will yield different test results.

The AIC method that we met in Section 16.2 is not really a decision test, but rather a method to rank different distributions. Other methods include the KS statistic and the AD statistic. The most popular is the KS statistic: however, AD is very important for tail fitting because it focuses on how the model fits at the low and high ends of the distribution.

In the following, we will use the KS test, which is summarised in Appendix A. As for the other methods, we will refer the interested reader to the many books on the subject, including the works of Klugman et al. (2008).

Let us now see how the KS test works for our case study. Let us first calculate the KS distance between the data points and the model, which has been calibrated using the data itself. By doing this, we obtain

$$d^{\mathrm{NORM}} = 3.58$$

which is well above any of the critical values listed in Appendix A, and corresponds to a probability of the Lognormal model being true of much less than 0.1%.

Actually, the parameters of the Lognormal distribution have been calculated using the data itself and therefore the critical values listed in Appendix A are too lenient in this case; however, in this case, this is not the problem as the model already fails the test under these more lenient conditions. Out of interest, however, let us try to split the data set into two equal parts DS-I and DS-II, calculate the parameters on DS-I and run the KS test on DS-II. Visually, this does not seem to make much difference (see Figure 16.8), but the normalised KS distance now becomes

$$d^{NORM} = 4.44 (\textit{using a training set and a validation set})$$

which means that the model is even less acceptable.

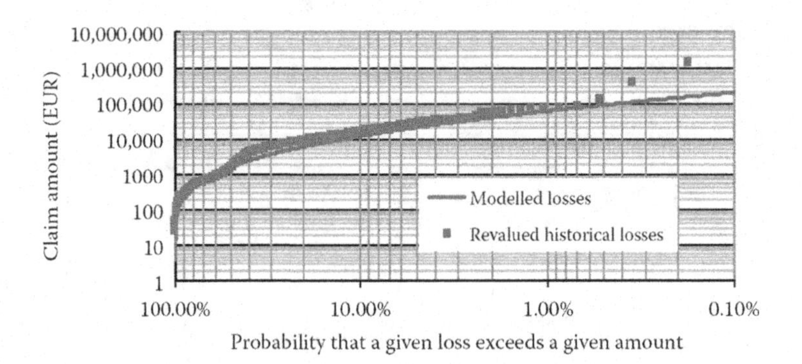

FIGURE 16.8
Visual comparison of a Lognormal model based on the 'training' data set DS-I against the validation set DS-II. Visually, it is not very different from Figure 16.7, but once we calculate the normalised KS distance, we see that it increases (from 3.58 to 4.44), as one would expect.

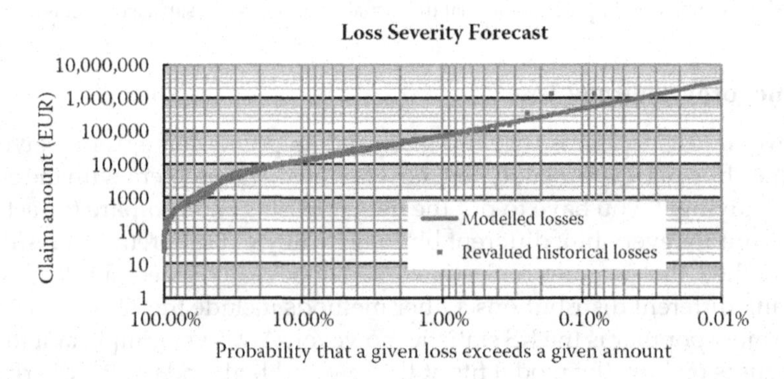

FIGURE 16.9
The loss data set (grey dots) against a spliced Lognormal/GPD model (black line). The splicing point is €10,000.

16.4.3 Moving Beyond the Lognormal Model

Because the Lognormal model seems to give a relatively poor fit, our simple approach suggests that we give up on it and use one of the following two methods:

1. Use a Lognormal model up to a certain threshold, and model the tail differently. This will sort out the fitting problem if it lies with the tail of the distribution
2. Use the empirical distribution instead up to that same threshold, and use a model for the tail (again, the same model as in Method 1).

Let us start with Method 1. Figure 16.9 shows how a spliced Lognormal/GPD model fits our data. The splicing point is at €10,000, which has been chosen by using the heuristic rules explained in Section 16.3.2, and is based on the graph of Figure 16.6.

The main difference with the Lognormal model is that this spliced model is closer to the last few large losses. This spliced distribution is, however, still not good enough in terms of fit because this is very much driven by the behaviour in the small loss region.

Therefore, let us try Method 2. This is illustrated in Figure 16.10, which uses an empirical/GPD model (rather than a Lognormal/GPD model), with the same tail. Naturally,

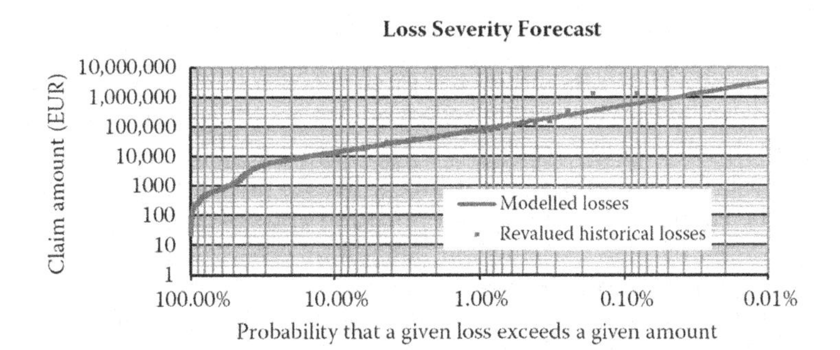

FIGURE 16.10
The loss data set (grey dots) against a spliced empirical/GPD model (black line). The splicing point is again €10,000. Note how the fit is by definition perfect at less than €10,000.

there is no problem with the fit below €10,000, as no modelling is attempted in that region. As for the tail, the normalised KS distance between the data points and the model above €10,000 is 0.52 (a very low value). This empirical/GPD model therefore seems satisfactory.[1]

16.5 Using Portfolio/Market Data

As we have seen in Section 16.3, tail modelling with extreme value theory has some limitations, such as the pricing horizon – the model is only good for modelling losses up to a certain amount. If you want to go beyond that, you need a larger data set, such as portfolio/market losses – and use a portfolio/market severity curve from a certain threshold on. (In the following, we just use 'market data' to refer to either portfolio data based on the insurer's other clients or market data based on data external to the organisation.)

A possible overall strategy for severity analysis is therefore as follows:

- Use Lognormal or empirical below a first threshold μ_1.
- Use client-based GPD between the first threshold μ_1 and a second threshold μ_2.
- Use market-based GPD above the second threshold μ_2.

Of course, using market data to model the 'extreme tail' has advantages and disadvantages.

1 Note, however, how the two largest observations are both well above the model. There are different possible interpretations for this. One explanation is that the model is right and that because of random fluctuations the data set happens to include losses that are expected to happen with lower frequency (about once every 3000 claims). Another explanation is that the model is not adequate to capture the extreme tail. Yet another explanation is that the value these losses (which are not settled yet) may still change, at which point they may well fall back into line. It is often impossible to be sure of which interpretation is the correct one based on the information we have, but it is always good to analyse the behaviour of the extreme tail. It should also be noted that this type of phenomenon – the last few claims being out of line with the model – is a common feature of severity modelling, and it ultimately means that the empirical distribution is never a good estimator of the tail.

16.5.1 Advantages of Using Market Data for Tail Modelling

- Using large market data sets moves the pricing horizon upwards, in the sense that it is possible to model the probability of larger losses and to price layers of reinsurance well beyond the client's experience.
- It provides a benchmarking opportunity – the severity curve for one risk (or one client) may be compared with that of a portfolio of risks or the whole market.
- For some classes of business, the tail behaviour can actually be proved not to be client-specific: for example, it may be possible to prove that, say, above £2M, the severity distribution of different clients are roughly the same. This is not surprising because the relative probability of having, say, a loss of £2M and a loss of £5M is unlikely to depend strongly on client-specific factors. This gives substance to the idea of using external information to model a client's loss behaviour above a certain level.

16.5.2 Disadvantages of Using External Data for Tail Modelling

- The market tail may not be fully relevant to the risk under analysis.
- It may be difficult to identify a 'market' that is sufficiently close to the risk we wish to analyse, for example, market curves for employers' liability for water companies may not be the same as that for oil companies.
- The client might feel misrepresented if it perceives that the analyst does not give sufficient weight to their specific characteristics.

16.6 Appendix A: Kolmogorov–Smirnov Test

The KS test is statistical decision test for goodness-of-fit based on the comparison of the cumulative distribution function of the model with the empirical cumulative distribution of the data, and relies on the calculation of the KS distance *d*. In general, the KS distance between two distributions *F* and *G* is given by

$$d = \sup_{x \in \mathbb{R}} \left| F(x) - G(x) \right| \tag{16.4}$$

where 'sup' is the least upper bound of a function.

When comparing a continuous distribution with a finite set of *n* data points $\{x_1, \ldots x_n\}$, this can be written more easily as

$$d = \max_{i=1,\ldots n} \left| F(x_i) - \hat{F}(x_i) \right| \tag{16.5}$$

where $\hat{F}(x_i)$ is the empirical cumulative distribution function of the data set. If the set $\{x_1, \ldots x_n\}$ is sorted in ascending order, that is, $\{x_1 < \ldots < x_n\}$, then one can write simply:

$$\hat{F}(x_i) = \frac{i}{n+1} \tag{16.6}$$

For example, in the case of a Lognormal, $F(x_i)$ can also be written simply as $F(x_i) = \exp\left(\Phi^{-1}(\log x_i - \mu)/\sigma\right)$, where $\Phi(x)$ is the cumulative distribution function of the Normal distribution.

Therefore, to check whether a model F is adequate to represent a data set, we have to:

1. Sort losses in ascending order
2. Calculate $\left|F(x_i)-\hat{F}(x_i)\right|$ for all data points x_i, as above
3. Calculate the maximum d of these $\left|F(x_i)-\hat{F}(x_i)\right|$
4. Check that d is less than a certain critical value (see below)

Critical values are provided for KS in many books for different significance levels, for example, $d < 1.22/\sqrt{n}$ (@ 10% s.l.), $d < 1.36/\sqrt{n}$ (@5% s.l.), $d < 1.63/\sqrt{n}$ (@ 1% s.l.) where n is the number of data points. It is sometimes simpler to speak in terms of the *normalised* KS distance, that is

$$d^{\text{NORM}} = \sqrt{n} \times d \tag{16.7}$$

in terms of which we can rewrite, more simply, $d^{\text{NORM}} < 1.22$ (@10% s.l.) etc.

However, one must be careful to note that these values are not valid if the parameters of the distribution are calculated from the data themselves – this is the usual problem with using the same set for training (finding the parameters) and validating the model. To use this method properly, we need to perform an independent validation, for example by

i. Splitting the data set into two subsets, randomly
ii. Calculating μ and σ on the first data set
iii. Comparing this model with the second data set

Alternatively, one can use a *modified KS test*, which can be calibrated by simulation – revised critical values for this test can be found in Porter et al. (1992).

16.7 Questions

1. Discuss the advantages and disadvantages of a *frequency/severity model* in comparison to a *burning cost approach* that includes adjustments for inflation, exposure, etc.
2. The following losses have been generated from a Lognormal distribution with parameters μ = 10 and σ = 1. Do the following:
   - Calibrate (i.e. find the parameters of) the Lognormal distribution based on these losses, pretending you do not know what the original parameters were.
   - Calculate the mean, the standard deviation and the percentiles 50%, 75%, 80%, 90%, 95%, 98%, 99%, and 99.5% of the severity distribution using the parameters calculated in (a) and compare these statistics with the same statistics calculated for the original Lognormal distribution (μ = 10, σ = 1). What is the percentage difference?
   - Calculate the KS distance between the Lognormal distribution with the parameters calculated as in (a) and the data points.
   - Compare the KS distance between the original Lognormal distribution and the data points.
   - Compare the KS distance as calculated in (c) and (d). Which one is the largest, and why?

3. In Section 16.4, we have justified the use of the empirical distribution to represent attritional losses on the basis of the Glivenko–Cantelli theorem and the Dvoretzky–Kiefer–Wolfowitz inequality. Why do we not use the empirical distribution to represent large losses as well?

4. (*) Show that the single-parameter Pareto, $F(x)=1-\left(\frac{\theta}{x}\right)^{\alpha}$, and the Type-II Pareto, $F(x)=1-\left(\frac{\theta}{x+\theta}\right)^{\alpha}$, are both special cases of the GPD, $F(x)=1-\left(1+\xi\left(\frac{x-\mu}{\sigma}\right)\right)^{-1/\xi}$.

5. (*) *Effect of claims inflation on various distributions.* Show that applying an inflation of r% to all the losses (i.e. $X'=X(1+r)$) modelled by:
   - a Lognormal distribution with parameters μ and σ leads to a Lognormal distribution with parameters $\mu'=\mu+\log(1+r)$, $\sigma'=\sigma$;
   - a single-parameter Pareto distribution $F(x)=1-\left(\frac{\theta}{x}\right)^{\alpha}$ while keeping constant the threshold θ(basically, this means assuming that the Pareto distribution continues to hold below θ, at least down to a certain value) leaves the Pareto distribution unchanged;
   - a GPD $F(x)=1-\left(1+\xi\left(\frac{x-\mu}{\sigma}\right)\right)^{-1/\xi}$ while keeping constant the threshold μ leads to a GPD with parameters $\xi'=\xi, \sigma'=\sigma(1+r)-\xi\mu r, \mu'=\mu$.

17

Aggregate Loss Modelling

In Chapter 6, we introduced the scientific basis for the stochastic modelling of risk, and we explored the implications of the two main types of aggregate loss models, the individual risk model, and the collective risk model. We then devoted several chapters (Chapters 7 through 16) to creating the building blocks for these aggregate loss models: a frequency model and a severity model.

Based on the frequency and the severity models that we have built, we already know how to calculate some simple statistics for the aggregate loss model: for example, the mean of the aggregate model is simply the mean claim count times the mean loss amount. However, to make sensible risk-management decisions, we need to know the full aggregate loss distribution so that we can answer questions such as, 'What is the amount the account is going to lose in a very bad year, for example, with a probability of 1%?'

This problem is normally impossible to solve exactly through analytical formulae. In this chapter, we will explain why this is the case, and how it is possible to get around the issue by using approximate solutions, the most famous of which is the Monte Carlo simulation. Figure 17.1 puts this step into the context of the risk costing process.

17.1 'Exact' Solution for the Collective Risk Model (Gross Losses)

Calculating the full distribution of the aggregate loss amount S is a tough problem because, except in a few cases, it cannot be solved analytically. The cumulative distribution function of S, $F_S(s)$, can be expressed as follows:

$$F_s(s) = \Pr(X_1 + \ldots + X_N \le s) \tag{17.1}$$

There are two difficulties here: first of all, we have a sum of random variables of which we know the distribution $F_X(x)$, and the distribution of the sum of random variables is a complicated function of the individual distribution, technically called a convolution. Second, the number N over which the sum must go is itself a random variable, and we need to take into account all possible values of N in the calculation.

DOI: 10.1201/9781003168881-19

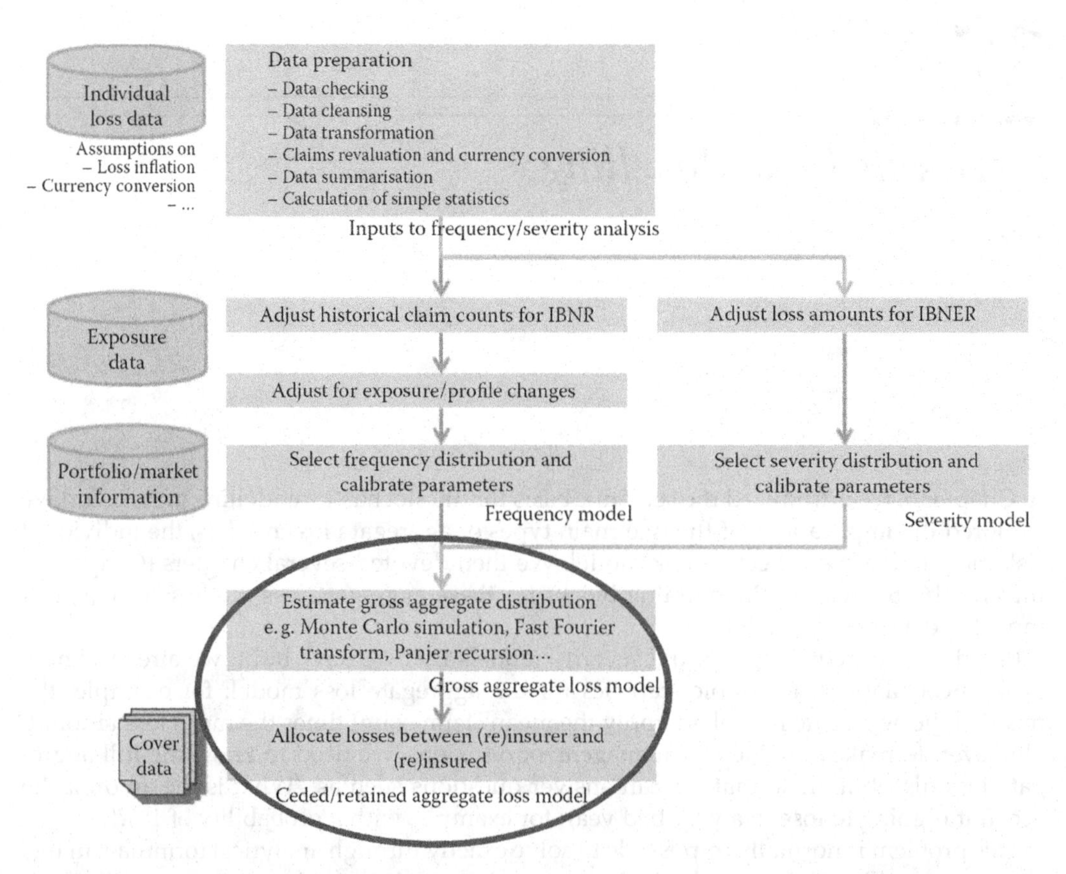

FIGURE 17.1
The Monte Carlo simulation (or one of the other methods outlined in this chapter) allows us to produce an aggregate loss model based on a frequency and severity model; it also allows us to calculate the distribution of ceded and retained losses based on the policy coverage.

Let us start by addressing the second problem first: because the different values of N correspond to mutually exclusive cases, we can write the probability as a weighted sum of the probability of mutually exclusive events:

$$F_s(s) = \Pr(X_1 + \cdots + X_N \leq s) = \sum_{n=0}^{\infty} \Pr(N = n)\Pr(N = n)\Pr(X_1 + \ldots + X_n \leq s) \quad (17.2)$$

Now the problem has been reduced to calculating convolutions of a *fixed* number of variables, and Equation 17.2 can be written as

$$F_s(s) = \sum_{n=0}^{\infty} \Pr(N = n)\left(F_X(s)\right)^{*n} \quad (17.3)$$

where $\left(F_X(s)\right)^{*n}$ is defined recursively as follows (see, for example, Klugman et al., 2008):

- $\left(F_X(s)\right)^{*0} = H(s)$, where $H(s)$ is the Heaviside step function, defined as 1 for $s \geq 0$ and 0 elsewhere

- $\left(F_X(s)\right)^{*n} = \int_{-\infty}^{\infty} \left(F_X(s-s')\right)^{*n-1} dF_X(s')$, which can be written as $\left(F_X(s)\right)^{*n} = \int_{0}^{\infty}\left(F_X(s-s')\right)^{*n-1} f_X(s')ds'$, when the severity distribution is continuous and with zero probability for negative losses[1]

In the case where the severity distribution is continuous, a formally identical relationship to Equation 17.3 holds for the probability density function:

$$f_S(s) = \sum_{n=0}^{\infty} \Pr(N = n)\left(f_X(s)\right)^{*n} \tag{17.4}$$

where $\left(f_X(s)\right)^{*n}$ represents here the n-fold convolution of $f_X(s)$ with itself, which is an integral in n dimensions.[2]

The important point to understand is that although Equation 17.3 provides a formal solution to the aggregate loss calculation problem as *an infinite sum of multidimensional integrals*, in general it will not be possible to write down this solution in closed form, and the problem needs to be tackled by numerical methods.

In the following few sections, we can see how this seemingly near-impossible problem can be broken down into something manageable and at what cost. It might be helpful to keep a very simple example in mind: that in which the frequency distribution is a Poisson distribution with Poisson rate $\lambda = 100$ and the severity distribution is a lognormal distribution with $\mu = 10$ and $\sigma = 1$ (corresponding to a mean of £36.3k and standard deviation of £47.6k).

17.2 Parametric Approximations

Parametric approximations work by approximating the aggregate loss distribution with a known distribution (such as Gaussian or lognormal) with mean and variance equal to the ones calculated using the formulae in Section 6.1.2. It should be mentioned that this method is in practice mainly used as a rough approximation in the context of burning cost analysis (Chapter 11) whereas in the context of frequency/severity modelling Monte Carlo simulation and FFT are the most common techniques.

1 Note that in the continuous case $\left(F_X(s)\right)^{*n}$ is defined as the convolution of $\left(F_X(s)\right)^{*n-1}$ with $f_X(s)$ rather than with $F_X(s)$, so we should not speak of $\left(F_X(s)\right)^{*n}$ as the n-fold convolution of $F_X(s)$, as it is sometimes done.

2 As for the cumulative distribution function, we can define $\left(f_X(s)\right)^{*n}$ recursively: $\left(f_X(s)\right)^{*n} = \left(f_X(s)\right)^{*n-1} * f_X(s) = \int_0^s \left(f_X(s-s')\right)^{*n-1} f_X(s')ds'$, with $\left(f_X(s)\right)^{*0} = \delta(s)$, $\delta(s)$ being the so-called Dirac delta function, a mathematical object (formally known as a generalised function) that can be thought of here as a way of representing a point mass of unit probability at $s = 0$, and as the derivative of the Heaviside (step) function. It was introduced informally by Paul Dirac in the 1920s as a function that was 0 everywhere except at $z = 0$ but somehow with a finite integral (a problematic definition from a mathematical standpoint), and was put on a rigorous footing by Sobolev and Schwartz in the context of distribution theory (see, e.g. Zemanian, 2003).

17.2.1 Gaussian Approximation

The first idea on how to solve Equation 17.3 in practice comes from the central limit theorem (CLT), which states that the sum $X_1 + \ldots + X_N$ of N independent, identically distributed variables (all of them with *finite* mean μ and variance σ^2) converges for $N \to \infty$ (but already rather well when $N \geq 30$) to a normal distribution with mean $N\mu$ and variance $N\sigma^2$.

At first sight, this seems to suit us just fine because our problem is indeed to find the distribution of the sum of N independent, identical variables. However, the CLT does not apply to this case. Before moving on, try answering the following question on your own.

> *Question*: Why is the aggregate loss distribution for a frequency distribution with N losses per year and a severity distribution with mean μ and variance σ^2 not simply a Gaussian with mean $N\mu$ and variance $N\sigma^2$, at least for large values of N?

The reason is quite simply that in our case N is itself a random variable. As a consequence, (i) the variance is not simply the sum of the variances of the severity distributions but also has a component due to the year-on-year volatility of the number of losses; (ii) also, and crucially, there is no result, under these changed circumstances, which guarantees that the resulting distribution is indeed a Gaussian.

Despite the fact that the CLT does not apply to this case, we may still hope that it provides a decent approximation of the aggregate loss distribution, perhaps after a modification of the variance to take the extra volatility into account. This will take care of point (i) above, and the mismatch between the true distribution and the Gaussian approximation will be due to (ii) only.

Figure 17.1 shows a comparison between the percentiles of the aggregate loss distribution calculated with two methods: the Monte Carlo simulation (which we will explain later, and that we use here simply as a benchmark) and the Gaussian approximation for a compound Poisson distribution with different rates ($\lambda = 10, 100, 1000$) and a lognormal distribution with $\mu = 10$ and $\sigma = 1$ as the severity distribution.

In the Gaussian approximation, we choose a mean equal to the theoretical mean and variance of the aggregate loss distribution for a compound Poisson/lognormal distribution:

$$\mathbb{E}(S) = \mathbb{E}(N)\mathbb{E}(X) = \lambda\mathbb{E}(X) = \lambda \times \exp\left(\mu + \frac{\sigma^2}{2}\right) \tag{17.5}$$

$$\mathbb{V}(S) = \mathbb{V}(X)\mathbb{E}(N) + \mathbb{E}(X)^2\,\mathbb{V}(N) = \mathbb{E}(N)\mathbb{E}(X^2) = \lambda \times \exp(2\mu + 2\sigma^2) \tag{17.6}$$

Equation 17.6 clearly shows what the effect of the year-on-year volatility of the number of losses is on the overall variance: if $\mathbb{V}(N) = 0$, $\mathbb{V}(S) = \mathbb{V}(X)\,\mathbb{E}(N)$; the frequency volatility adds a term equal to $\mathbb{E}(X)^2\mathbb{V}(N)$.

It is clear by looking at the percentage difference in the tail of the distribution (e.g. the 1 in 200 return period) that the approximation gets better and better as λ increases, as expected (Figures 17.2 through 17.4). Although this methodology has its appeal, it is clear that the error can be significant especially when the frequency is low, and what is most important is that it is difficult to calculate the upper bound for this error in a simple way. Another problem is that the aggregate loss distribution will not, in general, be symmetric and actually might have a significant right-hand tail, which the Gaussian simply does not capture.

17.2.2 Other Parametric Approximations

Other approximate methods, similar in principle to the Gaussian approximation, are the translated gamma approximation (Klugman et al., 2008), which models the aggregate

Return period (years)	Percentile	Gross losses (simulation)	Gross losses (Gaussian approximation)	Percentage difference
3 in 4	25%	230,133	235,448	2.3%
1 in 2	50%	330,364	363,155	9.9%
1 in 4	75%	459,801	490,862	6.8%
1 in 5	80.0%	496,880	522,506	5.2%
1 in 10	90.0%	601,405	605,802	0.7%
1 in 20	95.0%	705,981	674,589	−4.4%
1 in 50	98.0%	849,050	752,009	−11.4%
1 in 100	99.0%	971,408	803,623	−17.3%
1 in 200	0.995	1,093,126	850,859	−22.2%
1 in 500	0.998	1,265,393	908,102	−28.2%
1 in 1000	0.999	1,373,720	948,255	−31.0%
	Mean	363,001	363,155	0.0%
	Std Dev	189,192	189,339	0.1%

FIGURE 17.2
Aggregate loss distribution percentiles for a compound Poisson with $\lambda = 10$, severity distribution = lognormal with $\mu = 10$ and $\sigma = 1$.

Return period (years)	Percentile	Gross losses (simulation)	Gross losses (Gaussian approximation)	Percentage difference
3 in 4	25%	3,221,210	3,227,705	0.2%
1 in 2	50%	3,595,429	3,631,550	1.0%
1 in 4	75%	4,000,157	4,035,395	0.9%
1 in 5	80%	4,108,105	4,135,464	0.7%
1 in 10	90%	4,403,672	4,398,868	−0.1%
1 in 20	95%	4,670,789	4,616,392	−1.2%
1 in 50	98.0%	4,997,699	4,861,215	−2.7%
1 in 100	99.0%	5,215,945	5,024,431	−3.7%
1 in 200	99.5%	5,448,113	5,173,806	−5.0%
1 in 500	99.8%	6,026,667	5,354,825	−11.1%
1 in 1000	99.9%	6,253,777	5,481,800	−12.3%
	Mean	3,635,617	3,631,550	−0.1%
	Std Dev	598,354	598,741	0.1%

FIGURE 17.3
Aggregate loss distribution percentiles for a compound Poisson with $\lambda = 100$, severity distribution = lognormal with $\mu = 10$ and $\sigma = 1$.

loss distribution as a shifted gamma distribution, and the lognormal approximation, which matches the desired moments to those of a lognormal distribution. These methods have the added advantage of yielding a non-symmetric aggregate loss distributions – which might be more suitable than the Gaussian with low expected claim counts – but the other problems (difficulties in modelling the tail, difficulties in quantifying the errors) remain.

Return period (years)	Percentile	Gross losses (simulation)	Gross losses (Gaussian approximation)	Percentage difference
3 in 4	25%	35,088,519	35,038,433	−0.1%
1 in 2	50%	36,297,780	36,315,503	0.0%
1 in 4	75%	37,528,551	37,592,573	0.2%
1 in 5	80%	37,839,666	37,909,017	0.2%
1 in 10	90%	38,652,714	38,741,975	0.2%
1 in 20	95%	39,323,873	39,429,847	0.3%
1 in 50	98.0%	40,115,771	40,204,043	0.2%
1 in 100	99.0%	40,757,231	40,720,179	−0.1%
1 in 200	99.5%	41,255,871	41,192,543	−0.2%
1 in 500	99.8%	41,688,370	41,764,976	0.2%
1 in 1000	99.9%	42,147,794	42,166,507	0.0%
	Mean	36,320,099	36,315,503	0.0%
	Std Dev	1,804,736	1,893,387	4.9%

FIGURE 17.4
Aggregate loss distribution percentiles for a compound Poisson with $\lambda = 1000$, severity distribution = lognormal with $\mu = 10$ and $\sigma = 1$.

BOX 17.1 HOW WOULD YOU CALCULATE THE MULTIDIMENSIONAL INTEGRAL $(F_X(S))^{*N}$?

Calculating integrals numerically is rather simple. in one dimension! Under proper conditions, to calculate the integral $\int_a^b f(x)dx$, one simply needs to subdivide the domain *[a,b]* into small intervals $(a+n\delta x, a+(n+1)\delta x)$ and for each interval calculate the area between the function and the x axis, which is about $f(n\delta x)\delta x$. The integral can then be approximated as $\int_a^b f(x)dx \approx \sum_{n=0}^{(b-a)/' x} f(n' x)$ (Riemann's construction). If the domain is actually infinite (i.e. if $a=-\infty$ *or* $b=\infty$), special care needs to be taken in calculating the infinite summation, or approximating it with a finite sum.

This method can, in principle, be generalised to integrals over an arbitrary number of dimensions. However, when the number of dimensions is large, the problem becomes very difficult because of the 'dimensionality curse', by which the number of intervals necessary to achieve a certain degree of accuracy increases exponentially in the number of dimensions. In our specific case, we then have an *infinite* sum of these integrals, so the problem is also computationally very expensive.

It seems we are in a quandary ... unless, that is, we abandon this brute-force approach and look for smarter ways of doing the calculation, which are less general but exploit the specific features of the problem.

*17.3 Numerical Quasi-Exact Methods (Gross Losses)

Numerical methods for calculating the aggregate loss distribution simply attempt to calculate the convolution integrals directly using discrete mathematics, using the fact that

* BOX 17.2 ALGORITHM FOR CALCULATING THE AGGREGATE LOSS DISTRIBUTION USING FAST FOURIER TRANSFORM

1. Discretise the severity distribution $f_X(x)$ using steps of h, obtaining the vector $f = (f_0, f_1 \dots f_{M-1})$

   There are different ways of discretising the severity distribution. A good one, suggested in Embrechts and Frei (2009), is to set $f_i = F_X\left(jh + \frac{h}{2}\right) - F_X\left(jh - \frac{h}{2}\right)$ where $F_X(x)$ is the cumulative distribution function calculated at point x.
2. Calculate the DFT of $f = (f_0, f_1 \dots f_{M-1})$, obtaining the vector $\hat{f} = \left(\hat{f}_0, \hat{f}_1 \dots \hat{f}_{M-1}\right)$.
3. The DFT $\hat{f}^S$ of the aggregate loss distribution can be calculated for the Poisson case by Equation 17.15, which – once discretised – becomes $\hat{f}^S = \left(\hat{f}_0^S, \hat{f}_1^S \dots \hat{f}_{M-1}^S\right)$, where

   $$\hat{f}_j^S = \exp\left(\lambda\left(\hat{f}_j - 1\right)\right) \tag{17.8}$$

   In the negative binomial case, Equation 17.17 becomes

   $$\hat{f}_j^S = \left(1 + (\mathrm{v}-1)\left(1 - \hat{f}_j\right)\right)^{-\lambda/(\mathrm{v}-1)} \tag{17.9}$$

   Note that in both cases the calculations are straightforward because the $(j + 1)$th element of $\hat{f}^S, \hat{f}_j^S$ depends only on the $(j + 1)$th element of $\hat{f}, \hat{f}_j$.
4. Calculate the inverse Fourier transform of $\hat{f}^S, f^s = \left(f_0^S, f_1^S \dots f_{M-1}^S\right)$, which is a discretised version of the aggregate loss distribution.
5. *Basic statistics.* Once we have f^S in vector format, the calculation of the basic statistics is straightforward: for example, the mean of S can be approximated as $\bar{S} = \sum_{j=0}^{N} f_j^S jh$ and the quantiles can be approximated as $F_S^{-1}(p) \approx \max\left\{j^* \mid \sum_{j=0}^{j^*h} f_j^S \le p\right\} \times h$.

integrals can be approximated as a finite sum; for example, Riemann's construction mentioned in Box 17.1. All these methods require replacing both the severity distribution $f_X(x)$ and the aggregate loss distribution $f_X(s)$ with a discretised approximation using steps of h, obtaining vectors of the form $f = (f_0, f_1, f_2 \dots)$, where the value f_k is the value of $f(x) at\, x{=}hk$ after appropriate rounding (see Box 17.2, Step 1 for an example). The frequency distribution is already discrete and therefore does not require any further action.

More specifically, discretisation allows us to approximate the aggregate loss distribution, which can in general be written as $f_S(s) = \sum_{n=0}^{\infty} Pr(N = n)\left(f_X(s)\right)^{*n}$, as a finite sum:

$$f_{S,k} = \sum_{n=0}^{k} p_n f_{X,k}^{*n} \tag{17.7}$$

There are several different ways of calculating this sum. One option would be to do a straightforward calculation of Equation 17.7. This is computationally very demanding ($O(k^3)$ operations required where k is the number of points of the aggregate loss distribution that we wish to calculate). However, methodologies have been developed that drastically reduce the computational burden under certain conditions.

The fastest of these methodologies (Section 17.3.1) uses fast Fourier transform (FFT), and calculates the aggregate loss distribution with $O(k \log k)$ operations. However, it requires some artful tuning of the parameters involved. A valid alternative (Section 17.3.2) is the use of Panjer recursion, which is slower ($O(k^2)$ operations) but has a simpler implementation.

*17.3.1 Fast Fourier Transform Method

In a nutshell, the Fourier transform method works by transforming the distribution involved into functions in another space in which the operation of convolution becomes a simple product and therefore all the integrals become trivial; with this step complete, it is a simple case of returning to the original space.

This is how it goes. If the severity distribution is $f_X(x)$, we define its Fourier transform as the function

$$\hat{f}_X(\omega) := \int_{-\infty}^{+\infty} f_X(x) e^{-i\omega x} dx \tag{17.10}$$

That is, the Fourier transform takes a function over the variable x and, through an integral, transforms it into a complex function over ω. The result is, in general, a complex function because of the presence of the factor $e^{-i\omega x} = \cos(\omega x) - i \sin(\omega x)$, i being the imaginary unit.

Because $f_X(x)$ is a severity distribution, it will be zero for $x < 0$, and therefore the integral will reduce in practice to calculating $\hat{f}_X(\omega) = \int_0^{+\infty} f_X(x) e^{-i\omega x} dx$.

The first interesting fact about the Fourier transform is that, if we have $\hat{f}_X(\omega)$, we can obtain $f_X(x)$ back by an integral that is very much the same as that in Equation 17.10:

$$f_X(x) = \frac{1}{2\pi} \int_{-\infty}^{+\infty} \hat{f}_X(\omega) e^{i\omega x} dx \tag{17.11}$$

The two differences between Equations 17.10 and 17.11 are the sign in the exponential ($e^{i\omega x}$ rather than $e^{-i\omega x}$) and the factor $\frac{1}{2\pi}$ in front of the integral. Although the former is important, the latter is accidental and it would be easy to reformulate the definition of the Fourier transform so that this factor does not change in the inverse transformation; however, we have used the most common definition, which has the added advantage that if $f_X(x)$ is a probability distribution, the Fourier transform evaluated at $\omega = 0$ is $\hat{f}_X(0) = 1$.

The second interesting fact about the Fourier transform is that the Fourier transform of the convolution of two functions is the product of the Fourier transforms:

$$\hat{f}_{X_1+X_2}(\omega) = \hat{f}_{X_1}(\omega) \hat{f}_{X_2}(\omega) \tag{17.12}$$

Hence, the Fourier transform of the sum of n independent, identically distributed variables can be written as the product of individual transforms, that is,

$$\hat{f}_{X_1+\ldots+X_n}(\omega)=\prod_{i=1}^{n}\hat{f}_{X_i}(\omega)=\left(\hat{f}_X(\omega)\right)^n \quad (17.13)$$

The last equality derives from the fact that all variables are identically distributed as X. This allows rewriting Equation 17.4 in terms of the Fourier transform of both sides as

$$\hat{f}_S(\omega)=\sum_{n=0}^{\infty}p_n\left(\hat{f}_X(\omega)\right)^n \quad (17.14)$$

where p_n is the probability that there are exactly n losses in a given year. In the case where the frequency distribution is of a known form, this formula might be further simplified.

For example, for a Poisson frequency model with rate λ, p_n can be written as $p_n = e^{-\lambda}\lambda^n/n!$, and therefore

$$\hat{f}_S(\omega)=\sum_{n=0}^{\infty}\frac{e^{-\lambda}\lambda^n}{n!}\left(\hat{f}_X(\omega)\right)^n=e^{-\lambda\left(1-\hat{f}_X(\omega)\right)} \quad (17.15)$$

If the frequency model is a negative binomial with parameters (r,β), we can instead write:

$$\hat{f}_S(\omega)=\sum_{k=0}^{\infty}\binom{r+k-1}{k}\left(\frac{\beta}{1+\beta}\right)^k\left(\frac{1}{1+\beta}\right)^r\left(\hat{f}_X(\omega)\right)^k=\left(\frac{1}{1+\beta\left(1-\hat{f}_X(\omega)\right)}\right)^r \quad (17.16)$$

which can also be written more usefully in terms of the loss rate λ (i.e. the expected number of claims) and of the variance-to-mean ratio v as

$$\hat{f}_S(\omega)=\left(\frac{1}{1+(v-1)\left(1-\hat{f}_X(\omega)\right)}\right)^{\frac{\lambda}{v-1}} \quad (17.17)$$

Equations 17.15 and 17.17 demonstrate that for compound Poisson or negative binomial processes, we can express the aggregate loss model in a simple closed form as a function of the Fourier transforms of the severity distribution. This suggests a numerical method to calculate the aggregate loss distribution in practice using discrete Fourier transform (DFT), which is the discrete version of the Fourier transform.

17.3.1.1 Discrete Fourier Transform and Fast Fourier Transform

The DFT is defined as an operator that transforms a vector of complex numbers $f=\left(f_0,f_1\ldots f_{M-1}\right)\in\mathbb{C}^M$ (although we will only be interested in the case in which $f\in\mathbb{R}^M$) into another vector of complex numbers $\hat{f}=\left(\hat{f}_0,\hat{f}_1\cdots\hat{f}_{M-1}\right)\in\mathbb{C}^M$ where

$$\hat{f}_j=\sum_{k=0}^{M-1}f_k e^{i2\pi jk/M} \quad (17.18)$$

which can be inverted to yield

$$f_j = \frac{1}{M}\sum_{k=0}^{M-1} \hat{f}_k e^{-i2\pi jk/M} \tag{17.19}$$

Note that Equations 17.18 and 17.19 are the discrete versions of Equations 17.10 and 17.11, respectively.

The interesting things about the DFT from our point of view are that

i. A continuous distribution $f(x)$ can be approximated by sampling its values in a certain range $[0,Mh]$ at steps of length h and represented as a vector $f = (f_0, f_1 \dots f_{M-1})$. As long as h is small enough and M is large enough, the vector $= (f_0, f_1 \dots f_{M-1})$ provides a good approximation of $f(x)$.
ii. If we choose M to be a power of 2, $M = 2^m$, then a very efficient algorithm, FFT, can be used to calculate the DFT.[3]

17.3.1.2 Practical Issues

In practice, one also needs to make sure that the numerical errors are sufficiently small. For this purpose, one must carefully consider the choice of the parameters and the so-called aliasing problem.

17.3.1.2.1 Choice of the Parameters M, h

- h should be much smaller than the typical loss amount and, in any case, low enough that the statistics for the severity (such as the mean or a given percentile) calculated on the basis of that value of h does not differ significantly from the true one. The true value is not known but a good heuristic is that of comparing the statistics for increasingly smaller values of h until the difference is no longer significant (similar to the way in which one chooses the number of scenarios in a Monte Carlo simulation).
- The number of elements in the vector M should not be too large, to prevent the programme from taking too long to run.
- $M \times h$ should be chosen so that the probability of obtaining *total losses* larger than $M \times h$ is negligible, whatever we consider to be negligible – for example, $p^* = 0.1\%$. We can refer to the value p^* as the 'risk acceptance level' – that is, we do not care what happens with probability less than p^*. In other words, we want $F_S(M \times h) > 1 - p^* = 99.9\%$. Of course, calculating $F_S(M \times h)$ requires calculating the aggregate loss distribution with FFT or another method so this sounds (and is) circular. A heuristic approach to this problem is to start with a rough estimation of the order of magnitude of, for example, by running an FFT using $M \times h > F_X^{-1}(p^* \times \lambda)$, then trying it out and checking whether $M \times h$ is significantly larger than the estimated total losses at $1 - p^*$. If that is not the case, we must increase $M \times h$ by either increasing M or h, and again check if this leads to an acceptable result.

3 We will not get into the details of the algorithm here: suffice it to say that in its most common incarnation (the radix-2 Cooley–Tukey algorithm), the FFT is a recursive algorithm that breaks the problem of calculating the DFT of a vector with M elements into the problem of calculating the DFT of two vectors with $M/2$ elements. As a result, the algorithm takes a number of steps $T(M) \leq cM \log M$ (in theoretical computer science parlance, the algorithm has complexity $O(M \log M)$); that is, just a bit more than linear in the number of vector elements. The interested reader can find more details in many books: as in many other cases, a very good practical introduction with implementation details can be found in Press et al. (2007).

The catch is of course that you cannot get an arbitrarily large $M \times h$, an arbitrarily small h, and impose a tight constraint on M at the same time. The choice of M and h is therefore a delicate balancing act between conflicting criteria, and some trial and error is needed to find the correct parameters. Question 6 asks how to produce acceptable parameters in a simple case.

17.3.1.2.2 Aliasing Problem

A problem related to that of choosing the correct parameters for the numerical evaluation of the aggregate loss distribution is that of aliasing. This happens when the probability of the severity being above $M \times h, 1 - F_X(M \times h)$, is significant. This residual probability ends up being wrapped around and distributed over the range (0,...M − 1).

To get around this problem, a so-called *tilting procedure* has been suggested by Grübel and Hermesmeir (see Embrechts and Frei [2009] for full references). This works by fixing a value θ and replacing the original vector f with $T_\theta f = \left(f_0, e^{-\theta} f_1, e^{-2\theta} f_2, \ldots e^{-(M-1)\theta} f_{M-1}\right)$, where T_θ is the tilting operator. If the choice of θ is adequate, this causes the values $(T_\theta f)_j = e^{-j\theta} f_j$ to decrease to zero much faster than the values f_j. We can then calculate the DFT of $T_\theta f$, carry out the standard calculations for the aggregate loss distribution, and then 'untilt' at the end by applying the operator $T_{-\theta}$, producing $T_{-\theta} f^S = \left(f_0^S, e^{\theta} f_1^S, e^{2\theta} f_2^S, \cdots e^{(M-1)\theta} f_{M-1}^S\right)$.

As for the choice of θ, Embrechts and Frei (2009) suggest choosing a value as large as possible without producing overflows/underflows (the values of $e^{(M-1)\theta}$ and $e^{-(M-1)\theta} f_{M-1}$ can become too large and too small, respectively); specifically, they suggest a value of $\theta \sim M/20$ as adequate in most cases.

17.3.1.3 Implementation

Implementing an efficient FFT algorithm to estimate an aggregate loss distribution from scratch is not completely trivial. However, the problem has already been solved and the full code can be found in the book by Press et al. (2007); because the code is written in a pseudo-code very similar to C++, some minor adjustments need to be made if you are using VB. As usual, things are much simpler in R, and there are libraries that allow us to calculate an FFT in one line of code. Appendix B has the code for a basic implementation inclusive of tilting.

17.3.1.4 Output Example

In the following, we show some calculation examples with a Poisson distribution with $\lambda = 3$ as the frequency distribution and a lognormal distribution ($\mu = 10$ and $\sigma = 1$) as the severity distribution. In all cases, we have subdivided the amounts range in 65,536 subintervals, but we have used different discretisation steps.

As Figure 17.5 shows, the approximation becomes poor when h is relatively too small or too large compared with the losses being generated. The results of the FFT calculations are indeed quite close to the theoretical values for $h = 50$ and $h = 100$, although the difference becomes significant for $h = 10$ and $h = 1000$. The reason why the result for $h = 10$ is not accurate is that because h is rather small – 65,536 data points are not sufficient to capture the full range of loss amounts – the maximum generated loss is only £655,360, which is the 99.9654th percentile of the severity distribution. As for $h = 1000$, the problem is that the granularity of the severity distribution is not sufficient, and small losses are all approximated as £1,000.

*17.3.2 Panjer Recursion

Another popular method for calculating the aggregate loss distribution numerically is the so-called Panjer recursion, which we will touch on briefly here. We remember from the beginning of Section 17.3 that the aggregate loss distribution, which can in general be written as $g_k = \sum_{n=0}^{k} p_n f_k^{*n}$; this is the same as Equation 17.7 but with a slightly simplified notation. Panjer recursion hinges on the finding that g_k can be expressed with a simple recursion formula that does away with convolutions when the frequency distribution belongs to the $(a,b,0)$ class, that is, when it satisfies the equality $p_k = \left(a + \frac{b}{k}\right) p_{k-1}$ for $k \geq 1$, for example, a Poisson, binomial, or negative binomial distribution (see Chapter 14). The recursion formula is:

$$g_k = \frac{1}{1 - af_0} \sum_{j=1}^{k} \left(a + \frac{bj}{k}\right) f_j g_{k-j}, k = 1, 2 \cdots \tag{17.20}$$

where g_0, the probability that the aggregate losses $S = X_1 + \ldots + X_N$ are zero, is given by

$$g_0 = \Pr(S = 0) = \begin{cases} e^{-\lambda(1-f_0)} & \text{(Poisson case)} \\ \left(1 + q(f_0 - 1)\right)^m & \text{(Binomial case)} \\ \left(1 - \beta(f_0 - 1)\right)^m & \text{(NB case)} \end{cases} \tag{17.21}$$

The proof of Equation 17.20 can be found in the works of Panjer (1981) and in Klugman et al. (2008). Question 7 asks you to prove the much simpler Equation 17.21 about g_0.

FFT Gross Loss Model Results for Different Discretions Steps

Percentile	h = 100	h = 50	h = 10	h = 1000	Exact
50.0%	82,500	82,400	82,160	84,000	
75.0%	149,100	148,950	148,380	151,000	
80.0%	169,800	169,750	169,010	172,000	
90.0%	234,800	234,700	233,200	237,000	
95.0%	302,300	302,150	298,970	304,000	
98.0%	398,300	398,200	389,590	401,000	
99.0%	478,300	478,150	460,080	481,000	
99.5%	566,500	566,350	529,340	569,000	
99.8%	699,400	699,250	610,540	702,000	
99.9%	815,300	815,100	655,360	818,000	
Mean	109,096	109,017	106,872	110,446	**108,947**
Error on mean	0.14%	0.06%	−1.90%	1.38%	
SD	103,752	103,652	96,017	104,234	**103,705**
Error on SD	0.05%	−0.05%	−7.41%	0.51%	

FIGURE 17.5
Aggregate loss model calculated with FFT with different discretisation steps, but with the same number of discretisation points (65,536). The last column shows the theoretical value of the mean and the standard deviation, as calculated by Equations 6.3 and 6.4 (as to the percentiles, the exact value is unknown).

17.3.2.1 Practical Issues

The complexity of this method is $O(k^2)$, where k is the largest index for which we wish to calculate the aggregate loss distribution: $g_0, g_1, \ldots g_k$. The method is therefore slower than the FFT, which runs in little more than linear time.

On the positive side, there is no fine balance that needs to be achieved in the choice of the parameters, unlike the FFT approach, where several different constraints need to be imposed on the value of M (number of subdivisions), h (discretisation step), and θ (the tilting parameter).

In the case of Panjer recursion, the only tuning that needs to be done is the choice of the discretisation step h. Theoretically, any precision can be achieved by choosing a suitable small value of h. However, if the number of claims is large, we may face computational issues: in the Poisson case, $g_0 = e^{-\lambda(1-f_0)}$ might be lower than the smallest number that can be represented on your computer (what is technically termed an 'underflow'). See the articles by Shevchenko (2010) or Embrechts and Frei (2009) for a more in-depth discussion of this issue and possible workarounds.

These two articles also have discussions of how the discretisation error propagates over the recursion and of how instabilities occur in calculations under certain conditions.

17.3.2.2 Implementation

Implementing Panjer recursion in VBA is relatively straightforward, by using standard loops to represent the recursive nature of Equation 17.20. As for R, the package actuar has an implementation of the Panjer recursion, which works by discretising the severity distribution and then using aggregateDist, a function for calculating the aggregate loss distribution with five different methods, including a straightforward convolution, the Monte Carlo simulation and indeed the Panjer recursion (see Dutang et al. 2008). The example below is for a Poisson frequency with rate $\lambda = 10$ and a lognormal distribution with parameters $\mu = 7$ and $\sigma = 1$:

```
> fx <- discretise(plnorm(x,7,1),from = 0,to = 1000000,step = 500)
> Fs <- aggregateDist('recursive',model.freq = 'poisson',model.
sev = fx,
lambda = 10,x.scale = 500)
```

Once the aggregate loss distribution is calculated, it can be summarised in the usual way by calculating, for example, mean, standard deviation, and selected percentiles. For example, the function quantile gives some standard percentiles for the aggregate distribution calculated above:

```
> quantile(Fs)
25% 50% 75% 90% 95% 97.5% 99% 99.5%
9500 14000 20000 27000 32000 37500 44500 50500
```

17.4 Monte Carlo Simulation (Gross Losses)

The Monte Carlo simulation is probably the most popular way of calculating the aggregate loss distribution. This popularity is due mostly to the flexibility of the method and to

	No. of losses	Loss #1	Loss #2	Loss #3	Loss #4	Loss #5	Loss #6	Total losses
Simul #1	2	13,807	23,489					37,296
Simul #2	3	21,625	1863	20,397				43,885
Simul #3	1	20,589						20,589
Simul #4	4	28,254	81,563	11,780	60,441			182,038
Simul #5	2	43,171	16,310					59,481
Simul #6	1	10,124						10,124
Simul #7	0							–
Simul #8	2	8644	7337					15,981
Simul #9	2	20,549	50,082					70,631
Simul #10	6	98,877	32,575	8071	82,785	5401	69,620	297,329
Simul #11	3	14,476	5484	38,335				58,295
Simul #12	2	93,945	23,421					117,366
Simul #13	1	212,413						212,413
Simul #14	1	36,707						36,707
Simul #15	3	34,805	7755	46,916				89,476
Simul #16	5	11,107	56,779	172,196	251,647	21,481		513,210
Simul #17	5	3971	11,969	24,033	36,653	23,693		100,319
Simul #18	3	6400	2171	27,299				35,870
Simul #19	4	14,566	36,882	68,709	4588			124,745
Simul #20	2	21,672	10,976					32,648
							Average losses	102,920
							Standard deviation	122,544

FIGURE 17.6
Example of a Monte Carlo simulation. The frequency distribution is Poisson with rate $\lambda = 3$, whereas the severity distribution is lognormal with $\mu = 10$ and $\sigma = 1$.

the ease of its implementation. The idea is very simple, and it consists of simulating what could happen in one particular year by randomly choosing a number of losses (randomly, that is, based on the frequency distribution) and then, for each of these losses, choosing a severity based on the severity distribution. The result for one particular year is interesting, but it does not tell us whether the outcome we find is a typical one or not.

To have an estimate of the range of losses that could occur given a frequency and a severity distribution, we just need to repeat this procedure a large number of times. Figure 17.6 shows the result of repeating this procedure 20 times (typical values for the number of simulations are much bigger – 10,000 or more), for a case where the number of losses expected each year is small enough that full details of the simulation can be disclosed.

Even with such a small number of simulations, we can get a rough idea of what the average total losses and the volatility (as measured by the standard deviation) are going to be.

The exact values for the average losses and the standard deviation can be calculated by Equations 6.3 and 6.4 and are $\mathbb{E}(S) = 108,946.5$ and $\mathbb{V}(S) = 103{,}705.1$, not too far away from the approximate values shown in Figure 17.6, especially considering that we have used only 20 simulations.

Of course, if all we are interested in are the mean and the standard deviation, the formulae above do a better and quicker job than the Monte Carlo simulation. Thanks to the simulation, however, we can go a step beyond that and actually estimate the full aggregate loss distribution, that is, the probability that the total losses S will not be larger than an amount $s: F_S(s) = Pr(S \leq s)$. This can be done by sorting the total loss for the various

(a) Total losses	(b) Total losses (sorted)
37,296	–
43,885	10,124
20,589	15,981
182,038	20,589
59,481	32,648
10,124	35,870
–	36,707
15,981	37,296
70,631	43,885
297,329	58,295
58,295	59,481
117,366	70,631
212,413	89,476
36,707	100,319
89,476	117,366
513,210	124,745
100,319	182,038
35,870	212,413
124,745	297,329
32,648	513,210

FIGURE 17.7
(a) Results of the simulation shown in Figure 17.6. Each number represents the total loss amount for a given simulation. (b) The same simulation results, sorted in ascending order so that the different percentiles can be calculated.

Percentile	Total losses
10%	10,710
20%	23,001
30%	36,121
40%	39,932
50%	58,888
60%	81,938
70%	112,252
80%	170,579
90%	288,837
Average losses	102,920
Standard deviation	122,544

FIGURE 17.8
This shows the proportion of simulations of Figure 17.6 for which the total losses are below a certain level.

simulations in ascending order and then calculating the proportion of simulations for which $S \leq s$ (Figures 17.7 and 17.8).

Let us see now (Figure 17.9) what the results of a real Monte Carlo simulation look like. The accuracy of the results obviously depends on how many simulations have been performed to get to the results: for this reason, we show the percentiles of the aggregate loss distribution for four different runs, with 100, 1000, 10,000 and 100,000 simulations, respectively. The results in boldface at the right hand of the graph are the theoretical values.

Percentile	N = 100	N = 1000	N = 10,000	N = 100,000	Theoretical
50.0%	78,695	86,110	84,486	82,248	
75.0%	148,486	149,462	152,027	149,234	
80.0%	179,898	170,799	173,331	169,491	
85.0%	189,238	200,782	200,249	196,440	
90.0%	210,006	246,872	239,215	235,082	
95.0%	246,446	319,998	304,456	302,542	
98.0%	290,410	424,691	387,729	398,742	
99.0%	299,604	475,407	467,130	478,878	
99.5%	441,655	523,140	545,279	556,006	
99.9%	555,295	650,639	842,733	793,518	
Mean	101,305	111,038	110,581	108,959	**108,947**
SD	89,548	103,716	104,559	103,134	**103,705**

FIGURE 17.9
The larger the number of Monte Carlo simulations, the closer the mean and the standard deviation get to the theoretical results.

Note how the mean and especially the standard deviation take a while to converge to the correct theoretical value. This is mainly due to the effect of the tail of the lognormal distribution (occasional large losses will throw off the estimates of both mean and standard deviation, the latter being more strongly affected) and can be partially overcome by using more sophisticated Monte Carlo simulation methods, such as stratified sampling, including the Latin hypercube, which forces the algorithm to sample more evenly from different parts of the severity distribution.

17.4.1 Practical Issues

The Monte Carlo simulation is arguably the most flexible method for calculating the aggregate loss distribution. However,

- Achieving a good precision in the calculation of the quantiles requires a large number of simulations, making the methodology slow.
- What is worse, the number of simulations needed to achieve the desired level of accuracy cannot be known in advance (see, e.g. Shevchenko, 2010). In practice, one needs to continue performing simulations until the convergence of the desired quantiles is observed empirically.
- Also, calculation of the quantiles requires sorting all the simulated scenarios and therefore storing all the scenarios in the computer's memory, which may be challenging when the number of simulations is very large, although special techniques are available to deal with this problem (Shevchenko, 2010).

To be more specific on the issue of computational complexity, the number of operations T required for running a Monte Carlo simulation is proportional to the number of simulations n_{sim} and for each simulation to the expected number of losses, λ: $T(n_{sim}, \lambda) \propto n_{sim} \lambda$. Techniques to make convergence quicker, such as stratified sampling, will not change these relationships but will reduce the proportionality constant. The quantiles then need to be sorted, which require a number of operations proportional to $n_{sim} \times \log(n_{sim})$ with some algorithms.

One specific advantage of the Monte Carlo simulation from the computational point is that the generation of the n_{sim} different scenarios can be fully parallelised. Theoretically, *all* simulations could be run in parallel if one had n_{sim} different processors, and sorting could be performed in time proportional to $\log(n_{sim})$. With the availability of parallel cloud computing, this is a consideration that needs to be made in the development of professional simulation algorithms.

17.4.2 Implementation

The Monte Carlo simulation is straightforward to code in VBA, although making it more efficient with techniques such as stratified sampling requires more thought. In R, calculation of the aggregate loss distribution can be performed via the aggregateDist function in the actuar package (Dutang et al., 2008). However, given the importance of this technique and its widespread use amongst actuaries, we have thought it useful to present a simple implementation of the Monte Carlo simulation from scratch (Appendix A).

17.5 Coverage Modifications (Ceded and Retained Losses)

The methods above can be used to varying degrees to analyse not only the full losses but also the losses retained below a certain deductible, or the losses ceded above a certain deductible, or in general ceded to an insurance/reinsurance structure. In this section we will look at how we can calculate the *ceded* and *retained* aggregate loss distribution using these different methods.

First, however, it is useful to consider what can be said from a purely analytical point of view. As in the gross case, the elementary statistics can be calculated analytically, at least if there are no aggregate deductibles/limits.

As a common example, consider the simple structure in which the insured retains all losses between a deductible D and cedes the losses above D, up to a limit L on each loss. No aggregate features are considered in this example.

Assuming a collective risk model where N is the number of losses and X is the loss amount (see Section 6.1.2), the expected losses retained below the deductible are given by:

$$\mathbb{E}(S_{\text{ret}}) = \mathbb{E}(\min(X,D)) \times \mathbb{E}(N) \tag{17.22}$$

while the expected ceded losses are

$$\mathbb{E}(S_{\text{ced}}) = \mathbb{E}(X(D,L)) \times \mathbb{E}(N) \tag{17.23}$$

where $X(D,L)$ is shorthand for $\min(L, \max(X-D,0))$. The expected losses retained *above* the limit can be obtained simply as the expected gross minus the two components above.

The higher moments are also easy to express using the properties of the collective risk model (Section 6.1.2). For example, the variance for the retained below the deductible and the ceded losses are as follows:

$$\mathbb{V}(S_{\text{ret}}) = \mathbb{E}(\min(X,D))^2\, \mathbb{V}(N) + \mathbb{V}(\min(X,D))\,\mathbb{E}(N) \tag{17.24}$$

$$\mathbb{V}(S_{\text{ced}}) = \mathbb{E}(X(D,L))^2 \, \mathbb{V}(N) + \mathbb{V}(X(D,L))\mathbb{E}(N) \tag{17.25}$$

These formulae will normally be easy to calculate analytically although the process is tedious and the result long and unsightly. For example, in the simple case where N is given by a Poisson distribution with rate $\lambda (N \sim \text{Poi}(\lambda))$ and X comes from a Lognormal distribution $X \sim \text{LogN}(\mu, \sigma)$, the relations for the expected losses above become:

$$\mathbb{E}(S_{\text{ret}}) = \lambda \times \left(\exp\left(\mu + \frac{1}{2}\sigma^2\right) \Phi\left(\frac{\ln D - \mu - \sigma^2}{\sigma}\right) + D\bar{\Phi}\left(\frac{\ln D - \mu}{\sigma}\right)\right) \tag{17.26}$$

and

$$\mathbb{E}(S_{\text{ced}}) = \lambda \times \begin{pmatrix} \exp\left(\mu + \frac{1}{2}\sigma^2\right)\left(\Phi\left(\frac{\ln(D+L) - \mu - \sigma^2}{\sigma}\right) - \Phi\left(\frac{\ln D - \mu - \sigma^2}{\sigma}\right)\right) \\ +(D+L)\bar{\Phi}\left(\frac{\ln(D+L) - \mu}{\sigma}\right) - D\bar{\Phi}\left(\frac{\ln D - \mu}{\sigma}\right) \end{pmatrix} \tag{17.27}$$

where Φ is the normalised Gaussian cumulative distribution function, which we assume as a primitive function, and $\bar{\Phi}(x) = 1 - \Phi(x)$ is the survival function.

The formulae for the variance are even more unsightly and have been relegated to the solution to Question 9. Chapter 20 on non-proportional reinsurance has similar formulae where the severity distribution is a Pareto or a GPD – which are more useful in a reinsurance context.

The Downloads section of the book's website (Parodi, 2016) has the spreadsheet with the relevant calculations for the expected losses and the variance for the cases mentioned above. It also includes formulae for the case of a spliced Lognormal/Pareto or a Lognormal/GPD distribution, which are more useful ground-up severity distributions than a simple Lognormal (see Chapter 16 and Fackler (2013)).

To sum up, in the case where no aggregate features are present it is at least possible to calculate analytically the moments of the retained and ceded distribution. However, the calculation of the percentiles of a distribution is normally out of reach. The rest of this section shows how these can be calculated using various approximations and numerical methods.

17.5.1 Gaussian Approximation

We will limit ourselves to the case in which we have coverage modifiers that affect only the individual severity, without any constraints on the aggregate. The cases where there *are* constraints on the aggregate are much more difficult to tackle with the Gaussian approximation and ad hoc methods will need to be devised.

Assuming again the simple structure described at the beginning of Section 17.5, we can use Equations 17.22 to 17.25 to calculate $\mathbb{E}(S_{\text{ret}})$, $\mathbb{E}(S_{\text{ced}})$, $\mathbb{V}(S_{\text{ret}})$, and $\mathbb{V}(S_{\text{ced}})$. (If no model is available for the frequency and the severity, we can use the empirical means and variances.)

To estimate the percentiles of the aggregate retained/ceded loss distribution we can then calculate the percentiles of the normal distribution with mean and variance equal to those thus calculated. We can do similarly if we use a lognormal or translated gamma approximation.

*17.5.2 Fast Fourier Transform

Dealing with coverage modifiers is more complicated when using FFT technology. The difficulties arise especially if there are aggregate modifiers such as an annual aggregate deductible. For simplicity's sake, let us assume in this section that there are no aggregate coverage modifiers.

17.5.2.1 Aggregate Retained Losses

The retained severity distribution can be derived from the gross severity distribution by noting that $f_{\text{ret}}(x)$ is equal to $f(x)$ when x < EEL, to 0 when x > EEL and has a probability mass of value 1 − F(EEL) at x = EEL.

In DFT formalism, this can be approximated by a discrete probability vector whose first $\text{EEL}/h-1$ elements are equal to the elements of f, then an element of mass $S_{\text{EEL}} = 1 - \sum_{j=0}^{\text{EEL}/h-1} f_j$ the rest of the vector elements being zero.

$$f_{\text{ret}} = \left(f_0, f_1 \cdots f_{\text{EEL}/h-1}, S_{\text{EEL}}, 0, \cdots 0\right) \tag{17.28}$$

In the representation above, we have assumed that EEL is a multiple of h.

The aggregate retained distribution can be derived in the same way as in the gross case by taking the DFT of f_{ret}, $\hat{f}_{\text{ret}} = \left(\hat{f}_{\text{ret},0}, \hat{f}_{\text{ret},1} \cdots \hat{f}_{\text{ret},M-1}\right)$ and then compounding it with the frequency model according to Equation 17.8 in the Poisson case:

$$\hat{f}^S_{\text{ret},j} = \exp\left(\lambda\left(\hat{f}_{\text{ret},j} - 1\right)\right) \tag{17.29}$$

and according to Equation 17.9 in the negative binomial case:

$$\hat{f}^S_{\text{ret},j} = \left(\frac{1-p}{1-p\hat{f}_{\text{ret},j}}\right)^r \tag{17.30}$$

The aggregate loss distribution can then be obtained simply by inverting the DFT of $\hat{f}^S_{\text{ret}}$, obtaining f^S_{ret}.

17.5.2.2 Aggregate Ceded Losses

As for the ceded severity distribution, we need to translate the vector by EEL/*h* positions and then use the correct factor to make sure that the overall probability is 1:

$$f_{\text{ced}} = \left(\frac{f_{\text{EEL}/h}}{S_{\text{EEL}}}, \frac{f_{\text{EEL}/h+1}}{S_{\text{EEL}}}, \cdots \frac{f_M}{S_{\text{EEL}}}, 0, \cdots 0\right) \tag{17.31}$$

If there is also a limit L to the size of the individual losses that can be ceded, this can be taken into account as we did for the EEL deductible in the retained distribution, and f_{ced} becomes

$$f_{\text{ced}} = \left(\frac{f_{\text{EEL}/h}}{S_{\text{EEL}}}, \frac{f_{\text{EEL}/h+1,}}{S_{\text{EEL}}}, \cdots \frac{f_{\frac{\min(L,M)}{h}-1}}{S_{\text{EEL}}}, S_L, 0, 0 \cdots 0\right) \tag{17.32}$$

where $S_L = 1 - \sum_{j=0}^{(\min(L,M)/h - \text{EEL}/h - 1)} f_{\text{EEL}/h+j}$.

The case of the aggregate ceded distribution is similar, except that the rate is going to be lower because not all losses will make it to the cession threshold. Therefore, we first of all take the DFT of f_{ced}, $\hat{f}_{\text{ced}} = \left(\hat{f}_{\text{ced},0}, \hat{f}_{\text{ced},1} \cdots \hat{f}_{\text{ced},M-1}\right)$, then calculate the ceded frequency, $\lambda_{\text{ced}} = \lambda S_{\text{EEL}}$, and finally the DFT of the aggregate ceded distribution, which in the Poisson case reads:

$$\hat{f}^S_{\text{ced},j} = \exp\left(\lambda_{\text{ced}}\left(\hat{f}_{\text{ced},j} - 1\right)\right) \tag{17.33}$$

and in the negative binomial case:

$$\hat{f}^S_j = \left(1 + (v-1)\left(1 - \hat{f}_j\right)\right)^{-\lambda_{\text{ced}}/(v-1)} \tag{17.34}$$

17.5.2.3 More Complex Structures

We have mentioned that dealing with aggregate features is more complex with DFT formalism. Of course, it is easy to just impose a cap on the aggregate amount retained by adding a probability mass at AAD/h in vector $\hat{f}^S_{\text{ret}}$ and replace the rest of the vector with 0's. The calculation of the vector f_{ced} is more complex because what is now ceded does not only depend on the individual amount but also on whether the AAD has already been

	No. of Losses	Loss #1	Loss #2	Loss #3	Loss #4	Loss #5	Loss #6	Total losses
Simul #1	2	10,000	10,000					20,000
Simul #2	3	10,000	1863	10,000				21,863
Simul #3	1	10,000						10,000
Simul #4	4	10,000	10,000	10,000	10,000			40,000
Simul #5	2	10,000	10,000					20,000
Simul #6	1	10,000						10,000
Simul #7	0							–
Simul #8	2	8644	7337					15,981
Simul #9	2	10,000	10,000					20,000
Simul #10	6	10,000	10,000	8071	10,000	5401	10,000	50,000
Simul #11	3	10,000	5484	10,000				25,484
Simul #12	2	10,000	10,000					20,000
Simul #13	1	10,000						10,000
Simul #14	1	10,000						10,000
Simul #15	3	10,000	7755	10,000				27,755
Simul #16	5	10,000	10,000	10,000	10,000	10,000		50,000
Simul #17	5	3971	10,000	10,000	10,000	10,000		43,971
Simul #18	3	6400	2171	10,000				18,571
Simul #19	4	10,000	10,000	10,000	4588			34,588
Simul #20	2	10,000	10,000					20,000
							Average losses	23,411
							Standard deviation	13,892

FIGURE 17.10
This shows how to estimate the aggregate retained distribution based on a stochastic simulation. The individual losses in this table are obtained by imposing a cap of £10k (the deductible) on each loss of Figure 17.6. A cap of £50k is then imposed on the overall retained amount (e.g. in simulation no. 10).

	No. of Losses	Loss #1	Loss #2	Loss #3	Loss #4	Loss #5	Loss #6	Total losses
Simul #1	2	3807	13,489					17,296
Simul #2	3	11,625	–	10,397				22,022
Simul #3	1	10,589						10,589
Simul #4	4	18,254	71,563	1780	50,441			142,038
Simul #5	2	33,171	6310					39,481
Simul #6	1	124						124
Simul #7	0							–
Simul #8	2	–	–					–
Simul #9	2	10,549	40,082					50,631
Simul #10	6	88,877	22,575	–	72,785	–	59,620	247,329
Simul #11	3	4476	–	28,335				32,811
Simul #12	2	83,945	13,421					97,366
Simul #13	1	202,413						202,413
Simul #14	1	26,707						26,707
Simul #15	3	24,805	–	36,916				61,721
Simul #16	5	1107	46,779	162,196	241,647	11,481		463,210
Simul #17	5	–	1969	14,033	26,653	13,693		56,348
Simul #18	3	–	–	17,299				17,299
Simul #19	4	4566	26,882	58,709	–			90,157
Simul #20	2	11,672	976					12,648
							Average losses	79,510
							Standard deviation	112,929

FIGURE 17.11
This shows how to estimate the aggregate ceded distribution based on a stochastic simulation. The individual losses in this table are obtained by calculating the amount in excess of the deductible (if there is any excess amount) on each loss of Figure 17.6. The total ceded losses is calculated by subtracting the total retained losses (Figure 17.10) from the total gross losses (Figure 17.6).

exceeded. Dealing with this is possible but requires some code complications and a reduction of computational speed. Similar considerations arise for the annual aggregate limit on the ceded amount if more than one insurance layer is present.

*17.5.3 Panjer Recursion

Dealing with coverage modifications can be done with Panjer recursion in much the same way as for the FFT method, that is, by modifying the discretised severity distribution to account for individual policy modifiers and then applying the recursion using the modified severity distribution, and we will not expand on this here. The same difficulty with aggregate modifiers such as the AAD also applies to Panjer recursion.

17.5.4 The Monte Carlo Simulation

The main advantage of the Monte Carlo simulation is its complete flexibility: there is no conceivable insurance structure for which the calculations cannot be performed with a purpose-built simulation. To illustrate how the Monte Carlo simulation works as a method to calculate the aggregate retained and ceded distribution for a given insurance structure with EEL = £10,000 and AAD = £50,000, we use the results of the very simple stochastic simulation from Figure 17.6.

Every loss in the simulation is then replaced with the loss as modified by the insurance programme. Thus, for example loss no. 1 in simulation no. 1 is £13,807: as a consequence, the retained amount is £10,000 and the ceded amount is £3807 (Figures 17.10–17.12).

Gross, Retained and Ceded Losses for EEL = £10k, AAD = £50k

Percentile	Gross	Retained	Ceded
50%	82,361	26,803	53,691
75%	148,895	39,031	112,254
80%	169,812	40,000	131,235
85%	196,350	44,399	155,972
90%	234,381	50,000	192,071
95%	301,482	50,000	256,029
98%	398,365	50,000	351,318
99%	471,898	50,000	425,594
99.5%	563,486	50,000	514,782
99.9%	811,104	50,000	761,538
Mean	108,926	26,660	81,360
SD	103,264	14,275	93,731

FIGURE 17.12
The results of a larger simulation experiment (100,000 simulations) with the same gross loss model as in Figure 17.9. Note that the gross loss results are slightly different from those in Figure 17.9, as they originate from a different simulation.

17.6 A Comparison of the Different Methods at a Glance

It is perhaps useful to gather the advantages and disadvantages of the various aggregate loss calculation methods that we have explored in this chapter in a single table (Figure 17.13). A more detailed comparison of available methods (including some methods we haven't considered here) can be found in an article by Shevchenko (2010).

17.6.1 Conclusions

Overall, the Monte Carlo simulation should be preferred where the maximum amount of flexibility is needed, whereas FFT should be used where computational efficiency is paramount. Panjer recursion is a viable alternative to FFT where computational efficiency is less crucial while ease of implementation is needed. Gaussian approximation and similar approximation methods such as the translated gamma are mostly of historical and didactic interest.

17.7 Monte Carlo Simulation Results for our Case Study

We are now in a position to produce the aggregate loss distribution for our case study. This is normally shown in the form of a table that includes the basic statistics (mean, standard deviation, useful percentiles) for the total losses, the total number of losses, the losses retained below the deductible and the ceded losses, and the number of losses above the deductible. Specifically, the table in Figure 17.14 shows these statistics obtained through Monte Carlo simulation.

	Accuracy	Flexibility and implementation	Speed
Parametric approximations	Poor, especially when the yearly number of claims is small and the severity tail is fat. Also, accuracy cannot be improved beyond a certain limit.	Limited flexibility, e.g. when aggregate deductibles are present and when losses to a high layer need to be calculated. Implementation is trivial.	Since the only calculation required is finding the percentile of a Normal distribution or other simple statistics, results are obtained almost immediately.
Fourier transform	An arbitrary degree of accuracy can be achieved by calibrating the discretisation step and the number of points. The discretisation step may need to be adjusted depending on the number of claims and the severity range.	Some types of structures may be difficult to model, e.g. ceded losses when an aggregate deductible is present. Implementation requires some artful tuning of the parameters (such as discretisation step, tilting) depending on the frequency and severity distribution.	Very fast. Speed does not depend on the number of claims but only on the number of points sampled from the severity and aggregate loss distribution. A good degree of parallelism can be achieved.
Panjer recursion	An arbitrary degree of accuracy can be achieved by calibrating the discretisation step.	Same difficulty as Fourier transform with aggregate structures. Can be used only with certain frequency distributions Implementation is easier since the only parameter that needs to be tuned is the discretisation step. Computational issues (underflow/overflow) arise with large frequencies.	Reasonably fast. Speed does not depend on the number of claims but only on the number of points sampled from the severity and aggregate loss distribution.
MC simulation	An arbitrary degree of accuracy can be achieved, but this requires increasing the number of simulations, possibly encountering memory problems.	Completely flexible. Can include effects related to the order in which claims happen, dependencies, etc.	Slow when the number of claims is high and when high accuracy is needed. Note, however, that the simulations process lends itself to a high degree of parallelism – in principle, all different simulations can be run simultaneously.

FIGURE 17.13
A comparison between different methods to calculate the aggregate loss distribution.

17.8 Monte Carlo Simulation for the Individual Risk Model

Most of the material we have seen in this chapter refers to the collective risk model. The individual risk model works similarly but it is usually simpler. As an example, a Monte Carlo simulation to produce the aggregate loss distribution for the individual risk model can be carried out as follows.

Return Period (Years)	Percentile	Total Loss	Total Number	Retained Below Deductible	Ceded Frequency	Ceded Amount
1 in 2	50%	3.49	265	2.62	6	0.82
1 in 4	75%	4.31	300	3.03	9	1.44
1 in 5	80%	4.56	309	3.14	10	1.66
1 in 10	90%	5.29	335	3.20	30	2.31
1 in 20	95%	6.03	357	3.20	52	2.95
1 in 50	98%	7.09	379	3.20	81	4.11
1 in 100	99%	8.10	394	3.20	100	5.19
1 in 200	99.5%	9.23	411	3.20	113	5.41
	Mean	3.71	264.0	2.59	12.2	1.10
	Std Dev	1.33	36.3	0.48	19.6	0.97

All monetary amounts are in GBP million.

FIGURE 17.14
Main statistics for the gross, retained, and ceded aggregate loss distribution for our case study. The choice of the percentiles (and the corresponding return periods) to be shown is of course a matter of taste and need, but it is customary to focus on the top 50% of the distribution, as it is the most relevant to the insurer as well as the client. Larger return periods could be shown but the desire to focus on the extreme part of the distribution needs to be balanced against the uncertainty around those figures.

Assume we have R risks $r = 1, \ldots R$

1. For each simulated year i, scan through all risks r and decide whether they will 'fail' (e.g. die or bankrupt) based on the probability of failure p_r.
2. If risk r fails, sample from the 'loss-given-default' (LGD) distribution of r (if LGD is not fixed; otherwise, simply pick the unique LGD), obtaining $X_{r,i}$.
3. Total losses in year i:

$$S_i = \sum_{r \text{ that fail in year } i} X_{r,i}$$

4. Repeat for $i = 1, \ldots N_{\text{sim}}$ simulations to produce the aggregate loss distribution.

The risk costing process will look very much the same but with some simplifications. For example, IBNR is typically not a concern, and there is no need to choose a frequency distribution in the simple case in which each risk has a fixed probability of failing (the distribution is automatically a multinomial distribution).

17.9 Example Code

Examples of code for calculating the aggregate loss distribution (gross, retained, ceded) via FFT or Monte Carlo simulation are available in the Appendix to Chapter 17 in the book's website (Parodi, 2016).

17.10 Questions

1. *Basic statistics for the collective risk model.* An actuary has analysed company X's motor third-party liability losses for next year by modelling the number of losses as a negative binomial distribution with rate equal to 208.8 and a variance-to-mean ratio of 2.5, and the severity of losses as a lognormal distribution with $\mu = 10.5$ and $\sigma = 1.3$. Calculate the expected aggregate losses and the standard deviation of the aggregate losses for next year.
2. *Monte Carlo simulation (computer-based exercise).* Assume that losses related to public liability are generated according to this process: next year, the number of losses will come from a Poisson distribution with rate = 4.4 and severity coming from a lognormal distribution with $\mu = 11.2$ and $\sigma = 1.5$ (results are in GBP). Using the Monte Carlo simulation with at least 1000 simulations, produce an estimate of the total gross losses at the following percentiles: 50%, 75%, 80%, 85%, 90%, 95%, 98%, 99%, and 99.5%. Also, produce an estimate of the mean and the standard deviation, and compare the results with the theoretical values for the mean and standard deviation.
3. *Gaussian approximation (either computer-based or using numerical tables).* Suppose that losses can be modelled according to the frequency/severity model described in Question 2 (frequency: Poi(4.4), severity: LogN(11.2,1.5)). (a) Using the Gaussian approximation model, produce an estimate for the following percentiles: 50%, 75%, 80%, 85%, 90%, 95%, 98%, 99%, and 99.5%. (b) Also, produce an estimate of the same percentiles for the ceded loss model, given that there is an EEL deductible of £10,000. (*Hint: look up the formulae for the limited expected value of the lognormal severity distribution.*)
4. You are an actuary for a commercial lines insurance company. One of your jobs is to set the annual aggregate deductible (hence 'the aggregate') for commercial policies at a level that is satisfactory for clients *and* is breached with a likelihood of less than or equal to 5%. The policy you have been asked to price is a public liability policy with a straightforward structure – a (ranking) EEL deductible of £100k, an annual aggregate deductible (to be determined), and an EEL limit of £25M.
   1. Explain what the annual aggregate deductible is, and what its purpose is from the client's perspective.
   2. Explain how the aggregate works and show the calculations in practice with this simple numerical example: EEL = £100k, aggregate = £300k, and these four losses in a year: £80k, £120k, £70k, and £150k.

      You have produced a stochastic loss model (gross of any deductible/limit) based on the frequency/severity approach. The model is based on an analysis of, for example, 10 years of historical losses.
   3. Illustrate how you would use the stochastic loss model thus constructed to set the aggregate at an acceptable level.
   4. Describe how you would calculate by how much the expected ceded losses would increase as a result of imposing the aggregate as for Step iii.
   5. Illustrate how a burning cost analysis could be used for the purpose of setting the aggregate at a suitable level (i.e. AAD breached with a likelihood of $\leq$5%).
   6. Explain the *relative* merits of the stochastic (iii) and burning cost (v) approach *to calculating a suitable aggregate.*

5. You are a quantitative analyst for Company X. Company X sells cheap digital cameras, which are sold at €50 in the European market and have a production cost of €20. The cameras are offered with a 1-year warranty that offers a free replacement for failures during the warranty period. A new model (MB4) of camera with the same price as the rest of the range is going to be sold in the Christmas period in 2012–2013 and 25,000 sales across Europe are expected. A recent study of the probability of failure suggests that the failure rate for a similar camera model (MB2) was 5% during the first year, although some differences amongst different European territories were noticed.

   You can assume that:

   i. You can only get one replacement a year.
   ii. The overall replacement cost is €25 (= production cost + expenses including mailing expenses).
   iii. If a camera is replaced, the warranty expiry date remains the original one: it does not start again from the date when the camera is received.
      i. Calculate the expected cost of providing warranty cover for the lot of MB4 models sold during Christmas. State clearly all the assumptions you make.
      ii. Describe how you would calculate the 90th percentile of the statistical distribution of the costs (you do not need to carry out the actual calculations, only to set up the methodology clearly enough for one to follow). State clearly all the assumptions you make.
      iii. What are the *main* uncertainties around the expected cost calculated in (i)?

6. *R-based question.* Modify the R code in Appendix A so that it is able to deal with (a) a negative binomial distribution for the frequency; (b) a spliced lognormal/GPD severity distribution; (c) a generic severity distribution, entered as a vector of N percentiles $\left(\frac{1}{N+1},\frac{2}{N+1}\ldots\frac{N}{N+1}\right)$.

   Hints: The following function can be used to generate an N_sim claim counts from an negative binomial distribution with the correct mean and variance-to-mean ratio by a suitable choice of *a*, *b*:

```
claim_count = rpois(N_sim,rgamma(N_sim,shape =
                    a,scale = b))
```

   The following function creates a spliced lognormal/GPD distribution from two loss vectors analogous to indiv.losses in the code (one based on a lognormal and the other based on a GPD), the splicing point being GPD_mu:

```
Indiv.losses = ifelse(logn_vector<GPD_mu,
                      logn_vector, GPD_vector)
```

7. *R-based question.* Your aggregate loss model is a compound Poisson/lognormal model with parameters $\lambda = 5$ (Poisson rate) and $\mu = 10.2$, $\sigma = 1.2$. Using R and the FFT algorithm given in Appendix B, you want to produce aggregate loss models. Choose values of M and h so that (i) the running time is less than 10 s; (ii) the mean of the discretised severity distribution is within 1% of the true value; and (iii) the 99th percentile of the aggregate loss distribution is within 1% of its true value. (Hint: you

do not know the true value of the 99th percentile, but you can double the number of points until the difference between one run and the other is less than 1%.)

8. (*) Prove Equation 17.21. *Hint*: write $g_0 = \Pr(S = 0)$ as an infinite sum $\Pr(S=0) = \sum_{n=0}^{\infty} \Pr(N=n) \Pr(X_1 + X_2 + \cdots + X_N = 0 \mid N = n)$ and note that for a sum of positive-definite random variables to be zero, all of them must be simultaneously zero.
9. (*) Write down Equations 17.24 and 17.25 for the Poisson/Lognormal case.
10. (*) Write down Equations 17.22–17.25 for the case where the frequency model is Poisson and the severity is a spliced Lognormal/Pareto or a spliced Lognormal/GPD.

18

Identifying, Measuring, and Communicating Uncertainty

That's not right. That's not even wrong.

Wolfgang Pauli

In the previous chapters, we learned how to produce a stochastic model for our aggregate losses, which encapsulates our prediction for the losses in the next policy period. On the basis of this model, underwriters and clients may (if they trust us) take decisions on how much premium to charge (underwriters), whether to buy a policy and how much risk to retain (clients) and so on.

It is only fair to communicate how accurate our predictions are. A statement such as, 'the expected losses to this policy for the year 2012 are £1.62M' is neither right nor wrong if you are not also able, to some extent, to assess the error on your estimate: because saying that the expected losses are £1.62 ± £0.01M is very different from saying that they are £1.62 ± £1M!

Things get more confusing because we are not speaking of deterministic quantities such as your weight, but of the mean of a stochastic variable that fluctuates widely, so it may take several years to realise that an estimate may be materially wrong.

We have several types of uncertainty that we need to address in actuarial practice:

- Process uncertainty
- Parameter uncertainty
- Model uncertainty
- Assumption/data uncertainty
- Approximation errors in calculations

We are now going to look at each type of uncertainty in turn. After a few words of explanation, we will explain how to calculate the effect of that type of uncertainty on pricing and we will briefly comment on how that type of uncertainty can be communicated to a non-technical audience: a section on communicating uncertainty has been included because it is one of the things that the actuary is explicitly required to do in formal reports and because it is also something that is quite challenging.

In defining the various types of uncertainty, it is useful to keep the following simple examples in mind:

DOI: 10.1201/9781003168881-20

Example 1: Consider the yearly number of claims related to a risk. Assume that the number of claims follows a Poisson distribution and that the Poisson rate is estimated based on the number of claims observed over a past period.

Example 2: Consider the distribution of loss severities for a given risk. Assume that the loss amounts follow a lognormal distribution and that the parameters of the lognormal distribution are calculated based on a number of claims observed over a past period. Also, consider that the final loss amount of some of the claims is not known yet – only a reserve estimate is available.

Example 3: Consider the distribution of aggregate losses for a given risk, calculated combining the Poisson frequency model and the lognormal severity model as derived in Example 2 using a Monte Carlo simulation.

18.1 Process Uncertainty

Process uncertainty is the uncertainty that derives from dealing with inherently stochastic phenomena or phenomena that seem to be stochastic at the level we are able to look at them. This is intrinsic to the phenomenon and cannot be reduced.

In Example 1 above, even if we knew for sure that the number of claims for that risk could be modelled by a Poisson distribution and we knew the Poisson rate with infinite accuracy, the actual number of claims would exhibit random fluctuations from one year to another, and these fluctuations would be driven exclusively by the loss generation process itself.

18.1.1 How to Calculate Its Effect

That is what we have been doing until now! In the past few chapters, we have been showing how to produce models for the number of losses, for their severity, for the total amount of losses, etc. Each of these models provides an articulate view on process uncertainty because it gives the probability of obtaining a given number of claims or that a given claim has a severity larger than a certain amount and so on.

As an example, the aggregate loss distribution spells out the probability that the total losses for a given risk will be above a certain level next year. As such, it fully describes the process uncertainty for the overall loss amount.

18.1.2 How to Communicate It

Focus on things that your audience understands better first: the mean expected losses, the standard deviation …

Depending on the audience, it may be helpful to tweak the terminology a bit:

- Use 'average' instead of 'mean'
- Use 'volatility' instead of 'standard deviation'
- Use 'return period' or 'probability that losses are below/above a certain value' instead of 'percentile' or 'quantile'

18.2 Parameter Uncertainty

This is the additional uncertainty that derives from the fact that the parameters of our models are never known with 100% accuracy, even if the model is correct. To reduce this uncertainty, we need to use larger samples, but this is not always possible. In the case mentioned above, we may know for sure that our claims come from a Poisson process, but we will know what the rate is only with limited accuracy.

18.2.1 How to Calculate Its Effect

When you calculate the parameters with maximum likelihood estimation (MLE), there are mathematical results that give the 'standard error' (i.e. the volatility around the mean estimate) of the parameters – for example, Fisher's information matrix. This is usually provided in standard tools.

Calculating the effect on the price is more difficult: one way is to sample from the distribution of the parameters every time you run a simulation; for example, if you have a lognormal, every time you produce a lognormal variate, you use different parameters (μ,σ) sampling from the parameter distribution. You can do the same for the frequency distribution.

Here is an example of how you can take parameter uncertainty into account for a simple Poisson/lognormal model. Assume the Poisson rate is $\lambda = 100$ (calculated on $m = 10$ years of equal exposure, to make things simple) and the parameters of the lognormal are $\mu = 10.5$ and $\sigma = 1.3$ (calculated on n data points). The 'central' value of our expected losses estimate will be $\mathbb{E}(S) = \mathbb{E}(N)\mathbb{E}(X) = \lambda \times \exp\left(\mu + \frac{\sigma^2}{2}\right) = £8.45\text{M}$. Standard MLE results will give us the parameter uncertainty on each of these parameters, as measured by the standard error:

$$\text{s.e.}(\lambda) = \frac{\sqrt{\lambda}}{\sqrt{m}}, \quad \text{s.e}(\mu) = \frac{\sigma}{\sqrt{n}}, \quad \text{s.e}(\sigma) = \frac{\sigma}{\sqrt{2n}}$$

We can then simulate different values of the parameters from *independent*[1] normal distributions centred around λ, μ, σ, as in Figure 18.1, based on 100 simulations only (for illustration), of which we show the first 10.

As a result, we find a parameter uncertainty on the expected losses of about

$$\text{Parameter uncertainty} = \text{s.e.(expected losses)} = £1.23\text{M}$$

£1.23M. This is different from the process uncertainty, which is the standard deviation of the aggregate loss distribution and can be calculated by the standard formula for the variance of a compound Poisson distribution:

$$\mathbb{V}(S) = \lambda\mathbb{E}(X^2) = \lambda \times \exp(2\mu + 2\sigma^2)$$

Hence,

$$\text{Process uncertainty} = \text{SD}(S) = \sqrt{\lambda} \times \exp(\mu + \sigma^2) = £1.97\text{M}$$

1 The fact that the errors of the parameters μ, σ are independent is one of the advantages of the lognormal distribution.

Calculating Parameter Uncertainty with MC Simulation

λ	100	s.e.(λ)	3.16
μ	10.5	s.e.(μ)	0.106
σ	1.3	s.e.(σ)	0.075
No. of years	10		
No. of losses	150		

Expected losses (theor)	8,454,168
Expected losses (empir)	8,472,776
Process uncertainty (std dev, theor)	1,968,112
Parameter uncertainty (std error)	1,229,562
Overall uncertainty	2,320,622

Simulation ID	**μ**	**σ**	**λ**	**Expected losses**
1	10.45632664	1.204652	96.32733988	6,918,238
2	10.45440922	1.316405	101.0259179	8,337,278
3	10.5328109	1.302281	99.55598188	8,723,214
4	10.3775925	1.312729	98.02340736	7,455,236
5	10.50628158	1.460929	97.01579319	10,306,772
6	10.32770823	1.354988	100.8378806	7,719,183
7	10.60585124	1.374293	94.52575985	9,811,490
8	10.48887245	1.335992	102.2565536	8,964,609
9	10.64598033	1.36816	99.6742903	10,679,314
10	10.3841682	1.269511	91.76867193	6,644,292
...	...	...	...	...

FIGURE 18.1
Using Monte Carlo simulation, one can get a sense of what the uncertainty around our estimate of the expected losses is. In this case, the average uncertainty (as measured by the standard error on the expected losses) is approximately £1.23M. Note that this is different from the uncertainty around next year's outcome as a result of process uncertainty (£1.97M).

The total volatility around next year's outcome will be made of two independent components, the process uncertainty (£1.97M) and the parameter uncertainty (£1.23M). For independent random variables, the variance of the sum is equal to the sum of the variances:

$$\text{Total volatility} = \sqrt{(\text{£}1.97\text{M})^2 + (\text{£}1.23\text{M})^2} = \text{£}2.32\text{M}$$

which shows that one of the effects of parameter uncertainty is, quite naturally, to increase the overall uncertainty of next year's outcome.

Note that you may find that the effect of parameter uncertainty is not only that it causes an uncertainty around the mean and other statistics, but also – depending on the model – that it might introduce a *bias* in the calculated values. This is the case, for example, when you consider parameter uncertainty for tail modelling.

18.2.2 How to Communicate It

It is non-trivial for a non-technical audience to distinguish between parameter uncertainty and process uncertainty, and you may be asked: 'Do you not already have a distribution of values for next year'?

You should probably start explaining that even the average losses are only an estimate, let alone the large percentiles of the aggregate loss distribution. You can explain that this is especially true if you have scarce data.

18.3 Model Uncertainty

This is the additional uncertainty that comes from not being sure what the correct model is. In our claim count example of a Poisson process, we may not be fully sure that the process is Poisson: it might be a negative binomial distribution instead.

18.3.1 How to Calculate Its Effect

It is very difficult to calculate the effect of model uncertainty – or, to be more precise, it is impossible! To be able to calculate the error you make by using the wrong model, you need to know what the right model is; and knowing the right model is not different from knowing the right theory that describes a phenomenon. Theories are always only approximations of reality, and you can never be sure that your theory is correct under all circumstances.

The only case in which one can get a fair idea of model error is when the number of possible models is limited, in which case there are results from Bayesian analysis that guide us in the choice of the correct model and also allow us to calculate an estimate of the model error (Hastie et al., 2001). However, it should be noted that this is a very artificial situation and is unlikely to occur in the real world.

Rather than estimating the model uncertainty or assuming (unrealistically) that the number of possible models is limited, one can often achieve a much less ambitious, but still valuable, goal: consider a number of possible interesting models and compare the results of different models. The variability of the results will somehow be informative of the uncertainty brought in by the model choice.

18.3.2 How to Communicate It

It is not difficult to convey the idea that models are just simplified descriptions of reality and that as a result of this simplification there is some additional uncertainty in the outcome of your frequency, severity, or aggregate loss analysis. Also, you can explain that you may not be able to quantify this uncertainty exactly, but come to think of it, nobody else in the world can!

18.4 Assumption/Data Uncertainty

Data uncertainty is the additional uncertainty that comes from knowing the data only with limited accuracy. Assumption uncertainty – the uncertainty around assumptions such as claims inflation, future exposure, and others, can be seen as a type of data uncertainty. It is safe to say that in many actuarial studies (especially those that rely on limited data), assumptions uncertainty is the most relevant type of uncertainty affecting the results of the study.

In Example 1 above, not all claims may be reported immediately and therefore in every period, the number of claims will only be an estimate based, amongst other things, on our estimates of incurred but not reported (IBNR) claims.

In Example 2, there are at least two sources of data uncertainty:

i. The loss amount for *open claims* will not be the final settled amount, but a combination of (possibly) an amount already paid and an estimated reserve for the outstanding portion of the claim.
ii. For both *open and closed claims* because the severity model is built using the revalued losses, any uncertainty on the correct level of claims inflation to be applied to each claim will translate into an uncertainty on the revalued loss amount.

18.4.1 How to Calculate Its Effect

Generally speaking, the effect of assumption/data uncertainty on a model can be estimated by recalibrating the model under different assumptions on what the true values of the data really are. Let us see how this works in practice for Example 2.

18.4.1.1 Claims Inflation

To assess the effect of data uncertainty arising from claims inflation, which affects all claims simultaneously, one may simply want to undertake some *sensitivity analysis*: for example, if one's claims inflation assumption is 5%, one might want to repeat the analysis with different values of the claims inflation within a range of reasonable assumptions – for example, 3% to 7%.

18.4.1.2 Open Claims

To assess the effect of data uncertainty arising from claim amounts that have not yet been settled, you can use a *parametric bootstrap methodology*. This works by calculating the parameters of the severity distribution (e.g. a lognormal) repeatedly using different values of the loss amounts that are uncertain. The procedure is shown in Box 18.1, where we have chosen the case of a lognormal distribution for simplicity.

The case of loss data that may or may not end up being included in the data set can be dealt with in a similar way, by excluding (with a given probability) these data points from the loss data set. A simple illustration of this procedure can be found in Figure 18.2. Note that the effect of data uncertainty is in practice to increase the effect of parameter uncertainty.

Other strategies to deal with data uncertainty include the removal of data that is excessively uncertain, and the use of Bayesian analysis and fuzzy set theory. An overview of approaches can be found in Parodi (2012b).

18.4.2 How to Communicate It

The effect of assumption and data uncertainty is not difficult to explain: everyone understands the concept 'garbage in, garbage out'. What is more difficult is to explain,

BOX 18.1 PARAMETRIC BOOTSTRAP FOR A LOGNORMAL DISTRIBUTION WITH UNCERTAIN LOSS DATA

Inputs: k claims, some of which are open, each of which is represented by a random variable X_j $(j=1,...k)$ with a distribution of possible values $F(X_j)$ (such as a beta distribution, a triangular distribution, or a uniform distribution). In the case of closed claims, the distribution has a single value, the settled amount.

1. For each simulation $i = 1,\ldots N$
   a. For each claim $X_j\left(j = 1,\ldots k\right)$ in the data set, whose real value is assumed to come from a distribution of values centred around the best estimate and within a certain range:
      i. Sample a value $X_{j,i}$ for the loss amount from the distribution of possible values of $X_j, F_j\left(X_j\right)$
      ii. Calculate the parameters μ_i, σ_i of the lognormal distribution based on the loss data set $\{X_{1,i},\ldots X_{k,i}\}$
2. The volatility (and the whole distribution) of the possible values of μ_i, σ_i allows us to estimate the effect of data uncertainty on the severity distribution, and in turn on the expected losses and the overall distribution of the aggregate losses, much in the same way as for parameter uncertainty (Figure 18.1).

μ	10.37			10.45	10.44	10.38	10.32	10.38	...
σ	1.40			1.39	1.46	1.43	1.34	1.36	...
ID	**Best**	**Min**	**Max**	**Sim 1**	**Sim 2**	**Sim 3**	**Sim 4**	**Sim 5**	...
1	32,433	19,460	54,056	33,695	37,586	24,569	24,197	45,118	...
2	35,638	30,506	41,633	41,020	41,243	37,056	32,484	33,470	...
3	84,195	84,195	84,195	84,195	84,195	84,195	84,195	84,195	...
4	91,931	67,202	125,761	106,074	118,300	93,507	76,562	111,833	...
5	29,739	21,977	40,243	32,196	25,972	34,820	29,492	22,484	...
6	1814	1749	1882	1800	1865	1853	1804	1843	...
7	41,412	27,870	61,534	47,426	59,733	51,416	30,375	29,781	...
8	38,843	35,580	42,405	37,481	38,780	40,044	39,533	35,895	...
9	5554	3521	8761	7144	4859	4913	8088	8367	...
10	245,479	234,432	257,046	254,871	254,909	252,798	242,043	235,896	...

FIGURE 18.2
A simple illustration of how it is possible to take data uncertainty arising from the uncertainty on the settled claim amount. The table shows (column 'Best') the best estimate values of 10 claims, and (at the top of the same column) the parameters of a lognormal based on these values. For each claim, a minimum and maximum value is also given, and it is assumed (for simplicity) that the actual value of the value is equally likely to be any value within this range. Note that if a claim is already settled it will have Min = Max = Best (as the claim with ID = 3 in the table). What one does, then, is to run a number of simulations (the first five are shown here) and in each simulation, a value is simulated from the (min, max) range of each claim. One can then recalculate the parameters of the lognormal distribution based on these simulated values. By analysing the volatility of the lognormal parameters, it is then possible to understand the effect that data uncertainty has on the parameters of the lognormal distribution and hence on pricing.

in layman's terms, how you can calculate the effect of assumptions/data uncertainty on pricing.

The general idea, however, is not too difficult either: because some of the data are uncertain, the model must be recalculated under different assumptions on the data – such as different assumptions on claims inflation or different assumptions on each individual loss amount. This can be done by sensitivity analysis or by simulation.

18.5 Approximation Errors in calculations

'Approximation errors' is a heading under which we bring those errors that are not of a fundamental nature but are simply related to the approximate nature of the calculation methods we use in the pricing process. Although small approximation errors lurk everywhere in the pricing process, these are normally negligible, except for those errors that occur in the calculation of the aggregate loss distribution, whether we use simulation or other numerical methods such as fast Fourier transform (FFT).

Simulation Error in Monte Carlo Simulation. This depends on the fact that you can only use a limited number of simulations and every 'experiment' will produce a slightly different result. This can be reduced to some extent by increasing the number of simulations from, for example, 10,000 to 100,000 or 1,000,000. Better still, smarter simulation methods (such as Latin hypercube) can be used. In any case, there is an obvious trade-off between speed and accuracy.

Discretisation Error in FFT/Panjer Recursion. This depends on the fact that when using FFT/Panjer recursion, the severity distribution and the aggregate loss distribution need to be discretised and truncated: the larger the discretisation step, and the lower the truncation point, the larger the error introduced in the final estimate. Finding the right combination of discretisation step, truncation point, and number of subdivisions in the interval is an artful balancing act.

18.5.1 How to Calculate Its Effect

Simulation Error. One quick method to get an idea of the errors involved in simulation is to run the simulation several times and observe the variability of the result.

Discretisation Error. Upper bounds on discretisation errors can be estimated using methods from numerical analysis.

18.5.2 How to Communicate It

You seldom need to get into such details as simulation error in talking with a client, underwriter, or broker. One possible exception is when you present two different runs of a Monte Carlo simulation, and the results turn out to be slightly different – this may turn out to be confusing and you may need to explain where the difference comes from.

Apart from this, the key thing to communicate is that this type of uncertainty is negligible with respect to all other uncertainties in the costing process, such as parameter uncertainty and model uncertainty, so it is pointless worrying about this beyond making

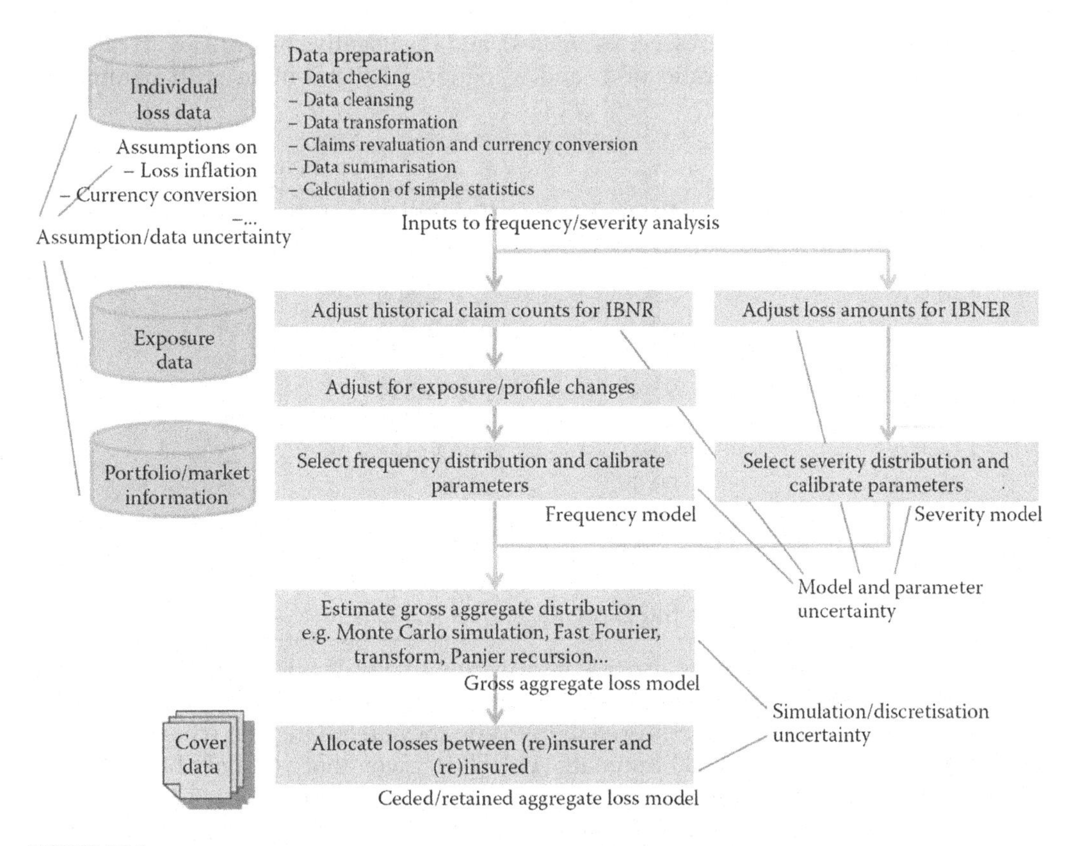

FIGURE 18.3
Data, parameter, model, and simulation uncertainty in the risk costing process.

sure you use a decent number of simulations in your simulation or a suitably small discretisation step in your FFT exercise.

18.6 Applications to the Risk Costing Process

Figure 18.3 shows where the various types of uncertainty become relevant in the various steps of the risk costing process.

18.7 Questions

1. As the consulting actuary for WWW, you have produced the following gross loss model for their public liability (PL) risks. The model is based on 1000 groundup losses (min = £150, max = £1,280,000) over the last 10 years (700 of which are fully

paid losses, and 300 are reserve estimates), and is a negative binomial with rate = 10 and variance-to-mean ratio of 2, and a lognormal distribution with parameters $\mu = 11$ and $\sigma = 1.6$.

		Gross losses	
Return period (years)	**Percentile**	**Total loss**	**Total number**
1 in 2	50%	1,480,167	9
1 in 4	75%	2,650,608	13
1 in 5	80%	3,008,045	14
1 in 10	90%	4,360,384	16
1 in 20	95%	5,791,894	18
1 in 50	98%	8,585,541	21
1 in 100	99%	11,320,404	23
1 in 200	99.5%	13,967,817	25
1 in 500	99.8%	20,279,446	27
1 in 1000	99.9%	24,251,589	28
	Mean	2,115,331	10.0
	Std Dev	2,451,803	4.6

The risk manager wants to use this information to decide on an aggregate limit for a PL policy. They would like to purchase insurance to cover for events with a return period of 1 in 200 years (0.5% probability), and they are asking you what confidence you have in your '1 in 200' estimate (£13.97M), given that you only had 100 data points to start with. Without actually doing any calculations,

a. Explain what types of uncertainty are at play here, and how you could go about estimating their effect.
b. Based on (a), comment on whether their concerns over the 1 in 200 estimate are exaggerated.
c. Describe two possible ways in which you could increase your confidence in the 99.5% figure, by using information from WWW only or by using external information.

19

Setting the Premium

The calculation of the expected losses (Chapters 6–18) is an essential component of pricing, but there are a few other elements that we need to consider before we translate this into the actual premium charged to policyholders (Figure 19.1). For one thing, we need to include loadings for expenses, commissions, reinsurance, target profit, and the like – allowing us to produce a 'technical premium', that is, the premium that should be charged on a purely technical basis, without taking into account supply-and-demand considerations (Sections 19.1 and 19.2).

We then need to consider that the insurer does not normally operate in a vacuum but participates in a market in which regulations and competitive pressure forces the company to deviate from the technical premium. Ultimately, the actual premium charged is a strategic commercial decision, and the company is forced to continually monitor not only its costs but also its sales and its position in the marketplace to ensure that it makes the optimal decisions and achieves its objectives (Sections 19.3).

Pricing actuaries are involved not only in the calculation of the expected losses but also in the determination of the *technical* premium, along with other corporate functions (Section 5.7), and are increasingly involved in the determination of the *competitive* premiums, especially in personal lines insurance, following the development of quantitative techniques such as price optimisation.

The main approaches for the calculation of the technical premium are the following (see Anderson et al. (2007a) for a more detailed classification):

- *Tariff-based calculations*, in which the regulator sets the premium or the rules around the calculations of the premium. Sometimes the scope of the term 'tariff system' is loosely extended to encompass de-regulated markets where premiums are set by conventional rules.
- *Cost plus pricing*, in which the premium is driven by the general formula

$$\text{Premium} = \text{Estimated costs} + \text{Target profit}$$

  where the (production) costs are given by the expected losses and the other expenses that the company incurs. These will be assessed quantitatively in a data-driven way wherever possible, ideally at policy level or at portfolio level when that is not possible. Note that cost plus pricing does not take into account supply and demand, and for this reason the actual premium charged will deviate from the formula above depending on business and market considerations. The formula is however still

DOI: 10.1201/9781003168881-21

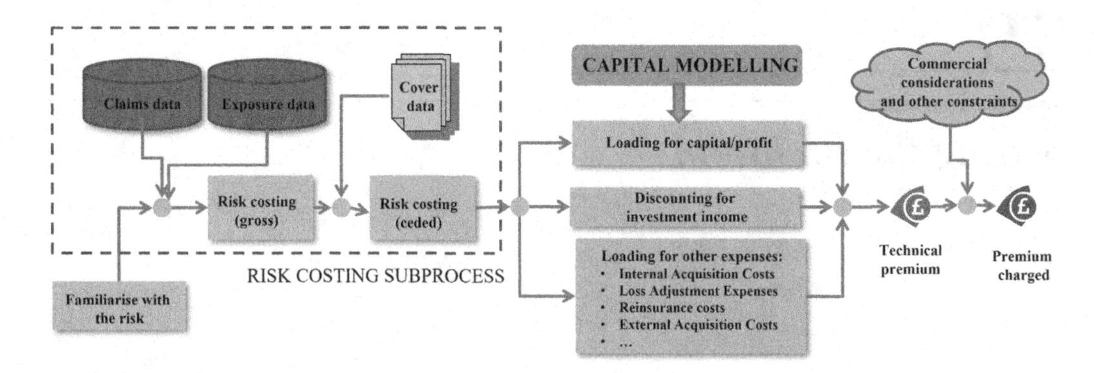

FIGURE 19.1
This shows where this chapter fits into the big picture of the risk pricing process that we have outlined in Chapter 1.

useful to provide a technical view of the premium as a guide to underwriters and as a reference for assessing performance. We will look at Cost Plus in Section 19.1, where we will describe its various components, while Section 19.2 will delve into the connection of the target profit with capital.

- *Pricing optimisation*, which builds on Cost Plus but also incorporates demand modelling, with the objective of setting the premium at a level that maximises profits. This methodology will typically be used by personal lines insurers with a large enough base of customers (see Anderson et al., 2007a, where it is goes under the name of 'distribution pricing'). Pricing optimisation is covered in Section 19.3.

19.1 Cost Plus Methodology in Detail

As explained in the introduction, the Cost Plus is a simple pricing methodology according to which the premium should be *driven* by the formula Premium = Estimated costs + Target profit. We say 'driven' because this way of setting the premium doesn't take supply and demand into account, and ultimately the actual premium charged will have to be a strategic decision that will need to take economic forces into account. So it is perhaps useful to break down the methodology as follows:

Cost Plus methodology for premium setting:

1. Calculate the 'technical' premium as

$$\text{Technical premium} = \text{Estimated costs} + \text{Target profit}$$

2. Set the actual premium charged taking into consideration the technical premium and other business/market considerations:

$$\text{Actual premium} = \text{Technical premium} +/- \text{Other considerations}$$

In an actuarial context, we can follow Anderson et al. (2007a) and break down the equation for the technical premium further as shown in Equation 19.1. This can be considered a particular incarnation of the *fundamental insurance equation* (Werner and

Modlin, 2016), which relates the technical premium to the various production costs of insurance and the desired profit.

$$\text{Technical premium} = \begin{array}{l} \text{Expected losses} \\ + \text{Allowance for uncertainty} \\ + \text{ Other costs}\left(\text{LAE, UW expenses, Reinsurance cost}\right) \\ - \text{ Investment income} \\ + \text{ Target underwriting profit} \end{array} \quad (19.1)$$

Note that by writing Equation 19.1 in this way we do not mean to provide an actual formula for the technical premium calculation: rather, we simply mean to illustrate the components of the technical premium and whether they contribute positively or negatively to the overall premium. The actual formula for the technical premium calculation need not be in the form of a sum (see, e.g. Equation 19.4). Let us look at each of the components in more detail.

19.1.1 Expected Losses

This is normally defined as the long-term average of the losses to the insurer – mathematically, the expected value of the aggregate *ceded* loss distribution, which was denoted as $\mathbb{E}(S_{\text{ced}})$ in Chapter 17.

The problem with this term is that it is not always easy to estimate. Expected losses will typically be estimated on the basis of some average over a limited number of years. Where this average is based on a burning cost analysis, it is customary to add a correction (positive or negative) for large losses and perhaps for natural catastrophe losses. However, these corrections are a means to improve the estimation of the expected losses (see Chapter 11), rather than terms to be added to the expected losses themselves.

19.1.2 Allowance for Uncertainty

Companies will sometimes include in the premium an allowance for uncertainty. The idea behind this is that the shakier the foundations on which you have built your expected loss estimate, the more you will want to load for extra safety – if you can.

Things that will normally make underwriters want to increase the premium are:

- Poor data (data uncertainty)
- Limited data (parameter uncertainty)
- Unconvincing model (model uncertainty)

Given the many sources of uncertainty, not all of which can be easily quantified, this allowance will inevitably be to some extent subjective.

19.1.3 Other Costs

19.1.3.1 Loss Adjustment Expenses (LAE)

This refers both to expenses allocated to specific losses such as lawyers' fees, claims management costs (allocated loss adjustment expenses, or ALAE), and unallocated costs such as policy administration, IT, accounting, cleaning, electricity (unallocated loss adjustment expenses, or ULAE). If ALAEs are already included in the loss amount (as is often the case), we only need to add an allowance for ULAE.

19.1.3.2 Underwriting Expenses

These can expenses related to the acquisition and servicing of business, and can be further subdivided into internal acquisition/servicing costs and external acquisition costs.

External acquisition costs (EAC) are the commissions and brokerage fees paid to external agents/brokers to procure business, and are typically proportional to the premium.

Internal acquisition/servicing costs (IASC) include activities carried out by the company to win business (e.g. marketing, salaries of sales representatives) and servicing it. They can be fixed (i.e. a monetary amount independent of the size of the policy), a percentage of the premium (e.g. premium taxes) or of the expected losses, or a more complicated function of the premium/expected losses.

19.1.3.3 Reinsurance Costs

This is the amount that we have to add to our premium to take account of the premium the company pay to the reinsurer (where the risk may be reinsured through a facultative or treaty arrangement). Note that reinsurance has the effect of removing a part of the risk (and therefore decreasing the expected losses) and replace it with a fixed amount (the reinsurance premium), much in the same way that insurance works.

The net reinsurance costs can therefore be written as:

$$\text{Net reinsurance cost} = \text{Reinsurance premium} - \text{Expected recoveries}$$

Alternatively, one can remove the expected recoveries from the expected losses and add the gross reinsurance cost (i.e. the reinsurance premium).

19.1.4 Investment Income

There is usually a delay between premium intake and claims payout, thus providing an investment opportunity for the insurer, especially for lines of business in which this delay is significant, such as liability.

This is so important that it has caused some to rethink what the underwriting business model should be. According to this view, because a significant portion of the income to the insurer comes in the form of investment income, the *operating ratio*[1] should normally be larger than 100%.

According to this model, insurance works by borrowing money (premiums) at a low interest rate and investing it a higher rate until losses need to be paid. As an example, consider a portfolio of policies with an overall premium of £1M and an operating ratio of 103% (so it loses on average 3% per year). Assume that the delay between premium receipt and claim payout is 1 year on average.

- The insurer 'borrows' money from the policyholders at the interest rate

$$\text{Interest rate} = \text{Operating ratio} - 1 = 3\%$$

1 The operating ratio is defined in several different ways: one of them is operating ratio = (incurred losses + expenses)/earned premium. The operating ratio is also called combined ratio or underwriting ratio.

- The insurer invests the money, obtaining an investment return larger than the interest rate above, for example, 5% per year
- At the end of the year, the insurer has paid the policyholder on average claims for an amount equal to the premium × operating ratio = £1.03M
- Also, the insurer has obtained investment returns equal to premium × (1 + investment return) = £1.05M
- Overall, the insurer has made a profit of £20,000 (2%)

This way of looking at insurance is interesting, although as a business model it has several faults. The main one is that if the operating ratio is not stable and profits are rather erratic. In any case, the main point we needed to make is that investment income is important and should be taken into account when calculating the premium, especially for long-tail classes of business such as liability.

How Investment Income Can Be Taken into Account

The effect of investment income is to decrease the break-even premium. Typically, this is considered by multiplying the expected losses by a discount factor. The discount factor can be calculated for example by an equation such as:

$$\mathrm{DF_L} = \sum_{t=1}^{M} \frac{x_t}{\left(1+r_{t-\frac{1}{2}}\right)^{t-\frac{1}{2}}} \tag{19.2}$$

where x_t is the percentage of losses expected to be paid during year t (measured from the inception date of the policy), r_t is the risk-free rate for bonds of duration t (which can be derived in general from the yield curve for the relevant risk-free investments, in which the yield for terms that do not correspond to any available bond are obtained by interpolation), and M is the maximum payment time (or a reasonable approximation of it). Note that the rate is calculated at the mid-point of the year $\left(t-\frac{1}{2}\right)$ on the assumption that losses are paid uniformly through each year.

Note that Equation 19.2 implicitly assumes that the full premium is paid at the inception date of the policy ($t = 0$). In many cases this assumption is good enough but where it isn't (because for example the premium is paid in instalments) it makes sense to also discount the premiums by a factor

$$\mathrm{DF_P} = \sum_{t=1}^{M} \frac{p_t}{\left(1+r_{t-\frac{1}{2}}\right)^{t-\frac{1}{2}}}$$

(p_t being the percentage of premium paid in year t), in which case we can use an effective discount factor of $\mathrm{DF} = \frac{\mathrm{DF_L}}{\mathrm{DF_P}}$.

A simplified version of Equation 19.2 often used in practice is this:

$$\mathrm{DF_L} = \frac{1}{(1+r_\tau)^\tau}$$

where τ is the *mean* delay between premium receipt and claim payout, and r_τ is the investment return of risk-free bonds of term τ.

This simplistic approach can be refined by considering the actual payment pattern and using different risk-free rates depending on when the various payments are expected, thus matching the portfolio of liabilities (the expected payments) with the corresponding assets (bonds of the relevant duration).

Why is the risk-free rate typically used rather than an investment return that considers the insurer's ability to invest more profitably? Two practical reasons are as follows:

1. Although it makes sense to discount the expected losses and consider the time value of money by using a risk-free rate, any gain beyond that is uncertain. It is the insurer who takes on the risk for the uncertain income, and it makes sense that it keeps the reward rather than passing it back to the policyholder by factoring it into the premium calculations in advance.
2. Also, this solution is the cleanest because it keeps the contribution of the underwriting team separate from the contribution of the investment team: the profit achieved by discounting the losses and costs by the risk-free rate can be fully credited to the underwriting team, whereas the investment income achieved in excess of the risk-free rate represents the valued added by the investment team.

19.1.5 Loading for Profit/Cost of Capital

This is an allowance for profit, based perhaps on company guidelines – for example, the desire that 10% of the overall premium contributes to the company's profit. One way to look at this is through the concept of return on capital (ROC), or – more commonly – the return on risk-adjusted capital (RORAC).[2] The shareholders of an insurer will require a certain return on the capital they invest in it, and this will have to be incorporated in the premium structure.

The profit required to achieve a given RORAC, considering the corporate tax rate paid by the company and the risk-free rate, is:[3]

2 The main difference between the ROC (return/capital ratio) and the RORAC (return/risk-adjusted capital ratio) is that the RORAC the denominator is the risk-adjusted capital. In other words, more capital is allocated to subportfolios that are more volatile, and therefore more return is demanded for those subportfolios to achieve the same RORAC.

3 The formula for the pre-tax profit % can be explained as follows. The standard definition of the RORAC is Net Income / Allocated Risk Capital (where the capital is allocated to various sub-portfolios on the basis of their riskiness). The net income (pre-tax) comes:

- partly from the profit of each individual policy (Pre-tax profit = Premium – Expected Losses – Expenses + Other Income), which will increase the existing capital (on average), and
- partly from investing the existing capital in safe assets (risk-free rate times the allocated risk capital).

Overall, therefore:

Net income (pre-tax) % = Pre-tax profit % + RiskFreeRate.

The RORAC is the after-tax net income: RORAC = (Pre-tax profit % + RiskFreeRate) * (1-TaxRate). By inverting this formula we obtain the Pre-tax profit % needed to achieve the desired RORAC.Note that the 'Other income' term mainly refers to the investment income that comes from the difference in timing between premium receipt and claim payout, and whose effect is separate from the income that one gets from the existing capital.

$$\text{Pre-tax profit\%} = \frac{\text{RORAC}}{1 - \text{TaxRate}} - \text{RiskFreeRate}$$

This target profit can be set at portfolio level and then apportioned to the individual policies either proportionally or by demanding a higher profit for policies that are more volatile on a stand-alone basis or that *contribute* more to the overall volatility of the portfolio (see Section 19.2.3).

Whatever the allocation method may be, the target profit for the policy will then be:

$$\text{Target profit} = \text{Pre-tax profit\%} \times \text{Capital allocated to the policy} \tag{19.3}$$

19.1.6 Technical Premium Formula

We have seen in the previous sections the components normally included as part of the technical premium, glossing over the exact algebraic structure of the formula.

Indeed, the formula for the technical premium can be written in many different ways, depending on the company's choice on how to apply the various loadings, but a typical one for the net technical premium (NTP) may look something like this:

$$\text{NTP} = \frac{\text{DF}_\text{L} \times (\text{Exp Loss} + \text{ALAE}) + \text{ULAE} + \text{IASC} + \text{Net R/I costs} + \text{Target profit}}{\text{DF}_\text{P}} \tag{19.4}$$

where

- DF_L and DF_P are the discount factors for losses and premium respectively, accounting for the effect of investment income;
- ALAE and ULAE are the allocated and unallocated loss adjustment expenses (Section 19.1.3.1). We are assuming here that only the ALAE should be discounted for investment income;
- IASC are the internal acquisition/servicing expenses (Section 19.1.3.2);
Net R/I costs are defined as in Section 19.1.3.3;
- the target profit is defined as in Equation 19.3.

Note that:

- we have applied the discount factor for losses only to the expected losses and the ALAE. Depending on how the various terms (e.g. the target profit) are defined and calibrated it could be applied to other terms as well;
- we have not included in Equation 19.4 the allowance for uncertainty, as this is not common and may be absorbed by the target profit/cost of capital component.

The gross technical premium (GTP) will typically be given by a formula like this:

$$\text{GTP} = \frac{\text{NTP}}{1 - \text{EAC\%}} \tag{19.5}$$

The external expenses (Section 19.1.3.2) are expressed here as a proportion EAC% to the premium. In practice, external expenses may be a fixed percentage of premium (e.g. a brokerage fee) but may also contain variable elements (e.g. profit commission, no claims bonus). The calculation of the variable elements may be complex and will in general require simulation; the technical premium can in this case still be expressed formally as Equation 19.5 but EAC% is not an input but a percentage estimated via simulation;

19.1.6.1 Calculation Example

Let us see an example calculation that uses Equations 19.7 and 19.8. Assume that for a given portfolio of general liability policies the company has these loading rules:

- the ULAE are 5% of the expected losses;
- the IASC are a fixed amount of USD 500 per policy plus 3% of the net premium plus 3% of the expected losses;
- the net reinsurance costs are set at 4% of the expected losses;
- the discount factor for losses is 0.9, and the discount factor for premiums can be ignored;
- the capital allocated is allocated uniformly across all policies as 70% of the net premium;
- the target profit is based on a RORAC of 10%, a tax rate of 25%, and a risk-free rate of 1%.

Also assume that the expected losses (EL) for this specific policy are USD 20,000, and that the ALAE are USD 1,200. The company pays 15% of brokerage fees.

The target profit percentage is therefore given by TP% = 10% / 75% – 1% ~ 12.3%, and the net technical premium (NTP) is

$$\begin{aligned}\mathrm{NTP} &= 0.9\times(\mathrm{USD}\,20{,}000+\mathrm{USD}\,1{,}200)+\mathrm{USD}\,500+0.05\times\mathrm{USD}\,20{,}000\\&+0.03\times\mathrm{USD}\,20{,}000+0.03\times\mathrm{NTP}+0.04\times\mathrm{USD}\,20{,}000+0.123\\&\times 0.7\times\mathrm{NTP}=\mathrm{USD}\,20{,}980+0.1163\times\mathrm{NTP}\end{aligned}$$

yielding

$$\mathrm{NTP}=\frac{\mathrm{USD}\,21{,}980}{1-0.1163}=\mathrm{USD}\,24{,}873$$

Finally, the gross technical premium (GTP) is given by:

$$\mathrm{GTP}=\frac{\mathrm{USD}\,23{,}742}{1-0.15}=\mathrm{USD}\,29{,}262$$

Note that we have assumed the simplest possible capital allocation rule for this example. An investigation of more sophisticated ways to allocate capital to policies is addressed in Section 19.2.

19.1.7 Actual Premium Charged

As mentioned many times, the technical premium is not the premium actually charged, but the premium calculated according to the actuarial analysis and the company guidelines on expenses, profit, and the like.

The actual premium charged is the result of a strategic decision, involving the consideration of:

- Regulatory constraints
- The premium charged by competitors
- The relationship with other products (for instance, selling an unprofitable product at a moderate loss may help sell other, more profitable, products)

This decision is – at least in the case of commercial lines and reinsurance – normally based on judgment and without involving actuaries. Having said that, large personal insurers have introduced techniques of price optimisation that have helped reduce the arbitrariness of the decision process. These techniques are illustrated in Section 19.3.

19.2 Capital Considerations in Pricing

To run their business effectively, insurers must have large amounts of capital: after all, what makes insurers able to fulfil their function is their ability to withstand a level of claims that would cripple or expose individuals and nonfinancial companies to too much financial volatility.

Given the fundamental social function of insurers, the financial regulator imposes minimum levels to the amount of capital that it must hold (*regulatory capital*) to remain solvent with high probability. In the United Kingdom, capital requirements are currently defined in the prudential sourcebook for insurers (INSPRU) issued by the Prudential Regulation Authority (PRA) and will eventually be defined by the Solvency II regulations that will be implemented throughout Europe.

The capital actually held by the insurer (*economic capital*) will in general be higher than the regulatory capital, and is driven by many considerations, including:

- The desire to protect the firm from insolvency
- The desire of the company to achieve and maintain a given rating with one of the major rating agencies such as S&P
- The desire to appeal to investors and generally ensure markets that the company is solid
- The desire to increase the number of clients, especially corporate clients, that are willing to buy insurance from the firm

Naturally, these drivers for the economic capital are strongly interdependent: for example, the desire to achieve/maintain a certain rating affects the ability to reach certain clients because insurance brokers and large corporate clients will often have policies that prevent them from purchasing insurance from participants that do not have a certain rating.

All this is very important, but *what does it have to do with pricing?*

The connection arises from the fact that ultimately, economic capital is provided by the investors in the company, who will require a certain return on that capital, above the level of risk-free (or almost risk-free) returns that they could get from government or corporate bonds. The exact level of return required will of course depend on the perceived riskiness of the business.

BOX 19.1 ONE POSSIBLE APPROACH TO CALCULATING THE CAPITAL LOAD FOR EACH POLICY

1. Determine the regulatory capital needed for the aggregate book
   This will depend on the regulatory framework.
2. Determine the economic capital for the aggregate book
   This is the capital that the company will in practice want to have in order to run its business effectively. As a minimum, it must be equal to the required regulatory capital.
3. Allocate the capital amongst the different subportfolios of risks (such as business units, lines of business – the actual segmentation will be very much company-dependent), in a risk-based fashion.
   The general idea is that riskier subportfolios and subportfolios that are strongly correlated with the rest of the portfolio should be allocated a larger capital. This is also what happens in financial theory when deciding which projects or which shares to invest on. This is expanded on in Sections 19.2.2 and 19.2.3.
   Also, since the ultimate volume of business written in a given underwriting year will be unknown until the year finishes, the capital is normally defined dynamically in terms of the ***capital intensity****, i.e. as the capital per unit of written premium (or other measures of volume).*
4. Consider the RORAC (return on risk-adjusted capital) demanded by the investors
5. Within each subportfolios, allocate capital to each policy in a risk-based fashion (see Section 19.2.3).
   Although allocating the capital in two steps (first to subportfolios and then to the individual policies) is arguably the most popular allocation approach, it is in principle possible to skip the intermediate step of allocating risk to the subportfolios first.
6. Determine the target profit for each policy $j = 1,\ldots n$ by the formula

$$(\text{Target profit})_j = \text{Pre-tax profit\%} \times (\text{Capital allocated})_j$$

Notice that we have used where in turn the pre-tax profit, because in order to guarantee a given return on capital to the shareholders, the revenues need to be grossed up to account for the tax rate paid by the company :

$$\text{Pre-tax profit\%} = \frac{\text{RORAC}}{1 - \text{TaxRate}} - \text{RiskFreeRate}$$

To achieve the desired return of capital, the portfolio of policies must achieve a certain profit or, in other words, a certain loss ratio. Also, it makes sense that different subportfolios have different target loss ratios, depending on how much the subportfolio contributes to the overall risk.

Hence, the need to provide a given level of profit translates into a loading towards the technical premium along with other items such as expenses and investment income, exactly as explained in Section 19.1.5.

This is the general idea behind how capital modelling considerations affect pricing. A more articulate outline of a methodology for determining the capital loads for each policy is given in Section 19.2.1. This methodology requires a method to allocate capital amongst different subportfolios and contracts based on some measure of risk: therefore, we look at risk measures in Section 19.2.2 and capital allocation methods in Section 19.2.3.

19.2.1 Overall Methodology

There is no unique protocol for adding a capital load on insurance policies, and many different approaches have been proposed (see, e.g. Cairns et al., 2008). However, the approach outlined in Box 19.1 is rather typical.

19.2.2 Risk Measures

The capital loading approach described in Box 19.1 is risk-based – that is, capital is allocated to each subportfolio based on some measure of risk. In this section, we will look at different risk measures and how suitable they are for capital modelling purposes. First of all, however, let us talk in general about the properties that a risk measure should have.

19.2.2.1 Coherent Risk Measures

There are good and bad risk measures. Technically, 'good' risk measures are called coherent, and a coherent risk measure is one that has a list of desirable properties. The definition we use here is slightly different from the one that is normally found in texts of financial economics because, in those texts, the random variable whose risk we want to measure is in that case related to profits, whereas we want to formulate it here in terms of the total losses S over a year.

So if S is a stochastic variable representing the total losses over a year (possibly net of cessions), and R is a function that maps loss variables into real numbers, we say that R is coherent if and only if:

a. R is subadditive: $R(S_1+S_2)\le R(S_1)+R(S_2)$: that is, we are never worse off if we seek diversification by combining two different portfolios.

   In practice, combining two different portfolios might mean writing a property portfolio alongside a liability portfolio, or even writing more liability policies.

b. R is monotonic, that is, if in all possible outcomes the losses from one portfolio S_1 are smaller than the losses from another portfolio S_2 (i.e. $S_1 \le S_2$), then[4] $R(S_1)\le R(S_2)$.

   If this were not the case, one could think of decreasing one's risk by writing more business!

c. R is positive-homogeneous: if $\alpha>0, R(\alpha S_1 1)=\alpha R(S_1)$, that is, doubling (halving) one's share in a portfolio of risks simply results in twice (half) the risk.

   Note that increasing the portfolio in this way does not bring any diversification benefit because the added risk is perfectly correlated with the existing one. An example is provided by quota share reinsurance: if a reinsurer takes 25% of all losses from a portfolio of 1000 properties, increasing its share to 50% exactly doubles the risk. This is completely different from the reinsurer taking the 25% from an additional portfolio of 1000 other similar properties, which would guarantee some diversification.

d. R is translational-invariant: if $k>0, R(S_1-k)=R(S_1)-k$.

   What this is trying to capture is that if you add cash to a portfolio, the risk decreases by the same amount. This is not surprising – if you have more cash, it

4 In most financial modelling books, this definition has $S_1 \ge S_2$ instead: this is because the emphasis there is on profits rather than on losses.

is more difficult to run out of it. This property also subtly (or not so subtly) tells us that the risk measure must be a monetary amount: hence, an object such as the probability of ruin (i.e. the probability of an insurer exhausting its capital) cannot be a risk measure.

Apart from being coherent, a good risk measure for use in capital modelling should give particular emphasis to tail risk (i.e. with relatively low likelihood but large effect), which is what worries insurers and regulators the most.

19.2.2.2 Examples of Risk Measures

Examples of risk measures that have been proposed to either determine the overall capital needed or allocate capital to different portfolios are shown in the table below.

Note that in what follows X is the aggregate loss distribution, net of anything that is ceded to other parties. We use 'X' rather than 'S' to avoid confusion with the survival distribution, $S(x)=1-F(x)$, which is used in the proportional hazard method.

Example spreadsheet calculations of the most common risk measures are available in the Downloads section of the book's website (Parodi, 2016).

Examples of Risk Measures	
Mean loss, $\mathbb{E}(X)$	This is the simplest possible risk measure and is trivially coherent. However, it is not a very good risk measure because it does not punish volatility in any way: two portfolios with the same mean losses but with very different volatility are seen as equally risky, which is unintended. Also, it is always strictly additive $(\mathbb{E}(X_1 + X_2)= \mathbb{E}(X_1)+ \mathbb{E}(X_2))$ rather than simply subadditive, and therefore never rewards diversification. Despite these shortcomings, the mean loss is the template for other, more interesting measures of risk, such as the mean of transformed loss distributions, proportional hazard, or Wang transform and TVaR.
Variance, $\sigma^2(X)$	The variance is sometimes proposed as a measure of risk although it is not quite clear why: because the variance has the dimension of a squared currency, it is not adequate to represent a monetary amount. This inadequacy is easily picked up by the 'coherence test': the variance fails both Property (c) because $\sigma^2(\alpha X_1)=\alpha^2\sigma^2(X_1)$ and Property (d) because $\sigma^2(X_1-k)=\sigma^2(X_1)$ for all k.
Standard deviation, $\sigma(X)$	Like the variance, the standard deviation is a measure of volatility, and therefore is a good starting point for assessing risk. It passes all tests of coherence except for (d), because $\sigma(X-k)=\sigma(X)$ for all k. This is not in itself a very serious shortcoming in the context of capital adequacy/allocation. A more subtle shortcoming of the standard deviation for the purposes of allocating capital is that it does not specifically focus on tail risk, but punishes volatility wherever it happens, whether for downside risk (the risk that the losses are much larger than the mean) or for upside risk (the 'risk' that the losses are much smaller than the mean).

Examples of Risk Measures	
Value at risk (VaR) at the pth percentile, VaR@p	The value at risk (VaR) is defined as the value of the loss distribution at a given quantile p. Mathematically, it can be defined as $$\mathrm{VaR}@p(X)=\inf\{x \mid F_x(x) \rangle p\}$$ where $F_X(x)$ is the CDF of the loss distribution. In the continuous case, this reduces to $\mathrm{VaR}@p(X)=F_X^{-1}(p)$. VaR is possibly the most widespread risk measure, embraced by insurers as well as regulators, possibly because it is very easy to understand and easy to compute (at least theoretically). However, it is not a coherent risk measure because it is not subadditive. This is much more than a mathematical annoyance: it means that diversification may not be taken into account correctly. Other shortcomings: • VaR@ p does not distinguish between two portfolios that have wildly different behaviours beyond p • For high values of p, any pretence of accuracy is purely illusory
Tail value at risk (TVaR) and excess tail value at risk (xTVaR)	Tail value at risk is a generalisation of the value at risk, which considers not the single value of the aggregate loss distribution at the percentile p but the average of all values above the value at risk at percentile p: it is therefore sometimes referred to as 'average value at risk'. As such, it overcomes one of the objections against VaR, which is that it is unable to distinguish between two distributions that have the same value at a given percentile but have significantly different values. It is defined as $$\mathrm{TVaR}@p(X)=\mathbb{E}(X \mid X \geq \mathrm{VaR}@p(X))$$ The TVaR can be estimated via Monte Carlo simulation in this way: a. Put the N simulated scenarios in descending order: $\{X_1^{\text{sorted}} \geq X_2^{\text{sorted}} \geq \ldots \geq X_N^{\text{sorted}}\}$, and pick the value corresponding to the $N\times(1-p)$-th value (k is k rounded down to the nearest integer). This is your estimate $\hat{V}$ of $\mathrm{VaR}@p(X)$. b. Take the average of all the simulated values that are larger than or equal to $\hat{V}$: $$\mathrm{TVaR}@p(X)\approx\frac{\sum_{\text{all } i \text{ such that } X_i^{\text{sorted}}\geq\hat{V}} X_i^{\text{sorted}}}{\#\text{ of simulated values} \geq \hat{V}}$$ TVaR is known to be coherent if the underlying loss distribution is continuous. If it is not (as is often the case, e.g., when pricing layers of (re)insurance), the tail value at risk can be replaced by the *expected shortfall*, which is coherent in general. An obvious shortcoming of TVaR is that, even more than in the case of VaR, its value is highly uncertain, especially for values of p close to 1. The additional difficulty with respect to VaR is that one must estimate not only the value of the loss distribution at the pth percentile but also at all percentiles above that.

Examples of Risk Measures

The TVaR is often found in the 'excess tail value at risk' variant, defined as:

$$\text{xTVaR}@p(X) = \text{TVaR}@p(X) - \mathbb{E}(X)$$

which has the same properties as $\text{TVaR}@p(X)$ but removes the expected losses from the calculation, thus focusing on what a risk measure needs to measure – the volatility of possible outcomes.

Expected shortfall (ES) and excess expected shortfall (xES)

Expected shortfall (also called conditional tail value at risk, CTVaR) is one of the most widespread risk measures amongst capital modellers because it is both relatively simple and theoretically satisfying (as it is coherent). It is conceptually similar to the TVaR, in the sense that it captures the average of all possible outcomes in the tail, not just one point. However, it is always coherent, even if the underlying distribution is not continuous.

It is defined as

$$\text{ES}@p(X) = \frac{1}{1-p}\int_p^1 \text{VaR}@q(X)\,dq$$

This definition may appear cryptic, but what it is saying is quite simple: one should take the average of all values at risk above p. And it becomes even clearer when you consider how the expected shortfall can be calculated via simulation: it simply means that you have to take the average of the top $N\times(1-p)$ simulations sorted in descending order, $\{X_1^{\text{sorted}} \geq X_2^{\text{sorted}} \geq \ldots \geq X_N^{\text{sorted}}\}$:

$$\text{ES}@p(X) \approx \frac{\sum_{i=\lfloor N\times(1-p)\rfloor}^{N} X_i^{\text{sorted}}}{N\times(1-p)}$$

The expected shortfall suffers from the same limitations in terms of uncertainty as the tail value at risk – the estimation error is necessarily larger than that of the VaR, especially for fat-tailed distributions.

It is easy to create realistic examples of Monte Carlo simulations for which the tail value at risk and the expected shortfall lead to different results when the cumulative distribution function is not continuous (see Question 4 at the end of the chapter).

As for the TVaR, the expected shortfall also comes in an 'excess' version, defined as:

$$\text{xES}@p(X) = \text{ES}@p(X) - \mathbb{E}(X)$$

which is a better risk measure for the same reason that xTVaR is more useful than TVaR: it only allocates capital to volatility and not for expected losses.

Note that the both $\text{ES}@p(X)$ and $\text{xES}@p(X)$ are coherent risk measures.

Examples of Risk Measures	
Mean of transformed loss, $\mathbb{E}^*(X)$	The idea behind transformed loss methods such as the proportional hazard transform and the Wang transform is to replace the original loss distribution $F_X(x)$ with a distribution $F_X^*(x) = g(F_X(x))$ that gives more emphasis to those losses that are of most concern for the insurer. Typically, this means giving more emphasis to the tail of the distribution.
Proportional hazard	Specifically, the *proportional hazard transform* works by replacing the original survival distribution $S_X(x) = 1 - F_X(x)$ with the distribution $S_X^* = (S_X)^{1-\lambda}$, with $0 \leq \lambda < 1$. The effect is, of course, to give increasingly more weight on the tail. For example, if $S_X(£1M) = 0.01$ and $\lambda = 0.5$, the transformed survival probability becomes $S_X^*(£1M) = 0.1$.
Wang transform	The Wang transform achieves a similar result – that is, giving more weight to the tail of the distribution – by shifting the percentiles of the loss distribution. The original distribution, $F_X(x)$, is replaced (in its most common incarnation) by $F_X^* = \Phi(\Phi^{-1}(F_X) - \lambda)$ where Φ is the standardised normal distribution. If $\lambda < 0$, this means that a given loss value x is now associated to a lower percentile and the new loss distribution is therefore more severe. Note that this transform maps old quantiles into new quantiles and the mapping is completely independent of the underlying distribution. For example, if $\lambda = 0.2$, the quantile 0.9 is mapped into the quantile $\Phi(\Phi^{-1}(F_X) - 0.2) \approx 0.86$ regardless of whether F_X is a lognormal distribution or a Pareto distribution.
Mean loss	Both the proportional hazard transform and the Wang transform can be used to define a simple risk measure: the mean loss calculated on the transformed distribution: $$\mathbb{E}^*(X) = \int_0^\infty x f_X^*(x)\,dx = \int_0^\infty S_X^*(x)\,dx$$ $\mathbb{E}^*(X)$ has the same properties of the mean loss (in particular, it is a coherent risk measure) but it is calculated on a distorted probability distribution and therefore emphasises the behaviour of certain parts of the loss distribution – typically, the tail. It therefore overcomes the standard objections to the use of the mean loss as a risk measure. Interestingly, VaR and TVaR can be seen as a special case of transform, in which the transformation $g(F_X(x))$ focuses even more drastically on one part of the distribution. The advantage of proportional hazard transform and Wang transform is that they take into account the whole loss distribution rather than just the tail (TVaR) or even a single point (VaR). However, they may be (or simply appear) too complicated for use in day-to-day capital modelling and communication to the management of an insurer. Also, the results depend crucially on the choice of the parameter λ. However, it might be difficult to justify a specific choice of this parameter, even more so than justifying the choice of the reference percentage for VaR or TVaR.

19.2.3 Capital Allocation Methods

The risk measures introduced in Section 19.2.2 can be used to determine the aggregate book capital and also to determine the capital allocated to each subportfolio and ultimately to determine the capital load for each policy. It is not necessary for the same risk measure to be used to determine the overall capital *and* the capital allocated to the different subportfolios. For example, the Solvency II requirement that capital should be held for the 1-in-200-year case is based on VaR. If the company decides to hold as economic capital the regulatory capital, it is using VaR for the overall capital. However, the company can then decide to allocate the target profit to the different lines of business using the expected shortfall because it is thought to be a better measure of the relative risk posed by the different lines of business.

There are many different ways to allocate capital (see, e.g. McNeil et al., 2005; Cairns et al., 2008). We will limit ourselves to the two methodologies that are perhaps the most popular: proportional spread and the allocation based on the Euler principle. Another methodology (game theory allocation) is also available as an online appendix in Parodi (2016).

19.2.3.1 Proportional Spread

The capital charge is allocated to risks (contracts or sub-portofolios) proportionally to a given risk measure such as one of those listed in Section 19.2.2, calculated on a stand-alone basis. Popular choices are the standard deviation, the VaR, and the expected shortfall. This method is quite common because of its simplicity; however, it does not allow for the correlation between different risks, and therefore it might overestimate/underestimate the capital charge.

An example of how proportional spread works with various risk measures is shown in Figure. Assume a simple portfolio with three risks (contracts, or subportfolios), and let Z_1, Z_2, Z_3 represent the relevant aggregate loss distributions. The insurer wants to calculate the overall capital needed to run the business and satisfy the regulator; it also wants to allocate capital to the various lines of business (and policies) equitably, according to two different risk measures (VaR@99.5% and expected shortfall @99%).

The basic statistics of interest for Z_1, Z_2, Z_3 and the various combinations of these three loss distributions are shown in Figure 19.2. You can assume that the overall economic capital is set equal to the regulatory capital, which in turn corresponds to the value at risk at the 99.5th percentile: hence,

$$\text{Overall capital} = \text{CCY}\,917.99\text{M}$$

where CCY stands here for a generic currency.[5]

We now look at how this capital can be allocated to the three different lines of business.

	Z_1	Z_2	Z_3	$Z_1+Z_2+Z_3$
Mean	9.32	16.98	11.73	38.03
St Dev	46.23	195.91	140.74	245.3
VaR@99.5%	221.24	506.7	403.52	918.99
ES@99%	405.39	1259.81	1026.21	1929.03
			All amounts in CCY M	

FIGURE 19.2
Basic statistics for the three risks in the portfolio.

5 To be pedantic, 'CCY' is just an abbreviation for 'currency' rather than the symbol for a generic currency. The orthodox symbol for a generic currency appears to be '¤' but it looks too confusing on the page.

	Based on value at risk		Based on exp shortfall	
	Capital allocation (%)	Capital allocation (CCY)	Capital allocation (%)	Capital allocation (CCY M)
Risk 1	19.6%	179.69	15.1%	138.42
Risk 2	44.8%	411.55	46.8%	430.17
Risk 3	35.7%	327.75	38.1%	350.4
Aggregate book	100.0%	918.99	100.0%	918.99

FIGURE 19.3
Allocation of CCY 918.99M of capital proportionally to the value at risk at the 99.5th percentile and to the expected shortfall at the 99th percentile.

The capital allocated to the different risks is proportional to their volatility, scaled so that the overall capital is equal to CCY918.99M.

For example, the capital allocated to risk 1 is

$$\text{Capital}(1;\text{VaR}) = \frac{221.24}{221.24 + 506.7 + 403.52} \times 918.99 = \text{CCY}179.69\text{M}$$

in the case that VaR is used, and

$$\text{Capital}(1;\text{ES}) = \frac{405.39}{405.39 + 1259.81 + 1026.21} \times 918.99 = \text{CCY}138.42\text{M}$$

in the case that expected shortfall is used.

Replicating these calculations for all three risks yields the results in Figure 19.3.

19.2.3.2 Allocation Based on the Euler Principle

There have been many attempts to allocate capital to contracts or subportfolio to reflect the correlations within the portfolio, but none of them is as elegant as the Euler allocation principle.

19.2.3.2.1 The Underlying Principle

The maths behind the Euler principle may look cryptic at first glance but the idea behind it is quite simple:

> *Allocate capital in direct proportion to how much a risk contributes to the overall volatility of the portfolio*

In the sentence above, the volatility may be assessed for example by the expected shortfall (Section 19.2.2), in which case 'contribution' means marginal increase in expected shortfall when the weight (volume) of a risk is increased.

The calculations are also in practice very simple to understand. We will therefore present these calculations and an example first, and leave the mathematical justification until later.

19.2.3.2.2 Practical Calculations (Expected Shortfall Case)

We will focus on the implementation that uses the expected shortfall as the measure of volatility, as it is the most popular (for good reasons). This allocation method is also called

BOX 19.2 ALGORITHM FOR ALLOCATING CAPITAL TO DIFFERENT RISKS USING THE EULER PRINCIPLE WITH EXPECTED SHORTFALL

Input: Aggregate loss models for risks $Z_1,\ldots Z_n$, and dependency structure.[6] We can assume that the policy structure is already applied.

1. Simulate a large number N of scenarios and for each scenario generate losses for all risks, respecting the underlying dependency structure (in the case of catastrophe losses, the losses for the various risks are usually already provided with the intended degree of correlation by the commercial catastrophe model).
2. For each scenario, calculate the total losses across all risks.
3. Select the top $(1-p)\%$ of scenarios according to total losses.
4. For each of the risks, calculate the average loss for the *selected* scenarios: $\bar{Z}_1, \bar{Z}_2 \ldots \bar{Z}_n$
5. Allocate the capital based on the formula: $AC_k = \frac{\bar{Z}_k}{(\bar{Z}_1 + \bar{Z}_2 + \ldots + \bar{Z}_n)} \times C.$

Assume that your portfolio is made of risks $Z_1,\ldots Z_n$, where Z_k is a random variable representing the aggregate losses to risk k. The overall losses are therefore represented by the random variable $Z = Z_1 + \ldots + Z_n$. Assume that the capital allocated to the overall portfolio (according to whatever method) is C.

The calculations for the case where the risk measure is the expected shortfall @$p\%$ go as in Box 19.1.

Figure 19.4 shows an example calculation for the simple case of three risks. For illustration purposes we have limited the number of simulations to 1,000 and as a risk measure we have chosen the expected shortfall at 99%. The allocation is therefore based on the top 1% (=10) simulations. The results are shown in Fig. 19.4.

Note that as for the case of proportional spread, the percentage in the last row of Figure 19.4 can be applied to the capital of the portfolio, however calculated (e.g. it could be the VaR at the 99.5th percentile as in Section 19.2.3.2.1), to obtain the monetary amount.

The procedure above can be easily amended to use the excess expected shortfall (expected shortfall above the mean) instead of the expected shortfall.

19.2.3.2.3 Dynamic Allocation

An obvious difficulty with the method described above is that it is static, in the sense that it requires the have the full view of the portfolio before allocating the capital to its risks, while what we need in practice is a dynamic allocation, with the allocated capital calculated at the point when every risk is priced (the problem is less important when we are trying to allocate capital between subportfolios).

Ideally, one would want to re-run the simulation of the full portfolio every time a new risk is added or renewed. However, that is computationally demanding, and various shortcuts can be adopted:

6 See Chapter 29 for an explanation of how to describe the dependency structure and how to incorporate it in the simulation.

Scenarios (in descending order)	Z_1	Z_2	Z_3	$Z_1+Z_2+Z_3$
Scenario 1	.02	21283.52	.29	21283.82
Scenario 2	2.89	.17	14805.64	14808.71
Scenario 3	.04	3595.8	.01	3595.84
Scenario 4	.	53.41	3168.79	3222.21
Scenario 5	.04	108.07	2579.92	2688.04
Scenario 6	1830.11	.5	.	1830.61
Scenario 7	.	.	1626.39	1626.39
Scenario 8	1.03	1520.34	.08	1521.45
Scenario 9	1511.08	.	.	1511.08
Scenario 10	.	1422.37	.32	1422.7
Scenario 11	2.23	1392.49	.	1394.73
Scenario 12	85.74	.	1231.93	1317.67
...	...	...	...	...
Scenario 1000	.	.	.	.
Overall average (scenario 1-1000)	19.06	39.02	36.95	95.03
Average top 1% (scenarios 1-10)	371.69	2951.31	2464.57	5787.57
Percentage	6.4%	51.0%	42.6%	100.0%

All amounts in CCY M

FIGURE 19.4
Allocation according to the Euler principle for the simple case of three risks (contracts/subportfolios). The percentages in the bottom line need to be multiplied by the overall capital for the portfolio to determine the actual capital allocated to each risk.

- If the full simulation for the portfolio is saved somewhere, one can remove the past simulation of the risk under investigation from the simulation matrix and produce a simulation array for the renewal risk, taking into account the dependency structure, then calculate the contribution to the expected shortfall of the added/renewed risk in the usual fashion.
- Simulate the overall portfolio with a single distribution and then add the new risk, taking into account the dependency with the overall portfolio. If the risk is a renewal of an existing contract and is large, this may not be a good approximation and it may be more useful to use the capital from the expiring contract adjusted, e.g., by the difference in exposure, expected losses, or stand-alone expected shortfall.

Also, the overall capital allocated to the portfolio will change depending on the amount of business written, and therefore the overall capital that is calculated at the beginning of an underwriting year will not be the same as the capital that will be needed at the end.

For this reason, it is customary to define the capital not in absolute terms but as a 'capital intensity', i.e. the amount of capital per unit of premium or expected losses. This can be defined at the beginning of the underwriting year and will lead to a good approximation of the necessary capital regardless of the exact amount of volume written.

We now look at the mathematical justification of the Euler principle. Those who are not interested can jump straight to the Properties section.

19.2.3.2.4 The Maths Behind the Euler Allocation Principle

To give a formal definition of the Euler principle, let us go back to our portfolio of risks and the aggregate losses $Z_1, \ldots Z_n$. However, we also assume that one can vary the weight of each of the risks by introducing weights w_k, and we call the case where all w_k are 1 the *basic portfolio*.

This is useful so that we can analyse what happens when we increase/decrease continuously the weight of a risk. The overall losses for the weighted portfolio become $Z(w_1,\dots w_n)=\sum_{k=1}^{n} w_k Z_k$

Assume as well that

- we have a risk measure $R(Z_1,\dots Z_n)$ that maps loss variables into positive real numbers;
- that this risk measure is positive-homogenous; namely, that $R(wZ_1,wZ_2,\dots wZ_n)=wR(Z_1,Z_2,\dots Z_n)$ for any $w>0$. As we saw in Section 19.2.2.2, mean, standard deviation, value at risk and expected shortfall are all examples of positive-homogeneous risk measures;
- that $R(w_1Z_1,w_2Z_2,\dots w_nZ_n)$ is differentiable with respect to all the weights $w_1,\dots w_n$

Since R is positive-homogeneous and differentiable, we can use Euler's homogeneous function theorem (Tasche, 2008) to write:

$$R(w_1Z_1,\dots w_nZ_n)=\sum_{k=1}^{n} w_k \frac{\partial R(w_1Z_1,\dots w_nZ_n)}{\partial w_k} \tag{19.6}$$

Since $R(w_1Z_1,\dots w_nZ_n)$ is a measure of the volatility of the portfolio, it is natural to interpret Equation 19.6 as breaking down the overall volatility into a weighted sum of contributions to that volatility from, with $w_k \frac{\partial R(w_1Z_1,\dots w_nZ_n)}{\partial w_k}$ being the contribution from risk k. Specifically, for the original portfolio the weights w_k are all 1 and Equation 19.6 becomes:

$$R(Z_1,\dots Z_n)=\sum_{k=1}^{n} \left.\frac{\partial R(w_1Z_1,\dots w_nZ_n)}{\partial w_k}\right|_{w_1=w_2=\dots=w_n=1} \tag{19.7}$$

If we write the capital allocated to the whole portfolio as $AC=c\times R(Z_1,\dots Z_n)$, where c is simply a scaling factor, the capital allocated to risk k for the basic (original) portfolio will be:

$$AC_k=c\times \left.\frac{\partial R(w_1Z_1,\dots w_nZ_n)}{\partial w_k}\right|_{w_1=w_2=\dots=w_n=1} \tag{19.8}$$

which immediately yields $AC=\sum_{k=1}^{n} AC_k$. Equation 19.8 formalises what we meant at the beginning when we said that the contribution to the volatility is the marginal increase to the volatility because of a change in weight.

19.2.3.2.5 Special Cases

Equation 19.8 provides the general solution for the allocation problem. Let's now look at what the solution looks like for specific choices of the risk measure.

Risk measure	*Allocation rule*	*Practical calculations with Monte Carlo simulation*
Expected loss	$AC_k = c \times \mathbb{E}(Z_k)$	Simulate a large number of scenarios and calculate the average for each risk across all simulations (no correlation).
Standard deviation	$AC_k = c \times \frac{cov(Z_k, Z)}{\sqrt{\mathbb{V}(Z)}}$	Calculate the covariance between the simulated array for each risk and the array for the overall losses, and divide by the standard deviation of the overall losses array.
Value at risk (VaR) at the p-th percentile, VaR@p	$AC_k = c \times \mathbb{E}(Z_k \mid Z = VaR@p(Z))$	Select the simulation corresponding to the p-th percentile for the total losses, and allocate capital on the basis of the losses for each risk for that specific simulation. Note that the practical implementation of this rule is highly numerically unstable, as it is based on a single scenario, and therefore it is normally not adopted.
Expected shortfall at thep-th percentile (ES@p)	$AC_k = c \times \mathbb{E}(Z_k \mid Z \geq VaR@p(Z))$	As above, but the capital is allocated on the basis of the top (1-p)% total losses (see details in Section 19.2.3.2.2).

FIGURE 19.5
Allocation rules according to the Euler principle using various popular (positive-homogenous) risk measures.

In the trivial case where the risk measure is the expected loss $(R(w_1 Z_1, \ldots w_n Z_n) = \mathbb{E}(w_1 Z_1 + \ldots + w_n Z_n) = w_1 \mathbb{E}(Z_1) + \ldots + w_n \mathbb{E}(Z_n))$, the partial derivative with respect to weight w_k will be $\left.\frac{\partial R(w_1 Z_1, \ldots w_n Z_n)}{\partial w_k}\right|_{w_1 = w_2 = \ldots = w_n = 1} = \mathbb{E}(Z_k)$ and the allocated capital will therefore be proportional to the expected losses for each risk: $AC_k = c \times \mathbb{E}(Z_k)$. The rule for the other popular risk measures are slightly more complicated and they are offered in Figure 19.5 without derivation. For the derivation, see for example Tasche (2000), McNeil et al. (2005).

For all the allocation rules in Figure 19.5 the practical calculations based on Monte Carlo simulation are straightforward.

19.2.3.2.6 Properties of the Euler Allocation Principle

The Euler allocation has various useful properties that makes it an ideal choice for capital allocation.

- It is easy to calculate numerically for most popular risk measures, including the expected shortfall (see table in Section 19.2.3.2.5, and detailed explanation for the expected shortfall case in Section 19.2.3.2.2).
- It is easy to understand, at least in terms of practical calculations.
- It is additive, in the sense that the sum of the capital allocated to each risk adds naturally to the overall capital (Equation 19.7).
- More volatile risks on a stand-alone basis will have more capital allocated to them.
- Diversification is taken into account naturally, as risks that are less correlated with the rest of the portfolio will have less capital allocated to them.
- Larger risks will naturally have more capital allocated to them.

One limitation of the Euler allocation principle based on the expected shortfall, which is the most popular method, is that it still has numerical instability problems (McNeil et al., 2005), albeit not as serious as the VaR-based allocation. Specifically, the presence of extremely large events in a particular simulation may lead to an allocation which is excessively skewed towards a particular risk.

19.3 Price Optimisation

Where data is sufficient, it is possible to move beyond purely loss-based pricing and attempt price optimisation – that is, setting the premium at a level that maximises expected profits, subject to a number of constraints. For this topic, we will follow mainly the works of Krikler et al. (2004) and Ruparelia et al. (2009), although information on the actual implementation of price optimisation is scant in that article as in most of current actuarial literature on the subject, probably because the topic is still commercially sensitive. We have tried here to outline the conceptual framework behind price optimisation, but the treatment will, by necessity, be quite simplified.

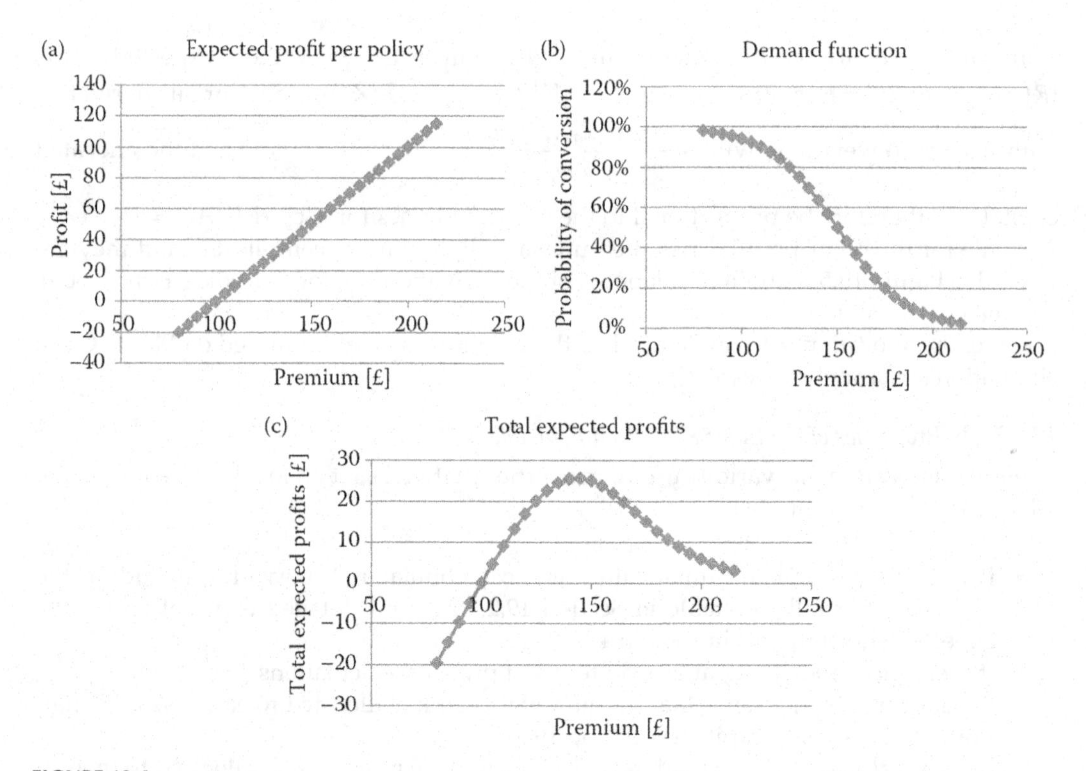

FIGURE 19.6
The total expected profits (a) can be obtained by multiplying the profit per policy (b) by the demand function (c). The price for which the total expected profits reach the maximum (in this case, this happens at approximately £145) is the optimal price.

To see how pricing optimisation works, first consider the case in which we have one product (the policy), which is sold at the same price P to all customers. As elementary price theory tells us, the profit per policy $\pi(P)$ increases linearly with price, whereas the demand curve, $D(P)$ (which describes the propensity of a customer to make a purchase) decreases and eventually goes to zero.

The total expected profits are therefore given by $TEP(P)=\pi(P)\times D(P)$, and there is typically going to be a price P^* for which $TEP(P^*)$ is maximum, as illustrated in Figure 19.6.

In general, the price of each policy will need to be different because different policyholders will have different values of the rating factors. For instance, the risk cost might be determined through a generalised linear modelling exercise. Also, the demand curve might be different depending on a number of factors – possibly overlapping with the rating factors – and might therefore be different for different policyholders. We can quite generally say that the premium for risk j will have the form

$$P_j = \text{Risk Premium}\left(\alpha_j\right) + \pi\left(P_j, \alpha_j\right) \tag{19.9}$$

In general, $\pi\left(P_j, \boldsymbol{\alpha}_j\right)$ depends on both the premium P_j and the vector $\boldsymbol{\alpha}_j = \left(\alpha_{1,j}, \ldots \alpha_{m,j}\right)$, which gives the value for risk j of the rating factors $\boldsymbol{\alpha} = (\alpha_1, \ldots \alpha_m)$ (for example, α_1 could represent the location, taking value 1 for urban locations and the value 0 for rural locations). This allows for the possibility of different categories of policyholders yielding different margins.

As for the demand function, this will also depend on the policyholder, although through factors $\boldsymbol{\beta} = (\beta_1, \ldots \beta_n)$ that are in general different from the rating factors $\boldsymbol{\alpha}$ that we used in Equation 19.9:

$$D_j = D\left(P_j, \boldsymbol{\beta}_j\right)$$

BOX 19.3 EXAMPLES OF POSSIBLE CONSTRAINTS FOR PRICING OPTIMISATION

- Keep the margin on each policy within certain limits, for example, $0.1\,P_j < \pi\left(P_j, \boldsymbol{\alpha}_j\right) \leq 0.2 P_j$.
- Keep the market share within a certain range.
- Keep the total revenue (or the number of policies) above a certain value.
- Limit the increase/decrease in price with respect to current policies within a certain range.
- Limit the difference between the prices of different policies in the portfolio.
- Keep the prices within a certain range of competitors' prices.
- The conversion rate across the portfolio, $\Sigma \frac{D_j}{(\#\text{offers})}$, must be above a certain level.
- Make sure the portfolio is well balanced, by requiring, for example, that the proportion of policies in different segments of the portfolio must be within a certain range.
- Avoid concentration of risk in the portfolio.

Where $\beta_j = (\beta_{1,j}, \ldots \beta_{n,j})$ is the value that the vector of factors $\boldsymbol{\beta}$ takes for risk j. The formula above should therefore be rewritten, for a portfolio of N policies each priced individually, as[7]

$$\text{TEP} = \sum_{j=1}^{N} D\left(P_j, \boldsymbol{\beta}_j\right) \times \pi\left(P_j, \boldsymbol{\alpha}_j\right)$$

The problem of pricing optimisation for a portfolio of policies can now be reformulated as the problem of maximising TEP by changing $\pi\left(P_j, \boldsymbol{\alpha}_j\right)$ subject to a number of constraints. Some examples of constraints are listed in Box 19.2.

We then have three problems to solve:

a. To calculate the profit $\pi\left(P_j, \boldsymbol{\alpha}_j\right)$, we need to be able to calculate the risk cost accurately enough. This is our usual risk costing problem, and most of this book is devoted to showing how this problem can be solved and what the uncertainties around it are (some of the techniques involving rating factors will be developed in Chapter 27).
b. We need to calculate the demand function. This is a new problem that we need to address separately (see Section 19.2.1).
c. We need to run an optimisation algorithm that gives us the value of $\pi\left(P_j, \boldsymbol{\alpha}_j\right)$ leading to the maximum total expected profits (Section 19.2.2).

19.3.1 Determining the Demand Function

Again, let us start with the simple case in which we have a homogeneous collection of policyholders all buying (or not buying) the same policy. To determine the demand function, we have several options. The most accurate is that of a price test.

19.3.1.1 Price Tests

A price test works by offering a randomly selected sample of policyholders a different price for the same policy and analysing the conversion rate at different levels of price. The conversion rate will be proportional to the demand curve. Krikler et al. (2004) calls this the 'demand probability'.

Because the conversion rate will be a number between 0 and 1, monotonically decreasing in price (at least typically), the demand function is often approximated as a logit function:

$$\text{logit}\left(D(P)\right) = \ln\left(\frac{D(P)}{1 - D(P)}\right) = -\frac{P - \text{shift}}{\text{scale}}$$

or, solving with respect to $D(P)$,

$$D(P) = \frac{1}{1 + \exp\left(\frac{\text{P} - \text{shift}}{\text{scale}}\right)}$$

7 Note that writing the total profit as the sum of the profit of individual policies is only a crude approximation that we have used for simplicity. In practice, total profit will depend not only on the sum of the profit of all underlying policies but also on fixed costs, which will be spread wider if the number of policies increases.

As an example, the illustrative demand function of Figure 19.6 has been obtained by imposing shift = £150, scale = £18. In practice, the logit will only be a rough approximation of the demand function. For example, it is unlikely that there exists a price at which the conversion rate is near 100%; some of the quotes will be sought simply for information purposes. In any case, the price test is never meant to test extreme scenarios, but rather to focus on premium levels around the current premium. For this reason, what we end up measuring is how the demand varies around a region where the demand function is roughly linear. The slope of this linear approximation of the demand function (its point derivative) is called *demand elasticity*.

19.3.1.2 Cost of a Price Test

As is typically the case when we measure something, price tests have a cost and an uncertainty, and there is a trade-off between the test cost and its uncertainty. The uncertainty of the test is, as can be expected, due mainly to the fact that only a small percentage of the portfolio is analysed, and different increments/decrements of the premium are applied to different customers. As a result, one particular increment/decrement δP is applied to an even smaller set of n quotes, and the error on the calculated demand function will be proportional to $\frac{1}{\sqrt{n}}$:

$$\text{Error} = \text{s.e}\left(\hat{D}(P^*)\right) \propto \frac{1}{\sqrt{n}}$$

where $\hat{D}(P^*)$ is the estimator of the demand probability $D(P^*)$.[8]

On the other hand, the cost of the price test for that particular increment δP, in the worst case (which occurs when the price is already optimal), is equal to the cost of having total expected profits $\text{TEP}(P^* + \delta P) \times n$ rather than $\text{TEP}(P^*) \times n$ on those n quotes [note that $TEP(P^* + \delta P) < TEP(P^*)$ because P^* is assumed to be optimal]. Using a second-order degree polynomial approximation around the maximum at $P = P^*$ where $\left(\left(\frac{dTEP}{dP}\right)_{P=P^*} = 0\right)$, we obtain

$$\text{Cost} = \text{TEP}(P^*) \times n - \text{TEP}(P^* + \delta P) \times n \approx -\frac{1}{2}\left(\frac{d^2\text{TEP}}{dP^2}\right)_{P=P^*} \times (\delta P)^2 \times n$$

(Note that the second derivative at $P = P^*$ is going to be negative because this is a maximum – hence, the cost is always positive).

There is therefore a trade-off between the cost of a price test (which increases linearly as the number of policyholders tested with a given price) and the uncertainty on the

8 The reasoning behind this formula is as follows: the demand probability at a given point P can be estimated by looking at the number k of policies that are purchased and dividing it by the number of quotes, n: $\hat{D}(P) = k/n$. The number k is the realisation of a binomial process with parameters n (the number of quotes) and $D(P)$ (the true demand probability). The expected value of k is therefore $\mathbb{E}(k) = n \times D(P)$ and its variance is $\mathbb{V}(k) = n \times D(P) \times (1 - D(P))$. Because the standard error is by definition the standard deviation of the estimator, we have $\text{s.e} = \sqrt{\mathbb{V}\left(\frac{k}{n}\right)} = \frac{1}{n}\sqrt{\mathbb{V}(k)} = \frac{1}{n}\sqrt{n \times D(P) \times (1 - D(P))} \propto \frac{1}{\sqrt{n}}$.

results of the test (which decreases more slowly, as the square root of the number of policyholders).

Krikler et al. (2004) suggest that the size of a price test should be chosen to maximise the expected net gain of the price test, defined as the expected additional profits deriving from the implementation of the results of the price test. As usual, the size of a price test is ultimately a business decision.

19.3.1.3 More General Case

In general, the demand function will depend not only on the price but also on a number of other factors as well, which include all rating factors but may also include additional factors such as competitors' prices. As an example, the demand for home insurance for middle-aged professionals may be less elastic than the demand for young students.

One possible way to model the demand function is to use a simple linear model linking the logit of the demand function to the factors $\boldsymbol{\beta} = (\beta_1, \ldots \beta_n)$:

$$\text{logit}\left(D\left(P_j, \boldsymbol{\beta}_j\right)\right) = a_0 + a_1\beta_{1,j} + \ldots + a_n\beta_{n,j}$$

More generally, one can use generalised linear modelling, which gives $\text{logit}\left(D\left(P_j, \beta_j\right)\right)$ as a linear combination of the values of functions $\varphi_k(\beta_1, \ldots \beta_n)(k = 1, \ldots K)$ of the same factors (see Chapter 27 for a more extensive treatment) – in this way, linearity in the factors is lost but it is retained in the functions $\varphi_k\,(\beta_1, \ldots \beta_n)$ (hence the term 'generalised linear modelling').

19.3.2 Maximising the Total Expected Profits

The problem of setting the profit per policy at a level that maximises the total expected profits in the general case of policies with many different cost and demand profiles is a numerical optimisation problem. This optimisation problem is straightforward to formulate but possibly tricky to solve because it runs over many dimensions and requires significant computing power.

Were the optimisation unconstrained, we could look at every policy, or every category of policy, individually, and find the optimal price for that policy. However, there are in general several global constraints that need to be satisfied, of which a selection are listed in Box 19.2. As a consequence, the optimisation needs to be run over many variables simultaneously.

19.3.3 Limitations

One should be aware of the limitations of pricing optimisation:

Uncertainties around Risk Cost. As we have seen, for example in Chapter 18, there are many uncertainties around the determination of the cost of risk. This uncertainty will be reflected in the estimate of the expected profit per policy, and therefore in the total expected losses. This problem will be less serious for personal lines insurance than for commercial lines insurance, but the large amount of data available will, in that context, be used to calibrate a number of rating factors as large as possible (to the limit of statistical significance), so that the problem of uncertainty remains relevant.

Uncertainty around the Demand Function. Determining the demand curve is also not trivial: as we have seen, the cost of price tests – through which the demand function can be probed – is inversely related to the accuracy that we are seeking for the demand function. As for the risk cost, trying to capture the presence of many factors affecting demand also makes it difficult to calculate the demand function accurately without a very large data set.

Regulatory Difficulties. Many countries will have regulations that severely constrain the use of pricing optimisation, for example, because it is illegal to carry out random price tests. Currently, pricing optimisation in insurance has been applied mainly in Europe and Israel, where the regulator's approach is more light-handed.

Changing Demand Function. The demand function changes with time, as the competition changes the landscape continually and macroeconomic conditions affect people's purchasing habits. This makes the price tests rapidly obsolete and requires them to be repeated frequently.

19.3.4 Other Considerations/Approaches

Lifetime Value. Price optimisation as we have described it thus far has a 1-year horizon. Attempts have been made (Ruparelia et al. 2009) to generalise this framework to a multiyear horizon introducing the concept of *customer lifetime value*, which captures the net present value of a policyholder over the years. This framework adopts a long-term strategy towards pricing; however, calculations over a multiyear horizon are much more uncertain and heavily reliant on assumptions about the future. It has also been pointed out that similar results can be obtained by adopting a 1-year horizon with 'exit constraints' (i.e. constraints on what the end-of-year position has to look like). According to the article by Ruparelia et al. (2009), no companies are currently using customer lifetime value as a means to optimise their prices – this, however, is a bold statement that is hard to verify.

Agent-Based Modelling. Price optimisation is an example of making decisions in an uncertain environment and can be theoretically solved in the framework of Markov decision processes (Parodi, 2012b). This is intrinsically a lifetime-value approach.

19.4 Questions

1. As a result of a pricing exercise, you produce the following estimates for a public liability risk:
   a. Expected losses = £850k
   b. Allowance for uncertainty = None
   c. Other costs = £150k
   d. Commissions = None
   e. Profit = 12% of premium
   f. Mean weighted payment time = 5.5 years

   Using current publicly available information for investment return information, produce an estimate of the technical premium.
2. You are asked to help your underwriter with an actuarial quote for an aviation liability policy. Your actuarial model produces yearly expected losses of £1.2M.

To produce an estimate of the technical premium, you use the Cost Plus methodology. You want to take these elements into account: expenses, investment income, and profit.

You have the following additional information:

a. The weighted payment time of the claims is 6 years
b. Current gilt yields (as found, for example, on the FT) are as follows: 0.22% (1 year), 0.91% (5 years), 1.43% (7 years), and 2.07% (10 years)
c. The target expense ratio (expenses/premium) is 28%
d. 10% of the premium is supposed to be for profit
e. No reinsurance will be purchased
f. The commission can also be ignored
    i. Based on the information above, write down a formula for calculating the technical premium explaining all its terms, and use it to estimate the technical premium

    After seeing your calculations, the underwriter is considering whether to buy excess-of-loss reinsurance in excess of £2M, and they ask you to estimate the losses that are expected to be ceded to the reinsurer as a result of that. You estimate that the expected losses in excess of £2M are going to be around £550k. In the meantime, the underwriter gets a quote from the reinsurer of £800k to cover that risk

    ii. Calculate the net reinsurance cost of the arrangement, and determine how the technical premium changes as a result of that
    iii. Explain why the technical premium might ultimately be different from the actual premium charged

3. Prove that TVaR is a coherent risk measure if the underlying probability density function of the loss distribution is continuous.
4. Suppose that your gross aggregate loss distribution for a contract can be well approximated by a lognormal distribution with $\mu = 13$, $\sigma = 2$ (outputs in USD). The contract covers losses to the layer USD 9M (in aggregate) xs USD 1M (in aggregate).
    a. Using Monte Carlo simulation, estimate the value at risk, the tail value at risk and the expected shortfall (all @ 99%) of the distribution of losses *ceded* to the layer.
    b. Produce a chart for the CDF of the ceded losses.
    c. Based on the chart produced in (b), explain where the difference between the tail value at risk and the expected shortfall comes from.

20

The Pricing Cycle and Rate Change Calculations

Everyone has a plan 'til they get punched in the mouth.

Mike Tyson

As we have already explained at some length in Section 5.7 (Pricing in the Corporate Context), pricing is not an isolated activity but has interactions with just about every other function in an insurer. Specifically, within most insurers, there is a cycle of monitoring, feedback, refinement, approval, audit, reporting and controls associated with pricing, and this cycle involves several corporate functions such as business planning, reserving, capital modelling, management information, and so on. This chapter aims at illustrating how this wider control cycle around pricing might look like – although of course its actual implementation may vary quite widely from company to company.

The most important part of the pricing cycle from an actuarial point of view is the calculation of the rate change for a portfolio. This will be addressed in Sections 20.2 to 20.8. More information on rate change calculations can be found in Bodoff (2008); Farr, Subasinghe, et al. (2014); and especially Lloyds' (2017, 2018).

20.1 The Pricing Control Cycle

20.1.1 Product Pricing

Before getting into the details of the pricing control cycle, it is perhaps useful to remember that pricing an insurance policy is simply one example of the more generic problem of pricing a product such as a car or a toothbrush. When pricing a generic product, one has to take into account the manufacturing costs and the related expenses, and at the same time consider the competitive landscape, and this is true for insurance. What sets insurance policies apart as a product is that the *manufacturing costs* (i.e. the cost of paying claims) are known only well after the policy is sold – actually, for some classes of business, they may not be completely known for many years.

Furthermore, the manufacturing costs are heavily influenced by the portfolio composition, that is, the client mix. This is different from a car manufacturer, which can to a large extent ignore who buys its vehicles as far as the assembly line costs are concerned.

DOI: 10.1201/9781003168881-22

20.1.2 Pricing as a Cycle

Pricing is not a one-off exercise but a process that goes (or should go) through continual improvement as more information becomes available. This process is normally referred to as the *actuarial control cycle* and distinguishes three phases in the actuarial approach to problems: (1) specifying the problem; (2) developing a solution; (3) monitoring the consequences (see Institute and Faculty of Actuaries (IFoA), 2010).

One way of looking at the actuarial control cycle is through the language of project management. The traditional approach to project management works as shown in Figure 20.1.

This formulation is perhaps more suitable to describe the pricing control cycle. In the case of pricing, 'planning' is the business planning by which the strategic goals of the insurers will be set. 'Executing' is the pricing process itself (costing and loading). 'Monitoring' is the process by which the consequences of pricing on sales and performance are analysed, and by which the assumptions are checked against actual outcomes.

In more detail, the process may be viewed as shown in Figure 20.2. As mentioned previously, the exact process is company-dependent, and this is only one possible way in which this cycle could be implemented. Also, note that the cycle will conceptually be roughly the same for personal lines insurers, commercial insurers, and reinsurers. However, the level of complexity is higher for personal lines insurers, where there is more information to be monitored (e.g. there are more distribution channels through which policies are sold). Finally, several fine details and interactions of pricing with other business functions, such as the role of claims management, have been ignored for simplicity of illustration.

20.1.2.1 Pricing Framework versus Pricing Decisions

It is worth pointing out that there is a difference between the pricing framework and the individual pricing decisions. The framework is set in the context of the company's strategic plan and will include parameters designed to promote coherent and consistent decision making. Individual pricing decisions are then made and monitored to track variance to plan. The flexibility allowed will vary depending on the risks involved; for big-ticket commercial risks, there is scope for commercial judgement and every pricing calculation will be subject to close scrutiny, but for high-volume personal lines, there will be limited discretion to move away from book rates. In any case, there will be the usual consideration of the effect on profit, cash flow, risk, market share, portfolio mix, and so on.

The Traditional Approach to Project Management

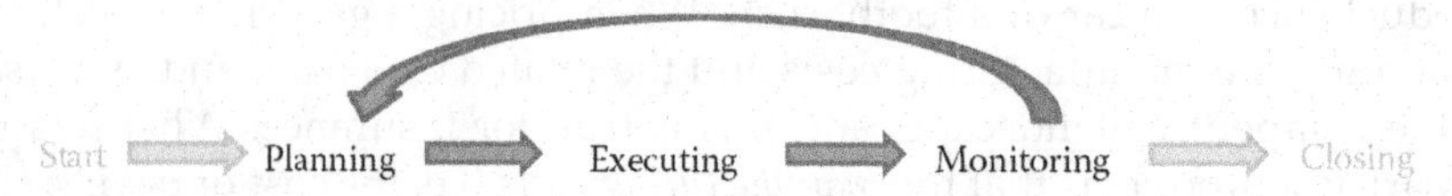

FIGURE 20.1
At the core of the traditional approach to project management is a control cycle (planning – executing – monitoring). Projects, however, are by definition one-off activities that always include a start and a closing point (the project basically closes when the monitoring activity shows sufficient alignment between the objectives of the project and its current status). In the case of pricing, start and closing will not be included because pricing can be viewed as an ongoing activity.

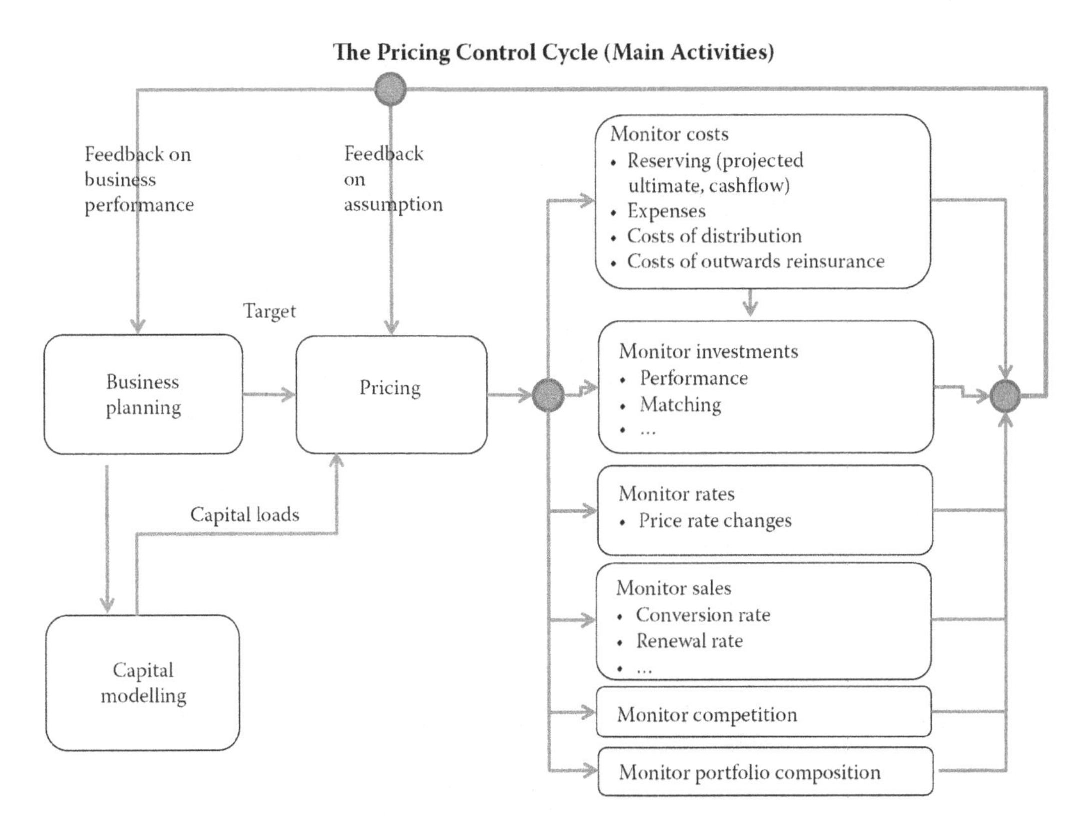

FIGURE 20.2
The inputs to the pricing process are ultimately coming from the business planning function of the company, which provides targets on profit, on the customer mix to achieve, and the like. Some of the inputs to pricing come from business planning through capital modelling, which has amongst other things the job of allocating capital loads to the different classes of business. A number of metrics are then monitored to (a) revise the assumptions underlying the pricing model (such as delays, frequency, or severity) as more experience accumulates and (b) give feedback to the management on sales and profitability so that the business plan can be revised.

20.1.2.2 Is the Pricing Control Cycle Really a Cycle?

One look at Figure 20.2 suggests that the pricing control cycle is more complex than a sequential cycle as one might imagine it to be. Many things happen at the same time in different parts of the cycle asynchronously and often in no particular order despite the fact that the general idea is indeed that of a continually improving process. Perhaps it would be more correct to describe this as a network rather than as a sequential cycle, an asynchronous network in which things happen at many different points in no particular order and possibly at the same time and ultimately a situation of temporary equilibrium is reached in which the experience becomes consistent with the pricing assumptions and with the business targets.

However, the situation of equilibrium will indeed be only temporary, as business planning may change targets, competition might change the market landscape, or regulation and the general environment change the risk.

20.1.2.3 Business Planning

The objective of business planning is to determine what parts of the business the company should focus on, and what profitability can realistically be achieved. The results of business planning then become guidelines for the underwriters.

The pricing function plays a crucial role during business planning by providing management with the necessary information on the expected profitability by business segment and under certain business scenarios. Apart from pricing, business planning sees the involvement of different functions of the company such as management (including underwriting management), reserving, and catastrophe modelling.

More in detail, the pricing function contributes to business planning by estimating a number of (prospective) profitability metrics, which will serve as guidance to underwriters for the year ahead. These profitability metrics may include, for example, a plan loss ratio (expected losses/earned premium), a plan underwriting ratio (plan loss ratio plus external expense ratio), and a plan combined ratio (underwriting ratio plus internal cost ratio), plus any other company-specific metric. All these metrics need to be calculated by business segment.

As far as the methodology is concerned, from a pricing perspective business planning is often akin to a portfolio-level burning cost exercise (Sections 11.1–4).

20.1.2.3.1 Inputs, Assumptions, and Sources

Typical inputs to business planning are:

- observed past losses, both paid and outstanding;
- estimated catastrophe (cat) losses (for property, in which case non-cat losses need to be excluded from the observed past losses to avoid double-counting);
- written premiums relevant to the planned year(s);
- IBNR factor assumptions.

We also need *assumptions* on

- IBNR factors (for both IBNYR and IBNER, typically combined);
- non-modelled cat losses;
- ILF curves (see Chapter 23) and other assumptions on non-working capacity, i.e. losses of a size for which experience is too scant or even null but for which cover is offered;
- estimated future rate changes and (prospective) *external* expense ratio;
- estimated *internal* expense ratio.

The assumptions above are needed for each business segment under consideration.

The observed losses and the written premiums are normally available through the technical accounting system. Estimated cat losses can be derived from a cat model and may already be available at contract level, or have to be run for the portfolio. IBNR factor assumptions will normally be provided by the reserving function. Non-modelled cat assumptions may come from the cat modelling team or the pricing team. ILF curves and other non-working capacity assumptions will normally come from the pricing team. Rate change/external expense ratio assumptions will come from underwriting management and others. Internal expense ratio estimates will normally come from Finance.

20.1.2.3.2 Calculations

The calculations are pretty much the same as for a burning cost exercise (Sections 11.1 and 11.4). In summary:

1. Losses that are for business that are not part of the company's appetite need to be excluded. Past non-excluded losses need to be revalued, and adjusted for IBNR. Large loss corrections must be introduced as needed, using ILF curves (liability business) or other non-working capacity assumptions. For property, nat cat loss results from cat models need to be incorporated (for modelled losses), as well as assumptions on non-modelled cat losses.
2. The exposures need on-levelled to make them relevant to the planned year.
3. The written premium needs to be made relevant to the planned year to reflect past and prospective rate changes. Earned premiums must be calculated from the written premium.
4. The loss ratio, underwriting ratio, combined ratio, etc. can then be calculated in the usual way.

These steps need to be replicated for each relevant business segment. This helps considering different scenarios including increasing/decreasing the amount of business written in a specific segment.

20.1.2.3.3 Outputs

As discussed above, the output of business planning (as far as pricing is concerned) is a number of profitability metrics that are useful to management, such as:

- Plan loss ratio (by segment and overall)
- Plan underwriting ratio (by segment and overall)
- Plan combined ratio (by segment and overall)

Since these outputs inform the decision process, business planning may be an iterative process in which different scenarios are examined, different assumptions are reconsidered, and so on.

20.1.2.4 Monitoring

The key feature of the pricing control cycle illustrated in Figure 20.2 – the feature that keeps the cycle going and allows the pricing process to improve and remain adequate – is monitoring. Some of the reasons why an insurer needs to monitor its business are

- To assess its performance against its targets
- To validate its pricing and reserving assumptions
- To report to management so that management can take affective decisions
- To report to regulators if required to do so
- To see how well it is faring against its competitors
- To gather information on the underwriting cycle

Because monitoring is so important, in this section, we are going to explain in more detail which variables are normally monitored and why. We will focus on those variables that are most related to pricing.

20.1.2.4.1 Monitoring 'Production' Costs

The pricing assumptions – such as the delay distribution, the frequency model, the severity model – can be monitored theoretically for each individual client. However, this is time consuming and the results are often not particularly meaningful because of their limited statistical significance: if the frequency model predicts 50 claims every year with a standard deviation of about 10, what are we to make of the fact that next year the number of claims is actually 70? It is more interesting to see how our assumptions have worked out collectively at the portfolio level as analysed by the reserving function. For example, the reserving function will give each year updated estimates of the loss development factors and will allow us to revise our assumptions on IBNR and IBNER.

The reserving function will also be able to estimate the timing of future cash flows, which would then inform investment choices because money must be invested in assets that are available when cash flows are expected.

Other costs that need to be monitored to be fed back into pricing are:

- The expenses (including the expected costs of distribution)
- The cost of reinsurance, both of which enter as loadings in the technical premium

20.1.2.4.2 Monitoring Investments

In general, the insurer's investments need to be monitored because

- Investment returns contribute to the technical profits of the company
- The matching of assets and liabilities must be validated periodically, as our estimation of outstanding liabilities and future cash flows change continually
- The regulatory position must also be validated periodically

Specifically for pricing, investment performance and investment mix need to be monitored because they are part of the pricing assumptions (Section 19.1.4).

20.1.2.4.3 Monitoring Rates

To assess whether premiums are adequate and business is profitable, insurers monitor premium rates. Premium rates can be defined in different ways, such as

a. Rate = premium income/exposure (also called 'rate on value'; the most popular)
b. Rate = premium income/limit (also called 'rate on line'; not applicable if policy is unlimited)
c. Rate = premium income/expected losses (requiring an actuarial model)

They can be calculated at the policy level or at the portfolio level, and they can be the gross or net of commissions.

Besides being a measure of profitability, premium rates (and especially premium rate changes) are normally required by reinsurers as part of the reinsurance submission (i.e. the information package that an insurer needs to provide to reinsurers to obtain cover), as this will allow the reinsurer to compare the exposure of different years when this is provided in terms of the original premiums (see Section 20.3.2).

There are two main aspects related to monitoring rates: one is the monitoring of rate adequacy (are the rates adequate to cover the expected losses, expenses, and desired profit?) and the other is the monitoring of how rates change over time as a result of changing market pressure. These two aspects of rate monitoring are obviously related since for example downward pressure on rates is likely to translate into negative rate changes – however, the relationship is not trivial because the portfolio itself changes over time, often because of deliberate actions by the underwriting management who may decide for example to off-load unprofitable segments of the portfolio or expand into more profitable areas.

Rate monitoring can be done in different ways and presents a number of subtleties. It is also the monitoring activity that actuaries are most involved with. For this reason, we have devoted a substantially portion of the chapter (Sections 20.2–8) specifically to it.

20.1.2.4.4 Monitoring Sales

Rates and the rating factor system (i.e. how the premium depends on the rating factors) affect sales as well as premium. Some of the variables related to sales that are typically monitored include the following:

Renewal rate	The renewal rate is the percentage (over a period) of expiring policies that is renewed. A low renewal rate may indicate that the premium rate is higher than that of the competitors.
Conversion rate (aka strike rate)	The conversion rate is the percentage of quotes that result into actual purchases. A low conversion rate also indicates that the premium rate is higher than that of the competitors.
New business volume	This is the number of new policies written over a period. Sudden changes in the new business volume may indicate changes in the competitive landscape.
Quote volume	This is the number of quotes received over a period. A high quote volume may indicate a successful marketing campaign or a successful distribution pipeline.

20.1.2.4.5 Monitoring the Portfolio Composition

Policyholders will be normally categorised depending on the characteristics of the risk. For example, in personal motor insurance, policyholders may be categorised based on their age, location, occupation, the car they own, and so on. In commercial lines insurance, holders of employers' liability policies will be categorised depending on the industry, the territory, the make-up of their workforce, and the like.

The portfolio composition (or business mix) can be loosely defined as the percentage of policyholders that fall into each of these categories. Changes in the portfolio composition need to be tracked as they will result into changes in the risk profile of the portfolio and ultimately in the profitability of the portfolio.

20.1.2.4.6 Monitoring the Competition

Insurers do not live in a bubble but operate in a market, and they need to know what competitors are doing to make effective decisions and revise their business plans and forecasts. For example, if your competitors offer lower prices than you, it may be a sign that they are trying to conquer a larger share of a market, possibly losing money in the short term, or that your expenses are too high. Either way, you need to choose the right response.

The problem of monitoring the competition is of course that the competition has no duty to provide you with detailed information on their operations, so you always see things through a haze. Some legitimate ways in which an insurer can track competition include the following:

- Engage in mystery shopping, that is, asking for quotes on a few standard policies to assess prices, terms and conditions, and the like – this is becoming quite easy with aggregators, which can give you prices from several different insurers at the same time. It is less easy, of course, with commercial insurance and reinsurance
- Ask your customers why they are leaving you and for whom
- Ask the competitions' customers why they are leaving them and joining you
- If your business comes through intermediaries, you might try and ask them about your competitors

Of course, none of these methods are perfect: mystery shopping will not enable you to determine the parameters of a 20-factor rating system (you would need millions of queries), the customers that respond to your queries may be unrepresentative of the whole lot and so on.

20.1.2.4.7 Features of a Good Monitoring System

A good monitoring system should have the following characteristics (see, e.g. IFoA, 2010):

- Its output should be concise and relevant to the strategic goals of the insurer … and should be consistent over time and with other analyses
- It should be accurate, both in terms of data and results
- It should be easy to use
- It should be well documented
- It should be able to produce an analysis quickly from data as-at date to results

20.2 Rate Adequacy Monitoring

Rates (or premiums, for that matter) are adequate to the extent that they cover the expected losses, the expenses, and the desired profit. There are different metrics that are useful to measure this. Some of these metrics are the following:

- The observed loss ratio (observed losses divided by premium actually charged)
- The priced (= expected) loss ratio (expected losses divided by premium actually charged)
- The underwriting loss ratio, based on either the observed or expected losses (observed/priced loss ratio plus expense ratio, including internal expenses)
- The combined loss ratio, based on either the observed or expected losses (observed/priced loss ratio plus expense ratio, including both external and internal expenses)
- The ratio between the actual premium **AP** (the premium actually charged, or the expected premium to be charged if variable features such as non-free reinstatements are included) and the technical premium **TP**.

Of these, the last metric is perhaps the most useful, as it compares directly the premium that should be charged to the premium that is actually charged. For this reason, we'll focus on this ratio, which we will refer to as premium adequacy index (PAI) – but goes under a variety of other names (price adequacy, premium adequacy, rate adequacy, etc.) – with 'index' included or not depending whether we wish to emphasise that the ratio can be used to track the performance of the portfolio over time.

$$\mathrm{PAI} := \frac{\mathbf{AP}}{\mathbf{TP}} \tag{20.1}$$

A PAI of 100% means that the actual premium covers exactly the expected losses, the various expenses (unallocated loss expenses, underwriting expenses, reinsurance costs, and external expenses if we are looking at it on a gross basis), the cost of capital/desired profit, minus any investment income.

Equation 20.1 can be calculated for a whole contract (or a portion of a contract, such as a layer of insurance), or a portfolio. For a portfolio, **AP** and **TP** represent respectively the sum of the actual premiums and the technical premiums at the insurer's share over all contracts in the portfolio.

Rate adequacy metrics – and specifically those that are based on pricing models – are of course only useful to the extent that the pricing models correctly reflect the risks. Therefore, the rate monitoring process must be accompanied by a regular assessment of how well the models reflect the reality of risk. For this reason, the modelled loss ratios and the modelled expenses must be regularly checked against the observed loss ratios and expenses. Also the way in which the expected actual premium is estimated should be revised periodically in case variable features such as reinstatements are included.

Tracking rate adequacy over time yields information on how the performance of the portfolio is changing, as long as the technical premium calculations have remained consistent or are appropriately rescaled for comparison purposes. However, it fails to distinguish changes in performance due to changes in portfolio composition from those due to changes in the rate of the renewed portfolio. The next section shows how to isolate the effect of rate changes.

An example of PAI calculation for a small portfolio is shown in Section 20.8.

20.3 Rate Change Calculations – Generalities and Non-Actuarial Approach

The rate change is a measure that attempts to capture the extent to which an insurer is charging more (or less) to cover *the same risk* from one year to the other.

As we mentioned in Section 20.1, an effective rate monitoring process is important to the success of an insurer and is an essential component of the pricing control cycle (Farr, Subasinghe, et al., 2014). Rate changes can be calculated by the underwriter based on simple calculations and expert judgment (see Section 20.3.3), or by the actuary based on a pricing model (Section 20.4), or by a mixture of the two (with cross-review/sign-off).

The rate change calculations are often presented along with a breakdown (a walk-through) of the different contributions (exposure, T&C, etc.) to the difference between the premium of a contract from one year to the next (Section 20.3.3).

You can speak of rate change of an insurance contract, or a portion of an insurance contract (what we call a 'contract unit' – e.g., a layer of (re)insurance). Companies are also interested, for monitoring purposes, in the average rate change of a portfolio (Section 20.3.5).

Before delving into the details of the rate change calculations, let us briefly discuss the uses of rate change (Section 20.3.1) and introduce some useful terminology (Section 20.3.2).

20.3.1 Uses and Perspectives

Rate change calculations may be used:

- For reporting purposes, e.g. to assess the market pressure and communicate to the management or the shareholders
- To carry out a market-wide study on rate behaviour
- To adjust the priced loss ratio during business planning
- To identify contracts that are under-/over-priced and to communicate to the underwriters at the time of pricing, as part of a pricing model.

Depending on the use, the definition of rate change may change slightly. It may also change on the perspective, i.e., if we are calculating it from the perspective of the whole market, of an insurance company, or the client. Use and perspective are of course to some extent overlapping categories.

20.3.2 Terminology and Notation

The following table shows some shorthand for the inputs and outputs of the rate change calculations.

Symbol	Name	Description
e, e'	Exposure	Exposure at year $n-1$ and n. This can be a simple measure of exposure such as sum insured (property) or vehicle years (motor), or a more complex measure such as a property schedule (with detailed information on different locations/ buildings) or a detailed fleet breakdown. In the latter case, the effect of assessing the impact of exposure on price involves not just scaling but recalculating the premium with the new schedule. As for loss inflation, we will consider it for simplicity as part of the exposure change, as is often done in practice – mainly to avoid too many indices. However, it is quite straightforward to treat the loss inflation index i as a stand-alone component of the premium change.
c, c'	Cover	Cover at year $n-1$ and n. This is an umbrella term for all relevant parameters defining the policy cover, e.g. deductibles, limits, aggregate deductibles/limits. Some of the parameters, e.g. local deductibles, might already be captured in the exposure definition and may therefore be excluded. The participation share is also excluded and captured by a separate parameter. Terms and conditions may be included here unless they are treated separately.

Symbol	Name	Description
s, s'	Share	The participation share at year $n-1$ and n for a given contract of part of a contract, e.g. a layer of (re)insurance – we need to keep this separate from the rest of the cover structure.
m, m'	Model	The rating models used at year $n-1$ and n to produce a technical premium along with the parameters that go with it (e.g. exposure curve parameters, expense ratio, cost of capital ...).
$\mathbf{TP}(e,c,s,m)$	Technical premium	This is the premium (also called 'office premium', 'modelled premium', or other) that the insurance company estimates to be adequate to cover expected losses, expenses/costs, and the desired level of profit. The technical premium is calculated (with pricing model m) for a given level of exposure e, cover (e.g. layer definition) c, co-insurance share s, and whatever other variable relevant to the situation at hand. This should be purely technical (unaffected by commercial/market considerations) and is in general different from what the company actually charges. When not indicated otherwise, TP will be net of any external expenses. When necessary to clarify, GTP will indicate the gross technical premium and NTP will indicate the net technical premium.
$\mathbf{AP}, \mathbf{AP}'$	Actual premium (expiring and current)	This is the premium actually charged (at year $n-1$ (expiring) and n (current)) by the company for a given contract, at the insurer's participation share. Where the premium includes variable features (e.g. paid reinstatements), AP can be replaced by the expected premium. The actual premium can either be net or gross of external expenses (e.g., brokerage). When not indicated otherwise, AP refers to the net premium. When necessary to clarify, GAP will indicate the gross actual premium and NAP will indicate the net actual premium.
$\mathbf{AP}_{\text{asif}}$	As-if actual premium	The actual premium that would have been charged in year $n-1$ if the characteristics of the risk (exposure, loss index, cover, share ...) had been as in year n.
$\mathbf{PAI}, \mathbf{PAI}'$	Premium adequacy index	This is an index that captures the profitability of a contract unit or contract or portfolio (at year $n-1$ and n). It is defined as the actual premium divided by the technical premium. See Section 20.2 for more details.

20.3.3 Non-Actuarial Rate Change Calculations

To produce a reliable estimate of the rate change, one should ideally be able to produce an actuarial estimate of the expected losses and of the technical premium. However, underwriters have for a long time been able to produce sensible rate change calculations without actuarial pricing. A typical definition of the percentage rate change is

$$\%R = \frac{\mathbf{rate}'}{\mathbf{rate}} - 1$$

where **rate** and **rate**′ are the premium rates at the expiring and current year respectively and may be calculated differently depending on the business written. This definition is useful if the risk and the cover haven't changed significantly from one year to the other.

More in general, the rate change percentage can be written as:

$$\%R = \frac{\mathbf{AP}'}{\mathbf{AP}_{\mathbf{asif}}} - 1$$

where $\mathbf{AP}_{\mathbf{asif}}$, $\mathbf{AP}'$ are the as-if premium and the current premium as defined in Section 20.2.2. Both premiums are net of external expenses (we'll come back to this point later).

The underwriter knows the expiring premium **AP**, the current premium **AP**′, and the characteristics of the expiring and current risk: exposure, share, cover, etc. Consequently, the underwriter will be able to estimate the premium that would have been paid last year had the risk characteristics been as in the current year.

If we assume for example that the risk is (roughly):

- proportional to the exposure e (this may include inflationary effects);
- proportional to the share s;

and that the cover c hasn't changed significantly, we can calculate the as-if premium as:

$$\mathbf{AP}_{\mathbf{asif}} \sim \mathbf{AP} \times \frac{e' \times s'}{e \times s}$$

and the rate change as

$$\%R \sim \frac{e \times s}{e' \times s'} \times \frac{\mathbf{AP}'}{\mathbf{AP}} - 1$$

If the terms and conditions do change, the underwriter might have to provide an estimate $(c'/c)_{est}$ of the impact of such changes (perhaps using loss curves – see Chapters 22 and 23), and then calculate the rate change ratio as

$$\%R \sim \frac{e \times s}{e' \times s'} \times \left(\frac{c}{c'}\right)_{\mathrm{est}} \times \frac{\mathbf{AP}'}{\mathbf{AP}} - 1$$

20.3.3.1 Example: Rate Change Calculations for Property Using Rate on Value and Rate on Line

Consider the simple case of property insurance where the actual premium charged is already expressed in terms of the exposure, specifically as rates on value (ROV), i.e. in premium per unit of insured value (see Section 21.8): $\mathbf{AP} = s \times \mathrm{ROV} \times \mathrm{TIV} \propto s \times \mathrm{ROV} \times e$, where the proportionality $\mathrm{TIV} \propto e$ is only an approximation, because 'exposure' is not just a number, and may depend on the type and location of the properties.

The inflation is normally disregarded (assuming that claims and the sums insured are revalued in the same way). Assuming no change in terms and condition (including attachment point/limit), the rate change ratio therefore simplifies to:

$$\%R \sim \frac{\text{ROV}'}{\text{ROV}} - 1$$

As a *numerical example*, consider an insurer who insures a supermarket chain in the UK. Last year the insurer charged £500,000 for a total insured value of £1BN. This year the total insured value has increased to £1.1BN, but the premium charged is now £605,000. This means that the premium rate has gone from 0.5‰ to 0.55‰, which corresponds to a rate change of about +10% (under the assumptions explained above).

Using the rate on line. The rate change is sometimes calculated based on changes in the rate on line, i.e. the premium divided by the amount of cover:

$$\%R \sim \frac{\text{ROL}'}{\text{ROL}} - 1$$

There is no sound basis for this definition. The actual premium can be written as $\mathbf{AP} = s \times \text{ROL} \times \text{Limit}$, but the limit is a very poor measure of exposure, and therefore the ROL-based definition of rate change is in general a poor approximation of the rate change. Despite this, rate change calculations based on the rate on line are sometimes used and may lead sensible results at a portfolio level.

20.4 Actuarial Rate Change Calculations (Single Contract or Contractual Unit)

The standard actuarial definition of rate change relies on having a sound pricing model, which allows to calculate the expected losses and the technical premium for the expiring and current risk.

We'll focus here on the rate change for the smallest contractual unit (which could be, for example, a full contract or a layer of (re)insurance), leaving aggregation considerations to later (Section 20.7).

The rate change ratio can then be defined as the ratio between the actual premium at year n, $\mathbf{AP}'$, and the 'as-if' actual premium at year $n-1$, $\mathbf{AP}_{\text{asif}}$. This is the premium that (in the analyst's estimation) would have been charged at year $n-1$ if the various characteristics of the contract and the risk (e.g. exposure, cover …) had been the ones relevant for year n.

$$\rho_R := \frac{\mathbf{AP}'}{\mathbf{AP}_{\text{asif}}} \tag{20.2}$$

AP_{asif} can be formally defined from an actuarial point of view in terms of the change in the actuarially calculated technical premium:

$$\mathbf{AP}_{\text{asif}} := \frac{\mathbf{TP}(e',c',s',m')}{\mathbf{TP}(e,c,s,m')} \times \mathbf{AP}$$

Note that the model m' appears both at the numerator and at the denominator. The idea is that the model according to which the technical premium is calculated should be the one that provides the most up-to-date and trusted view of the risk. See Section 20.4.3 for a more in-depth discussion around this point.

Hence, Equation 20.2 can be written in expanded form as:

$$\rho_R := \frac{\mathbf{TP}(e,c,s,m')}{\mathbf{TP}(e',c',s',m')} \times \frac{\mathbf{AP}'}{\mathbf{AP}} \tag{20.3}$$

The percentage rate change is then given by

$$\%\delta R = \rho_R - 1 = \frac{\mathbf{AP}'}{\mathbf{AP}_{\mathbf{asif}}} - 1$$

Finally, the change in absolute monetary amount is given by

$$\delta R = \mathbf{AP}' - \mathbf{AP}_{\mathbf{asif}} = \%\delta R \times \mathbf{AP}_{\mathbf{asif}}$$

20.4.1 Other Sources of Change

The factors included in the technical premium are e,c,s,m. Obviously, other factors subject to change may contribute to the technical premium and might therefore be looked at when considering the rate change. An example is given by changes to the *breadth of cover*, i.e. terms and conditions that are not captured by the pricing model, for example changes in contract wording. No pricing model is ever able to incorporate the impact of *all* aspects of a risk or of a contract.

Changes of this type can be incorporated by using a judgment-based 'other effects' factor. In the following, we will mostly assume that these effects can be ignored or included as part of the exposure factor.

20.4.2 A Note on Model Changes

As we mentioned above, according to Equation 20.9 the same model should be used to calculate the technical premium for the expiring and current contract. In other words, if the model has changed, either in structure or parameters, the technical premium for the expiring contract must be re-calculated with the new model before calculating the rate change.

The reason is that the rate change is trying to capture whether the extent to which more or less premium is charged for the same risk, and the model is an assessment of the risk. If the risk remains the same but our assessment of the risk changes because of a model improvement, that shouldn't impact the rate change.

This might be understood through a *thermometer analogy*. Alice measures the temperature outside her house on Saturday with a cheap old thermometer and the reading is 18°C. Since it feels warmer than that, she throws away the thermometer. On Sunday, she goes to visit her neighbour Bob[1] to borrow his expensive thermometer, and the reading this time

1 Everyone has heard about Alice and Bob, two obsessive experimentalists. During the week they exchange encrypted messages (Rivest et al., 1978) and run experiments on quantum entanglement (Flarend & Hilborn, 2022). During the weekend they're relaxing at home measuring the temperature to decide what to wear for their day trip – they just can't help themselves.

is 20.5°C. Has the temperature risen from Saturday? Difficult to say, because the thermometer she used on Saturday was rubbish. It also doesn't feel like it's warmer anyway. The story may end with this question mark except that – Alice being Alice and Bob being Bob – Alice calls Bob to ask him if he had taken the temperature outside on Saturday as well with his fancy thermometer, and it turned out that he had, and that the reading was 20.5°C. So now, based on this new information, has there been a temperature change? Most probably not.

The obvious moral of the story is that you don't want to say that the temperature (the risk) has risen only because the thermometer (the pricing model) has changed: you want to say it if it's actually higher! So whenever possible you want to compare the temperature (the risk) with the best thermometer (most recent pricing model) that you've got.

Disregarding model changes also has the advantage that it establishes a continuity between the underwriters' view of the rate change and the actuarial rate change. The underwriter will have a 'feel' for whether the risk has changed, exactly as Alice has her own 'feel' for whether it's warmer or not, even if that perception is only an approximation. To the extent that this perception is correct, you want the conclusions based on the underwriter's perception to be the same as those based on the pricing model – and that's only possible if you use the same model for the expiring and current contract.

20.4.2.1 Exceptions to The Rule of Disregarding Model Changes: Time-Dependent Parameters

However, the rule above – that model change should be disregarded – is only true for those changes that change our assessment of risk for *both* the expiring *and* the current policy. However, if the model has changed because the risk has genuinely changed from one year to the other, that could be considered by the company part of exposure change and not of rate change. The best way to take this into account is in the latter case to consider time (e.g. the underwriting year) a parameter of the model.

Here are a couple of examples that show how to reason about model change.

Example 1. Assume that a property risk has remained identical over the last year (except perhaps for an increase to the sum insured due to inflationary pressure on property values), and the premium charged is the same (1‰ of the insured value). However, the model has changed because a new historical loss data analysis has revealed that the base rates should double, causing a doubling of the technical premium. *This change should apply to all years*. In this case, the rate change is 0%.

Example 2. Assume that a fleet of cars is insured at CCY 500 per vehicle. Assume 0% claims inflation for simplicity. The model has changed to reflect the decreasing frequency of car accidents leading to at TP reduction of 3%. *This change should apply only to the current year*. If the actual premium remains the same, that could be interpreted as a rate increase of 3%.

In both examples, Equation 20.3 is still valid, but in Example 1 the model m' replaces m for all years, whereas in Example 2 $m' = m$ but has a parameter (the frequency of car accidents) that is time dependent.

20.4.3 A Note on Loss History

How should emerging loss experience be considered in the rate change calculations?

For example, suppose that a client has an expiring policy with a premium of $50,000 and the exposure and the cover have not changed from one year to the other, but it has experienced a large loss of $100,000. However, you renew the policy at the same premium.

Different companies may take a different approach to this, but in my opinion what you do with claim history should depend on how you respond to this question:

> *Has the occurrence of that loss changed your opinion on the risk of that client, or do you think that the client has simply been unlucky?*

- If the emerging loss experience has changed your view of the risk, then this is like a model change (see previous section), and therefore, in the case you're considering, you'd have a negative rate change.
- If the emerging loss experience has not changed your view of the risk (and it is therefore due to process randomness), then your rate change should be 0%.

What one should *not* do is always to consider the emerging loss experience as evidence of a change in risk, unless one has a consistent system in place that punishes/rewards all accounts that perform differently from the average. Otherwise, if one just does that for clients who've had a loss, one could have this paradox: imagine a portfolio that has remained the same in all aspects of risk (exposure, inflation = 0%, cover ...) and the actual premium has renewed at the same level: **AP'**= **AP** for all contracts. The loss experience has remained the same at portfolio level. The contracts that have no loss will have an estimated rate change of 0%, while the ones who had loss will have negative rate change. So overall the portfolio will have a negative rate change even though the overall loss experience has not changed, the exposures have not changed, and the premiums have not changed!

20.4.4 A Note on Loadings

The technical premium includes loadings for internal expenses and profit (we'll discuss external expenses later). What should we do if these loadings change from one year to the other?

As a general rule, the same loadings should be used for both expiring and current year. This is because the rate change focuses on risk, not on the structure of the premium.

This is especially obvious for the profit loading – you don't want to have the paradox that you increase the required return on capital to increase your profits but you don't consider that as a rate change. That is exactly how a rate change happens: by increasing your profit loading!

As to what loadings one should use (expiring or current), one can choose either. Using the current ones is more in line with the prescription of using the newest model. However the difference is usually minimal and other considerations (e.g. whatever is easier to do technically) may help one decide.

A company may also decide not to use loadings at all and use the expected losses instead of the technical premium in Equation 20.3 and all formulae derived from it.

20.4.5 Gross vs Net Rate Change Calculations

The rate change formula can be written both for the gross and the net actual premium AP. The rate change calculations shown so far are for the net technical premium (see also Section 20.3.2 for notation), i.e. the technical premium net of the external expenses.

However, the rate change calculation can also be calculated on the gross basis:

$$\rho_R^{\text{gross}} := \frac{\mathbf{GTP}(d,e,c,s,m')}{\mathbf{GTP}(d,e',c',s',m')} \times \frac{\mathbf{GAP}'}{\mathbf{GAP}}$$

where d stands for deductions (commissions, broker's fees, etc.)

If the external expenses are (as is typically the case) a proportion of the premium ($\mathbf{GAP} = \mathbf{AP}/(1-d)$, where d is the percentage of external expenses), the two definitions are equivalent: $\rho_R^{gross} = \rho_R^{net} = \rho_R$, since the factor $(1-d)$ divides both actual and technical premium.

However, some organisations (e.g. Lloyd's, 2018) calculate two views of the rate change – one from the insurer's perspective (the one we have discussed so far) and one from the insured's perspective, on the assumption that, from the point of view of the insured, an increase in the external expenses (brokers' fees) will be indistinguishable from change in the underwriting company's rates:

$$\rho_R^{\text{insured's view}} := \frac{\mathbf{TP}(e,c,s,m')}{\mathbf{TP}(e',c',s',m')} \times \frac{\mathbf{GAP}'}{\mathbf{GAP}}$$

where **TP** is the *net* technical premium. Note that if the gross premium is related to the net premium by the relationship $\mathbf{GAP} = \mathbf{AP}/(1-d)$ mentioned above, then $\rho_R^{\textit{insured's view}}$ is related to ρ_R by the relationship:

$$\rho_R^{\text{insured's view}} = \frac{1-d}{1-d'} \times \rho_R$$

20.4.6 Relationship to the Premium Adequacy Index

As we saw in Section 20.2, the premium adequacy index (PAI) for a contract tries to capture the extent to which the actual premium is adequate, using the technical premium as the benchmark, and is defined as $\text{PAI} / \mathbf{AP} / \mathbf{TP}(e,c,s,m)$.

The change in profitability can be expressed as the ratio between the premium adequacy calculated at two different points in time. This is in turn equal to the rate change ratio:

$$\frac{\text{PAI}'}{\text{PAI}} = \frac{\mathbf{TP}(e,c,s,m')}{\mathbf{TP}(e',c',s',m')} \times \frac{\mathbf{AP}'}{\mathbf{AP}} = \rho_R \tag{20.4}$$

Note that if the contract is made of different contract units (for example, different layers) each of which has a different participation share, Equation 20.4 cannot be used directly to calculate the rate change of the contract. Section 7 explains among other things how to combine the rate change of different contract units.

20.5 Breaking Down the Premium Change into Its Components

Many companies are interested in determining not only the rate change but also in explaining the impact of the various factors (exposure, model, share, rate change ...) on how one goes from **AP** to **AP**′ for a particular contract unit. This 'walk-through' depends in general on the order with which we apply the changes. The following is therefore only an example, in which we assume that any change in inflation is incorporated in the exposure change. The following chain is a description of the various changes that lead from **AP** to **AP**′:

$$\mathbf{AP}\overset{e\to e'}{\to}\mathbf{AP}^{[e]}\overset{s\to s'}{\to}\mathbf{AP}^{[es]}\overset{c\to c'}{\to}\mathbf{AP}^{[esc]}\equiv\mathbf{AP}_{\mathbf{asif}}\overset{R\to R'}{\to}\mathbf{AP}'$$

This walk-through can be best understood in terms of the ratio between **AP**′ and **AP**, and by assuming that the effect of the various changes on the technical premium is multiplicative:

$$\frac{\mathbf{AP}'}{\mathbf{AP}}=\frac{\mathbf{AP}_{\mathbf{asif}}}{\mathbf{AP}}\times\frac{\mathbf{AP}'}{\mathbf{AP}_{\mathbf{asif}}}=\frac{\mathbf{TP}(m',e',s,c)}{\mathbf{TP}(m',e,s,c)}\times\frac{\mathbf{TP}(m',e',s',c)}{\mathbf{TP}(m',e',s,c)}\times\frac{\mathbf{TP}(m',e',s',c')}{\mathbf{TP}(m',e',s',c)}\times\rho_R$$

By noting that $\mathbf{TP}(m',e,s,c)=s\times\mathbf{TP}(m',e,1,c)$, where $\mathbf{TP}(m',e,1,c)$ is the technical premium at 100% share, this can be re-written as

$$\frac{\mathbf{AP}'}{\mathbf{AP}}=\frac{\mathbf{TP}(m',e',1,c)}{\mathbf{TP}(m',e,1,c)}\times\frac{s'}{s}\times\frac{\mathbf{TP}(m',e',1,c')}{\mathbf{TP}(m',e',1,c)}\times\rho_R$$

This can be written in a more compact form by giving a name to the ratios in the formula:

$$\rho_e=\frac{\mathbf{TP}(m',e',1,c)}{\mathbf{TP}(m',e,1,c)},\quad\rho_s=\frac{s'}{s},\quad\rho_c=\frac{\mathbf{TP}(m',e',1,c')}{\mathbf{TP}(m',e',1,c)}$$

Hence, the walk through the different changes from **AP** to **AP**′ is simply:

$$\frac{\mathbf{AP}'}{\mathbf{AP}}=\rho_e\times\rho_s\times\rho_c\times\rho_R \tag{20.5}$$

Equation 20.5 has all the information we need about the different contributions to the premium change. However, it is often more useful to express these contributions in terms of percentage changes and monetary amounts.

Percentage changes can be obtained simply by subtracting 1 to the ratios above and expressing these numbers as percentages:

$$\%\delta e=\rho_e-1$$

$$\%\delta s=\rho_s-1$$

$$\%\delta c=\rho_c-1$$

$$\%\delta R=\rho_R-1$$

As for the monetary changes, these depend on the order with which the contributions are applied at (to some extent, this is also true for the percentage changes and the ratios). For one particular order, the monetary changes can be written as:

$$\delta e = (\rho_e - 1) \times \mathbf{AP}$$

$$\delta s = (\rho_s - 1) \times \rho_e \times \mathbf{AP}$$

$$\delta c = (\rho_c - 1) \times \rho_s \times \rho_e \times \mathbf{AP}$$

$$\delta R = (\rho_R - 1) \times \rho_c \times \rho_s \times \rho_e \times \mathbf{AP}$$

where δe, δs, δc and δR are the changes (in absolute monetary amounts) due to changes in exposure, share, cover and rate, in this order.

The walk-through (see Equation 20.5) can then also be written down additively as:

$$\mathbf{AP}' = \mathbf{AP} + \delta e + \delta s + \delta c + \delta R \qquad (20.6)$$

The premium walk is often visualised with the help of a waterfall chart as in Figure 20.3.

We do not address here the case of the rate change for non-proportional treaty reinsurance. Suffice it to say that it is mostly the same as the rate change for direct, but the need arises to split the overall rate change into a primary premium rate change and the reinsurance premium rate change.

20.5.1 Taking External Expenses into Account in the Walk from AP to AP′

Equation 20.6 gives the walk between the expiring and current *net* actual premium. What about the walk for *gross* actual premium?

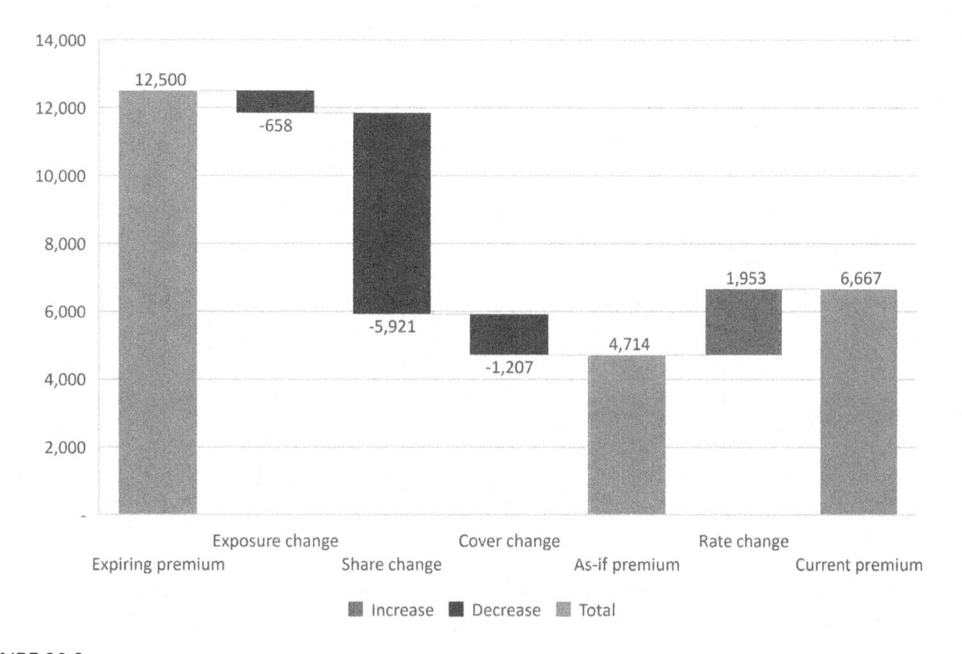

FIGURE 20.3
Example of waterfall chart with notional figures, which shows Equation 20.6 in graphical format. For clarity, the intermediate calculation of the as-if premium is shown, so that the equation is broken down into two parts: $\mathbf{AP}_{\mathbf{asif}} = \mathbf{AP} + \delta e + \delta s + \delta c$ and $\mathbf{AP}' = \mathbf{AP}_{\mathbf{asif}} + \delta R$.

Regardless of whether the insurer is interested in providing the insured's view (see Section 20.4.6), it will be interested in seeing the full walk from **GAP** to **GAP′**, isolating the effect of the changes in the external expenses. We can indeed write

$$\mathbf{GAP}' - \mathbf{GAP} = \delta e + \delta s + \delta c + \delta d + \delta R$$

where δd encapsulates the effect of the change in the external expenses. The articulation of the full sequence of formulae for this case is analogous to the net case and is left to the reader.

20.6 The Loss Ratio Walk

Apart from the premium walk, the company might also be interested – e.g. for business planning purposes – to produce a *loss ratio walk*, i.e. to break down the changes that explain the difference of the expected priced loss ratio (expected losses divided by actually charged premium) from one year to the other.

This can be done at contract level but is important especially at portfolio level.

The loss ratio walk works roughly like the premium walk but in this case we *normally* also need to take the contribution of the model change into account, as the expected losses depend on the model and any difference in the model is part of the explanation for a difference in the expected loss ratio.[2]

In this case, it is more straightforward to define the change ratios in terms the expected losses (which are already part of the definition of the loss ratio) rather than of the technical premium:

$$\rho_e = \frac{\mathbf{EL}(m',e',1,c)}{\mathbf{EL}(m',e,1,c)}, \quad \rho_s = \frac{s'}{s}, \quad \rho_c = \frac{\mathbf{EL}(m',e',1,c')}{\mathbf{EL}(m',e',1,c)}$$

The rate change ratio is defined as before ($\rho_R = \mathbf{AP}' / \mathbf{AP}_{\text{asif}}$) where however we define $\mathbf{AP}_{\text{asif}}$ in terms of the change in expected losses:

$$\mathbf{AP}_{\text{asif}} = \frac{\mathbf{EL}(m',e',s',c')}{\mathbf{EL}(m',e,s,c)}$$

And the model change ratio is simply:

$$\rho_m = \frac{\mathbf{EL}(m',e,1,c)}{\mathbf{EL}(m,e,1,c)}$$

Since $\mathbf{LR} = \mathbf{EL}(m,e,s,c) / \mathbf{AP}$ and $\mathbf{LR}' = \mathbf{EL}(m',e',s',c') / \mathbf{AP}'$, we can write:

2 As always, this actually depends on the use we are making of the loss ratio walk. We are assuming here that the loss ratio is a figure that needs to be presented to the management. Therefore, any change in it from the previous year needs to be explained, whether it comes from changes in modelling or in the actual risk.

$$\mathbf{LR}' = \mathbf{LR} \times \frac{\rho_m}{\rho_e \times \rho_s \times \rho_c \times \rho_R} \tag{20.7}$$

Note that, unlike the premium walk, all the ratios (except for the modelling change ratio) are now at the denominator, for the obvious reason that the premium also appears at the denominator. The modelling change ratio goes at the numerator as an increase in the estimated expected losses leads to an increase in the estimated priced loss ratio, while an increase, e.g., in the rates leads straightforwardly to a reduction in the priced loss ratio.

As we did in the case of the premium walk (Equations 20.5 and 20.6), we can turn Equation 20.7 into an additive expression, where the actual values of the addenda depend on the order in which they are defined:

$$\mathbf{LR}' = \mathbf{LR} + \Delta m - \Delta e - \Delta s - \Delta c - \Delta R \tag{20.8}$$

We have prefixed the addenda in Equation 20.8 (which are pure numbers) with 'Δ' rather than with 'δ' not for any profound mathematical reason but as a reminder that these are not the same as the addenda that appear in the premium walk in Equation 20.6 (which were monetary amounts), and the '–' has been added as a reminder that the effect is the opposite as the one in the premium walk.

The relationship between $\Delta m, \Delta e, \Delta s, \Delta c, \Delta R$ and the ratios $\rho_m, \rho_e, \rho_s, \rho_c, \rho_R$ can be worked out similarly to Section 20.5 and is left as an exercise to the reader.

20.7 Rate Change at Programme and Portfolio Level

The rate change calculations described in Section 20.4 are for the 'atomic' level – a single contract unit or layer of (re)insurance. In this section we are looking at how the rate change of different units/layers can be combined to calculate the rate change at programme level (i.e. the overall contract) and at portfolio level. In the following we will normally speak of 'portfolio', but the aggregation method is the same when you aggregate the layers of a contract.

The objective is to calculate the rate change for a portfolio from underwriting year $n-1$ to underwriting year n, where each contract is allocated to the underwriting year based on its inception date.

To calculate the rate change for the portfolio, we need to limit ourselves to the contracts that are included in both year $n-1$ and year n. Lapses and new contracts do however have an impact on profitability, of course, and therefore should be considered when calculating the premium adequacy index.

20.7.1 Combining the Rate Change of Different Contract Units

In the following, we assume that corresponding contracts and contract units are mapped explicitly, e.g. through a common identification number. $\mathbf{AP}_j$ denotes a contract unit incepting in underwriting year $n-1$, $\mathbf{AP}'_j$ one incepting in underwriting year n.

The overall actual premium of the portfolio can be written as

$$\mathbf{AP}^{\wp} = \sum_{\text{all matching units } j} \mathbf{AP}_j$$

$$\mathbf{AP}'^{\wp} = \sum_{\text{all matching units } j} \mathbf{AP}'_j$$

where the sum is extended over all contract units that are explicitly mapped between year $n-1$ and year n. In other words, contracts or layers that are dropped or new from one year to the other are not included in the calculations.

The monetary changes across a programme (or portfolio) $\wp$ can be written as the sum of the changes for each mapping layer:

$$\delta e^{\wp} = \sum_{\text{all matching units } j} \delta e_j$$

$$\delta s^{\wp} = \sum_{\text{all matching units } j} \delta s_j$$

$$\delta c^{\wp} = \sum_{\text{all matching units } j} \delta c_j$$

$$\delta R^{\wp} = \sum_{\text{all matching units } j} \delta R_j$$

The change percentages can then be calculated as

$$\%\delta e^{\wp} = \frac{\delta e^{\wp}}{\mathbf{AP}^{\wp}}$$

$$\%\delta s^{\wp} = \frac{\delta s^{\wp}}{\mathbf{AP}^{\wp} + \delta e^{\wp}}$$

$$\%\delta c^{\wp} = \frac{\delta c^{\wp}}{\mathbf{AP}^{\wp} + \delta e^{\wp} + \delta s^{\wp}}$$

$$\%\delta R^{\wp} = \frac{\delta R^{\wp}}{\mathbf{AP}^{\wp} + \delta e^{\wp} + \delta s^{\wp} + \delta c^{\wp}}$$

Note that this can be re-written also as

$$\%\delta R^{\wp} = \frac{\delta R^{\wp}}{\mathbf{AP}^{\wp}_{\mathbf{asif}}} = \frac{\mathbf{AP}'^{\wp}}{\mathbf{AP}^{\wp}_{\mathbf{asif}}} - 1$$

where $\mathbf{AP}^{\wp}_{\mathbf{asif}} = \sum_{\text{all matching units } j} \mathbf{AP}_{\mathbf{asif},j}$.

Finally, the portfolio change ratios can be calculated as:

$$\rho^{\wp}_{e} = \%\delta e^{\wp} + 1$$

$$\rho^{\wp}_{s} = \%\delta s^{\wp} + 1$$

$$\rho^{\wp}_{c} = \%\delta c^{\wp} + 1$$

$$\rho^{\wp}_{R} = \%\delta R^{\wp} + 1$$

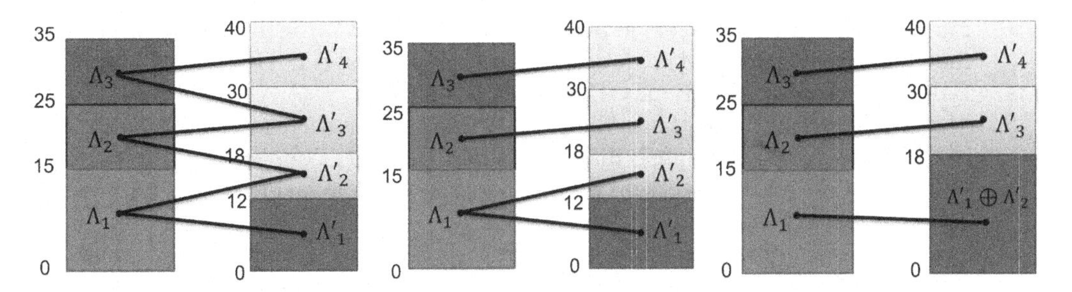

FIGURE 20.4
(Left) The layers for years $n-1$ and n are connected between them with an edge if there is some overlapping: e.g. Λ_1 is connected with Λ'_1 and Λ'_2 but not with Λ'_3 because the attachment point of Λ'_3 is higher than the exit point of Λ_1. (Centre) After removing the edges for which the overlapping is at most 50% *both ways*, the graph only has three connected components: $(\Lambda_1;\Lambda'_1,\Lambda'_2)$, $(\Lambda_2;\Lambda'_3)$, and $(\Lambda_3;\Lambda'_4)$. (Right) Layers Λ'_1 and Λ'_2 are part of the same connected component $(\Lambda_1;\Lambda'_1,\Lambda'_2)$ and are therefore merged to form a composite layer $\Lambda'_1 \oplus \Lambda'_2$.

*20.7.2 Layer Mapping

When a programme has more than one layer, it is necessary to map the layers at time $n-1$ and n. The mapping may not always be obvious, as layers may overlap only partly and may split or merge from one year to the other. The mapping may be done by the underwriters manually (which is time consuming and requires a suitable user interface) or automatically.

Once the mapping is done, specific rules for the rate change calculation must be specified if layer merging or splitting is present.

In the following we describe a possible methodology for automating layer mapping (subject possibly to overriding by the underwriters). We are assuming for both years a tower of layers with finite (non-zero) share with no gaps. A visual illustration of this methodology is provided in Figure 20.4.

1. Build a *bipartite graph*[3] for which the nodes of the left-hand side of the graph are the layers $\Lambda_j = \left(L_j, L_j + D_j\right)$ at year $n-1$, and the nodes of the right-hand side of the graph are the layers $\Lambda'_k = \left(L'_k, L'_k + D'_k\right)$ at year n.
2. Draw an edge between two nodes Λ_j and Λ'_k if and only if the two layers are overlapping, i.e. if $\min\left(L'_k + D'_k, L_j + D_j\right) > \max\left(L_j, L'_k\right)$. Note that a layer for year $n-1$ can be connected to multiple layers for year n and vice versa.
3. For each edge, measure these two quantities:

$$\mathrm{OLR}_{n-1} = \frac{\min\left(L'_k + D'_k, L_j + D_j\right) - \max\left(D_j, D'_k\right)}{L_j}$$

$$\mathrm{OLR}_{n} = \frac{\min\left(L'_k + D'_k, L_j + D_j\right) - \max\left(D_j, D'_k\right)}{L'_k}$$

Where OLR_{n-1} (overlapping ratio at $n-1$) represents the length of intersection of the two layers as a percentage of layer Λ_j, and OLR_n the ratio between the length of the

3 That is, a set of nodes and edges (i.e. connections between nodes) where the nodes are split into two separate subsets (the left-hand side and the right-hand side) and edges are possible only between nodes of different subsets.

intersection of the two layers as a percentage of layer Λ'_k (the indices $n-1, n$ refer to the fact that layers Λ_j and Λ'_k are at year $n-1$ and n respectively).[4]
4. Delete all edges for which $\max(OLR_{n-1}, OLR_n) \leq 50\%$.
5. Merge the left-hand nodes and the right-hand nodes in each connected component of the graph (in plain English that means that for every connected group of nodes you can merge all the layers on the left-hand and all the layers on the right hand). The symbol for merging two layers is '$\oplus$'.
6. The output is a bipartite graph in which certain nodes (layers) will be connected (matched) from one year to the next, some will be on their own, and some other will be connected after grouping (e.g. this will happen when layers are split, or merged).

How does layer matching affect the rate change calculations?
Layers that are on their own will not contribute to the rate change calculation, exactly as contracts that are not renewed or new contracts.
The case of a single layer matching another single layer requires no further explanation as it was the subject of Section 20.4 (actuarial rate change between two contract units).
The most interesting case is the calculation of the rate change between merged layers. The rate change can be defined without any major difficulty by using Equation 20.3 where the technical and actual premiums are the sum of the premiums for the various layers calculated at the company's share. In the general case where the merged layer $\Lambda_1 \oplus \Lambda_2 \oplus \ldots \oplus \Lambda_n (n \geq 1)$ is matched to merged layer $\Lambda'_1 \oplus \Lambda'_2 \oplus \ldots \oplus \Lambda'_{n'} \ (n' \geq 1)$, the rate change is given by

$$\rho_R = \frac{\sum_{i=1}^{n} \mathbf{TP}(m', e, c_i, s_i)}{\sum_{j=1}^{n'} \mathbf{TP}(m', e', c'_j, s'_j)} \times \frac{\sum_{j=1}^{n'} \mathbf{AP}'_j}{\sum_{i=1}^{n} \mathbf{AP}_i}$$

The complication arises when one wants to calculate the premium walk, as we need to tread lightly on how to define the share change, which in turns depends on how we define the *average* share for a merged layer. If we define the average share in this way:

$$s^* = \frac{\sum_{i=1}^{n} \mathbf{TP}(m', e, c_i, s_i)}{\sum_{i=1}^{n} \mathbf{TP}(m', e, c_i, 1)}, \quad s^{*\prime} = \frac{\sum_{j=1}^{n'} \mathbf{TP}(m', e', c'_j, s'_j)}{\sum_{j=1}^{n'} \mathbf{TP}(m', e, c'_j, 1)}$$

a natural way of defining the other contributions of the premium walk is the following:

$$\rho_e = \frac{\sum_{i=1}^{n} \mathbf{TP}(m', e', c_i, s_i)}{\sum_{i=1}^{n} \mathbf{TP}(m', e, c_i, s_i)}$$

$$\rho_s = \frac{s^{*\prime}}{s^*}$$

$$\rho_c = \frac{\sum_{j=1}^{n'} \mathbf{TP}(m', e', c'_j, 1)}{\sum_{i=1}^{n} \mathbf{TP}(m', e', c_i, 1)}$$

4 The value of OLR_{n-1} and OLR_n defines the type of relationship between two layers: e.g. $\mathrm{OLR}_{n-1} \times \mathrm{OLR}_n = 0$ means that the two edges are not overlapping; $\mathrm{OLR}_{n-1} = \mathrm{OLR}_n = 1$ means that the two layers are identical; $\mathrm{OLR}_{n-1} = 1$, $\mathrm{OLR}_n < 1$ means that Λ_j is fully included in Λ'_k; $\mathrm{OLR}_{n-1} < 1$, $\mathrm{OLR}_n = 1$ means that Λ'_k is fully included in Λ_j.

A little algebra should convince you that these definitions are internally consistent, i.e. that $\sum_{j=1}^{n'} \mathbf{AP}_j{}' = \rho_e \times \rho_s \times \rho_c \times \rho_R \times \sum_{i=1}^{n} \mathbf{AP}_i$.

Note that the formula for the cover change includes the terms $\sum_{i=1}^{n} \mathbf{TP}(m', e', c_i, 1)$ and $\sum_{j=1}^{n'} \mathbf{TP}(m', e', c'_j, 1)$, where the sum is over all layers in each merged layer. This is different (although normally by not much) than the technical premiums calculated for the union of the layers for the merged layers in the expiring and current year respectively – the difference is that the loadings for the combined layer may not be the sum of the loadings for the layers being merged.[5]

20.8 A Worked-Out Example

This section shows examples of how to calculate the rate change and premium walk for a single contract unit and for a portfolio, and the premium adequacy index at two different times.

A simple spreadsheet for rate change calculations that allows to run this and more complicated examples is made available in the book's website (Parodi, 2016).

The table below shows two snapshots of a company's portfolio at two consecutive underwriting years. For each contract the table specifies the participation share of the company.

Portfolio at year n-1

Contract ID	Exposure index	Technical premium (@100%)	Actual premium charged (@100%)	Company share
0001	100	100	90	10%
0002	100	250	150	5%
003	100	15	5	100%

All monetary amounts are in kCCY

Portfolio at year n

Contract ID	Exposure index	Technical premium (@100%)	Actual premium charged (@100%)	Company share
0001	110	120	80	10%
0002	110	250	50	1%

All monetary amounts are in kCCY

5 This may cause the slightly counterintuitive effect that simply combining two layers from one year to the other causes a spurious (if small) contribution to the rate change. This is no more paradoxical than the fact that the premium of the combined layer may be smaller than the two layers purchased separately if the risk loading is sub-additive– a desirable property of risk measures.

You can assume that no model change was put into effect between years $n-1$ and n, and that any difference in the technical premium @100% per unit of exposure can be explained by changes in cover, such as different retentions and limits.[6] You can also assume that the effect of exposure is purely multiplicative, namely that doubling the exposure causes the technical premium to double.

20.8.1 Rate Change Calculations and Premium Walk for Each Contract

The rate change calculations can be calculated for Contracts 0001 and 0002 only, as they are the only two renewed contracts.

Contract 0001

- $\mathbf{AP}$ = CCY 9k, $\mathbf{AP}'$ = CCY 8k
- $\mathbf{AP}$ = CCY 10k, $\mathbf{TP}'$ = CCY 12k

The rate change is given by $\%\delta R = (\mathbf{AP}' \times \mathbf{TP})/(\mathbf{TP}' \times \mathbf{AP}) - 1 = 80/108 - 1 = -25.9\%$. This can also be written as $\%\delta R = \frac{\mathbf{AP}'}{\mathbf{AP}_{\mathbf{asif}}} - 1 = \frac{8}{10.8} - 1 = -25.9\%$.

The exposure change is $\%\delta e = e'/e - 1 = +10\%$. The cover effect is given by $\%\delta c = (\mathbf{TP}'@100\%/e')/(\mathbf{TP}@100\%/e) - 1 = +9.1\%$. The participation share remains at 10% and therefore $\%\delta s = 0$.

The full premium walk from **AP** to **AP**′ (which also provides the monetary amount for each change) is given by:

	AP	δe	δs	δc	AP_asif	δR	AP
CCYk	9.00	0.9	0	0.9	10.80	–2.80	8.00
%		*10.0%*	*0.0%*	*9.1%*		*–25.9%*	

Contract 0002

- **AP** = CCY 7.5k, **AP′** = CCY 0.5k
- **TP** = CCY 12.5k, **TP′** = CCY 2.5k

The rate change is given by δR = (**AP′** × **TP**) / (**TP′** × **AP**) – 1 = 6.25/18.75 – 1 = –66.7%.

The exposure change is $\%\delta e$ = e′/e – 1 = 10%. The cover effect is given by $\%\delta c$ = (**TP′**@100%/e′)/(**TP**@100%/e) – 1 = –9.1%. The share change is $\%\delta s$ = 1%/5% – 1 = –80%.

The full premium walk from **AP** to **AP′** is given by:

	AP	δe	δs	δc	AP_asif	δR	AP′
CCYk	7.50	0.75	–6.6	–0.15	1.50	–1.00	0.50
%		*10.0%*	*–80.0%*	*–9.1%*		*–66.7%*	

6 This will allow us to calculate the impact of cover without using exposure curves, ILFs and so on that will be addressed later in the book.

20.8.2 Portfolio Rate Change Calculations

The only contracts that matter are 1 and 2, because Contract 3 has lapsed in year n. The full walk-through can be obtained by simply adding the monetary amounts for Contracts 1 and 2. For example, the overall rate change will be

$$\delta R^{\wp} = \delta R(0001) + \delta R(0002) = -2.80 - 1.00 = -\text{CCY}3.80\text{k}$$

$$\%\delta R^{\wp} = \frac{\delta R^{\wp}}{\mathbf{AP}_{\text{asif}}^{\wp}} = \frac{\delta R^{\wp}}{\mathbf{AP}_{\text{asif}}(0001) + \mathbf{AP}_{\text{asif}}(0002)} = -30.9\%$$

The full results for the portfolio are given in the table below.

	AP	δe	δs	δc	AP_asif	δR	AP′
CCYk	16.50	1.65	–6.6	0.75	12.30	–3.80	8.50
%		*10.0%*	*–36.4%*	*6.5%*	*–30.9%*		

20.8.3 Premium Adequacy Calculations

To calculate the premium adequacy index for both years, we need to include Contract 3 back in the calculation of the PAI for year $n-1$.

$$\mathbf{PAI}^{\wp} = \frac{\mathbf{AP}(0001) + \mathbf{AP}(0002) + \mathbf{AP}(0003)}{\mathbf{TP}(0001) + \mathbf{TP}(0002) + \mathbf{TP}(0003)} = 57.3\%$$

$$\mathbf{PAI}'^{\wp} = \frac{\mathbf{AP}'(0001) + \mathbf{AP}'(0002)}{\mathbf{TP}'(0001) + \mathbf{TP}'(0002)} = 58.6\%$$

The percentage difference in the PAI from one year to the other is +2.2%.

20.8.4 Interpretation of Results

The rate change and the PAI change give two different messages: the rates are reducing but the portfolio performs better in terms of expected profitability. The main reason for this apparent contradiction is that the PAI of year $n-1$ includes Contract 0003 – which has an AP/TP ratio of 33.3% and therefore brings the average PAI down. The removal of that contract in year n causes the PAI to increase.

This is a (very simplistic) example of how a company may compensate the reduction in profitability due to reducing rates by reducing (or removing) participation in unprofitable contracts and increasing (or adding) participation in more profitable contracts. Needless to say, this is easier said than done.

Also note that we are talking here about *expected* profitability. Even if PAI < 100% for both years, that doesn't immediately mean that the company is not making money: the observed loss ratio may be much lower than the expected loss ratio, because, for example, the portfolio may be heavily veered towards cat risks, that may not materialise in benign seasons.

A more systematic comparison of the difference between the rate change results and the PAI results can be made using the formula that Question 2 asks to prove.

20.9 Questions

1. You are monitoring a small portfolio of marine hull policies. The following table gives a snapshot of the portfolio for contracts written during 2019 (left) and 2020 (right).

	Year 2019		Year 2020	
Contract ID	**Technical Premium**	**Premium charged**	**Technical premium**	**Premium charged**
1	10,000	12,000	11,000	12,500
2	8,000	9,000	9,000	9,500
3	12,000	15,000	Not renewed	Not renewed
4	Not existing	Not existing	7,000	10,000
5	6,000	9,000	6,500	10,000

All monetary amounts are in FRD

   i. Determine whether the premium adequacy has improved or deteriorated between 2019 and 2020. Explain your workings.
   ii. Calculate the rate change for the portfolio, by explaining your reasoning.
   iii. Briefly comment on the consistency (or lack thereof) of the results in (i) and (ii).

2. (*) The Venn diagram below shows symbolically a portfolio at two successive underwriting years. The shaded area (intersection) $\wp' \cap \wp$ is made of the contracts that have been renewed; the area $\wp - \wp'$ is made of the contracts that have lapsed; $\wp' - \wp$ is made of new business.

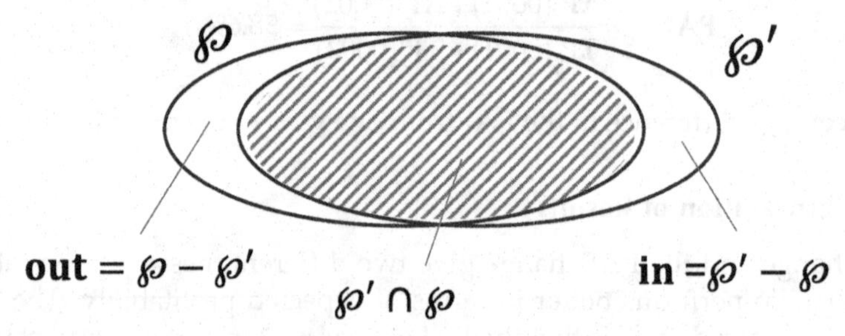

   i. Prove that the change in premium adequacy between $\wp$ and $\wp'$ can be approximately written as:

$$\delta\mathbf{PAI}(\wp \to \wp') \approx \%\mathbf{RC} + \delta\mathbf{PAI}'(\wp \cap \wp' \to \mathbf{in}) \times \%\mathbf{in} - \delta\mathbf{PAI}(\wp \cap \wp' \to \mathbf{out}) \times \%\mathbf{out}$$

where:

- $\delta\mathbf{PAI}(\wp \to \wp') = \mathbf{PAI}'(\wp') - \mathbf{PAI}(\wp)$ is the difference in premium adequacy between the two portfolios;
- **%RC** is the percentage rate change for the matching portfolio;
- $\delta\mathbf{PAI}'(\wp \cap \wp' \to \mathbf{in}) := \mathbf{PAI}'(\mathbf{in}) - \mathbf{PAI}'(\wp \cap \wp')$ is the difference in premium adequacy between the new business and the common portfolio;
- $\delta\mathbf{PAI}(\wp \cap \wp' \to \mathbf{out}) = \mathbf{PAI}(\mathbf{out}) - \mathbf{PAI}(\wp \cap \wp')$ is the difference in premium adequacy between the lapsed portfolio and the common portfolio;

- $\%\mathbf{in} = \mathbf{AP'(in)} / \mathbf{AP'}(\wp')$ is the percentage (in premium) of the current portfolio that is due to new business;
- $\%\mathbf{out} = \mathbf{AP(out)} / \mathbf{AP}(\wp)$ is the percentage (in premium) of the expiring portfolio that has lapsed.

ii. Give an interpretation of the formula in (i) and explain why it is not an exact relationship.

3. *(For some readers, this problem might be better tackled after reading Chapter 22.)* Petrochemical company X insures its single oil refinery with insurer Y and its property damage insurance has just been renewed and will incept 1 May 2023 (2023 policy year). Traditionally Y's pricing guidelines prescribed from-the-ground-up premium rate of 0.1% of the total insured value, but a recent recalibration with a richer data set has found that a rate of 0.12% better reflects the risk (which itself hasn't changed over recent years).
Assuming that
   - the full value of the refinery is insured for both 2022 and 2023;
   - the insured value has been revised from $800M (2022) to $850M (2023);
   - the actual premium (AP) charged was $900k in 2022 and has been renewed for $1.05M;

   i. Calculate the rate change between the 2022 and the 2023 policy years (both incepting 1 May), explaining your strategy to deal with the change to the recommended premium rate.
   ii. Calculate the premium walk from AP(2022) to AP(2023) breaking down the contribution (where applicable) of exposure change, policy structure change, and rate change.
   iii. Explain how your answer to parts (i) and (ii) would change if the increase in the premium rate from 0.1% to 0.12% were due to the deployment of new, less consolidated technology in policy year 2023 (ignore exact timing) which is expected to increase the overall frequency of losses by 20% and revise your calculations accordingly.

   Your risk management department undertakes a pricing review for the property business, and notes that using premium rates rather than expected loss rates could fail to differentiate clients with less volatility (e.g. chains of supermarkets) from clients with more volatility (e.g. single refineries like in this case), for which a higher loss ratio should be required.

   iv. Explain how you could modify the pricing methodology in order to take that into account, giving an example of a formula that you could use to calculate the technical premium for a generic client with k different locations with insured values $IV_1, IV_2 \ldots IV_k$.

Part III

Business-Specific Pricing

21

Experience Rating for Non-Proportional Reinsurance

The objective of this section is to show how we can adapt the risk-costing process that we developed in the previous chapters for the purpose of pricing non-proportional reinsurance, especially in the context of treaty reinsurance. Note that proportional reinsurance, because it just cedes a proportional amount of each risk to the reinsurer, will not be significantly different from direct insurance and therefore does not require a separate treatment.

Non-proportional treaty reinsurance pricing[1] has some peculiarities we need to cater to. For example,

i. The policy basis may be 'losses occurring during' (LOD; not different from the standard occurrence policy of direct insurance), 'claims made' (as in direct insurance) and 'risk attaching during' (RAD; which has no equivalent in direct insurance)
ii. Only losses above a certain agreed threshold, called the *reporting threshold* (e.g. £500,000) will be reported to the reinsurer to price layers well above that threshold (e.g. £3M xs £2M). This is sometimes also true for commercial insurance, but the threshold is much lower and less of a concern for the analysis
iii. Specific exposure alignments and adjustments are needed to take into account the difference between 'written premium' and 'earned premium' and also to take into account rate changes of the original direct policy
iv. It is more common in reinsurance to receive information on the historical development of each claim
v. The presence of the indexation clause leads to the need to consider the settlement pattern (London Market clause) and the payment pattern (continental clause) to estimate the effect of the changes in layer definition
vi. The presence of long payment delays for large liability losses also leads to the need to consider the payment pattern to set premiums that take investment income into account (as is also the case for direct insurance but to a lesser degree)

Let us look at each of these peculiarities in the context of the risk-costing process and see how they lead to redefining the risk-costing process as in Figure 21.1.

1 Some of the points below (i, iii, v) do not normally apply to facultative reinsurance.

DOI: 10.1201/9781003168881-24

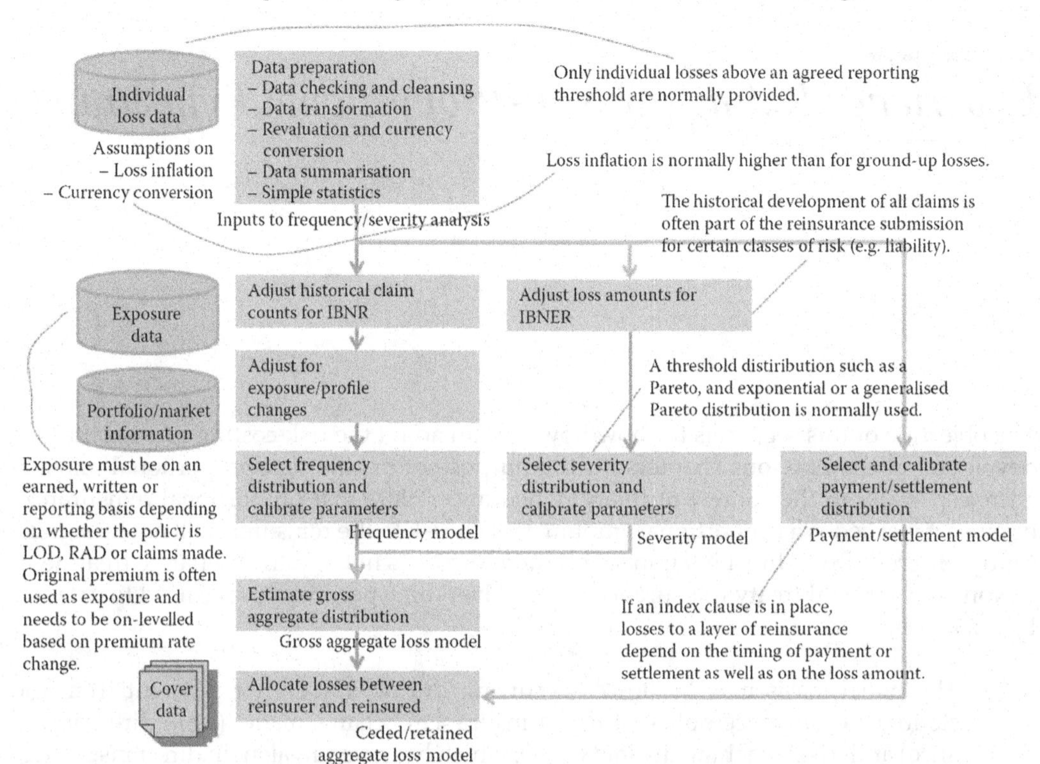

FIGURE 21.1
The revised risk-costing (sub)process for reinsurance pricing. The main difference with the process of Figure 6.1 is the presence of a third branch for the creation of a payment/settlement pattern model, which is used to deal with the index clause when this features in the reinsurance contract (as is normally the case in treaty reinsurance for liability classes). However, the process needs to be tweaked almost everywhere to accommodate the idiosyncrasies of reinsurance pricing.

21.1 Types of Contract

As we have seen in Section 3.3, there are several types of contract in reinsurance, depending on which losses are covered by the policy. As a reminder, the most widely used types of contract are

- LOD, which covers all losses occurring during the specified period, regardless of the original policy's inception date and of the loss reporting date
- RAD, which covers all losses attaching to policies that have incepted during a specified period, regardless of occurrence and reporting date
- 'Claims made', which covers all losses reported *to the reinsured* during the reinsurance policy period, regardless of the original policy's inception date and the loss occurrence date

Claims made is probably the least used of the three. It is an important basis in direct insurance for some lines of business such as directors' and officers' liability (D&O) and

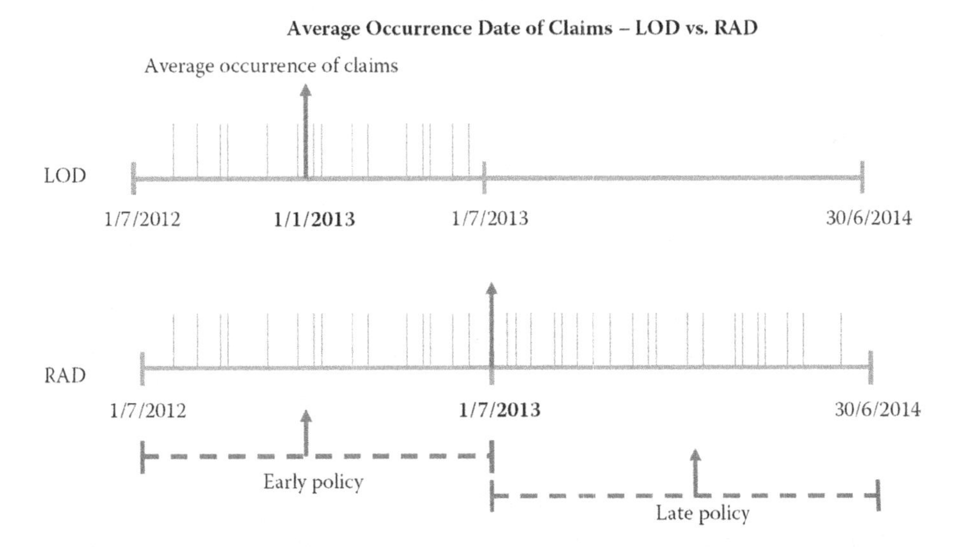

FIGURE 21.2
The average occurrence of a claim for a risk attaching policy is shifted by about 6 months from that of a claim for a losses-occurring policy with the same policy period. This is a pictorial explanation of why this is so.

professional indemnity (PI). We will look at it again in Chapter 24, where we will review professional indemnity insurance in more detail.

LOD and RAD are both popular policy bases, and we will focus here on the difference between these two. The distinction between LOD and RAD has consequences, for example, on the revalue-at date: for RAD, it should be approximately 6 months ahead (at least under the assumption that losses occur uniformly over the year, and that policies also incept uniformly over the year), as illustrated in Figure 21.2.

21.2 Inputs

The input to the reinsurance costing process is similar to direct insurance: claims data, exposure data, and insurance coverage data. However, there are some differences in the format of these data sets.

21.2.1 Policy Information

The most important data for treaty reinsurance policies is as follows:

- Policy period
- Type of reinsurance (such as quota share, excess of loss, aggregate excess of loss, catastrophe excess of loss)
- Basis of cover (RAD, LOD, claims made, etc.)
- Treaty limits and excess points, individual and on aggregate (including information on the reinstatements structure) and reinsurer's share of losses (where applicable, as in quota share and surplus treaties)

- Details on indexation clause (where applicable)
- Details on hours clause and sunset clause (catastrophe cover)

21.2.2 Claims Data

When pricing non-proportional reinsurance, and especially in the case of Risk XL, in which only losses above a certain attachment point/deductible trigger a payment, only claims data above a certain level, called the *reporting threshold* (or the reporting level), are communicated by the reinsured to the reinsurer.

More specifically, if the reporting threshold is RT (e.g. £500,000), the data set will be made of all claims $X_1, \ldots X_N$ such that $X_j > \text{RT}$ at some point during the reserving history of the claim (note: the claim *must* be reported if *at some point* it was reserved as an amount larger than RT. It does not matter that the claim later fell below the threshold or even to zero: the reinsurer will want to see it anyway).

The existence of a reporting threshold has an effect on the level above that we can carry out a frequency and severity analysis, the *analysis level*. To see why, let us make an example.

Assume that your reporting threshold is RT = £500,000 and that you have claims over a period of 10 years (2002–2011). You are asked to price a Risk XL reinsurance policy for year 2012. You normally use a claims inflation of 8.5%.

To produce a severity model, you might be tempted simply to revalue all claims using 8.5% claims inflation and then fit a curve through those points. However, if you do that, you introduce a bias in the analysis.

The reason why this is the case is that all losses below the reporting threshold of £500,000, which *would* be above the threshold once they have been revalued to current terms, are excluded from the analysis data set. For example, a loss of £350,000, which occurred in 2002 (let us say halfway through the policy year), would now be revalued to £350,000 × 1.085^{10} ~ £791,000, which is well above £500,000. At the same time, a loss of £510,000 in 2011 would be included in the data set although it will only be revalued to £553k, well below the loss (£791k), which has been excluded.

Overall, this results in losses between £500k and £500k × 1.085^{10} ~ £1.13M, the amount to which a loss that just made it into the data set in the earliest year (2002) would be revalued, being underrepresented in the severity distribution.

The solution to this problem is to set the analysis level such that this problem disappears, and only analyse (for both frequency and severity purposes) losses that are above that analysis level. In the example above, this is any level above £1.13M because a loss that in current terms would be larger than that would certainly appear in the data set.

In general terms, this is how we calculate the analysis threshold: if

RT = reporting threshold
r = claims inflation
RY = renewal year
FY = first year from which losses are available

Then the analysis threshold AT must be set for *losses occurring* policies at a monetary level such that:

$$\text{AT} \geq \text{RT} \times (1+r)^{\text{RY}-\text{FY}+\frac{1}{2}} \tag{21.1}$$

where the 1 / 2 was included because a loss occurring *at the beginning* of the first policy year available would need to be revalued *to the midpoint* of the renewal policy year (assuming that claims occur uniformly over the policy year).

In the case of risk attaching policies, the mean occurrence date (after the usual assumption of claims occurring uniformly over the period and that the original policies are written uniformly over the period) would be *at the end* (rather than at the midpoint) of the renewal policy year and therefore Equation 21.1 should be modified as follows:

$$\mathrm{AT} \geq \mathrm{RT} \times (1+r)^{\mathrm{RY}-\mathrm{FY}+1} \tag{21.2}$$

Box 21.1 deals with a situation that you often encounter in reinsurance, where it is often the case that the reporting threshold is set at 50% of the value of the attachment point of the policy (the deductible of the lowest layer), exactly to deal with this bias. However, 50% may be too high if claims inflation is substantial, as the following worked problem shows.

BOX 21.1 WHAT TO DO WHEN THE REPORTING THRESHOLD IS TOO HIGH

Let us assume as before that the reporting threshold is £500k, that you have losses for the period 2002 to 2011 and that you are pricing a £1M xs £1M layer on a losses occurring basis renewal year 2012. Assuming that you believe that the correct claims inflation is 8.5% per annum, you realise that Equation 21.1 requires an analysis level of at least £1.13M. What are you going to do?

SOLUTION 1: REDUCE THE ANALYSIS PERIOD

One solution is to reduce the analysis period (e.g. considering losses from 2004 only) – so that the difference between RY and FY is reduced and Equation 21.1 becomes valid. This workaround has the unhappy consequence that you have to discard some of the data, and reinsurance losses are already sparse enough as they are. However, the resulting analysis is at least unbiased.

SOLUTION 2: 'EXTRAPOLATE BACK' THE SEVERITY DISTRIBUTION

Another solution is to carry out the analysis for more than £1.13M and then 'extrapolate back' the severity distribution (and as a consequence, gross up the frequency) to start from £1M, assuming that the shape of the distribution remains the same. If you are only rolling back by a small amount (e.g. £130,000 in our case) you are unlikely to lose much accuracy.

How do you roll back a severity distribution in practice? For the generalised Pareto distribution (GPD; the most important distribution for reinsurance), things work as follows

μ = the minimum AT at which *I can* do the analysis, in our case £1.13M
ξ = the shape parameter calculated using all revalued losses above μ
σ = the scale parameter calculated using all revalued losses above μ
μ' = the threshold at which *I would like* to do the analysis ($\mu' < \mu$), in our case £1M
ξ' = the new, 'extrapolated back' shape parameter
σ' = the new, 'extrapolated back' scale parameter

The parameters for the extrapolated-back distribution are related to the distribution above μ by the following relationship (see Question 1):

$$\xi' = \xi$$

$$\sigma' = \sigma + \xi \times (\mu' - \mu)$$

And as a consequence, the frequency should be increased like this:

$$\nu' = \nu \times \left(\frac{\sigma}{\sigma'}\right)^{1/\xi}$$

(note that $\sigma' < \sigma$ if $\mu' < \mu$).

Similar formulae (actually, simpler formulae) can be obtained for other threshold distributions such as the single-parameter Pareto and the exponential (see Question 1).

SOLUTION 3: DECREASE THE REPORTING THRESHOLD

The best solution of all is probably to go back to the reinsured and ask for all claims in excess of a lower reporting threshold, say RT = £400k. This may not be feasible when you have already agreed on a reporting threshold with the client, but you can ask it for next year so that next year you will not have the same problem.

21.2.3 Exposure Data

In the case of reinsurance, the reinsurer has two options as to what measure of exposure to use to model a given class of business:

1. Use the original exposure of the reinsured's clients (such as number of employees or wageroll for employers' liability, vehicle years for motor, turnover for public liability and sum insured for property), after aggregating it at the reinsured's level (that is, summing over all the relevant clients of the reinsured).
    a. For example, if reinsurer *X* is reinsuring the motor portfolio of insurer *Y*, and insurer *Y* has clients $Z_1, \ldots Z_k$ each with exposure $VY_1, \ldots VY_k$, then the exposure of *Y* is $VY_1 + \ldots + VY_k$
2. Use the overall premium originally charged by the reinsured as a proxy for the exposure, because the reinsured has an interest in pricing its clients proportionally to the risk they pose.
    a. For example, if reinsurer *X* is reinsuring the EL portfolio of insurer *Y*, and insurer *Y* has clients $Z_1, \ldots Z_k$ to which it has charged premiums $P_1, \ldots P_k$, then the exposure of *Y* is $P_1 + \ldots + P_k$

The main advantage of using original premium as a measure of exposure is that if the direct insurer has priced its business consistently across its portfolio and over time, the original premium may be a better measure of exposure than any direct measure of exposure because it already incorporates the factors that make one risk more expensive than another.

There are, however, several disadvantages as well:

- There is no guarantee that the insurer is pricing its business consistently. Pricing is a commercial decision that depends on the negotiation with each client (especially for large contracts), on competition, and in particular on the underwriting cycle.
- Whenever two insurers merge and seek to purchase reinsurance for the joint portfolio, simply summing the total average premium for the two insurers may not be sufficient as the two insurers might have different pricing methods and commercial attitudes.
- Because original premium is a monetary amount, it needs to be corrected for rate changes (if the reinsurer decides to halve the price of its portfolio, the reinsurer will not halve the price of reinsurance if it considers that the risk has remained exactly the same and that the insurer is only trying to hold on to its market share).
- What makes one risk more expensive from the ground up (and is hence reflected in the original premium) may not be the same as what makes one risk more expensive in the large-loss region – this is, arguably, only a fine point, but it is worth keeping in mind.

21.2.3.1 Examples

- In the United Kingdom, vehicle years is the most common exposure measure for motor reinsurance
- Total maximum possible loss (MPL) or total sum insured are common exposure measures for property
- For employers' and public liability, it is more common to use original premium as a measure of exposure, as there is a large variation between companies as to the risk they pose in proportion to turnover or payroll, or any other likely monetary measure of exposure. Also, in this way, we take into account the fact that companies buy different limits.

21.3 Frequency Analysis

Frequency analysis is performed much in the same way as for direct insurance. The main difference is that we are looking here to the number of claims above the analysis threshold, rather than the number of claims from the ground up. Another (less important) difference is that the treatment of exposure is slightly more complicated, especially when original premiums (i.e. the premiums charged by the direct insurer to its clients) are involved.

As usual, before exposure can be used meaningfully to compare the number of claims coming from different policy/accident years, adjustments need to be made. One such adjustment is to make sure that the exposure used is adequate for the contract basis (Section 21.3.1). Another such adjustment, which is necessary when the exposure is a monetary amount, is to 'on-level' the exposure to reflect the current environment. We have learnt how to do this earlier on but, in the case of reinsurance, there are some peculiarities when the exposure is the original premium (Section 21.3.2).

21.3.1 Exposure and Contract Basis

Regardless of whether the exposure is the original premium or the original exposure that the direct insurance uses, the exposure needs to be properly matched to the claims experience, and how this is done depends on the contract basis.

For RAD policies, the correct exposure is the premium (or, e.g. the number of vehicle years) *written* during the reinsurance policy period n, WP(n). This can be matched to the claims *attaching* to policy year n, that is, the claims that are covered by the policy in that year.

For LOD policies, the correct exposure is the premium *earned* during the reinsurance policy period n, EP(n). This can be matched to the claims *incurred* during the policy period n, that is, the claims that have occurred during that time.

It is often the case that we are not given the earned premium but that we are only given the premium written by the reinsured during a certain period, WP(n). Under the assumption that the premium is written uniformly over the year and that the earning pattern is also uniform – that is, claims occur uniformly over the period of the policy – we can then write the earned premium as the arithmetic average of the written premium of two consecutive years (see also Section 11.4):

$$\mathrm{EP}(n)=\frac{\mathrm{WP}(n-1)+\mathrm{WP}(n)}{2} \tag{21.3}$$

The same principle applies for other types of exposure, such as vehicle years for motor: we can replace 'WP' with 'WVY' (written vehicle years) and 'EP' with 'EVY' (earned vehicle years) in Equation 21.3.

21.3.2 Incorporating Rate Changes When Using Original Premium as Exposure

When the exposure is a monetary amount (such as wageroll, turnover, or original premium), the historical exposure must be adjusted to current terms, or 'on-levelled', to make a like-for-like comparison between the number of claims of different years possible. We have already seen how to on-level exposure of the monetary variety in previous chapters.

In the case of original premiums, this is most usefully obtained by taking into account the rate changes that the original insurer (the reinsured) has implemented over the years; these are usually well documented because they are part of the reinsured's strategy.

21.3.2.1 Worked-Out Example

Assume that in 2011, the insurer was writing £1.25M of insurance and the year after (2012), it was writing £1M and that the claims attached to these years are shown in Figure 21.3.

The temptation would be to say that the risk has gone down from 2011 to 2012 because of the decrease in exposure. However, it might well be the case that the insurer has exactly the same clients with the same risks but has applied an across-the-board 20% decrease in its premiums.

Having this information, we realise that taking 20% of £1.25M away from £1.25M, we have exactly £1M; therefore, the exposure has not changed at all. The risk may have remained exactly the same but the insurer just gets less money for it.

To take this into account, we adjust for the rate change by creating a rate index that starts at 100 and changes depending on the rate change information. This rate index works

Policy year	Original premium	Rate change	Number of claims
2011	£1.25M	N/A	8
2012	£1M	-20%	7
2013 (est.)	£1M	0%	N/A

FIGURE 21.3
Example of input to reinsurance burning cost with original premium as an exposure measure.

Policy year	Original premium	Rate change	Rate index	On-levelled premium	Claims attached
2011	£1.25M	N/A	100	£1M	8
2012	£1M	-20%	80	£1M	7
2013 (est.)	£1M	0%	80	£1M	N/A

Frequency unit of exposure (2011–12) = (8 + 7)/(£1M + £1M) = 7.5 claims/£M

FIGURE 21.4
Example of frequency calculations for a RAD policy with original premium as an exposure measure.

exactly like any other inflation index and it allows us to on-level all exposures to renewal terms in the following way:

$$\text{Onlev Exp}(2011) = \text{Exp}(2011) \times \frac{\text{rate index}(2013)}{\text{rate index}(2011)} = £1.25\text{M} \times \frac{80}{100} = £1\text{M}$$

which reveals that the exposure hasn't changed at all since 2011.

The calculations for all years are shown in Figure 21.4. The same reasoning can be applied for a burning cost calculation of the total loss experience (rather than the number of claims) – see, for instance, Question 4.

21.3.3 IBNR Analysis

IBNR analysis works very much like in direct insurance. However, in Risk XL pricing, we count the number of claims from a particular cohort (such as policy year, or occurrence year) that, on a revalued basis, are exceeding the threshold.

Therefore, if the reserve for a specific claim is revised downward to the point that it drops below the threshold, it is also excluded from the claim count. This is analogous to the case of claims settling as zero after being included with a non-zero reserves that we dealt with in Chapter 13.

21.3.4 Selection of a Frequency Model

The selection of a frequency model works much in the same way as in the direct case (Chapter 14).

21.4 Severity Analysis

The approach to severity analysis is not too different from the standard risk-costing process, with the main differences coming mostly from the fact that we are dealing here mostly with large losses.

21.4.1 IBNER Analysis

Theoretically, IBNER analysis is not any different from the direct insurance version. However, in reinsurance, it is the norm and not the exception to look at how individual claims develop, for example, with Murphy–McLennan's method (see Section 15.2.3). Because there are relatively few losses for a Risk XL policy, it is customary to provide detailed information on the claims in the form of development triangles or similar structures.

You might remember that in Method 3a in Section 15.2.3, we only calculated the IBNER factors between successive loss values that were both above zero because the case of losses dropping to zero or going from zero to a finite value already contributed to the IBNR calculations for non-zero claim counts.

In excess-of-loss reinsurance, we do something analogous, except that 'above zero' is replaced with 'above the analysis threshold': therefore, in the calculation of the IBNER factors, we only consider development between loss values that are both above the analysis threshold. Movements between below-threshold and above-threshold loss values will increase or decrease the claim count, and will therefore be taken into account by the IBNR analysis of above-the-threshold claim counts. This is illustrated in Figure 21.5.

21.4.2 Selecting and Calibrating a Severity Model

We are interested only in losses above a certain threshold, and therefore we normally use 'threshold' distributions only. There is no formal definition of a threshold distribution, but the idea is that a threshold distribution assumes that only losses above a certain value are possible, and that the probability density is decreasing (which makes intuitive sense). Examples of threshold distributions are the exponential, Pareto (single- or two-parameter), and GPD. Occasionally, a shifted lognormal distribution, that is, a distribution such that $X - \mu$ (μ being the threshold) is a lognormal distribution, is used, but this is not a recommended choice because the decreasing-probability property is not satisfied.

	Development year, n						
	0	1	2	3	4	5	6
Revalued loss amount	15,000	245,000	**650,000**	**700,000**	**720,000**	**450,000**	450,000
Paid	0	0	**0**	**100,000**	**100,000**	**100,000**	450,000
IBNER factor ($n - 1 \rightarrow n$)	N/A	16.33	2.65	**1.08**	**1.03**	**0.63**	1.00
Contribution to claim count	0	0	1	1	1	0	0

FIGURE 21.5
Example of calculation of the IBNER factors for an individual claim, in the case of an analysis threshold of £500,000. The IBNER factor between two successive development years $n - 1$, n contributes to the overall IBNER calculations only if the incurred amount is above the analysis threshold both in year $n - 1$ and in year n. This is consistent with the fact that the claim contributes to the claim count only when the incurred claim amount is more than £500,000, that is, in development years 2, 3, and 4, which affects the IBNR analysis.

Despite the fact that, historically, many different distributions have been used to model large losses, the GPD is becoming the standard choice. As we have seen in Section 16.3, if the threshold is large enough, extreme value theory *prescribes* that we should use a GPD for modelling large losses. If a GPD does not work over the full range of severities, we might want to use a GPD/GPD model (i.e., two different GPDs at two different thresholds), or an empirical/GPD model, much in the same way as we did for ground-up losses.

Given the scarcity of data for large losses, it is often the case that market data need to be used to price layers above a certain level because of the 'pricing horizon' phenomenon (Section 16.3).

21.5 Payment and Settlement Pattern Analysis

A novelty with respect to direct insurance pricing is the index clause, which makes it important to calculate when a claim is actually paid/settled. Note that we speak of 'payment/settlement' because whether one needs the payment pattern or the settlement pattern to model the index clause depends on the type of cause (such as London Market or continental). Furthermore, the payment pattern is often used to determine the settlement pattern and is used independently to determine investment income.

21.5.1 Settlement Pattern

When using the London Market clause, the reference date for a lump sum compensation is the settlement date. Because we have the history of development for each claim (which is shown in Figure 15.3), we also have the information on what percentage of each claim was paid in each development year. Based on this information, we can build a settlement pattern such as that in Figure 21.6, which gives the percentage of claims settled by development year (development year = number of years from reporting). We will use this later for simulation purposes.

Note that there is a difficulty here: we normally look at claims over a period of up to 10 years, but the maximum settlement time for a claim may be 20 years or more. For this reason, it is important to have some model (possibly based on a large collection of claims, and over a longer period) that tells us what the general shape of the curve looks like, and possibly fit it with a simple model such as a log-logistic:[2]

$$F(x)=\frac{(x/\theta)^{\gamma}}{1+(x/\theta)^{\gamma}} \tag{21.4}$$

Note that the parameter θ is also the median of the distribution. The mean of the distribution is $\mathbb{E}(X)=\theta\frac{\pi/\gamma}{\sin(\pi/\gamma)}$ $(\gamma>1)$ and the value corresponding to the p-th percentile is

$$F^{-1}(p;\theta,\gamma)=\theta\left(\frac{p}{1-p}\right)^{\frac{1}{\gamma}}$$

2 As usual, using a sensible distribution based on some prior idea of what the distribution should look like is better than picking a distribution from a distribution-fitting tool that just happens to fit your specific data set.

Note also that there is no need to correct for the bias arising out of the limited observation window (as we did in Chapter 13 for IBNR analysis) because the statistics are based on reported claims and therefore include *all* reported claims within the observation window. (Enlarging the observation window does however improve the accuracy of the calculation because we will have a larger set of reported claims to base our statistics on.)

21.5.2 Payment Pattern

This is more complex because the payment pattern will in general be different depending on the settlement time. We may need a payment/settlement matrix that gives a payment pattern for each settlement time. Calibrating this matrix, however, requires plenty of data and is normally impossible to obtain at the client level – analysis of portfolio data will be necessary.

Alternatively, we can merge all settlement times together and produce a payment pattern on the basis of the estimated ultimate incurred amount. This is perhaps a more common approach and would also benefit a portfolio-level analysis. The payment pattern will have a shape similar to that of Figure 21.6. The payment pattern is important not only for the continental index clause but also because it allows us to take investment income into account properly in pricing. This is especially important in excess-of-loss liability reinsurance given that the average settlement time of liability losses can be approximately 7 to 8 years.

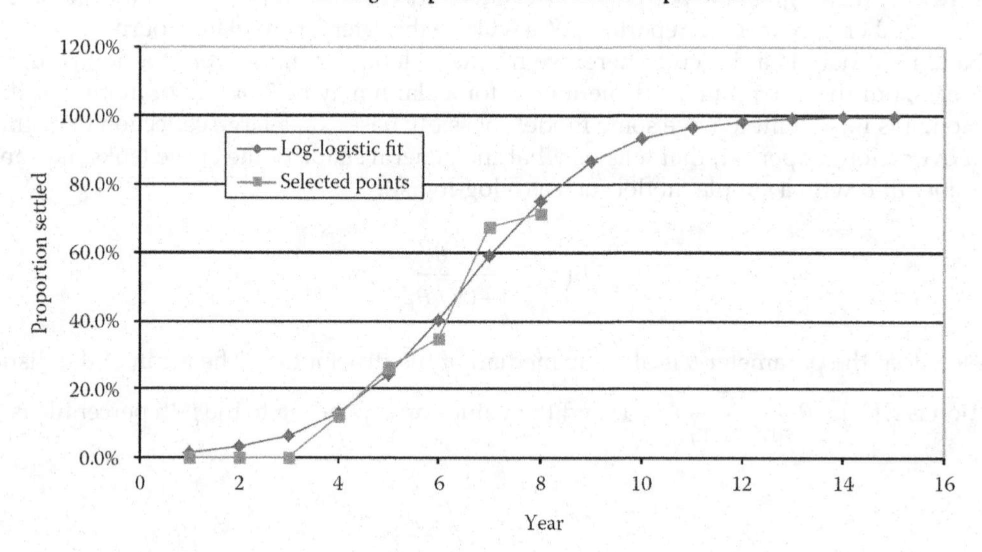

FIGURE 21.6
An example of a settlement pattern. The x axis has the number of years from reporting, while the y axis gives the proportion of claims already settled, both observed and modelled: so for example we see that the out of all claims with reporting year 6 (i.e. all claims reported within 6 years from inception) the percentage settled observed was about 35%.

21.6 Aggregate Losses to an Excess Layer

After looking at how frequency and severity modelling need to be adjusted to the needs of reinsurance modelling, we are now in a position to produce an aggregate loss model. Before looking at how this can be achieved in practice using the Monte Carlo simulation or other numerical techniques, let us see what we can achieve analytically, that is, based on formulae.

In the case of a ground-up collective risk model, we saw that the expected losses could be written as $\mathbb{E}(S) = \mathbb{E}(N)\,\mathbb{E}(X)$, where N, X, and S are random variables representing the number of ground-up losses, the loss amount, and the overall losses, respectively.

How does the collective risk model help us when we are pricing a layer of reinsurance L xs D, or $(D, D + L)$ *without any aggregate deductible/limit*? Let us first consider the case where there is no index clause. Given a loss X, the loss to the layer $(D, D + L)$ is

$$X(D,L) = \min(X, D+L) - \min(X, D) \qquad (21.5)$$

The expected loss to the layer if an individual loss has occurred is therefore

$$\mathbb{E}(X(D,L)) = \mathbb{E}(\min(X, D+L)) - \mathbb{E}(\min(X, D)) \qquad (21.6)$$

If N is the number of losses from the ground up, the expected *total* losses to the layer are

$$\mathbb{E}(S(D,L)) = \mathbb{E}(X(D,L)) \times \mathbb{E}(N) = (\mathbb{E}(\min(X, D+L)) - \mathbb{E}(\min(X, D)) \times \mathbb{E}(N) \qquad (21.7)$$

This formula, which is the standard formula for the expected losses to a layer of reinsurance, is not used in practice because the expected claim count from the ground up ($\mathbb{E}(N)$) is not normally known. Because losses are reported only above a certain threshold (the *reporting threshold*), frequency can be estimated only above another, higher threshold called the *analysis threshold*. Equation 21.7 it therefore more usefully rewritten as

$$\mathbb{E}(S(D,L)) = \frac{(\mathbb{E}(\min(X, D+L)) - \mathbb{E}(\min(X, D)))}{\Pr(X > a)} \times \mathbb{E}(N_a) \qquad (21.8)$$

where a is the analysis threshold, $\mathbb{E}(N_a) = \mathbb{E}(N)\Pr(X > a)$, is the number of claims expected to exceed the analysis threshold each period, and the ratio $(\mathbb{E}(\min(X,D + L)) - \mathbb{E}(\min(X,D)))/\Pr(X > a)$ is the average severity conditional on the severity being above the analysis threshold. Crucially, this conditional average severity can be calculated without knowing $\Pr(X > a)$ (which we do not normally know), based on the severity model above a:

$$\mathbb{E}(S(D,L)) = (\mathbb{E}(\min(X, D+L) | X > a) - \mathbb{E}(\min(X, D)) | X > a) \times \mathbb{E}(N_a) \qquad (21.9)$$

Section 21.6.1 has examples of Equation 21.9 calculated analytically for the most popular severity distributions used in XL reinsurance (Pareto, GPD). It also provides formulae for the variance of losses to a layer based on Equation 6.4. The full distribution cannot be obtained analytically even for these simple cases and numerical techniques need to be used (Section 21.6.2).

Contracts with an indexation clause

Things get a bit more complicated if the contract has an index clause. A first approximation of this case can be obtained by simply assuming that all losses will be settled after τ_s years (the mean settlement time) and that the index to which the layer is linked increases as compound interest does with an average wage inflation index equal to $w{:}I(\tau_s) = I(0)\times(1+w)^{\tau_s}$, where $t = 0$ is the time at which the policy incepts. Under this approximation, $\mathbb{E}(S(D,L))$ can be written as

$$\mathbb{E}(S(D,L))\,|_{\text{index clause}} \approx \mathbb{E}(X(D\times I(\tau_s), L\times I(\tau_s))|X>a)\times\mathbb{E}(N_a) \tag{21.10}$$

A more accurate estimation of the expected losses to the layer in the case where an indexation clause is present can be obtained through simulation (Section 21.6.2).

21.6.1 Analytical Calculation of Expected Losses and Variance for a Layer

For some selected severity distributions such as the commonly used Pareto distribution and its generalisation used in EVT, the GPD, the expected losses of Equation 21.9 can be calculated analytically using the following formulae. (Note again that we are assuming that no aggregate deductible/limit is present.)

For the GPD case with parameters μ (threshold), ξ (shape) and σ (scale), Equation 21.9 becomes:

$$\mathbb{E}(S(D,L)) = \mathbb{E}(N_\mu)\times\begin{cases} \dfrac{\sigma}{1-\xi}\times\left(\left(1+\dfrac{\xi}{\sigma}(D-\mu)\right)_+^{1-\frac{1}{\xi}} - \left(1+\dfrac{\xi}{\sigma}(D+L-\mu)\right)_+^{1-\frac{1}{\xi}}\right) & \xi\neq 0,1 \\ \sigma\times\left(e^{-\frac{D-\mu}{\sigma}} - e^{-\frac{D+L-\mu}{\sigma}}\right) & \xi = 0 \\ \sigma\times\ln\left(\dfrac{1+\dfrac{D+L-\mu}{\sigma}}{1+\dfrac{D-\mu}{\sigma}}\right) & \xi = 1 \end{cases} \tag{21.11}$$

where $\mathbb{E}(N_\mu)$ is the number of claims above μ and $(x)_+$ is shorthand for $\max(0,x)$. Note that Equation 21.11 is valid for all values of ξ, and for $D>\mu$ (the layer must be above the threshold).[3] The proof of Equation 21.11 can be found in the solution to Question 3.

In the case of the single-parameter Pareto (defined above θ, for all $\alpha>0$) this simplifies to:[4]

$$\mathbb{E}(S(D,L)) = \mathbb{E}(N_\theta)\times\begin{cases} \dfrac{\theta}{\alpha-1}\times\left(\left(\dfrac{\theta}{D}\right)^{\alpha-1} - \left(\dfrac{\theta}{D+L}\right)^{\alpha-1}\right) & \alpha\neq 1 \\ \dfrac{\theta}{\alpha}\times\ln\left(\dfrac{\theta(1-\alpha)+\alpha(D+L)}{\theta(1-\alpha)+\alpha D}\right) & \alpha = 1 \end{cases} \tag{21.12}$$

3 Although we have given the formula for all values of ξ, only the $\xi\neq 0,1$ case is needed in practice, as the cases $\xi = 0$ and $\xi = 1$ can be approximated as accurately as needed by using $\xi = \varepsilon$ and $\xi = 1+\varepsilon$ with a very small ε.

4 Remember that a GPD reduces to a single parameter Pareto if you set $\mu = \theta$, $\xi = 1/\alpha$, $\sigma = \theta/\alpha$.

where $\mathbb{E}(N_\theta)$ is the number of claims above θ.

Note that these formulae are valid independently of the frequency model. The frequency model has an impact on the volatility of the losses to the layer, but not on the mean.

For the Pareto and the GPD distribution it is also possible to calculate the variance of the losses to the layers by using Equation 6.4 for the variance $\mathbb{V}(S(D,L))$ of a collective risk model with frequency N_μ (GPD) or N_θ (Pareto) and severity $\mathcal{L}_{D,L}(X)$.

The formula for the simplest case, that of a Pareto severity with a Poisson claim count distribution, is given by:

$$\mathbb{V}(S(D,L)) = 2E(N_\theta)\theta^2 \times \begin{cases} \dfrac{1}{1-\alpha}\left(l(l+d)^{1-\alpha} + \dfrac{d^{2-\alpha} - (d+l)^{2-\alpha}}{2-\alpha} \right) & \alpha \neq 1,\ \alpha \neq 2 \\ d\ln\left(\dfrac{d}{d+l}\right) + l & \alpha = 1 \\ \ln\left(\dfrac{d+l}{d}\right) - \dfrac{l}{d+l} & \alpha = 2 \end{cases} \quad (21.13)$$

where we have written $l = L/\theta, d = D/\theta$ for increased legibility.

The more general case of a GPD severity with a generic frequency distribution is shown in the text of Question 6.

21.6.2 Aggregate Loss Distribution to a Layer via Monte Carlo Simulation

The Monte Carlo simulation to produce a gross (i.e. before reinsurance) aggregate loss model by combining the frequency and severity model works very much like the standard risk-costing process, except that the model is normally not ground-up but above a certain analysis threshold.

The calculation of the losses ceded to the reinsurer, however, may be complicated (in the case of liability reinsurance) by the presence of an indexation clause, which creates a dependency between the layer definition and the time of settlement or payment. In the following, we are going to assume that we need to price the losses to a layer of reinsurance $(D, D + L)$, possibly unlimited $(L = \infty)$.

Therefore, in the case of Risk XL we need these inputs to the Monte Carlo simulation:

1. A *frequency model*, such as a Poisson or a negative binomial distribution representing the number of claims *in excess of the analysis threshold*. The analysis threshold needs to be below the attachment point of the layer, D.
2. A *severity model*, such as a GPD representing the distribution of losses above the analysis threshold.
3. A *payment/settlement pattern model* (a settlement pattern model will suffice in the case of the London Market clause), such as a log-logistic distribution giving the percentage of claims likely to be settled by a given year of development.
4. An assumption on the *future values of the index* used in the indexation clause. Typically a flat future increase assumption ($I(t) = I(0)(1+r)^t$, where $t = 0$ is the time at which the base index is calculated) is used.

Therefore, when we simulate, for example, 100,000 scenarios with the London Market clause, we need for each scenario to:

1. Sample a number n of losses from the frequency model
2. For each of these n losses, sample a random amount from the severity model: $x_1,\ldots x_n$
3. For each of these n losses, x_k $(k = 1,\ldots n)$:
   a. Sample a random settlement time t from the settlement model (t will in general be different for each loss).
   b. Calculate the (expected) value of the attachment point and limit at time t, using the index clause for that date: $D(t) = D \times I(t)/I(0)$, $L(t) = L \times I(t)/I(0)$.
   c. Calculate the loss to the indexed layer: $x_k^{D,L} = \min\left(L(t), \max\left(x_k - D(t), 0\right)\right)$
4. Sum over the n losses to find the aggregate loss for each scenario: $s^{D,L} = \sum_{k=1}^{n} x_k^{D,L}$
5. Repeat Steps 1 through 4 for each of the 100,000 scenarios

The outcome of the 100,000 simulated scenarios, put in ascending order, provides an approximation of the distribution of the aggregate losses to the layer. The mean across of the scenarios provides (theoretically) a better approximation of the expected losses to the layer than Equation 21.11, while obviously the simulation doesn't improve on Equation 21.10 when there is no indexation clause.

A similar algorithm will yield of the distribution of the aggregate losses to the layer for the continental version of the index clause (Question 7). The extension to pricing a *tower* of layers rather than a single layer is trivial. The extension to layers with aggregate deductibles/limits is also straightforward.

21.7 Technical Premium Calculations

Technical premium calculations are more complicated in excess-of-loss reinsurance than for direct insurance for several reasons. Here are some examples:

- For higher layers of reinsurance, there is little activity, and the results of the models are very unreliable and as a result qualitative underwriting considerations play a larger role than in personal and commercial lines insurance
- Cost of capital considerations – that is, the marginal cost of setting capital aside for the extra treaty contract written – are important, especially for higher layers
- In the case of liability business, the payment lag may be very long and therefore the role of investment income is crucial in setting the correct price
- In some classes of business (especially property), there may be reinstatement premiums, and therefore the premium paid over the course of the contract is itself a stochastic variable. The upfront premium needs to be calculated based on the expected value of this stochastic variable and a correction factor driven by the estimated number of reinstatements that are likely to be used.

21.7.1 No Reinstatement Premiums, Index Clause

If no allowance for uncertainty is included, and no retrocession (reinsurance of reinsurance) is included either (therefore no 'Net R/I costs term is necessary), and the London

Market clause is assumed, the technical premium can be written approximately (using a variation of Equation 19.4) as

$$\text{Tech Prem} = \frac{\left(\mathbb{E}\left(\min\left(X,(D+L)\times(1+i)^{\tau_s}\right)\right)-\mathbb{E}\left(\min\left(X,D\times(1+i)^{\tau_s}\right)\right)\right)\times\mathbb{E}(N)\times(1+\text{LAE\%})}{(1-\text{EAC\%})\times(1-\text{Profit\%})\times(1+r)^{\tau_p}} \tag{21.14}$$

where:
LAE% is the percentage of loss allocated expenses
EAC% is the percentage of external acquisition costs
τ_s is the mean settlement time
τ_p is the mean delay between premium receipt and claim payout
i is the assumed future value of the wage inflation index used in the treaty contract
r is the discount rate associated with risk-free bonds of duration approximately equal to τ_p
If the continental clause is used, in Equation 21.14 τ_s should be replaced with τ_p

21.7.2 Reinstatement Premiums, No Index Clause

In certain types of reinsurance, such as property XL, it is common to cap the amount of money that can be recovered from the reinsurer through the system of reinstatements, which has been explained in Chapter 3. The Monte Carlo simulation will have no problem determining the expected losses to a layer no matter how complicated the structure is, for example, if a limited number of reinstatements are present.

However, we also need to take into account that in certain circumstances (especially property), a premium must be paid for the layer to be reinstated. Therefore, the premium that will ultimately be paid is not fixed in advance but is itself a stochastic variable that needs to be simulated.

Analytical formulae have been developed based on making it possible to calculate what premium should be charged upfront depending on the desired reinstatement structure. Note that reinstatements are typically used for short-tail risks only so it is not necessary to consider layer indexation.

If we have a layer L xs D without any aggregate deductible and without reinstatements, and we have losses $X_1, \ldots X_n$, these will yield aggregate losses to the layer $S(D,L)=\sum_{i=1}^{n}X_i(D,L)=\sum_{i=1}^{n}(\min(X_i,D+L)-\min(X_i,D))$ and expected losses of $\mathbb{E}(S(D,L))=\sum_{i=1}^{n}\mathbb{E}(X_i(D,L))$.

If we now introduce the constraint that there will only be k reinstatements to the layer (i.e. an aggregate limit AL = $(1+k)L$), and that for each reinstatement j there will be a reinstatement premium equal to $P_j=\beta_j\times P_0$, where P_0 is the upfront premium and β_j is a number normally between 0 and 1 (extremes included) the expected losses become (see Antal, 2003):

$$\mathbb{E}(S(D,L;0,(k+1)L))=\mathbb{E}\left(\min\left(\sum_{i=1}^{N}X_i(D,L),0\right)\right) \tag{21.15}$$

and the expected overall premium is:

$$\mathbb{E}(P) = P_0\left(1+\sum_{j=0}^{k-1}\frac{j}{L}\mathbb{E}\left(\min\left(\sum_{i=1}^{N}X_i(D,L),(j+1)L\right)-\min\left(\sum_{i=1}^{N}X_i(D,L),jL\right)\right)\right) \quad (21.16)$$

Equation 21.16 can be inverted to calculate P_0 based on the expected overall premium, and a slight modification of Equations 21.15 and 21.16 also allow us to deal with the case where an aggregate deductible is present.

Antal (2003) provides a derivation of these formulae, and also a number of simplified formulae for special cases (such as unlimited numbers of reinstatements with the same reinstatement premiums).

In practice, however, it is more common (and even simpler) to calculate the expected losses and premiums through the Monte Carlo simulation, which also provides the full distribution of losses to the layer.

This can be done as follows: for every simulated scenario, the number of reinstatements that are needed is calculated and the additional reinstatement premium is calculated, so that for a total loss S_i and a total premium P_i is associated with every simulated scenario, which can be expressed as a multiple (in general, a non-integer multiple) of P_0: $P_i = \rho_i \times P_0$ ($\rho_i \geq 1$, because the premium will never be less than the upfront premium). At the end of the simulation, we will therefore have a large number of scenarios, from which an estimate of the expected losses $\mathbb{E}(S(D, L;0,(k + 1)L))$ and of the expected premium $\mathbb{E}[P]$ can be derived from the upfront premium as $\mathbb{E}[P] = \mathbb{E}[\rho] \times P_0$. Figure 21.7 provides a simple illustration of this process.

Note that what the simulation gives is only an estimate of $\mathbb{E}[\rho] = \mathbb{E}[P]/P_0$, not of the total premium or the upfront premium individually.

Technical premium	1,202,809
Upfront premium	575,423

To determine the upfront premium (in the simple case where the premium is based only on the expected losses to the layer and not on the full aggregate loss distribution), the process is like this:

1. Estimate the *expected losses* to the layer via a Monte Carlo simulation.
   - In the example of Figure 21.7, this would be £730,707.
2. Estimate the *technical premium* based on a formula such as the following, which is a simplified version of Equation 21.12 after stripping out the effect of the indexation clause:

$$\text{Technical premium} = \frac{\left(\mathbb{E}(\min(X,D+L)) - \mathbb{E}(\min(X,D))\right)\times\mathbb{E}(N)\times(1+\text{LAE\%})}{(1-\text{EAC\%})\times(1-\text{Profit\%})}$$

   - If, in our example, we assume LAE% = 25%, EAC% = 10%, Profit = 10%, we conclude that the technical premium should be equal to £731K×125%/(90% × 90%) ≈ £1.13M.
3. Estimate the *upfront premium* based on the value of ρ obtained through the simulation.
   - In our example, ρ = 2.09 and therefore the (technical) upfront premium is equal to $TP_0 = TP/\rho$ = £1.13M/2.09 = £540k. Note that we are speaking here of TP_0 and

Determining the Total Premium as a Function of the Upfront Premium (P_0) When Reinstatement Premiums Are Present

Layer limit [£]	500,000
Max # of reinstatements, k	3
Aggregate limit AL=(1+k) L	2,000,000
First reinstatement premium (% of P_0)	100%
Second reinstatement premium (% of P_0)	75%
Third reinstatement premium (% of P_0)	50%

Simulation ID	Total losses to layer before aggregate limit [£]	Total losses to layer, capped at $AL = (1+k)L$[£]	No. of reinstatements	Total premium paid [£]
1	625,058	625,058	1.25	$2.19 \times P_0$
2	0	0	0.00	$1.00 \times P_0$
3	1,108,098	1,108,098	2.22	$2.86 \times P_0$
4	231,979	231,979	0.46	$1.46 \times P_0$
5	2,774,671	2,000,000	3.00	$3.25 \times P_0$
6	1,029,764	1,029,764	2.06	$2.78 \times P_0$
7	0	0	0.00	$1.00 \times P_0$
8	540,318	540,318	1.08	$2.06 \times P_0$
9	1,745,266	1,745,266	3.00	$3.25 \times P_0$
10	26,583	26,583	0.05	$1.05 \times P_0$
Average	808,174	730,707	1.31	$2.09 \times P_0$

FIGURE 21.7
This simple 10-scenario simulation illustrates how the relationship between upfront premium and total premium can be worked out. In each simulation, a certain number of reinstatements is required, and the total premium is calculated as the sum of the upfront premium plus the reinstatement premiums used in that simulation. As an example, in the third simulation the layer (which has a limit of £500,000) needs to be reinstated fully twice (the first time at 100% of the upfront premium, and the second time at 75% of the upfront premium) and partially (22%) the third time, at 50% of the upfront premium. The total premium paid is therefore $P = P_0 + P_0 + 0.75\ P_0 + 0.22 \times 0.50\ P_0 = 2.86\ P_0$. On average, a total premium of $\rho = 2.09$ times the upfront premium needs to be paid.

TP to make it clear that we are speaking about the technical premium and not the premium that will ultimately be charged to the reinsured.

21.8 Questions

1. (*) Prove the formulae given in Box 21.1 to extrapolate back the GPD. By remembering that the single-parameter Pareto distribution and the exponential distribution are but special cases of the GPD, adapt these formulae so that they can be used straightforwardly for these two distributions.
2. Simplify the risk-costing process for Risk XL reinsurance so that it can be used for costing property reinsurance, stripping out all the features that are necessary only for liability reinsurance.
3. (*) Prove analytically Equations 21.11, and prove that it reduces to Equation 21.12 in the case of a single-parameter Pareto.

4. (*) Generalise Equation 21.12 to the case where instead of a single parameter Pareto distribution above θ your distribution is made of a spliced Pareto distribution, i.e. a distribution with this cumulative distribution function:

$$F(x)=\begin{cases} 1-\left(\frac{\theta}{x}\right)^{\alpha} & \theta \le x < \theta' \\ 1-\left(\frac{\theta}{\theta'}\right)^{\alpha}+\left(\frac{\theta}{\theta'}\right)^{\alpha}\left(1-\left(\frac{\theta'}{x}\right)^{\alpha'}\right) & x \ge \theta' \end{cases}$$

Also, prove that this generalisation reduces back to Equation 21.11b when $\alpha' = \alpha$.

5. (*) Prove Equation 21.13 and generalise it to the case where (a) the severity distribution is Pareto but the claim count distribution is not Poisson; (b) the severity distribution is a GPD and the claim count distribution is Poisson; (c) the severity distribution is a GPD and the claim count distribution is not Poisson.
6. (*) Prove that the variance of losses to a layer for the case of a GPD severity with a generic claim count distribution is as follows. The formula is quite complex so we'll need to define a few symbol for shorthand purposes:

$$A_x(\xi)=\left(1+\xi\frac{x-\mu}{\sigma}\right)_+$$

$$\lambda := \mathbb{E}(N_\mu);\ v=\mathbb{V}(N_\mu)$$

It's also a good idea to break down the formula into two terms:

$$\mathbb{V}(S(D,L))=\lambda\sigma^2 V_{\text{Poi}}(D,L)+(v-\lambda)\sigma^2 V_{\text{NonPoi}}(D,L)$$

where the second terms disappears in case the frequency distribution is Poisson. We can now write the two components as:

$$V_{\text{Poi}}(D,L)=\begin{cases} \frac{2L}{\sigma(\xi-1)}A_{D+L}(\xi)^{-\frac{1}{\xi}+1}-\frac{\left(A_{D+L}(\xi)^{-\frac{1}{\xi}+2}-A_D(\xi)^{-\frac{1}{\xi}+2}\right)}{(\xi-1)\left(\xi-\frac{1}{2}\right)} & \xi \ne 0,\frac{1}{2},1 \\ \frac{-4L}{\sigma A_{D+L}\left(\frac{1}{2}\right)}+8\ln\left(\frac{A_{D+L}\left(\frac{1}{2}\right)}{A_D\left(\frac{1}{2}\right)}\right) & \xi=\frac{1}{2} \\ 2\left(\frac{L}{\sigma}\left(\ln\left(e^2A_{D+L}(1)\right)\right)-\ln\left(\frac{\left(eA_{D+L}(1)\right)^{A_{D+L}(1)}}{\left(eA_D(1)\right)^{A_D(1)}}\right)\right) & \xi=1 \\ 2\left(-\frac{L}{\sigma}e^{-\frac{D+L-\mu}{\sigma}}+\left(e^{-\frac{D-\mu}{\sigma}}-e^{-\frac{D+L-\mu}{\sigma}}\right)\right) & \xi=0 \end{cases}$$

$$V_{\text{NonPoi}}(D,L) = \begin{cases} \dfrac{1}{(\xi-1)^2}\left(A_{D+L}(\xi)^{-\frac{1}{\xi}+1} - A_D(\xi)^{-\frac{1}{\xi}+1}\right)^2 & \xi \neq 0, \dfrac{1}{2}, 1 \\ 4\left(\left(A_{D+L}\left(\dfrac{1}{2}\right)\right)^{-1} - \left(A_D\left(\dfrac{1}{2}\right)\right)^{-1}\right)^2 & \xi = \dfrac{1}{2} \\ \left(\ln\left(\dfrac{A_{D+L}(1)}{A_D(1)}\right)\right)^2 & \xi = 1 \\ \left(e^{-\frac{D-\mu}{\sigma}} - e^{-\frac{D+L-\mu}{\sigma}}\right)^2 & \xi = 0 \end{cases}$$

7. Modify the algorithm in Section 21.6.2 for the Monte Carlo simulation for a Risk XL contract layer so that it works for a European Indexation Clause.
8. Company C has purchased risk excess of loss (XL) cover for its property portfolio. A reinsurer R offers a £3M xs £5M layer of reinsurance with these characteristics:
   a. Losses occurring
   b. One-year policy incepting 1 April 2011
   c. Reinsurance premium = £400k
   d. Two reinstatements only, the first is a free reinstatement, and the second is a 50% premium reinstatement
   e. No index clause
      i. Explain what 'losses occurring' means
      ii. Explain what we mean by 'reinstatement' and why reinstatements may be a desirable feature for both the reinsurer and the reinsured
      iii. What is the maximum amount that can be recovered by C from R (show your reasoning)?
      iv. During the policy year, company C has experienced only two losses of more than £5M: £7.5M and £11.7M. Calculate the total amount of reinsurance premium paid by C to R, and the total amount of claims recovered by C from R
      v. Why do you think the index clause is not used in this case? In replying to this, you should mention the purpose of the index clause, but you need not explain how the index clause works
9. You want to adapt the risk-costing process introduced in Section 6.2 so that it works for experience rating of excess-of-loss reinsurance of a liability portfolio, such as employers' liability
   a. Which feature would you need to add so that this process is adequate to deal with the indexation clause, and why?

   Another feature that changes in the case of an XL reinsurance treaty has to do with the individual claims data provided: typically, only claims above a certain threshold R are reported to the reinsurer.

   Assume that a reinsurer is pricing a £4M xs £1M layer for an XL employers' liability treaty for a losses occurring annual policy starting 1 January 2013. The reinsurer asks for all losses that *occurred* over a period of 10 years (let us say, from 1 January 2003 until the data collation date), and estimates a large-loss claims inflation of 10%.
   b. Propose an acceptable reporting threshold, and explain your reasoning.

10. An insurance company X has been writing a mixture of EL and PL business (roughly in the same proportion) over the last 10 years. The summary statistics (for the last 6 years only) are as follows:

Policy year (incepting at 1/1)	Premium *written* during the year (all annual policies)	Rate change with respect to the previous year	Claims incurred during policy year
2006	£1.45M	N/A	£1.53M
2007	£1.50M	+3%	£1.15M
2008	£1.80M	+5%	£1.25M
2009	£1.75M	+7%	£1.67M
2010	£1.90M	-2%	£1.05M
2011	£1.95M	3%	£0.41M

Source: Summary statistics as of 31 December 2011.

The premium expected to be written in 2012 is £2M, and there are no planned rate changes. Based on your experience, you expect that as a result of IBNR, 2011 is only 30% complete, 2010 is 80% complete, 2009 is 95% complete, and the previous years are 100% complete.

What is the expected (incurred) loss ratio for policy year 2012? What are the expected losses?

Remember that the loss ratio is defined as

$$\text{(Incurred)loss ratio} = \text{claims incurred} / \text{premium } \textit{earned}$$

Also note that the loss ratio is defined in terms of the earned premium, whereas the table above gives the written premium. You can assume that the claims inflation is 5%. State all other assumptions you make.

22

Exposure Rating for Property Insurance

Exposure rating is a method to price an insurance policy when the data for a specific client is not sufficient to produce a reliable severity model. It relies on the use of so-called 'exposure curves', which are reengineered severity curves and are usually based on losses from a large number of clients, as collected by large institutions such as Lloyd's of London or Swiss Re. Ultimately, however, the origins of many of the exposure curves used by underwriters and actuaries in the London market remain mysterious.

Exposure rating was initially developed in the context of treaty property reinsurance; however, its use is now common in both reinsurance and direct insurance.

We will start by looking at the inadequacy of standard experience rating when applied to property risks (Section 22.1). We will then look at exposure curves in some detail: how they naturally arise in the context of pricing a layer of (re)insurance (Section 22.2), what relationship they have with severity curves (Section 22.3), what their main properties are (Section 22.4). The standard Bernegger (MBBEFD) curves are introduced in Section 22.5. Section 22.6 explains how one can derive exposure curves from scratch when sufficient data is available.

Having laid the foundational aspects of exposure rating, Sections 22.7 and 22.8 describe at length the exposure rating process in reinsurance and direct reinsurance respectively. The issue of combining experience rating with exposure rating and that of incorporating cat risks are also addressed, along with more advanced issues such as MPL uncertainty and the presence of non-scalable losses.

Although much of the chapter assumes that the insured interest are buildings, plants, and so on, the same principles apply to any property-like line of business such as Marine/Aviation Hull or Fine Art, where there is a schedule of insured items each of which has an insured value and the maximum possible loss that it may incur.

22.1 Inadequacy of Standard Experience Rating for Property Risks

One way to price property business for a portfolio of property is of course to use experience rating: collect all losses over a certain period, produce a frequency model and a severity model, and combine them with a Monte Carlo simulation or a similar method. However, experience rating is typically inadequate for property. There are several reasons for this, the most important of which is this:

DOI: 10.1201/9781003168881-25

Past property losses for a client's portfolio of locations are a poor guide to future losses unless the profile of insured values of that portfolio of locations has remained broadly unchanged.

As an example, if the property schedule (i.e. the list of properties in the portfolio) for a client included in the past a property of £100M that had a total loss in the past, and now the top property in the portfolio has a value of £20M, a model based on past losses will predict a loss of £100M with finite probability, but that loss is not possible with the new profile.

On top of this, if one looks at the typical property loss experience of a client, one will notice that loss data sets tend to be characterised by many small losses with (possibly) a few large spikes corresponding to buildings being completely destroyed or severely damaged. As a result, even if the profile of the client's portfolio hasn't changed it is very difficult to produce a reliable severity model, unless a very large data set is available.

A better approach to property modelling involves looking in more depth at the nature of property risk. First of all, property risk is inherently limited: because property insurance will repay only the damage to the insured's property (and possibly some related expenses), the loss amount is limited by the value of the property, which normally corresponds to the *insured value* (or *sum insured*, which we will use interchangeably). Often (especially in the case of large commercial/industrial properties), it will be possible to limit this loss even further: for example, if a large plant is made of a number of well-separated buildings, we may be able to assume that a fire will normally be confined to one of these buildings. In this case, we say that although the insured value is x, the *maximum possible loss* (MPL)[1] is some fraction of x.

As a consequence of this, severity curves for an individual property risk can be thought of as curves that give the probability that the damage is less than a given percentage of the insured value or the MPL, rather than the probability that the damage is less than an absolute monetary amount. This is of course not possible for risks that involve liability because there is no natural cap for liability losses.

The second fact about property risk is that there is normally a finite (rather than infinitesimal) probability that the loss is a total loss, that is, that as a result of the damage the property must be written off completely. In the case of a building, that means that it must be torn down and rebuilt from scratch. In the case of a plane, that means that the plane is damaged so extensively that repairing it would result in a craft which would be unsafe to fly.

Taking these facts into account, the severity curve for an individual property might look something like Figure 22.1. This figure shows that, for example, the probability that a loss costs less than 20% of the value of the maximum possible loss for the property is approximately 70%. The probability then increases smoothly as the damage percentage approaches 100%, until it gets to approximately 85%. Then it jumps suddenly to 100%. The height of this jump is the probability of having a total loss.

All this is perhaps interesting but certainly rather theoretical; however, how are we to derive a severity curve for an individual property? We cannot expect to have many examples of properties that have so many claims that a full severity curve can be fitted through these claims. Also, if the first loss is a total loss, we are done collecting losses

1 This is one of many acronyms with roughly the same meaning: maximum probable loss (MPL again), maximum foreseeable loss (MFL), expected maximum loss (EML), possible maximum loss (PML) ... different practitioners will use different names for the same thing, and many will mean slightly different things by these different names, for example, trying to capture losses at different levels of probability, or losses involving different types of failure in the risk control mechanisms, by using different names. In this book, by MPL we'll mean maximum possible loss and by maximum possible loss we'll simply mean the largest possible loss due to 'FLEXA' (fire, lightning, explosion, aircraft collision), not including therefore so called 'catastrophic' losses (earthquake, flood, windstorm, etc.).

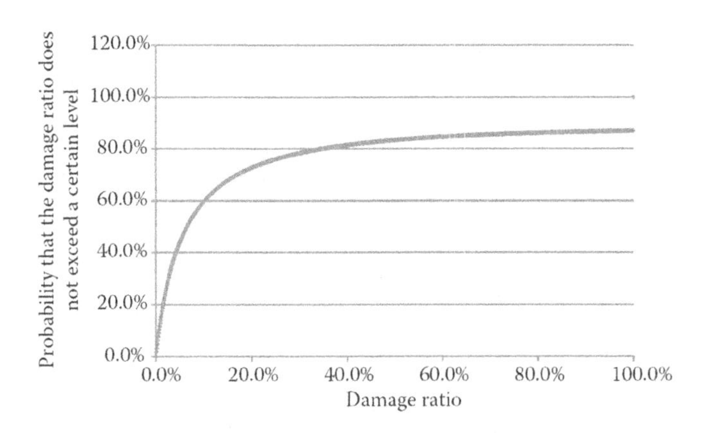

FIGURE 22.1
An example of a severity curve for an individual property. The damage ratio is the ratio between the loss and the maximum possible loss (MPL). Note how the graph approaches a probability of approximately 85%, but then jumps to 100% when the damage ratio is 100%. This means that there is a probability of roughly 15% of having a total loss (100% loss).

for that property! So we cannot work out the severity distribution from the losses from a particular property. We need an extra assumption.

The third fact, or rather reasonable assumption, is the scalability of property losses.

> **Scalability assumption**: *similar properties – i.e. properties with the same occupancy type (residential houses, paper mills, warehouses, refineries…) – will have a similar severity curve once you scaled it down in terms of the maximum possible loss.*

This is the crucial assumption on which all of exposure rating is built, and it gives us a recipe to construct empirical severity distributions:

1. Collect individual losses.
2. Divide each loss by the MPL of the relevant property, obtaining the damage ratio (a number from 0 to 1).
3. Sort the damage ratios in ascending order, obtaining an empirical severity distribution.

We can then fit a model to the empirical severity distribution and use that as our relative severity model, which can then be used for rating.

Note that a single client (whether a non-insurance company or an insurer) will not in general have enough losses in its portfolio to allow one to build a robust severity curve in this way; for this reason, standardised severity curves (or rather, their reengineered version, the exposure curves) are normally the result of large-scale studies.

22.2 How Exposure Curves Arise

We are now just one step away from the concept of exposure curves, although exposure curves look rather different, at first sight, from severity curves. The most important

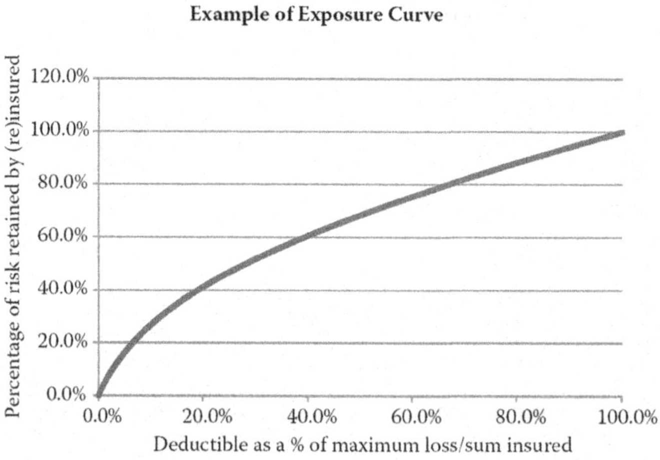

Example of Exposure Curve

Percentage of risk retained by (re)insured

120.0%
100.0%
80.0%
60.0%
40.0%
20.0%
0.0%

0.0% 20.0% 40.0% 60.0% 80.0% 100.0%

Deductible as a % of maximum loss/sum insured

FIGURE 22.2
An example of an exposure curve. The curve is to be interpreted like this: if for example a deductible of the 40% of the insured value is imposed, this will reduce the expected claims to the (re)insurer by approximately 60%.

difference is in the information they show: an exposure curve gives the percentage of risk that is retained by the (re)insured (and therefore cast off by the (re)insurer) if a given deductible is imposed. The other differences are in the way they look: as in the example of Figure 22.2, they are always continuous, they are always concave, and they always go up to 100%, regardless of the probability that a total loss occurs.

It may not be immediately clear why this is so, but it turns out that this representation is much more helpful than a standard normalised severity curve in the context of excess-of-loss pricing, so let us start from there. Assume you are pricing a layer of reinsurance L xs D, or (D, D + L). Because we are dealing with property reinsurance, we can also safely assume that there is no index clause.

Given a loss X, the loss to the layer $(D, D + L)$ is

$$X(D,L) = \min(X, D+L) - \min(X, D) \quad (22.1)$$

The expected loss to the layer if an individual loss has occurred is therefore

$$\mathbb{E}(X(D,L)) = \mathbb{E}(\min(X, D+L)) - \mathbb{E}(\min(X, D)) \quad (22.2)$$

If N is the number of losses from the ground up, the expected *total* losses to the layer are

$$\mathbb{E}(S(D,L)) = \mathbb{E}(X(D,L)) \times \mathbb{E}(N) = \left(\mathbb{E}(\min(X, D+L)) - \mathbb{E}(\min(X, D))\right) \times \mathbb{E}(N) \quad (22.3)$$

This is the standard formula for the expected losses to a layer of reinsurance (Klugman et al., 2008).

Assume now that X represents a loss to a particular property, and that the maximum possible loss to that property is M. Because the loss cannot exceed M, it is quite natural to describe any loss occurring to that property as a percentage of its maximum value M, $x = X/M$.

We can then write

$$\mathbb{E}(X) = \mathbb{E}(X/M) \times M = \mathbb{E}(x) \times M, \tag{22.4}$$

and similarly,

$$\mathbb{E}(\min(X,D)) = \mathbb{E}(\min(x,d)) \times M \tag{22.5}$$

where $d = D/M$.

The expected losses to the layer $(D, D + L)$ can then be written as

$$\begin{aligned}\mathbb{E}(S_{D,L}) &= \mathbb{E}(\min(X,D+L)) - \mathbb{E}(\min(X,D)) \times \mathbb{E}(N) = \\ &= \frac{\mathbb{E}(\min(X,D+L)) - \mathbb{E}(\min(X,D))}{\mathbb{E}(X)} \times \mathbb{E}(X) \times \mathbb{E}(N) = \\ &= \frac{\mathbb{E}(\min(x,d+l)) - \mathbb{E}(\min(x,d))}{\mathbb{E}(x)} \times \mathbb{E}(S) = \\ &= (G(d+l) - G(d)) \times \mathbb{E}(S)\end{aligned} \tag{22.6}$$

where

$$G(u) = \frac{\mathbb{E}(\min(x,u))}{\mathbb{E}(x)} \tag{22.7}$$

is a function from [0,1] to [0,1].

Now, would it not be good if we had the function $G(u)$ already tabulated for us so that we could simply plug that in into the equation and find the expected losses to the layer given the overall expected losses?

This is indeed what exposure rating is about: providing curves that give $G(d)$ as a function of d as illustrated in Figure 22.1. What we haven't yet explained is why $G(d)$ can be interpreted as the percentage of risk retained by the (re)insured after the imposition of a deductible $D = d \times M$ (where M represents the MPL). The reason is that if a loss X is experienced, the loss retained by the (re)insured is the minimum of the loss itself and the deductible, $X_{\text{ret}} = \min(X, D)$, which can also be written as $X_{\text{ret}} = M \times \min(x, d)$. The average expected retained loss is therefore $\mathbb{E}(X_{\text{ret}}) = M \times \mathbb{E}(\min(x, d))$. If no (re)insurance were there, the average expected loss to the property would be $\mathbb{E}(X) = M \times \mathbb{E}(x)$. The percentage of loss retained by the (re)insured is therefore the ratio between these two:

$$\%\text{ of risk retained by the (re)insured} = \frac{\mathbb{E}(X_{\text{ret}})}{\mathbb{E}(X)} = \frac{M \times \mathbb{E}(\min(x,d))}{M \times \mathbb{E}(x)} = G(d) \tag{22.8}$$

22.3 Relationship with Severity Curves

As we mentioned previously, exposure curves are in one-to-one correspondence with severity curves, so they can be considered market severity curves that are reengineered to look like Figure 22.2. To make this clearer, we show in Figure 22.3 a normalised severity curve and the exposure curve that corresponds to it. How does one calculate the exposure curve corresponding to a given severity curve, and vice versa?

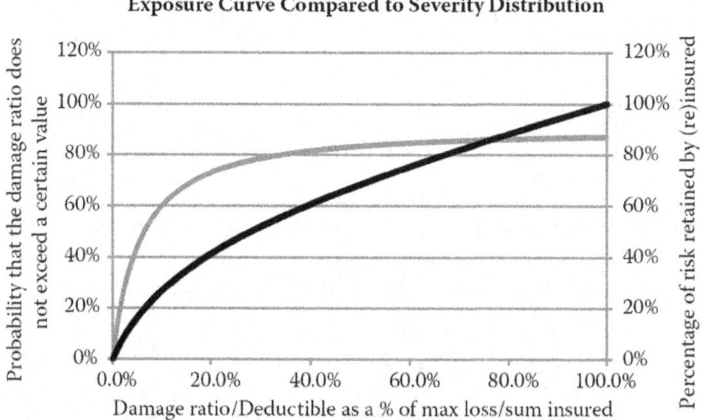

FIGURE 22.3
A comparison of a normalised severity curve (grey line) and the exposure curve (black line) that corresponds to it. The comparison shows that although the severity curve has a discontinuity at 100% (the probability that a given loss does not exceed a given percentage of damage jumps from approximately 85% to 100%), the corresponding exposure curve increases to 100% continuously.

Equation 22.7 provides a recipe for calculating the exposure curve based on the severity curve. To obtain the severity curve from the exposure curve, notice first that the numerator in Equation 22.7 can be written as

$$\mathbb{E}(\min(x,u)) = \int_0^1 \min(x,u) f(x) dx = \int_0^u x f(x) dx + u(1 - F(u)) = \int_0^u \bar{F}(x) dx \tag{22.9}$$

where the last equality was derived using integration by parts and setting $\bar{F}(x) = 1 - F(x)$.

We can now rewrite Equation 22.7 as

$$G(u) = \frac{\int_0^u \bar{F}(x) dx}{\mathbb{E}(x)} = \frac{\int_0^u \bar{F}(x) dx}{\int_0^1 \bar{F}(x) dx} \tag{22.10}$$

and by taking the derivative of both sides, we obtain

$$G'(u) = \frac{1 - F(u)}{\mathbb{E}(x)} \tag{22.11}$$

Notice in particular that because $F(0) = 0$, the derivative of G at 0 is $G'(0) = 1/\,\mathbb{E}(x)$.

We can therefore write the cumulative distribution function $F(x)$ as

$$F(x) = \begin{cases} 1 - \dfrac{G'(x)}{G'(0)} & 0 \le x < 1 \\ 1 & x = 1 \end{cases} \tag{22.12}$$

22.4 Properties of Exposure Curves

Equations 22.7 and 22.12 also allow us to prove certain properties of the exposure curves that we have mentioned in passing in the previous sections.

Continuity. $G(x)$ is continuous over its whole range of values, [0,1]. This is because $G(x)$ is, apart from a proportionality constant, equal to the integral of a function, $1 - F(x)$, whose value is limited between 0 and 1.

Concavity. $G(x)$ is concave in [0,1]. In mathematical terms, this means that for any two points $a, b \in [0,1]$, $G(x) \geq G(a) + \frac{G(b)-G(a)}{b-a}(x-a)$; that is, the function always lies above (or on) the straight line connecting two of its points. In the case where the function $G(x)$ has both the first and the second derivative over the whole interval [0,1] (which in our case is true if $F(x)$ is continuous and its first derivative exists), demanding concavity is equivalent to demanding that $G''(x) \leq 0$ for all x in [0,1]. Because $G''(x) = -f(x)/\mathbb{E}(x)$ and $f(x)$ is a probability density (always ≥ 0), this is automatically the case for exposure curves. In the case where $F(x)$ does not have these properties, the proof is slightly less immediate (see Question 4) but the property still holds.

What does the degree of concavity tell us? If the exposure curve is strongly concave, as in the example in Figure 22.4b, even a small deductible (e.g. 10% of the MPL) achieves a strong reduction in the risk ceded to the reinsurer, and the probability of a total loss is negligible. At the other extreme, if the exposure curve has no concavity at all, which is achieved by $G'(u) = 1$ for all u, then all losses are total losses (the severity curve will be a flat line up to 1 and then suddenly jump to 1) and a deductible of d% will achieve a reduction in the loss of exactly d%. The case shown in Figure 22.4a is not so extreme but nonetheless shows a situation (curves for residential buildings) in which the exposure curve is much nearer to the diagonal and deductibles are far less effective, and there is a significant probability of a total loss (<20%).

Probability of total loss. The probability of a total loss is by definition the probability that the loss is exactly $x = 1$. This is (by definition):

$$\Pr(x=1) = \Pr(x \leq 1) - \Pr(x < 1) = 1 - \lim_{x \to 1^-} F(x) \tag{22.13}$$

Using Equation 22.12, this may be also written as

$$Pr(x=1) = 1 - 1 + G'(1)/G'(0) = G'(1)/G'(0) \tag{22.14}$$

which shows that the only exposure curves for which the probability of a total loss equals zero are those that end up completely flat ($G'(1) = 0$). Note that the possibility that $\Pr(x = 1)$ is zero because $G'(0^+) = \infty$ (vertical tangent at the beginning of the graph) is not to be entertained, because this would also mean that the average loss for a property is zero[2] (see Equation 22.15 below).

Mean loss. By setting $u = 0$ in Equation 22.11, we see that $G'(0) = 1/\mathbb{E}(x)$. Therefore the average damage ratio is $\mathbb{E}(x) = 1/G'(0)$ and the mean monetary loss is:

2 Despite this shortcoming, exposure curves with vertical tangent at zero are sometimes used in practice. Riegel (2010) calls them 'quasi-exposure curves' and analyses their use for his burning cost-adjusted exposure rating method.

(a)

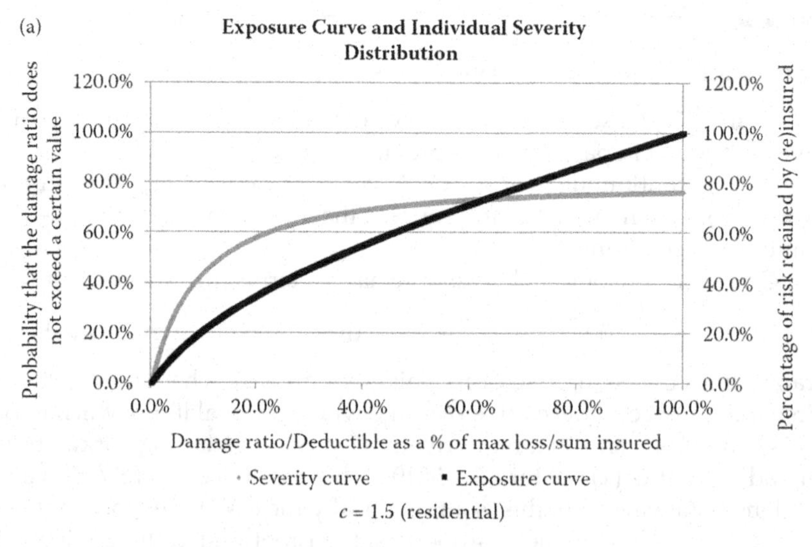

(b)

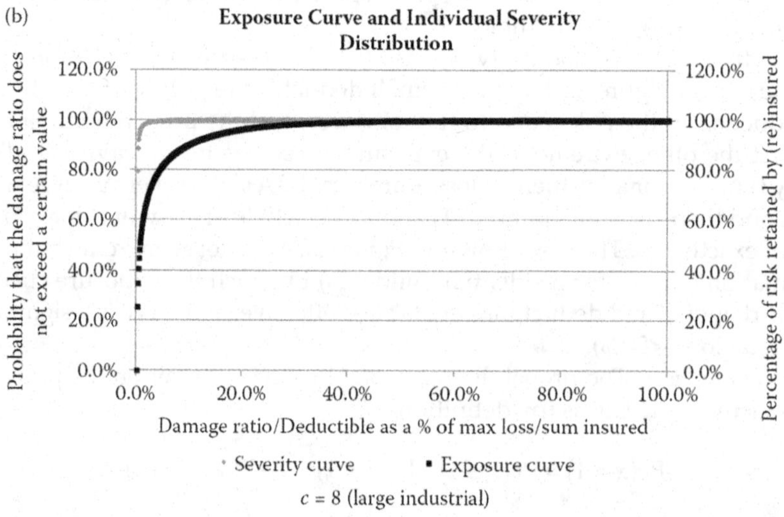

FIGURE 22.4
Two examples of exposure curves (with their corresponding severity curves) modelling fire risk for different types of properties: residential properties (a), which have a significant probability of a total loss and the exposure curve is not too far away from a straight line, and large industrial properties (b) for which the probability of a total loss due to fire is minimal and deductible of, for example, 20% of the MPL will get rid of most of the risk for the reinsurer. The meaning of the parameter c is discussed in Section 22.5.

$$\mathbb{E}(X) = \frac{M}{G'(0)} \tag{22.15}$$

That is, the steeper the exposure curve is at the beginning (which is often the case with large industrial plants), the smaller the average loss is relative to the value of the property. On the other hand, if the exposure curve is the bisector of the first quadrant, then $G'(u) = 1$ for all u including $u = 0$, and therefore $\mathbb{E}(x) = 1$ (all losses are total losses).

22.5 Commonly Used Curves and Bernegger's Parameterisation

Exposure curves were derived empirically from collections of historical losses by institutions such as the ISO in the United States and by corporations, some of which (such as Lloyd's of London and Swiss Re) have also made them publicly available.

A famous example is given indeed by the Swiss Re curves, which are exposure curves for fire risk with a single parameter c that is related to the concavity of the curve (the letter 'c' was not originally meant to stand for 'concavity', but it is a good mnemonic[3]). A curve with a small value of c will be close to a straight line (the 45° line through the origin), whereas a curve with a large value of c will be very concave, as in the two examples in Figure 22.4. The value $c = 0$ corresponds to the case with zero concavity, that is, the case in which the exposure curve is a straight line and all losses are total losses. In a publicly available technical report, Swiss Re has provided a table (see Figure 22.5) to guide the choice for the correct value of c for different types of property.

Note that Figure 22.5 also provides the basis for 'total loss' to use for different types of property: sum insured (= insured value) or MPL. As a matter of fact, the basis is almost always 'MPL', and the only cases for which this is not true are those cases for which the MPL is equal to the sum insured anyway. For more details, the interested reader is referred to Swiss Re's technical publication 'Exposure rating' (2004) for more details.

What should one do if one wants to use the table in Figure 22.5 but only has the insured value, not the MPL, for the properties in the portfolio? A rough method is to set the MPL equal to a certain proportion (e.g. 70%) of the insured value, and use the relevant value of c. The proportion can be chosen depending on the underwriter's or the actuary's experience and will depend on the type of property. Another (possibly even rougher) method that is sometimes used is to increase the value of c to make up for the fact that a total loss based on insured value is more unlikely to happen than a total loss based on MPL because the insured value is larger than the MPL.

Two problems with this approach are that the tables that describe the various exposure curves must be stored somewhere – in a software tool or in an old-fashioned booklet – and

Type of property	Value of c	Basis
Personal lines	1.5	Sum insured
Commercial lines (small-scale)	2	Sum insured
Commercial lines (medium-scale)	3	Sum insured
Captive business interruption	3.1	MPL
Captive PD/BI	3.4	MPL
Captive PD	3.8	MPL
Industrial and large commercial	4	MPL
Lloyd's (Y5): Industry	5	Top location
Large-scale industry/multinational companies	Up to 8	MPL

FIGURE 22.5
Recommended values of c for different types of property.
Source: **Swiss Re, Exposure rating, 2004.**

3 The letter 'c' actually stands for 'continuous family of curves' because at a later stage (see the end of this section) it allowed to vary continually the shape of the Swiss Re curves, originally available as a discrete set.

that they will be available only for a small number of parameters. This problem could be overcome if there was a simple way to express these curves analytically.

22.5.1 Bernegger (MBBEFD) Curves

This is exactly what Stefan Bernegger, a former physicist working as an actuary for Swiss Re, set out to do in his widely read article (Bernegger, 1997). Bernegger found a useful parameterisation for exposure curves and the corresponding severity curves. Specifically, he noted that the distribution of energy levels for a gas of classical or quantum-mechanical particles could be used to approximate the severity curves underlying the exposure curves used by underwriters and actuaries. Because of this, Bernegger's curves are also called MBBEFD curves: depending on the values of their parameters they are in a formal correspondence to the Maxwell–Boltzmann distribution, the Bose–Einstein distribution, and the Fermi–Dirac distribution.[4] The correspondence is not normally made explicit in actuarial papers (which is perhaps unsurprising) but the reader interested in exploring the connection may want to have a go at Question 8, which provides the necessary background.

As it is made clear in Bernegger (1997), the analogy is purely formal, and there is no suggestion that buildings on fire should behave like gases of quantum-mechanical particles. Despite this, this parameterisation is in practice very useful because:

- it is flexible enough to capture most curves derived empirically;
- it allows for the building of more compact exposure rating tools, without the need of a large number of look-up tables and interpolation rules;
- it allows for easy Monte Carlo simulation because Bernegger's CDF is analytically invertible (see below).

The general form of the exposure curves is:

$$G(x)=\begin{cases}\dfrac{\ln\left(\dfrac{(g-1)b+(1-bg)b^x}{1-b}\right)}{\ln(bg)} & b>0,\ b\neq 1, bg\neq 1, g>1\\[2ex] \dfrac{1-b^x}{1-b} & bg=1,\ g>1\\[2ex] \dfrac{\ln(1+(g-1)x)}{\ln(g)} & b=1,\ g>1\\[2ex] x & g=1 \text{ or } b=0\end{cases} \tag{22.16}$$

which corresponds to this cumulative distribution function for the severity:

4 A more intimidating line-up of fine minds could hardly have been co-opted in connection to insurance rating! Maxwell was the founding father of theoretical electromagnetism; Boltzmann, of statistical mechanics; Einstein, of the theory of relativity, and (amongst several others) of quantum mechanics; Fermi was one of the founding fathers of nuclear physics; Dirac managed to combine quantum mechanics and relativity. Bose, the least known of the lot, was a self-taught physicist who contributed to laying the foundations for quantum statistics.

$$F(x)=\begin{cases}\dfrac{b(g-1)(1-b^{x})}{b(g-1)+(1-bg)b^{x}} & x<1, b>0,\ b\neq 1, bg\neq 1, g>1\\ 1-b^{x} & x<1,\ bg=1,\ g>1\\ 1-\dfrac{1}{1+(g-1)x} & x<1, b=1,\ g>1\\ 0 & x<1,\ g=1 \text{ or } b=0\\ 1 & x=1\end{cases} \tag{22.17}$$

Finally, the probability density function is given by:

$$f(x)=\begin{cases}\dfrac{(b-1)(g-1)\ln(b)b^{1-x}}{\left((g-1)b^{1-x}+(1-bg)\right)^{2}} & x<1, b>0, b\neq 1, bg\neq 1, g>1\\ -\ln(b)\times b^{x} & x<1, bg=1, g>1\\ \dfrac{g-1}{\left(1+(g-1)x\right)^{2}} & x<1, b=1, g>1\\ 0 & x<1, g=1 \text{ or } b=0\\ \dfrac{1}{g}\times\delta(x-1) & x=1\end{cases} \tag{22.18}$$

where $\delta(z)$ is the Dirac's delta, a mathematical object that for the purpose of this book can be visualised as a mass point of finite probability at point $z = 0$ (see footnote in Chapter 17 for more details).

By varying the variables b and g, we can simulate Swiss Re curves with different values of c, by using the formula:

$$b=e^{\alpha+\beta c(1+c)},\quad g=e^{(\gamma+\delta c)c} \tag{22.19}$$

with $\alpha = 3.1$, $\beta = -0.15$, $\gamma = 0.78$ and $\delta = 0.12$ for the Swiss Re curves.

22.5.2 Properties of Bernegger's Curves

The probability of a total loss for Bernegger's curves is (for all values of b, g) given by:

$$\Pr(\text{total loss})=F(1)-F(1^{-})=\frac{1}{g} \tag{22.20}$$

The mean damage ratio can be calculated using Equation 22.15:

$$\mathbb{E}(x)=\begin{cases}\dfrac{\ln(bg)(1-b)}{\ln(b)(1-bg)} & b>0,\ b\neq 1, bg\neq 1, g>1\\ \dfrac{b-1}{\ln(b)} & bg=1,\ g>1\\ \dfrac{\ln(g)}{g-1} & b=1,\ g>1\\ 1 & g=1 \text{ or } b=0\end{cases} \tag{22.21}$$

BOX 22.1 SIMULATING LOSSES FROM A BERNEGGER (MBBEFD) CURVE

Inputs

- Maximum possible loss (MPL)
- Parameters b, g of the Bernegger curve

a. Generate a pseudo-random number u uniformly between 0 and 1

b. If $u < 1 - 1/g$, set $y = \ln\left(\dfrac{b(g-1)(1-u)}{b(g-1)+(1-bg)u}\right) / \ln(b)$; else, set $y = 1$

c. Return from-the-ground-up loss: Loss = $y \times \text{MPL}$

22.5.3 Simulating Losses Using a Bernegger Distribution

One of the advantages of Bernegger severity curves (Equation 22.17) are that they are easily invertible and can therefore be used for simulation purposes using **inverse transform sampling** (Devroye, 1996), i.e. by generating a pseudo-random number uniformly between 0 and 1 and then calculating the corresponding normalised loss as $y = F^{-1}(u)$. To be more specific, to produce a random loss between 0 and MPL we can use the simple algorithm in Box 22.1. Note that this algorithm produces a loss equal to MPL with finite (i.e. non-zero) probability.

The simple algorithm in Box 22.1 is at the basis of the simulation of losses for a portfolio for properties and will be used in the applications to reinsurance and direct insurance that we will discuss in Sections 22.7 and 22.8.

22.6 Build and Calibrate Your Own Exposure Curve

In some cases, you may have sufficient data to derive your own exposure curve for a portfolio of properties. We show below how to produce your own severity curve and calibrate it using Bernegger's parameterisation.

The first method (Box 22.2) is arguably the most popular and relies on finding the maximum likelihood estimate of the parameter of Bernegger's *cumulative distribution function* (Equation 22.17) rather than fitting the exposure curve directly.

A simple spreadsheet tool for the calibration of a Bernegger curve based on the method above can be found in the downloads section of Parodi (2016).

The second method (Box 22.3) is based on fitting the exposure curve directly – specifically, choosing a Bernegger curve that minimises the Kolmogorov-Smirnov distance between the empirical exposure curve and the model. This might appear to be theoretically the same thing, but it has important practical differences. As was shown in Parodi & Watson (2019), it is possible to create an example of a pair of distributions for which the distance between the CDFs is arbitrarily large (i.e. KS distance arbitrarily close to 1) while the distance between the exposure curves is arbitrarily small (i.e. KS distance arbitrarily close to 0). Question 9 explores how this can be done. The underlying reason is that very small losses are picked up by MLE and have a high impact on the value of the parameters,

BOX 22.2 BUILDING AND CALIBRATING AN EXPOSURE CURVE USING MAXIMUM LIKELIHOOD ESTIMATION

Inputs

- A large number of claims for a portfolio of reasonably similar properties (such as residential, commercial offices, industrial plants)
- The MPL for the property involved in each claim

1. Divide each claim by the corresponding MPL and obtain damage ratios $y_1, y_2 \dots y_N$, where $0 < y_k \le 1$ for each k. These losses can be sorted to form an empirical (relative) severity curve. For simplicity (and obviously, without loss of generality) we can assume that the losses are already sorted: $y_1 \le y_2 \le \dots \le y_N$. We will also assume that the first N' $(0 \le N' \le N)$ damage ratios are partial losses and the remaining $N - N'$ are total losses: $y_1 \le y_2 \le \dots \le y_{N'} < 1, y_{N'+1} = \dots = y_{N-1} = y_N = 1$.
2. Fit a Bernegger severity curve in the form of Equation 22.17 to the empirical severity curve by maximum likelihood estimation, obtaining the 'best' values of the parameters b and g. The log likelihood to maximise is the following:

$$LL = \ln\left(\prod_{j=1}^{N'} f(y_j) \times \left(\frac{1}{g}\right)^{N-N'}\right) = \sum_{j=1}^{N'} \ln f(y_j) - (N - N') \times \ln g \qquad (22.22)$$

   where $f(y)$ is the probability density function for Bernegger's curves, as defined by Equation 22.18. Note that the problem does not have a closed form solution and requires numerical optimisation.
3. The exposure curve is then that defined by Equation 22.16 with the values of b^* and g^* found in Step 2.

Output

An exposure curve $G(x)$ with Bernegger parameterisation and parameters b^*, g^*.

Note

If you are happy to use a Swiss Re curve for your portfolio and simply want to check which value of c to use, you can write Equation 22.22 in terms of c using the replacement rules given in Equation 22.19. Finding the best value of c is in this case much easier as it requires only solving a one-dimensional optimisation problem.

but have a much smaller impact on the exposure curve. The method in Box 22.2 therefore risks focusing too much on the smaller losses, which are the least relevant from a cost perspective – the exposure curve is by construction more directly related to the monetary amount that will be paid out for a given layer than the CDF.

22.6.1 How Can You Calibrate an Exposure Curve if You Don't Have the MPLs?

In case MPL information is not available, there are two sub-cases:

(a) you do not have the MPL information, but you know the insured value relevant to each loss;
(b) you have nothing.

In case (a), you could try to make some broad assumption on the relationship between the MPL and the IV in your portfolio. For example, you can assume that MPL is roughly equal to, say, 70% of the IV across all properties (the actual number will depend on what you are pricing exactly). You could also use a more complex relationship, that recognises that smaller properties often have MPL = TIV, but larger properties will have a more marked difference. You can of course make exceptions for losses that are already above your assumed MPL. The calculations then proceed as in Box 22.2 or Box 22.3.

BOX 22.3 BUILDING AND CALIBRATING AN EXPOSURE CURVE BY MINIMISING THE KS DISTANCE BETWEEN THE EMPIRICAL EXPOSURE CURVE AND THE MODEL

Inputs

- A large number of claims for a portfolio of reasonably similar properties
- The MPL for the property involved in each claim

1. Produce N sorted damage ratios as in Box 22.1, Step 1: $y_1 \le y_2 \le \ldots \le y_{N'} < 1,\ y_{N'+1} = \ldots = y_{N-1} = y_N = 1$.
2. Calculate the empirical survival distribution $\hat{S}(y_k)$ in the usual way as the percentage of damage ratios strictly above y_k.
3. Calculate the empirical exposure curve $\hat{G}(y_k)$ with, e.g., the following formula:

$$\hat{G}(y_k) = \frac{\sum_{i=0}^{k-1}\left(\hat{S}(y_i)+\hat{S}(y_{i+1})\right)\times\left(y_{i+1}-y_i\right)}{\sum_{i=0}^{N'}\left(\hat{S}(y_i)+\hat{S}(y_{i+1})\right)\times\left(y_{i+1}-y_i\right)} \quad (22.23)$$

where y_{i+1} is conventionally set to 1 if $N' = N$, in which case $\hat{S}(y_{N'}) = 0$ and the last addendum of the denominator is 0. This is an application of the trapezium rule for the calculation of the integrals in Equation 22.10.
4. Fit a Bernegger exposure curve in the form of Equation 22.16 to the empirical exposure curve by finding the parameters b^* and g^* that minimise the KS distance between the Bernegger model $G(b,g;y)$ and the empirical curve $\hat{G}(y_k)$:

$$d_{KS}\left(G,\hat{G}\right) = \max_{k\in\{1,\ldots N'\}}\left|\hat{G}(y_k) - G(b,g;y_k)\right| \quad (22.24)$$

As in Box 22.2, solving this problem requires numerical optimisation.

Output

An exposure curve $G(x)$ with Bernegger parameterisation and parameters b^*, g^*.

Note

As in Box 22.2, if you are happy to use a Swiss Re curve for your portfolio and simply want to check which value of c to use, the optimisation problem above becomes one-dimensional.

In case (b) – a very common case, unfortunately – the recommendation is: don't. Formally, you can certainly assume an MPL profile for your properties based on some sensible assumption, and you can write down the likelihood leaving the MPL as a variable; in practice, the uncertainty around the parameters that you will find is likely so high that the exercise is purely academic and you'll be much better off just assuming a sensible exposure curve from Figure 22.5 (or a similar source).

22.7 Exposure Rating Process in Reinsurance

In the previous section, we have seen the reasons behind the introduction of exposure curves, we have become familiar with their properties, and we have learnt to parameterise them and fit them to a claims data set. We are now going to look at how exposure curves are used in practice as a tool for rating, and we are going to start with excess-of-loss reinsurance, in which context the technique was first developed. The application of exposure rating to direct insurance is described in Section 22.8.

22.7.1 Basic Process

The basic process for rating excess-of-loss reinsurance for a portfolio of properties hinges on Equation 22.6, which gives the expected losses to a layer L xs D for an individual property with MPL equal to M. According to Equation 22.6, the expected losses to a layer, $\mathbb{E}(S_{D,L})$, can be calculated by multiplying the expected losses, $\mathbb{E}(S)$, by the difference in the value of the exposure curves at points $d = D/M$ and $d + l = (D + L)/M$, that is, $G(d + l) - G(d)$.

Assuming we have already decided what exposure curve to use and we therefore know $G(d + l) - G(d)$, the main question for us is how to determine $\mathbb{E}(S)$: because we are probably using exposure curves due to the lack of sufficient loss information or because we want an independent view on rates, using experience rating to determine $\mathbb{E}(S)$ is out of the question.

The method normally used by reinsurers to determine $\mathbb{E}(S)$ is, instead, to use the premium originally charged by the insurer for the portfolio (the *original* premium). Because, by definition, the expected loss ratio is related to the expected losses by this simple equality:

$$\mathbb{E}(\text{LR}) = \frac{\mathbb{E}(S)}{\text{Premium}} \tag{22.25}$$

any estimate of the expected loss ratio $\mathbb{E}(\text{LR})$, for example, the actual loss ratio LR* achieved by the insurer over a number of past years,[5] can be used as an estimate of the expected losses:

$$\mathbb{E}(S) = \mathbb{E}(LR) \times \text{premium} \approx \text{LR}^* \times \text{premium} \tag{22.26}$$

We can now rewrite Equation 22.6 for a single property as

$$\mathbb{E}(S_{D,L}) = (G(d+l) - G(d)) \times \mathbb{E}(\text{LR}) \times \text{premium} \tag{22.27}$$

5 As you can imagine, care must be taken in accounting for the underwriting cycle and other effects that might bias the estimation of the loss ratio over a limited number of years.

To generalise this to a portfolio of K properties, we simply sum Equation 22.21 over all K properties in the portfolio:

$$\begin{aligned}\mathbb{E}(S^{D,L}) &= \sum_{k=1}^{K}\mathbb{E}(S_k^{D,L}) = \\ &= \sum_{k=1}^{K}(G_k(d_k+l_k)-G_k(d_k))\times\mathbb{E}(S_k) = \\ &= \sum_{k=1}^{K}(G_k(d_k+l_k)-G_k(d_k))\times\mathbb{E}(\mathrm{LR})\times P_k\end{aligned} \tag{22.28}$$

where

- $d_k = D/M_k$, $l_k = L/M_k$ are the limits of the layer expressed in terms as a percentage of M_k, which in turn is the MPL of the kth property.
- P_k is the original premium paid to the insurer for insuring the kth property from the ground up.
- $\mathbb{E}(\mathrm{LR})$ is the expected loss ratio for the client.
- $G_k(x)$ is the exposure curve for the kth property.

Note that although we have allowed for the exposure curve to be different for each property, in practice each property will be classified as belonging to one of a certain numbers of occupancies (such as residential, office, industrial plant) and on the basis of this, the curve for that occupancy will be chosen.

If the portfolio of property is homogeneous in terms of occupancy so that we can assume the same exposure curve for all properties, this formula simplifies to:

$$\mathbb{E}(S^{D,L}) = \mathbb{E}(\mathrm{LR})\times\sum_{k=1}^{K}(G(d_k+l_k)-G(d_k))\times P_k \tag{22.29}$$

where the dependence on k is limited to the different values of MPL/SI (incorporated in d_k, l_k) and to the premium paid for each property.

The accuracy of the exposure rating calculation by the reinsurer, therefore, also depends on how accurate the rating process of the insurer is. A more formal treatment of the material in this section can be found in Riegel (2010).

22.7.2 How This Is Done in Practice

Even if we have a portfolio of millions of houses, it is straightforward, with current computing powers, to estimate the expected losses to a layer of reinsurance by a direct application of Equation 22.29. In practice, however, the reinsurer may not be able (and often may not need, nor want) to obtain from the insurer the full list of properties of the portfolio for which reinsurance is sought. It is much more common that the reinsurer will require the insurer to send the details of the portfolio to be insured in aggregate form, as in the table shown in Figure 22.6, further specifying of course the type of properties in the portfolio, so that the reinsurer may choose adequate exposure curves.

In a table like this, it is not possible to know the individual values of MPL/insured value because all properties with similar such values have been put together in an MPL/insured value band (e.g. 1,000,001 → 1,500,000). For each band, the total premium written is specified.

Exposure Rating – Input Example

Sum insured (lower limit)	Sum insured (upper limit)	Total premium	Original loss ratio	Total sum insured	No. of risks
0	500,000	734,874	60%	274,593,762	1546
500,001	1,000,000	192,912	60%	79,510,325	113
1,000,001	1,500,000	237,273	60%	119,516,504	96
1,500,001	2,000,000	211,035	60%	119,537,850	69
2,000,001	2,500,000	194,928	60%	114,476,022	51
2,500,001	3,000,000	161,752	60%	112,416,894	41
3,000,001	3,500,000	186,473	60%	97,451,400	30
3,500,001	4,000,000	164,826	60%	96,066,251	26
4,000,001	4,500,000	82,589	60%	59,858,354	14
4,500,001	5,000,000	144,744	60%	75,537,922	16
5,000,001	6,000,000	184,771	60%	176,025,332	32
6,000,001	7,000,000	144,440	60%	143,631,814	22
7,000,001	8,000,000	257,513	60%	158,346,925	21
8,000,001	9,000,000	148,573	60%	126,442,425	15
9,000,001	10,000,000	224,765	60%	188,811,487	20
10,000,001	12,500,000	439,826	60%	321,599,576	29
12,500,001	15,000,000	229,173	60%	217,847,911	16
15,000,001	17,500,000	181,344	60%	149,875,850	9
17,500,001	20,000,000	188,916	60%	261,859,399	14
20,000,001	25,000,000	260,115	60%	249,267,460	11
25,000,001	30,000,000	446,295	60%	375,702,369	14
30,000,001	35,000,000	156,364	60%	134,461,423	4
35,000,001	40,000,000	371,903	60%	372,272,606	10
40,000,001	45,000,000	121,245	60%	170,303,070	4
45,000,001	50,000,000	2,003,079	60%	924,738,609	19

FIGURE 22.6
An example of input data for a reinsurance submission. The portfolio of properties is assumed to be homogeneous enough to be used in this way (e.g. all properties may be commercial offices or industrial plants).

Note that the loss ratio included in the table is only an estimate, based on the past performance of the reinsured or the future performance expectation.

Based on the information in Figure 22.6, it is possible to calculate an approximate version of Equation 22.29:

$$\mathbb{E}\left(S^{D,L}\right) \approx \mathbb{E}(LR) \times \sum_{\beta=1}^{B}\left(G\left(d_\beta + l_\beta\right) - G\left(d_\beta\right)\right) \times P_\beta \tag{22.30}$$

where $d_\beta = D/M_\beta$, $l_\beta = L/M_\beta$, M_b in turn being the average MPL/SI in a band: $M_\beta = (M_{\beta,\max} + M_{\beta,\min})/2$. The premium P_β is simply the sum of all the premiums paid for properties in

BOX 22.4 EXPECTED LOSS CALCULATION USING EXPOSURE RATING IN PROPERTY REINSURANCE

Inputs

a. Total premium by band (*source: insurer*),
b. Expected loss ratio (*source: insurer, or past performance*),
c. Total sum insured or MPL values by band (*source: insurer*)

i. Select the desired exposure curve G for the occupancy type in the insurer's portfolio. (If there is more than one occupancy, repeat the procedure for each occupancy and sum over the results.)
ii. For each band, calculate the minimum and maximum of the layer $(d_\beta, d_\beta + l_\beta)$ in terms of the average MPL/SI of that band.
iii. For each band, calculate the expected percentage of loss to the layer, $G(d_\beta + l_\beta) - G(d_\beta)$ and the expected losses from the ground up for that band, $P_\beta \times \mathbb{E}(\text{LR})$. The expected losses to the layer are then given by $P_\beta \times \mathbb{E}(\text{LR}) \times (G(d_\beta + l_\beta) - G(d_\beta))$.
iv. Calculate the overall expected losses to the layer by summing all contributions in Step 4 over the different bands $\beta = 1,\ldots B$, yielding Equation 22.30.

Output: estimated expected losses to the layer L xs D.

band b: $P_\beta = \sum_{\text{all properties } k \text{ in band } \beta} P_k$. (The use of the letter 'β' to stand for 'band' does not arise from a disorderly passion for Greek letters but from the need to avoid confusion with the parameter b in Bernegger's formalisation of exposure curves.)

The basic algorithm for calculating the expected losses to a layer $(D, D + L)$ through exposure rating in property reinsurance is therefore as in Box 22.4.

Figure 22.7 shows how these calculations are done in practice with the use of a spreadsheet, using the data set of Figure 22.6. You will have noticed, actually, that we haven't used all the information that was provided in Figure 22.6; this is because for the limited scope of this exercise (finding the expected losses to the L xs D layer of reinsurance) the information on the number of risks and the total sum insured is not strictly necessary. We will see later, however, how this information can be used to assess the volatility around this estimate.

22.7.2.1 Sources of Uncertainty

The success of the method outlined above depends crucially on a number of assumptions, and where there is an assumption there is a source of uncertainty, that we need to keep in mind and communicate properly when presenting the results of an exposure rating exercise.

1. The correctness of the results will depend on choosing the correct *exposure curves*. Given the fact that each property in a portfolio probably has unique features, the exposure curve for a given occupancy will always be just an approximation.
2. The *loss ratio* assumption is also critical, and it is subject to many uncertainties. When based on past performance of the insurer, it depends in turn on assumptions

Example of Exposure Rating Exercises									
A	B	C	D	E	F	G	H	I	J
Sum insured (lower limit)	Sum insured (upper limit)	Average in band, M	Total premium	Original loss ratio	Expected losses	G(d)	G(d+l)	G(d+l) − G(d)	Loss cost to layer
0	500,000	250,000	734,874	60%	440,925	99.9%	99.9%	0%	0
500,001	1,000,000	750,001	192,912	60%	115,747	99.9%	99.9%	0%	0
1,000,001	1,500,000	1,250,001	237,273	60%	142,364	99.9%	99.9%	0%	0
1,500,001	2,000,000	1,750,001	211,035	60%	126,621	99.9%	99.9%	0%	0
2,000,001	2,500,000	2,250,001	194,928	60%	116,957	93.5%	99.9%	6%	7,578
2,500,001	3,000,000	2,750,001	161,752	60%	97,051	83.6%	99.9%	16%	15,831
3,000,001	3,500,000	3,250,001	186,473	60%	111,884	76.3%	99.9%	24%	26,403
3,500,001	4,000,000	3,750,001	164,826	60%	98,895	70.7%	99.9%	29%	28,950
4,000,001	4,500,000	4,250,001	82,589	60%	49,554	66.0%	99.9%	34%	16,798
4,500,001	5,000,000	4,750,001	144,744	60%	86,846	62.2%	99.9%	38%	32,743
5,000,001	6,000,000	5,500,001	184,771	60%	110,863	57.4%	94.7%	37%	41,314
6,000,001	7,000,000	6,500,001	144,440	60%	86,664	52.4%	86.3%	34%	29,340
7,000,001	8,000,000	7,500,001	257,513	60%	154,508	48.4%	79.7%	31%	48,412
8,000,001	9,000,000	8,500,001	148,573	60%	89,144	45.1%	74.5%	29%	26,211
9,000,001	10,000,000	9,500,001	224,765	60%	134,859	42.3%	70.2%	28%	37,607
10,000,001	12,500,000	11,250,001	439,826	60%	263,896	38.2%	64.1%	26%	68,133
12,500,001	15,000,000	13,750,001	229,173	60%	137,504	33.9%	57.4%	24%	32,417
15,000,001	17,500,000	16,250,001	181,344	60%	108,806	30.5%	52.4%	22%	23,799
17,500,001	20,000,000	18,750,001	188,916	60%	113,349	27.7%	48.4%	21%	23,427
20,000,001	25,000,000	22,500,001	260,115	60%	156,069	24.5%	43.7%	19%	29,970
25,000,001	30,000,000	27,500,001	446,295	60%	267,777	21.3%	38.7%	17%	46,822
30,000,001	35,000,000	32,500,001	156,364	60%	93,818	18.9%	35.0%	16%	15,151
35,000,001	40,000,000	37,500,001	371,903	60%	223,142	17.0%	32.1%	15%	33,705
40,000,001	45,000,000	42,500,001	121,245	60%	72,747	15.5%	29.6%	14%	10,237
45,000,001	50,000,000	47,500,001	2,003,079	60%	1,201,847	14.2%	27.5%	13%	160,488
						Overall cost to layer			**755,338**

FIGURE 22.7
In this exposure rating calculation, the layer is defined as D = £2M, L = £3M; the exposure curve is a Swiss Re curve with $c = 4.0$; the expected loss ratio is 60%. The expected losses (column F) are calculated from the premium (column D) and the expected loss ratio (column E) as $F = D \times E$. To obtain the value of d and $d + l$ to be used in columns G and H for the calculation of $G(d)$ and $G(d + l)$ (G is the exposure curve), the value of the deductible and the exit point (= deductible + limit) is divided by the average sum insured [column C; note that $C = (A + B)/2$], and capped to 1. Therefore, for example, $G(d) = 100\%$ every time that the deductible is larger than the average sum insured in the relevant band. The expected loss cost to the layer (column J) is calculated for each band as $J = (H - G) \times I$. The overall cost to the layer (£755,838) is the sum of all the (other) values in column J. Note that in the table above, the inputs are on a white background, while the calculated figures are on grey background.

on the underwriting cycle and on the consistency of the insurer's pricing strategy. Furthermore, the loss ratio might have fluctuated quite a bit in the past due to the presence or absence of large losses or out of sheer randomness, with the usual consequences on parameter uncertainty.

3. The *property schedule* needs to be fully up-to-date for the exercise to reflect the true risk. Specifically, the information on the MPL and the sum insured must be correct and up-to-date.
4. We must always make sure that the *perils covered by the policy* match those for which the exposure curves have been derived. Any mismatching adds uncertainty to our expected loss estimate.

As an example, these curves were normally derived for non-catastrophe perils – mostly fire – so applying them to include catastrophe risks is always problematic. Catastrophe risks should be analysed independently (see Section 22.8.2.6).

22.7.3 Simulation of the Aggregate Loss Distribution

In Section 22.7.2, we saw how exposure rating could be used to produce an estimate of the expected losses to a layer of reinsurance. As we've done for experience rating, would like to be able to also produce an estimate of the full aggregate loss distribution; and as for experience rating, we can do this by using a frequency/severity model. Interestingly, an approximate frequency/severity model can indeed be produced in exposure rating with a suitable use of the information contained in the property schedule smartly.

BOX 22.5 ESTIMATING THE AGGREGATE LOSS DISTRIBUTION FOR A LAYER L XS D FROM A BANDED PROPERTY SCHEDULE

Inputs: the exposure curve G, the property schedule by band, the premium per band, the expected loss ratio.

STEP 1 – PRODUCING A SEVERITY MODEL FOR EACH BAND:

1. For each band b, approximate the severity curve assuming that all properties have MPL equal to M_b. Given the exposure curve G, we can derive (Equation 22.12) the normalised severity curve $F(x)$, which once scaled to M_b represents the monetary amounts.
2. Based on the severity model thus produced, we can also produce an estimate of the average individual loss amount for each band, $\mathbb{E}(X_b)$.

STEP 2 – PRODUCING A FREQUENCY MODEL FOR EACH BAND:

3. For each band, we also have an estimate of the expected ground-up losses based on the premium written for that band, P_b, and the expected loss ratio for that band $\mathbb{E}(LR_b)$ (normally, this will be the same for all bands): $\mathbb{E}(S_b) = \mathbb{E}(LR_b) \times P_b$.
4. Remembering the collective model (Equation 6.3) we can derive an estimate for the expected number of losses for that band: $\mathbb{E}(N_b) = \mathbb{E}(LR_b) \times P_b / \mathbb{E}(X_b)$.
5. We can assume that the number of claims in band b follows a Poisson process with rate $\lambda_b = \mathbb{E}(N_b)$.

STEP 3 – PRODUCING AN AGGREGATE LOSS MODEL ACROSS ALL BANDS:

6. The aggregate loss distribution across all bands can be obtained from the frequency and severity models for each band by Monte Carlo simulation using a large number N of scenarios. For each scenario:
   a. Sample at random the number of losses for band b, N_b
   b. For each of the N_b losses for band b, sample at random a loss amount based on the severity curve for each band: $X_{b,1}, X_{b,2}, \dots X_{b,N_b}$ (see, e.g., Section 22.5.3 for the case of Bernegger curves)
   c. The total loss for simulation k is the sum of all losses generated at random across all bands: $S = \sum_{b=1}^{B} \sum_{j=1}^{N_b} X_{b,j}$
   d. The total loss to the layer $(D, D + L)$ for simulation k is, analogously, the sum of all losses to the layer across all bands: $S(D,L) = \sum_{b=1}^{B} \sum_{j=1}^{N_b} \min\left(\max\left(X_{b,j} - D, 0\right), L\right)$

Sort the values of S and $S(D,L)$ in ascending order to obtain the empirical aggregate loss distribution for both the ground-up losses and the losses to the layer.

Note that this simulation method implicitly assumes that the loss processes for properties of the different bands are independent: the method above is equivalent to simulating first separately for each band, and then summing the results scenario by scenario – the numerical equivalent of performing the convolution of the aggregate loss distributions of each band.

Output: the aggregate loss distribution and the ceded loss distribution for layer $(D, D + L)$, in the same format as the output of an experience rating exercise.

The underlying idea is this: we already know that the exposure curve provides a severity model, and therefore we can estimate the average severity $\mathbb{E}(X_b)$ of an individual loss for each band. Furthermore, we know the expected ground-up losses $\mathbb{E}(S_b)$ for each band (column F). Using the standard formula $\mathbb{E}(S_b) = \mathbb{E}(N_b) \times \mathbb{E}(X_b)$ for the collective model (see Equation 6.3), this provides us with an estimate of the expected number of claims for each band: $\mathbb{E}(N_b) = \mathbb{E}(S_b) / \mathbb{E}(X_b)$. This is not yet a frequency model, but it is commonly assumed that the number of claims follows a Poisson distribution with Poisson rate $\mathbb{E}(N_b)$.

The details of how to simulate an aggregate loss distribution based on banded property exposure information is shown in Box 22.5. We are looking at the pricing of a layer L xs D by way of example but the extension to more complex structures is trivial.

Note that the algorithm above can also be used in a simplified version in which an overall severity distribution is first produced by sampling from all bands in proportion to the relevant expected claim count. This might be quite a large sample with, for example, 100,000 different loss amounts. We can then simulate using for the number of losses a process which has an average claim count of $E(N) = \sum_{b=1}^{B} E(N_b)$, and as a severity model the 100,000 values derived above.

22.8 Exposure Rating in Direct Insurance

Although property exposure rating was born in the context of reinsurance, it is routinely used in a direct insurance as well. The core difference between reinsurance and direct insurance is that in direct insurance there is obviously no such thing as an 'original premium' that one can use to calculate the ground-up expected losses for each property (or each band). A direct insurer will need an independent estimate of the ground-up expected losses for each property. This independent estimate typically comes in the form of a 'rate on value', a prior estimate of the expected losses expressed as a percentage of the insured value of the property.

In the following, we will look at the main ingredients and the methodology typically used for pricing. We will limit ourselves to the pricing of what are traditionally referred to as FLEXA losses (fire/lightning/explosion/aircraft collision), ignoring the treatment of natural catastrophe losses, which are normally not addressed by exposure rating but require the use of catastrophe models.

We will consider the case where we have separate rates and curves for property damage and business interruption, which is what normally happens in more sophisticated models.

22.8.1 Ingredients to Pricing

A pricing exercise requires certain inputs about the client's properties, such as the value of the properties and their location, and on the coverage (Section 22.8.1.1). Pricing guidelines (typically incorporated into a pricing model) will then use such information to determine a suitable rate on value (ROV) for the specified programme (Section 22.8.1.2) and ultimately the premium to be charged (Section 22.8.1.3).

22.8.1.1 Contract-Specific Inputs

The two main inputs that we need for pricing purposes are:

a) Cover details (attachment point, limit, co-insurance share ...)
b) A property schedule

The cover details require no further explanation. As for the property schedule, we should note that unlike in reinsurance, where the number of properties is very large and it is common to aggregate properties into bands by insured value or MPL, in direct insurance the full list of properties is typically available for pricing and underwriting purposes.

The property schedule will typically specify for each location:

- the type of **occupancy** (refinery, paper mill, supermarket ...), geographical location (used especially for cat modelling) and possibly additional information such as construction material;
- the **insured value (IV)** – normally split into a property damage (PD) and business interruption (BI) value;
- **maximum possible loss (MPL)** – unlike the insured value, which is a contractual item, this is typically estimated by the underwriter based on the information that they receive from the client or from risk engineers. This will also be separate for PD and BI;

ID	*Occupancy*	*IV (PD)*	*IV (BI)*	*MPL (PD)*	*MPL (BI)*	*LD (PD)*	*LD (BI)*	*IP*
1	Refinery	$800M	$500M	$400M	$200M	$1M	30d	18M
2	Office	$50M	$10M	$25M	$10M	$100k	5d	12M
3	Warehouse	$10M	$10M	$10M	$10M	$50k	3d	6M

FIGURE 22.8
An example of a property schedule with three locations.

- **local deductible (LD)** – this is the underlying deductible that is removed from the gross losses before applying the terms of the cover, and varies in general depending on the individual location (hence the qualifier 'local'). This can be a monetary amount or (in the case of BI) a number of days. A combined PD/BI monetary deductible is sometimes used;
- **indemnity period (IP)** – this is the maximum period over which BI is covered. It's normally expressed in months.

An example of a property schedule is shown in Figure 22.8.

As an example of how the local deductible works, consider a contract with attachment point $5M and limit $5M, with the property schedule shown above. Assume that your office (ID=2) has experienced a fire loss which has caused a property damage loss of $7.3M, and a business interruption of 16 days, which translates roughly (assuming that each day of interruption causes a loss of $10M /365 days ~ $27.4K ($10M is the MPL for BI)) into a monetary loss of $438K.

To calculate the ceded loss, you first remove the deductible from the PD and BI loss: the PD loss net of LD is therefore $7.2M, and the net BI loss is 16–5=11 days of loss, corresponding roughly to[6] $300K, for a total loss of $7.5M net of local deductibles. This translates into a ceded loss of $2.5M to the $5M xs $5M layer of insurance.

22.8.1.2 Portfolio-Level Pricing Guidelines

For each type of occupancy (e.g. Crude Processing – Refineries), the following will typically be specified in the pricing guidelines for a property portfolio:

- A table of **base rates on value (base ROV**, or simply **BR)**, which give the expected losses (or, in traditional underwriting practice, the premium)[7] per unit of insured value assuming a **standard deductible** (SD), i.e. a reference level of local deductible. Normally expressed in ‰ (per mille) of the insured value.
- A table of **exposure curves parameters**. These could be for example the values of b and g for the Bernegger curves (Section 22.5). The curves may either be from the ground up or be provided in excess of the SD. Unless otherwise specified, we will

6 This is only an approximation – in practice, BI loss calculations will be based on the actual loss of profit minus the variable costs.

7 In traditional underwriting practice, the ROV is the premium per unit of insured value. In actuarially informed pricing models, however, the ROV is the expected loss ('pure premium') per unit of insured value, which allows a separate treatment of the loadings. Unless otherwise specified, in this book we will stick to the actuarial definition of ROV, but the reader should be aware of the ambiguity when looking at the output of pricing models.

assume that the exposure curve $G(d)$ is defined in terms of the MPL, and that the likelihood of a loss above MPL is zero.[8]

- A table of **BI-specific parameters**, e.g. the factor by which the base rate should be multiplied for changes to the indemnity period.
- If the exposure curves are provided in excess of the underlying deductible, a **deductible change impact table or function** must be provided that spells out the impact on the rates of using a local deductible that is different from the standard deductible (see Section 22.8.2.1 for examples).
  Note: If the exposure curve is given from the ground up, such table is implicitly incorporated in the exposure curve.
- Rules on how much premium or expected losses should be allocated between MPL and IV.

22.8.1.3 Company-Level Information

These include rules (along with the relevant parameters) on how to calculate the technical premium based on expected losses, expenses, investment income and target profit, in a 'Cost+' fashion (see Chapter 19). Since these rules depend critically on the financials of the company (including such information as the cost of capital) they may change regularly.

Note that when the rates on value are already premium rates this step is not necessary.

22.8.2 Pricing Methodology

Using the ingredients specified in Section 22.8.1, it is possible to tackle various pricing-related issues.

Note that in the following calculations we will normally assume that for BI we can translate the (local) deductible expressed in days into a monetary amount using the rough approximation:

$$LD[\text{monetary amount}] = LD[\text{days}] \times \frac{MPL}{IP} \tag{22.31}$$

Local deductible	Multiplication factor
0% of standard deductible	× 4.0
25% of standard deductible	× 2.0
50% of standard deductible	× 1.5
100% of standard deductible	× 1.0
150% of standard deductible	× 0.8
200% of standard deductible	× 0.6

FIGURE 22.9
An example of a property schedule with three locations.

8 This is correct theoretically based on the definition of the MPL, but the MPL is only an estimate and because of the uncertainty around this estimate losses above the MPL are indeed possible. See discussion in Section 22.8.4.1.

22.8.2.1 ROV and Expected Losses for the Full Value of a Property (Single Location)

By 'full value' we normally mean the whole value of the property (TIV) in excess of the local deductible. Since we have a table that gives us the standard deductible for each type of occupancy, finding the rate at the local deductible is straightforward. The actual method depends on whether the exposure curve used by the company is from the ground up or in excess of the local (underlying) deductible.

22.8.2.1.1 ROV Using Exposure Curves in Excess of the Underlying Deductible

In this case, the rate on value can be simply obtained from the base rate using the rates vs local deductible table. This table can be something as simple as a table like this:

The table above allows us to calculate the rate on value by multiplying the base rate by the relevant multiplication factor: e.g. if the standard deductible is £100,000 and the base rate is 0.5‰, the rate on value for a deductible of £150,000 is going to be 0.5‰ × 0.8 = 0.4‰. Obviously, if the value of the local deductible sits between different rows in the table above some interpolation is necessary.

A less crude approach would allow for the multiplication factor to depend on the actual monetary value of the standard deductible *and* on the MPL (a £1M deductible will of course have less impact on a £1bn property than on a £10M property!).

In general, what we need is something that can be formally described as a transformation matrix $\mathbb{T}(SD, LD, MPL)$ that gives the multiplication factor to go from the standard deductible to a specific local deductible for a property with a given MPL:

$$\text{ROV}(LD) = \text{BR} \times \mathbb{T}(SD, LD, MPL) \tag{22.32}$$

where BR is the base rate (on an expected losses basis or a premium basis). The transformation matrix can be given in the form of a table or as an algebraic relation, such as (see Lee & Frees, 2016):

$$\mathbb{T}(SD, LD, MPL) = \left(\frac{\theta + \frac{LD}{MPL}}{\theta + \frac{SD}{MPL}} \right)^{-\alpha+1} \tag{22.33}$$

For example, if $\theta = 1$, $\alpha = 10$ and MPL=£10M, going from a standard deductible of £100K to a local deductible of £50k causes a 4.5% increase in the rates. Note that θ is a pure number and that this formula is currency-independent.

22.8.2.1.2 ROV Using from-the-Ground-Up Exposure Curves

In this case, the rate on value can be obtained from the base rate using the properties of the exposure curves.

Let BR be the base rate (either on a premium or expected losses basis), and FGU_Rate the rate that would be applied if there were no deductible. Then (using the definition of exposure curve):

$$BR = \text{FGU\_Rate} \times \left(1 - G\left(\frac{SD}{MPL} \right) \right) \tag{22.34}$$

and

$$ROV(\text{LD}) = \text{FGU\_Rate} \times \left(1 - G\left(\frac{LD}{MPL}\right)\right) \tag{22.35}$$

Hence,

$$\text{ROV(LD)} = BR \times \frac{1 - G\left(\dfrac{LD}{MPL}\right)}{1 - G\left(\dfrac{SD}{MPL}\right)} \tag{22.36}$$

(Remember that we have assumed for simplicity that no premium should be allocated between MPL and TIV.)

22.8.2.1.3 Expected Loss Calculation for the Full Value of the Property

Once we have the ROV for the local deductible the expected losses for the full value of a single location are given by

$$\mathbb{E}(S) = \text{ROV}(LD) \times IV \tag{22.37}$$

22.8.2.1.4 Expected Losses for PD and BI Combined

The formulae above hold for both PD and BI separately. The overall expected losses for a given location are obtained by simply summing over the two contributions:

$$\mathbb{E}(S) = \text{ROV}^{PD}(LD^{PD}) \times IV^{PD} + \text{ROV}^{BI}(LD^{BI}) \times IV^{BI} \tag{22.38}$$

22.8.2.2 Expected Losses and Premium for a Layer of Insurance

Again, this calculation proceeds differently depending on whether we have a from-the-ground-up exposure curve or our exposure curve is on top of a standard deductible. Note that the calculations in this section are exclusively for modelling PD-only losses, or when a single exposure curve is used to capture combined PD/BI losses approximately. The general treatment of losses with a PD and a BI component requires stochastic modelling (Section 22.8.2.3).

22.8.2.2.1 Using Exposure Curves in Excess of an Underlying Deductible

The expected losses to a layer L xs D under a local deductible equal to LD can in this case be written as

$$\mathbb{E}(S(D,L)) = \text{ROV}(LD) \times \left(G\left(\frac{D+L}{MPL-LD}\right) - G\left(\frac{D}{MPL-LD}\right)\right) \times IV \tag{22.39}$$

where $\text{ROV}(LD)$ is calculated as in Section 22.8.2.1. Note that we have subtracted LD from MPL since MPL is a ground-up quantity and the exposure curve is defined above LD. (Honestly this is a bit precious given the fact that the local deductible should normally be very small with respect to the MPL, and that the MPL is an estimate anyway – so in practice this will often be ignored and it will be assumed that the MPL is above the local deductible whatever value that has.)

22.8.2.2.2 Using from-the-Ground-Up Exposure Curves

The expected losses to a layer L xs D, with $L+D<MPL$ and with a local deductible equal to LD can in this case be written as

$$\mathbb{E}(S(D,L)) = \text{FGU\_Rate} \times \left(G\left(\frac{D+L+LD}{MPL} \right) - G\left(\frac{D+LD}{MPL} \right) \right) \times IV \tag{22.40}$$

22.8.2.3 Simulation of the Aggregate Loss Distribution (Single Location)

The exposure curve approach can also be used for stochastic modelling, using a frequency/severity approach to produce not only an estimate of the expected losses for the full value or for a layer (which can only be done analytically for simple structures with no aggregate features), but also an estimate of the volatility and of the full aggregate loss distribution. We will start by looking at the case of a single location. In Section 22.8.2.4 we will build on this to consider contracts with multiple locations (which is the most common case).

We will look at the simulation algorithm for the case where the exposure curve is defined from the ground up. The other case requires only simple modifications and is given in the solution to Question 10.

22.8.2.3.1 Severity Model

The severity model is uniquely determined by the exposure curve $G(x)$, which is in one-to-one correspondence with the normalised severity distribution $F(x)$ through Equation 22.10. We will in general need two severity models, one for the PD component and the other for the BI component, each with their own parameters. We will assume that both of them can be approximated as Bernegger curves.

We will also need to consider that the severity of the PD and BI component are correlated: under standard terms and conditions you can have a PD loss without a BI component but not vice versa; also, a large PD loss is more likely than a small one to cause a large BI loss. Chapter 29 addresses the problem of modelling correlated variables, e.g. through copulas; in this chapter we'll simply assume that we can simulate correlated variables.

22.8.2.3.2 Frequency Model

As for frequency, the most commonly adopted frequency model in a property insurance context is the Poisson distribution, which should normally be sufficient if systemic effects can be disregarded (see Chapter 14), a reasonable assumption in property insurance.[9] We will again need two models, one for PD and one for BI. The two models are related because under standard cover a business interruption claim requires some degree of property damage. We will use therefore need:

- a Poisson model for PD claims: $N^{PD} \sim \text{Poi}(\lambda^{PD})$, where λ^{PD} is the expected number of PD claims for that location;
- a Poisson model for BI claims: $N^{BI} \sim \text{Poi}(\lambda^{BI})$, where λ^{BI} is the expected number of BI claims for that location. We can assume that $\lambda^{BI} \leq \lambda^{PD}$;
- rules that ensure that we do not generate BI-only losses (requiring that $\lambda^{BI} \leq \lambda^{PD}$ is not enough).

9 Of course, there are systemic elements affecting multiple locations when catastrophes are involved, but as we said catastrophes will need to be dealt with separately using catastrophe models, not standard exposure rating.

BOX 22.6 SIMULATING THE AGGREGATE (GROSS) LOSS DISTRIBUTION FOR A SINGLE LOCATION

Inputs

- Location features: type of occupancy, IV^{PD}, IV^{BI}, MPL^{PD}, MPL^{BI}
- Parameters of the Bernegger's curves for both PD and BI: b^{PD}, g^{PD}, b^{BI}, g^{BI}; base rates BR^{PD}, BR^{BI}; standard deductibles SD^{PD}, SD^{BI} (all for the relevant occupancy)

a. Calculate the ground-up frequency for both PD and BI: $\lambda^{PD}, \lambda^{BI} \leq \lambda^{PD}$ using Equation 22.41. Calculate the ratio $r = \lambda^{BI} / \lambda^{PD}$. Note that $0 \leq r \leq 1, r = 0$ corresponding to a PD-only cover.
b. For each simulated scenario, sample a number of PD losses: $N^{PD} \sim \text{Poi}(\lambda^{PD})$
c. For each PD loss $j = 1, \ldots N^{PD}$, sample a severity amount X_j^{PD} $(0 < X_j^{PD} \leq MPL^{PD})$ for the Bernegger distribution with parameters b^{PD}, g^{PD} using the algorithm in Box 22.1
d. For each PD loss $j = 1, \ldots N^{PD}$, generate a pseudo-random uniform number u between 0 and 1 and if $u < r$ also generate a BI loss X_j^{BI} from a Bernegger distribution with parameters b^{BI}, g^{BI}. If $u \geq r$ simply set $X_j^{BI} = 0$ (meaning that this is a PD-only loss)
e. For each loss $j = 1, \ldots N^{PD}$, calculate the total loss: $X_j = X_j^{PD} + X_j^{BI}$
f. For each loss $j = 1, \ldots N^{PD}$, calculate the total loss net of the local deductible(s): $X_j(\text{net of } LD) = \max(0, X_j^{PD} - LD^{PD}) + \max(0, X_j^{BI} - LD^{BI})$. Note that in the case of a combined PD/BI deductible, that will become $X_j(\text{net of } LD) = \max(0, X_j^{PD} + X_j^{BI} - LD^{\text{combo}})$
g. Calculate the aggregate loss for each scenario by summing over the J losses: $S(\text{net of } LD) = \sum_{j=1}^{N^{PD}} X_j(\text{net of } LD)$
h. Repeat for N different scenarios
i. Sort the values of $S(\text{net})$ in ascending order to produce an empirical aggregate (gross) loss distribution

The Poisson rate for the number of claims can be easily determined from the knowledge of the expected aggregate losses (from the ground up) for a given location, $\mathbb{E}(S) = \text{FGU\_Rate} \times IV$, the application of the usual equality for the collective risk model, $\mathbb{E}(S) = \lambda \times \mathbb{E}(X)$, and formula for the expected severity from the ground up (Equation 22.15), $\mathbb{E}(X) = MPL / G'(0)$.

The frequency from the ground up is then:

$$\lambda = \frac{BR \times IV \times G'(0)}{\left(1 - G\left(\frac{SD}{MPL}\right)\right) \times MPL} \tag{22.41}$$

Equation 22.41 can be used for both PD and BI: note that for BI we may also have to translate the standard deductible into a monetary amount as per Equation 22.31.

The simulation then proceeds as in Box 22.6.

22.8.2.3.3 Aggregate ceded loss distribution – An example

The outcome of this simulation is an empirical aggregate loss distribution on a gross basis. If we are interested in losses for a specific policy structure (e.g. a layer of reinsurance) we can as usual (see Chapter 17) apply to each of the losses X_j (net of LD) generated by the process the relevant policy modifier, and then apply the relevant aggregate modifiers to the aggregate loss distribution.

For example, to price a layer of insurance L xs D with an aggregate limit AL we need to calculate for each scenario

$$S(\text{ceded}) = \min\left(\sum_{j=1}^{N^{PD}} \min\left(L, \max\left(0, X_j\left(\text{net of } LD\right) - D\right)\right), AL\right)$$

and sort the values in ascending order to produce the empirical aggregate *ceded* loss distribution.

22.8.2.4 Contracts with Multiple Locations

The formulae in Section 22.8.2.2 are for contracts with a single location. Most contracts, however, include multiple locations. The generalisation of these results to multiple locations is trivial: if $S_k\left(D, L\right)$ are the losses for location $k = 1, \ldots K$, the expected losses to a L xs D layer for the whole contract for the case of the ground-up exposure curve are simply given by the sum of the expected losses over all locations:

$$\mathbb{E}\left(S(D,L)\right) = \sum_{k=1}^{K} \mathbb{E}\left(S_k\left(D,L\right)\right) = \text{ROV}\left(LD_k\right) \times \left(G_k\left(\frac{D+L+LD_k}{MPL_k}\right) - G_k\left(\frac{D+LD_k}{MPL_k}\right)\right) \times IV_k \quad (22.42)$$

The case where the exposure curves are in excess of the underlying deductibles is analogous. As for the single-location case, this formula holds for the PD-only case or for the case where the exposure curve is for combined PD and BI, while the general case has to be addressed with stochastic modelling (Box 22.7).

22.8.2.4.1 Simulation of the aggregate loss distribution for contracts with multiple locations

The full aggregate loss distribution for a multi-location contract can be estimated easily enough if we assume that losses arising from the different locations are independent (a safe enough assumption for non-catastrophe losses).

The algorithm is similar to that for a single location, the main difference being that once the number of losses is generated from a Poisson distribution (whose mean is equal to the sum of the mean for each of the K individual properties) the losses will need to be matched to the locations in proportion to the expected claim count for each location. The severity of each claim (both PD and BI components) will then need to use the exposure curve parameters for the occupancy relevant to those locations. The algorithm is described in some detail in Box 22.7.

As explained at the end of Section 22.8.2.3, this algorithm for the gross loss distribution can be easily adapted for the calculation of the distribution of ceded or retained losses.

22.8.2.5 Premium Calculation

The technical premium for a property contract can be calculated based on the expected losses, the whole aggregate distribution, and the other company-level guidelines on expenses, cost of capital, etc. using a formula such as that discussed in Section 19.1.6.

BOX 22.7 SIMULATING THE AGGREGATE (GROSS) LOSS DISTRIBUTION FOR A CONTRACT WITH MULTIPLE LOCATIONS

Inputs

- Location features: occupancy type, $IV^{PD}(k)$, $IV^{BI}(k)$, $MPL^{PD}(k)$, $MPL^{BI}(k)$ for all locations $k = 1,\ldots K$
- Parameters of the Bernegger's curves for both PD and BI, base rates; standard deductibles for all the occupancy types

a. Calculate the ground-up frequency for both PD and BI: $\lambda^{PD}(k), \lambda^{BI}(k)$ for all locations using Equation 22.41. Calculate the ratio $r(k) = \lambda^{BI}(k)/\lambda^{PD}(k)$. Note that $0 \le r(k) \le 1, r(k) = 0$ corresponding to a PD-only cover.
b. For each simulated scenario, sample a number of PD losses: $N^{PD} \sim \text{Poi}\left(\sum_{k=1}^{K} \lambda^{PD}(k)\right)$
c. For each PD loss $j = 1,\ldots N^{PD}$, allocate the loss to one of the K locations in proportion to the ground-up frequencies $\lambda^{PD}(1), \lambda^{PD}(2)\ldots\lambda^{PD}(K)$. Let $\text{loc}(j) \in \{1,2\ldots K\}$ be the location selected for loss j
d. For each PD loss $j = 1,\ldots N^{PD}$, sample a severity amount X_j^{PD} $(0 < X_j^{PD} \le MPL^{PD}(\text{loc}(j))$ for the Bernegger distribution with parameters relevant to the occupancy type for that location, using the algorithm in Box 22.1
e. For each PD loss $j = 1,\ldots J$, generate a pseudo-random uniform number u between 0 and 1 and if $u < r(\text{loc}(j))$ also generate a BI loss $X_j^{BI} \le MPL^{BI}(\text{loc}(j))$ from the Bernegger distribution relevant to that occupancy. If $u \ge r(\text{loc}(j))$ simply set $X_j^{BI} = 0$ (meaning that this is a PD-only loss).
f. For each loss $j = 1,\ldots J$, calculate the total loss: $X_j = X_j^{PD} + X_j^{BI}$
g. For each loss $j = 1,\ldots J$, calculate the total loss net of the local deductible(s) for that location: $X_j(\text{net}) = \max\left(0, X_j^{PD} - L^{PD}(\text{loc}(j))\right) + \max\left(0, X_j^{BI} - L^{BI}(\text{loc}(j))\right)$. Note that in the case of a combined PD/BI deductible, that will become $X_j(\text{net}) = \max\left(0, X_j^{PD} + X_j^{BI} - L^{\text{combo}}(\text{loc}(j))\right)$.
h. Calculate the aggregate loss for each scenario by summing over the J losses: $S(\text{net}) = \sum_{j=1}^{J} X_j(\text{net})$
i. Repeat for N different scenarios
j. Sort the values of $S(\text{net})$ in ascending order to produce an empirical aggregate loss distribution

22.8.2.6 Dealing with Natural Catastrophe Losses and Other Complications

The pricing methodology that we have discussed here is the basic form that can be used for exposure rating in property insurance. In practice, more bells and whistles need to be added in practice: as an example, the system may need to deal not only with FLEXA losses but also with machinery breakdown losses, with Contingent Business Interruption (CBI), and especially with natural catastrophe losses.

Natural catastrophe modelling is indeed an integral part of most property pricing models, and the canonical way of dealing with it is by using catastrophe modelling tools. These are dealt with in some detail in Chapter 25. For now, the important thing to notice is that catastrophe losses coming from cat modelling tools such as RMS or EQECAT need to be added to the non-cat losses and also merged in the simulation process – which is relatively straightforward as cat and non-cat losses can be assumed to be independent and therefore simulations from non-cat losses and cat simulations can be added scenario by scenario (which achieves what in mathematics is called a convolution).

An extra layer of complexity comes from the fact that not all relevant cat losses are modelled by the current cat modelling tools available in the market or to the company. Non-modelled catastrophe losses therefore will need to be treated separately. This is normally done by introducing 'cat rates' analogous to our base rates described above, which depend on the combination of peril and territory (see Chapter 26).

BOX 22.8 AN EXAMPLE OF HYBRID EXPERIENCE/ EXPOSURE RATING METHODOLOGY FOR PROPERTY

Inputs

- The property schedule at the most recent date available
- Loss/claim experience over a number of years and possibly over a threshold
- Historical exposures, such as total sum insured over the same period as the loss data set

STEP 1 – PRODUCING A FREQUENCY MODEL (BASED ON EXPERIENCE): Produce a frequency model, much in the same way as explained in Chapter 13. This will typically be Poisson model with a given Poisson rate.

STEP 2 – PRODUCING A SEVERITY MODEL FOR THE WHOLE PROPERTY PORTFOLIO (BASED ON EXPOSURE):

i. Select the desired exposure curve G for the type of properties in the insurer's portfolio (different exposure curves may be used for different locations, depending on the occupancy type).
ii. Let us assume that we have K properties, each with its MPL. We build a 'semiempirical' severity distribution of N points (N should be large, e.g. 10,000 or 100,000 points) by this simple intermediate simulation:
    a. Pick a property k at random ($k = 1, \dots K$) with probability proportional to the insured value of that location (this criterion can be refined if we have some prior assumption around the base rate relativities between different occupancies).
    b. Draw at random a value of the relevant MPL-normalised severity for the selected property k, then multiply by MPL to get the absolute severity value.
    c. Repeat this N times.
    d. Sort all the losses thus obtained in ascending order.
    e. The output is a sample of losses that we can use as a severity model (admittedly a rather ragged one) for the portfolio of losses.

STEP 3 – PRODUCING AN AGGREGATE LOSS MODEL ACROSS ALL BANDS:

i. Because we now have a frequency model (Step 1) and a severity model (Step 2), we can combine them to produce an aggregate loss model in the usual way, as shown for example in Chapter 6 ('The scientific basis for pricing').
ii. *Output*: the aggregate loss distribution and the ceded/retained loss distribution as needed, in the same format as the output of an experience rating exercise.

22.8.3 Hybrid Experience/Exposure Rating for Property: Basic Algorithm

As discussed at the beginning, stand-alone experience rating does not generally work for property pricing. However, a combination of experience and exposure rating might be useful: the attritional loss experience may provide a better guide to future attritional losses than the small losses coming out of exposure rating, and the past observed claim count may provide a better frequency estimate than that based on base rates.

However, the number of losses experienced in the past may give valuable information on the frequency of losses expected in the future, and this added information might make the premium information unnecessary. Having these things in mind, let us see how the algorithm above can be modified to rate property damage and business interruption using exposure rating.

A hybrid experience/exposure rating approach might therefore use observed attritional loss experience for frequency modelling and exposure curves with the current MPL profile could be used to produce an overall severity model. A possible methodology that achieves this is outlined in Box 22.8. For simplicity of exposition, we will assume PD losses only, but the extension should be obvious based on the material in Section 22.8.2.

Note that if enough client losses are available, it may make sense to go one step further use the client's losses up to a certain threshold and use the severity derived by exposure rating to model the tail of the severity distribution. We will leave the necessary modification of the methodology described in Box 22.8 to the reader's initiative. Note that this is similar to the general approach to severity modelling (Chapter 16) where one only trusts the client's severity distribution up to a certain threshold and uses a portfolio or market curve above that threshold; it is also similar to what we saw in Chapter 11 when discussing how burning cost analysis could be enhanced with ILF curves to tackle the problem of large losses.

*22.8.4 Non-Scalable Losses and MPL Uncertainty: Relaxing the Assumptions Behind Exposure Rating

In this section we'll look critically at two assumptions that we have used for exposure rating: (a) loss scalability; (b) absence of losses above the MPL; and point to what can be done when these assumptions are not satisfied. The reader interested in exploring this topic further is referred to Parodi (2020), from which most of the material below is taken.

*22.8.4.1 *Scalable vs Non-Scalable Losses*

As we have seen in Section 22.1, property exposure rating is predicated on the assumption of loss scalability: for a given occupancy type, we normally assume that the probability that the damage ratio is less than or equal to a given percentage is independent of the MPL. This allows us to build *normalised* severity curves, i.e. severity curves that depend only on the damage ratio and not on the absolute monetary amount of the loss. From these severity curves we can build *exposure curves*, which are also normalised by construction.

However, this assumption is only true to the extent that the properties are indeed similar. The fact that each property is unique and that two properties never share exactly the same exposure curve, is, however, only the most obvious problem with exposure rating. Another, subtler one is that some of the losses may not actually be scalable, i.e. they may remain the same (or scale non-proportionally) when 'transferred' to a property of different value. Possible examples of this are content of fixed value or contained electrical damage.

It is tempting to identify non-scalable losses with the attritional losses, and the scalable losses with the large losses. This is not completely correct as scalable losses can also be small and some non-scalable losses can be large, but it is useful as a first approximation and is actually reflected in the practice of considering exposure curves in excess of the local deductible rather than from the ground up, as we saw in Section 22.8.2, and have a separate table that quantifies the impact of changing the local deductible on the rates.

This deserves expanding a little. If one uses a single MPL-independent exposure curve, especially with large industrial properties where MPLs in excess of $100M or even $1bn are not unusual, one will notice that the effect of changing the local deductible from a standard value of, say, $1M to a value of $500k is negligible, while the market will charge a significantly higher rate (let's say, +10%). This might be partly for commercial reasons, but many underwriters argue that the main reason for the rate increase is that the local deductible allows to get rid of attritional losses that are not scale-dependent and that represent a non-negligible monetary amount.

We may therefore need a mechanism to deal with non-scalable losses. One solution is to build separate frequency and severity models for scalable losses (standard portfolio exposure curves) and non-scalable losses (using monetary severity curves, perhaps derived for each client via experience rating).

Another solution (Parodi, 2020) is to use *generalised exposure curves* that combine scalable and non-scalable losses and that are therefore MPL-dependent. This can be achieved by assuming that losses are a mixture of the non-scalable losses (with cumulative distribution function $F_A(x)$, and maximum possible loss M_A) and scalable losses (with CDF $F_L(x)$, and maximum possible loss MPL). The combined CDF can be written as:

$$F(x) = \frac{\lambda_A / \lambda_L}{1 + \lambda_A / \lambda_L} F_A(x) + \frac{1}{1 + \lambda_A / \lambda_L} F_L(x) \tag{22.43}$$

where λ_A / λ_L is the ratio between the frequency of non-scalable (attritional) and scalable (large) losses, which we can often assume is an invariant across properties of a given occupancy type. The corresponding exposure curve can then be calculated using Equation 22.10, leading again to a convex combination of the exposure curves $G_A(d), G_L(d)$ corresponding to $F_A(x)$, $F_L(x)$:

$$G(d) = w_A G_A(d) + (1 - w_A) G_L(d) \tag{22.44}$$

where $w_A = \lambda_A \mathbb{E}(X_A) / (\lambda_A \mathbb{E}(X_A) + \lambda_L \mathbb{E}(X_L))$, and in turn $\mathbb{E}(X_A) = M_A \times \mathbb{E}(x_A)$, $\mathbb{E}(X_L) = MPL \times \mathbb{E}(x_L)$ are the expected absolute monetary amounts. The weight to be given to the two exposure curves $G_A(d)$, $G_L(d)$ will therefore vary depending on the frequency ratio λ_A / λ_L, the expected damage ratio $\mathbb{E}(x_A)/\mathbb{E}(x_L)$, and the ratio of the maximum possible losses M_A / MPL.

A special case of practical relevance is where the idea that non-scalable losses are the same as attritional losses and scalable losses are the same as large losses is taken literally and we assume that all losses below M_A are attritional and all losses above that threshold are scalable. The threshold M_A should be picked so that the local deductible will typically be below that value. See Parodi (2020), Section 4 for further details.

The generalised exposure rating approach adds a layer of complexity during the construction and especially the calibration of the pricing model (see Section 4.4 of Parodi (2020)). From a user's point of view, however, working with it is as straightforward as the standard exposure rating methodology.

*22.8.4.2 MPL Uncertainty

Another practical problem that we address here is that the MPL is not a contractual feature but an underwriter's estimate (perhaps based on an engineering assessment), and therefore it is subject to uncertainty. This means that one needs to allow for the possibility that a loss between MPL and IV (assuming that the latter is larger) occurs.

Companies might deal with this problem in different ways – for example, they may impose a minimum rate on line or rate on value for any layer or part of layer which is above the largest MPL in the client's portfolio. Another simple approach is to impose that a certain percentage of the expected losses be allocated to the region between MPL and IV, for example through a simple rule like this:

$$\%EL(>MPL) = \begin{cases} 0 & \text{if } MPL = IV \\ \%EL_{\max} \times \left(1 - \dfrac{MPL}{IV}\right) & \text{if } MPL < IV \end{cases} \quad (22.45)$$

A more robust way of addressing the problem of MPL uncertainty is to incorporate this uncertainty directly into the exposure curve. The cleanest way of taking this into account is arguably that of allocating a fixed probability p(say, 1%) of exceeding MPL, and then assuming a simple conditional distribution for that region which then translates into the correct behaviour for $G(d)$.

Let's see how this works in practice with a simple example. Let us assume that $F_0(MPL_{est};x)$ is the (non-normalised) severity and exposure curves for losses up to MPL_{est} (the *estimated* maximum possible loss). Let us also assume that the *actual* MPL may be any value between MPL_{est} and IV with equal probability. The overall (non-normalised) severity distribution from 0 to IV can then be written as:

$$F(IV;x) = \begin{cases} (1-p) \times F_0(MPL_{est};x) & \text{if } 0 \le x < MPL_{est} \\ 1-p+p \times (x - MPL_{est})/(IV - MPL_{est}) & \text{if } MPL_{est} < x \le IV \end{cases} \quad (22.46)$$

We can then derive the exposure curve $G(d)$ in the standard way through Equation 22.10[10] (see Question 11).

10 Note that selecting a value of p uniquely entails a value for %EL and vice versa, so one can set either value at the start and derive the other (Parodi, 2020).

This methodology can obviously be refined using more sophisticated probability density functions such as a decreasing exponential to ensure that it is more likely for the actual *MPL* to be near the estimated value rather any random value between *MPL* and *IV*. We could also allow for the actual *MPL* actually to fall either above or *below* the estimated MPL, and use, say, a Gaussian distribution centred around the estimated *MPL*. However, this type of overthinking very quickly leads to spurious accuracy.

Note that we have been assuming that $MPL \leq IV$. In practice, this is not always the case (when, for example, the maximum possible loss for a property allows for the possibility of losing neighbouring properties), but we can safely ignore this case, as in that case we can assume that the severity curve goes all the way up to *MPL*, and *IV* is used for the calculation of the overall expected losses and the expected number of losses but doesn't show in the curve.

A final note of caution: whatever method one decides to use to incorporate MPL uncertainty into the exposure curve, one should always start by extending the cumulative distribution function beyond MPL and *then* calculating the corresponding exposure curve rather than attempting to tamper with the exposure curve itself, as it is very easy in that case to end up with an exposure curve that is not mathematically consistent – e.g. a curve that may lead to negative prices for some layers! Starting with a well-defined CDF (basically: piecewise continuous, monotonically increasing, with finite support), instead, ensures that exposure curve has all the right properties, including continuity and concavity.

22.9 Questions

1. Assume that you have a gunpowder factory which is so badly protected against fire that, once *any* fire starts, all the gunpowder blows up. (i) Draw the severity curve and the exposure curve. (ii) If you were to represent this as a Swiss Re curve with parameter c, what value of c would you need to choose?
2. According to Equation 22.15, the average loss only depends on the derivative at point 0, regardless of what the curve does afterwards. Therefore, very different exposure curves will give the same average loss as long as they start in the same way. How do you explain that?
3. Consider the curve $y = \frac{x^3}{4} - \frac{3}{4x^2} + \frac{3}{2x}$. Can this be an exposure curve? If so, calculate: (a) the probability of a total loss and (b) the average loss.
4. Prove that an exposure curve $G(x)$ is concave in [0,1] in the general case (the one in which $F(x)$ is not necessarily continuous and with defined first derivative over [0,1]).
5. Your house is insured for £300,000 and you pay £500 for your household insurance, which mainly insures your house against fire events. On average, the insurer expects to achieve a loss ratio of 60% on its portfolio of houses. Your insurance company goes to the reinsurer to buy facultative reinsurance for your house (an unusual circumstance) and purchases a reinsurance policy with a deductible of £90,000. Using the standard Swiss Re curve for the relevant occupancy, what is the expected cost to the reinsurer?

6. As the actuary for a Lloyd's syndicate specialising in property reinsurance, you are asked to quote a £3M xs £2M layer for a Risk XL policy, which is covering a portfolio mostly of large commercial properties. You are given the following risk profile for an insurer who insures its property risk.

MPL (lower limit)	MPL (upper limit)	Average in band, M	Total premium
0	500,000	250,000	3,484,440
500,001	1,000,000	750,001	9,009,920
1,000,001	1,500,000	1,250,001	7,848,230
1,500,001	2,000,000	1,750,001	6,995,310
7,000,001	8,000,000	7,500,001	7,011,480
8,000,001	9,000,000	8,500,001	7,512,650
9,000,001	10,000,000	9,500,001	3,170,790

All amounts are in GBP

You decide to use an exposure curve, some of whose values are tabulated below. You can assume that the original loss ratio is 66%. Determine the expected losses to the layer, stating any additional assumption you need to make and interpolating the values in the table as necessary.

Priority/SI	Swiss Re 3.0	Priority/SI	Swiss Re 3.0	Priority/SI	Swiss Re 3.0
2.0%	15.7%	36.0%	68.9%	70.0%	87.7%
4.0%	24.8%	38.0%	70.3%	70.0%	87.7%
6.0%	31.3%	40.0%	71.6%	72.0%	88.6%
8.0%	36.4%	42.0%	72.9%	74.0%	89.5%
10.0%	40.6%	44.0%	74.2%	76.0%	90.4%
12.0%	44.1%	46.0%	75.4%	78.0%	91.2%
14.0%	47.3%	48.0%	76.5%	80.0%	92.1%
16.0%	50.1%	50.0%	77.7%	82.0%	92.9%
18.0%	52.6%	52.0%	78.8%	84.0%	93.7%
20.0%	54.9%	54.0%	79.9%	86.0%	94.6%
22.0%	57.1%	56.0%	80.9%	88.0%	95.4%
24.0%	59.1%	58.0%	82.0%	90.0%	96.2%
26.0%	60.9%	60.0%	83.0%	92.0%	96.9%
28.0%	62.7%	62.0%	84.0%	94.0%	97.7%
30.0%	64.4%	64.0%	84.9%	96.0%	98.5%
32.0%	66.0%	66.0%	85.9%	98.0%	99.2%
34.0%	67.5%	68.0%	86.8%	100.0%	100.0%

7. Given the exposure curve tabulated in Question 6, and knowing that it is a Swiss Re curve with $c = 3.0$:
   i. Using Bernegger's formulae in Section 22.5, verify that the values of the exposure curve for percentage of sum insured equal to 26%, 58% and 74% are as shown in the table.
   ii. Calculate the probability of having a total loss.
   iii. Calculate the mean loss for a property with sum insured equal to £5.5M.
8. (*)Show the connection between the following probability distributions from statistical mechanics: $f(\varepsilon) = A/(\exp(\varepsilon/kT)-1)$ (Bose–Einstein), $f(\varepsilon) = A/(\exp(\varepsilon/kT)+1)$

(Fermi–Dirac), $f(\varepsilon) = A\exp(-\varepsilon / kT)$ (Maxwell–Boltzmann) and the Bernegger distribution, and identify the values of b, g for which you have the three cases.

Hint: Compare the probability distributions above not with the pdf of Bernegger distribution but with its survival probability $1 - F(x)$.

9. (*) Show that it is possible for the KS distance between the CDF of two cumulative distribution functions $F(x)$ and $F^*(x)$ with domain [0,1] can be made arbitrarily close to 1 while at the same time making their corresponding exposure curves $G(x)$ and $G^*(x)$ arbitrarily close.

   Hint: Assume that $F^*(x)$ is related to $F(x)$ as below and ensure that ε is as small possible while k is as close as possible to 1.

$$F^*(x) = \begin{cases} (1-k)F(x) & 0 \le x < \varepsilon \\ k + (1-k)F(x) & \varepsilon \le x \le 1 \end{cases}$$

10. (*) Adapt the algorithm in Section 22.8.2.3 for the simulation of losses for a single location to the case where the exposure curve is defined in excess of the underlying deductible.
11. (*) Calculate the exposure curve corresponding to the cumulative distribution function that incorporates MPL uncertainty (Equation 22.39).
12. Consider a food manufacturing plant with the following characteristics:
    - TIV (PD) = £200M, TIV (BI) = £200M, MPL (PD) = £130M, MPL (BI) = £200M
    - Local deductible (LD) for PD = £1M, LD(BI) = 45d
    - Indemnity period (IP) = 12 months (from the accident date)

The underwriting guidelines for food manufacturing plants are as follows:

- The standard deductible is £1M for PD and 30 days for BI.
- Assume no allowance for losses between MPL and TIV.
- The exposure curves in use are the Swiss Re curve with c=3.8 for PD, c=3.1 for BI. In both cases, these are defined above the standard deductible.
- The base rate is equal to 0.7‰ (PD), 1‰ (BI)
- The premium breakdown is as follows: 80% to cover the expected losses, 20% to cover costs and profit.
- The company also has a deductible impact table as below. The company uses it for both PD and BI.

	LD/SD=50%	LD/SD=100%	LD/SD=150%	LD/SD=200%
MPL=£100M-	125%	100%	90%	85%
MPL=£150M	120%	100%	92%	87%
MPL=£200M+	115%	100%	93%	88%

Where 'MPL=£100M-' means an MPL of £100M or less, and 'MPL = £200M+' means an MPL of £200M or more.

Calculate:

i. The rate at the local deductible for both PD and BI
ii. The expected losses to the layer £90M xs £10M (all in excess of the local deductible)

13. Consider the same food manufacturing plant as in Question 12. The underwriting guidelines for food manufacturing plants are also as before, with the following exceptions:
    - The exposure curves in use are the Swiss Re curve with c=3.8 for PD, c=3.1 for BI. In both cases, these are defined from the ground up.
    - Since the curves are from the ground up, there is no need for a deductible impact table

    Calculate:
    i. The rate at the local deductible for both PD and BI
    ii. The expected losses to the layer £90M xs £10M (all in excess of the local deductible)

23

Liability Rating Using Increased Limit Factor Curves

We have seen how exposure rating works with property type of cover. In this chapter we consider what you could loosely call exposure rating for liability cover. The method is based on the use of the so-called increased limit factor (ILF) reference curves (Miccolis 1977; Palmer 2006; Riegel 2008). The key idea behind ILF curves is that they should help you determine how the expected losses increase by increasing the limit purchased, and ultimately that helps you determine the cost of a specific layer of (re)insurance based on the rate per unit of exposure at a given limit.

We'll start by looking at how ILF curves arise (Section 23.1), and then we'll look at how they can be applied to pricing in practice (Section 23.2), with instruction on how ILFs need to be modified to take inflation into account (Section 23.3). In Section 23.4 we'll look at how ILF curves can be derived from either a severity model (e.g. a lognormal/GPD spliced distribution) or from the observed losses based on the ISO methodology. Some peripheral issues around ILFs are discussed briefly in Section 23.5. In Section 23.6 we'll look at one popular type of ILF curves, the Riebesell curves. We'll finish by noting that in many circumstances it's probably better not to use ILFs at all and relying on market severity curves enhanced with the knowledge around the frequency of losses above a certain threshold (Section 23.7).

23.1 How ILF Curves Arise

The main difference between liability and property in the construction of reference curves is that for liability, there is no natural upper limit to the size of the loss such as the sum insured or the maximum possible loss (MPL) for property. True, liability is usually purchased up to a given limit L, but this limit does not provide in itself information on the potential loss severity: you can buy £5M of cover for employers' liability but still get a £50M claim.

One of the consequences of this difference is that the ILF curves – the reference curves for liability business – inevitably deal with absolute monetary amounts and not relative damages – and unless the ILF curve has a very specific shape that is invariant at different

DOI: 10.1201/9781003168881-26

scales (such as a single-parameter Pareto), then the shape of the curve will depend on the basis value, that is, the lowest amount for which the curve is provided.

23.1.1 Assumptions Underlying ILF Curves

There are two main assumptions behind ILF curves (Palmer 2006):

- The ground-up loss frequency is independent of the limit purchased
- The ground-up severity is independent of the number of losses and of the limit purchased

These two assumptions are quite reasonable although not unquestionable: the second one is basically one of the assumptions of the collective risk model, and we have already discussed in Chapter 6 how this can easily be breached. As usual, they are quite useful hypotheses that should be abandoned only if someone had good empirical evidence that they are not true – and a more sophisticated model to replace that created on these assumptions.

If the assumptions above are satisfied, the shape of an ILF curve depends only on the underlying severity distribution – in other words, there is a 1:1 correspondence between severity curves and ILF curves, as for property exposure curves.

23.1.2 Definition of ILF Curves

Suppose that we know the expected losses to a liability risk with a policy limit b and that we want to know the expected losses to a policy with a limit u. Using the familiar results on the expected losses to a layer of insurance (in this case, a ground-up layer with limit u) from Chapter 20, we can then write

$$\begin{aligned}\mathbb{E}(S_u) &= \mathbb{E}(\min(X,u))\times\mathbb{E}(N)\\ &= \frac{\mathbb{E}(\min(X,u))}{\mathbb{E}(\min(X,b))}\times\mathbb{E}(\min(X,b))\times\mathbb{E}(N)\\ &= \mathrm{ILF}_b(u)\times\mathbb{E}(S_b)\end{aligned} \tag{23.1}$$

where

$$\mathrm{ILF}_b(u)=\frac{\mathbb{E}(\min(X,u))}{\mathbb{E}(\min(X,b))}=\frac{\int_0^u\bar{F}(y)dy}{\int_0^b\bar{F}(y)dy} \tag{23.2}$$

where in turn $\bar{F}(y)=1-F(y)$ and $F(y)$ is the cumulative distribution function (CDF) of the severity. As we did in Chapter 20, we ask: would it not be good if we could have $\mathrm{ILF}_b(u)$ already tabulated for us, so that we could simply plug that information into Equation 23.1 and find the expected losses for any limit we want? This is what casualty rating with ILF curves is all about. Indeed, ILF curves are often provided as tables that give the value of the ILF as a function of a monetary amount (in a specified currency, and at a given point in time), as in Figure 23.1. An example of an ILF curve in graphical format is given in Figure 23.2.

x	ILF(x)
100,000	1.00
200,000	1.37
300,000	1.58
400,000	1.72
500,000	1.83
600,000	1.92
700,000	1.99
800,000	2.05
900,000	2.10
1,000,000	2.15
1,100,000	2.19
1,200,000	2.23
1,300,000	2.26
1,400,000	2.30
1,500,000	2.33
1,600,000	2.35

FIGURE 23.1
Example of ILF curve in tabular format. Note that the currency unit of x must be specified along with the context to which this curve can be applied and the point in time at which this curve was valid.

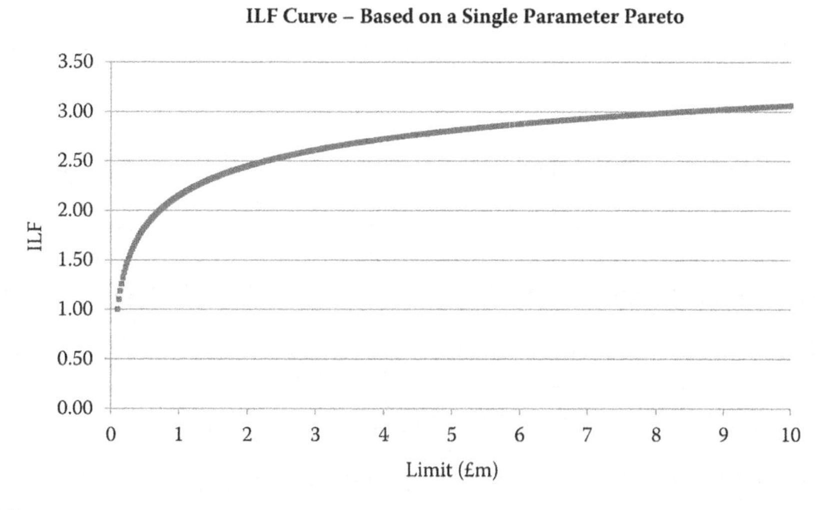

FIGURE 23.2
Example of ILF curve in graphical format. This was drawn based on the (simplistic) assumption that the underlying severity curve is a single-parameter Pareto with parameter $\alpha = 1.1$.

23.1.3 Relationship with Severity Curves

As is the case with exposure rating curves for property, there is a one-to-one correspondence between severity curves and ILF curves. Given the cumulative distribution function (CDF) $F(x)$ – and the corresponding survival distribution $\bar{F}(x)$ – of a severity curve, we can obtain the ILF using Equation 23.2.

Conversely, if we know the ILF curve $\mathrm{ILF}_b(x)$, it is easy to invert Equation 23.2 and obtain an expression for $S(x)$ that holds for $x > b$:

$$\bar{F}(x) = \mathrm{ILF}_b'(x)\int_0^b \bar{F}(y)dy \quad (23.3)$$

Note that Equation 23.3 seems to require the knowledge of $\bar{F}(x)$ over the interval $[0,b]$ in the right-hand side to derive $\bar{F}(x)$ above b on the left-hand side. However, what we are really interested in is the conditional survival probability above b, $\bar{F}_b(x) = \Pr(X \le x \mid X \ge b)$, which can be written (after some elementary algebraic manipulations – see Question 2) as

$$\bar{F}_b(x) = \frac{\mathrm{ILF}_b'(x)}{\mathrm{ILF}_b'(b)} \quad (23.4)$$

Finally, can we obtain the ILF curve above b, $\mathrm{ILF}_b(x)$, if we have only the conditional severity distribution above b, $F_b(x)$? The short answer is 'no': Equation 23.4 can be inverted to give

$$\mathrm{ILF}_b(x) = 1 + \mathrm{ILF}_b'(b)\int_b^x \bar{F}_b(y)dy \quad (23.5)$$

which can also be written as

$$\mathrm{ILF}_b(x) = 1 + \frac{\int_b^x \bar{F}(y)dy}{\int_0^b \bar{F}(y)dy} = 1 + \frac{\int_b^x \bar{F}_b(y)dy}{\int_0^b \bar{F}_b(y)dy}$$

where $\int_0^b \bar{F}(y)dy$ is unknown unless we have the CDF from the ground up. Therefore all we can say is that

$$\mathrm{ILF}_b(x) = 1 + c\int_b^x \bar{F}_b(y)dy \quad (23.6)$$

where the constant c depends on the ground-up loss distribution $F(x)$. In other words, we know $\mathrm{ILF}_b(x) - 1$ only up to a multiplicative constant.

On the positive side, the value of c can be estimated if we have extra information. For instance, Riegel (2008) outlines a method for determining a full ILF model based on (a) the knowledge of the severity distribution in excess of a threshold u; and (b) the knowledge of the ratio between typical premiums at two different limits (both above u). The viability of this method depends, of course, on the reliability of the estimated ratio in (b).

23.1.3.1 Properties of ILF Curves

Note that Equation 23.3 can also be used to prove that ILF curves are necessarily nondecreasing and concave, in that it can be inverted to yield $\mathrm{ILF}_b'(x) = (1 - F(x)) / \left(\int_0^b (1 - F(y)) dy \right)$, which is always positive, and by taking the derivative of this to yield $\mathrm{ILF}_b''(x) = -f(x) / \left(\int_0^b (1 - F(y)) dy \right)$, which is always negative.

23.2 Applications to Pricing

The application of ILF curves to pricing is straightforward. Given the expected losses for a from-the-ground-up policy with a limit of l, $\mathbb{E}(S_l)$, it is possible to calculate the expected losses for a from-the-ground-up policy with another limit l' (an immediate application of Equation 23.1).

$$E(S_{l'}) = \frac{\mathrm{ILF}_b(l')}{\mathrm{ILF}_b(l')} \times \mathbb{E}(S_l) \tag{23.7a}$$

where b is the starting point of the ILF curve.

More in general, we can calculate the expected losses for a layer of (re)insurance L xs D without aggregate features as:

$$\begin{aligned} \mathbb{E}(S_{D,L}) &= \mathbb{E}(\min(X, D+L)) - \mathbb{E}(\min(X, D)) \times \mathbb{E}(N) \\ &= \frac{\mathbb{E}(\min(X, D+L)) - \mathbb{E}(\min(X, D))}{\mathbb{E}(\min(X, b))} \times \frac{\mathbb{E}(\min(X, b))}{\mathbb{E}(\min(X, l))} \times \mathbb{E}(\min(X, l)) \times \mathbb{E}(N) \\ &= \frac{\mathrm{ILF}_b(D+L) - \mathrm{ILF}_b(D)}{\mathrm{ILF}_b(l)} \times \mathbb{E}(S_l) \end{aligned} \tag{23.7b}$$

Note that $\mathbb{E}(S_l)$ can be estimated either based on past experience or can be given in a pure exposure rating fashion as a base rate applied to a given measure of exposure, such as wage roll, turnover, or whatever is used for a particular liability product:

$$\mathbb{E}(S_l) = BR@l \times \text{Exposure}$$

Even when the expected losses at a given limit l are not available, we can estimate the expected losses to a layer L' xs D' if we have the ILF curve and the expected losses to another layer L xs D:

$$\mathbb{E}(S_{D',L'}) = \frac{\mathrm{ILF}_b(D'+L') - \mathrm{ILF}_b(D')}{\mathrm{ILF}_b(D+L) - \mathrm{ILF}_b(D)} \times \mathbb{E}(S_{D,L}) \tag{23.7c}$$

Equations 23.7a–23.7c can be used in the context of exposure rating as shown in the example below. They can also be used as part of experience rating, e.g. to price layers outside the range for which we trust our loss-calibrated models.

An Example of Application of ILF Curves to Exposure Rating

Let us consider a simple example of an employers' liability policy – a similar example to that described in Chapter 1. We want to calculate the losses to a layer £15M xs £10M.

Let us assume that the (loss) base rate per unit of exposure for a limit of £5M is $BR@£5M = 0.2\%$, and let us assume that the wage roll is £100M. That means that the expected losses will be given by $\mathbb{E}(S_l) = \frac{0.2}{100} \times £100M = £200k$

Let us also assume (as in Chapter 1) we have a ILF curve defined as[1]:

$$ILF_b(u) = (1+\rho)^{\log_2\left(\frac{u}{b}\right)}$$

where $b = £500k$ and $\rho = 0.2$.

The expected loss to the £15M xs £10M can then be calculated as:

$$\mathbb{E}\left(S_{£5M,£5M}\right) = \frac{\mathrm{ILF}_{£500k}(£25M) - \mathrm{ILF}_b(£10M)}{\mathrm{ILF}_{£500k}(£5M)} \times £200k = £15.9k$$

23.3 Effect of Claims Inflation on ILF Curves

Because ILF curves are expressed in monetary terms, the question arises about the extent to which they remain valid as time goes by and inflation pushes up the cost of the claims. As it turns out, ILF curves *do* in general need to be adjusted for inflation. First of all, note that if the claims inflation applies uniformly over all loss sizes, so that $X = (1 + r)X$ where r is the claims inflation per annum, then

$$\mathbb{E}(\min(X',u)) = (1+r) \times \mathbb{E}\left(\min\left(X, \frac{u}{1+r}\right)\right)$$

As a result, the revised ILFs are given by

$$\mathrm{ILF}_b^{(\mathrm{rev})}(u) = \frac{\mathbb{E}(\min(X',u))}{\mathbb{E}(\min(X',b))} = \frac{\mathbb{E}\left(\min\left(X, \frac{u}{1+r}\right)\right)}{\mathbb{E}\left(\min\left(X, \frac{b}{1+r}\right)\right)} = \mathrm{ILF}_{\frac{b}{1+r}}\left(\frac{u}{1+r}\right) = \frac{\mathrm{ILF}_b\left(\frac{u}{1+r}\right)}{\mathrm{ILF}_b\left(\frac{b}{1+r}\right)} \tag{23.8}$$

This allows us, amongst other things, to calculate the effect of claims inflation on an XL layer $(D, D + L)$. If the cost in year y is

$$C_L = \left(\mathbb{E}(\min(X,D+L)) - \mathbb{E}(\min(X,D))\right) \times \mathbb{E}(N)$$

Then the cost in year $y + 1$ is

$$C_L^{(\mathrm{rev})} = (1+r) \times \left(\mathbb{E}\left(\min\left(X, \frac{D+L}{1+r}\right)\right) - \mathbb{E}\left(\min\left(X, \frac{D}{1+r}\right)\right)\right) \times \mathbb{E}(N)$$

1 As we'll see in Section 23.6, this is called a Riebesell curve.

and the increase in cost is

$$\begin{aligned}\frac{C_L^{(\text{rev})}}{C_L} &= (1+r)\times\frac{\mathbb{E}\left(\min\left(X,\frac{D+L}{1+r}\right)\right)-\mathbb{E}\left(\min\left(X,\frac{D}{1+r}\right)\right)}{\mathbb{E}(\min(X,D+L))-\mathbb{E}(\min(X,D))}\\ &= (1+r)\times\frac{\text{ILF}\left(\frac{D+L}{1+r}\right)-\text{ILF}\left(\frac{D}{1+r}\right)}{\text{ILF}(D+L)-\text{ILF}(D)}\end{aligned} \tag{23.9}$$

23.4 Derivation of ILF Curves

ILF curves are normally derived by considering a large amount of portfolio/market losses and building a severity distribution and then an ILF curve by using the relationship $\text{ILF}(x) = \mathbb{E}(\min(X, x))/ \mathbb{E}(\min(X, b))$. Therefore, the issues that arise in the derivation of ILF curves are the same that arise in the derivation of a severity model (Palmer, 2006):

1. Some losses may be censored because of policy limits (e.g., if a claim is £7M and the policy limit is £5M, we may only see the claim recorded as £5M).
2. Some losses may be net of a deductible, or even if they are gross of the deductible, they may originate from policies that have a deductible and for which the claims below the deductible are absent, therefore creating a bias.
3. Losses must be revalued to current terms (and claims inflation itself needs to be estimated and may not be the same over the whole range of losses), which brings significant uncertainty.
4. Many claims are still open and as such, there is going to be IBNER, which also adds significant uncertainty.

We will now look at two methods for deriving ILF curves and deal with the issues above. The two methods differ mainly in the way they deal with the issue of IBNER. The simpler method in Section 23.4.1 uses both closed and open claims, adjusting where possible the open claims for the effect of IBNER (Chapter 15); the method in Section 23.4.2 is the methodology in use by the Insurance Services Office (ISO) in the United States as illustrated by Palmer (2006), and uses closed claims only, with a procedure to correct the bias that arises from the fact that larger claims take longer to settle.

23.4.1 Derivation of ILF Curves from a Severity Model

The first method that we present simply assumes that we have proper ground-up data to build our severity model with, and that our revaluation method (Chapters 9 and 10) and our IBNR adjustments (Chapter 15) are sufficiently correct. A severity model is built as explained in Chapter 16, using for example a lognormal/GPD model or an Empirical/GPD model. The ILF curve is then derived by using Equation 23.2, which is shown here again for convenience:

$$\mathrm{ILF}_b(u)=\frac{\mathbb{E}(\min(X,u))}{\mathbb{E}(\min(X,b))}=\frac{\int_0^u \bar{F}(y)dy}{\int_0^b \bar{F}(y)dy} \tag{23.2}$$

In the following, we give some examples of the application of Equation 23.2 using models that should be familiar from Chapter 16: a simple lognormal distribution (Section 23.4.1.1), a spliced lognormal/Pareto (Section 23.4.1.2), a spliced lognormal/GPD (Section 23.4.1.3), and a spliced Empirical/GPD (Section 23.4.1.4).

The formulae for the various special cases are given here without proof. See online solutions to Question 3 for the derivation (Parodi, 2016).

23.4.1.1 Lognormal Distribution

In the simple case where the severity model is a lognormal distribution, the corresponding ILF curve is as follows.

$$\mathrm{ILF}_b(u)=\frac{\exp\left(\mu+\frac{\sigma^2}{2}\right)\Phi\left(\frac{\ln u-\mu-\sigma^2}{\sigma}\right)+u\,\bar{\Phi}\left(\frac{\ln u-\mu}{\sigma}\right)}{\exp\left(\mu+\frac{\sigma^2}{2}\right)\Phi\left(\frac{\ln b-\mu-\sigma^2}{\sigma}\right)+b\,\bar{\Phi}\left(\frac{\ln b-\mu}{\sigma}\right)} \tag{23.10}$$

where $\Phi(x)$ is the CDF of the normalised Gaussian distribution (aka as the error function) and $\bar{\Phi}(x)=1-\Phi(x)$ is the corresponding survival function.

23.4.1.2 Lognormal/Single-Parameter Pareto

Another popular severity model is the lognormal/Pareto model. We will only consider the case where the splicing point θ is smaller than b (the basis limit). This is the most interesting case as ILFs are mainly used to investigate policy limits that are beyond the region of attritional losses, that captured by the lognormal part.

For completeness we show the curve for all values of u; however, the most interesting part of the curve is again that above the basis limit b, which is assumed to be above θ.

$$\mathrm{ILF}_b(u)=\begin{cases}\dfrac{\exp\left(\mu+\frac{\sigma^2}{2}\right)\Phi\left(\frac{\ln\theta-\mu-\sigma^2}{\sigma}\right)+\bar{\Phi}\left(\frac{\ln\theta-\mu}{\sigma}\right)\left(\theta+\frac{\theta}{\alpha-1}\left(1-\left(\frac{\theta}{u}\right)^{\alpha-1}\right)\right)}{\exp\left(\mu+\frac{\sigma^2}{2}\right)\Phi\left(\frac{\ln\theta-\mu-\sigma^2}{\sigma}\right)+\bar{\Phi}\left(\frac{\ln\theta-\mu}{\sigma}\right)\left(\theta+\frac{\theta}{\alpha-1}\left(1-\left(\frac{\theta}{b}\right)^{\alpha-1}\right)\right)} & u\geq\theta\\[2ex] \dfrac{\exp\left(\mu+\frac{\sigma^2}{2}\right)\Phi\left(\frac{\ln u-\mu-\sigma^2}{\sigma}\right)+\bar{\Phi}\left(\frac{\ln u-\mu}{\sigma}\right)u}{\exp\left(\mu+\frac{\sigma^2}{2}\right)\Phi\left(\frac{\ln\theta-\mu-\sigma^2}{\sigma}\right)+\bar{\Phi}\left(\frac{\ln\theta-\mu}{\sigma}\right)\left(\theta+\frac{\theta}{\alpha-1}\left(1-\left(\frac{\theta}{b}\right)^{\alpha-1}\right)\right)} & u<\theta\end{cases} \tag{23.11}$$

where:

- $\Phi(x)$ and $\bar{\Phi}(x) = 1 - \Phi(x)$ are the CDF and the survival function of the normalised Gaussian distribution;
- μ, σ are the parameters of the lognormal;
- the single-parameter Pareto has CDF $F(x) = 1 - (\theta / x)^{\alpha}$.

23.4.1.3 Lognormal/GPD

Slightly more general than the lognormal/Pareto model, the lognormal/GPD model uses the lognormal distribution to model the attritional losses and the GPD to model the tail, as per the teachings of Extreme Value Theory.

As in the previous section, we will only consider the case where the splicing point θ is smaller than b, but we provide the curve for all values of u.

$$\mathrm{ILF}_b(u) = \begin{cases} \dfrac{\exp\left(\mu + \frac{\sigma^2}{2}\right)\Phi\left(\frac{\ln\theta - \mu - \sigma^2}{\sigma}\right) + \bar{\Phi}\left(\frac{\ln\theta - \mu}{\sigma}\right)\left(\theta + \frac{\nu}{1-\xi}\left(1 - \left(1 + \xi\frac{u-\theta}{\nu}\right)^{1-1/\xi}\right)\right)}{\exp\left(\mu + \frac{\sigma^2}{2}\right)\Phi\left(\frac{\ln\theta - \mu - \sigma^2}{\sigma}\right) + \bar{\Phi}\left(\frac{\ln\theta - \mu}{\sigma}\right)\left(\theta + \frac{\nu}{1-\xi}\left(1 - \left(1 + \xi\frac{u-\theta}{\nu}\right)^{1-1/\xi}\right)\right)} & u \geq \theta \\ \dfrac{\exp\left(\mu + \frac{\sigma^2}{2}\right)\Phi\left(\frac{\ln u - \mu - \sigma^2}{\sigma}\right) + \bar{\Phi}\left(\frac{\ln u - \mu}{\sigma}\right)u}{\exp\left(\mu + \frac{\sigma^2}{2}\right)\Phi\left(\frac{\ln\theta - \mu - \sigma^2}{\sigma}\right) + \bar{\Phi}\left(\frac{\ln\theta - \mu}{\sigma}\right)\left(\theta + \frac{\nu}{1-\xi}\left(1 - \left(1 + \xi\frac{u-\theta}{\nu}\right)^{1-1/\xi}\right)\right)} & u < \theta \end{cases} \tag{23.12}$$

where:

- $\Phi(x)$ and $\bar{\Phi}(x) = 1 - \Phi(x)$ are the CDF and the survival function of the normalised Gaussian distribution;
- μ, σ are the parameters of the Lognormal;
- the GPD has CDF $F(x) = 1 - (1 + \xi(x - \theta)/\nu)^{-1/\xi}$.

23.4.1.4 Empirical/GPD

As we have seen in Chapter 16, it is often sensible to focus the modelling efforts on the tail and use the empirical distribution for the body of the distribution. It is easy to adapt our methodology to incorporate this case. All we need to do is to choose a splicing point θ and calculate $\int_0^{\theta} \bar{F}(y)dy$ numerically, e.g. as a Riemann sum:

$$\int_0^{\theta} \bar{F}(y)dy \approx \sum_{i \text{ s.t. } y_i \leq \theta} \bar{F}_{\text{emp}}(y_i)(y_i - y_{i-1}) = \sum_{i=1}^{K} \frac{n-i+1}{n+1}(y_i - y_{i-1})$$

where $\{y_1, y_2 \ldots y_N\}$ are the observed losses in ascending order and $\{y_1, y_2 \ldots y_K\}$ is the subset of the losses that are less than θ.

Limiting ourselves again to the case where the splicing point θ is smaller than b (the basis limit) and u, we can write the ILF curve as follows:

$$\mathrm{ILF}_b(u) \approx \frac{\sum_{i=1}^{K} \frac{n-i+1}{n+1}(y_i - y_{i-1}) + \left(1-\frac{K}{N}\right)\left(\theta + \frac{\nu}{1-\xi}\left(1-\left(1+\xi\frac{u-\theta}{\nu}\right)^{1-1/\xi}\right)\right)}{\sum_{i=1}^{K} \frac{n-i+1}{n+1}(y_i - y_{i-1}) + \left(1-\frac{K}{N}\right)\left(\theta + \frac{\nu}{1-\xi}\left(1-\left(1+\xi\frac{b-\theta}{\nu}\right)^{1-1/\xi}\right)\right)} \tag{23.13}$$

23.4.2 ISO Methodology for the Derivation of ILF Curves

The ISO methodology has special interest for us not only because it shows how ILF curves are derived in practice but also because it provides an example of how a severity distribution can be built based on closed claims only, something we mentioned in Chapter 15. The ISO's attitude to IBNER has indeed been that adjusting for IBNER leads to excessive uncertainties and that should be avoided, and that closed claims are much more trustworthy. The issue, as we have seen in Chapter 15, is that considering only closed claims may cause a severe bias towards smaller claims because larger claims typically take longer to settle. The ISO methodology addresses this issue by analysing claims separately depending on the time it takes for them to settle (or more accurately, depending on the weighted payment time).

The ISO methodology can be outlined as follows. The reader is referred to Palmer (2006) for a more thorough account.

Input. All *closed* losses, each with information on the time at which each intermediate payment was made.

1. Revalue all closed losses to the midterm of the policy.
2. Subdivide the loss data according to the *payment lag*. Payment lags above a certain value (e.g. 4–6 years) are normally grouped together as the severity distribution has been seen not to depend significantly on the payment lag if the payment lag is long enough.

   The payment lag is defined as

   $$\text{Payment lag} = \text{Payment year} = \text{Occurrence year} + 1$$

   where the payment year is the year to which the payment date belongs, and the payment date is the *money-weighted average date of the indemnity payments*. As an example, if a claim is settled at £42,000 through three payments of £10,000 on 20 February 2010, £24,000 on 15 July 2011, and £8000 on 6 March 2012, the weighted payment time is 29 April 2011. The mechanism of the calculation is illustrated in Figure 23.3.
3. Calculate the empirical distribution of closed claims for each payment lag.

   If there is neither truncation (claims below or above a certain level are excluded from the data set) nor censoring (claims above the limit are recorded as the value of the limit), this exercise is straightforward: you simply need to put all relevant losses in ascending order and the empirical cumulative distribution is defined as $F_{PL=k}^{\mathrm{emp}}(x) = (\#\text{of losses} \le x \text{ with payment lag} = k)/(n+1)$.

   If there is censoring, a method based on the Kaplan–Meier estimator (see, for example, Klugman et al., 2008) is used. The interested reader is referred to the article by Palmer (2006), in which the procedure is explained and a simple numerical example is provided. In this chapter, we will ignore the issue.

Loss ID	Payment date	No. of days from first payment	Paid amount
1	2/20/2010	0	10,000
2	7/15/2011	510	24,000
3	3/6/2012	745	8000
	Weighted average	433.33	
	Weighted payment date	4/29/2011	
	Payment year	2011	

FIGURE 23.3
The weighted payment time of the three payments in the table is calculated as follows: first of all, the number of days from the first payment is calculated. Second, the money-weighted average of the number of days from the first payment is calculated, where the weight is the paid amount: in our case, the weighted average is WA = (10,000 × 0 + 24,000 × 510 + 8000 × 745/(10,000 + 24,000 + 8000) = 433 (we can ignore the decimals). The weighted payment date is then calculated as 433 days from the payment date, which takes us to 29 April 2011. The payment year is then the calendar year of the payment date, in this case 2011.

4. Estimate the percentage of losses that are settled at different payment lags.
   Based on historical data, we can estimate the payment lag distribution, that is, what percentage of losses is settled with a payment lag of 1, 2, 3, and so on. This can be denoted as $p_k = \Pr(\text{payment lag} = k)$.
5. Combine the empirical severity distributions corresponding to each payment lag into an overall empirical severity distribution.

$$F(x) = \Pr(X \le x) = \sum_{k=1}^{\infty} \Pr(X \le x \mid PL = k) p_k \tag{23.14}$$

where $p_k = \Pr(\text{PL} = k)$ is, as explained in Step 4, the probability that a given claim has a payment lag equal to k. In practice, we only have an empirical estimate $F_{PL=k}^{emp}(x)$ of the distribution $\Pr(X \le x \mid \text{PL} = k)$ for each payment lag and an empirical estimate $\hat{p}_k$ of the probability p_k of a loss having a given payment lag; furthermore, all payment lags above a certain value k are combined to produce a statistically significant estimate. As a consequence, Equation 23.14 can be rewritten less ambitiously as

$$\hat{F}(x) \approx \sum_{k=1}^{\bar{k}} F_{PL=k}^{\text{emp}}(x)\hat{p}_k + F_{PL>\bar{k}}^{\text{emp}}(x) \times \left(1 - \sum_{k=1}^{\bar{k}} \hat{p}_k\right) \tag{23.15}$$

6. Smooth the tail of the combined empirical distribution, $\hat{F}(x)$.
   The empirical distribution will inevitably be a bit bumpy in the large-loss region. Therefore, it makes sense to replace the empirical distribution with a parametric distribution above a certain threshold. The results of extreme value theory (see Chapter 16) suggest that a generalised Pareto distribution (GPD) should be used (however, other choices have traditionally been made as well). The parameters of the distribution can be determined, for example, by percentile matching.
7. Fit a mixed exponential distribution to the empirical distribution with a smoothed tail.

This step is to replace the empirical distribution (with a smoothed tail) with something with which it is easier to perform calculations. Traditionally, this has been done by using a mixed exponential distribution, that is, by approximating the CDF with a weighted sum of exponential severity distributions: $F(x) \approx 1 - \sum_{i=1}^{r} w_i \exp(-x / \mu_i)$, with $\sum_{i=1}^{r} w_i = 1$. In practice, up to approximately 10 exponential distributions can be used to reproduce the behaviour of the empirical severity distribution.

You might be wondering what to make of a method like this, which flies in the face of every principle of statistical parsimony that we have insisted on through the whole book. One way to look at this is that it is simply a practical method that makes calculations easier, but it is not really attempting to 'model' the severity distribution – it is more like a harmless interpolation method. Any other method of interpolation that you can think of – such as linear interpolation or spline interpolation, or simply be happy with resampling from empirical distribution – would work equally well. The advantage of using a combination of exponential distributions is that exponential distributions tend to go to zero quite quickly and therefore different exponents will have their effect in different regions and then they will 'disappear'. The only region where the method may not be so harmless is in the tail, as the distribution may tail off too quickly for practical purposes.

8. Derive the ILF curve based on Equation 23.2.

   Using Steps 1 through 7, we have estimated the cumulative distribution function $F(x)$ of the loss severities. By applying Equation 23.2 to the severity distribution thus constructed, we can now obtain the ILF curve, which can then be used, amongst other things, to price excess of loss layers through Equation 23.7b.

23.5 Other Issues

23.5.1 Indexation Clause

The effect of the index clause can be roughly taken into account by considering the weighted payment time (continental clause) or the mean settlement time (London Market clause), assuming a given wage inflation index (or whatever index the layer endpoints are contractually linked to), and then calculating the new endpoints of the layer based on these numbers.

23.5.2 Discounting

Discounting for investment, and the like – this can be done as with experience rating (Chapter 20), that is, by dividing the expected losses to a layer by a discount rate $(1 + r)^{\tau}$, where r is the risk-free investment return and τ is the weighted payment time.

23.5.3 Dealing with Expenses

ILF curves can be provided on an indemnity-only basis or may include expenses and other loadings. In the former case, ILF are defined as in Equation 23.2. In the latter case, ILFs are defined as

$$\mathrm{ILF}_b^{\text{plus costs}}(u) = \frac{\mathbb{E}(\min(X,u)) + \text{Expenses}(u) + \text{Risk load}(u)}{\mathbb{E}(\min(X,b)) + \text{Expenses}(b) + \text{Risk load}(b)}$$

The 'expenses' term in the equation above includes in general both allocated and unallocated expenses. One of the problems with dealing with expenses is that they may inflate differently from indemnity losses. Refer to the article by Palmer (2006) for more details on how to deal with expenses.

23.5.4 Per Claim versus Per Occurrence Basis

ILFs can be provided on a per claim basis or on a per occurrence basis. In the former case, the limit is the maximum amount that will be paid to an individual claimant, whereas in the latter case, the limit is the maximum amount that will be paid for all claims arising from the same event. You need to be aware of what the ILF is actually showing before using it for your own purpose.

23.6 Riebesell (ILF) Curves

We have already seen various examples of ILF curves:

- curves built from popular severity models (Section 23.4.1.1–4);
- ISO curves for a variety of casualty business (Section 23.4.2).

Riebesell curves are another popular example of ILF curves, used often in European casualty treaty business. Riebesell curves are heuristic curves based on the assumption that each time that the sum insured doubles, the expected losses increase by a constant factor, $1+\rho$, with $0<\rho<1$ (in formulae: $\mathbb{E}(\min(X,2L)) = (1+\text{Á}) \times \mathbb{E}(\min(X,L))$). For the more general case where the limit increases by a generic factor a, this translates into the Riebesell's rule:

$$\mathbb{E}(\min(X,aL)) = (1+\rho)^{\log_2 a} \times \mathbb{E}(\min(X,L)) \tag{23.16}$$

Note that since $x^{\log(y)} = y^{\log(x)}$ in any base, 23.16 can equivalently be written as $\mathbb{E}(\min(X,aL)) = a^{\log_2(1+\rho)} \times \mathbb{E}(\min(X,L))$, as it is sometimes done. The ILF curve is therefore given by:

$$ILF_b(u) = (1+\rho)^{\log_2\left(\frac{u}{b}\right)} = \left(\frac{u}{b}\right)^{\log_2(1+\rho)} \tag{23.17}$$

As shown by Mack and Fackler (2003), Riebesell's rule is consistent with the assumption that the tail of the severity distribution has a Pareto tail above a certain threshold θ, that is, a CDF of the form $F(x) = 1 - (\theta/ x)^{\alpha}$. This is in turn consistent with extreme value theory, according to which Pareto-like behaviour is one of the three admissible behaviours of tail losses (Pareto-like, exponential, and beta). However, the value of α that results from the application of the Riebesell curves is $\alpha = 1 - \log_2(1+\rho)$, which is between 0 and 1, leading to the apparently absurd consequence that the expected losses for an unlimited policy are infinite. This is normally not a serious problem if the policy limit always remains finite (as is typically the case) but undermines the approach in classes of business such as motor in the United Kingdom where liability is unlimited by law and it ultimately points to the fact that Riebesell's rule is not realistic, at least for very high limits. Despite these issues, Riebesell curves remain popular because they are easy to use and are scale-invariant: no correction for claims inflation or for currency changes are needed.

23.7 Do We Really Need ILF Curves?

We have seen in Section 23.1.3 that if you only have information on losses above a certain threshold, the ILF curve built based on those losses is only defined up to a constant. Therefore, you can build ILF curves only if you have ground-up loss experience.

However, this is only because of the way ILF curves are engineered. If instead of using reengineered severity curves (the ILF curves) for pricing purposes, you used the market severity curves themselves, combined with the information on the frequency above a certain threshold, you would be able to price a layer that falls within the range of definition of the market severity curve. This is how (re)insurance pricing should normally work, data permitting.

Furthermore, ILF curves cannot be used directly to derive other statistics beyond the expected losses to a layer. For example, they provide no information on the volatility of the layer loss (whether expressed by the standard deviation, VaR, expected shortfall, or other statistics), which might be used as a risk loading to calculate the technical premium. Using a straightforward severity model would instead allow us to obtain these statistics via analytical formulae or simulation.

Therefore, by collating market/portfolio losses and transforming them into ILF curves rather than straightforward market severity curves, we may, in some cases, be creating an unnecessary problem for ourselves!

23.8 Questions

1. Assume that the correct severity distribution for a public liability portfolio is a lognormal with $\mu = 10.5$ and $\sigma = 1.2$. Derive the analytical form of the corresponding ILF curve above £100,000 and draw it using a spreadsheet.
2. (*) Prove that the CDF of the severity curve above b (conditional on being above b) corresponding to an ILF curve with basis point b is $F_b(x) = 1 - \mathrm{ILF}_b'(x)/\mathrm{ILF}_b'(b)$.

Hint: Start from the conditional survival distribution above b.

3. (*) Prove that the ILF curves corresponding to the following severity models: (a) lognormal, (b) lognormal/Pareto, (c) lognormal/GPD, (d) empirical/GPD are as in Sections 23.4.1.1–4.
4. (*) Prove Equation 23.9 directly for the case of the lognormal distribution, starting from Equation 23.2.
5. Your company uses the following table of ILFs.

Policy limit (£)	ILF
95,238	0.953
100,000	1.000
105,000	1.046
476,190	1.773
500,000	1.785
525,000	1.796
952,381	2.281
1,000,000	2.285
1,050,000	2.289

a. Define ILFs, and show (with a formula) how they are related to the underlying loss severity distribution.
b. Explain what ILFs are used for, and in what context.
c. Explain how ILFs differ from exposure curves.
d. Briefly describe how ILFs can be derived, and what the data issues involved in such derivation are.
e. Calculate the updated ILFs based on 1 year of inflation at 5% for the limits of £500k and £1M, stating any assumption you need to make to update the ILF in this way.
f. Assuming 5% inflation, calculate the expected percentage increase in losses for the layer £500k xs £500k for a policyholder who purchases the same limits of insurance 1 year later.

24

Pricing Considerations for Specific Lines of Business

This chapter is a quick gallery of pricing techniques for specific lines of business. It is neither comprehensive (many specialist lines of business have been left out) nor exhaustive (if you are a specialist in one of these lines of business, you will inevitably find the treatment of that class of business rather shallow).

Rather, the lines of business included here have been selected to illustrate specific pricing techniques:

- Professional indemnity (Section 24.1) to illustrate the pricing peculiarities of the claims-made basis and of round-the-clock-reinstatements.
- Weather derivatives (Section 24.2) as an example in which rating is based on an index rather than the policyholder's experience.
- Credit risk (Section 24.3) as an example in which the role of systemic risk is paramount and pricing is based on the individual risk model rather than the collective risk model.
- Extended warranty (Section 24.4) as an example in which the focus is on the failure of products and the probability of a loss is a function of age of the product.
- Aviation hull and liability (Section 24.5.1) as an example in which losses are modelled based on publicly available information of losses, and there is tail correlation between different components of the same loss.
- Business interruption (Section 24.5.2) as an example in which one may find it advantageous to use a compound severity distribution.
- Commercial motor insurance (Section 24.5.3) as another example in which each claim has a property and liability component and a way of introducing the concept of periodic payment orders (PPOs).
- Product liability (Section 24.5.4) as an example to illustrate the pricing peculiarities of the integrated occurrence basis.
- Construction (Section 24.5.5) as an example of cover with unusual duration and unusual earning pattern.

DOI: 10.1201/9781003168881-27

24.1 Professional Indemnity Cover

24.1.1 Legal Framework

The legal framework for professional indemnity policies (and similar policies) is tort law, as it is for all liability insurance. Specifically, however, these policies deal with negligent advice given by professional people. It therefore applies to all professionals, whether they are medical doctors, actuaries, engineers or lawyers, and this cover is typically bought by firms and sole practitioners that are in the business of providing advice or services to clients. Examples are law firms, engineering firms, actuarial firms, medical centres, and medical consultants.

24.1.2 Policy

24.1.2.1 Claims-Made Basis

Professional indemnity policies are normally sold on a claims-made basis (occurrence-based policies are however also common). This means that the policy will pay for losses that have been *discovered/reported* (rather than occurred) during a certain period. The difference between 'discovered' and 'reported' is important because once the insured discovers that negligent advice has been given, whether or not a claim has been made, it has to report the 'discovery' to the insurer within a certain period. Each policy normally has a window at the end of each policy year allowing for, say, a month's delay in reporting after the policy has expired.

24.1.2.2 Retroactive Date

Claims-made policies can be 'mature' or at the retroactive date of inception (RDI). A mature policy may apply to all reported claims regardless of date of occurrence, or to all claims that occur after a given date (the 'retroactive date'), depending on the contract. In an RDI policy, the retroactive date is the same as the inception date.

24.1.2.3 Reinstatements

One interesting feature of claims-made policies is that they often include a provision for reinstatements, like many reinsurance policies do. There are two types of reinstatements available for professional indemnity policies: direct reinstatements and round-the-clock reinstatements, a rather exotic type of structure.

Direct reinstatements work as we explained in Section 3.3.1, that is, once a layer of insurance is exhausted, it is 'reinstated' as often as specified in the contract, achieving the same effect as an aggregate limit on the compensation. A separate reinstatement arrangement will be specified for the different layers of insurance; therefore, the compensation recovered through the different layers of insurance can be worked out for each different layer in isolation.[1]

With *round-the-clock reinstatements*, a dependency between the various layers of insurance is created. Whenever a layer of insurance is exhausted, the layer above it drops down

1 As an example, a loss of £3.7M will trigger a recovery of £1.7M from a £3M xs £2M Layer of (re)insurance (assuming no indexation clauses, aggregate deductibles/limits, or reinstatements), regardless of whether or not the (re)insured has also bought a £1M xs £1M layer.

	(Ground-Up)	(Retained below the Deductible)	(Primary)	(Excess)	(Excess)	(Retained above Overall Limit)
	xs 0	0.5 xs 0	0.5 xs 0.5	1 xs 1	3 xs 2	xs 5
Loss no. 1	2.5	0.5				
			0.5			
				1		
					0.5	
Loss no. 2	5	0.5				
					2.5	
			0.5			
				1		
					0.5	
Loss no. 3	7.2	0.5				
					2.5	
			0.5			
				1		
					0.5	
						2.2

All monetary amounts are in £M

FIGURE 24.1
An example of how round-the-clock reinstatements work in the case of unlimited reinstatements and the following insurance structure: an each-and-every-loss deductible of £500k (the '0.5 xs 0' column), a primary layer of £0.5M xs £0.5M and two excess layers (£1M xs £1M and £3M xs £2M). The figure should be read as follows: Loss no. 1 (ground-up amount of £2.5M) is retained for the first £500k, then exhausts both the primary and the excess layers and then burns £500K of the £3M xs £2M layer. Loss no. 2 is again retained for the first £500k, then goes straight to the £3M xs £2M Layer, exhausting it. It then goes back to the £0.5M xs £0.5M layer (hence the expression 'round the clock'), then to the £1M xs £1M layer and then again to the £3M xs £2M layer). The following loss (Loss no. 3) keeps going through the cycle: however, only an amount equal to the width of the full tower of insurance (£4.5M) can be recovered, and the rest (£2.2M) is retained.

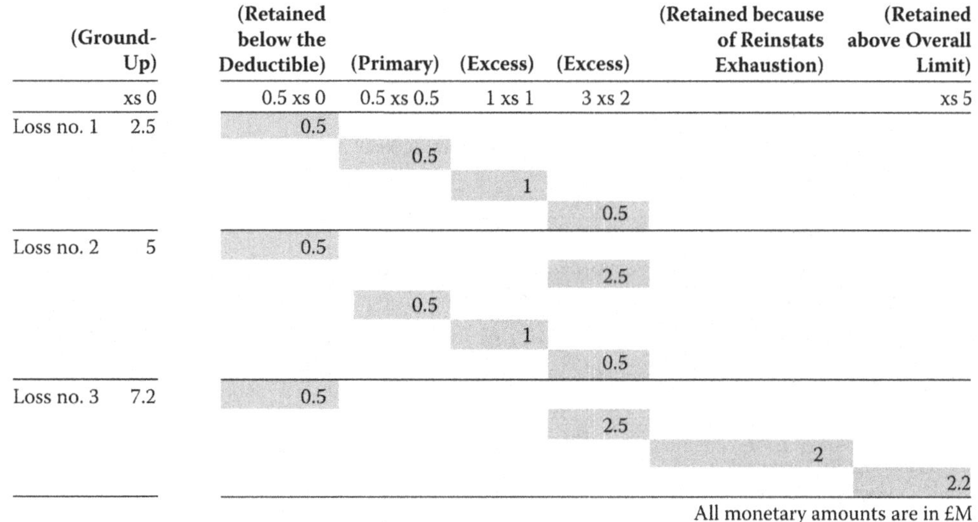

	(Ground-Up)	(Retained below the Deductible)	(Primary)	(Excess)	(Excess)	(Retained because of Reinstats Exhaustion)	(Retained above Overall Limit)
	xs 0	0.5 xs 0	0.5 xs 0.5	1 xs 1	3 xs 2		xs 5
Loss no. 1	2.5	0.5					
			0.5				
				1			
					0.5		
Loss no. 2	5	0.5					
					2.5		
			0.5				
				1			
					0.5		
Loss no. 3	7.2	0.5					
					2.5		
						2	
							2.2

All monetary amounts are in £M

FIGURE 24.2
An example of how round-the-clock reinstatements work in the case where all layers have one reinstatement only.

to cater to the losses that cannot be recovered from the first layer and so on for every other layer on top of that. When all layers are exhausted in that order, the cycle starts again beginning from the first layer if a reinstatement is available. Figures 24.1 and 24.2 provide an illustration of how round-the-clock reinstatements work with a simple case that involves three losses.

24.1.3 Calculating Reporting Year Exposure

Professional indemnity policies are normally sold on a claims-made basis, although occurrence-basis policies are not uncommon. There is no consistent measure of exposure for claims-made policies because losses may arise from many different past years. To deal with this, we need information on the distribution of lags between occurrence and reporting, and we build the exposure based on this lag information.

Assume that the reporting pattern is similar to the one shown in Figure 24.3. Therefore, the actual exposure for the renewal year (2013) comes from different years. An example calculation for the 'reporting year exposure' is shown in Figure 24.4.

Often, however, we will not have a detailed reporting pattern but only a rough indication on the average delay between the loss trigger (e.g. legal advice or provision of a medical service) and the reporting of a potential claim. We can then use, for example, an exponential delay approximation to calculate the contribution of the different years of exposure, as described in Section 13.3.

Year	n	$n+1$	$n+2$	$n+3$	$\geq n+4$
% Reported	10%	30%	40%	20%	0%

FIGURE 24.3
Example of a reporting pattern for professional indemnity claims. The table should be read as follows: of all claims originating from professional activity performed in year n, 10% are expected to be reported in year n, 30% in year $n + 1$, etc. This can also be expressed by saying that 10% of the exposure of year n is 'earned' in year n, 30% is earned in year $n + 1$, etc.

Year	Turnover (Revalued)	Reporting Year 2010	2011	2012	2013	RY Exposure in 2013
2009	£0.6M	30%	40%	20%	0%	0% × £0.6M
2010	£0.7M	10%	30%	40%	20%	+20% × £0.7M
2011	£0.9M	0%	10%	30%	40%	+40% × £0.9M
2012	£1.1M	0%	0%	10%	30%	+30% × £1.1M
2013	£1.4M	0%	0%	0%	10%	+10% × £1.4M
						= £0.97M

30% of the 2011 turnover is 'earned' in 2012

FIGURE 24.4
The reporting year exposure for the renewal year (2013) can be calculated as a weighted average of the turnover figures of different years. Here, we have used turnover as a measure of exposure. Other measures, such as the number of professionals, are also acceptable.

Delay (Years)	Factor	Cumulative Earned	Incremental Earned	Reporting Year	Turnover (Revalued), £M	Turnover (Reporting Year Basis), £M
1	8.68	11.5%	11.5%	<=2008	600.0	**600.0**
2	3.22	31.1%	19.6%	2009	600.0	**600.0**
3	2.16	46.3%	15.2%	2010	700.0	**611.4**
4	1.72	58.2%	11.9%	2011	900.0	**653.9**
5	1.48	67.5%	9.2%	2012	1100.0	**731.3**
>=6		100.0%	32.5%	2013	1400.0	**847.3**

FIGURE 24.5
The reporting year exposure for the renewal year (2013) is calculated as a weighted average of the turnover figures of different years, based on the assumption that the average delay between trigger and reporting is 4 years. The cumulative earned percentage is calculated by Equation 24.1 with $a' = 1$, a = delay in year, $\tau = 4y$. The reporting year exposure (last column) is calculated as in Figure 24.4: as an example, the reporting year for the year 2012 is 11.5% × 1100 + 19.6% × 900 + 15.2% × 700 + (11.9% + 9.2% + 32.5%) × 600 = £731.3M.

Specifically, Equation 13.5, which provides the projected number of claims given the reported number of claims, can be manipulated to yield the percentage of claims that is expected to be reported:

$$\frac{\text{reported}_a(a')}{\text{projected}_a(a')} = \frac{a' - \tau\left(\exp\left(\frac{a'-a}{\hat{\tau}}\right) - \exp\left(-\frac{a}{\hat{\tau}}\right)\right)}{a'} \tag{24.1}$$

As an example, assume we have an annual professional indemnity cover, and that the average delay between trigger and reporting is 4 years, and that the exposure (the turnover, in this case, already on-levelled) is as in Figure 24.4. Also assume that the on-levelled turnover before 2009 is £600M. A detailed calculation of the reporting year exposure is shown in Figure 24.5.

24.1.4 Frequency Analysis

It is often said that there is no pure IBNR for claims-made policies because the policy covers only reported claims. What this actually means is that the time between an error being discovered and the time when the claim is communicated to the insurer is very small. A small time window is normally available to the insured, to report a potential claim. For example, policies may give a month after the expiry date for the insured to report claims that occurred in the previous year. However, there are other delays on account of which the final position for each year may be known only with a significant delay.

As is clear from Figure 24.6, there are different sources of delay between the moment when the loss 'occurs' (that is, the moment when negligent advice is given, which is the *trigger* for the claim) and when it appears in the underwriter's books with a figure attached to it. Losses can therefore be 'triggered but not discovered' (TBND), 'discovered but not reported' (DBNR), 'reported but not quantified' (RBNQ), and at each stage some delay is accumulated. Of course, there is also the possibility that losses may be triggered and never discovered, discovered but never reported (unlikely), or reported but never quantified (this will be common for errors that are discovered but no complaint is received).

How is this relevant to frequency analysis? TBND delay is relevant to the way we calculate the 'reporting year exposure'. The cumulative delay between discovery and quantification

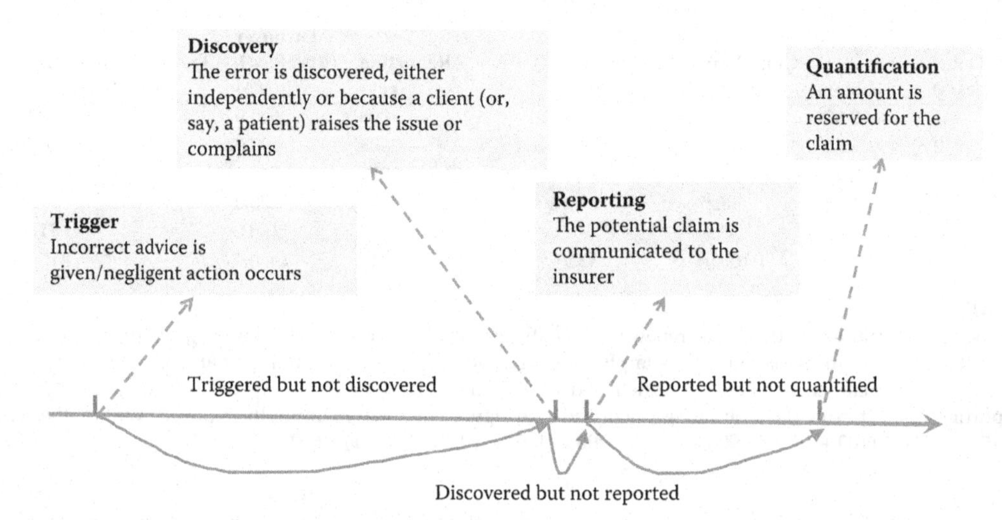

FIGURE 24.6
A simplified claims reporting process for professional indemnity claims. For simplicity, we have considered here only the case of a claim that is eventually quantified.

(DBNQ = DBNR + RBNQ) is instead relevant to the projection of the number of quantified claims (i.e. reported as non-zero) to ultimate.

The 'RBNQ' delay is normally quite small for standard lines of business such as motor or employers' liability; however, in the case of PI claims, there is a tendency to associate a figure to the claim only when it is pretty certain what the claim amount is going to be,[2] which could happen 6 months to 1 year after the claim is actually reported. Therefore, you should not expect that just because a policy is claims-made, the figures from very recent years are already developed to ultimate.

All this may sound a bit unfamiliar and daunting, but what this means in practice is that normally we will have as an input a triangulation of reported non-zero claims counts that we will need to project to ultimate according to the usual methodology.

24.1.5 Severity Analysis

24.1.5.1 IBNER

There may be IBNER, which may be partly compensated by the general practice of putting a figure on the claim only when the claim amount becomes reasonably certain.

24.1.5.2 Severity Distribution

- Because the occurrence date (i.e. the date in which professional advice was given) has no bearing on the claim amount, and in any case, it is not normally available, it is customary to revalue the claims from the reporting date to the midterm of the renewal policy.

2 The reason is at least partly the fact that the plaintiff's lawyers may gain access to the reserved amounts and use them as lower bounds for any compensation request.

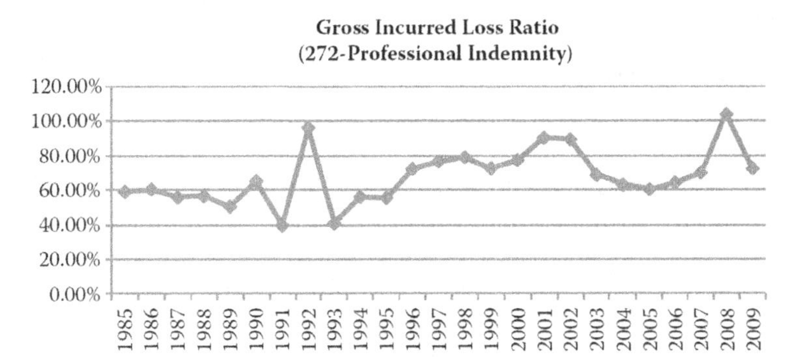

FIGURE 24.7
Loss ratio statistics for professional indemnity cover. Note the spikes in 1992 and 2008.
(Source: 2010 FSA returns).

- The loss severity is very sensitive to things such as court inflation and jurisdiction.
- Very large losses – $100M or more – are possible, especially in the United States.

24.1.6 Systemic Effects

Systemic effects due to the economy, such as the 2008 recession, can lead to an increase of frequency and severity (Figure 24.7); therefore, this is a line of business that is difficult to rate without knowledge of the business environment. There is also scope for underwriters to adjust the rates (or even withdraw capacity) on the basis of external factors such as the state of the economy. In terms of modelling, the dependency of frequency and severity on the state of the economy can be addressed using common shock modelling (see Sections 6.1.3.1, 14.4.3.2.2, and 24.3.2.3).

24.1.7 Aggregate Loss Analysis

We mentioned that professional indemnity is often sold with round-the-clock reinstatements. This must be taken into account when running the Monte Carlo simulation to obtain the distribution of losses ceded to the insurer.

24.1.8 Recap on Professional Indemnity

- PI is normally sold on a claims-made basis, either 'mature' or 'RDI'
- Exposure needs to be adjusted for TBNR to align it to a 'reporting year' basis
- The number of reported claims needs to be adjusted for RBNQ (also true for burning cost)
- Frequency and severity very dependent on
    - Systemic effects due to the economy
    - Territory
- Type of profession is a crucial rating factor
- During simulation, we may need to take account of round-the-clock reinstatements

More information on the pricing of claims-made policies can be found in Marker and Mohl (1980).

24.2 Weather Derivatives

Weather derivatives are risk management tools that allow organisations to protect themselves against the financial consequences of the vagaries of weather. The reason why we include a section on weather derivatives here is that actuaries are often involved in the pricing of these financial instruments and such pricing differs substantially from our standard pricing process. This is not surprising because the product itself is intrinsically different and attempts to do different things from a standard insurance policy. Before looking at how the pricing of weather derivatives works, we should perhaps start by considering what problems weather derivatives are trying to solve.

24.2.1 What Needs Weather Derivatives Respond To

Imagine you own a jewellery shop in the high street. You notice that on cold and snowy days during wintertime, the number of purchases is significantly reduced, presumably because fewer people go out for their shopping. You therefore realise that if next December is very harsh, you will end up selling much less jewellery for Christmas. Can you not insure your loss of revenues?

You can see how this may be difficult with a standard insurance policy. First of all, no damage needs to be done to your shop by the cold weather or the snow for you to have a loss of revenue – just the misery of it will suffice – so a property damage/business interruption policy would not be of help. Insurers also would need to make sure that it was actually the weather that caused your slump in revenues, and not the fact that you have hired a grumpy shop assistant who scares customers away, or that another jewellery shop has just been opened on the other side of the street. The risk of moral hazard would also be significant: in the presence of bad weather, you may over-rely on your insurance and not bother about making all efforts for your business to be successful.

Another, certainly more common, example is that of a power company, which has the opposite problem: if a winter is too mild, they will not sell enough electricity or gas and will not make the profits that have been budgeted. Again, various business risks muddle the picture when you try to prove the link between a mild winter and reduced revenues.

Yet another example is that of a farmer who wants insurance to protect him from the possibility that winter frost will destroy or severely damage their harvest the following spring – again, it will be difficult to discriminate between things that the farmers has done wrong and the genuine effect of low temperatures.

The list goes on and on: golf courses that have to close down for excess of rain, zoos that have a decrease in ticket sales because of generally bad weather, cinemas that on the other hand sell less tickets because the sun is shining … the need to cover for the whims of the weather is widespread and very diverse – which in itself should make it attractive for the market who can hedge its risks by, for example, insuring both low and high temperatures.

24.2.2 What Weather Derivatives Are

The key idea behind weather derivatives is to shift the attention from the loss of profits to the weather itself. The payout is not related to the actual loss of profits of the organisation but is simply triggered by the value of an index, which is based on objective, uncontroversial weather data obtained by a weather station whose independence is recognised by both buyer and seller.

The most common example is that of a 'heating degree days' (HDD) or 'cooling degree days' (CDD) contract. In an HDD contract, the difference between a base temperature T_b and the average temperature during day t is calculated and set to zero if negative:

$$\mathrm{HDD}(t) = \max\left(0, T_b - \frac{T_{\min}(t) + T_{\max}(t)}{2}\right) \tag{24.2}$$

In a typical structure, the paid amount may depend on the cumulative number of HDD over a given period (t_0 to t_1), for example, after applying an excess (strike) D and a limit L:

$$\text{Payout} = \min\left(\max\left(\sum_{t=t_0}^{t_1} \mathrm{HDD}(t) - D, 0\right), L\right) \times \text{tick value} \tag{24.3}$$

where the tick value is the payout for a HDD. For example, if the cumulative HDDs between October and February is 6000, the strike value is 5500, the limit is 1000 and the tick value is £10,000, the company will receive a payout equal to min(max[6000 – 5500,0], 1000) × 10,000 = 500 × 10,000 = £5M.

The terminology may be a bit confusing: a HDD contract pays when the weather is cold. The name, however, is not absurd: if the temperature goes below a certain value, a building needs to be heated and the colder it is, the more it is to be heated; therefore, the number of HDDs over the winter season gives you a fairly reasonable idea of how much energy you will need to use to heat your house. To give you an idea, the typical HDDs for New York City during the heating season is approximately 5000 (source: http://en.wikipedia.org/wiki/Heating_degree_day).

A CDD contract will work similarly but based on the difference between the average temperature and the base temperature, as long as this difference is not negative, it is otherwise set to zero. Similar indices can be worked out for wind, rain, snow, or even – as we will see later on – for a combination of weather variables. The important thing is that the data comes from an independent source and therefore involves no negotiation between the buyer and the seller.

24.2.3 Market for Weather Derivatives

As mentioned previously, the main buyers of weather derivatives are energy companies and other companies or even individuals who need to smooth their business results against adverse weather ('adverse' for the business, not for a picnic in the park). However, because weather derivatives are not indemnity based but index based, no insurable interest needs to be proved and they can therefore, in principle, be purchased by anybody regardless of the reason.

As for the sellers, some standardised products such as HDDs/CDDs linked to a limited number of weather stations can be bought *off-the-shelf* from exchanges such as the Chicago Mercantile Exchange. *Over-the-counter* derivatives with non-standard indices or non-standard weather stations can be bought from specialist insurers and hedge funds.

24.2.4 Basis Risk

Because of the way weather derivatives work, the payout is simple and uncontroversial. However, because the focus is not on the buyer's losses but on an index based on weather data coming from stations near the area that affects the buyer's business results, there is a potential basis risk: that is, the index is a not a perfect hedge for the insured's risk and the payout is not fully aligned to the losses incurred by the insured.

This may cut both ways: the insured may receive payout when the business goes smoothly and may not receive it (or not receive enough of it) when it is needed. The risk transfer is therefore imperfect. Many things concur to posing a basis risk. Here are a few examples:

- Weather may not matter so much as expected. There may still be a statistically significant correlation between weather and business results, but the effect may not be important as initially thought.
- Weather may matter but the index may be inadequate – for example, the insured may be buying a temperature-based product whereas its business results are affected by a combination of temperature, rain, and other elements; or the index is built on the basis of insufficient past statistics and suffers from model/parameter uncertainty.
- Weather may matter, and the index would in itself be adequate but there is no weather station for which the weather is sufficiently correlated with the weather in the area that matters to the insured – for example, the yield of a farm may depend on a microclimate that is not captured by any available weather station, or combination of weather stations.

24.2.5 Valuation of Weather Derivatives

Several methods have been proposed for valuing weather derivatives: see, for example, articles from Hamisultane (2008), Brody et al. (2002) and Alaton et al. (2007). We will consider two methods here that are based on option pricing techniques and the actuarial method. However, our main focus will be on the actuarial method, which is arguably the most common approach to the problem.

24.2.5.1 Option Pricing Method

The name 'weather derivative' itself suggests that option pricing techniques similar to those used to value call/put options on shares could be used for weather derivatives. We are not in a position to explain in detail how these techniques work, but the gist of it is that to determine the price of a derivative, you must find a way to express the derivative as a combination of the underlying security and a risk-free asset in such a way that the payoff is exactly replicated. As an example, a call option can be replicated as a suitable combination of shares and a risk-free asset ('cash').

The price of the option must then be equal to that of the replicating portfolio; otherwise, there is the opportunity of an arbitrage – that is, making money without taking any risk. The market will eventually catch up and remove the possibility of the arbitrage by converging towards the correct price.

24.2.5.1.1 Issues with the Option Pricing Method

There are, however, difficulties in applying this pricing method to weather derivatives. One of the assumptions of arbitrage-free pricing when applied to stocks is that both stocks and cash can be bought and sold in any amount (including fractional amounts). The problem with this in the weather derivatives case is that the underlying meteorological index (such as temperature), unlike stocks, is not traded.

Possible workarounds have been proposed.

- Geman (1999) has proposed the use of an energy contract instead of the meteorological index as the underlying variable used to price the weather derivative. The problem with this approach is that the prices of energy contracts are poorly correlated with the meteorological index and depend significantly on demand.
- Brix et al. (2002) have proposed the use of correlated weather future contracts instead of the meteorological index. However, these futures contracts have limited liquidity.

Because of the limitations of these solutions, the hedge is not perfect and the price thus determined is not unique – there is more than one price associated with an arbitrage-free replication strategy.

If the situation is difficult with temperature, for which there are at least traded substitution assets such as energy contracts and weather future contract, it becomes desperate for over-the-counter products for rainfall, snow, wind, or a combination of several weather variables.

These pricing difficulties, and the fact that many insurers are not comfortable with option pricing methods, open up a space for actuarial methods, which are arguably the most popular for pricing these products.

24.2.5.2 Actuarial Method

As we have discussed through the book, the standard way of pricing a commercial/reinsurance policy is to look at historical losses and create a frequency/severity model, possibly complementing such information with exposure curves or other market benchmarks. In personal lines insurance (and in some commercial lines risks), one will normally look at a portfolio of varied risks and identify the rating factors that allow us to discern which policyholders are more or less risky.

The pricing of weather derivatives is different because of what the product is trying to achieve: hedging a company's loss of revenues with a financial instrument that pays under the same meteorological conditions that cause the revenue dip.

As a consequence, the work around a weather derivative can be divided in roughly two stages:

i. Finding the meteorological index that best (or well) hedges the losses of revenue of the company. The techniques used at this stage range from trial and error to complex machine learning techniques.
ii. Once this index is identified, use weather data to model the index and structure/price a product around this index. The techniques used in this stage are by and large the standard burning cost analysis and stochastic modelling that we have described in previous chapters.

An overview of the actuarial valuation method is shown in Figure 24.8 and is expanded upon in Section 24.2.6.

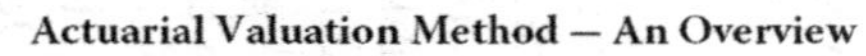

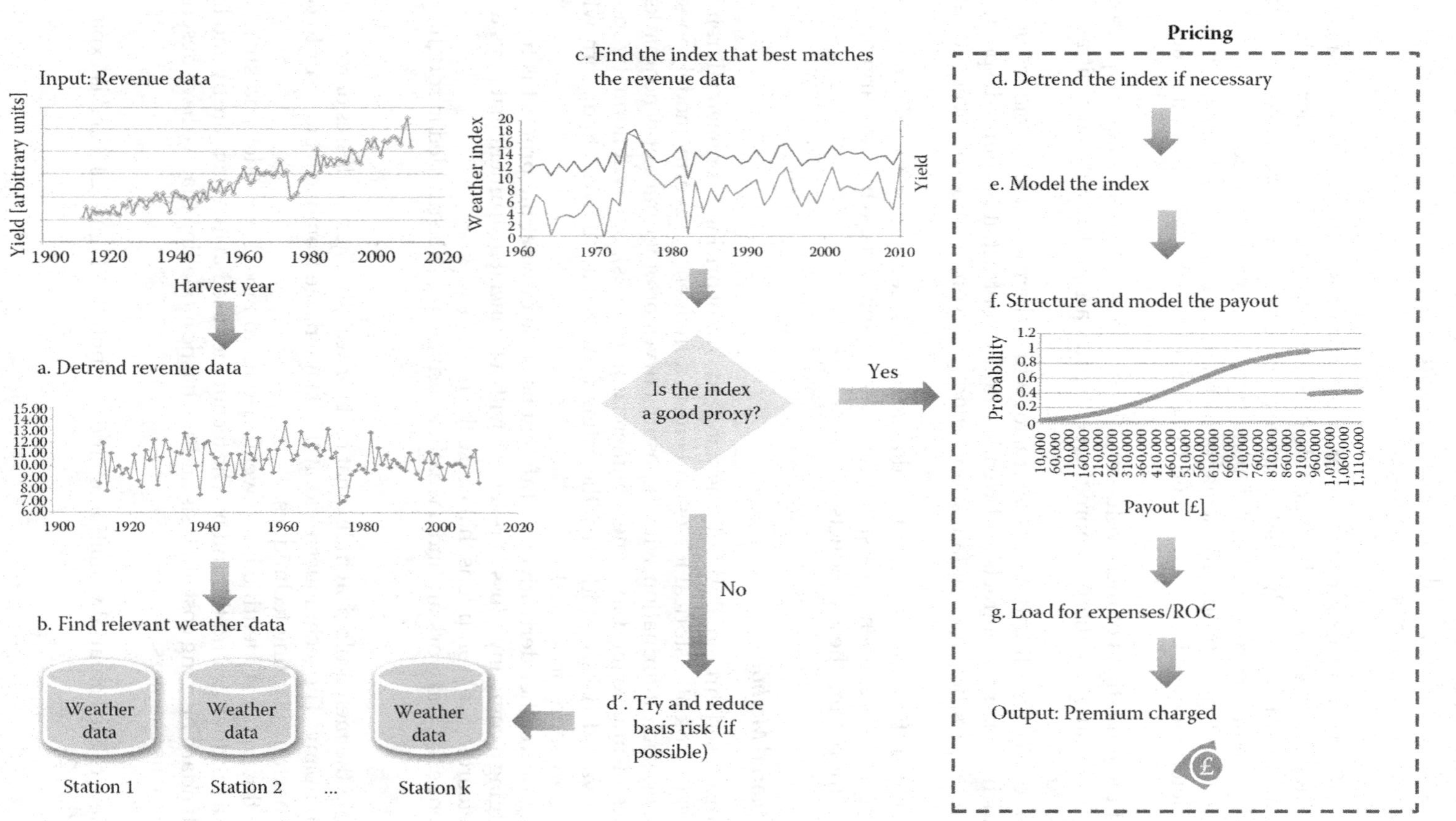

FIGURE 24.8
An overview of the actuarial method of valuing a weather derivative. The process can be roughly broken down into two phases: an exploratory analysis that seeks to identify an index which is a good proxy for revenues/costs/volumes (and which can therefore provide the basis for a hedge), and the actuarial valuation itself based on weather data.

24.2.6 Actuarial Valuation Method in Detail

We will now show how the actuarial method works using anonymised data from various real-world cases.

24.2.6.1 Phase 1: Finding the Index

To explain things more concretely, let us look at the different steps with a specific example in mind: a farm that produces an agricultural product for which growth and harvest are affected by weather.

24.2.6.1.1 Input: Revenue-Related Data

The main input to this phase is given by the *revenue* data or by another variable that is related to the revenues but is the one that is more closely influenced by the weather. In Figure 24.9, we use yield,[3] which is a good proxy for the revenues of a farmer selling an agriculture product and is actually more closely related to the weather than the revenues themselves. Other examples of revenue-related data are the tickets sold in an entertainment park or the amount of gas sold for heating purposes.

It is immediately evident that yield goes up and down from one year to the other; however, there is also a general trend, in all likelihood due to the technological improvements in farming. This increasing trend is obviously extraneous to what we are trying to capture and needs to be removed if we are to isolate the effect of the weather on crop production.

24.2.6.1.2 Detrending Revenue-Related Data

To isolate the effect of the weather, we have to remove trends due to non-weather factors, such as the time value of money or trends due to technological improvements.

In the case of the crop yield shown in Figure 24.9, the trend can be modelled by assuming an exponential increase with an average $r = 1.3\%$ year-on-year increase.[4] Once this trend is

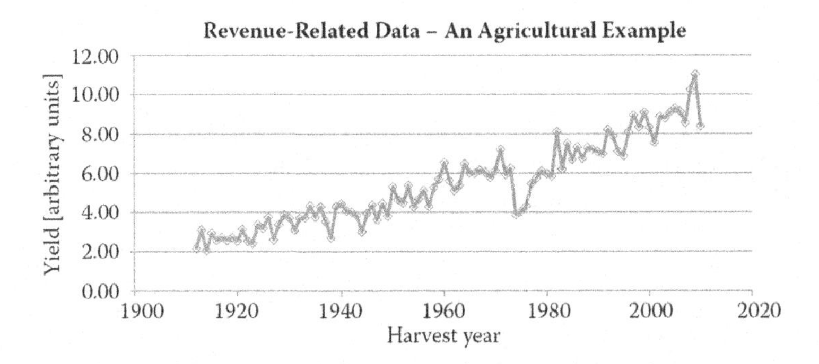

FIGURE 24.9
Yield of a given crop over the years – a more-than-linear increasing trend can be discerned even by visual inspection.

3 Crop produced per unit of cultivated area.
4 Ideally, the technological trend should be assessed independently of the actual yield series, because this may also be influenced by other trends, such as global warming; alternatively, we can attempt to model all trends affecting revenues (including weather) at the same time. We ignore these issues as a first approximation, perhaps changing the detrending assumptions later on in the exercise when we have determined the dependency of yield on the chosen meteorological index.

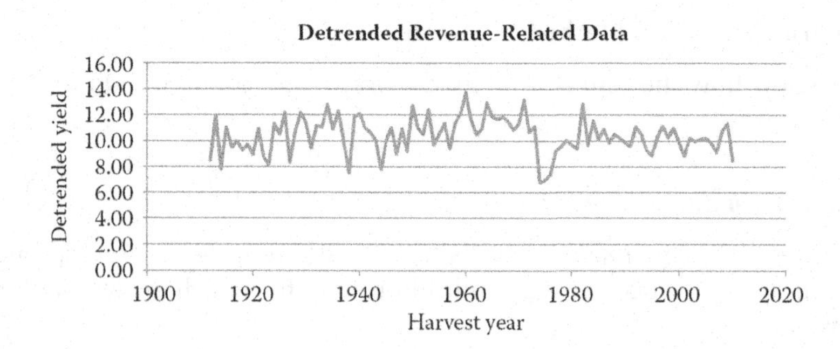

FIGURE 24.10
Yield of a given crop over the years after removing the effect of technological improvements on yield. Meteorological trends have been ignored at this stage.

removed by creating a new series $Y'(t) = Y(t) \times (1+r)^{t_0 - t}$ (t_0 being the year for which insurance is sought), the time series should appear to fluctuate randomly around a central value as in Figure 24.10.

24.2.6.1.3 Find Relevant Weather Data

Once we have accounted for the obvious trend in revenue/volume/cost due to technological improvements and other factors, we seek to explain the remaining variations in terms of the weather. We then need to look for relevant weather data coming from stations near the farm.

At this stage, 'relevant' really means 'anything thought to have a potential effect on the business results', erring on the side of caution and leaving it until later to discard variables that are not statistically significant.

Weather data provided by weather stations normally include basic statistics on these variables (amongst others):

- Temperature (high, low, average, etc.)
- Pressure (high, low)
- Wind strength (60-min average, 120-min average, gust, etc.)
- Wind strength at high altitude (some weather stations)
- Rainfall (total in a day, 60-min maximum, etc.)
- Humidity
- Snow (some weather stations)

Weather stations guarantee an independent source of information of the weather to which both the buyer and the seller can refer. However, some care is needed in the choice of the weather station. You must make sure that the weather station:

- Has a long history of recording covering at the very least the period over which revenue/cost/volume data are collected and more to identify certain long-term trends
- Has a consistent method of reporting data
- Has not been moved and the conditions around the weather station haven't changed significantly (e.g. if buildings have been constructed around the weather station, the weather station will be partially shielded from wind)
- Will continue to operate in the foreseeable future

- Will report frequently enough (e.g. it is no good if the weather station records temperatures daily but reports them twice a year)

Also, data cleansing of recordings from a weather station by a specialised company is normally necessary.

24.2.6.1.4 Find the Index That Best Matches the Profit Data

This is the most difficult problem around the valuing of weather derivatives. Some of the problems we need to solve to construct an index that predicts profits in the most effective way are the following:

a. *Which weather stations should we use?*
   We need to find the weather station whose measurements are most correlated with the weather in the farm. Normally (but not always), this will be the weather station that is nearest to the farm. However, sometimes the area affected by the weather is quite large and we need to use a combination of weather stations. We then have the additional problem of how to combine the results of different weather stations.
b. *What are the weather variables that have a significant effect on revenues?*
   Aided by physical knowledge and other expert knowledge, or simply by statistical analysis, we need to determine which weather variables have the most significant effect on revenues. In the case of our agricultural product, temperature and rainfall are obvious candidates, but we may need to consider all the other variables provided by a weather station. It should be noted that standard, off-the-shelf products have just one variable, the most common one being temperature.
c. *What function of the selected variables actually has the most effect on revenues?*
   Choosing the functional form in which the weather variables 'enter the equation' cannot normally be done separately from selecting the relevant variables themselves (Step b), but here we discuss it separately for clarity's sake.

In the case of our farm example, temperature might enter the equation in the form of HDDs (the cumulative number of degrees below the threshold over a given period). If that is the case, we need to fine-tune the period and the threshold to optimise the match with revenues. However, more complicated functions may be needed, because 2 days in a row of frost might have more effect than 2 days of frost over the period, and HDD will not take this into account.

Also, other weather variables may play a role in complicated ways. For example, rainfall might play a role in two different ways: drought (periods without rain) in the growing season may damage the yield, but excessive rainfall may also damage the yield, so we may need to have separate components for drought and rainfall in our model. Each of these components will have its own parameters – days of drought, threshold above which rainfall has an effect, compounding factors for successive days of rain … the possibilities are endless.

Finally, each factor may not have a sufficient effect when considered individually, but an index built as a combination of several factors (such as cold weather and rainfall) may match revenues much more closely. The meteorological index could be anything from a simple HDD to a complex multifactor model.

How do we then go about, therefore, finding a good index?

- Trial and error.
- Think of the physical reality behind the numbers – what experts think about the factors that affect revenue is a good starting point although not necessarily the last word: for example, farmers will normally know what types of weather are to be feared the most: long spells of cold weather? Drought followed by rainfall?
- Start with well-tested indices, such as HDD/CDD for temperature, and move on to indices more specifically tailored to your problem.
- Use a 'statistical learning' approach, for example, use generalised linear modelling or regularised regression with a model selection mechanism, such as starting from the simplest model with one variable to more complex combinations. All this is explained in Chapter 27 ('Rating factor selection and GLM').

In our farm example, we have used a 'frost index' loosely based on cumulative HDD over the winter period but with an allowance for the compounding effect of a sequence of cold days, by giving a larger weight to the HDD of day t if the HDD value for $t-1$ was also positive.

24.2.6.1.5 Index Validation

This quest will yield an index that is the best we have been able to find. The interesting thing is that finding a good index may be quite difficult and proving that it is the *best* possible index may well be impossible, but once an index is proposed, it is quite straightforward to determine whether it is a good proxy of the revenue-related variable, using standard model validation techniques.[5]

It is a bit like the travelling salesman problem: finding a tour through a large number of cities that is shorter than, for example, 1500 km is extremely difficult because of the large number of combinations that need to be tested, but if someone proposes a specific tour, it takes a simple sum to check if that person is right.

Successful indices have actually been found for a number of situations:

- HDD has a strong correlation with power usage.
- Rainfall has a strong correlation with the power output of hydroelectric stations.
- Windfall has a strong correlation with the power output of wind farms.

In many other cases, an index that has a statistically significant effect on revenues can be found, but the correlation may still be rather poor (e.g. $\leq$50%) because a lot of volatility remains unexplained. This of course results in basis risk. There are many reasons as to why this might be the case, including that we haven't been good enough at finding an index. However, the most frequent explanation is probably that weather is only one of the causes of the revenue slump. Note that this does not immediately mean that the index is not useful – an imperfect hedge may still be part of a wider risk management strategy.

A case in point is given by our farm and as illustrated in Figure 24.11: the correlation between the frost index and yield is obvious but the match is far from perfect, and part of the reason for this is the presence of non-weather factors affecting the yield. A better index could be created if past statistics of these factors were available.

5 See Chapter 12 for the general concept of model validation, and Chapter 27 for an application to rating factors selection.

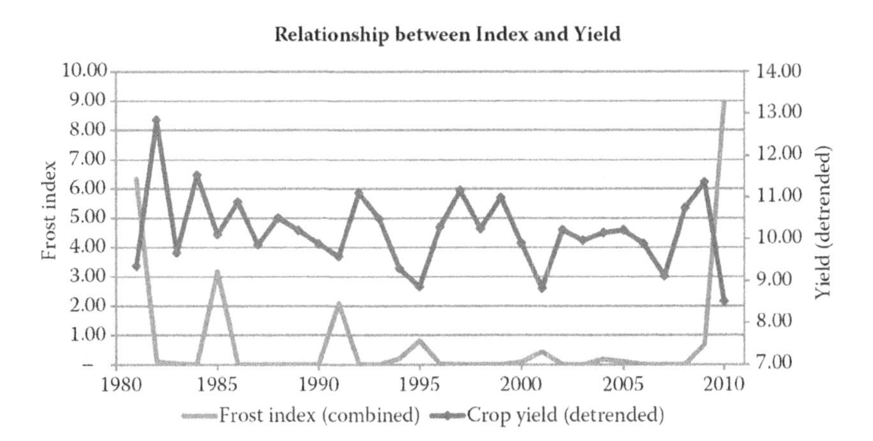

FIGURE 24.11
The chart shows the existence of an obvious correlation between the value of our frost index and the crop yield: in most years in which there has been a strong dip in yield, the value of the frost index is high or at least non-zero. However, the correlation is not perfect: a few dips are not captured (e.g. 1983) and others (such as 1995) are captured but the value of the index for 1995 is not high enough considering how bad the dip was. A likely explanation for the 1995 anomaly is the presence of pests, which damaged the harvest.

If no satisfactory index is found, the choice is between widening the scope of the analysis – for example, looking at data from different weather stations, or looking at different variables or different overall models, and giving up on the idea of a weather product because the impact of weather turns out to be less important than expected.

If a good index is found, we can basically forget about the revenue/cost/volume data and about how well they match the index and move on to pricing the weather derivative. The good news is that in the actuarial framework this is much simpler and less controversial than finding the correct index.

24.2.6.2 Phase 2: Modelling the Payout

We now abandon the example of our farm and (for the issues to be clearer) move on to an example in which the indices are HDDs based not on the average temperature but on the minimum temperature in each day.

24.2.6.2.1 Detrending the Weather Variables

In standard actuarial pricing, one looks for a model for future losses based on past loss statistics. Analogously, in the pricing of weather derivatives, one looks for a statistical model for the future behaviour of the index based on past index statistics.

To build such a model, the effect of any long-term trend in weather such as global warming needs to be removed, bringing past statistics in current weather terms, much in the same way as the effect of claims inflation is neutralised. Figure 24.12 shows the effect of detrending the minimum temperature (averaged over each year), which shows an increasing trending over the period 1950 to 2012.

After detrending, each day's temperature reading is replaced with an amended reading. The historical values of the index can then be calculated based on the amended readings and used for modelling.

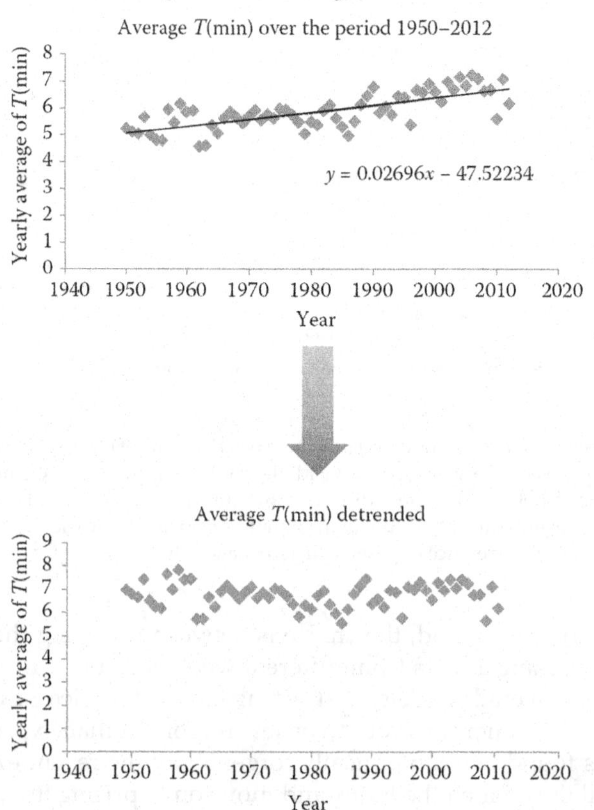

FIGURE 24.12
The year-on-year increase of the average minimum temperature for a given weather station in the United Kingdom is roughly linear.

24.2.6.2.2 Modelling the Index

After detrending, the index (in this case, the number of HDDs over a certain period and below a certain threshold) can be modelled with a statistical distribution. Very often, a normal distribution will be sufficient. This is because weather indices tend to display a degree of volatility that is much lower than, for example, the total losses for a portfolio of property risks or employers' liability risks.

In the case illustrated in Figure 24.13, the index was modelled as a normal distribution with mean = 1400 HDD and standard deviation = 129 HDD, and the fit was (as it often happens when dealing with weather that has limited tail risk) very impressive.

24.2.6.2.3 Estimating the Payout

The payout can now be calculated exclusively based on our model for the weather index, the payout per unit ('tick value') and the payout structure; for example, the strike value (the point at which the derivative starts paying) and the exit value (the point at which it stops).

In our example, we have chosen the strike point at 1360 HDD (slightly below the mean value of the index), and the exit point at 1100 HDD. The tick value is £10,000/HDD.

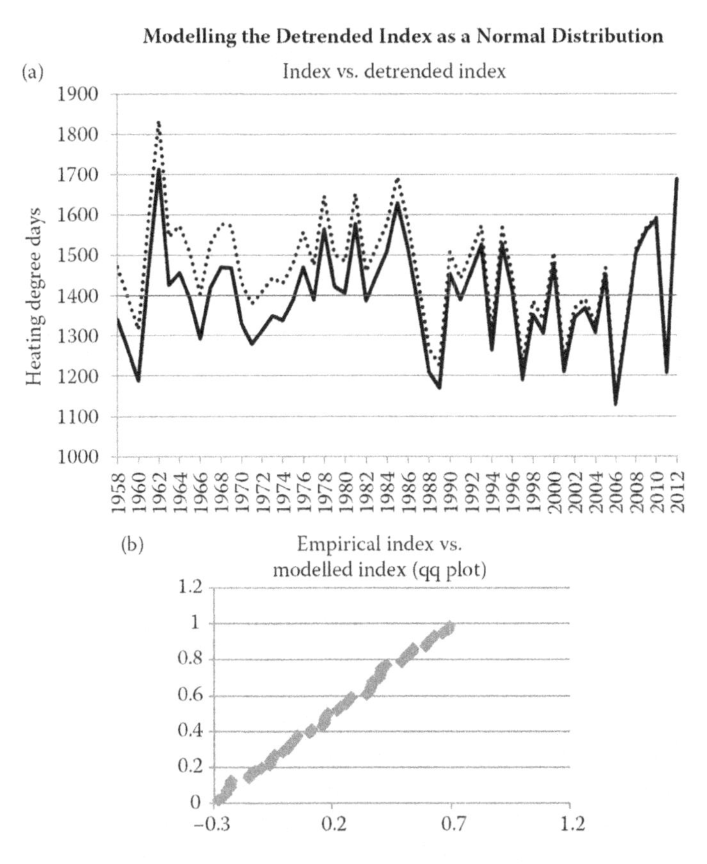

FIGURE 24.13
(a) Value of the index (cumulative number of HDD over a period) for a number of years, before (dashed line) and after (continuous line) detrending. (b) qq plot comparing the empirical values of the index with a Gaussian model. All the points lie near the bisector, which is an indication that the normal distribution is a good fit for the year-on-year volatility.

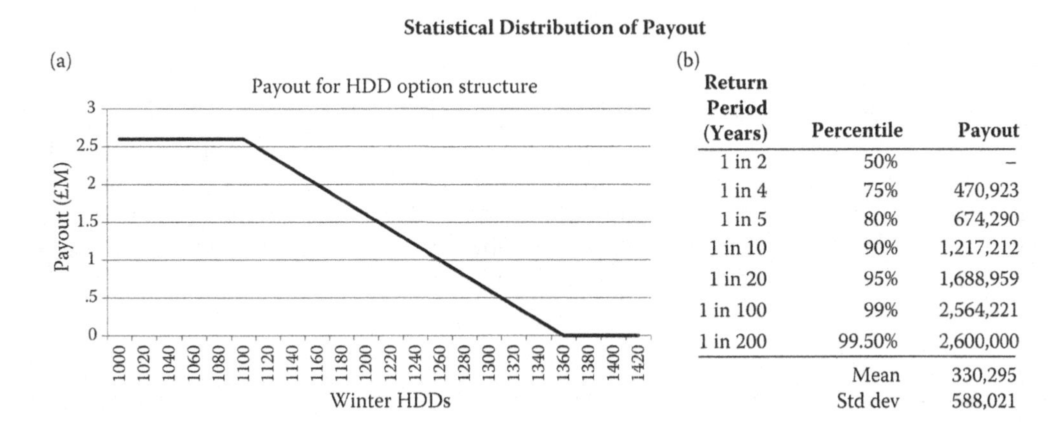

Return Period (Years)	Percentile	Payout
1 in 2	50%	–
1 in 4	75%	470,923
1 in 5	80%	674,290
1 in 10	90%	1,217,212
1 in 20	95%	1,688,959
1 in 100	99%	2,564,221
1 in 200	99.50%	2,600,000
	Mean	330,295
	Std dev	588,021

FIGURE 24.14
(a) Payout structure of the derivative: the derivative starts paying if the cumulative HDDs are below 1360 and increases as the cumulative number of HDDs decreases, up to a maximum value of £2.6M, achieved when the number of HDDs is ≤. (b) Output of the Monte Carlo simulation, shown as a standard percentile table.

Year	HDD	As-If Payout
...	...	...
1992	1455	0
1993	1526	0
1994	1265	952,032
1995	1527	0
1996	1411	0
1997	1191	1,692,360
1998	1351	85,886
1999	1307	529,475
2000	1478	0
2001	1212	1,484,714
2002	1344	159,490
2003	1369	0
...	...	...

Return Period (Years)	Percentile	Payout
1 in 2	50%	–
1 in 4	75%	481,710
1 in 5	80%	652,788
1 in 10	90%	1,515,290
1 in 20	95%	1,749,274
1 in 100	99%	N/A
1 in 200	99.50%	N/A
	Mean	337,847
	Std dev	598,861

FIGURE 24.15
In this method, the payout is calculated on an as-if basis for each year in the observation period (i.e. on the assumption that the derivative has been in place for the whole observation period), and then some simple statistics (mean, standard deviation, a few percentiles) are calculated.

The expected payout at different levels of probability can then be obtained by the Monte Carlo simulation (although, in this case, they could also be obtained by direct calculation). The payout structure and the results of the Monte Carlo simulation are shown in Figure 24.14.

An alternative way to produce the statistical distribution of payouts is to bypass modelling and do a *burning cost analysis* using the values of the index that actually occurred over the observation period (after detrending). The results are shown in Figure 24.15.

In the specific example, the burning cost does a good job of estimating mean and volatility (standard deviation), but is inadequate in calculating, for example, the losses expected with a 1 in 10 return period and has simply nothing to say on higher percentiles such as 1 in 100, because there are not enough elements in the sample (in this case, there were only about 60 elements in the sample).

24.2.6.2.4 Determining the Premium

Once the statistical distribution of the payouts is estimated, it is then possible for the seller to determine the premium that it should charge for the derivative. This will be based on the expected payout (or on other features of the payout distribution) with loadings for process/model/parameter uncertainty, expenses, and return on capital – much in the same way as for any other policy, as illustrated at length in Chapter 19.

24.3 Credit Risk

We already mentioned credit risk in our example in Chapter 7 on the individual risk model. We saw that in credit risk, we have a number of loans that may or may not default, and default is a binary state, just like being alive or dead. If a loan defaults, the loss is equal to the loss given default (LGD), which is a percentage of the overall loan amount.

Different pricing methodologies have been proposed for credit risk modelling. According to the article by McNeil et al. (2005), which has a good review of credit risk techniques, the two main classes of credit risk models are structural (firm value) models and reduced-form models. In structural models (examples of which are the Merton model and Moody's widely used Kealhofer–McQuown–Vasicek model), default happens when the difference between assets and liabilities (represented by stochastic variables) falls below a given threshold. In reduced-form models, the reason for the default happening is unspecified, and the default time is modelled by a stochastic variable, which in turn depends on economic factors. For a comparison of these approaches, see the reports by McNeil et al. (2005).

Most of these methods are rather complex and are beyond the scope of this book. In this section, we are going to illustrate a simple ratings-based approach, in which the probability of default depends on preassigned ratings. This approach can be viewed as a special case of a reduced-form model.

Examples of credit insurance products are Trade Credit and Mortgage Indemnity Guarantee.

24.3.1 Data Requirements

To price a portfolio of loans, we need:

a. Current exposure information
b. Historical losses (if available) and
c. Information on the wider economic context

The information in (b) may not be used directly for pricing but serves as a sanity check and helps understand the mechanism by which defaults happen for the portfolio under consideration and how loss amounts are related to the size of the loan.

It should be stressed, as done by McNeil et al. (2005), that 'data problems are the main obstacle to the reliable calibration of credit models'. This is because the big difficulty in this type of exercise is calibrating the model with the correct probabilities of default, the loss given default percentages and the way in which all this depends on systemic economic factors.

24.3.1.1 Current Exposure Information

The current exposure information is essentially a description of each loan in the portfolio. For each loan we will ideally need:

- Loan start date
- Loan duration
- Some measure of asset quality (such as the credit rating). The source of this measure may be internal (such as a bank assigning a probability of default to each loan) or public (such as S&P, Moody's, Fitch, and others)
- Loan amount
- Loan repayment schedule
- Territory
- Other useful descriptors, such as industry

Note that information on territory, industry and other useful descriptors of the loan are important to establish what systemic risks may be at play for the portfolio under consideration.

24.3.1.2 Historical Loss Information

Historical losses do not matter for credit risk to the same extent that they do for, say, employers' liability or motor losses because the adage that 'the past is not necessarily a good guide to the future' is especially relevant here. They are, however, an important indication of what to expect in years with similar economic conditions and similar profiles of loans as in the past. We therefore require:

- Past losses over a number of years
- Past exposure (number of loans, loan amounts, loan description over a number of years)

24.3.1.3 Information on the Wider Economic Context

This is information to be found in the public domain and which is relevant to the probability and effect of default, such as gross domestic product (GDP) growth rates.

Prob ID	Probability of Default
Q1	0.1%–0.2%
Q2	0.2%–0.7%
Q3	0.7%–2%
Q4	2%–5%
Q5	5%–20%

FIGURE 24.16
Some buyers of credit insurance have internal models for determining asset quality, based on empirical loss data.

	AAA	AA	A	BBB	BB	B	CCC/C	D
AAA	91.33%	7.88%	0.55%	0.06%	0.08%	0.03%	0.06%	0.00%
AA	0.60%	90.51%	8.10%	0.56%	0.06%	0.09%	0.03%	0.03%
A	0.04%	2.14%	91.50%	5.61%	0.42%	0.17%	0.03%	0.08%
BBB	0.01%	0.16%	4.14%	90.24%	4.28%	0.74%	0.17%	0.26%
BB	0.02%	0.06%	0.21%	5.87%	83.87%	7.99%	0.89%	1.10%
B	0.00%	0.06%	0.17%	0.30%	6.45%	82.97%	4.93%	5.12%
CCC/C	0.00%	0.00%	0.27%	0.40%	1.13%	13.77%	54.60%	29.85%
D	0.00%	0.00%	0.00%	0.00%	0.00%	0.00%	0.00%	100.00%

FIGURE 24.17
Example transition matrix, after the fashion of those produced by S&P, although actual tables will include some other complications, such as the probability of making a transition from a given rating to a non-rated state. Note that state D denotes 'default' and that by definition the probability of moving out of the default state is 0%. Note that although only the central estimate for each migration probability is estimated here, a measure of volatility is normally also provided.

24.3.2 Pricing Methodology

There are three elements to an actuarial analysis of the credit risk for a portfolio of loans: the probability of default of each loan (Section 24.3.2.1), the LGD for each loan (Section 24.3.2.2), and the correlation between the different loans (Section 24.3.2.3) and how this affects the probability of default and the LGD. Based on these three elements, one can calculate expected losses, volatility, and the full distribution of credit risk losses (possibly conditioned on a specific economic scenario).

24.3.2.1 Probability of Default

Based on the chosen measure of asset quality, we associate a probability of default with each asset. This could be the annual probability of default or – in the (common) case of multiyear loans – the probability of default over the loan period. The probability of default may come as a simple table such as the one in Figure 24.16.

It may come in the form of a transition matrix in which not only the probability of default but also the probability of a rating event (transitioning from one rating to another) is given, similar to the credit migration matrix in Figure 24.17.

Formally, we can define the transition matrix $T(r,s;1) = \Pr(r \rightarrow s;1)$, where $\Pr(r \rightarrow s;1)$ is the probability of going from state r (i.e. credit rating) to state s within a year. We can then determine the probability of going from state r to state s in 2 years by multiplying the matrix T by itself: $T(r, s;2) = T^2(r, s;1) = \sum_q T(r, q;1)T(q, s;1)$; and we can generalise that to k years: $T(r,s;k) = T^k(r,s;1)$. Specifically, the probability of defaulting within k years (where k may be the duration of the loan) is $T(r,D;k)$. Note that $T(D,r;k) = 0$ whenever $r \neq D$ and $T(D,D;k) = 100\%$.

As a concrete example, if we have a loan-rated BBB, the probability that it defaults within 1 year is

$$\Pr(BBB \rightarrow D;1) = 0.26\%$$

whereas the probability that it defaults within 2 years is

$$\begin{aligned}\Pr(\text{BBB} \rightarrow D;2) &= \sum_{r=\text{AAA}}^{D} \Pr(\text{BBB} \rightarrow r;1)\Pr(r \rightarrow D;1) \\ &= 0.01\% \times 0.00\% + 0.16\% \times 0.03\% + \ldots + 0.26\% \times 100\% \\ &= 0.63\%\end{aligned}$$

Therefore, this means that the probability of going from BBB to D (default) in 2 years is the probability of going from BBB to any other state (including D) in the first year and then from that state to D in the second year.

Based on the credit migration matrix, and under the assumption of the stationarity of the probabilities of default (although this assumption can be relaxed, as we will see), we will be able to calculate the probability that a loan defaults in the first year, the second year, or in the third year, and so on.

24.3.2.2 Loss Given Default

Once a default happens, the loss will not necessarily be the full amount of the loan, but a percentage of it – this is called the LGD. If the loan is repaid periodically, the LGD will be lower than or equal to the present value of the outstanding amount to be repaid (plus an allowance for interests lost).

Furthermore, a portion of the outstanding amount may still be recovered.

- In some cases, it is possible to assume that a fixed percentage will be recovered. This is the case when, for example, the recovery of the debt is outsourced to specialist credit recovery firms that charge the creditor a fixed percentage of the outstanding amount (e.g. 20%–30%).
- In general, however, the LGD is a stochastic variable that can be modelled, for example, as an MBBEFD distribution.

24.3.2.3 Correlation between Defaults

The main consideration around credit risk is probably that most of the loss volatility comes from the correlation between individual defaults rather than from random fluctuations in the number of loans defaulting independently: the overall volatility of a portfolio of independent loans is indeed minimal. For this reason, credit risk products for a portfolio of loans are often referred to as 'correlation products'.

Correlation between the defaults of a portfolio of loans emerges as a result of two different causes (McNeil et al., 2005):

- First (in the case of commercial loans), different loans may be purchased by firms with direct economic links between them, so that the default of one brings about the default of the other.
- Second (for all loans), the default of individual loans may depend on systemic factors such as the health of the economy: during an economic recession, the general level of economic activity falls and many firms fall with it.

In a reasonably diversified portfolio, the first cause has a negligible effect, whereas the second is paramount. One way to take systemic factors into account is by conditioning the probability of default on economic health, which is a way of producing correlation through a common correlation shock, very much in the same way that we will illustrate in Section 29.7 ('Common shocks'). Note that in this approach *all loans are assumed to be independent given the value of the underlying economic factors.*

As for what economic variables are relevant drivers behind the default probabilities, Duffie and Singleton (2003) show evidence that empirical default rates are strongly correlated with fluctuations in the GDP growth rate: a good starting point is therefore to use GDP growth rates to provide the systemic element upon which the probabilities of default depend. This is called a one-factor model for the obvious reason that the probabilities are driven by the value of a single underlying stochastic variable.

A simple example is the Bernoulli model, in which one can model the probability of default of each loan, p_i, as a random variable that is dependent on an external factor ψ (such as the GDP growth rate, or a function of the growth rate) distributed according to some underlying distribution such as a beta, a probit-normal, or a logit-normal (McNeil et al., 2005). In the case of the beta distribution, the probability of default has a distribution $f(q) \alpha\ q(\psi)^a (1 - q(\psi))^b$ where a, b may vary depending on the asset quality of the loan (normally a categorical variable) and $q(\psi)$ is a suitable function of the underlying economic variable.

An even simpler example is that in which the underlying variable is discrete, that is, the probability of default has only a few different values depending on the economic scenario:

$$p_i(\text{AQ}) = \begin{cases} p_i^{\text{boom}}(\text{AQ}) & \text{under stressed economic conditions}(\Psi = 1) \\ p_i^{\text{normal}}(\text{AQ}) & \text{under normal economic conditions}(\Psi = 0) \\ p_i^{\text{bust}}(\text{AQ}) & \text{under bad economic conditions}(\Psi = -1) \end{cases} \tag{24.4}$$

where AQ represents the asset quality of the loan. The probability of the three different states is given by $q(\psi)$: $q(1) = q_1$, $q(0) = q_0$, $q(-1) = q_{-1}$, with $q_1 + q_0 + q_{-1} = 1$.

The dependency of the default probabilities on an economic factor ψ can be generalised to the credit migration probabilities: in general, migration towards lower (respectively, higher) ratings will be more likely during an economic recession (respectively, expansion). Because the credit migration probabilities are normally given with some measure of historical volatility (such as the standard deviation, or a range of past values), as a first approximation, one may associate various values of the economic factor with different tables, each picking a different point of the range.

For example, in the simple case where ψ has only three discrete values as in Equation 24.4, if the range of historical values for Pr(A → BBB) is between 3% and 9% with a central point of 5.6%, such as that in Figure 24.17, we may assume $\Pr(A \to BBB \mid \psi = 1) = 3\%$, $\Pr(A \to BBB \mid \psi = 0) = 5.6\%$, $\Pr(A \to BBB \mid \psi = -1) = 9\%$.

We have focused above on the probability of default. However, in general, the LGD, too, might depend on the economic conditions. For example, if LGD is modelled as a beta distribution on the interval [0,1], we can have the mean and the standard deviation of the beta distribution to depend on the economic conditions ψ through a continuous variable or through a discretised condition such as Equation 24.4.

24.3.2.4 The Monte Carlo Simulation

A model for the aggregate losses can be obtained based on the methodology outlined in Section 17.7 on individual risk models. Some modifications are, however, required to take into account the systemic element and the multiperiod nature of the risk.

Let us see how this works in the case where we want to estimate the distribution of aggregate losses in the next K periods for a portfolio of L loans $l = 1, \dots L$, with varying start dates (start dates may be before or during the period analysed) and expiry dates (expiry dates may be during or after the period analysed). We make some simplifying assumptions for illustration purposes.

24.3.2.4.1 Simplifying Assumptions

i. The payments for each loan are made at the end of each period (in practice, this hypothesis is always true as long as we choose a sufficiently short period).
ii. Ratings are available for all loans, and probabilities of default and LGD are available for each rating.
iii. We have a single systemic economic factor ψ, which is assumed to remain constant over the K periods.

Let N_{sim} be the number of simulated scenarios, $i = 1, \dots N_{\text{sim}}$.

1. For each simulated scenario i, simulate a value of the economic factor ψ (this could be a discrete or a continuous value, as shown in point c).
2. For that same scenario, scan through all L loans and for each loan l decide whether it will default based on the probability of default $p_l = p_l(\psi)$. Note that if a credit migration table such as that in Figure 24.16 is provided (with credit migration probabilities depending in general on ψ), this probability can be written as $p_l(\psi) = T(R_l,D;K)$ where R_l is the initial rating for loan l.
3. If loan l defaults at time t_l ($t_l = 1,\ldots K$), sample a value $x_{l,i}$ from the LGD distribution of l, possibly dependent on ψ (if LGD is fixed, simply pick the unique LGD) and multiply it by the outstanding amount $O_l(t_l)$ of debt at time t_l, obtaining $X_{l,i} = x_{l,i} \times O_l(t_l)$. If loan l does not default in the ith scenario, we just set $X_{l,i} = 0$.
4. The total losses over K periods for the ith scenario are $S_i = \sum_{l=1}^{L} X_{l,i}$.
5. Repeat for $i = 1, \ldots N_{\text{sim}}$ simulations to produce the aggregate loss distribution, then calculate the various statistics (mean, standard deviation, percentiles and others) in the usual way.

The algorithm above has one important simplification: the economic factor ψ is assumed to remain constant over the K periods (iii). This assumption, however, is there only to simplify the description of the algorithm, and taking the possibility that ψ may depend on time does not add significant conceptual difficulties once a time series $\{\psi_t\}$ of values of ψ is available.

The difficult part is to be able to generate values of $\{\psi_t\}$ that make economic sense. The time series needs to take into account the fact that the value of ψ_t depends to some extent on $\psi_{t-1}, \psi_{t-2}\ldots$ (technically, this is called autocorrelation): for instance, a period of GDP expansion is more likely to be followed by another period of GDP expansion than by a period of GDP contraction. Also, if there is more than one economic factor concerned, that is, ψ is a vector with several components (such as GDP growth rate, interest rates, inflation, equity returns, etc.), it is important to take into account the dynamic relationships between the different economic factors. All this requires, in practice, is an economic scenario generator, which simulates future economic scenarios.

An economic scenario generator would feed into the Monte Carlo simulation by creating, for each simulation, a sequence of K values of ψ_t: $\psi_1, \psi_2\ldots\psi_K$, one for each period. Step 2 needs to be amended to take into consideration the fact that the transition probabilities change over the period (see Question 5). Note that, as a consequence of the need to simulate random sequences $\psi_1, \psi_2\ldots\psi_K$ (one for each simulation), a much higher number of simulations will be needed for the estimated aggregate loss distribution to be stable.

24.3.2.5 Output

The output of an aggregate loss model produced with the pricing methodology outlined in Sections 24.3.2.1 through 24.3.2.4 is the cumulative distribution function, which can be represented in a standard way, such as a percentile table or a graph. In some cases, it might also make sense to present output models for different fixed scenarios, for example, for an expansion, normal, or recession economic scenario, similar to Figure 29.7.

24.3.2.6 Limitations

The ratings-based methodology outlined above is simple and reasonably comprehensive. However, the devil is in the calibration! The methodology assumes that

a. A rating is associated with every loan, reflecting the relative riskiness of the loans.
b. The future probability of default and the LGD for each rating can be somehow derived from past empirical default rates or another source.
c. The relationship between the economic factor(s) ψ and the probability of default/LGD is known.
d. The economic scenario generator produces realistic future scenarios.

As you can easily appreciate, the number of parameters (such as ratings) and distributions (such as the LGD distribution) is impressive, and a massive data set is required for a meaningful calibration. As usual, it is pointless to try and make your credit risk model more 'realistic' (i.e. incorporating more features) if the added features cannot be calibrated reliably. In practice, therefore, it may well turn out that a much more basic model than the one described above is more useful for your particular endeavour.

Even if you use a very basic model, note that rating agencies are slow to adjust the credit rating of a firm, and therefore you should not trust a table such as that in Figure 24.17 to fully reflect current economic conditions.

24.4 Extended Warranty

As mentioned in Chapter 3, extended warranty allows the insured to extend the warranty that typically comes with a purchased product (such as a car, computer, or camera) by a number of years (or, in some cases, in perpetuity), in exchange for a premium.

Note that extended warranty may be provided by the retailer (point-of-sale extended warranty), an independent insurer, or the manufacturer itself. If the retailer or the manufacturer offers the product, they may then buy reinsurance from a professional insurer. For rating purposes, the most important entity of extended warranty risk is the failure rate, that is, the probability that a device will need repair or substitution within a certain time. As for the severity, for most products, an average cost of repair/replacement will be sufficient, although of course nothing prevents the actuary from analysing the distribution of claim amounts in more detail.

An interesting feature of extended warranty is that failure rates are not constant in time. Actually, they behave according to the so-called 'bathtub' curve – a different behaviour for the infantile, mature, wear-out and end-of-warranty phases (Figure 24.18).

24.4.1 Data Requirements

To calculate a failure rate, we need both exposure and claim information. The exposure information is typically given by the number of sales per day/month, subdivided by type of product, model, territory of sale, and other relevant factors. An example is provided in Figure 24.19. The claims data will include items such as claim ID, territory, model,

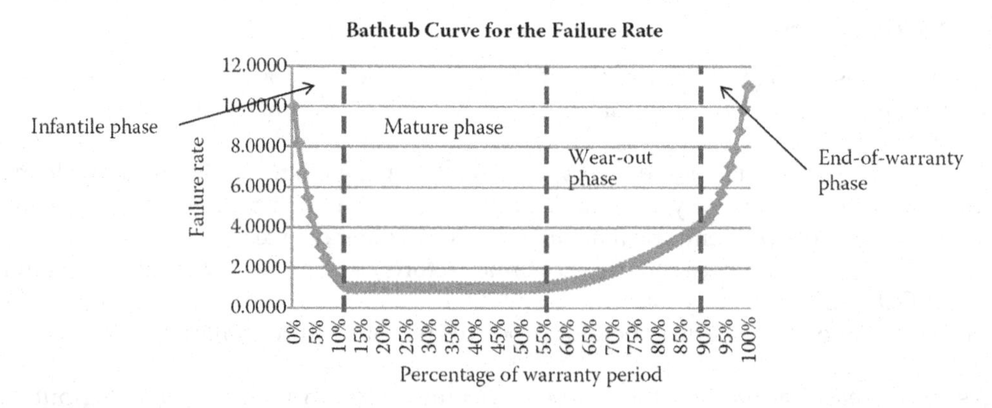

FIGURE 24.18
The failure rate for a product, such as an electronic device, follows a bathtub-shaped curve, which is reminiscent of a mortality table (mortality is higher for infants and after a stable period, increases with age) and borrows part of its terminology. However, devices (unlike humans) also have an 'end-of-warranty' phase during which some policyholders rush to make claims before the policy expires. This may be partly fraudulent and partly simply an effect of warranty holders postponing genuine claims for devices rarely used until the eleventh hour.

Exposure (Sales) Data for Extended Warranty

Month	Country 1	Country 2	Country 3	Country 4	Country 5
2007/06	872	144	45	1317	461
2007/07	1800	395	1826		6404
2007/08	14,144	3614	8842	4406	16,380
2007/09	14,204	10,244	9107	37,852	29,156
2007/10	13,296	6010	6836	20,294	25,188
2007/11	15,960	6405	8396	16,332	44,217
2007/12	21,025	7078	7290	22,711	35,352
2008/01	13,807	4546	7971	24,937	17,184
2008/02	23,548	6626	19,161	25,442	42,108
2008/03	25,306	12,217	13,797	44,104	65,771
2008/04	24,457	11,613	20,451	57,635	59,154
2008/05	23,674	10,644	10,234	55,725	36,611
...	...	...	...	...	...

FIGURE 24.19
Number of products sold in each month and in each country. The same information must be provided for different models and other risk factors.

purchase date, notification date, repair date, cost of repair/replacement, symptom, type of defect, and type of repair.

24.4.2 Failure Rate Analysis

The ultimate objective of failure rate analysis is being able to predict the number of failures over the timeframe of the extended warranty, that is, after the original warranty expires. This prediction will often be based on the failure rate observed during the period of the original warranty.

Calculation of the Base Failure Rate

		Number of Months from Purchase												
Month	**Sales**	**0**	**1**	**2**	**3**	**4**	**5**	**6**	**7**	**8**	**9**	**10**	**11**	**...**
Jul-08	22,936	183	176	61	21	43	31	29	26	26	29	26	20	...
Aug-08	25,815	207	90	41	39	32	39	36	32	26	25	32	43	...
Sep-08	21,747	187	75	78	40	36	35	37	37	37	36	33	53	...
Oct-08	14,571	207	175	124	55	59	66	63	55	53	72	82	88	...
Nov-08	10,645	249	146	78	39	40	35	39	33	46	48	42	41	...
Dec-08	13,983	212	118	64	38	26	21	30	23	33	33	36	48	...
Jan-09	9157	216	98	43	28	18	19	43	35	32	36	40	26	...
Feb-09	6427	203	55	48	12	22	25	39	21	16	31	27	22	...
Mar-09	7870	148	74	66	21	34	44	32	22	36	24	20	33	...
Apr-09	5083	146	68	59	38	32	30	22	26	30	17	18	34	...
May-09	6543	142	63	65	43	26	28	33	43	38	23	30	36	...
Jun-09	11,717	107	71	71	29	14	41	22	24	29	24	22	19	...
Jul-09	16,016	180	132	59	31	29	32	29	37	17	29	22	27	...
Aug-09	15,747	165	83	45	42	27	29	34	22	25	29	26	54	...
...	...	...	...	...	...	...	...	...	...	...	...	...	...	...

FIGURE 24.20
At its most granular level, this table allows us to calculate the empirical failure rate for a particular cohort after a given number of months from purchase: for example, products sold in November 2008 have at the fifth month from purchase (April 2009) an annualised failure rate of $35/10{,}645 \times 12 = 3.95\%$.

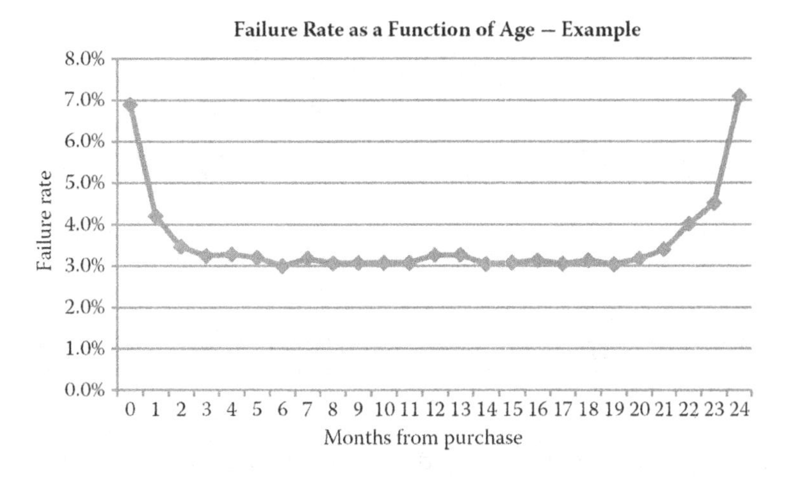

FIGURE 24.21
This graph is obtained by aggregating by row the information in Figure 24.20, that is, by dividing the number of claims for a specific age (months from purchase) by the number of products sold and multiplying by 12 to produce an annualised figure. As a result, it seems clear that the mature phase has kicked in at around age 3. The base failure rate can then be calculated as an average over a sufficiently long period within the mature phase, for example, from ages 3 to 11. Notice that these particular failure rates have been calculated based on claims for products which didn't have an extended warranty but had a 2-year original warranty – hence the end-of-warranty hike at the end of the 2 years, probably before the wear-out phase has started. Also note that the calculation spans between month 0 and 24 (rather than month 0 and 23) – this is because age has been defined discreetly by calculating the month of purchase, the month of claim, and then subtracting one from the other – leaving open the possibility of products purchased in November 2008 to have a claim in November 2010: this rough methodology is sufficient for the purpose of calculating the base failure rate.

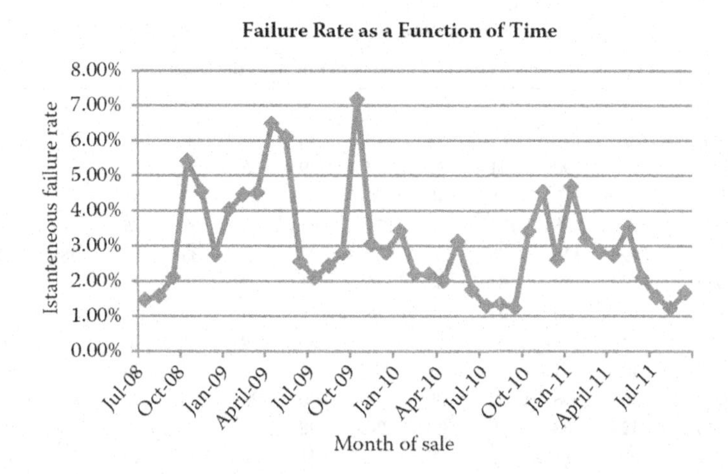

FIGURE 24.22
This graph shows how the failure rate depends on the cohort and is obtained by calculating the base failure rate over ages 3 to 11 for each individual month of purchase: for example, for products sold in November 2008, the (annualised) base failure rate is $(39 + 40 + 35 + \ldots + 41)/10{,}645 \times 12/9 = 4.55\%$. Much of the month-on-month volatility can be explained by normal statistical fluctuations. However, pronounced peaks such as that in October 2009 may be a sign that a new model with some teething problem has been introduced. This type of graph therefore allows us to keep an eye on how each individual model pans out. Typically, each new model has a higher failure rate at the beginning, which then subsides as the issues with components, software and others, are corrected.

Adjustment to the Base Failure Rate

Year	Basis Failure Rate	Infantile Gross-Up	Wear-Out Gross-Up	End-of-Warranty Gross-Up	Estimated Failure Rate
1	3.05%	43%	0%	0%	**4.37%**
2	3.05%	0.0%	0%	0%	**3.05%**
3	3.05%	0.0%	12%	0%	**3.42%**
4	3.05%	0.0%	25%	0%	**3.81%**
5	3.05%	0.0%	44%	14%	**5.01%**

FIGURE 24.23
The estimated failure rate can be calculated using the base failure rate and making adjustments for the different phases. For example, the estimated failure rate for year 5 is given by $3.45\% \times 1.44 \times 1.14 = 5.01\%$. These correction factors cannot normally be determined for every single product and model but are based on portfolio estimates.

One possible way to go about this is to determine the *base failure rate*, that is, the failure rate during the mature phase, and to make the necessary adjustments to project the failure rate during the extended warranty period.

The basis failure rate can be calculated, for example, by subdividing all sales in 1-month cohorts and, for each month of sale, calculate the number of claims received within the same month, within 2 months, 3 months, etc., as in Figure 24.20. This representation allows us to estimate, amongst other things:

- The 'instantaneous' failure rate for a particular cohort and for a given number of months from purchase.
- The exact shape of the bathtub curve (at least for the first period), and specifically the time it takes for the mature phase to set in.

- The base failure rate for a specific cohort, based on the mature phase derived above.
- How the base failure rate changes with time.

By aggregating the information in Figure 24.20 along different dimensions, we can derive several important bits of information: as an example, by summing over all rows, we can derive the empirical shape of the failure rate as a function of age, that is, time from purchase (Figure 24.21), and identify the mature phase – which in turn allows us to calculate the base failure rate.

By aggregating the information in Figure 24.20 along columns, we can determine whether the failure rate is changing as new models are introduced in the market (see Figure 24.22).

One difficulty is that when we first introduce extended warranty for a product, we were often trying to determine the failure rate of the product at an advanced age, for example, 3 or 4 years, based on the failure rate as measured during the original warranty period. This is relevant because the extended warranty will normally start during the mature phase and will be affected by the wear-out phase. As for the end-of-warranty hike, this will happen regardless of the duration of the extended warranty and might actually precede the wear-out phase when the duration is short.

For the wear-out phase, market/portfolio information might need to be used because information for new products will not be available. Figure 24.23 shows an example of how portfolio information can be used to derive failure rates across the whole period.

Another difficulty is that new models are continually introduced with slightly (or not so slightly) different risk profiles, and by the time the failure rate is known, the premium has already been paid. The premium is paid when the model is sold and by the time enough experience has been accumulated about the post-warranty failure rate, the model may have already been replaced by newer models.

Finally, note that the failure rate calculated in this way is not a controlled statistical calculation of the failure rate of a device: there is no guarantee, for example, that the cohort on which we are basing the calculation remains the same in time. People may get rid of the product or replace it before it fails and therefore never use part of the cover, they may forget about the extended warranty, they may decide not to repair their product because they've lost interest in the product, and by their behaviour, they may make their item more likely to fail and so on. However, from an underwriting point of view, this hardly matters, as long as we are able to associate a number of claims with a number of sales.

24.4.3 Frequency Analysis: From Failure Rate to Expected Claim Count

The expected claim count for a cohort of policies can be simply calculated by multiplying the cumulative failure rate over the extended warranty period by the number of sales. Considering, for example, Figure 24.23 and assuming that the original warranty period is 1 year, the total failure rate will be 3.05% + 3.42% + 3.81% + 5.01% = 15.29% (we are assuming that all products are kept until the end of the extended warranty, and that more than one claim per product is possible). If 10,000 extended warranties are sold, we expect 1529 claims on average. Under the assumption that more than one claim per item is possible, a Poisson or negative binomial distribution will be adequate to model the frequency.

24.4.4 Severity Analysis

Given the large number of claims, and the fact that the distribution of loss amounts is capped by the value of the item under warranty, the production of a severity model is

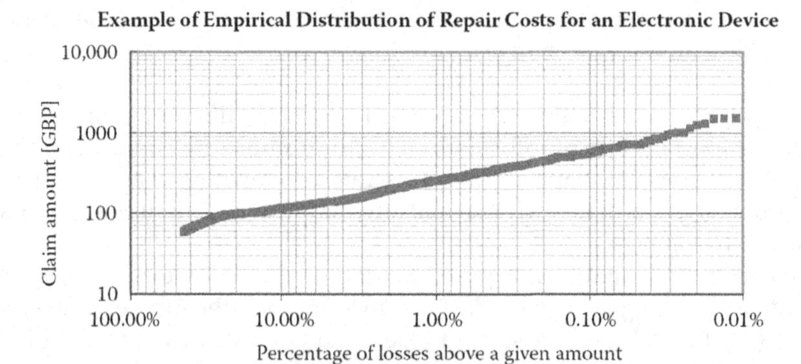

FIGURE 24.24
An example of an empirical distribution of (revalued) repair costs for a specific product, which seem to have a natural cap at approximately £1800, the current value of the product. For this particular product, the small number of claims corresponding to the top values provides evidence that repair – rather than replacement – was the favoured strategy of the manufacturer.

normally unnecessary; resampling from the empirical distribution (Figure 24.24) will be sufficient.

24.4.5 Aggregate Loss Model

Often, the underwriter will be content with a failure rate estimate (the cumulative failure rate over the period of the extended warranty), multiplied by an average claim amount: the volatility of the average claim amount is so low that any additional modelling effort may not be justified. Most of the uncertainty, as explained below, comes from systemic risk.

If aggregate loss modelling is necessary, however, the expected claim count (frequency model) over the relevant period can be combined with the severity model via the Monte Carlo simulation to produce the total losses at different levels of probability, in the usual way.

24.4.6 Systemic Factors

Under normal circumstances, the number of failures and their severity will be reasonably stable in time, and underwriters will have a reasonable idea of how products in general wear out over the years. The first few months of a product's life will provide a reasonable idea of how a new model is performing against its predecessors.

The real unknown, however, is the systemic risk. The most important systemic element is arguably that of epidemic failure: occasionally, a model will introduce a new component or some other innovation that turns out to have some fundamental flaw, which leads to a hike in the failure rate. This is sometimes referred to as 'epidemic failure'. If the extended warranty is sold by the retailer or by the manufacturer, and there is a reinsurance contract with an insurer, the reinsurance contract may explicitly exclude epidemic failures.

Another systemic element to consider is the territory where the product is going to be sold. Different countries will exhibit very different failure rates, and underwriters will be keen to make sure that the territorial profile remains relatively constant, or that different rates are charged to different territories.

24.4.7 Rating Factor Analysis

Extended warranty lends itself quite naturally to ratings factor analysis because the number of claims is quite large (sometimes hundreds of thousands, for very popular products). Rating factors will include territory of sale, make, model, year of make, duration of the warranty, exclusions, plus any product-specific factors (such as engine size for cars).

24.4.8 Pricing

As usual, the pure risk cost calculated above needs to be loaded for expenses and profit in the usual way. One specific observation to make is that investment income may play a significant role because the time between the premium for the extended warranty is paid (often up-front) and the payment may be substantial: for example, if a camera has an original warranty of 1 year and a 4-year extension is purchased at point of sale, the average time between premium receipt and claims payment will be (very roughly) 3 years.

24.5 Miscellaneous Classes of Business

We now look briefly at a number of different classes of business, focusing on their distinguishing features rather than attempting to provide a full-blown pricing methodology for any of them.

24.5.1 Aviation Insurance

The most important types of cover for aviation insurance are hull and liability. Other types of coverage include product liability (the liability of providers of aircraft or their components), airport liability, and ground handling. It includes in its scope both airplanes and helicopters.

As mentioned in Section 2.4, aviation liability is international in nature and is regulated by the 2005 Montreal Convention, which provides for unlimited liability to airlines and allows claimants ample discretion in the choice of jurisdiction, with a general trend towards using US courts wherever possible. As to aviation hull, the payout is obviously limited to the value of the hull. However, the value of the hull can be very high: up to approximately $400M in 2013.

From the pricing point of view, this line of business is characterised by a very small frequency of large claims, whose severity can be extremely large: the collision of two large airplanes (such as in the Tenerife accident in 1977) is often taken as a worst-case scenario, with a magnitude that can run into billions of dollars.

A good feature of this class of business, from a pricing point of view, is that most if not all the large losses are in the public domain because it is difficult to keep this information confidential. It is therefore possible to build a database of large losses for the whole aviation market and use the frequency and severity information from that database to help in pricing individual carriers.

24.5.1.1 Exposure

Because most aviation accidents causing large losses occur during take-off and landing, a good measure of exposure for aviation liability for an airline is the number of departures. As to aviation hull, the value of the planes also plays an obvious role, and therefore the average fleet value is more commonly used. Other measures commonly used are revenue passenger miles, number of passengers, and number of planes. None of these measures are perfect – that is, the number of departures does not take into account whether most departures refer to small or large planes (which is obviously relevant to aviation liability risk) and the average fleet value does not take into account which planes (high- or low-valued) have more departures (obviously relevant to aviation hull risk).[6] Correlation analysis can help find the exposure measure that is most correlated to the frequency of losses or the total losses per year; however, this study must be accompanied by an analysis of the statistical significance of the results.

24.5.1.2 Frequency Analysis

The frequency of attritional losses for both aviation liability and hull can be based on the individual carrier's experience. As for large losses (e.g. \$1M–\$10M), it makes more sense to rely on market experience, by scaling down the experience of the overall aviation market (or the relevant peer group) using the market share of the carrier. For example, if there are on average x losses per year for the whole market, and a carrier has a 2% share of the overall market, a rough measure of the expected frequency is $0.02x$. For larger airlines, it may make sense to use a credibility approach (see Chapter 26 for details), which combines the frequency estimate based on the airline's data with the estimate based on the market share.

24.5.1.3 Severity Analysis

For aviation liability, the severity distribution can be modelled as usual by using the airline's data for the main body of the distribution and market losses for the tail. For aviation hull, we can either adopt a similar approach to that for aviation liability, making sure, however, that our model does not produce losses above the largest value in the fleet.

An alternative, better method is probably to use an exposure rating approach such as that illustrated in Section 22.8.1: the maximum probable loss is, in this case, simply the value of the aircraft. As a matter of fact, and despite historical practice, exposure rating is probably more suitable when applied to aircraft than to buildings because each building is unique whereas aircraft have broadly similar features, at least once they are classified into their main categories such as jets, turboprops, narrow body, wide body, and the like.

The difficulty with exposure rating, as usual, is that suitable exposure curves may need to be developed in-house. This is conceptually straightforward (see Section 22.6), but it requires knowing for each loss the value of the aircraft at the time of loss (the 'agreed value' in the aircraft schedule), which takes into account not only the model characteristics but also depreciation. Because there is a degree of subjectivity in determining the agreed

6 Also, you will remember from Section 8.3.2 that AFV may be a good measure of exposure for burning cost analysis but less so for frequency analysis, because the frequency of losses is not affected in an obvious way by the value of the planes (a more expensive, bigger plane could actually be safer than a small turboprop).

value, and all the losses are transformed into ratios between the loss and the agreed value, this might reduce the accuracy of the exposure curve.

24.5.1.4 Aggregate Loss Modelling

The frequency and severity model can be combined to produce an aggregate loss model in the usual way (Chapter 17).

24.5.1.5 Rating Factor Analysis

Aviation risk lends itself quite well to rating factor analysis (see Chapter 27), thanks to the detailed information available on each loss. Common rating factors include general type of aircraft (jet, turboprop, rotor wing, etc.), characteristics of the aircraft (narrow body, wide body, hybrid, etc.), broad territory of operation (US, Europe – this affects the jurisdiction where claims are likely to be litigated), use of aircraft (such as passenger/cargo), type of ownership (private, commercial, military), loss history, and so on.

24.5.1.6 Pricing

Aviation has long been an odd class of business, for which the long-term expected losses often exceed the premium that the markets are able to charge, and the markets seem content to make a profit in most years while taking the occasional hit that wipes out their profits. One of the reasons is that historically most of the profit has not come from underwriting but from investment income, exploiting the relatively long delay between premium receipt and the payment of losses, especially for liability, and especially for the largest liability losses. It is therefore important to take into account investment income. Other reasons are more commercial, such as the downward pressure on premiums of several years without large losses.

24.5.2 Business Interruption

Business interruption, as we have seen in Chapter 3, covers the loss of profit arising as a result of property damage. For this reason, it is also called *consequential loss* and, in the context of energy risk, *loss of production income*. The cover will normally be limited to a maximum period of inactivity, for example, 12 or 18 months (cover period).

This cover is normally offered as part of a 'property damage and business interruption' cover rather than as a stand-alone cover. One difficulty of this type of cover lies in the fact that whereas the deductible and limit of the property component is normally expressed as a monetary amount, the business interruption component is often expressed as a number of days, for example, a deductible of 30 days, with a limit of 12 months. As a consequence of the different nature of deductibles and limits, the property section and the business interruption section of the policy may need to be modelled separately.

24.5.2.1 Frequency/Severity Analysis

Business interruption is one of those risks that cries out for being modelled with a *compound severity distribution*: in this case, this means separately modelling the number of days of interruption and the loss per day of interruption, and then combining the two models much in the same way as we combine the number of losses and their severity in the collective risk model. This will work, of course, if there are enough losses to produce these models reliably. One way in which this could work is as follows.

Assume that as input we have a property schedule. For each property, the schedule indicates not only the insured value (IV) and/or the maximum possible loss (MPL) for property damage (PD) but also the IV and MPL for business interruption (BI). Assume that the number of days during which business is interrupted is independent of the site and follows a cumulative distribution function (CDF) $F_T(t) = \Pr(T \leq t)$.

Then assume that whenever business is interrupted in site j, the monetary loss per day is given by a fixed amount X_j, which can be typically extracted from the property schedule because each site will often have a maximum probable loss per year, which we can divide by 365.25 to obtain an approximate average amount per day.[7] Note that, because we assumed that X_j is a fixed amount, X_j is automatically independent of the delay t.[8] Also assume that each site has the same probability to suffer from business interruption.

Given these assumptions, one could model business interruption as follows:

1. Produce a frequency model in the usual way. A good measure of exposure is the total sum insured for business interruption, or the total sum insured for both property damage and business interruption (although that includes elements of severity as well as frequency, so it is only partly useful for frequency analysis).
2. Produce a model for the number of interruption days, in exactly the same way as we produce a severity model.
3. Produce an aggregate loss model through a Monte Carlo simulation as follows:
   a. For each simulated scenario, simulate the number of losses N from the frequency model
   b. For each loss $k = 1, \ldots N$, assign it to a specific property in the property schedule, j...
   c. ... and simulate a random number of days $d(k)$ from the interruption mode. In general, we may need to keep a tab of the number of days of interruption for each property, so that a single property cannot have an interruption of more than 365 days
   d. The severity of the loss is $X(k) = d(k) \times \text{MPL per day}(j)$
   e. The total losses for that simulation is therefore $S = X(1) + \ldots + X(N)$
   f. This is repeated for all simulations. The relevant statistics (mean, standard deviation, percentiles, etc.) are then extracted in the usual way.

Note that this is just one possible way in which business interruption can be modelled and, depending on the information available, the procedure above can be made more sophisticated or simplified further.

When information is too scant for such an exercise to be carried out – and this is sadly almost always the case! – exposure rating (Chapter 22) can be used. Exposure curves are available for pure BI risk, and for mixed PD/BI risks. In the case of direct insurance, the

7 Theoretically, we could do better than this, modelling the monetary amount of loss given interruption as a stochastic variable (e.g. with a beta distribution, from 0 to a maximum value of loss), and taking seasonality of production into account – but one has to be realistic as to how much useful information can be extracted from a limited number of historical losses! If one has sufficient losses, it may be useful to look at the actual amount paid for BI losses and divide it by the number of days of BI, and see how this compares with the simple rules MPL per day = MPL(year)/365.25.

8 However, if we intend to boldly go through the route of modelling X_j as a stochastic variable, this assumption becomes important.

exposure curves can be translated into severity curves and the rate per unit of exposure, combined with the exposure curve, can be used to derive the frequency (see, e.g. Equation 22.41). In the case of reinsurance, the frequency can be derived from the combination of original premium, estimated loss ratio and the exposure curve (Box 22.5).

24.5.2.2 Dependency Analysis

The remaining issue is that of the dependency between the PD and BI component of the severity. Naturally, large property damage losses will tend to be associated with large business interruption losses, and if we have gone along the route of modelling PD and BI separately, we need to take into account this dependency. This can be done, for example, by coupling the PD and BI component with a t-copula, which has the advantage of taking into account tail correlation (the correlation between large losses). It is unlikely that the calibration of such a copula can be achieved based on the information of a single policyholder – hence, portfolio/market information needs to be used as a proxy. For more information on how to produce an aggregate loss model for two correlated severities, see Section 29.5.1.

24.5.3 Commercial Motor Insurance

The main sections of a motor insurance policy are third-party liability (TPL) and own damage (OD), also called accidental damage (AD). TPL covers policyholders against legal liability to third parties for bodily injury (TPBI) or property damage (TPPD) caused by negligent driving. OD covers policyholders for own damage to the owner's vehicle.

TPL is compulsory from the ground up with unlimited liability for bodily injury in the UK (liability for property damage must be provided up to at least £2.5M), whereas in the European Union there is a minimum policy limit of €5M.

There are two types of commercial motor insurance: fleet insurance and commercial vehicles. A commercial motor policy will often cover both.

Fleet insurance insures the cars owned and operated by an organisation. It would be a management nightmare to deal with each vehicle's insurance separately, so a fleet policy is purchased, based on an annual declaration of the number of vehicles on the basis of which a premium is charged. For small fleets (less than around 10 vehicles) an individual rating based on the actual vehicles may be used. For larger fleets, a flat premium will be charged for all vehicles to minimise underwriting costs.

Experience rating is used rather than NCD, as it increases flexibility.

There are four main types of *commercial vehicles* insurance:

a. Goods-carrying vehicles – such as vans and lorries (damage to goods is excluded)
b. Passenger-carrying vehicles – such as taxis/coaches (risk of heavy liability claims)
c. Agricultural and forestry vehicles
d. Special types – such as construction plant working vehicles registered for the road, ice cream vans, mobile chip vans, ambulances, hearses, and others

Motor trade vehicles are a category of their own ('trade plates', customers' vehicles, demonstration vehicles, testing, courtesy cars, etc.). The underwriting of motor trade policies is quite complex.

24.5.3.1 General Modelling Considerations

It is customary to model TPL and OD claims separately: this is because the profile of these two types of claims are quite different, and also because they will be subject to different policy deductibles. TPL is compulsory in the United Kingdom from the ground up, whereas OD can be retained by the policyholder (fully or in part). If necessary, third-party property damage and third-party bodily injury can also be modelled separately.

As usual, when you model various components separately, you have to consider the question of dependency between these components (Section 29.1); therefore, this component splitting should be kept to a minimum. However, in practice, OD and TPL can be safely modelled separately, and although there will be some correlation between the severity of the OD component and that of the TPL component, this correlation can, in the first approximation, be ignored with little quantitative consequences.

24.5.3.2 Exposure

The most common exposure measure is vehicle years (by which each vehicle counts only the fraction of the year in which it was in the fleet). When the vehicle years are not available, use the number of vehicles at a specified point in time.

24.5.3.3 Frequency Analysis

Frequency analysis can be carried out in the standard way for both OD and TPL. The number of claims is normally quite high; thus, for large fleets, there will be a good degree of stability. The delays involved in liability claims can be significant and therefore this risk can be considered long-tail, although most claims are reported quickly – within a month or so.

There are systemic elements at play – for example, a bad winter can cause an increase in the number of accidents. Also, there is a recognised decreasing trend in the number of motor accidents, although this is offset by an increased propensity to claim.

24.5.3.4 Severity Analysis

For OD, the main thing to remember is that the loss amount needs to be capped at the value of the most expensive vehicle in the fleet. Theoretically, this could be a textbook case for using exposure rating, by focusing on the damage ratio (loss divided by current value of the vehicle). However, the effort of building appropriate exposure curves and of having to calculate updated values for each vehicle in the fleet may not be justified for a risk that is experience-rich and reasonably stable.

For TPL, the usual decomposition of attritional losses/large losses can be used. Ideally, for very large losses (e.g. £500,000 or more) a market severity curve should be used if access to reinsurance-level losses is available. The largest motor losses are in the region of several tens of millions of pounds and are normally driven by the bodily injury components of the third-party liability (TPBI). The largest motor claim in the UK is the Selby claim, in which 10 people were killed and two trains were heavily damaged.

A specific feature of the severity of motor claims is that many large claims are now PPOs. We therefore devote a brief section to explaining what PPOs are. For more in-depth information, see, for example, the works of Cockcroft et al. (2011) and Leigh (2013).

24.5.3.5 Periodic Payment Orders

A periodic payment order (PPO) is a way of compensating victims of accidents alternative to lump sum payments. It is designed to cover long-term cost of care and loss of earnings. Periodic payment orders were introduced with the 2003 Courts Act in England and Wales, to give more certainty to claimants, who might otherwise find that the lump sum accorded to them is insufficient because of erroneous medical assessment of because the sum is misinvested/misspent by them or by their family, falling back on the health care system. For this reason, PPOs can be imposed by the court even against the wish of both claimant and plaintiff.

Periodic payment orders are increasingly used as a means of compensation for liability claims, *especially for bodily injury claims in motor*, and especially for large claims involving traumatic brain injuries and for victims such as children for which the uncertainties are greater.

Initially, PPOs were indexed using retail price inflation (RPI), which was however widely considered inadequate because medical cost care inflation normally outstrips general inflation. After the *Thompstone v Thameside & Glossop Acute Services NHS Trust 2008* appeal, RPI was abandoned as an indexation method, and two separate indices are now used instead:

- The wage inflation index for the loss of earnings component of the claim
- The 75th percentile (most commonly) of Subsection 6115 of the Annual Survey of Hours and Earnings (ASHE 6115), which is related to the salary of health care personnel

Note that the periodic compensation may be 'stepped up' over and above any indexation by a court decision if circumstances change. After this 2008 ruling, PPOs took off and have been increasing steadily ever since.

PPOs increase the uncertainty for insurers and especially reinsurers (which then feeds back as additional costs for insurers). Specifically, these risks are transferred from the claimant to the (re)insurer.

- Longevity risk (claimant living longer than expected)
- Investment risk (investment not achieving as predicted for lump sums)
- Future inflation risk (medical cost of care running ahead of expectations)

We will not look into the details of the consequences of PPOs on pricing, which are the subject of ongoing research are and are still insufficiently understood. Suffice it to say that the risks listed above make it difficult to price and reserve PPOs accurately. Some of these risks (longevity and possibly future inflation) might be hedged by adequate life and capital markets products, but the market for these products is not yet fully developed. The interested reader can find more details about pricing and reserving issues arising from periodic payment orders in Cockcroft et al. (2011).

24.5.3.6 Aggregate Loss Modelling

The frequency and severity model can be combined in the usual way to produce an aggregate loss model. Because the number of claims is large and the experience is reasonably stable, a simple burning cost analysis will often suffice to give a fair forecast of the loss experience; however, an allowance for the possibility of infrequent large claims needs to be made.

24.5.3.7 Rating Factor Analysis

Because the number of motor claims is normally very large, some rating factor analyses can be attempted. Common rating factors that can be considered are type of fleet/fleet composition (e.g. commercial vehicles vs cars), type of industry, type of cover third-party liability with fire and theft [TPF], comprehensive, etc.), territory, drivers' details (such as age or specific occupation), number of vehicles (for economy of scale), loss experience, exclusions (especially excess), use of vehicle (such as business-only, mixed), postcode, vehicle make/model/etc. (especially for commercial vehicles [CVs]), driving restrictions (such as named drivers).

24.5.4 Product Liability

Product liability is often sold along with public liability in the United Kingdom. For many organisations, the product liability component of the package is a minor consideration. However, for some organisations, such as pharmaceutical companies, a faulty product can elicit losses of several billion dollars (the use of the currency 'dollar' is not accidental because the largest losses originate from cases brought before US courts). For these organisations, special product liability cover needs to be arranged, normally in excess of a very high threshold. It is on this type of excess cover that we focus in this section, as it gives us the chance of introducing the integrated occurrence basis, a basis first introduced by Bermudan insurers.

It should be noticed that although large individual claims may be made for faulty products, for example, if the product causes serious bodily injury or death, the largest effect on an organisation is the systemic element, that is to say, a product causing harm to a large number of customers.

From the pricing point of view, there are a few things to keep in mind. First of all, product liability excess cover is often sold on an integrated occurrence basis. Second, and partly as a consequence of the integrated occurrence basis, it makes sense to calculate the frequency and severity of all claims related to the product as if they were a single claim. Third, product liability is hugely affected by the territories where the products are sold, and the inclusion or exclusion of the United States amongst these territories is crucial because of the higher propensity to claim, the possibility of class action lawsuits and the potentially very large payouts (including punitive damages) accorded by US courts.

24.5.4.1 Integrated Occurrence Basis

Some product liability covers are sold on an 'integrated occurrences' basis. This basis allows the policyholder to aggregate individual claims related to a given product in a batch and to have it treated by the insurer as a single occurrence, which is attached to the policy year in which the integrated occurrence was declared. Note that the status of integrated occurrence is not automatic but needs to be declared by the insured, and (depending on the wording) may be activated only when the number of individual claims exceeds some threshold, that is, when there is evidence of an epidemic failure.

This policy has two main advantages for the policyholder:

- When the policy has a high deductible, the individual claims may not reach the deductible, but the aggregate amount may.
- Once the integrated occurence (IO) is declared for a product, the policy limit and other exclusions remain frozen for all claims that can be legitimately associated with that product, regardless of when the claim was made and whether the policy has been renewed (this should be contrasted, for example, with the standard claims-made policy, which requires the claim to be made during the period of the policy).

If the policyholder has more certainty, so does the insurer, because the policy will normally have a specified annual limit, and because all integrated occurrence claims are attached to the year in which an integrated occurrence was declared, once the annual limit is reached, the insurer will have to make no further payments for that product, even if claims keep coming in.

24.5.4.2 Frequency Analysis

One consequence of the integrated occurrence basis is that the natural 'unit' for the analysis is the batch of claims rather than the individual claim. For consistency, it is better to look at batches of claims related to the same product even when an integrated occurrence is not declared – the idea being that batches that are not epidemic in nature will not attach to the excess cover anyway. Each integrated occurrence will of course be allocated to the year in which it was declared, whereas the undeclared batches may be tentatively assigned to the year in which the first claim was made.

Therefore, the frequency analysis will focus on the number of batches, and this can be expected to be quite a small number because the number of different products for an organisation will be limited. Incurred but not reported losses are not a concern here because the policy focuses on the batches reported (declared) during each year. As for the most natural exposure measures for product liability, these will relate to the number of sales, or the total revenues from sales.

24.5.4.3 Severity Analysis

Similarly, for severity analysis, it makes sense to look at the overall severity of the batches rather than at the individual claims. Historical batches can be revalued to current terms in the same way as individual claims, using a claims inflation index that is adequate for large liability losses. Batches will of course be made of individual claims that may be filed over the years, but the usual principle of stationarity holds (see Section 10.2.1), by which we can assume that future batches will be similarly spaced in time, and it therefore makes sense to revalue the whole batch as a single lot.

Note also that to have a more substantial number of data points, it may make sense to batch all claims related to the same product whether they are declared as integrated occurrences or not. In any case, it will only be the ones that are declared as integrated occurrences that are likely to exceed the deductible. IBNER may be a concern for integrated occurrence claims because reserve development is affected by an estimate of the number of future individual claims that will be received for a product more than by the development of the individual claims, and the number of future claims is harder to predict. Conceptually, however, we may still analyse the way in which estimates develop in time along the lines of Section 15.2.2.

We can then fit a distribution through the historical batches revalued and IBNER-adjusted, much in the same way as we do for individual claims. The focus of the analysis should of course be on the tail, because the cover we are pricing is normally an excess cover. A Monte Carlo simulation can then be used to produce an estimate for the aggregate losses to a layer of insurance, L xs D.

24.5.5 Construction Cover

As discussed in the online Appendix to Chapter 2 (Parodi, 2016), there are two main types of policies that normally go under the heading of 'Construction': Contractors (or Construction) All Risks (CAR), which covers the property and liability risks arising from construction projects, and Engineering (or Erection) All Risks (EAR), which covers the property and liability risks arising more specifically from the construction and setting up of plants/machinery/industrial processes. The cover may be extended to cover the maintenance/testing phase after the completion of the project. Delay in start-up (DSU) cover may also be purchased for both CAR and EAR to protect the contractor against delays in the completion of the project. For simplicity, the following discussion will focus on the property ('Works') component of CAR, which is perhaps the most interesting (the discussion for EAR would be similar, and the liability component is more traditional).

24.5.5.1 Pricing Characteristics

Three characteristics of construction policies immediately set construction pricing apart from other lines of business:

- Projects are one-off endeavours and therefore client-based experience rating is typically not feasible. Furthermore, certain exercises such as rate change calculations cannot be done in the standard way because there is no renewal.
- Projects have in general non-annual duration and claims can occur at any point during the project.
- The earning pattern for the premium is not uniform over the duration of the project. Construction projects may start with the proverbial 'hole in the ground' and end up with an expensive building. It therefore makes sense that the risk of property damage increases with time.

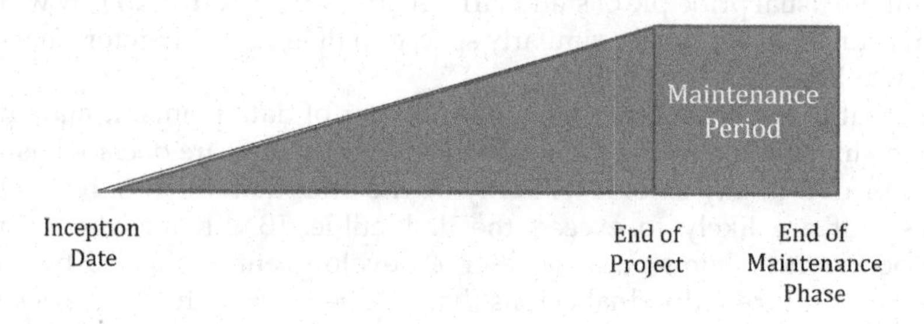

FIGURE 24.25
Earning pattern for a construction project. The risk is zero at the beginning and maximum at the end of the project. If one wants to emphasise that the increase in value is faster in the middle of the project and then it stabilises, one can use an S-shaped curve.

Because of these features, construction pricing normally uses exposure rating. In its simplest form, this may include – similarly to what discussed in Chapter 22 – the following:

- A number of **base rates** (one for each main type of construction, such as building, bridge, dam, etc.) which apply to the total contract value (TCV) of the project
- **Risk factors** relevant to the project (e.g. the number of floors for a building, or whether a bridge is on the sea or on a river).
- **Exposure curves**, to calculate the effect of deductibles and to allocate losses to layers between 0 and the maximum possible loss,[9] or if we need to sample losses to understand the volatility of an account.

Assuming for simplicity that the base rates are from the ground up, and that there is a deductible d, the expected losses for a layer L xs D of insurance could then be written with a formula of the form:

$$\text{Exp losses} = f_1 \times f_2 \times \ldots \times f_k \times \text{BR} \times \text{TCV} \times \left(G\left(\frac{D+L+d}{\text{MPL}}\right) - \times G\left(\frac{D+d}{\text{MPL}}\right)\right)$$

where BR is the base rate, TCV the total contract value, $f_1, f_2 \ldots f_k$ the risk factors, MPL the maximum possible loss, and G the exposure curve.

The devil, as is often the case, is in the calibration. And as is also often the case, the most difficult thing to calibrate are the exposure curves and the risk factors. In the absence of a large portfolio these may be chosen based on underwriting judgment. As for the base rates, it will normally be possible to calibrate the general level of the base rates at portfolio level or by type of project, depending on the size of the data set (more on this in Section 31.7, for a general pricing model). This can be done by a burning cost exercise based on a portfolio of relevant past projects (see Section 11.4), using the standard techniques to match claims and exposures.

The only complication around this calculation for Construction is that as mentioned above the exposure is not earned uniformly but increases over the time of the project. A common assumption (because it's simple, and because it's not that easy to do provably better) is to assume that the risk is zero at the beginning of the project and then increases linearly with time, reaching its maximum at the end of the project, and staying there if there is a maintenance period, as in Figure 24.25. The details of the calculations for an arbitrary earning pattern are discussed in Section 11.4.

It should be noted that the increasing earning pattern is more of a severity effect than a frequency effect.

24.5.5.2 Pricing Cat Risks

Catastrophe risks might also need to be considered. This is difficult to model since standard cat models do not normally cater for construction projects (which have peculiarities not only in terms of the increasing exposure, but also of varying vulnerability to catastrophes at different points in the project), although there are specialist models for construction for certain geographies (Mitchell-Wallace et al., 2017). When specialist models are not

9 Of course one can be more sophisticated than that – since the value of the works increases with time, the maximum possible loss can also be seen as a function of time, and therefore if the simulation takes timing into consideration, we need to take that into account. If not, one can simply use an exposure curve that applies to the overall loss regarding of time. In the latter case, the exposure curve will be more concave.

available, one option is to use the standard models and adapt the outputs by scaling them to obtain a proxy for cat risk; alternatively, one can use partly judgment-based, partly data-driven cat rates and hazard maps (Section 25.4).

24.5.5.3 Rate Change Calculations

Since the construction policies discussed here are not subject to renewal, the rate change calculations techniques (Chapter 20) do not apply to them. The calculation of rate change for a portfolio of construction policies can therefore only be measured approximately by considering variations in the premium adequacy, inevitably mixing the effect of actual rate changes with the changes in the business written.

24.6 Questions

Note that the questions below are either on material presented in this chapter or an invitation to reflect on what the quirks of pricing are for classes of business that we haven't directly addressed here.

1. Professional indemnity, claims-made policies
   i. Explain the difference between a claims-made basis policy and an occurrence-basis policy in direct insurance.
      X is a law firm that buys professional indemnity. Assume that of the claims originating from advice given in year y,
      - 25% of the claims are notified to the insurer in year y
      - 35% in year $y + 1$
      - 40% in year $y + 2$

   The claims reported in each year are shown below (note that both the exposure and the claims reported have already been revalued).

Year	Exposure (Turnover, Revalued), $M	Claims Reported (Already Revalued), $M
<2008	110	
2008	110	1.80
2009	100	1.72
2010	80	1.66
2011	50	1.07
2012	50	1.13
2013 (est)	30	

   ii. Explain what adjustments need to be made to exposure before being used for a burning cost exercise, and why.
   iii. Assuming that:
      - X buys PI insurance from the ground up
      - the claims in the table above are from the ground up
      - the retroactive date is 2005

estimate the expected losses for year 2013 based on a simple burning cost analysis.

iv. Explain what other assumptions you needed to make in (iii) besides those already stated in the text.

v. X is considering whether to move to an occurrence basis in 2013 as an alternative to renewing its claims-made policy. Explain what the main problem in doing this is going to be, and describe a possible solution for that problem.

2. Answer the following questions about periodic payment orders (PPOs).
   i Explain what a PPO is.
   ii Explain why they were introduced, and who is ultimately responsible for deciding whether a PPO rather than a lump sum should be granted.
   iii Explain the main advantages and disadvantages of PPOs compared with lump sums from the point of view of claimants and insurers.
3. Extended warranty (and common shocks)

   You are a quantitative analyst for Company X, which sells cheap digital cameras at €50 in the European market at a production cost of €20. The cameras are offered with a 1-year warranty that comes with a free replacement for failures during the warranty period.

   A new model (MB4) of camera with the same price as the rest of the range is going to be sold in the Christmas period in 2014 and 25,000 sales across Europe are expected.

   A recent study of the probability of failure suggests that the failure rate for a similar camera model (MB2) was 5% during the first year, although some differences amongst different European territories were noticed.

   You can assume that:
   i. You can only get one replacement a year.
   ii. The overall replacement cost is €25 (= production cost + expenses including mailing expenses).
   iii. If a camera is replaced, the warranty expiry date remains the original one: it does not start again from the date when the camera is received.
      i. Calculate the expected cost of providing warranty cover for the lot of MB4 models sold during Christmas. Clearly state all the assumptions you make.
      ii. Describe how you would calculate the 90th percentile of the statistical distribution of the costs (you do not need to carry out the actual calculations, only to set up the methodology clearly enough for one to follow). Clearly state all the assumptions you make.
      iii. What are the *main* uncertainties around the expected cost calculated in (i)?
4. Personal accident
   i. Describe the main coverage of personal accident insurance and give an example of a benefit schedule.
   ii. Mention two composite personal lines product to which personal accident is normally attached and explain why personal accident is relevant to these products.
   iii. Explain why a fixed benefit schedule is typically used for PA.
   iv. It is 5 February 2014. The following statistics are available for the policy years 2008 to 2012 for a portfolio of personal accident policies offering *identical death benefits* as specified in the table below (the amount of benefit is changed at the beginning of the year for all policies incepting in that year).

Policies Incepting on Year	No. of Deaths	Amount Offered for Death Benefits across All Policies [£M]	Total Loss to the Insurer [£M]
2008	2500	1.20	3000
2009	2250	1.20	2700
2010	2610	1.30	3393
2011	2405	1.50	3608
2012	2550	1.60	4080
2013	N/A	1.60	N/A

Calculate the total losses related to the death benefits that you expect for the portfolio of policies incepting in 2013, stating all assumptions you make.

v. What frequency model would be adequate for the number of losses in 2013? Why?

5. Credit risk

Modify the algorithm in Section 24.3.2 (b) to take into account changes in the value of the economic factor over the K periods. Assume the values of ψ over the period, ψ_t, are provided by an economic scenario generator. Discuss what effect this might have on the number of simulations that you need to produce a reliable aggregate loss model.

25

Catastrophe Modelling in Pricing

And then came Katrina.

Michael Lewis

One of the key risks that organisations face is exposure to catastrophes – for example, a production plant may be destroyed by an earthquake, a tsunami, or a hurricane. Therefore, modelling catastrophes is an integral part of pricing property insurance.

Standard actuarial techniques do not work for modelling catastrophes: catastrophes are very rare events at the client's level and if one looks at the experience over a short period of 5 to 10 years, as is customary for other risks such as motor or employer's liability or fire-related property damage, one normally sees no loss activity at all. If a client has been particularly unlucky, they may have seen one catastrophe, in which case the client's loss experience is again not representative of future losses. (Even at portfolio level, experience is not sufficient.)

Therefore, the client's experience in catastrophes is *always* inadequate and, in the language of Chapter 26, has near-zero credibility. The solution to this problem is an altogether different approach to modelling – that of catastrophe modelling. Catastrophe models do not use the client's experience – they use:

- A database of non–client-specific historical events for a country or the whole world and
- A scientific understanding of phenomena such as hurricanes and earthquakes, which allows us to produce new events, some of which may have never occurred historically to that extent to build scenarios that are then combined with the knowledge of the exposures of the client (its buildings, plants, etc.) to determine possible losses that the client may incur, and the corresponding probabilities.

Catastrophe modelling is a fast-developing discipline that has changed much in recent decades (see Figure 25.1). More than anything else, it has changed because specific events such as Hurricane Andrew (1992) have forced the issue of modelling catastrophes on insurers, and because other events, such as Hurricane Katrina (2005), have exposed the inadequacies of the first generations of catastrophe models. In a way, we can say that catastrophe models are constantly playing catch-up with reality: every large catastrophe forces a rethink of the way we model natural hazards.

Catastrophes can be either natural or man-made (see Section 25.3). This chapter is mainly concerned with natural catastrophes, for which modelling is more advanced. Natural

DOI: 10.1201/9781003168881-28

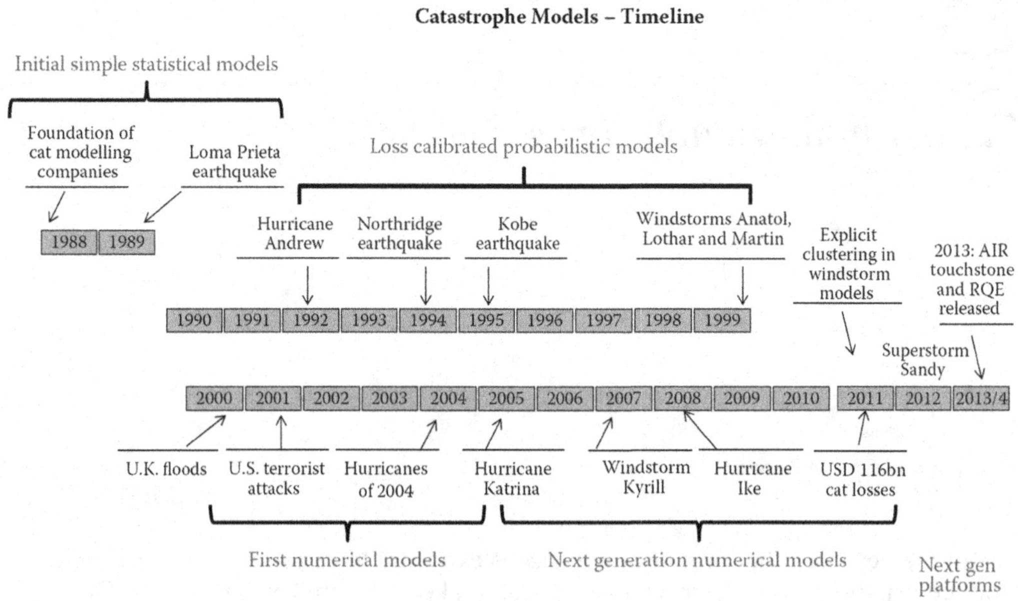

Catastrophe Models – Timeline

Initial simple statistical models
Foundation of cat modelling companies
Loma Prieta earthquake
1988 1989
Loss calibrated probabilistic models
Hurricane Andrew
Northridge earthquake
Kobe earthquake
Windstorms Anatol, Lothar and Martin
1990 1991 1992 1993 1994 1995 1996 1997 1998 1999
Explicit clustering in windstorm models
2013: AIR touchstone and RQE released
Superstorm Sandy
2000 2001 2002 2003 2004 2005 2006 2007 2008 2009 2010 2011 2012 2013/4
U.K. floods
U.S. terrorist attacks
Hurricanes of 2004
Hurricane Katrina
Windstorm Kyrill
Hurricane Ike
USD 116bn cat losses
First numerical models
Next generation numerical models
Next gen platforms

FIGURE 25.1
The short, bumpy story of catastrophe models. (Courtesy of Willis Re Analytics.) Although the scope of some of the models has widened, there hasn't been much methodological advance since 2012, with most of the development from then on being on the technological side.

catastrophe modelling is a large and rich subject, and this short chapter cannot do justice to it. Readers interested to explore this subject in more depth are referred to the book by Mitchell-Wallace et al. (2017). They would also do well to read the short but enlightening history of the subject written by Michael Lewis (Lewis, 2007), the financial writer known for several best-sellers including *Liar's Poker*, *The Big Short*, *Moneyball*, and many others.

25.1 Structure of a Catastrophe Model

A catastrophe model is *typically* a sophisticated piece of software produced by a specialist company. There is nothing in principle that prevents a company, especially an insurance company, from producing an in-house tool for catastrophe modelling. However, the investment necessary to create a catastrophe modelling tool is massive and this has favoured the creation of specialised companies such as RMS, AIR, and EqeCat. This has also allowed the creation of a number of state-of-the-art tools that are recognised across the industry and therefore make it easier for market participants to agree on the price of a risk. For example, a large insurance company or its broker and a reinsurance company are likely to be able to run the same accepted software on the same portfolio of properties.

A catastrophe model can be described as being made up of three components: a hazard module, which models the peril; a vulnerability module, which estimates the likely damage/ injury that structures/people are likely to suffer because of the hazard; and

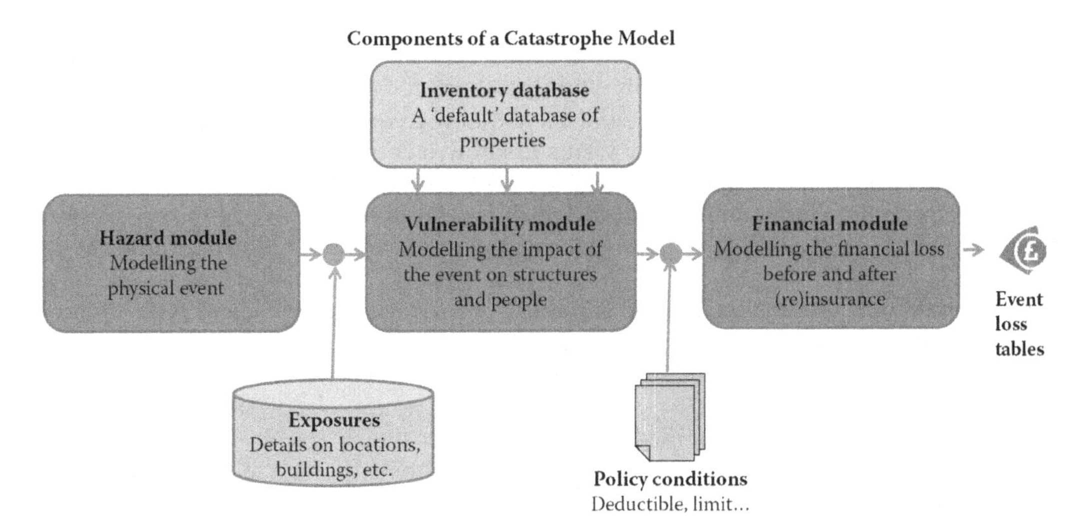

FIGURE 25.2
The three components of a catastrophe model (the hazard model, the vulnerability model, and the financial model) and the necessary inputs (location details, policy conditions). Note that different views of the process are possible.

a financial module, which estimates the financial loss consequential to these damages/injuries. A simple graphical representation is shown in Figure 25.2. A more detailed explanation of these modules is given below.

25.1.1 Hazard Module

The hazard module does two things:

1. *It generates a set of potential catastrophic events* – each with an event ID, an occurrence rate, a severity, and a geographic area affected (plus possibly other features). For example, an earthquake event might include the likelihood of occurring, the magnitude according to the Richter scale, and the fault (such as San Andreas Fault in California) at which the earthquake happens. This catalogue of events is based on an analysis of historical events (e.g. actual past earthquakes) and on a scientific analysis of the hazard.
2. *It assesses the level of physical hazard for each event* – this is one step beyond measuring the intensity of the event but one step short of actually estimating the damage caused: the level of physical hazard is calculated for each event in (1) taking into account the characteristics of the location. For earthquakes, the level of physical hazard may be given by the level of ground motion, and for hurricanes, by the strength of the wind after terrain and buildings are taken into account.

As an example of what the level of physical hazard means in practice, the Mexico City earthquake in 1985 had a huge death toll and resulted in significant property damage, despite its epicentre being 350 km off the coast, because the old city was built on a drained lake and the soft terrain was prone to propagating and amplifying seismic shakes.

25.1.2 Vulnerability Module

The hazard module is concerned only with the characteristics of the catastrophic event itself – where it will hit, how strong it will be, and how frequently we expect it to happen. However, what an underwriter really wants to know is more specific, that is, how much damage the event is likely to cause to the properties of their clients. To respond to this query, we need the vulnerability module, which translates the hazard characteristics into the expected physical damage after considering the characteristics of the properties in the portfolio.

The *input to the vulnerability module* is the list of properties of the client with location details and information for each property on occupancy type (such as residential, multiple occupancy, industrial ...), number of storeys, construction type, year of construction, and so on, where available.

The vulnerability module provides *vulnerability curves*, which give the degree of damage consequential to an event as a function of the hazard intensity, such as the magnitude of an earthquake, the wind speed of a hurricane or – as in the case shown in Figure 25.3 – the depth of a flood.

Note that:

- Vulnerability modules also model to some extent the uncertainty around the degree of damage, providing, for example, both a mean damage ratio (mean damage as a function of the replacement value) and a standard deviation (SD).
- Different vulnerability curves are normally available for damage to the building, to its contents, and for business interruption.

25.1.2.1 Inventory Database

In many cases, the company that has commissioned a catastrophe model report (either an insurer seeking reinsurance or a non-insurance company seeking insurance) will not have thorough information on the characteristics of its portfolio of properties but will only know certain features such as the occupancy type. To cater to this, modern catastrophe models are normally equipped with an additional module, the inventory database, which provides a mixture of buildings with a suitable variety of characteristics, such as height

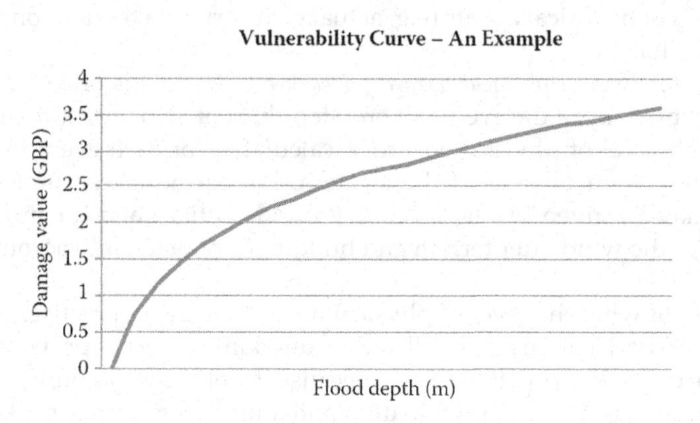

FIGURE 25.3
A vulnerability curve for flood, with arbitrary values on the axes.

and construction material, that can be used to produce realistic results even when the exposure information available is not detailed enough.

25.1.3 The Financial Module

As the vulnerability module translates hazard into physical damage, the financial module translates physical damage into financial loss. The input to the financial module is given by the *policy conditions* (such as deductibles and limits), which are necessary to calculate the retained and ceded amounts to a policy.

The main outputs of the financial module (and hence of the catastrophe model) are the event loss tables (ELTs), which give a detailed list of events and the related losses. These can be calculated at a gross level (no insurance), or they can provide the ceded losses for a specific insurance structure: so there is a different ELT for each insurance structure, plus a gross ELT.

The gross ELT typically shows the following information:

- Event ID and description
- Annual rate of occurrence for each event
- Expected gross losses for each event
- SD of the gross losses for each event (also called 'secondary uncertainty'). This is often split into an independent standard deviation (SDI) and a correlated standard deviation (SDC). The split is made with the purpose of helping to combine event loss tables for the same peril but different contracts/locations together while preserving the existing correlation (at least approximately). More details are given in Section 25.2.5 but in simple terms, the SDIs of different event loss tables for the same ID combine by summing the variances (fully independent events), while the SDCs combine by summing the standard deviations (fully correlated events).
- Amount of exposure to each event – i.e. the maximum amount that can be lost.
- Other fields that may be shown are a description of the event and the size of the event in physical term (e.g. earthquake magnitude).

An example of an ELT is given in Figure 25.4.

Ceded ELTs will have the same format as Figure 25.4, but the expected loss and SD for each loss will be those for the ceded amount rather than for the gross amount.

Event Loss Tables – An Example

Event ID	Rate	Expected loss	Correlated standard deviation (SDC)	Independent standard deviation (SDI)	Overall standard deviation (SD)	Exposure
1	0.015	2,300,000	322,000	1,288,000	1,610,000	11,500,000
2	0.008	4,000,000	560,000	2,240,000	2,800,000	28,000,000
3	0.010	5,500,000	880,000	3,520,000	4,400,000	27,500,000
4	0.025	7,200,000	864,000	3,456,000	4,320,000	43,200,000

FIGURE 25.4

A standard format for the ELT. Each event (Event ID in Column 1) causes a loss with a certain frequency (Rate: Column 2), and a loss amount between 0 and the exposure (Column 7), with a mean value equal to the expected loss (Column 3) and a standard deviation equal to SD (Column 6). The standard deviation SD can be split into an independent and correlated part (Columns 4, 5): this information can be used when combining events with the same ID from different ELTs. *The monetary amounts are in a generic currency CCY.*

25.2 Other Outputs of a Catastrophe Model

Based on the ELT, which is the basic output of a catastrophe model, a number of other outputs can be produced, which are listed below.

25.2.1 Occurrence Exceedance Probability

This is the probability that a single occurrence will exceed a loss of a given size in any one year. It is *not* the severity distribution as we know it from actuarial modelling because it is related to both the severity *and* the frequency distribution.

Formally, OEP(x) can be defined as follows:

$$\text{OEP}(x) = \Pr\left(\exists \text{X occuring during a given period such that } X > x\right) \tag{25.1}$$

As a consequence, if the frequency is higher, the OEP(x) is also higher. This would be the same as the severity distribution $F(x)$ only if the number of losses in each year was exactly one. As we mentioned, the exceedance probability OEP(x) is related to both the frequency and the severity distribution. The exact relationship can easily be found by noticing that 1 – OEP(x) (the probability that all losses are $\leq x$) can be written as

$$\begin{aligned} 1-\text{OEP}(x) &= \sum_{k=0}^{\infty}\Pr\left(\max\left(X_1,\ldots,X_k\right) \leq x\right)\Pr(N=k) \\ &= \sum_{k=0}^{\infty}\Pr(X_1 \leq x)\Pr(X_2 \leq x)\ldots\Pr(X_k \leq x)\Pr(N=k) \\ &= \sum_{k=0}^{\infty}\left(F(x)\right)^k \Pr(N=k) \end{aligned} \tag{25.2}$$

Hence,

$$\text{OEP}(x) = 1-\sum_{k=0}^{\infty}\left[\text{F}(x)\right]^k \Pr(N=k) \tag{25.3}$$

In the special case where the frequency distribution is a Poisson, $\Pr(N = k) = \exp(-\lambda)\lambda^k/k!$ and after some simple algebra, Equation 25.3 becomes:

$$\text{OEP}(x) = 1-\exp\left[-\lambda S(x)\right] \tag{25.4}$$

where $S(x) = 1 - F(x)$ is the survival probability.

Also, if the frequency distribution is a negative binomial, $\Pr(N=k) = \binom{k+r-1}{k} p^k (1-p)^r$ and Equation 25.3 becomes

$$\text{OEP}(x) = 1-\frac{(1-p)^r}{\left[1-pF(x)\right]^r} = 1-\frac{(1-p)^r}{\left[1-p+pS(x)\right]^r} \tag{25.5}$$

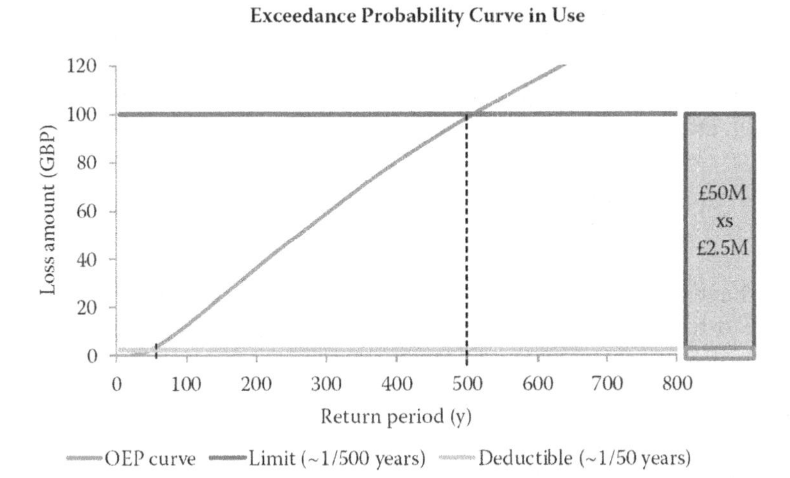

FIGURE 25.5
An example of an OEP and of how it can be used to calibrate an insurance structure. This particular representation helps to see immediately the relationship with the insurance structure, and the use of the RP rather than the probability in the x axis helps to show a number of different orders of magnitude of probability and therefore has the same purpose as the logarithmic scale in our severity curves. In this case, the deductible of the policy (£2.5M) has been chosen to be close to the 1 in 50 RP for the OEP and the limit of the policy (£100M) to be close to the 1 in 500 RP.

An example of how the OEP can be used to calibrate an insurance structure is shown in Figure 25.5.

25.2.2 Aggregate Exceedance Probability

This is the probability that aggregated event losses in a year will exceed a threshold. This is basically the same as our aggregate loss distribution in actuarial models, but it is expressed in terms of the return period.

25.2.3 Return Period

This is the number of years between occurrences of a loss event of a given size in the region. In general, the larger the loss event, the longer the return period (RP). It represents the likelihood that a loss will equal or exceed the loss amount displayed.

25.2.4 Average Annual Loss

This is an estimate of the amount of premium required to balance catastrophe risk over 1 year, also referred to as pure modelled premium and equivalent to expected loss on an annual basis. This is what we call 'expected losses' in our actuarial models.

For the example in Figure 25.4, the average annual loss (AAL) would be given by

$$\text{AAL} = \sum_{j=1}^{\text{no. of events}} \text{Rate}_j \times \text{Exp loss}_j = 0.015 \times 2.3\text{M} + \ldots + 0.025 \times 7.2\text{M} = \text{CCY } 0.302\text{M}$$

where CCY is the symbol for a generic currency.

25.2.5 Standard Deviation

This is the square root of the variance of the aggregate loss distribution (Figure 25.5). In the case of a Poisson distribution the relationship $\mathbb{V}(S)=\mathbb{E}(N)\mathbb{E}(X^2)$ holds (see Chapter 6) and the standard deviation can be calculated as follows:

$$\text{SD}^2 = \sum_{j=1}^{\text{no. of events}} \text{Rate}_j \times \left(\text{SD}_j^2 + \text{Exp Loss}_j^2\right) = \left(\text{CCY } 1.602\text{M}\right)^2$$

where SD_j is the *overall* standard deviation for event j. If the frequency distribution is not Poisson we need to use the relationship $\mathbb{V}(S)=\mathbb{E}(N)\mathbb{V}(X)+\mathbb{V}(N)\mathbb{E}(X)^2$(Chapter 6) and make an assumption on the variance-to-mean ratio for the number of events, as this quantity is not given in the ELT.

25.2.6 Period Loss Tables (PLTs) and Year Loss Tables (YLTs)

The vendor of cat modelling system may also provide along the so-called period loss tables (PLTs), which are simulated scenarios derived from the ELTs using proprietary methodologies that preserve the correlation between events, and so they can be used without further processing for pricing purposes, as shown in Section 25.5.3.

An Example of a PLT

Scenario ID	No of losses	Loss ID	Event ID	Rep ID	Loss Date	Loss Amount (CCY)
4	1	1	1488497	1	26/05/2023	1,096,726
10	1	1	1488497	1	10/10/2023	69,206,000
14	1	1	1488498	1	26/11/2023	7,430
19	1	1	1488497	1	15/07/2023	1,029,865
21	1	1	1488498	1	23/02/2023	9,431
22	1	1	1488497	1	21/04/2023	37,632
27	1	1	1488498	1	24/01/2023	17,564
31	1	1	1488498	1	07/12/2023	101,105
32	2	1	1488498	1	28/03/2023	7,119,118
32	2	2	1488498	2	19/07/2023	629,966
34	1	1	1488498	1	13/09/2023	5,763,452
35	1	1	1488497	1	10/02/2023	8,206,597
37	1	1	1500256	1	12/06/2023	38,127,336
47	1	1	1488498	1	02/01/2023	52,719,656
51	1	1	1504304	1	12/01/2023	4,861
54	2	1	1488498	1	12/08/2023	4,636,739
54	2	2	1488498	2	17/08/2023	1,309,822
71	1	1	1488415	1	06/09/2023	13,223
…	…	…	…	…	…	…
99999	2	2	1488498	1	24/05/2023	2,227,344

FIGURE 25.6
An example of a period loss table with 100,000 simulated scenarios, of which only the ones with more than loss are shown. Each entry shows (from left to right): the scenario identification number; the number of losses for that scenario; the loss identification number within each scenario, ordered chronologically; the identification number of the event from which the loss originates; the identification number of the loss to differentiate different losses arising from the same event; the loss date (a crucial piece of information when the peril has some seasonality); and, finally, the loss amount (in a generic currency CCY as the amounts are fictional).

More specifically, a PLT provides a list with a large number of scenarios (each scenario representing a fictional realisation of the catastrophe losses in a given period). Each scenario includes a list of one or more individual losses (scenarios with no losses are excluded), with loss amount and occurrence date. The occurrence date is important as many catastrophes are seasonal (tropical cyclones in the North Atlantic region – which occur between June and November – being an obvious example).

An example of a PLT with an explanation for each field is shown in Figure 25.6. Note that typically a different PLT will be produced for each region/peril. Independent PLTs (i.e. PLTs with different region or peril) with the same number of simulations can be combined in a straightforward way.

Note that a PLT that covers a period of one year is called a YLT.

25.3 Key Perils Modelled

Unlike actuarial modelling, which is often (although not necessarily) carried out on an all-risk basis, that is, looking at the loss experience without regard to the specific cause of each loss, catastrophe modelling is by its nature dependent on the peril modelled.

You therefore have models for hurricanes, for earthquakes, for floods, and for several other perils, although you sometimes find combined models for weather perils that are related or derive from a common cause (e.g. a combined tornado, hail and straight-line wind model has been produced by AIR). Very often, these models will be developed specifically for the territories that are most affected by a specific peril.

Catastrophes are commonly subdivided into natural and man-made catastrophes. Natural catastrophes are those caused by natural events. Examples of these are:

- Tropical cyclones
- Tornados
- Floods
- Hail
- Volcano eruptions

Man-made catastrophes, on the other hand, are caused not by natural causes but by human activity and negligence. Examples of man-made catastrophes are:

- Terrorist attacks
- Nuclear explosions
- Chemical spillages
- Collapse of human-built structures (dams, bridges …)
- Fires caused by arsonists

but the list of past disasters caused by human activity and potential future ones is as long as the list of things that can go wrong because of human malevolence (think terror attack, or arson) or negligence (think lack of maintenance in order to reduce expenditure).

The distinction is not always clean – perils of a biological nature such as a pandemic or wildfire could be classified as either natural or man-made depending on what caused them.

Figure 25.7 gives a list of perils commonly modelled. These include mostly natural perils of a geophysical nature, but as the list shows, models are also being developed for man-made catastrophes. A much more detailed view of model coverage by territory can be found in Mitchell-Wallace et al. (2017).

Peril	Territories affected (examples)	Key parameters of the hazard model	Observations
Hurricane/ windstorm/ storm surge	United States Japan Northern Europe Caribbean Australia/New Zealand Far East Asia	Track (hurricane path) Maximum speed Rate of decay from storm centre Storm radius Ground friction Coastal terrain (for storm surge) ...	Hurricane models are (together with earthquake models) the most developed of catastrophe models, although the quality of the models will vary depending on the territory.
Earthquake	United States Japan Southern Europe Far East Asia Australia/New Zealand	Magnitude Fault line Soil conditions Focal depth ...	Earthquake models are also very well developed, especially for the United States and Japan. Earthquake models are available for property damage but also (in some territories) for loss of life, which is relevant to workers' compensation/ employers' liability policies and also to group life policies.
Tornado	United States South America Parts of Europe (e.g. Germany) South-Eastern Asia Australia/New Zealand	Occurrence rate Intensity Path length and width Motion direction	The science of tornadoes is not as developed as that for hurricanes. As a consequence, a lot of hazard modelling is done by reverse-calibrating past claim information (to a larger extent than for, say, hurricanes and earthquakes).
Hail	Europe Africa Most territories above (to some extent)	`Occurrence rate Size of hailstones	Africa has been included as hail is an issue in the continent but the territory is hardly covered by existing cat models because the market for it would be very small.
Flood (river and pluvial)	UK Europe (esp. Germany) United States	*River:* Inundation (i.e. elevation of water level) Tangential velocity (normally not included) *Pluvial:* Amount of rain in mm	Most countries have a flood risk but only models for a few territories have been developed, because models are very localised and computer-intensive and require specific platforms to run. E.g. in the UK flood modelling has been developed because flood insurance for houses is compulsory.
Wildfire	United States Canada	Fire intensity (temperature) Spread rate	Models mostly developed for Nort America, especially the westernmost states of the United States.
Epidemics	All countries can be affected		Based on statistical epidemiology models. Models are in their infancy.
Terrorism	Different countries will be affected most heavily in different historical periods		Models are in their infancy. The scale of the attacks can vary widely (from a vehicle bomb to a nuclear a bacteriological attack).

FIGURE 25.7
A list of perils commonly modelled, at different degrees of sophistication, by catastrophe models.

25.4 Pricing Using Cat Rates and Hazard Maps

The most basic type of cat pricing does not make full use of catastrophe models but uses maps or tables that assign a risk level to each point in the map. Naturally, hazard maps will be different for each peril, so that for example a zone may be classified as a low risk for earthquake but as a high risk for winter storm. An example of hazard map for earthquake is shown in Figure 25.8. A hazard table might be an even simpler object with a list of zones (CRESTA zones for earthquakes are a popular examples), each with an associated risk level.

A pricing model will then assign a **catastrophe base rate** to each risk level, and the expected losses to a property of given insured value will be

$$\text{Expected Losses} = \text{Cat Rate} \times \text{Insured Value}$$

The pricing model will also use **exposure curves** as for non-catastrophe property business.

Apart from these generic observations, the specifics of the model will vary depending on the company and on how the cat rates have been derived. For example, the various locations may be grouped by CRESTA zone, city, postcode, or other aggregation criteria, and the maximum possible loss (MPL) may be taken as equal to a percentage of the total insured value for the grouped locations, to account for the fact that a whole area is not likely to have a 100% loss. Alternatively, one may simply decide to model each property separately and take the MPL equal to the insured value, on account of the fact that it is generally easier to have a loss equal to the sum insured in a catastrophe.

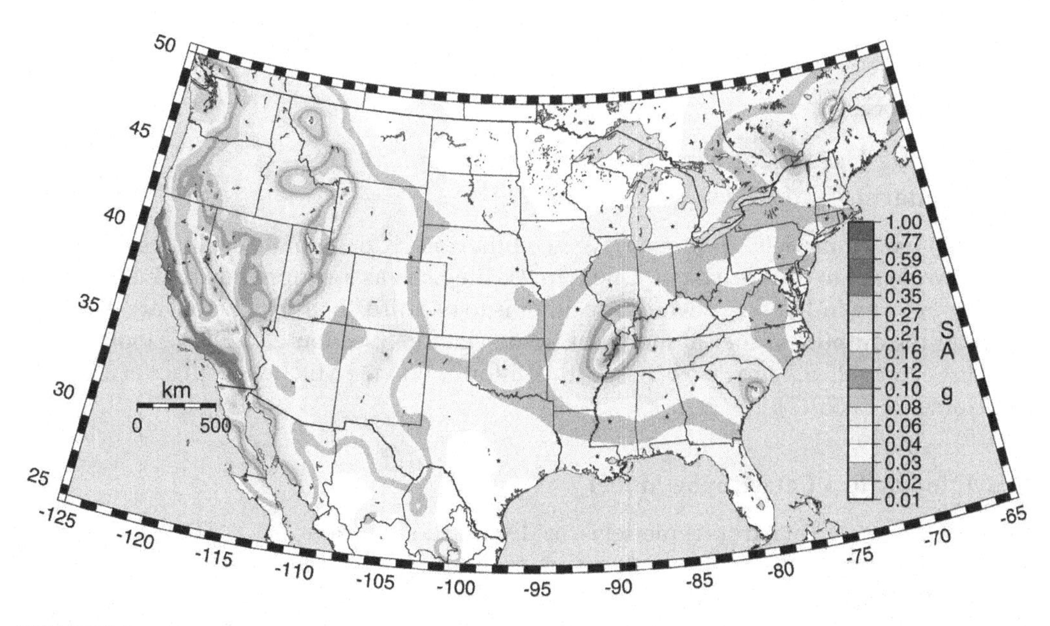

FIGURE 25.8
An example of a hazard map for the earthquake peril in the United States.
Source: **flickr.**

The exposure curve $G(x)$ will in general depend on the peril and will in general be different from the exposure curves used for FLEXA risk for the same location.

If simulation is needed, this can be easily carried out by using for example a Poisson model for the frequency and a Bernegger (MBBEFD) curve for the severity. For simplicity of illustration, we will consider the case where exposure curves are applied location by location and MPL is equal to the insured value. The Poisson model will then have a rate λ for each location equal to:

$$\lambda = \frac{\text{Cat Rate} \times \text{Insured Value}}{\text{Average Damage Ratio} \times \text{MPL}} = \text{Cat Rate} \times G'(0)$$

where we have assumed MPL = Insured Value and have used Equation 21.15 for the average damage ratio. As for the severity model, simulation from a Bernegger curve can be carried out as explained in Section 22.5.2.

It should now be clear the sense in which this is a simplified approach: you have a very basic form of hazard module (a map or a table), but there's no vulnerability model that attempts to determine the amount of damage on the basis of the property features, and the financial model is simply the application of a rate and an exposure curve to the insured value.

Note that even when pricing is carried out with cat modelling software, there will be region/peril combinations that are not modelled by the software and for which a price needs to be determined. In this (very common) case, the analyst can use the cat modelling software for the modelled perils and the cat-rate-based method for the non-modelled perils.

Combining non-catastrophe and catastrophe losses is straightforward, as one can simply treat each region/peril combination for a given location as an additional separate location (attempts to model correlations would almost certainly be spurious).

25.5 Pricing Using Cat Models: Simulation of Cat Loss Scenarios and PLTs

When we use cat models for pricing, we need the ability to simulate cat losses. There are mainly two ways to produce them. One is to use the ELTs as the input and derive scenarios from them using in-house software. The other is to use directly the PLTs that are provided as one of the outputs of the cat model by some vendors (Section 25.2.6). We look at both cases in Sections 25.5.2 and 25.5.3. For further information see Mitchell-Wallace et al. (2017) and Homer & Li (2017).

25.5.1 Inputs to a Catastrophe Model

The input data to a catastrophe model is, as discussed in Section 25.1.2, a list of properties with location details (geocoding), occupancy type, and features such as number of storeys, construction type, year of construction, and so on.

Needless to say, ensuring good data quality is crucial in order to have reliable outputs of catastrophe models. It is also needless to say that data quality varies considerably depending on the territory being modelled, and this has to be considered when assessing the limitations of the model outputs.

An extensive discussion of exposure data and issues around data quality can be found in Mitchell-Wallace et al. (2017).

25.5.2 Simulating Cat Losses Using ELTs

Based on the ELT, we can calculate not only the expected losses on a yearly basis but also a full aggregate loss model, given some assumptions on the frequency and severity distributions. There are several ways of calculating an aggregate loss model. The one described below should sound familiar to actuaries.

Let us assume first that we have a single ELT to simulate from.

One possible way to produce an aggregate loss model is to model the number of losses with a Poisson distribution with a rate λ equal to the sum of all rates from the ELT:

$$\lambda = \sum_{j=1}^{\text{no. of events}} \text{Rate}_j$$

For example, in the case of Figure 25.4, this yields:

$$\lambda = 0.015 + 0.008 + 0.01 + 0.025 = 0.058$$

We can assume (as is typically done) that the loss amount for each event will come from a beta distribution with minimum value equal to 0, maximum value equal to the exposure (Column 7 in Figure 25.4), and mean and standard deviation as in Columns 3 and 6, respectively. We can then sample from each of the beta distributions in proportion to the rate given in Column 2.

Combining the frequency and severity information, we can create a Monte Carlo simulation in the usual way. Box 25.1 shows one possible algorithm.

The AEP can then be calculated by a change of coordinates, using RP instead of probability.

Note that although this simulation is necessary to calculate the full aggregate loss distribution, some basic statistics such as the AAL (and, obviously, the annual frequency) can be calculated immediately from the ELT using the formulae already mentioned and listed again here for your convenience.

$$\text{AAL} = \sum_{j=1}^{\text{no. of events}} \text{Rate}_j \times \text{Exp Loss}_j$$

$$\lambda = \sum_{j=1}^{\text{no. of events}} \text{Rate}_j$$

25.5.2.1 Combining Multiple ELTs

Note that the treatment above is for a single ELT. If you have more than one ELT for different locations/policies/etc., care must be taken about the correlation between the ELTs when these include the same event ID. We assume that the ELTs come from compatible versions of the cat model (Mitchell-Wallace et al., 2017).

BOX 25.1 ALGORITHM TO PRODUCE AN AGGREGATE LOSS MODEL FROM A SINGLE ELT

Input: the event loss table

i. Set $k = 1$, where k represents the scenario index.

ii. Pick a random number $n(k)$ from a Poisson distribution with rate $\lambda = \sum_{j=1}^{\text{no. of events}} \text{Rate}_j$. $n(k)$ represents the number of claims in simulated scenario k.

iii. For each of these claims, pick at random the event which has caused the claim, in proportion to the event rate.
(For example, assume that Event 3 from the ELT in Figure 25.4 has been chosen.)

iv. For the event chosen, pick at random the loss amount from a beta distribution with: left endpoint = 0, right endpoint = the exposure, mean = the expected loss, and standard deviation = SDC + SDI for that event. The output of this step is $n(k)$ claims with values $X[k; 1],\ldots X[k; n(k)]$.
(For example, if we have picked Event 3, simulate a random beta variate between 0 and CCY 27.5M, with mean = CCY 5.5M and SD = CCY 4.4M.)

v. Sum over all claims simulated for that year, obtaining $S(k) = X[k; 1] + \ldots + X[k; n(k)]$

vi. Move to the next simulation ($k \leftarrow k + 1$) and repeat Steps ii through v, for a large number of scenario, for example, 100,000.

vii. Once the MC simulation has finished, calculate the desired statistics (mean, SD, percentiles …) for the aggregate loss distribution.

This correlation is only a concern when two ELTs are for the same region and peril. If two ELTs are for different regions, or different perils, the event IDs will certainly be different, and it will be possible to assume independence. In that case we simply add the rows of one ELT to the other and apply the algorithm described in Box 25.1.

Let us focus on the most interesting case, the one where we have two ELTs – let's call them ELT(a) and ELT(b) – for the same region and peril. This is where the distinction between correlated standard deviation (SDC) and independent standard deviation (SDI) becomes useful – this split is indeed provided by the vendor so that you can combine events from different ELTs considering – at least approximately – the correlation between the two events.

The rules for combining ELT(a) and ELT(b) into a common table ELT(a+b) are as follows:

1. For each event whose ID that appears only in one of the tables, add it to the common table ELT(a+b) as it is.
2. For each event whose ID appears in both ELT(a) and ELT(b), add a new event to ELT(a+b) with these parameters:
   i. Event ID: remains the same
   ii. $\text{Rate}(a+b) = \text{Rate}(a) + \text{Rate}(b)$
   iii. $\text{ExpLoss}(a+b) = \text{ExpLoss}(a) + \text{ExpLoss}(b)$
   iv. $\text{Exposure}(a+b) = \text{Exposure}(a) + \text{Exposure}(b)$

v. $SDC(a+b) = SDC(a) + SDC(b)$ (i.e. add the two standard deviations as you would do for the standard deviations of two perfectly correlated variables)
vi. $SDI(a+b) = \sqrt{SDI(a)^2 + SDI(b)^2}$ (i.e. add the two variances as you would do for the variances of two independent variables)
vii. $SD(a+b) = SDC(a+b) + SDI(a+b)$

Once ELT(a) and ELT(b) are combined, one can simulate cat losses by using the algorithm in Box 25.1 to ELT(a+b).

Needless to say, the method above can be easily generalised to an arbitrary number of ELTs.

What if the split of the standard deviation into a correlated and independent component in the ELT is not provided? A commonly used solution (and the most pessimistic, or simply cautious, one) is to consider that the ELTs are fully correlated, and sum the standard deviations of events with the same ID together (Mitchell-Wallace et al., 2018). This will of course tend to exaggerate the likelihood of very large cat losses.

25.5.3 Simulating Using the Period Loss Tables (PLTs)

We introduced period loss tables (PLTs) in Section 25.2.6. PLTs are simulated scenarios made in some cases available by the cat model. They are also based on the relevant ELTs so they do the same as the algorithm described in Box 25.1, but taking full advantage of the proprietary methodology of the vendor, and likely to account more accurately of the correlation between events.

They are therefore the most straightforward solution for the calculation of the aggregate loss distribution, as long as you believe that the process by which they are constructed is reliable.

25.5.4 Combining Non-Catastrophe and Catastrophe Losses

Combining catastrophe and non-catastrophe losses in simulation is relatively straightforward as we can normally assume that they are uncorrelated.

In case we generate cat losses from an ELT (Section 25.3.c.1), we can simply create one simulated scenario for cat losses for each simulated scenario for non-cat losses. The total losses for each particular scenario is then the sum of the two. In other terms, you do a convolution of the aggregate distribution of the cat and non-cat losses.

If, on the other hand, the cat losses are taken directly from a PLT (Section 25.3.c.2), and the PLT simulates N scenarios, the easiest way to operate is to generate the same number N of non-cat scenarios through simulation and pair up the scenario keeping the same order. If, however, the number of simulated scenarios in the PLT is lower than the number of simulated scenarios needed for stable non-cat results,[1] one simple solution is to use full or partial replicas of the PLTs until you match the number of non-cat scenarios.

This can be done for both the gross losses and the losses after the policy structure is applied. Note that the deductibles and (sub)limits to be applied to the cat losses may be different from those to be applied to non-cat losses.

1 We can safely disregard the case where the number of PLT scenarios is larger than the number of non-cat scenarios: in this case the simplest solution is just to increase the number of simulations.

BOX 25.2 ALGORITHM TO PRODUCE A SEVERITY DISTRIBUTION

i. $k = 1$, k represents the simulation index
ii. For each k, pick at random the event that has caused the claim, in proportion to the event rate; for example, assume that event 3 (Hayward 7.0) has been chosen
iii. For the event chosen, pick at random the loss amount from a beta distribution with limits 0 and exposure and, mean and SD as in the ELT. The output of this step is a claim $X(k)$; for example, if we have picked Hayward 7.0, simulate a random beta variate between 0 and 50 m, with mean = 6.5 m and SD = 5 m
iv. Move to the next simulation ($k \leftarrow k + 1$) and repeat Steps ii through v, for a large number N of simulations
v. Sort the losses $X(k)$ and calculate the severity distribution in the usual way

25.6 Calculation of the Severity Distribution

Note that a straightforward modification of the algorithm in Box 25.1 allows us to produce a severity distribution with a given number of points, for example, $N = 100{,}000$. The idea is simply to simulate random loss amounts from the events of the ELT in proportion to the rate of each event.

This also suggests an alternative way of calculating the aggregate loss model, by sampling directly from the frequency distribution and the severity distribution calculated in this section.

25.7 What Role for Actuaries?

Actuaries need to know about catastrophe models because consideration of catastrophes must inform their activity in pricing, reserving, and capital modelling. Specifically, an allowance for catastrophes needs to be made in *pricing*, especially in the pricing of property risk, because the possibility of a catastrophe obviously affects the total losses expected for a policy. Catastrophes will also be one of the main causes of volatility of the results for an insurer's property portfolio and this will have an impact on the *capital requirements*, which in turn will affect the capital allocated to each policy (Chapter 19).

But do actuaries actually do catastrophe modelling? Actuaries do have a role to play in catastrophe modelling, but this role is by no means central. The development of catastrophe models is, in practice – as we have already mentioned – mostly the reserve of specialised companies that employ scientists, software developers, statisticians, engineers, and financial professionals (including actuaries). The running of catastrophe models and the interpretation of their outputs is also normally done by professionals who have a specific background and training in the relevant sciences and the relevant software. An ideally balanced catastrophe modelling team in an insurer, an insurance broker or a consultant should include:

- Analysts with a background in the physical sciences (such as geology, geophysics, geography) who start from a vantage point in understanding the hazards.
- Analysts with a background in engineering (such as structural engineering) who will be comfortable with the vulnerability component.
- Mathematicians and actuaries who will be comfortable in running and interpreting the financial component, especially when it is necessary to test unfamiliar structures or combine the outputs of a catastrophe model with other more traditional actuarial models, for example, in pricing complex structures involving many different lines of business.

Despite the different specialisations, however, professionals in a catastrophe modelling team will be familiar with all the components of a catastrophe model and will be able to run a standard exercise from start to finish.

Whether or not they are embedded in a catastrophe modelling team, it should be clear that actuaries need to acquire some familiarity with the outputs of catastrophe models especially if modelling property risk and need to be able to use these outputs to run simulations and assess the effect of catastrophes in pricing.

25.8 Questions

1. A large industrial company X wants to insure its properties against the risk of earthquakes, because many of its plants are in earthquake-prone regions.
   a. Explain why a simple burning cost analysis or a frequency/severity analysis over the last 10 years of experience may very well not provide a good indication of the cost of insurance.
   b. Explain how a catastrophe model allows one to overcome the difficulties in (a).
   c. State what the components of a catastrophe model for earthquakes are, and briefly describe each of them.
   d. A simple example of an event loss table is given in Figure 25.4 and is repeated here:

Event ID	Rate	Expected loss	Correlated standard deviation (SDC)	Independent standard deviation (SDI)	Overall standard deviation	Exposure
1	0.015	2,300,000	322,000	1,288,000	1,610,000	11,500,000
2	0.008	4,000,000	560,000	2,240,000	2,800,000	28,000,000
3	0.010	5,500,000	880,000	3,520,000	4,400,000	27,500,000
4	0.025	7,200,000	864,000	3,456,000	4,320,000	43,200,000

Briefly describe the meaning of each column.

Calculate the *overall* yearly rate of catastrophes for the simple example set out in Question (d), and briefly explain how you would simulate losses based on that table. Briefly explain the difficulty of building a catastrophe model for terrorist attacks relative to a natural catastrophe model.

Part IV

General-Purpose Techniques

26

Credibility Theory

The trick is to find a balance.

Cliché

The experience-based calculation of the 'expected loss' (basically, the pure cost of risk per unit of exposure, without any allowance for expenses or profit) for an insurance policy is affected by several sources of uncertainty, the most obvious – and perhaps the best understood – of which is the limited size of the historical database of losses of the client. Other important uncertainties include data uncertainty and model uncertainty.

To make up for such uncertainties, the analyst may use benchmark information from the portfolio or the market (the portfolio/market expected loss[1]) to replace or complement the client expected loss. The problem with this is that the portfolio benchmark is never fully relevant to a particular client. This is usually captured by the spread, or heterogeneity, of the different client rates around the portfolio benchmark.

The standard way to combine client and portfolio information is through credibility theory. The credibility rate (expected loss per unit of exposure) is the *convex combination*[2] of the client rate and the portfolio benchmark rate:

$$\text{Credibility rate} = Z \times \text{Client rate} + (1 - Z) \times \text{Portfolio benchmark} \tag{26.1}$$

where Z is a real number between 0 and 1, reflecting the 'credibility' that you give to the client's experience: 0 means that you give it no credit at all, and 1 means that you fully trust it and you don't need to look elsewhere

Typically, the credibility factor will be given by an expression like this:

$$Z = \frac{\text{Accuracy of the client's rate}}{\text{Accuracy the of client's rate} + \text{Relevance of the portfolio benchmark}} \tag{26.2}$$

Where the accuracy of the client's rate is in turn inversely proportional to the uncertainty around that estimate:

1 The main difference between the portfolio and the market is that 'portfolio' typically refers to data held by the company on a specific segment of business – e.g. all the employers' liability clients – whereas 'market' refers to external data, i.e. data held by third parties and made available through published studies or subscriptions.

2 'Convex combination' is a mathematical term for a linear combination of points or functions with non-negative real coefficients adding up to 1: $z = tx + (1-t)y$ with $t \in [0,1]$. It might sound a bit fancy but it's a useful shorthand.

DOI: 10.1201/9781003168881-30

$$\text{Accuracy of the client's rate} \propto \frac{1}{\text{Uncertainty of the client's estimate}} \tag{26.3}$$

And the relevance of the portfolio benchmark is deemed to be inversely proportional to the heterogeneity of the portfolio, i.e. the degree to which the rates of the different clients making up the portfolio are close to one another – the idea being that the more spread out they are, the less information that gives us about the rate that we should apply to our client:

$$\text{Relevance of the portfolio benchmark} \propto \frac{1}{\text{Heterogeneity of the portfolio rates}} \tag{26.4}$$

So that the credibility factor can also be expressed as:

$$Z = \frac{\text{Heterogeneity of the portfolio rates}}{\text{Uncertainty of the client's estimate} + \text{Heterogeneity of the portfolio rates}} \tag{26.5}$$

Ultimately, the credibility approach can be seen as an example of a Bayesian approach to pricing, in which a prior estimate (the portfolio benchmark) is corrected by the observation of client data (observations), producing a posterior estimate (the credibility estimate): the main difference is that in general a Bayesian approach will not yield a simple convex combination as Equation 26.1 except in special cases.

While the explanations above give the general idea of credibility (around which we'll put some more meat in Section 26.1 and that we'll apply to a number of simple circumstances in Section 26.2), credibility theory takes many different forms (Section 26.3), some of which are rather sophisticated: for example, in their book *A Course in Credibility Theory and its Applications* (2005), Bühlmann and Gisler discuss not only the now-standard approaches but also introduce hierarchical credibility, evolutionary credibility (Kalman filtering) and multidimensional credibility. However, the credibility approach has in practice severe limitations, which we will address in Section 26.4.

26.1 The General Idea of Credibility

In the introduction to this chapter we have seen that the general idea of credibility is that we can improve on our estimate of the client rate by using a benchmark coming from our portfolio of clients (or a market estimate), and that we can do that by choosing a rate which is between the client rate and the benchmark (Equation 26.1), with the weight given to the client rate depending on both the accuracy of the client rate estimate and the relevance of the portfolio benchmark to the client. In this section we will look at how the calculations can be carried out in practice.

Before giving the calculation details, let's sharpen up the terminology a bit.

26.1.1 Expected Loss Rate

The *expected loss rate* R for an insurance contract is given by

$$R = \frac{\mathbb{E}(S)}{w} \tag{26.6}$$

where $\mathbb{E}(S)$ is the expected aggregate loss in a given year and w is the expected exposure in that same year.[3] (For simplicity of description we assume one-year contracts, although this is obviously not necessary.)

Using the collective risk model assumption, $\mathbb{E}(S)$ can be written as $\mathbb{E}(S) = \mathbb{E}(N)\mathbb{E}(X)$ where $\mathbb{E}(N)$ is the expected number of claims and $\mathbb{E}(X)$ is the expected individual claim amount. To derive $\mathbb{E}(N)$ and $\mathbb{E}(X)$, we need to know the underlying frequency and severity distributions with their exact parameter values.

For example, if N follows a Poisson distribution, $N \sim \text{Poi}(\lambda w)$, and X follows a lognormal distribution, $X \sim \text{Log}\,N\,(\mu,\sigma)$, then[4] $\mathbb{E}(S) = \mathbb{E}(N)\mathbb{E}(X) = \lambda w \exp\left(\mu + \frac{\sigma^2}{2}\right)$.

26.1.2 Client's 'True' Expected Loss Rate

Let R_c be the 'true' expected loss rate of the client (or simply *client rate*). This is given by $R_c = \frac{\mathbb{E}(S_c)}{w_c}$ where S_c is the aggregate loss in a year and w_c is the exposure in the same year. According to the collective model, $\mathbb{E}(S_c)$ can be written as $\mathbb{E}(S_c) = \mathbb{E}(N_c)\,\mathbb{E}(X_c)$, where N_c is the number of claims and X_c is the claim amount. In practice, we will have only an estimate of $\mathbb{E}(S_c)$ and therefore of the expected loss rate, and not its exact value.

26.1.3 Client's Estimated Expected Loss Rate, and Its Uncertainty

Let $\hat{R}_c$ be an estimate of the true expected loss rate R_c of the client. This will typically be obtained by

- A simple burning cost approach applied to the aggregate losses.
- Estimating the average frequency and severity and calculating their product.
- Estimating the parameters of the frequency and severity distribution and calculating the average frequency and severity based on those estimates.
- Hybrid approaches.

This estimate will be affected by several sources of uncertainty: the models for frequency and severity will not replicate reality perfectly (model uncertainty); the values of the model parameters will only be known approximately because the data sample is always limited in size (parameter uncertainty); and the data themselves are often reserve estimates incorporating various assumptions (for example, on claims inflation) rather than known quantities (data and assumption uncertainty).

Data uncertainty has the effect of increasing parameter uncertainty, and so could be incorporated into parameter uncertainty. Model uncertainty is generally ignored in the treatment of credibility but that's only because it is difficult – if not impossible – to

3 Note that we speak about 'expected loss rate' because we are interested in the expected losses per unit of exposure.

4 However, reality is usually not so straightforward because it is not always possible to express $\mathbb{E}(S)$ in a simple analytical form. This may be because of policy modifications (reinstatements, indexation clauses …), or simply the fact that the severity distribution is available only as an empirical collection of points rather than a model. Therefore, $\mathbb{E}(S)$ will often be estimated by a stochastic simulation or by an approximate formula.

estimate; however, it is definitely a contributor to the uncertainty of an estimate of the expected losses.

The standard error (i.e. the standard deviation of an estimator) is often used as a measure of uncertainty of the client rate. In general, the standard error of the expected loss rate will depend on the process by which the expected loss rate is estimated. Notice that the standard deviation of the expected loss rate estimator should not be confused with the standard deviation of S/w, the aggregate loss per unit of exposure.

We can therefore consider that the client rate estimate will be affected by a standard error se_c, whose empirical estimate we call $\widehat{\mathrm{se}}_c$.

In formulae, we can say describe the estimated client rate $\hat{R}_c$ given the true expected loss rate, R_c, by a random variable with mean $\mathbb{E}(\hat{R}_c \mid R_c) = R_c$ and variance $\mathbb{V}(\hat{R}_c \mid R_c) = \mathrm{se}_c^2$. In other terms,

$$\hat{R}_c \mid R_c = R_c + \mathrm{se}_c \varepsilon_c \tag{26.7}$$

where ε_c is a random variable with zero mean and unit variance: $\mathbb{E}(\varepsilon_c) = 0$, $\mathbb{E}(\varepsilon_c^2) = 1$. No other assumption is needed on the distribution of ε_c, although if the error is purely due to parameter uncertainty this can be approximated as Gaussian noise. As we will see below, R_c is also a random variable.

26.1.4 Portfolio Benchmark

We can also define the benchmark rate $R_{\mathcal{P}}$ for the portfolio (or market). The portfolio benchmark rate $R_{\mathcal{P}}$ may be obtained in a similar way to $\hat{R}_c$ but using data from all clients in a portfolio (including or not the data used to calculate $\hat{R}_c$) or it could be derived by market surveys that look at the average rate charged for a given type of policy.

Whatever the case, the rates of the clients making up the portfolio are assumed to be centred around the benchmark $R_{\mathcal{P}}$, with a spread σ_h (also called *heterogeneity*), whose empirical estimate we can call s_h, and which represents the heterogeneity of the different clients around the mean portfolio rate. The spread will typically be determined empirically.

(To be completely general, we should also consider the uncertainty on $R_{\mathcal{P}}$ as well when the portfolio benchmark comes from statistical analysis of the portfolio data (Parodi and Bonche 2010), but for simplicity, we will ignore this second-order effect here and assume that $R_{\mathcal{P}}$ is *not* a random variable.)

We therefore can express the client *true* rate R_c (before any observation is made) as a random variable whose mean is $\mathbb{E}(R_c) = R_{\mathcal{P}}$ and whose variance is $\mathbb{V}(R_c) = \sigma_h^2$:

$$R_c = R_{\mathcal{P}} + \sigma_h \varepsilon_h \tag{26.8}$$

where ε_h is a random variable such that $\mathbb{E}(\varepsilon_h) = 0, \mathbb{E}(\varepsilon_h^2) = 1$. No assumption is needed on the shape of the distribution of ε_h.

Note that this allows us to rewrite Equation 26.7 as follows:

$$\hat{R}_c = R_{\mathcal{P}} + \sigma_h \varepsilon_h + \mathrm{se}_c \varepsilon_c \tag{26.9}$$

26.1.5 Credibility Rate

Let us now make the crucial assumption that the credibility estimate for the expected loss rate R_c has the form:

$$\hat{R} = Z \times \hat{R}_c + (1-Z) \times R_{\mathcal{P}} \tag{26.10}$$

We want to choose a value Z(and therefore of $\hat{R}$) that makes $\hat{R}$ as close as possible to the true rate R_c. To do this, we need an error function (also called a loss function, a concept we encountered already in Chapter 12) – i.e. a function that measures the error we make by choosing $\hat{R}$ as an estimate for R_c. The problem of finding Z then becomes the problem of minimising this error function.

The most common error function used in credibility theory is the mean squared error $E_{c,h}\left[\left(\hat{R} - R_c\right)^2\right]$, where the expected value is taken on the joint distribution of ε_c, ε_h, which are assumed uncorrelated ($\mathbb{E}(\varepsilon_c \varepsilon_h) = 0$). It is straightforward[5] to prove that the value of Z that minimises the error function is:

$$Z = \frac{\sigma_h^2}{\sigma_h^2 + se_c^2} \tag{26.11}$$

Note that this formula can be rewritten in a form that is perhaps easier to understand intuitively and is easier to generalise to more than two sources of information:

$$Z = \frac{\dfrac{1}{se_c^2}}{\dfrac{1}{\sigma_h^2} + \dfrac{1}{se_c^2}} \tag{26.12}$$

where $\dfrac{1}{se_c^2}$ can be interpreted as a measure of accuracy and $\dfrac{1}{\sigma_h^2}$ as a measure of relevance of portfolio/market information, and therefore Equation 26.12 has now the same form as Equation 26.2.

Note that σ_h, se_c are in practice not available but can be approximated as s_h, $\widehat{se}_c$, we can only actually calculate an approximate (empirical) version of the credibility factor:

$$\hat{Z} = \frac{s_h^2}{s_h^2 + \widehat{se}_c^2} \tag{26.13}$$

This credibility approach is illustrated pictorially in Figure 26.1.

26.1.5.1 Notes on Using the Credibility Estimate

In practice we don't often have a good estimate of the standard error, and even more frequently the standard error calculated in the usual way (e.g. using Fisher's matrix) will underestimate the uncertainty around the client estimate. For example, using a burning

5 The result is indeed straightforward once we express $\hat{R} - R_c$ as a function of ε_c, ε_h only and remember that $\mathbb{E}(\varepsilon_h^2) = \mathbb{E}(\varepsilon_c^2) = 1, \mathbb{E}(\varepsilon_h \varepsilon_c) = 0$. The mean squared error is then given by $\mathbb{E}_{c,h}\left[\left(Z \times \hat{\varphi}_c + (1-Z) \times \varphi_m - \varphi_c\right)^2\right] = \mathbb{E}_{c,h}\left(\left[(Z-1)\sigma_h \varepsilon_h + Z\sigma_c \varepsilon_c\right]^2\right) = (Z-1)^2 \sigma_h^2 + Z^2 \sigma_c^2$. By minimising with respect to Z, one obtains the expression for Z.

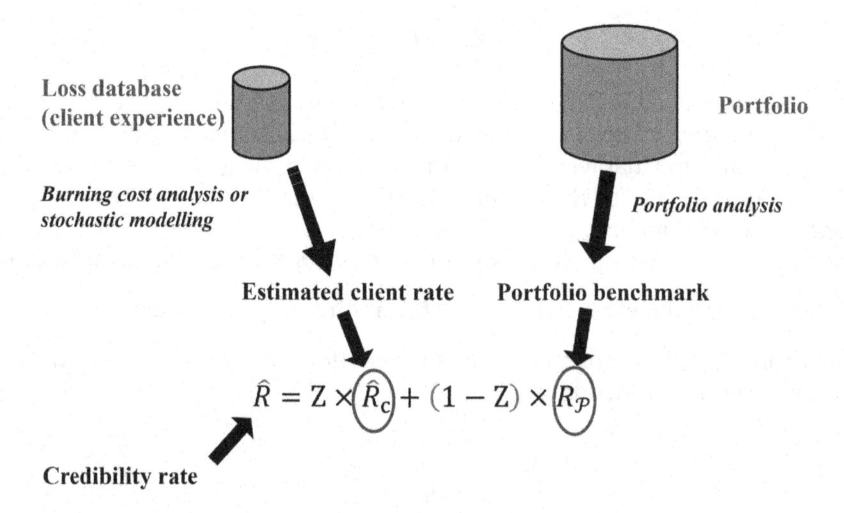

FIGURE 26.1
An illustration of how the credibility premium may be calculated by combining the client rate with a portfolio benchmark (as determined by the analysis of a portfolio or of external market data).

cost analysis might require making assumptions around the frequency and severity of large losses (see Chapter 11), the possibility of systemic years, the relevance of past historical losses (especially the farthest-back years) and so on, which will add to the uncertainty of the client rate estimate in a way that is not trivial to estimate.

Very often, therefore, the client uncertainty in Equation 26.13 may need to be assessed in a looser, less mathematical way, based on judgment or on the basis of a scoring system where the various components of uncertainty (data, assumption, model …) are assessed by means of heuristic rules.

As for the spread of the portfolio benchmark, it may also often not easy to estimate accurately on the basis of a careful portfolio analysis (e.g. analysis of the different expected loss rates of a number of clients), so this may be replaced by proxies such as the different premiums found in the market with broad assumptions on the loss ratios.

26.1.5.2 Other Loss Functions

Equation 26.11 (and related formulae) have the form they do as a consequence of choosing the mean squared distance as the error function. This is a popular choice but it is not the only one and it is not always the most sensible one depending on the circumstances. For example, squaring the quantities tends to exaggerate the impact of the difference between the uncertainty and the spread, so that very often we'll have credibility factors very close to 0 and to 1, 'binarising' the credibility results.

Various alternatives to the mean squared loss can be used. An obvious choice is the mean absolute loss $\mathbb{E}_{c,h}\left[\left|\hat{R} - R_c\right|\right]$, leading (under suitable assumptions about the noise) to the following form for the credibility factor, which is less sensitive to differences between the portfolio spread and the client uncertainty:

$$Z = \frac{\sigma_h}{\sigma_h + \mathrm{se}_c} = \frac{\dfrac{1}{\mathrm{se}_c}}{\dfrac{1}{\sigma_h} + \dfrac{1}{\mathrm{se}_c}} \tag{26.14}$$

26.1.5.3 Extension to Any Number of Sources of Information

Equations 26.10 and 26.12 can be easily generalised when you wish to combine more than two sources of information, typically a data-driven client rate estimate $\hat{R}_c$ and several different external benchmarks $R_{\mathcal{P}_1}, R_{\mathcal{P}_2} \ldots R_{\mathcal{P}_k}$, each with their own spread $\sigma_{h,j}$:

$$\hat{R} = Z_c \times \hat{R}_c + \sum_{j=1}^{k} Z_{\mathcal{P}_j} \times R_{\mathcal{P}_j} \tag{26.15}$$

$$Z_c = \frac{\frac{1}{se_c^2}}{\sum_{j=1}^{k} \frac{1}{\sigma_{h,j}^2} + \frac{1}{se_c^2}}, \quad Z_{\mathcal{P}_j} = \frac{\frac{1}{\sigma_{h,j}^2}}{\sum_{j=1}^{k} \frac{1}{\sigma_{h,j}^2} + \frac{1}{se_c^2}} \tag{26.16}$$

Notice that $Z_c + \sum_{j=1}^{k} Z_{\mathcal{P}_j} = 1$ by construction.

26.2 Examples and Applications

Let us now look at various different examples and applications of the credibility approach.

26.2.1 A Simple Example

Let's look at how the credibility approach works for a simple example. Imagine that you have carried out a burning cost analysis for a professional indemnity policy for law firm G. The firm employs 550 lawyers, and this number has remained broadly constant over the last 12 years.

The burning cost analysis was based on 10 years of experience. As a result of the burning cost analysis, you found that the average of the total losses over the years (after revaluation and all other necessary corrections for exposure, large losses, risk profile changes etc.) is FRD 2,070k, and the sample standard deviation is FRD 630k (FRD = Freedonian dollar).

In order to improve on this estimate, a credibility approach is adopted. The premium information for six other clients (A, B, C, D, E, F) whose risk have recently been underwritten by your company is shown in Figure 26.2.

Firm	Premium charged (FRDk)	Number of lawyers	Premium per unit of exposure (FRDk)
A	2,100	300	7.00
B	1,000	180	5.56
C	650	120	5.42
D	330	50	6.60
E	2,500	400	6.25
F	520	80	6.50

FIGURE 26.2
Portfolio information for the professional indemnity credibility example. To calculate the portfolio the premium per unit of exposure (in this case, number of lawyers) needs to be determined.

Assuming a loss ratio for this line of business of 65%, you want to determine the premium that the insurer should charge for firm G based on a credibility approach.

Solution. For comparison purposes we need to calculate the expected losses per unit of exposure, where the exposure is in this case the number of lawyers in the firm. The client estimate for the expected loss rate is:

$$\hat{R}_c = \frac{\text{FRD}2{,}070\text{k}}{550} = \text{FRD}3.76\text{k}$$

As a measure of uncertainty around this estimate we can take the standard error of the estimate (which is roughly equal to the sample standard deviation divided by the square root of the number of years) and divide it by the exposure:

$$\widehat{\text{se}}_c = \frac{\text{FRD}630\text{k}}{\sqrt{10}\times 550} = \text{FRD}0.36\text{k}$$

The portfolio estimate can be obtained by taking a *straight* average of the premium per unit of exposure across different clients and then multiplying by the loss ratio:

$$R_p = \text{FRD}6.22\text{k}\times 65\% = \text{FRD}4.04\text{k}$$

The heterogeneity can be measured by calculating the sample standard deviation of the premium per unit of exposure across different clients and multiplying by the loss ratio:

$$s_h = \text{FRD}6.19\text{k}\times 65\% = \text{FRD}0.40\text{k}$$

The credibility factor can then be calculated using Equation 26.13:

$$\hat{Z} = \frac{0.40^2}{0.40^2 + 0.36^2} = 0.556$$

The credibility expected loss rate is then:

$$\hat{R} = 0.556\times\text{FRD}3.76\text{k} + 0.444\times\text{FRD}4.04\text{k} = \text{FRD}3.88\text{k}$$

The credibility (technical) premium for firm G is then:

$$\text{TP} = \frac{550\times\text{FRD}3.88\text{k}}{65\%} = \text{FRD}3.29\text{m}$$

26.2.2 Credibility of a Frequency/Severity Model

We have seen in Chapter 18 an example of how you can take parameter uncertainty into account for a simple Poisson/lognormal model. We had then assumed that the Poisson rate was $\lambda = 100$ (calculated on $m = 10$ years of equal exposure, to make things simple) and the parameters of the lognormal were $\mu = 10.5$, $\sigma = 1.3$ (calculated on 150 data points).

We saw that the 'central' value of our expected losses estimate was

$$\mathbb{E}(S) = \mathbb{E}(N)\mathbb{E}(X) = \lambda\times\exp\left(\mu + \frac{\sigma^2}{2}\right) = £8.45\text{M}$$

Standard MLE results gave us the parameter uncertainty on each of these parameters, which was then propagated by MC simulation as in Figure 18.1, yielding a parameter uncertainty of £1.23M. As a result, we find a parameter uncertainty on the expected losses of approximately £1.23M, which is what we need for our credibility estimate, except that it is expressed as an absolute amount and not per unit of exposure. This can be easily taken care of if we know the exposure measure.

Let us therefore assume that this loss model is for a third-party liability motor insurance for a fleet of 10,000 cars, so that the expected loss rate is actually

$$\hat{R}_c = \frac{\text{Estimated expected losses}}{w_c} = \frac{£8.45\text{M}}{10,000} \sim £845.4$$

With an estimated standard error of

$$\widehat{\text{se}}_c = \frac{£1.23\text{M}}{10,000} \sim £123.0$$

Let us also assume that we have carried out a market analysis on our base of 25 clients (Figure 26.3).

Based on this market analysis, we find that the market premium is

$$R_{\mathcal{P}} = £981.8$$

with a measured heterogeneity (spread) equal to

$$s_h = £285.7$$

The credibility factor can be calculated as

Market analysis

Mean (weighted)	981.8	(Market risk premium)
Std. dev. (weighted)	285.7	(Heterogeneity)

Client ID	No. of vehicles	Risk premium
1	6234	1040.6
2	21,683	957.1
3	7804	926.2
4	21,988	421.1
5	24,610	1086.5
6	25,222	1359.0
7	25,973	908.8
8	20,984	682.6
9	27,019	1056.0
10	19,604	743.0
...	...	...

FIGURE 26.3
The derivation of the market expected loss rate based on an analysis of our client base for motor fleet insurance (25 clients, 10 shown). Note that in this example, both the mean (= market expected loss rate) and the standard deviation of the expected loss rate across different clients (= market heterogeneity) are calculated on a weighted basis (i.e. by giving more weight to clients with a larger fleet), unlike our toy example about salaries.

$$\hat{Z} = \frac{s_h^2}{s_h^2 + \widehat{se}_c^2} = \frac{(£285.7)^2}{(£285.7)^2 + (£123.0)^2} = 0.844$$

And the credibility premium is

$$\hat{R} = 0.844 \times £845.4 + 0.156 \times £981.8 = £866.7$$

As you will notice, high credibility has been given to our client's estimate. This is a consequence of the facts that it is a large fleet, and that the spread amongst different clients is quite large.

26.2.3 Mixing Experience and Exposure Rating

Although we have thus far spoken about expected loss rates and market information based on a study of relevant clients, credibility theory can be applied to all situations in which there is a quantity, for example, the expected loss rate, the frequency, or the average severity, among others, calculated based on the client's data and any portfolio/market benchmark of that same quantity that is considered relevant to the client, so long as it is possible to estimate the spread around that benchmark that different clients will have.

One such application is mixing experience and exposure rating. Experience rating is the determination of the risk cost based on the client's own experience, whereas exposure rating has the same objective but is based on the original premiums (or base rates) and exposure curves.

The idea is to use experience rating as the client's estimate and exposure rating as the market benchmark. The estimation of $\hat{R}_c$ is no different from that of any other credibility exercise, so we will just assume that we have R_c and $\widehat{se}_c$.

As for $R_{\mathcal{P}}$, we assume that this comes from an exposure rating exercise based on the knowledge of the client's portfolio of properties and (in the case of reinsurance) of its loss ratio.

For illustration purposes, let us assume that the exposure rating estimate for the expected loss rate is $R_{\mathcal{P}}$ and, to make things more concrete, let us assume that this was derived based on the original premium P, a loss ratio LR, and the Swiss Re curve with parameter c. The missing bit is the 'spread' σ_h around $R_{\mathcal{P}}$.

How do we estimate σ_h? The two pieces of information on which $R_{\mathcal{P}}$ is based – the loss ratio LR and parameter c – are only estimates themselves. The LR is an *assumed* loss ratio, which is probably based on the historical performance of the account and will therefore have some uncertainty, $\sigma_{h,LR}$. The value of the c parameter is also based on the type of properties in the portfolio, but it will never be determined with total accuracy; more likely than not, there will be some uncertainty around that as well. For example, for different clients with a similar profile, we might have chosen slightly different values of c. Sometimes, the value may have been determined empirically based on the losses themselves, and in this case, we will have a collection of different values of c that were appropriate for different clients. That will give us a spread $\sigma_{h,c}$.

Whatever the sources of the uncertainties, we can write

$$R_{\mathcal{P}} = R_{\mathcal{P}}(\mathrm{LR}, c)$$

where LR and c are themselves random variables with certain distribution and standard deviations $\sigma_{h,LR}$ and $\sigma_{h,c}$, respectively. The heterogeneity σ_h for $R_{\mathcal{P}}$ can therefore be derived,

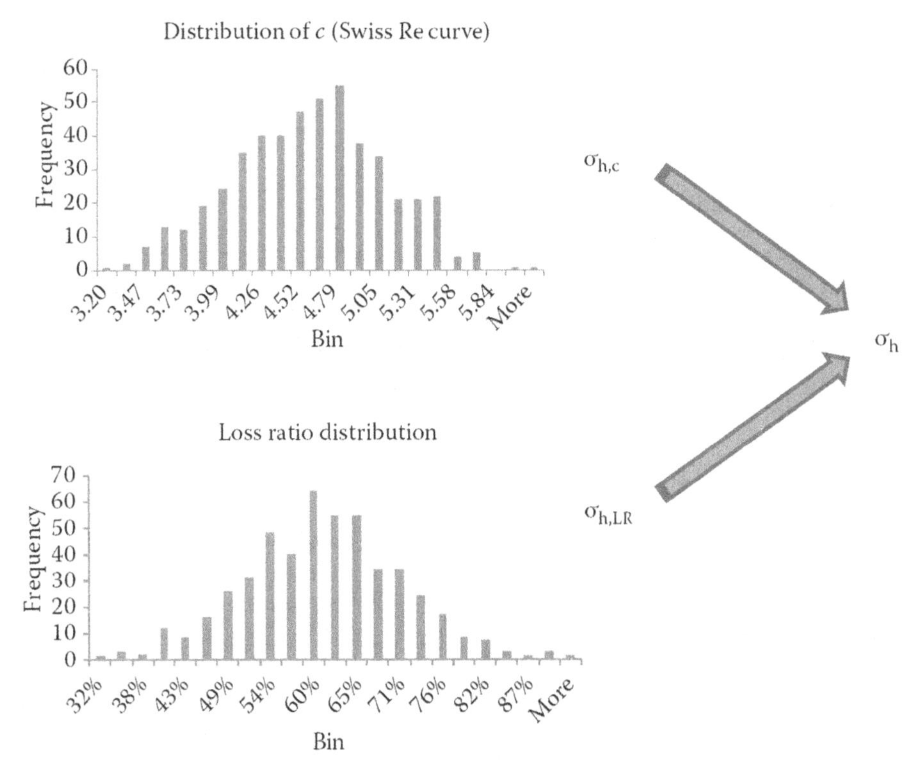

FIGURE 26.4
The heterogeneity for exposure rating can be derived from the heterogeneity of LR and c.

by simulation or straightforward error propagation, from the distribution of c and LR around the assumed value, much in the same way as the uncertainty on experience rating can be derived by calculating the expected losses repeatedly for different values of the parameters (Figure 26.4).

Once we have an estimate of σ_h, we can calculate the credibility factor Z and the credibility estimate in the usual way. It is important to notice that the procedure outlined above is not the standard approach to mixing experience and exposure rating that *needs* to be followed in every circumstance: it is more important to grasp the principles behind this approach and adapt them to the specific circumstances and information at our disposal.

26.3 Specific Approaches to Credibility

We have used the framework of uncertainty-based credibility (as developed, for example, by Boor (1992), and Parodi and Bonche (2010)) as a relatively straightforward way of illustrating the concept of credibility as the attempt to achieve a trade-off between the accuracy of the client estimate and the relevance of the portfolio benchmark. However, many different approaches to credibility have been developed in the past, all of which (except perhaps the standard for full credibility used in classical credibility) achieve this trade-off between accuracy and relevance in their own different way.

We will not go into the technical details and the proofs of these various methods. However, a good introduction to the various types of credibility can be found in the work by Klugman et al. (2008). The reader who needs a more in-depth treatment can instead refer to the book by Bühlmann and Gisler (2005).

26.3.1 Classical Credibility

Historically, the concept of credibility was born at the beginning of the twentieth century in the context of workers' compensation insurance in the United States. The first question that classical credibility was trying to answer (see, for example, Mowbray 1914) was a simple one: How much client data does one need to calculate the pure premium with sufficient accuracy, without resorting to market information? A data set that satisfies this criterion is said to have 'full credibility'.

26.3.1.1 Criterion for Full Credibility

The main idea behind full credibility is this. Assume we are trying to estimate the expected value of a random variable Y, of which we have a certain number (possibly as low as 1) of distinct measurements. Y could (and typically will) be the number of claims over a period ($Y = N$), or the average loss amount [$Y = (X_1 + \ldots X_n)/n$], or the total losses over a period ($Y = S$).

Also assume that Y can be approximated as belonging to a normal distribution: $Y \sim N(\mu_Y / \sigma_Y^2)$ and call $CV_Y = \mu_Y/\sigma_Y$ the coefficient of variation of Y.

We want to explore conditions under which Y is within a certain proportion α of the true amount μ_Y with probability at least p:

$$\Pr(-\alpha\mu_Y \leq Y - \mu_Y \leq \alpha\mu_Y) \geq p \tag{26.17}$$

Because we have assumed that Y follows a normal distribution, this can be rewritten as

$$2\Phi\left(\frac{\alpha\mu_Y}{\sigma_Y}\right) - 1 \geq p \tag{26.18}$$

where $\Phi(\cdot)$ is the cumulative distribution function of the standard normal distribution (that with mean equal to 1 and variance equal to 1). Equivalently,

$$CV_Y = \frac{\sigma_Y}{\mu_Y} \leq \frac{\alpha}{\Phi^{-1}\left(\frac{1+p}{2}\right)} \tag{26.19}$$

The other idea of credibility is that to keep the coefficient of variation of the estimator Y low, we must make sure that we have sufficient data. This takes different forms depending on whether we are dealing with frequency, average loss amount, or average total losses.

26.3.1.1.1 Full Credibility of Claim Count

Consider the case of frequency, where $Y = N$, N being the number of losses that occurred over a period T. The larger the number of losses we expect over a period (the 'rate'), the more the number of losses *recorded* over that period is representative of the rate. If the claim count process is Poisson, we know that if the Poisson rate λ is sufficiently large, the number of losses can be approximated as a normal distribution with mean and variance equal to the Poisson rate: $N(\lambda,\lambda)$. The coefficient of variation is therefore equal to $CV_N = 1/\sqrt{\lambda}$, which becomes very small for large Poisson rates. The condition for full credibility therefore becomes (by manipulating Equation 26.19):

$$\lambda \geq \left(\frac{\Phi^{-1}\left(\frac{1+p}{2}\right)}{\alpha} \right)^2 \tag{26.20}$$

Because the only estimate that we have of λ is the number of claims actually observed n_{obs}, Equation 26.20 can be rewritten replacing λ with n_{obs}.

$$n_{obs} \geq \left(\frac{\Phi^{-1}\left(\frac{1+p}{2}\right)}{\alpha} \right)^2 \tag{26.21}$$

It will be useful for future reference to give a name to the right-hand side of Equation 26.20:

$$n_0 = \left(\frac{\Phi^{-1}\left(\frac{1+p}{2}\right)}{\alpha} \right)^2 \tag{26.22}$$

The number n_0 represents the standard of full credibility for the number of claims from a Poisson process.

If the claim count process is not Poisson but a process with variance-to-mean ratio equal to $1 + r$ (where r could possibly be negative), Equation 26.21 becomes

$$n_{obs} \geq (1+r) \times \left(\frac{\Phi^{-1}\left(\frac{1+p}{2}\right)}{\alpha} \right)^2 = (1+r) \times n_0 \tag{26.23}$$

26.3.1.1.2 Full Credibility of Average Severity

In the case of average severity, Y is the average severity over a number n_{obs} of claims: $Y = \left(X_1 + \ldots + X_{n_{obs}}\right) / n_{obs}$. Each of the n_{obs} claims is assumed to be sampled from the same distribution $X \sim F_X$. Under very broad assumptions, this is a textbook example of the application of the central limit theorem, and for large values of n_{obs} the variable Y will be approximately distributed as a normal distribution: $Y \sim N\left(m_X, s_X^2 / n_{obs}\right)$ where m_X and s_X are the mean and standard deviation, respectively, of the severity distribution. By replacing $\sigma_Y = \sigma_X / \sqrt{n}$ and $\mu_Y = \mu_X$ in Equation 26.19, we find the following condition for the full credibility of the average severity:

$$n_{\text{obs}} \geq \left(\frac{\sigma_X}{\mu_X} \frac{\Phi^{-1}\left(\frac{1+p}{2}\right)}{\alpha} \right)^2 = \left(\frac{\sigma_X}{\mu_X}\right)^2 \times n_0 = CV_X^2 \times n_0 \tag{26.24}$$

26.3.1.1.3 Full Credibility of Aggregate Losses

In the case of the aggregate losses, Y is the total losses over a given period, $Y = S = X_1 + \ldots + X_N$ where N is the number of claims (itself a random variable). We will assume here, as in the case of frequency and severity, that S can be approximated as a normal variable, as we have also done, for example, in Section 7.2.2.1. Remembering that in the collective risk model, where the number of claims and their severity are independent, the mean and the variance of S can be calculated as (see Chapter 7): $\mathbb{E}(S) = \mathbb{E}(N)\,\mathbb{E}(X)$, $\mathbb{V}(S) = \mathbb{V}(N)\,\mathbb{E}(X)^2 + \mathbb{E}(N)\,\mathbb{V}(X)$, then the coefficient of variation is given by

$$\text{CV}_s = \frac{\sqrt{1+r+\text{CV}_X^2}}{\sqrt{\lambda}} \tag{26.25}$$

where $\lambda = \mathbb{E}(N)$ represents the mean claim count, and $1 + r = \mathbb{V}(N)/\,\mathbb{E}(N)$ represents the variance-to-mean ratio of the claim count. By substituting Equation 26.25 into Equation 26.19 and stating everything in terms of the number of observations n_{obs}, we obtain:

$$n_{\text{obs}} \geq (1+r+\text{CV}_X^2) \times n_0 \tag{26.26}$$

which reduces, in the case of a Poisson distribution ($r = 0$), to $n_{\text{obs}} \geq (1+\text{CV}_X^2) \times n_0$. Interestingly, the standard for full credibility can be written as the sum of the standards for full credibility of the frequency $[(1 + r) \times n_0]$ and of the severity $(\text{CV}_X^2 \times n_0)$: in a way, it is like saying that we need to have enough data to satisfy both the frequency and severity criterion separately. Note that the normal approximation for the aggregate losses is not as good as the same approximation for either frequency or, especially, severity, as we already found out in Section 7.2.2.1.

26.3.1.2 Partial Credibility

Although Mowbray (1914) did not attempt to answer it himself, the question arises of what one should do when the data at one's disposal does not have full credibility. The solution proposed in the early days of credibility is the one that still stands today, that is, write the credibility premium as a linear combination of the client's pure premium and the market's pure premium: CP = Z × *P*(client) + (1 − Z) × *P*(market). The problem therefore becomes how to calculate Z.

The recipe for classical credibility is to write Z as

$$Z = \min\left(1, \sqrt{\frac{n_{\text{obs}}}{n_{\text{FC}}}}\right) \tag{26.27}$$

where n_{obs} is the number of losses over which the random variables (frequency, severity, and aggregate losses) are calculated, and n_{FC} is the number of losses necessary for full credibility (FC). Equation 26.27 has the advantage of being simple and of behaving reasonably (Z is less than 1 if n_{obs} is not sufficient to guarantee full credibility and is 1 if $n_{obs} > n_{FC}$). Its exact form is justified by the fact that it is the one which ensures that the variance of the credibility estimate is limited to that of a sample with full credibility (see Klugman et al., 2008).

Despite its simplicity and reasonableness, classical credibility suffers from a few major drawbacks:

- The choice of α and p in producing a standard for full credibility is rather arbitrary.
- There is no underlying model behind the choice of P(cred) as the linear combination of P(client) and P(market) – it is a purely heuristic rule.
- The calculation of Z is simple but has one major flaw, as discussed in Klugman et al. (2008): Equation 26.27 is designed to limit the variance of the credibility premium (CP), rather than trying to minimise the expected error made by using CP as a proxy for the true (unknown) expected loss. This error can be measured, for example, by the mean squared error. Examples of how the mean squared error can be used to this end are shown in Section 26.3.2 (Credibility) or in Section 26.5 (Uncertainty-Based Credibility).

26.3.2 Bayesian Credibility

Credibility has a natural interpretation in the context of Bayesian theory: the 'market premium' can be interpreted as a prior value for the premium, and one can say that the premiums of different clients are positioned around the market premium according to a *prior distribution* of the premiums.

Also, for each policyholder, we will have past losses $X_1, X_2 \ldots X_n$ (X_j is a random variable that may represent the number of losses in year j, the revalued amounts of all losses, the revalued total claims in each year/period, or the loss ratio in each year/period, etc.; the actual values are denoted as $x_1, x_2 \ldots x_n$). The past losses $X_1, X_2 \ldots X_n$ are independent, identically distributed conditional on the value of a parameter θ, which measures the risk propensity of each policyholder and is a realisation of a random variable Θ, and this common conditional distribution (that we will call the *likelihood function*) can be written as $f_{x|\Theta}(x \mid \theta)$.

Based on the likelihood function and prior distribution, we will then be able to produce a *posterior distribution* for Θ given the observed data, and ultimately, produce an estimate for the expected loss for next year given the past observations, $\mathbb{E}(X_{n+1} \mid X_1 = x_1 \ldots X_n = x_n)$. A simple example of how this works is when both the likelihood distribution and the prior distribution of the market expected loss are normal.

26.3.2.1 Normal/Normal Case

Assume that the aggregate claims S of the client can be modelled as a normal distribution $S \mid \Theta \sim N(\Theta, \sigma_c^2)$ where σ_c^2 is known and Θ is a random variable that has a normal prior distribution $\Theta \sim N(\mu, \sigma_m^2)$ (μ, σ_m^2 are also assumed to be known). This is called the *likelihood function*.

Also assume that n values of S, $s_1 \ldots s_n$, have been observed through the years, from which a client estimate can be produced as $\bar{s} = (s_1 + \ldots + s_n)/n$.

We can then produce a posterior distribution for Θ given $\bar{s}$, and it is easy to prove (see Question 1) that the posterior distribution is itself a normal distribution:

$$\Theta \mid \bar{s} \sim N\left(\frac{\mu\sigma_c^2 + n\sigma_m^2\bar{s}}{\sigma_c^2 + n\sigma_m^2}, \frac{\sigma_c^2\sigma_m^2}{\sigma_c^2 + n\sigma_m^2}\right) \tag{26.28}$$

The mean $\mathbb{E}(\Theta \mid \bar{s})$ of the posterior distribution is our credibility estimate of the expected losses for next year, and it should be noted that it can be expressed as a convex combination of $\bar{s}$ and μ:

$$\mathbb{E}(\Theta \mid \bar{s}) = Z\bar{s} + (1 - Z)\mu \tag{26.29}$$

where

$$Z = \frac{n\sigma_m^2}{\sigma_c^2 + n\sigma_m^2} = \frac{n}{n + \sigma_c^2 / \sigma_m^2} \tag{26.30}$$

and is therefore in the standard 'credibility estimate' format. Note that when n approaches infinity, Z approaches 1.

Observe that if we divide both the numerator and denominator of Equation 26.30 by n, the credibility factor Z is in the format of Equation 26.11 by noting that σ_c^2 / n can be taken as a rough measure of the standard error of the client premium and σ_m^2 is a measure of market heterogeneity. Another popular way of writing Equation 26.30 is as $Z = n/(n + k)$, where $k = \sigma_c^2 / \sigma_m^2$.

26.3.2.2 Empirical Bayesian Approach

Although the problem of finding a credibility estimate has been expressed above in the pure language of Bayesian credibility, in practice, an *empirical* Bayesian approach may be adopted, in which the values of the parameters μ, σ_m^2, σ_c^2 are estimated from the data themselves: for example, σ_c^2 can be estimated by looking at the volatility of the aggregate losses of the client over the years, and μ, σ_m^2 can be estimated by an analysis of a portfolio of clients.

26.3.2.3 Poisson/Gamma Case for Claim Counts

An interesting example of the Bayesian approach to credibility is that which attempts to predict the expected number of claims for a risk given a benchmark frequency and the observed number of losses for a client.

The theory for this was developed in the context of the bonus-malus calculations for motor insurance claims in Switzerland and was based on the observation that the main information to consider when assessing the goodness of a risk was the number of past claims and not the severity of the claims, as the latter is too volatile (Bühlmann and Gisler, 2005).

Let us assume that:

- N_j is the number of claims for a particular risk policyholder in year j;
- The number of claims N_j for year $j = 1, 2 \ldots n$ is Poisson distributed with rate $\Theta = \theta$ (observed data);

- Θ is Gamma-distributed with mean $\mathbb{E}(\Theta) = \alpha / \beta$ and variance $\mathbb{V}\mathrm{r}\,(\Theta) = \alpha / \beta^2$ (prior distribution).

Under these assumptions, the credibility estimate for the number of claims for that client is given by:

$$\mathbb{E}(\Theta \mid \bar{N}) = Z\bar{N} + (1 - Z)\frac{\alpha}{\beta} \tag{26.31}$$

Where $\bar{N} = \sum_{j=1}^{n} N_j$ is the average number of losses and

$$Z = \frac{n}{n + \beta} \tag{26.32}$$

Note that $1 / \beta = \mathbb{V}(\Theta) / \mathbb{E}(\Theta)$ is a measure of the heterogeneity of the portfolio and therefore β is a measure of the relevance of the portfolio benchmark. The total number of years n is very straightforwardly a measure of accuracy for the Poisson rate of the policyholder.

A significant advantage of this approach with respect to, e.g. using Equation 26.12 for the number of claims is that this approach works well even when the number of claims observed over n years is zero.

26.3.2.4 Pros and Cons of the Bayesian Approach to Credibility

The Bayesian approach is probably the 'cleanest' approach to credibility from a theoretical point of view, and one may wonder why the expression of credibility in this language is not more widespread. The main reason is that the Bayesian approach gives the standard convex combination of the market premium and the client premium for only a few selected distributions. We were able to express the credibility estimate in the form of Equation 26.17 only because we have chosen a normal distribution both for the likelihood function and the prior distribution. Under less fortunate circumstances, and for generic prior and client distributions, all we can say is that given observations $X = (x_1 \ldots x_n)$, the expected value of another observation (in the example above, the value of the expected losses for next year) can be written as (Klugman et al., 2008):

$$\mathbb{E}(X_{n+1} \mid X_1 = x_1 \ldots X_n = x_n) = \int \mu_{n+1}(\theta)\, \pi_{\Theta|X}(\theta \mid x_1 \ldots x_n)\, d\theta$$

In the integral above, $\mu_{n+1}(\theta)$ is the hypothetical mean – that is, the mean for the client premium if the unknown parameter were θ, which in our case above is simply θ itself – and $\pi_{\Theta|X}(\theta \mid x_1 \ldots x_n)$ is the posterior distribution of θ given the data $(X_1 \ldots X_n)$. Also note that if Θ is discrete, the integral can be replaced by a sum. This is a perfectly legitimate outcome of our quest for the credibility estimate of the premium but does not conform to the simple convex combination recipe and is not therefore commonly used.

26.3.2.5 Other Interesting Cases

It should be noted that the Normal/Normal case and the Poisson/Gamma case are not the only ones that yield a credibility estimate in the form of Equation 26.29 (i.e. a convex combination).

There is actually a general rule to establish when the Bayesian credibility approach can be used to yield a credibility estimate in the form of a convex combination: the likelihood function must belong to the exponential family of distribution (the same that we will meet again when addressing generalised linear modelling) and the prior probability must be its conjugate prior, which always exists for the exponential family. A more extensive treatment of this can be found, for example, in Bühlmann and Gisler (2005).

26.3.3 Bühlmann Credibility

One of the most widespread and successful methods to deal with credibility is perhaps that by Bühlmann (1967), later extended by Bühlmann and Straub (1970). The main assumptions behind Bühlmann's credibility are as follows:

- For each policyholder, the past losses $X_1, X_2 \ldots X_n$ (X_j may represent the number of losses in year j, or the revalued amounts of all losses, or the revalued total claims in each year/period, or the loss ratio in each year/period, etc.) are independent, identically distributed variables conditional on the value of a parameter θ, which measures the risk propensity of each policyholder and is a realisation of a random variable θ. (This is exactly as in Section 26.4.2.)
- The credibility premium – that is, the expected losses for next year – is a linear function of the past losses $X_1, X_2 \ldots X_n$: $\mathrm{CP}(\theta) = a_0 + \sum_{j=1}^{n} a_j X_j$, where θ is the (unknown) value of θ for that particular policyholder. The values of the parameters $a_0, a_1 \ldots a_n$ are chosen so as to minimise the mean squared loss $\mathbb{E}\left(\left(\mathrm{CP}(\theta) - a_0 - \sum_{j=1}^{n} a_j X_j\right)^2\right)$.
- The mean and the variance of the losses for each policyholder depend only on the parameter θ: we can therefore define $\mu(\theta) = \mathbb{E}(X_j \mid \Theta = \theta)$ and $v(\theta) = \mathbb{V}(X_j \mid \Theta = \theta)$.

Given these assumptions, we can easily prove (Klugman et al., 2008) that the credibility premium can be written as $\mathrm{CP} = Z\bar{X} + (1-Z)\mu$, where $\bar{X} = \sum_{j=1}^{n} X_j / n, \mu = \mathbb{E}[\mu(\theta)]$ and

$$Z = \frac{n}{n + \mathbb{E}[v(\theta)]/\mathbb{V}[\mu(\theta)]} \tag{26.33}$$

which is in the same format as Equation 26.18 for Bayesian credibility. This time, however, the result does not depend on specific choices for the aggregate loss distribution and the prior distribution but is much more general. This generalisation was made possible by imposing that the credibility premium be a linear combination of the past losses.

26.3.4 Bühlmann–Straub Credibility

One limitation of the Bühlmann model is that it assumes that all years have the same weight when calculating the credibility premium, regardless of possibly having different exposures and therefore different degrees of individual credibility, so to speak. Bühlmann and Straub (1970) presented a variation to Bühlmann's method that deals with this issue.

As before, the past losses $X_1 \ldots X_n$ are assumed to be independent, conditional on the parameter θ, and with mean $\mu(\theta) = \mathbb{E}(X_{j|\Theta} = \theta)$ as before. However, the variance is assumed to depend on both θ and a measure of exposure, m_j (e.g. m_j might represent the number of vehicles in a car fleet), as follows:

$$\mathbb{V}\left(X_j \mid \Theta = 0\right) = \frac{v(\theta)}{m_j}$$

The credibility premium can still be written as $\text{CP} = Z\bar{X} + (1-Z)\mu$, where $\mu = \mathbb{E}[\mu(\theta)]$ as before but the client estimate $\bar{X}$ is the average of the past losses, weighted by the exposure:

$$\bar{X} = \sum_{j=1}^{n} \frac{m_j}{m} X_j$$

where $m = m_1 + \ldots + m_n$, and

$$Z = \frac{m}{m + \mathbb{E}\left[v(\theta)\right] / \mathbb{V}\left[\mu(\theta)\right]} \tag{26.34}$$

26.4 Limitations of the Credibility Approach

Credibility provides an intuitive and simple way of balancing client information with a portfolio benchmark. However, despite the fact that much academic ink has been spent on this topic,[6] credibility estimates should not be taken *too* seriously (that's perhaps true for the results of models in general!). This explains why this chapter is relatively short and focuses only on the main ideas behind credibility despite the long and illustrious story of the subject.

The basic idea is of course useful – mixing various sources of information can allow you to have a more balanced view of the risk – however, the credibility estimate is only a broad heuristic rule to provide some audit trail behind a pricing decision. Any added accuracy coming from complex applications of the concept is, however, likely to be spurious.

Here are some of the problems that credibility estimates tend to suffer from:

- Standard credibility formulae tend to place excessive focus on parameter uncertainty as a source of client uncertainty, because this is the easiest source of uncertainty to capture. However, data/assumption/model uncertainty and other soft factors also play a large role in the accuracy of the client estimate and the relevance (and accuracy) of the portfolio benchmark.
- Many credibility formulae place an excessive focus on the variance as a measure of uncertainty/spread, which is hard to justify when the underlying uncertainties do not come from a Gaussian distribution – leading to an exaggerated impact

6 Having added to the literature with the umpteenth paper about credibility in my first years as an actuary I am also guilty of this, so I am hardly in a position to vent.

of the differences between the portfolio spread and the uncertainty of the client estimate.

- The credibility framework relies on the idea that the best estimate is a convex (linear) combination of the client rate and the portfolio benchmark. Using a Bayesian framework (which is arguably the soundest theoretical framework for credibility), however, shows that the posterior estimate given the prior benchmark and the additional observations from the client, is in general not a linear combination of the client and the benchmark. This does not mean that we should deploy an even more powerful mathematical arsenal in concrete situations (see observation above on spurious accuracy), but that we should take any credibility estimate with caution.

26.5 Questions

1. (*) Using the Bayesian approach to credibility, prove that the posterior distribution of Θ given s (see Section 26.3.2 for the meaning of these parameters) is itself a normal distribution with these parameters:

$$\Theta \mid \bar{s} \sim N\left(\frac{\Theta\sigma_c^2 + n\sigma_m^2 \bar{s}}{\sigma_c^2 + n\sigma_m^2}, \frac{\sigma_c^2\sigma_m^2}{\sigma_c^2 + n\sigma_m^2}\right)$$

2. (*) Prove that Bayesian credibility estimate – when the number of claims of a client follows a Poisson distribution and the prior distribution for the Poisson rate follows a gamma distribution – can be expressed in the standard format of Equation 26.29.
3. You are the actuary for a reinsurance company. An insurance company, A, wants to buy a layer of excess-of-loss reinsurance (£10M × £5M) for its property portfolio from your company. The underwriter has asked you to produce a price for this product by loading the expected losses to this layer by 40%.

   You decide to price this excess-of-loss policy based on a combination of experience rating and exposure rating and combine them with a credibility approach

   a. Describe what data you would require company A to provide to perform experience rating and exposure rating. Make a separate list for the two rating methods, although some items may be the same.

   As a result of the experience rating exercise, you estimate that the expected losses to the layer will be £100,000 ± £30,000, where £30,000 is the standard error, which you obtained by error propagation methods based on the parameters of the frequency and severity distribution.

   As to your exposure rating exercise, you reckon that the portfolio of company A is made of approximately similar types of property whose loss curve can be approximated as a Swiss Re curve with the same value as parameter c.

   You estimate that the value of c is distributed as a normal distribution centred around c = 5.0 with a standard deviation $\sigma(c)$ = 0.5 (all properties have the same value of c, but the exact value of c is not known). As a result of this, your exposure rating calculations yield the following values for the exposure-based losses: £150,000 with a standard deviation of £70,000 (the standard deviation of £70,000 is related to the uncertainty on the parameter c).

b. Write a credibility formula that combines the experience rating and the exposure rating result, and a formula you would use to calculate the credibility factor Z. Explain what the terms in Z are and why it makes sense to apply the formula to this case.

c. Use the formula in (b) to produce a credibility estimate for the premium.

27

Rating Factor Selection and Calibration: GLMs, GAMs, and Regularisation

> Plurality is not to be posited without necessity.
>
> **Ockham's razor**

> An explanation should be as simple as possible, but not simpler.
>
> **Albert Einstein (abridged)**

In situations where a large amount of data is available, such as in personal lines insurance (and, in some cases, in commercial lines insurance or reinsurance), it is possible to rate policyholders according to their individual characteristics rather than with a one-size-fits-all approach.

We are all familiar with this mechanism: the premium that you pay for your car insurance will depend on your age, location, profession, and so on. Your household insurance premium will depend on whether you live along a river or on the top of a hill, and on whether your roof is made of tiles or of straw; and for commercial insurance, it obviously makes a difference to your public liability policy if you have an accounting firm or a nuclear power station.

The characteristics that make a policyholder more or less risky are called 'risk factors'. This definition includes everything, whether it is measurable or not. For example, your riskiness as the owner of a car insurance policy will depend on how aggressive you are as a driver: however, insurers are unable to measure your aggressiveness. When they give you a questionnaire, they ask you plenty of questions, but what they would really like to know is if you are aggressive, reckless, incompetent, and inclined to fall asleep at the wheel. Because there is no point in asking you those questions, they ask you other, seemingly more innocent questions about other measurable factors that are hopefully a proxy for your riskiness: age, sex, profession, type of vehicle, and the like.

Therefore, you have two types of factors: the factors that really drive the risk (risk factors) and the factors according to which you are going to be rated (rating factors), as in the table below.

Risk factors	Risk factors are the characteristics of the policyholder (individual or company) that make the policyholder more or less risky, that is, more or less expensive to cover. Examples: A driver's recklessness, the position of a house in relation to water hazards

DOI: 10.1201/9781003168881-31

Risk Factors and Rating Factors	
Rating factors	Rating factors are the factors that are actually used for pricing a policy, and they are normally measurable proxies for risk factors (or genuine risk factors when these are measurable). Examples: A driver's age, the distance of a house from the river and its altitude

27.1 Why Rating Factors Are Useful

Before explaining how insurers rate their policies using rating factors, let us consider the effect of properly performed rating-factor selection in the non-life insurance industry.

A company that can use *effectively* a larger number of rating factors has a competitive advantage on its peers. Consider this very simple example in which you know exactly in advance what the risk factors are. For example, suppose you write scooter insurance and you can make the following simplifying assumptions:

- the only factor that matters to the level of risk a policyholder poses is location[1] (and specifically, whether the owner lives in an urban vs rural area);
- we have only two companies in the market place: Company A and Company B;
- both A and B use exactly the same premium structure, which for simplicity consists in setting the technical premium equal to the expected losses (we ignore expenses and profits). This is just to avoid distracting complexities but it doesn't have an impact on the argument;
- Company A uses location as a rating factor and is able to calculate correctly the technical premium, while Company B charges a flat premium to all clients based on the portfolio expected losses divided by the number of clients. In both cases, we assume that the expected losses estimates are 100% accurate – in other terms, both A and B calibrate their models perfectly, although they use different levels of granularity. Again, this is unrealistic but doesn't affect the argument;
- all policies are bought at the beginning of each year and held for the full year;
- customers have a flat demand curve, i.e. they are extremely price sensitive – they will always switch to the lowest price. Such things as customer inertia and service levels can be ignored.

To make our life easier, we will assume that at the beginning of Year 1, both A and B have the same numbers of clients by category, as in Figure 27.1.

Because of our assumptions, Company A charges the appropriate technical premium for each category, whereas Company B charges £1,050 to everyone. On average, things will work out fine for both Company A and Company B during Year 1.

However, what do you think will happen once it is time to renew, at the beginning of Year 2? Since customers are price sensitive, all urban customers will migrate from A to B

1 The reason why it is necessary to assume that this is the only risk factor at play is that when we are going to compare different companies we want to ensure that there are no other unknown factors differentiating the risk profile of the two companies.

Year 1	Age	No of customers (A)	No of customers (B)	Tech Premium	Company A	Company B
	Urban	100	100	1,500	1,500	1,050
	Rural	100	100	600	600	1,050
			Total expected profit		0	0

All monetary amounts are in GBP

FIGURE 27.1
Company A is able to charge every customer the right price. We will assume (conveniently) that at the beginning of year 1 it has exactly the same exposure (same number of customers) for each combination of factors. Note that for simplicity of illustration we are assuming (unrealistically) that both companies are pricing at the breakeven premium ignoring costs – hence the zero expected profits.

Year 2	Age	No of customers (A)	No of customers (B)	Tech Premium	Company A	Company B
	Urban	0	200	1,500	1,500	1,050
	Rural	200	0	600	600	1,050
			Total expected profit		0	-90,000

All monetary amounts are in GBP

FIGURE 27.2
The migration of rural customers to Company A and to urban customers to Company B is not a problem for Company A but causes Company B an expected loss.

as they will be charged £1,050 instead of £1,500; and all rural customers will migrate from to B to A as they will be charged £600 instead of £1,050. The situation in Year 2 will be as in Figure 27.2.

Now the problem with Company B is that the average cost is £1,500, but it is charging £1,050. As a consequence, Company B will make an expected loss of (£1,500 − £1,050) × 200 = £90,000. How has this happened? Simply because the customers that have migrated to B were the riskier ones, whereas the safer customers have fled. At this point Company B increases their prices to £1,500 and the situation is stable, but the past economic damage is done, and B is now limited to one type of customer only.

If there are more rating factors and more categories, the situation doesn't change, but it may take more time to reach the stable situation: at each step, some of the bad risks will migrate to B and some of the good risks will migrate to A, and no matter what the starting point is, you end up with one category only for Company B (or more categories with the same right price) that is priced correctly – for the customers in this/those categories, staying or going is a matter of indifference.

The moral of the story is simple: companies that are able to differentiate better between risks using more rating factors will end up attracting the good risks while the companies that use blunter models will end up attracting the bad risks and under-pricing them.

This is the main reason why it is necessary to use as many rating factors as possible (as long as these rating factors make the model more predictive) and calibrate them well. The situation we have depicted, however, is very idealised and reality is much fuzzier. In practice:

- We don't know the right model and specifically we don't know the complete list of *risk* factors – only some *rating* factors that seem to work.

- Even if we knew the right form of the model, we wouldn't know the parameters with perfect accuracy and therefore the technical premium would be known with uncertainty.
- Different companies have different premium structures with different expenses, cost of capital, etc.
- Although we have assumed that decisions are made at the end of each year, in reality, reactions can be far quicker, with personal lines insurers changing the way policies are priced monthly or in some cases even daily.
- We have ignored other factors affecting whether customers may or may not change carrier, such as customers' inertia and level of service.

27.1.1 Some Facts about Rating Factors in Practice

Building on the limitations mentioned above, the following facts affect how rating factor selection is done in practice:

- Twenty rating factors or more are common in some lines, such as motor, and certain factors (such as age) are much more granular than in our toy example.
- Changes to the pricing structure occur more frequently based on the conversion rate. (The conversion rate is the percentage of quotes that turn into actual purchases: if the conversion rate is too low, that might be an indication of overpricing, and if it is too high, it might be an indication of underpricing.)
- Statistically based rating factor selection is only the first part of the story: strategic commercial considerations are paramount. For example, the pricing structure may have to satisfy the need for portfolio balancing, and this will be achieved with cross-subsidies. As an example, a company may not want to have a portfolio disproportionately made of old rural customers, because a competitor effectively targeting this segment might cause a significant drop in the company's revenues. To keep the portfolio balanced, we may need (for example) artificially to decrease the price of young urban customers while increasing to some extent that of old rural customers (cross-subsidy).
- The price elasticity of different categories may be different: not only are customers 'stickier' than our simplified model suggests, but different categories may be stickier than others, for example, middle-aged professionals may fret less about a £20 increase than young students or retired people.
- The introduction of new rating factors is delicate because we may not have enough data to establish their statistical significance.
- The statistical aspects are far subtler than indicated in our simple example. For one thing, you cannot really perform a controlled experiment because your population keeps changing: you see only your own customer base and not that of your competitors, so you do not know exactly which risks you are going to attract. In practice, you have rating actions affect statistics and may lead to instabilities, and you may never converge to a stable situation where you have found the optimal price for everybody.
- Ethical/legal implications: you cannot use all rating factors that come to your mind. Factors such as race cannot in practice be used. Now this seems obvious today, but it has not been long since the race of the policyholder was actually used in rating. A more recent example is that related to sex: although many territories are still using sex as a rating factor, a recent European Court ruling has made it illegal to do so.

- Other reasons why it may not be possible or practical to use more factors are: system limitations, pressure by brokers or others to keep the models simple, keeping the quoting system fast, avoiding undesired side-effects on policyholder behaviour when certain rating factors are used.

27.2 Non-Actuarial Approach to Models with Rating Factors

Models with rating factors are very familiar to underwriters. They normally include formulae that allow them to calculate the technical premium for a customer based on a multiplicative structure like this:

$$\text{Technical Premium} = \text{Base Premium} \times f_1 \times f_2 \times \ldots \times f_n$$

where the Base Premium is the technical premium for the base case (the one without any adjustments), and f_j is an adjustment factor for different values of variable X_j. For example for motor insurance X_j might be a variable 'location' with two possible levels: 0 = rural, 1 = urban, and the value of the factor might be $f_j = 1$ for rural and $f_j = 1.8$ for urban, meaning that urban customers are charged 80% more all the rest being equal. The value of X_j that corresponds to $f_j = 1$ is called the *base level*, as it does not modify the base premium.

Slightly more sophisticated models would distinguish those adjustment factors $f_1, f_2 \ldots f_n$ affecting frequency from those factors $h_1, h_2 \ldots h_m$ affecting severity, and would distinguish the cost and profit part of the technical premium from the expected (i.e. average) total losses:

$$\text{Expected No of Losses} = \text{Base Frequency} \times \text{Exposure} \times f_1 \times f_2 \times \ldots \times f_n$$

$$\text{Expected Severity} = \text{Base Severity} \times h_1 \times h_2 \times \ldots \times h_m$$

$$\text{Expected Total Losses} = \text{Expected No of Losses} \times \text{Expected Severity}$$

$$\text{Technical Premium} = \text{Expected Total Losses} + \text{Loadings}$$

In terms of structure, they are not very different from the generalised linear models (GLMs) treated at length in this chapter, at least GLMs with a multiplicative structure and without interacting factors.

Where these non-actuarial rating models differ from GLMs is mainly in the way in which we decide which rating factors to use, and in the way we calibrate these factors. Specifically:

- the rating factors are typically those factors that have an impact on the risk according to the underwriters, rather than through a data-driven selection process;
- the calibration of the factors is also typically based on underwriters' judgment and is not data-driven.

As it can be imagined, there is no guarantee that models built in this way will have sufficient predictive power. Underwriters will certainly have a good 'feel' as to which rating factors are relevant: however, quantifying their effect (calibration) is a much more difficult problem which involves probability assessments, for which human intuition is famously ill-equipped. Also, exactly because underwriters often have an in-depth knowledge of the risks they are underwriting, they will tend to try to make their models more realistic by including minor factors whose effect is swamped by the uncertainty around the more significant factors and therefore only give the illusion of improvement (this is normally called *spurious accuracy*).

The next section looks into what methods can be used to develop a rating factor model that has the right number of factors which are also well-calibrated.

27.3 How Do We Develop a Rating-Factor Model in Practice? From One-Way Analysis to Generalised Linear Models

Having looked at why (relevant) rating factors are useful, we now turn our attention to the question of *how* we can select and calibrate the right factors. In the previous section we have already seen that traditional rating models include a number of factors chosen by the underwriters on the basis of their knowledge of what has an impact on the risk and quantified on the basis of judgment. In this section, we present actuarially sound methodologies for rating factor selection and calibration.

The simplest and most intuitive methods, and possibly one of the earliest to be used, were one-way analysis and multi-way analysis.

27.3.1 One-Way Analysis

The idea behind one-way analysis is to look at each factor independently to determine which factors are relevant and how the can be calibrated. The way this works is very simple. Assume that you want to build a model for the number of losses for motor insurance, and that you want to use two binary rating factors, age (young vs old) and location (urban vs rural). The model would look like this:

$$\text{Expected No of Losses} = \text{Base Frequency} \times \text{Exposure} \times f_{\text{age}} \times f_{\text{location}}$$

Where the exposure is given by vehicle-yearsos, and the base frequency is that for old clients living in rural locations. The following statistics are available for age and location.

One-way analysis for the Age variable

Age	Exposure	# Losses	Frequency
Old	250	35	0.14
Young	300	140	0.47
	Estimated Young/Old Factor Ratio		3.33

The table suggests that the frequency for young drivers is 3.33 times that for old drivers. Analogously we can analyse the relativities for Location with the following table:

One-way analysis or the Location variable

Location	Exposure	# Losses	Frequency
Rural	300	40	0.13
Urban	250	135	0.54
Estimated Urban/Rural Factor Ratio			4.05

The table above suggests that the frequency for urban clients is 4.05 larger than that for rural clients.

Based on one-way analysis, therefore, we could be tempted to conclude that the age and location factors should have these values: $f_{\text{age}}(\text{old}) = 1, f_{\text{age}}(\text{young}) = 3.33, f_{\text{location}}(\text{rural}) = 1, f_{\text{location}}(\text{urban}) = 4.05$. Such a model, however, would predict that the frequency for young urban clients is 13.5 time higher than the frequency for old rural clients.

This may or may not be correct. However, when we break down the data further, we may find out the following statistics:

Age	Location	Exposure	# Losses	Frequency
Old	Rural	200	20	0.10
Old	Urban	50	15	0.30
Young	Rural	100	20	0.20
Young	Urban	200	120	0.60

The data is perfectly consistent with the two one-factor tables above, in the sense that these tables are obtained from the two-factor table by aggregating the results over age and location. However, this table reveals that the relativity between (young, urban) and (old, rural) is only 6, not 13.5; and the model that explains the two-factor table best is one that has $f_{\text{age}}(\text{old}) = 1, f_{\text{age}}(\text{young}) = 2, f_{\text{location}}(\text{rural}) = 1, f_{\text{location}}(\text{urban}) = 3$. What went wrong?

The problem can be traced to *exposure correlation*: some combinations of age and location are more likely than others (have larger exposure). Specifically, young clients seem to live mainly in urban areas, and old clients seem to live mainly in rural areas, creating spurious correlations between the variables.[2]

If this example doesn't convince you, try an even more extreme situation, where the exposures of age and location are perfectly correlated – so that you have zero exposures for old/urban and for young/rural – and see what the one-way analysis suggests (Question 1 at the end of t-he chapter).

There seems to be a simple solution to this problem: do a two-way analysis instead. However, the problem repeats itself: there might be other relevant factors, for example 'occupation' for which the exposures are correlated with age and/or location (as it seems likely), and ignoring these factors will distort our estimates for f_{age} and f_{location}. So why stop?

27.3.2 Multi-Way Analysis

Based on the example above, it looks like there might be a simple solution to the problem of exposure correlation: instead of doing one- or two-way analysis, do a multi-way analysis that includes all the relevant factors.

2 Note that this has nothing to do with actual interaction between the two variables – i.e. the possibility that young urban clients are a bigger risk than the two facts that they are both young and urban can separately explain. This will be dealt with in Section 27.4.4.

This, however, runs into an obvious problem. The number of observed losses is only a proxy of the expected number of losses for a given combination of factors. So for example if the number of observed losses is 100, and the underlying process is Poisson, we expect (roughly) a standard error equal to the standard deviation, which is equal to $\sqrt{100} = 10$, or 10% of the observed losses; that becomes ~ 30% for 10 observed losses, and 100% when you observe only one loss. When you observe no losses, the error is indeterminate and the model predicts that you will not have losses for that particular combination of factors, which would be a nice find but it's not realistic.

Also how do we judge if factors are relevant or not, in the presence of such errors? Maybe we could start adding factors once combinations with low numbers of losses start to appear, but that is a very strict criterion and will probably results in business-relevant factors to be excluded.

The upshot is that multi-way analysis might work where we have plenty of data and a large number of data points, with exposures distributed uniformly enough. In real-world situations, however, we may easily have 20 factors, with an average of 3 levels each, leading to $3^{20} \sim 3.5$ millions of combinations, each of which needs to have at least 100 data points for the error to remain below 10%. This means 350 million losses at the very minimum, but probably more around 1 billion (10^9) if we allow for the exposure distributed non-uniformly. And the number of customers will be larger than that. No insurance company in the world has that type of data set.

In abstract terms, the reason for this failure is as follows. Multi-way analysis implicitly attempts an *independent* model calculation for each combination of factor levels, but the large number of factor levels used in practice quickly leads to a combinatorial explosion of independent calculations with increasingly smaller exposures. This is called *dimensionality curse* – some problems seem simple enough to solve with few variables but become unmanageable when the number of variables increases.

So what do insurers use in practice? The industry standard for rating-factor selection is *generalised linear modelling*, although the industry is increasingly experimenting with other machine learning methods, of which generalised linear models should be seen not just as precursors but as legitimate examples. Before generalised linear models (GLMs) came into use, other methods were used; some of them (minimum variance methods) may be seen as a special case of GLMs (Mildenhall, 1999).

The way in which GLMs and other methodologies overcome the problems of multi-way analysis is by fitting a model simultaneously (and not independently) to all factors, using a limited number of parameters (roughly equal to the *sum* of the number of levels of each factor, at least in the case where there are no interacting variables[3]).

27.4 Generalised Linear Modelling

The use of GLM for rating factor selection has been a great success story for actuarial science. Actuaries have of course done rating factor selection almost from the start of their involvement in pricing, using multi-way analysis, minimum bias methods, and others, but

3 Interestingly, a fully interacting GLM – i.e. a model in which each factor is assumed to interact with every other factor – would face the same combinatorial explosion as a multi-way analysis. However, interacting factors are used sparingly in GLMs.

it is with GLM that large-scale analysis of rating factors has been put on a firm footing and proper diagnostic tools have been developed.

The machinery of GLM looks daunting at first, and classic mathematics books such as those by McCullagh and Nelder (1990) inevitably include more mathematical formalism than the practicing actuary will need. However, the ideas behind GLM are really simple and most of what actuaries do in practice with GLM is based on a clear understanding of these ideas rather than on a detailed knowledge of the formalism and the algorithms involved.

We have therefore tried here to summarise the main ideas behind GLMs and we have glossed over many of the technicalities. The reader who would like to explore further the treatment of GLM as used in pricing is referred to Anderson et al. (2007b), Iwanik (2011), Dean (2014) and the book by Ohlsson and Johansson (2010).

27.4.1 Least Squares Regression and Multivariate Linear Modelling

To understand GLM, one could start with simple (multivariate) linear modelling (without the 'generalised' bit). The basic idea behind multivariate linear regression is actually the same as that of one-dimensional linear modelling, but with some algebraic complications. So let us take one further step back and remind ourselves briefly of how linear modelling works.

In the one-dimensional case illustrated below, we have n pairs of data points (x_i, y_i) where the y_is (outputs) represents realisations of Y (the dependent variable), and the x_is (inputs) represent realisations of X (the independent variable). The underlying idea of linear modelling is that the true relationship between X and Y is a linear one, but is corrupted by noise (such as random measurement errors) ε:

$$Y = aX + b + \varepsilon \tag{27.1}$$

The noise ε is itself a random variable, assumed to be normally distributed with the same variance for each realisation of X: $\varepsilon \sim N(0, \sigma^2)$. (The same-variance assumption is not critical.) As a consequence of Equation 27.1, each pair of data points can be written as $y_i = ax_i + b + \varepsilon_i$, where ε_i is a realisation of the Gaussian noise process above.

The basic problem of linear modelling is then reduced to finding the values of a and b that best explain the observed pairs (the 'best fit line'). The classic way of solving this

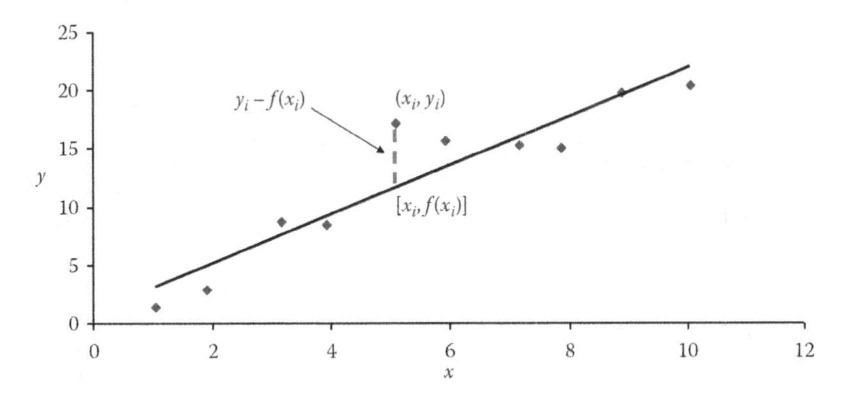

FIGURE 27.3
A pictorial illustration of least squares regression: the goal is to find the 'best' line through all the dots.

problem (which we owe to Gauss) is through least squares regression: choose a and b that maximise the likelihood that the n pairs (x_i, y_i) come from the linear model of Equation 27.1, as illustrated in Figure 27.3. This is equivalent to minimising the following *squared loss function*.[4]

$$L(a,b)=\sum_{i=1}^{n}\left(y_i-ax_i-b\right)^2 \tag{27.2}$$

The values of the parameters that minimise $L(a, b)$ are

$$\hat{a}=\frac{\sum x_i y_i-\dfrac{\sum x_i \sum y_i}{n}}{\sum x_i^2-\dfrac{\left(\sum x_i\right)^2}{n}}, \hat{b}=\frac{\sum y_i-a^* \sum x_i}{n}$$

27.4.1.1 Noise Estimation

Note that the variance of the noise, σ^2, is not something that needs to be provided beforehand but can be estimated as part of model fitting once the other parameters have been determined: $\sigma^2=\sum_{i=1}^{n}\left(y_i-\hat{a}x_i-\hat{b}\right)^2/n$.

27.4.1.2 Note on Loss Functions

Despite the fact that in the case of Gaussian noise Equation 27.2 is the 'canonical' loss (as it is linked to the maximum likelihood estimation), other loss functions (for example, $L(a,b)=\sum_{i=1}^{n}\left|y_i-ax_i-b\right|$) could be used as well, which would lead to different estimates of the parameters. In machine learning, for example, it is quite customary to decide on a loss function on its own merits, without necessarily attempting to connect it with the noise function.

By way of terminology, we will sometimes use $L(Y,f(X))=\|Y-f(X)\|^2$ as shorthand for the squared loss function and $L(Y,f(X))=\|Y-f(X)\|$ for the absolute loss function.

Let us now see what happens if, instead of having a single dependent variable, we have k different variables so that the model in Equation 27.1 becomes

$$Y=a_0+a_1X_1+\ldots a_kX_k+\varepsilon$$

which translates for the individual pairs into $y_i=a_0+\sum_{r=1}^{k}a_r x_{i,r}+\varepsilon_i$.

The best-fit function is, in this case, a hyperplane whose parameters are those that minimise the squared loss function

4 If the variance is different for each point (heteroscedasticity), $\varepsilon_i \sim N(0,\sigma_i^2)$, it is easy to see that the best-fit line is that which minimises $L(a,b)=\sum_{i=1}^{n}\left(y_i-ax_i-b\right)^2/\sigma_i^2$.

$$L(a_0, a_1, \ldots, a_k) = \sum_{i=1}^{n} \left(y_i - a_0 - \sum_{r=1}^{k} a_r x_{i,r} \right)^2$$

where

The solution to this minimisation problem is straightforward once expressed in matrix terms:

$$\beta = \left(\boldsymbol{X}^T \boldsymbol{X}\right)^{-1} \boldsymbol{X}^T \boldsymbol{Y}$$

where $Y = \{y_1, \ldots y_n\}^T$ is the vector of observations, $\boldsymbol{\beta} = \{a_0, a_1, \ldots a_k\}^T$ is the vector of the parameters, and

$$\boldsymbol{X} = \begin{pmatrix} 1 & x_{1,1} & \cdots & x_{1,k} \\ 1 & x_{2,1} & \cdots & x_{2,k} \\ \cdots & \cdots & \cdots & \cdots \\ 1 & x_{n,1} & \cdots & x_{n,k} \end{pmatrix}$$

is the input vector ($x_{i,j}$ represents the value of the ith variable in the jth observation, and the '1s' are included as a trick to produce the constant term a_0).

The multivariate linear model is quite powerful; however, the assumptions of a linear dependency and of Gaussian noise are too restrictive in many real-world situations. Generalised linear models provide a response to these limitations.

27.4.2 Beyond the Linear Model: Generalised Linear Models

There are three directions in which GLM extends the linear model:

i. Replacing the simple linear combination of inputs to describe the expected value of the dependent variable Y:

$$\mathbb{E}(Y) = a_0 + a_1 X_1 + \ldots + a_n X_n$$

with a *linear* combination η of a wider dictionary of functions $\psi_j (X_1, \ldots X_n)$:

$$\eta = \sum_j \beta_j \Psi_j (X_1, \ldots X_n)$$

Examples of admissible functions are

- $\psi_1 (X_1, \ldots X_n) = X_1$
- $\psi_2 (X_1, \ldots X_n) = X_1 X_2$
- $\psi_3 (X_1, \ldots X_n) = \log (X_1 + X_5)$ (odd but theoretically possible)
- $\psi_4 (X_1, \ldots X_n) = 1$ if X_1 = 'urban', 0 if X_1 = 'rural' (we will, however, introduce a more convenient shorthand for variables that can only assume a finite set of values)

ii. Introducing a further link function g, which transforms the responses so as to enforce certain constraints on the output (such as positivity). This allows to write the expected value of Y as $\mathbb{E}(Y) = g^{-1}[\eta]$ or, in extended form:

$$\mathbb{E}(Y) = g^{-1}\left[\sum_j \beta_j \Psi_j(X_1,\ldots X_n)\right] \tag{27.3}$$

iii. Replacing the Gaussian noise with a more general type of noise (also known as 'error structure') belonging to the so-called exponential family. This means that the observed values of Y, whose expected value is given by Equation 27.3, are assumed to distributed according to the following probability density distribution:[5]

$$f(y;\theta,\varphi) = \exp\left[\frac{\theta y - b(\theta)}{a(\varphi)} + c(y,\varphi)\right]$$

The *exponential family* includes (amongst others) Gaussian, binomial, Poisson and Gamma noise.

A distribution belonging to the exponential family has the following two properties (see Anderson et al. 2007b):

- The distribution is specified by its mean and variance.
- The variance of an output y_i Y_I is a function of its mean (in contrast to the basic linear model, where the variance is the same for each output). This dependence can be written as

$$\mathbb{V}(y_i) = \frac{\phi V\left(y_i^{\text{fitted}}\right)}{\omega_i}$$

where $V(x)$ is the so-called variance function, ω_i is a constant that assigns a weight to observation i (more on this later), and ϕ is a scale factor (also called dispersion parameter, and sometimes denoted as φ) for the variance.

Figure 27.4 shows how some of the best-known distributions can be parameterised as members of the exponential family by a suitable choice of the functions $a(\phi)$, $b(\theta)$ and $c(y,\phi)$.

Note that the distributions in Figure 27.4 also contain a variable called ω. This is the prior weight given to a particular observation. The reason why in general we need this term is that observations based on a larger number of exposures are more reliable. For example, if we are modelling the average severity of claims, the weight we give to each observation should reflect the number of claims over which that average has been calculated, and for this reason we might be using a Gamma noise with $a(\varphi) = \varphi/\omega$, where ω is the number of claims.

As in the case of classical least squares regression, the parameters of a given model $f(X) = \sum_j \beta_j \Psi_j(X_1,\ldots X_n)$ can be calculated by maximum likelihood estimation or, in other words, by minimising a log-likelihood loss function:

5 Informally, one can use (and we will do that at times) the notation $Y = g^{-1}\left[\sum_j \beta_j \Psi_j(X_1,\ldots X_n)\right] + \varepsilon$, with noise ε belonging to the exponential family. This notation doesn't mean that ε is *added* to the expected value $\mathbb{E}(Y)$, but simply that Y is sampled from an exponential family distribution with mean $\mathbb{E}(Y)$, and ε represents the residual (see later on).

Examples of Exponential Family Distributions

	$a(\phi)$	$b(\theta)$	$c(y, \phi)$
Normal	ϕ/ω	θ^2	$-\frac{1}{2}\left(\frac{\omega y^2}{\phi} + \ln(\frac{2\pi\phi}{\omega})\right)$
Poisson	ϕ/ω	e^{θ}	$-\ln(y!)$
Gamma	ϕ/ω	$-\ln(-\theta)$	$\left(\frac{\omega}{\phi} - 1\right)\ln\left(\frac{\omega y}{\phi}\right) - \ln(\Gamma(\omega/\phi))$
Binomial (m trials)	ϕ/ω	$\ln(1 + e^{\theta})$	$\ln\binom{m}{y}$
Exponential	ϕ/ω	$\ln(\theta)$	0

FIGURE 27.4
Parameterisation of a few popular distributions as members of the exponential family. Other distributions that are part of this family and that are sometimes used by practitioners are the Inverse Gaussian and the Tweedie distribution (see Anderson et al., 2007b for more details).

Linear model vs. Generalised Linear Model

	Linear model	Generalised linear model
The model	$\mathbb{E}(Y) = a_0 + \Sigma a_j X_j$	$\mathbb{E}(Y) = g^{-1}(\Sigma a_j \psi_j(X_1, X_2 \ldots X_n) =$
		$(\text{e.g.}) = \exp(a_1 X_1 + a_2 X_2 + a_3 X_1 X_2)$
The loss function	$L(Y, f(X)) = \lVert Y - f(X) \rVert^2$	$L(Y, f(X)) = -2\log \text{Pr}_{f(X)}(Y)$
The noise	Gaussian	Exponential family (Gaussian, Poisson, Gamma)

FIGURE 27.5
The main differences between the linear model and the GLM.

$$L[Y, f(X)] = -2\log \text{Pr}_{f(X)}(Y) \tag{27.4}$$

where the factor '2' has been introduced so that in the case of Gaussian noise, the loss function reduces to the standard squared loss of least squares regression (Figure 27.5).

In general, there will be no simple closed formula such as that for least squares regression to calculate these parameters. However, numerical methods such as the Newton-Raphson iterative algorithm for function optimisation can be (and are) used to achieve the same goal.

27.4.3 Frequently Used Models for Pricing

GLMs allow a large amount of flexibility in terms of functional form, link function and noise structure but for pricing purpose there are a couple of models that are most frequently used.

Claims frequencies (and claim counts) are usually modelled by a GLM with a logarithmic (or simply log) link function and Poisson noise. One good reason to use Poisson to model claims frequency is that the results of the model do not depend on the time interval chosen (e.g. days or years). The log link function has the effect that all outputs are positive (which is a desirable property) and that the effect of the factors is multiplicative, which has been observed to be a realistic assumption in the real world. The prior weights are normally given by the exposures (such as vehicle-years, number of employees, etc.) in the case of claim frequencies, and are set to 1 in the case of claim counts, for which the exposures are factored out (see discussion on offset terms in Section 27.4.4.3).

	Claim frequencies	Average claim amounts	Probability of renewal
Link function g(x)	$\ln(x)$	$\ln(x)$	$\ln[x/(1-x)]$
Error	Poisson	Gamma	Binomial
Scale parameter φ	1	Estimated	1
Variance function V(x)	x	x^2	$x(1-x)$
Prior weights ω	Exposure	No. of claims	1

FIGURE 27.6
An at-a-glance view of some standard GLMs for general insurance.

Claims amounts are usually modelled using Gamma noise. One good reason to use Gamma noise is that the results of the model do not depend on the currency used for the monetary amount. The log link function is chosen for claims amounts as well, for much the same reasons as above (positivity, multiplicativity). Note that because the Gamma distribution does not allow for zero claims, zero claims are removed from both the frequency and the severity and are modelled at a later stage if needed for expense purposes. Also, note that claims are normally capped as large claims can significantly distort the modelling process (Iwanik, 2011).

See Figure 27.6 for a quick review of some standard GLMs used in general insurance.

27.4.4 Model Structuring

We now look at how we can structure a GLM in practice, assuming we already know its variables and functional dependencies (the issue of selecting the right variables will be addressed in Section 27.4.5).

27.4.4.1 Data Preparation and Exploratory Analysis

Before delving into the actual process of model structuring and rating factor selection, it is useful to perform some exploratory analysis. That includes:

Data checking and cleansing. This in turn includes the standard checks that we discussed in Chapter 10 (missing data, unusual data points, extreme values …), with an eye to items that could cause problems for GLMs, e.g. losses with zero amounts when using a Gamma model for the severities.

Feature engineering (continuous vs categorical, number of levels, etc.). This is the point in which we decide, e.g. how many different levels we need to model age: shall we use a continuous function? Or a discrete function rounded by year? And can we group all years between 40 and 49 years in a single bucket?

One/two-way analysis. While one-way analysis is an insufficient tool for modelling, it is a great tool for running checks. It allows us to check whether there is enough data for each level after feature engineering, or whether there is enough information around a given factor in general. It also provides a first indication about the relevance of each factor – e.g. if we observed no difference in output for different levels of one factor we might suspect (albeit not conclusively) that that factor is not really relevant. Two-way analysis can also be run in some cases.

Analysis of the correlation of exposures (Cramer's V test). We saw how correlated exposures create spurious correlations between factors. The correlation between the exposures of a pair of categorical variables can be analysed through, e.g. the Cramer's V test. This is not directly used in the analysis – the GLM will work its way around that correlation – but it is useful to see why, for example, a one-way analysis is not sufficient. This test is based on the calculation of the following statistic:

$$V(X_1, X_2) = \sqrt{\frac{\chi^2}{N \times \min(\#\text{levels}_1 - 1, \#\text{levels}_2 - 1)}} \tag{27.5}$$

where N is the number of data points, $\#\text{levels}_1$ and $\#\text{levels}_2$ are the number of levels for variables X_1 and X_2 respectively, χ^2 (the chi-square statistic) is equal to:

$$\chi^2 = \sum_{i,j} \frac{\left(n_{i,j} - \frac{n_{i,\cdot} \times n_{\cdot,j}}{N}\right)^2}{\frac{n_{i,\cdot} \times n_{\cdot,j}}{N}}$$

and where in turn $n_{i,j}$ is the total number of exposures for which $X_1 = i$ and $X_2 = j$, $n_{i,\cdot} = \sum_k n_{i,k}$, $n_{\cdot,j} = \sum_h n_{h,j}$. Figure 27.8 will show an example of this calculation.

27.4.4.2 Modelling Continuous and Categorical Variables

GLMs can deal with both continuous and categorical variables. Examples of *continuous variables* are the annual mileage and the driver's age. These may or may not be related monotonically to the modelled variable; for example, claims frequency is likely to increase monotonically as the annual mileage increases, but the relationship with the driver's age is obviously more complicated (Ohlsson and Johansson, 2010).

Examples of how continuous variables may enter a GLM are analytical functions such as $\psi_1(X_1,\ldots X_n) = X_1$, $\psi_2(X_1,\ldots X_n) = X_4 X_7$, $\psi_3(X_1,\ldots X_n) = \log(X_1 + X_n)$, $\psi_4(X_1,\ldots X_n) = \sin(X_3 + \exp(X_4))$. The difficulty, of course, is that the functional form in which continuous variables should enter the model is not normally known a priori but needs to be discovered by expert judgment or simply by trial and error. For example, one can try many different functional forms and discard the unsuited by model selection: what remains will hopefully be a decent approximation of the true functional form.

If the continuous variable is not related monotonically to the modelled variable, or is likely to have an otherwise complex shape, finding the correct functional form is particularly challenging, and the standard approach is to transform the continuous variable into a discrete variable by grouping the continuous values into intervals. For example, age could be grouped into the following intervals: 17 to 20, 21 to 25, 26 to 35, 36 to 49, 50 to 64, 65 to 74, 75+, and treated as a categorical variable (more on categorical variables later). A more advanced approach is that of using *generalised additive models*, which approximate the functions with splines or other 'smoothers' (see Section 27.5).

Categorical variables are variables with a limited number of possible values (called 'levels') and they have *in general* no inherent order. Examples are sex, occupation, car type, and geographical location. As we have seen above, continuous variables are often transformed into categorical variables, and categorical variables are by far the most common type of variables for applications of GLM to general insurance (Anderson et al. 2007b).

Without loss of generality, we can assume that a categorical variable X with L levels is a variable that can take one of the integer values from 0 to $L-1$: $X \in \{0,1 \ldots L-1\}$. If X has no interactions with other variables, it can be included in the model as the linear combination

$$\Psi(X) = \sum_{j=1}^{L-1} \beta_j \delta_j(X)$$

where δ_j: N $\rightarrow$ {0,1} is the function such that $\delta_j(X) = 1$ if $X = j$, 0 otherwise.[6] Note that $\Psi(X) = 0$ if $X = 0$ (base level), and that this is equivalent to writing:

$$\Psi(X) = \beta_j \text{ if(and only if) } X = j$$

where $\beta_0 = 0$. (For example, if X represents occupation, with $X = 0$ standing for 'clerical' and $X = 1$ for 'manual', $\Psi(X)$ will have the value $\beta_0 = 0$ for a customer employed in an office and β_1 for a customer doing manual work.)

Note that there are L levels associated with the variable X but only $L - 1$ non-zero parameters $\beta_1 \ldots \beta_{L-1}$. As mentioned above, the level $X = 0$ represents a 'base level'. Therefore, we can say that variable X contributes $L - 1$ parameters (or degrees of freedom) to the overall model.

If categorical variables $X \in \{0,1 \ldots L-1\}$, $X' \in \{0,1 \ldots M-1\}$ are interacting, and they do not have interactions with other variables, the general form in which they enter the model is:

$$\Psi(X, X') = \sum_{\substack{i \in \{0,1 \ldots L-1\} \\ j \in \{0,1 \ldots M-1\} \\ i+j>0}} \beta_{i,j} \delta_i(X) \delta_j(X') \tag{27.6}$$

It is easy to see that the variables X, X' contribute amongst themselves $L \times M - 1$ parameters to the overall model.

Equation 27.6 can be easily generalised to interactions involving an arbitrary number of factors.

27.4.4.3 Bringing All Pieces Together

The typical way of writing an overall model with a log link function will be by adding terms like the ones described above plus an *intercept term* β_0 and an *offset term* $\Xi(X_{n+1}, \ldots X_{n+m})$, which were implicit in Equation 27.3 but are here stated explicitly:

$$\mathbb{E}(Y) = \exp\left[\beta_0 + \sum_j \beta_j \Psi_j(X_1, \ldots X_n) + \Xi(X_{n+1}, \ldots X_{n+m})\right]$$

27.4.4.3.1 The Intercept Term

Each term $\Psi_j(X_1, \ldots X_n)$ is built to be equal to zero for a particular combination (the same for all j) of variable values $X_1 = x_1, \ldots X_n = x_n$ (the *base levels*). Without loss of generality, we can assume that $x_j = 0$ for the categorical variables.

6 This is basically a slightly different way of writing Kronecker's delta.

The process by which we isolate the intercept term is an example of *(intrinsic) aliasing*. A treatment of aliasing is beyond the scope of this chapter: the interested reader is referred to Anderson et al. (2007b), on which the following brief summary is based on, or to McCullagh and Nelder (1990).

In a nutshell, aliasing is a process by which one removes parameters that are redundant because of linear dependencies among the independent variables, either because of dependencies inherent in the definition of the independent variables (*intrinsic aliasing*) or because one level of a given factor is perfectly correlated with one level of another factor (*extrinsic aliasing*). Standard GLM routines will perform intrinsic and extrinsic aliasing automatically. When the correlation between levels in extrinsic aliasing in large but not perfect we speak of '*near aliasing*', which can be dealt with heuristically.

27.4.4.3.2 The Offset Term

The term $\Xi(X_{n+1},\ldots X_{n+m})$ is again a term which is implicitly contained in Equation 27.3 but is introduced here to factor out the known effects of a variable. The classical example is when a variable, say X_{n+1}, represents the exposure and the claim count Y is known – or assumed – to be proportional to that exposure, and therefore we introduce a term $\Xi(X_{n+1}) = \ln(X_{n+1})$ which ensures that $\mathbb{E}(Y)$ is proportional to X_{n+1}.

27.4.4.4 Examples of Models

Consider the example of a model for the number of motor losses with three explanatory variables:

- X_1 (type of location, with two levels: 0, rural; and 1, urban)
- X_2 (type of vehicle, with five levels, 0–4)
- X_3 (annual mileage, continuous)
- X_4 (population density in the owner's postcode)

Assume Poisson noise and a logarithmic link function. Also assume that the number of losses is proportional to the annual mileage, and to an unknown power of the population density. Finally, assume that there is no interaction between the variables. The general form of such a model would be

$$\mathbb{E}(Y) = \exp\left[\beta_0 + \beta_1^{(1)}\delta_1(X_1) + \sum_{i=1}^{4}\beta_i^{(2)}\delta_i(X_2) + \ln(X_3) + \beta_1^{(4)}\ln(X_4)\right]$$

where β_0 is the intercept term, corresponding to the value of Y for $X_1 = 0$, $X_2 = 0$, and $X_3 = 1$, and δ is the function defined above.

The number of parameters of the overall model is therefore equal to 7: one for the *intercept term*, one additional parameter for *type of location*, four additional parameters for *type of vehicle*, one additional parameter for the *population density* (the exponent of the power law). No additional parameter is needed for the *annual mileage*, which can be interpreted as an offset term as the effect of this variable is fully known.

If we wanted to allow for the interaction between X_1 and X_2, the model would become

$$\mathbb{E}(Y) = \exp\left[\beta_0 + \sum_{\substack{i\in\{0,1\}\\ j\in\{0,1\dots4\}\\ i+j>0}} \beta_{i,j}\delta_i(X_1)\delta_j(X_2) + \ln(X_3) + \beta_1^{(4)}\ln(X_4)\right]$$

The number of parameters is now 1 (intercept parameter) + 9 (additional parameters for X_1, X_2) + 1 (additional parameter for X_4) = 11.

Note that if in the two examples above the noise model had been a two-parameter model such as Normal or Gamma, then the dispersion of the model should have also been calculated from the data, and the number of parameters would be 8 in the non-interacting and 12 in the interacting case.

More examples of model structuring are shown in the problems section, including examples with hybrid continuous/categorical interacting terms.

27.4.5 Model Selection

We have now spoken at some length about how moving from linear models to GLMs has considerably extended our modelling flexibility. However, we still haven't addressed the main problem, which is how to select the right model. For example, should we use as many variables as possible, so that the experimental data can be replicated almost perfectly, or should we limit ourselves to a simple model?

Before looking at answers to this question, it is perhaps useful to mention what is possibly the simplest example of model selection.

A Trivial Example of Model Selection: Univariate Feature Selection

This is based on applying a statistical test to each factor separately to assess its significance (hence the term 'univariate'). An example of such statistical test is that of calculating the p-value of each parameter based on an estimate of its standard error. This test is included in most GLM packages. Only the features whose p-value is below a certain value (e.g. 0.05) are kept.

This method is as simple as it gets: when applied to GLM, it simply requires running the GLM algorithm on the full model (with all variables and levels included), and only the significant variables/levels are kept. Once the rest are discarded, the model can be re-fitted to obtain a better calibration of the remaining parameters.

This method is quick but is also quite rough. One of the reasons is that by definition the method doesn't take into account the combined effect of variables – e.g. two or more variables may appear non-significant in isolation but may have explanatory power in combination.

To overcome the limitations of such a simple approach, it pays to take a look at the science behind model selection: machine learning.

Model Selection and Machine Learning

As we have already mentioned in Chapter 12, questions around model selection are better answered in the context of *machine learning*, whose objective is to build models that can predict the behaviour of a system based on experimental data (what we would call 'historical experience'). The central idea of machine learning is probably that there is a trade-off between complexity and prediction accuracy. Many people – especially in a business environment – hold (more or less implicitly) the view that the more complex a model is, the more accurate it is. Now this seems quite an extraordinary view that no actuary could hold. However, this view often comes in a more subtle version: if some factor is perceived as relevant to a phenomenon, it *must* be included in the model lest the model be too naïve a representation of reality. Is this still something which sounds unfamiliar to actuaries?

Most of us will remember Ockham's razor, which can be stated as saying that amongst a number of models (explanations) that explain a given phenomenon, one should always choose the simplest. And here is one quote from Albert Einstein, which partly counterbalances Ockham's principle: 'It can scarcely be denied that the supreme goal of all theory is to make the irreducible basic elements as simple and as few as possible without having to surrender the adequate representation of a single datum of experience' (usually rendered as 'everything should be made as simple as possible, but not simpler').

Interestingly, machine learning is able to put these philosophical ideas into mathematical form by considering the prediction error of a model as a function of the complexity of the model itself. This is well depicted in Figure 12.4. One key message of Figure 12.4 is that there is a complexity (the optimal complexity) at which the prediction error is minimal.

A sound model selection approach typically needs two ingredients:

a. a criterion to assess whether a model is better than another (validation against a hold-out sample, cross-validation, AIC, BIC...) – several of these criteria are approximations of the expected prediction error;
b. a mechanism to navigate the space of models (exhaustive search, greedy approach, regularisation ...).

We will look at these ingredients in detail in the next couple of sections, and then we'll look at examples of how these can be used in combination to select a good model.

27.4.5.1 Model Ranking Criteria

The first ingredient of a model selection methodology is a ranking criterion in the space of models, which tells us when one model is better than another. We already know that a loss function evaluated on the training set is not enough as a ranking criterion, because the unadulterated loss function (e.g. negative loglik) will tend to choose models that are too complex, without any real limiting principle. Examples of ranking criteria are discussed below.

27.4.5.1.1 Hold-out Sample Method

As we explained in Chapter 12, the cleanest way of selecting a model is by splitting the available data into a training set and a test/selection set (and possibly even a validation test). We can then navigate the model space with one of the techniques explained in Section 27.4.5.2, calibrate each model using the training set and calculating the loss function using the selection set. We can then select the model with the smallest loss function and assess its absolute performance against the validation set.

The main problem with the hold-out sample method is that it requires a large data set, large enough that we can afford to use only a subset for it for training.

27.4.5.1.2 Deviance Tests for Nested Models

Deviance tests are tests for comparing *nested models* based on the calculation of the *deviance* of the model. The deviance of the model is a measure of the distance between the model and the observations. It is a generalisation of the concept of the sum of squared errors as we find it in least squares regression.

Deviance comes in two flavours:

- the *scaled deviance*, which is simply defined as:

$$D^*_M = 2\times\left(\text{loglik}_{M_{\text{sat}}} - \text{loglik}_M\right)$$

  i.e. as the difference between the log likelihood of the saturated model M_{sat} (the model with as many parameters as the number N of data points, and therefore able to fit the data perfectly) and the log likelihood of model M, multiplied by 2;
- the *unscaled deviance* of model M, defined as

$$D_M = 2\times\hat{\phi}\times\left(\text{loglik}_{M_{\text{sat}}} - \text{loglik}_M\right) = \hat{\phi}\times D^*_M$$

  i.e. the scaled deviance multiplied by the estimated dispersion parameter (scale factor): $D_M = \hat{\phi}\times D^*_M$.

The strategy for comparing two nested models is different depending on whether the dispersion parameter needs to be estimated (e.g. in the case of Gaussian noise) or is already known (e.g. in the case of Poisson noise, where the dispersion parameter is 1).

If the dispersion parameter is known, it can be shown that the difference in the deviance between two nested models M_1 and $M_2 \supset M_1$ (the symbol $\supset$ is here to be interpreted as meaning that M_2 includes all variables of M_1 and at least one more) follows a chi-square distribution with degrees of freedom equal to the difference Δp in the number of parameters between the two models:

$$D^*_{M_1} - D^*_{M_2} \sim \chi^2_{\Delta p}$$

We can then check whether the additional parameters Δp have brought a significant reduction to the deviance to justify the added complexity, by checking whether the probability that

$$\Pr\left(\chi^2_{\Delta p}\right) > \text{observed difference in deviance}$$

is above a given significance level (say, 5%).

If the dispersion parameter is not known but has to be estimated (e.g. Gaussian noise, Gamma noise), one can instead use the F-test. The F-test is again a test for comparing *nested models* based on the F-distribution.[7]

The F-statistic for the comparison of models M_1 and $M_2 \supset M_1$ (the symbol $\supset$ is here to be interpreted as meaning that M_2 includes all variables of M_1 and at least one more) is then calculated as the following quantity:

7 The F-distribution ('F' stands for Fisher, the famous statistician) is the distribution of the ratio of the random variable $F = \dfrac{U_1/v_1}{U_2/v_2}$, where U_k is a random variable with a chi-square distribution with v_k degrees of freedom (see Freund, 1999).

$$F = \frac{D_1 - D_2}{\Delta p \times \hat{\phi}_2}$$

where $\Delta p = p_2 - p_1$ is the number of new parameters introduced by going from M_1 to M_2. If the value of F is larger than the value $F^{\alpha}_{\Delta p, N-p_2}$ of the F-distribution with Δp and $N - p_2$ degrees of freedom, at a chosen significance level α (say, 5%), then we can judge that the new variable(s) introduced are significant and we can retain them. In the case of categorical variables, note that the addition of a variable translates into the addition of $L-1$ new parameters, where L is the number of levels of the categorical variable.

27.4.5.1.3 Analytical Criteria: AIC and BIC

A simpler way is to use the Akaike information criterion (AIC), which we have also introduced in Chapter 12. This is given by AIC = −2 loglik + 2d, where d is the number of parameters in the model. We have already mentioned that thanks to the penalty term '+2d', the model with the smallest AIC achieves a compromise between fit and complexity – and hence AIC can be used as a selection criterion. One of the reasons why this works is that (for a certain class of models that include linear models) the prediction error is related asymptotically to the AIC (Hastie et al., 2001):

$$\text{Error} \to -\frac{2}{N}\log\text{lik} + 2\frac{d}{N} = \frac{\text{AIC}}{N} \text{ for } N \to \infty$$

where N is the number of data points. In other terms, the prediction error can be approximated by the AIC for large values of N. In practice, the fact that this holds only for large values of N is frequently disregarded, and this criterion is used for all values of N. An easy way to improve on this is to use a small sample correction on AIC as advocated, for example, in the book by Burnham and Anderson (2002). The corrected criterion is AICc = AIC + $2k(k+1) / (N-k-1)$, where k is the number of parameters in the model and N is the number of data points.

The AIC is an example of an *analytical criterion* – i.e. a criterion that uses a simple function to compare two models. Another example of analytical criterion is to choose the model that minimises the Bayesian Information Criterion (BIC), which is defined as

$$\text{BIC} = -2\text{loglik} + (\log N)d$$

Where N is defined as above and log is the natural logarithm. Note that the penalty for the BIC is more severe (apart from the case where N is less than 8) than for the AIC, so this criterion will tend to select models that are simpler. There is an asymptotic result for this criterion as well – it can be proved (Hastie et al., 2001) that the probability that the correct model will be chosen (among a finite set of models *which includes the true one*) goes to 1 as the number of data points goes to infinity. This is to be contrasted with AIC, which tends to select models that are overly complex as the number of data points goes to infinity (Hastie et al., 2001).

Despite this theoretical result that seems to give a clear edge to BIC with respect to AIC, in practice the relative performance of the two methods is not straightforward: first of all, the result above is only valid when the set of models considered includes the true model (which is almost never the case); secondly, the result is indeed asymptotic, and BIC tends to choose models that are overly simple for relatively small samples.

Two other examples of analytical criteria are the MDL (briefly discussed in Chapter 12), Mallow's C_p (equivalent to the AIC under certain circumstances) – however, the list of possible criteria is long, as a simple web search will reveal.

27.4.5.1.4 Cross-Validation

Cross-validation is one of the standard techniques – if not *the* standard technique – for model selection in machine learning. It is a variant of the hold-out sample method, but it uses just one sample, which it splices and dices it dynamically so that different subsets play the role of training set and test (selection) set – as a result, it allows to use all data points in turn as the training set and requires less data to work.

More in detail, cross-validation, or more specifically K-fold cross-validation, estimates the prediction error by dividing at random the data set into K different subsets (typical values of K are $K = 5$ or $K = 10$). Each subset $k = 1, \ldots K$ is in turn removed from the data set and the model is fitted to the remaining $K - 1$ subsets. The kth subset is used as a test set. The process is repeated for all K subsets and the cross-validation estimate of the prediction error is given by the average log-likelihood over the K different subsets:

$$\mathrm{CV}^{(K)} = \frac{1}{K}\sum_{k=1}^{K} L_k$$

where L_k is calculated using Equation 27.4 for the kth subset. For model selection purposes, the cross-validation estimate of the prediction error, $\mathrm{CV}^{(K)}$, is then calculated for different models of different complexities (in our case, the number n of parameters in the GLM), selected with forward selection, and the model with the lowest value of $\mathrm{CV}^{(K)}(n)$ (which corresponds to the minimum of the prediction error curve shown in Figure 12.4) is selected.

Cross-validation is a more general technique than AIC and although it also overestimates the prediction error, it does so to a much lesser extent than the AIC and similar criteria (Hastie et al. 2001).

27.4.5.2 Navigating the Model Space

This section looks at different ways in which different models are taken up for consideration. Note that these methods are indeed only methods to navigate the space of models – they assume that we have a way to compare models and decide which one is better (Section 27.4.5.1).

27.4.5.2.1 Exhaustive Search

When the number of candidate models is finite, the most obvious way of navigating the space of models is through exhaustive search. As the name suggests, this simply means trying all the models and for each of them calculating the ranking criterion. This method has the obvious advantage of ensuring that we will produce a global solution, because no model is left behind. The main disadvantage is also obvious, and it's called 'combinatorial explosion'.

For example, if the only possible functions are $\psi_1(x_1), \ldots \psi_n(x_n)$ and the link function is the identity function, so that all models are straightforward linear combinations of some of the functions $\psi_j(x_j)$, there are still 2^n possible models to try, ranging from $\psi(x) \sim 1$ to $\psi(x) \sim$

$\psi_1(x_1) + \ldots + \psi_n(x_n)$ – an exponential function of the number of functions. And this is before we even consider the possibility of interacting variables and other complications.

27.4.5.2.2 Greedy Approach (Forward/Backward Selection)

In GLM, the specific model for a set of data is usually selected using a so-called 'greedy approach'.[8] This comes in two main varieties, *forward* and *backward* selection. Forward selection starts from the simplest model, Y = constant, and adds at each step the function $\psi_j\,(X_1,\ldots X_n)$, which reduces the loss function (or another suitable distance) by the largest degree.[9] Backward selection starts from the most complex model and removes functions one by one: however, this method can only be applied when the dictionary of functions is finite.

To decide whether the selected additional function in forward selection is actually an improvement, one needs to measure the prediction error. This is difficult! To do this properly, we need an independent data set (the validation set) against which to test the various models parameterised using the training set. By calculating the loss function repeatedly on the validation set, we can actually get the empirical curve that gives the prediction error in Figure 12.4.

Note that hybrid forms of greedy selection are possible. While forward selection starts from the simplest model up and backward selection starts from the fullest model down, one can start for example for example from the legacy model, i.e. the model currently in use by the company (or, say, the model derived using a univariate feature selection approach), and try to move up with forward selection and down using backward selection until finding the model smallest error.

The method of forward/backward selection is straightforward but has an important drawback, which is a common feature of greedy algorithms: *it does not necessarily yield the global optimum* because it does not explore the full model landscape but follows only the most promising lead by adding at each step the variable that for example reduces the AIC by the largest amount, possibly ending up in a *local optimum point*. In some cases, however, finding the best solution would require adding a variable that is not the one that decreases the AIC by the largest amount, but that eventually takes you to the global optimum.

27.4.5.2.3 Sparsity-Inducing Regularisation

Regularisation (which we introduced in Chapter 12) replaces the problem of minimising the loss function (say, $L\left[Y,f(X)\right] = -2\text{loglik}$) with the problem of minimising a regularised loss function:

$$L_{reg}\left[Y,f(X)\right] = -2\ \text{loglik} + \lambda g(\beta)$$

where β is the set of possible parameters of the model $f(X)$, $g(\beta)$ is a function of these parameters and λ is called the smoothing (or regularisation) parameter. The net effect

8 A greedy approach, or algorithm, is one that subdivides a problem into stages and makes the locally optimal choice for that stage, although this may not result in a global optimum. The Wikipedia entry for greedy algorithm (http://en.wikipedia.org/wiki/Greedy_algorithm) has some good examples.

9 Assuming that the dictionary of functions is finite. When the dictionary is (infinitely) countable, we need to impose constraints on how we add functions. For example, we may demand that at each step we only add one extra variable of the n variables allowed.

of regularisation is finding a solution which loses something in terms of being able to replicate the behaviour of the data but smoothes out some of the noise and is hopefully better at producing a more predictive model. The higher the value of λ is, the smoother the model.

By choosing a suitable function $g(\beta)$ carefully, we can perform feature selection at the same time as parameter calibration. For example, by using lasso regularisation $\left(g(\beta)=\|\beta\|\right)$ or elastic net regularisation $\left(g(\beta)=\|\beta\|+c\|\beta\|^2\right)$, some of the parameters in the optimal solution are set to zero. Therefore we talk about 'sparsity-inducing regularisation'. The advantage of this is that for a specific value of λ, the solution can be found by numerical techniques such as gradient descent, and feature selection is performed in one swoop without the need for explicitly trying different combinations of variables.

27.4.5.3 The Model Selection Process

By combining a model ranking criterion (Section 27.4.5.1) with a method for navigating the space of possible models (Section 27.4.5.2) we now have a selection process. Three classical examples of processes used in practice are:

- Forward selection using a deviance test to compare successive (nested) models
- Forward selection using AIC to compare models
- Lasso regularisation using cross-validation

However, virtually any combination of ranking criteria and navigation methods will produce a valid model selection process.

27.4.6 Model Structure Validation

The model selection process should yield the model with the lowest prediction error, and the size of this prediction error gives a first level of validation, especially if it's done with the proper approach as recommended by machine learning. Ideally, in a data-rich situation the prediction error should be measured by using a validation set separate from both the training set and the selection set; in other cases we'll have to make do with the best selected model.

Apart from this, there are other diagnostics for validating the structure of the model and the data used to build it. Examples of these diagnostics are mentioned in McCullagh and Nelder (1990) and Anderson et al. (2007).

27.4.6.1 Checking the Noise Function

When we build a GLM we have to make assumption about the noise function (Gaussian, Poisson, Gamma ...). To check that this assumption is appropriate, the most natural approach is to check whether the (raw) residuals $y_i^{\text{observed}}-y_i^{\text{fitted}}$ (y_i^{observed} being the i-th observed data points and y_i^{fitted} being the value predicted by the model based on the value of the independent variables for observation i: $y_i^{\text{fitted}}=\mathbb{E}\left(Y|\,X_1=x_1^{(i)},\ldots\right)$) are indeed distributed according to the selected noise structure.

In practice, analysts do not use the raw residuals above but a normalised version of the residuals that allows to compare observations with different variances, weights, and relevance of the observation. One popular normalisation is achieved by using the standardised Pearson residuals,[10] which is basically the raw residual divided by the adjusted standard deviation:

$$r_i = \frac{y_i^{\text{observed}} - y_i^{\text{fitted}}}{\sqrt{\frac{\varphi}{\omega_i} V\left(y_i^{\text{fitted}}\right)}}$$

Now if the model form is suitable the standardised deviance residuals should be distributed approximately as a Gaussian distribution with zero mean and unit variance (as can be expected, this approximation breaks down for low values of y_i^{fitted}). If the noise structure is incorrect – if, for example, we use Gaussian noise when we have an underlying Poisson process – the residual will show a strong departure from that distribution: e.g. the residual may not be distributed symmetrically around zero but skewed. This may already be noticeable by looking at a residual plot, which shows the residual for each fitted value. Figure 27.12 shows examples of residual plots.

27.4.6.2 Checking the Variance Function

This can also be done by using the residual plots and statistics – specifically, by inspecting the way that the standardised deviance residuals depend on the fitted value. If the variance function is the correct one, the residual plot should show look uniform across the range of fitted value, with variance approximately constant. If, for example, we choose a constant variance where the underlying noise structure is Poisson (and therefore with variance linearly increasing with the fitted value) the residual plot will show a dispersion that increases with the square root of the fitted value.

Note that this check overlaps with that described in the previous section but not completely, as it is perfectly possible, for example, to build a model with Gaussian noise and linearly increasing variance if we so desire.

27.4.6.3 Checking the Link Function

To examine the appropriateness of the link function we can use the Box-Cox transformation. This means assuming that the link function has this functional form:

$$g(\lambda; x) = \begin{cases} \dfrac{x^\lambda - 1}{\lambda} & \lambda \neq 0 \\ \ln(x) & \lambda = 0 \end{cases}$$

and repeating model fitting for different values of λ. If the highest likelihood is achieved for $\lambda \sim 0$, this points to a log link function; if it is achieved for $\lambda \sim 1$, this points to an identity link function with a base level shift; for $\lambda \sim -1$, this points to an inverse link function with a base level shift.

10 Anderson et al. (2007) have a slightly different formula, which gives different weights depending on how much each data point influences the final outcome ('leverage'). We will ignore this complication here.

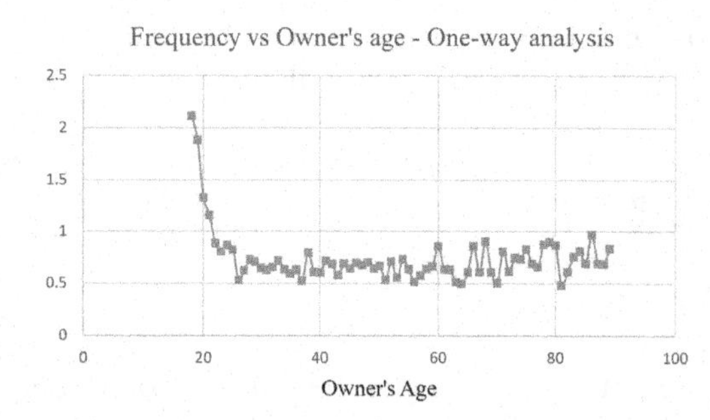

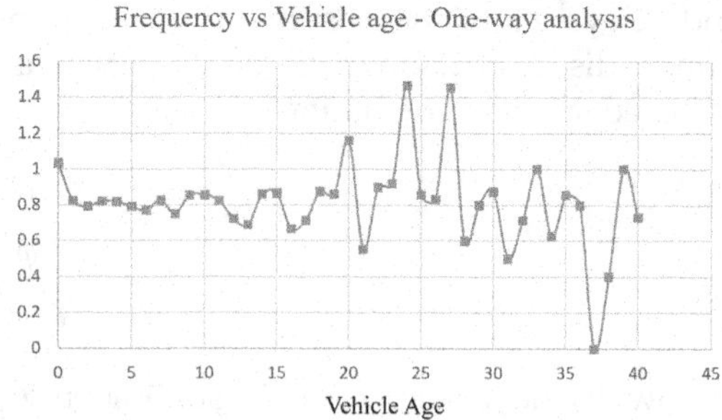

FIGURE 27.7
One-way analysis for the variables Owner's Age (left) and Vehicle Age (right). Note the erratic behaviour of Vehicle Age towards the end of the chart, a consequence of scarcity of data points with large vehicle ages.

27.4.6.4 Checking for the Outsized Influence of Individual Data Points

Methods (such as leverage checks, consistency checks, influence checks) are also available to identify isolated data points that have a dramatic influence on the outcome of the model. Once identified, these points may be removed from the data set and the GLM may be re-calibrated. We do not get into the details of this but the interested reader is referred to the usual sources (McCullagh and Nadler, 1983; Anderson et al., 2007).

27.4.6.5 Final Validation (Actual vs Expected)

Ideally the checks outlined in this section are best performed during the model selection process, so that the selection of the best model may be redone with the necessary corrections (noise, variance, link function, and reduced data set).

Once the final selection has been made, the selected model can be validated based on its predictive ability against – ideally – a test set separate from the training and selection sets (see Chapter 12).

27.4.7 A Practical General Insurance Example: Modelling Claims Frequency for Motor Insurance

In this section, we will look at a simple but common example of how GLMs are used: the modelling of claim count data. We will use synthetic data, i.e. artificially generated data. Synthetic data are useful in certain circumstances because they are a first test for a modelling technique: if that technique works on synthetic data, it *might* also work on real-world data. If it doesn't even work on synthetic data, there is basically no hope that it will be useful in real-world situation.

Section 27.4.7.11 has a summary description of the 'hidden' model from which the synthetic data was produced. The details of the model and the data set are available in the Downloads section of the books's website (Parodi, 2016) for the readers who wish to run the analysis by themselves and make changes to the parameters.

27.4.7.1 The Data Set

The data set records the number of losses for a relatively small number of policies (20,000). It also records the value for each policy of the following variables (this particular choice of factors is not necessarily the wisest, but it will do for illustration purposes):

	Rural	Urban		Electric/Diesel	Petrol
Age Band 1	772	6,164	Age Band 1	2,805	4,131
Age Band 2	1,085	5,917	Age Band 2	2,823	4,179
Age Band 3	7,170	18,456	Age Band 3	10,202	15,424
Age Band 4	2,196	2,788	Age Band 4	1,988	2,996
Age Band 5	2,328	3,124	Age Band 5	2,210	3,242

FIGURE 27.8
The table on the left shows the joint distribution of exposures for the pair of variables Age Band and Location Type, for which V=23.6% (which with 50,000 data points amounts to a significant exposure correlation) – this makes sense because younger drivers are overrepresented in an urban setting. The table on the right shows negligible exposure correlation (V=0.6%).

- Owner's Age at the start of the policy (an integer or floating number ≥18)
- Location Type (urban or rural)
- Region (one of five regions in which the territory is subdivided)
- Vehicle Age at the start of the policy (an integer or floating number ≥0)
- Fuel type (Electric/Diesel or Petrol)
- Number of no-claim years at the beginning of the policy (0, 1, 2, 3, 4+)
- Dummy1 (0 or 1)
- Dummy 2 (D1, D2, D3)
- Exposure (i.e. fraction of the year for which the policy was on the book)

The name of the variables Dummy1 and Dummy2 gives away the game that these are variables that are not actually correlated with the output data but are introduced to check whether the selection algorithm prunes them out.

Note that the average number of losses across all policies was chosen not to be necessarily realistic but to ensure that meaningful statistics could be produced even with a small number of policies.

27.4.7.2 Feature Engineering

Two of the variables (Owner's age and Vehicle age) can be considered continuous, and we may want to transform them into categorical variables. An exploratory one-way analysis can help us decide what the best way of categorising is.

Based on the graphs above, we decide to create these categories for Owner's Age:

Age Band	*Range*
AgeBand1	18–21
AgeBand2	22–26
AgeBand3	27–60
AgeBand4	61–75
AgeBand5	76+

and these categories for Vehicle Age:

Vehicle Age Band	*Range*
VehAgeBand	00
VehAgeBand	11
VehAgeBand	22+

27.4.7.3 Analysis of Exposure Correlation

We use the Cramer's V statistic (Equation 27.5) to identify the presence of exposure correlation between pairs of categorical variables. As already mentioned, this has no direct impact on the selection and calibration of the model but will enable us to see to what extent one-way analysis is sufficient to capture the behaviour of the model. Figure 27.8 shows Cramer's V statistic for two pairs of variables: Age Band vs Location Type and Age Band vs Fuel Type. The exposure correlation is significant for the first pair, showing that one-way analysis of both the Age Band variable and the Location Type variable is unlikely to tell the whole story.

27.4.7.4 The Model Structure

We are now in a position to build a model with up to 8 categorical variables. Since the model will represent claim counts, we are going to use a multiplicative model (log link function) with Poisson noise. We will use an offset function to take account of the exposure. For simplicity, we will assume no interactions between terms.

The simplest model with the structure above is

$$\mathbb{E}(Y) = \exp[\beta_0]$$

while the most complex model (without interaction terms) is

$$\begin{aligned}\mathbb{E}(Y) = \exp\Bigg[\beta_0 &+ \beta_1^{(\text{LocType})}\delta_1\left(X_{\text{LocType}}\right) + \sum_{i=1}^{4}\beta_i^{(\text{AgeBand})}\delta_i\left(X_{\text{AgeBand}}\right) \\ &+ \sum_{i=1}^{4}\beta_i^{(\text{Region})}\delta_i\left(X_{\text{Region}}\right) + \sum_{i=1}^{2}\beta_i^{(\text{VehAgeBand})}\delta_i\left(X_{\text{VehAgeBand}}\right) \\ &+ \beta_1^{(\text{FuelType})}\delta_1\left(X_{\text{FuelType}}\right) + \sum_{i=1}^{4}\beta_i^{(\text{NCB})}\delta_i\left(X_{\text{NCB}}\right) + \beta_1^{(\text{Dummy1})}\delta_1\left(X_{\text{Dummy1}}\right) \\ &+ \sum_{i=1}^{2}\beta_i^{(\text{Dummy2})}\delta_i\left(X_{\text{Dummy2}}\right) + \ln(\text{Exposure})\Bigg]\end{aligned} \tag{27.7}$$

Note that the shorthand way of writing the two models above (adopted for example by the R package glm) is to write them as:

```
lnY ~ LocationType
```

and

```
lnY ~ LocationType + VehAgeBand + AgeBand + NCB + Region +
FuelType + Dummy1 + Dummy2
```

This way of writing misses a lot of important information about the models (about, e.g. whether one variable is categorical or continuous), but it is indeed a convenient shorthand that allows to represent the output information in a compact way.

27.4.7.5 Model Selection Using Univariate Feature Selection

The most straightforward way of selecting the model is that of starting from the complete model (Equation 27.7) and remove the factors (or some of the categories of the factor) for which the probability that they are different from zero by chance is above a certain threshold, e.g. 5%.

In our case (see box below), all the coefficients are significantly different from zero except for:

- Dummy1
- Dummy2

We can therefore exclude these three variables and re-run the model with the remaining variables, and keep that as our best model.

```
glm(formula = y ~ LocationType + VehAgeBand + AgeBand + NCB + Region
+ FuelType + Dummy1 + Dummy2, family = poisson(),data = input_
data, offset=log(exposure))

Deviance Residuals:

     Min        1Q    Median       3Q       Max
 -3.0281   -0.9751   -0.3529   0.4791    4.2232
```

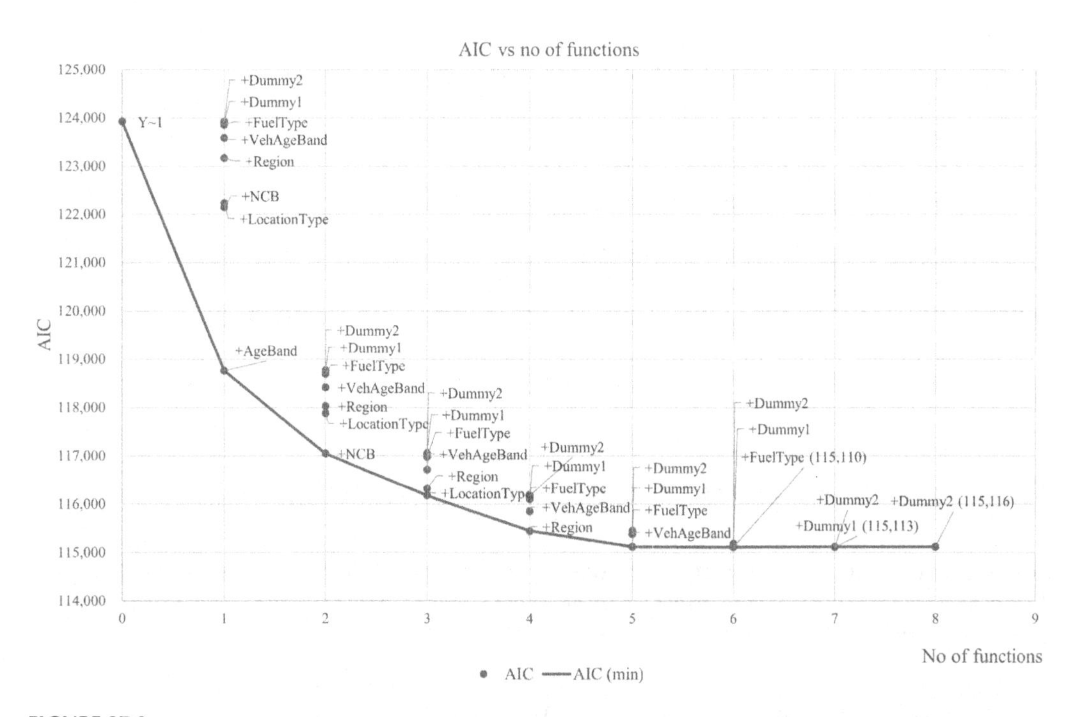

FIGURE 27.9
To illustrate the process of model selection, this chart shows the AIC value for all models tested via forward selection. The models are grouped by the number of factors included (regardless of the number of levels) and the models with the lowest AIC for every number of factors are connected with a line. All the models in each group are obtained from the winner from the previous group by adding a new factor. The winning model is that corresponding to six factors.

```
Coefficients:
                    Estimate  Std. Error z value  Pr(>|z|)
(Intercept)         0.960555  0.028685    33.486  < 2e-16 ***
LocationTypeUrban   0.333788  0.011791    28.308  < 2e-16 ***
VehAgeBandVA1      -0.298134  0.020117   -14.820  < 2e-16 ***
VehAgeBandVA2      -0.348395  0.018647   -18.684  < 2e-16 ***
AgeBandA2          -0.566290  0.014773   -38.332  < 2e-16 ***
AgeBandA3          -0.768750  0.011400   -67.435  < 2e-16 ***
AgeBandA4          -0.722469  0.018464   -39.128  < 2e-16 ***
AgeBandA5          -0.606583  0.017071   -35.533  < 2e-16 ***
NCBC2               0.163235  0.015926    10.249  < 2e-16 ***
NCBC3               0.351325  0.013553    25.922  < 2e-16 ***
NCBC4               0.492034  0.017415    28.253  < 2e-16 ***
NCBC5               0.592794  0.016976    34.920  < 2e-16 ***
RegionB             0.119910  0.012478     9.610  < 2e-16 ***
RegionC            -0.071656  0.011902    -6.021  1.74e-09 ***
RegionD             0.194415  0.015307    12.701  < 2e-16 ***
RegionE            -0.271800  0.018412   -14.762  < 2e-16 ***
FuelTypePetrol      0.071112  0.009267     7.674  1.67e-14 ***
Dummy1              0.006053  0.009461     0.640    0.522
Dummy2D1            0.002173  0.011910     0.182    0.855
Dummy2D2           -0.002954  0.013015    -0.227    0.820

---
Signif. codes: 0 '***' 0.001 '**' 0.01 '*' 0.05 '.' 0.1 ' ' 1
(Dispersion parameter for poisson family taken to be 1)
Null deviance: 59288 on 49999 degrees of freedom
Residual deviance: 50446 on 49980 degrees of freedom
AIC: 115116
```

Note how the software automatically sets to zero the coefficient of the first category for each factors (AgeBand1, NCB1 …).

27.4.7.6 Model Selection Using Exhaustive Search and Any Ranking Criterion

As discussed earlier, a simple way to select the right model among a finite set that is to try *all* possible combinations of factors. However, for our example this means trying $2^8 = 512$ different models,[11] and many more if interactions terms are included. We will therefore ignore this approach and use a variety of heuristics instead.

27.4.7.7 Model Selection Using the Greedy Approach (Forward Selection) with the Akaike Information Criterion (AIC)

A more scientific method is to use the forward selection methodology with the AIC. This means that we start from the simplest model ($\ln \mathbb{E}(Y) = \beta_0$) and at each step we add the

11 Every non-interacting model can be written down as an 8-digit binary number where the k-th digit is 1 if factor k is included, 0 otherwise.

variable that leads to the model with the smallest AIC. Figure 27.9 shows the process in summarised format.

This chart is reminiscent of the bias-variance trade-off chart (Chapter 12) – however, the number of factors is not an exact measure of complexity, as different categorical factors have different numbers of levels.

At the end of the process the winning model is `lnY ~ LocationType + VehAgeBand + AgeBand + NCB + Region + FuelType`, which includes all factors except for Dummy1 and Dummy2, and whose parameters are as follows:

```
Call:
glm(formula = y ~ LocationType + VehAgeBand + AgeBand + NCB +
Region + FuelType, family = poisson(), data = input_data,
offset = log(exposure))

Deviance Residuals:

    Min         1Q     Median        3Q       Max
-3.0222    -0.9751    -0.3528    0.4786    4.2212

Coefficients:

                      Estimate  Std. Error  z value   Pr(>|z|)

(Intercept)           0.964784  0.026513     36.389   < 2e-16 ***
LocationTypeUrban     0.333748  0.011791     28.306   < 2e-16 ***
VehAgeBandVA1        -0.298215  0.020115    -14.826   < 2e-16 ***
VehAgeBandVA2        -0.348482  0.018645    -18.691   < 2e-16 ***
AgeBandA2            -0.566274  0.014773    -38.331   < 2e-16 ***
AgeBandA3            -0.768775  0.011399    -67.440   < 2e-16 ***
AgeBandA4            -0.722531  0.018464    -39.132   < 2e-16 ***
AgeBandA5            -0.606629  0.017071    -35.536   < 2e-16 ***
NCBC2                 0.163231  0.015926     10.249   < 2e-16 ***
NCBC3                 0.351301  0.013552     25.922   < 2e-16 ***
NCBC4                 0.492160  0.017414     28.262   < 2e-16 ***
NCBC5                 0.592959  0.016974     34.933   < 2e-16 ***
RegionB               0.119940  0.012477      9.613   < 2e-16 ***
RegionC              -0.071666  0.011901     -6.022   1.73e-09 ***
RegionD               0.194511  0.015307     12.707   < 2e-16 ***
RegionE              -0.271741  0.018411    -14.759   < 2e-16 ***
FuelTypePetrol        0.071137  0.009267      7.677   1.63e-14 ***

---
Signif. codes: 0 '***' 0.001 '**' 0.01 '*' 0.05 '.' 0.1 ' ' 1
(Dispersion parameter for poisson family taken to be 1)

    Null deviance: 59288 on 49999 degrees of freedom
Residual deviance: 50446 on 49983 degrees of freedom
AIC: 115110
```

So we can see that *in this case* the simple method of removing statistically not significant coefficients at the 5% significance level yields the same model as the more sophisticated forward selection method. This is not the case in general.

27.4.7.8 Model Selection Using the Greedy Approach with a Deviance Test

A variant of the method in the previous section is using forward selection with a deviance test. Since for the case of Poisson noise the dispersion factor is known the test to apply is the chi-square test.

This works similarly to forward selection with AIC, except that in this case the test is valid only for nested models: therefore, every time we add a variable there is no natural way of selecting the best variable (among the ones for which the deviance test shows that the addition of such variable reduces the deviance significantly). Naturally, one can still use one of various heuristics so as to proceed without ambiguities, such as using the variable that reduces the deviance by the largest amount, either as an absolute amount or divided by the number of parameters)

We will not provide here a fully worked-out example of this approach. However, to illustrate how it works, let's assume we have already chosen the first three variables:

```
model_a: y ~ AgeBand + NCB + LocationType
```

which has scaled deviance $D^* = 51,569$, and we need to decide which variable (if any!) among `Region, VehAgeBand, FuelType, Dummy1, Dummy 2` we should add.

The options are therefore six, which are listed below in ascending order of deviance:

```
y ~ AgeBand + NCB + LocationType
                 Df Deviance
+ Region          4    50830
+ VehAgeBand      2    51248
+ FuelType        1    51506
+ Dummy1          1    51568
+ Dummy2          2    51568
+ <none>          0    51569
```

Since the model that causes the largest reduction in deviance (both in absolute terms and divided by the number of parameters) is

```
model_b: y ~ AgeBand + NCB + LocationType + Region,
```

we can run a chi-square test that compares this model and the one without the variable Region:

```
anova(model_a,model_b,test='Chisq')
```

The results are as follows:

```
Analysis of Deviance Table

Model 1: y ~ AgeBand + NCB + LocationType
Model 2: y ~ AgeBand + NCB + LocationType + Region
```

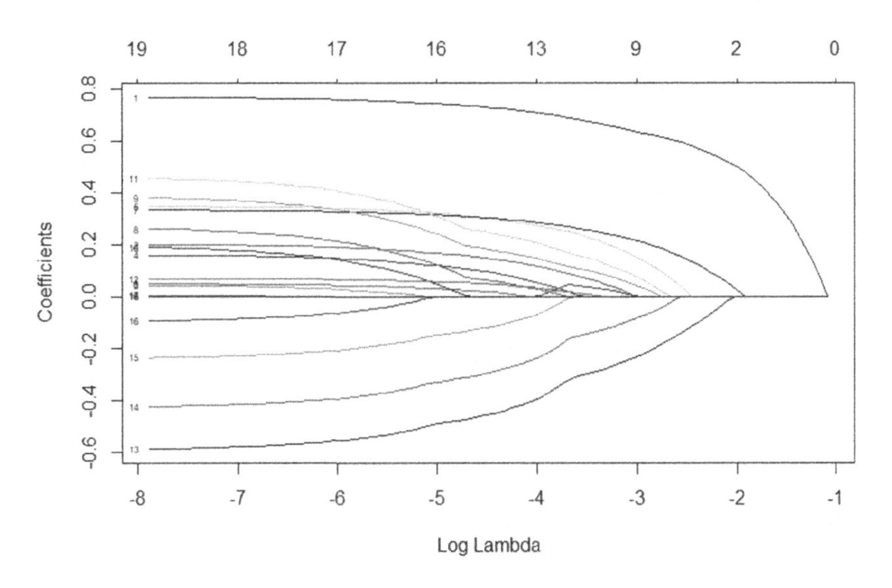

FIGURE 27.10
With lasso regression, the number of factors selected depends on the value of the regularisation parameter λ. New factors (or components of factors) are introduced gradually as the value of λ is reduced, from right to left. For low-enough values of λ all factors are part of the selected model.

```
Resid. Df Resid. Dev Df Deviance Pr(>Chi)
1       49990       51569
2       49986       50830  4  739.18 < 2.2e-16 ***
---
Signif. codes: 0 '***' 0.001 '**' 0.01 '*' 0.05 '.' 0.1 ' ' 1
```

The results show that the additional variable is significant, with the probability that the difference in deviance be so high by chance estimated to be less than 2.2×10^{-16}. Note that if instead of the Region we had chosen one of the variables VehAgeBand, FuelType we would still have obtained similar results, with different probabilities. However, if we chose the model with the Dummy1 variable and we ran a chi-square test:

```
model_c: y ~ AgeBand + NCB + LocationType + Dummy1
anova(model_a,model_c,test='Chisq')
```

then the chi-square test would give the following results:

```
Model 1: y ~ AgeBand + NCB + LocationType
Model 2: y ~ AgeBand + NCB + LocationType + Dummy1
  Resid. Df Resid. Dev Df Deviance Pr(>Chi)
1       49990         51569
2       49989         51568  1  0.79309   0.3732
```

showing that the probability of obtaining the observed difference in deviance is way above a 5% significance level. This indicates that the variable is unlikely to have explanatory power.

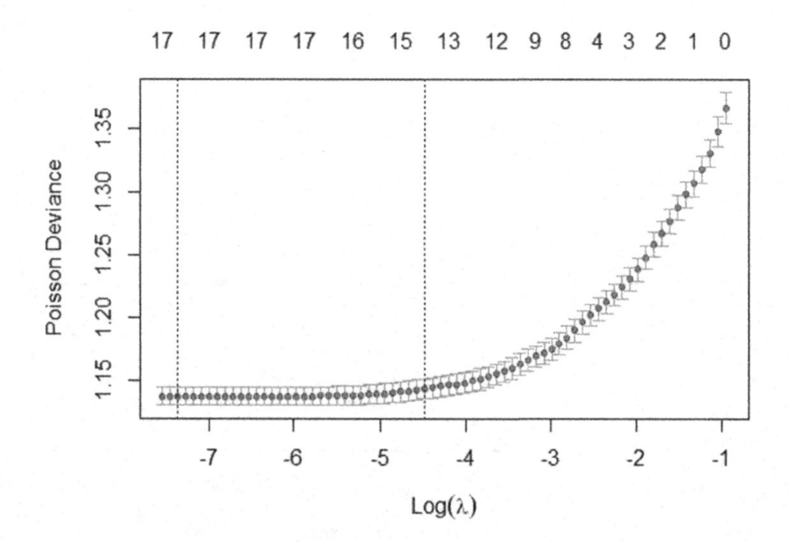

FIGURE 27.11
The bias-variance trade-off approximated via lasso selection for different values of the regularisation parameter λ. Complexity here goes from low (right) to high (left) rather than in the usual representation, because a high level of λ forces the elimination of more factors and therefore entails a lower complexity. The selected value of $\lambda\left(\lambda \sim \exp(-4.43)\right)$ corresponds to the largest value of λ for which the expected prediction error (Poisson deviance) is within one standard deviation of the minimum.

It is easy to show that this approach leads – at least for our practical case – to choosing the same overall model as the greedy approach with AIC.

27.4.7.9 Model Selection Using Lasso Regularisation

Unlike forward/backward selection, this estimates the global minimum in an automated, and does so without the combinatorial explosion of exhaustive search. The method uses cross-validation based on regularisation. That is, the method attempts to minimise the following regularised loss function:

$$L\left[Y, f_{\text{FULL}}(X)\right] = -2\log \Pr_{f(X)}(Y) + \lambda\beta_1$$

where $\log \Pr_{f_{\text{FULL}}(X)}(Y)$ is the log-likelihood that is optimised with maximum likelihood estimation in standard GLM regression; $f_{\text{FULL}}(X)$ is the full model, i.e. the model with all factors included; β is the set of parameters; β_1 is the l_1 norm for the parameters (so for example if the parameter set can be written as $\beta = \left(\beta^{(0)}; \beta_0^{(1)}, \beta_1^{(1)} \ldots \beta_{n_1-1}^{(1)}; \ldots\right)$ then $\beta_1 = \left|\beta^{(0)}\right| + \left|\beta_0^{(1)}\right| + \left|\beta_1^{(1)}\right| + \ldots + \left|\beta_{n_1-1}^{(1)}\right| + \ldots$); and λ is the regularisation parameter, which controls how large is the penalty for having non-zero coefficients.

The example here uses the R package glmnet, developed by Friedman et al. (2022).

As we learned in Chapter 12, what makes the lasso interesting for model selection is that for a given value of λ the optimal solution may be one where some of the coefficients are set to zero. The higher λ is, the fewer factors are selected, as shown in Figure 27.10.

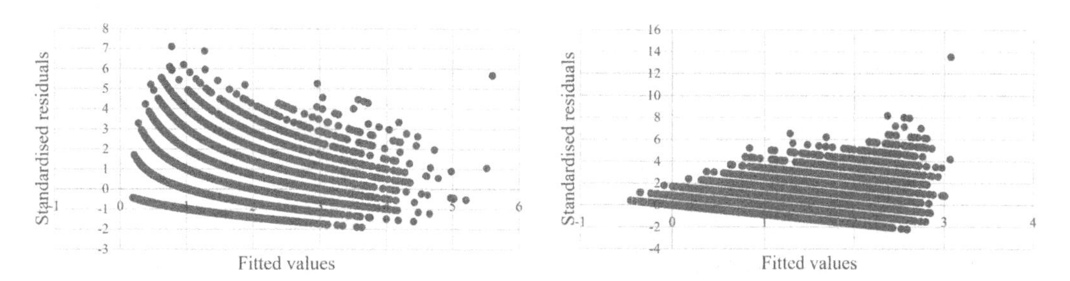

FIGURE 27.12
Standardised residual plots for Poisson noise (left) and Gaussian noise (right). The residual plot for the correct noise should show that the points are distributed (roughly) normally with mean 0 and standard deviation 1. This is indeed roughly the case for the left chart, but only for the residuals with high enough fitted values (see text). For low values of the fitted values we see that the normal approximation breaks down and we see plenty of very high standardised residuals. On the other hand, the chart on the right shows a steadily increasing variance, which is what one would expect with Poisson noise. Therefore this provides some indication that the Poisson noise is a good choice.

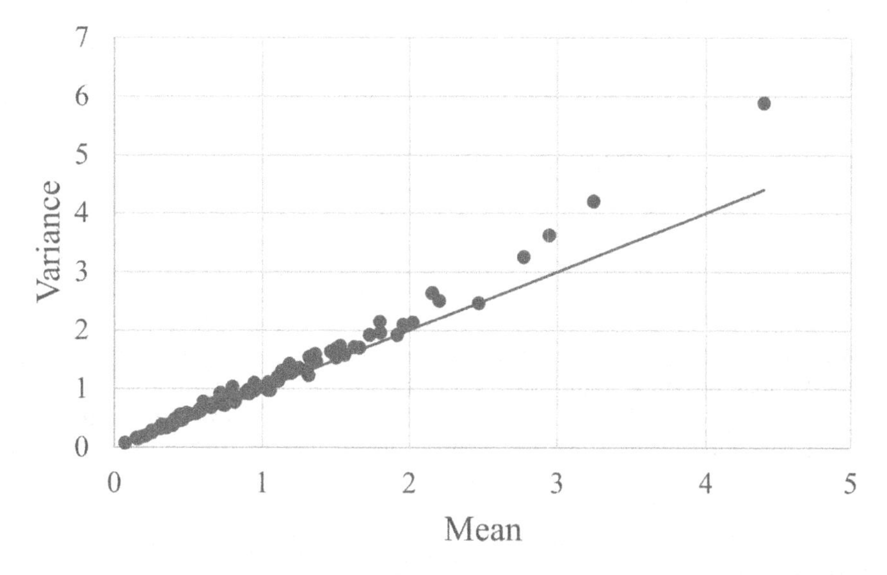

FIGURE 27.13
The variance vs mean chart for our data set, segmented into 100 groups with the same number of points. The chart shows that by and large the points lay on the line variance=mean, except for the last few points. This chart is not consistent, for example, with a Gaussian noise with uniform standard deviation.

The best model is theoretically the one corresponding to the value λ_{min} of the regularisation parameter λ that minimises the prediction error (Poisson deviance) calculated through cross-validation (Figure 27.11). In practice, the chart often flattens below a certain value of λ, and choosing λ_{min} may cause the inclusion of irrelevant variables. It might therefore be better to select the value λ_{1se} that is within one standard error (1se) of the minimum – the idea being that results within one standard error are not statistically different in a significant way. This is the dashed vertical line in Figure 27.11 corresponding to $\log \lambda \sim -4.43$.

The model's coefficients are as below. It can be noticed that this model penalises complexity slightly more than the one obtained by forward selection, as it gets rid not only of the dummy variables but also of some of AgeBand4 and RegionC, because these categories don't have a meaningful impact above the base level (AgeBand3 and RegionE respectively) for their variables.

```
(Intercept)     -0.36906191
ageband1         0.72659264
ageband2         0.15065467
ageband4         .
ageband5         0.09655918
vehage0          0.30735048
vehage1          0.01570875
locurban         0.30800286
regiona          0.06213316
regionb          0.18548272
regionc          .
regiond          0.23836055
petrol           0.05596774
ncb_1           -0.45087787
ncb_2           -0.30574607
ncb_3           -0.12226721
ncb_4            .
dummy1           .
dummy2_0         .
dummy2_1         .
```

Notice also that the parameters are different from those obtained via forward selection. This is because the base value for each of the factor has been chosen differently by glmnet: e.g. the glm package has selected AgeBand1 as the base level (lexicographic order) while the glm package selects AgeBand3 (lowest level). The results in terms of predictive values are however consistent.

27.4.7.10 Model Structure Validation

Through the analysis of residuals we can check the appropriateness of the **noise function** and of the **variance function** (Section 27.4.6). As Figure 27.12 shows, the analysis of the results is often not trivial.

As a further check that the variance function is correct we can for example (Iwanik, 2011) order the observations in ascending order of fitted value, subdivide them in, say, 100 different groups of equal size, and calculate the mean and variance for each group. The results for our data set are shown in Figure 27.13, and confirm that the variance behaves as expected for Poisson noise (variance = mean).

We will leave the checks for the link function and the outsized influence of individual data points to the reader, along with the final validation of actual vs expected. Since our model is based on artificial data and therefore is ultimately known, we can do something even better that is in general not possible: we can compare the models we've produced against the true model. We'll do that in the next section.

27.4.7.11 The Big Reveal

And finally, as we mentioned at the beginning, the data we based our example on is synthetic so the true model from which the data set was produced is known, and it can be revealed now.

The claim count for each contract was generated using a Poisson distribution. The *expected claim count* $\mathbb{E}(Y)$(Poisson rate) for each contract during a given underwriting year is given by this simple multiplicative model with no interacting factors:

$$\mathbb{E}(Y) = \beta_0\ \mathcal{E}\ R(\text{Age})\ R(\text{Region})\ R(\text{VehAge})\ R(\text{LocType})\ R(\text{FuelType})\ R(\text{NCB})$$

where $\mathcal{E}$ is the exposure of the contract (assumed to be a number between 0 and 1, and representing the time in years for which the contract overlaps with the underwriting year), and:

$$\beta_0 = 0.904837$$

$$R(\text{Age}) = \frac{2\exp\left(-\frac{\text{Age}-18}{2.5}\right) + 1 + 0.1\exp\left(\frac{\text{Age}-70}{20}\right)}{1.0196}$$

$$R(\text{Region}) = \begin{cases} 1.3 & \text{Region} = \text{A} \\ 1.5 & \text{Region} = \text{B} \\ 1.2 & \text{Region} = \text{C} = \exp\left(\ln(1.3)\,\delta_A(\text{Region}) + \ln(1.5)\,\delta_B(\text{Region}) + \ldots\right) \\ 1.6 & \text{Region} = \text{D} \\ 1 & \text{Region} = \text{E} \end{cases}$$

$$R(\text{VehAge}) = 1 + 0.4\exp\left(-\frac{\text{VehAge}}{0.5}\right)$$

$$R(\text{LocType}) = \begin{cases} 1 & \text{LocType} = \text{Rural} \\ 1.4 & \text{LocType} = \text{Urban} \end{cases}$$

$$R(\text{FuelType}) = \begin{cases} 1 & \text{FuelType} = \text{Electric / Diesel} \\ 1.1 & \text{FuelType} = \text{Petrol} \end{cases}$$

$$R(\text{NCB}) = \begin{cases} 0.7 & \text{NCBCategory} = \text{C1} \\ 0.85 & \text{NCBCategory} = \text{C2} \\ 1 & \text{NCBCategory} = \text{C3} \\ 1.15 & \text{NCBCategory} = \text{C4} \end{cases}$$

Note that the variables Dummy1 and Dummy2 do not appear in the model. Also note that the parameters found with the methods described in Sections 27.3.6.5–9 do not tie up *at first glance* with those of the true model, even after taking parameter uncertainty into account. The reason is that the same model admits (infinitely many) different representations depending on the value of the overall base factors β_0, on which level the algorithm chooses as the base level for each categorical variable,[12] and on how we discretise the continuous variables.

More details about the hidden model, including information on the relative exposures of the various levels for each category, the correlation between the exposures, and information on the binning of the continuous variables (Age, VehAge) can be found in the downloads section of the book's website (Parodi, 2016).

27.5 Generalised Additive Models (GAMs)

Generalised linear models are not the only way in you the multivariate linear model can be generalised. Another popular methodology is that of Generalised Additive Models (GAMs), which were introduced by Hastie and Tibshirani in 1990 and are described, for example, in Hastie et al. (2001). Their use specifically for insurance is explored for example in Ohlsson and Johansson (2010) and in Wüthrich and Buser (2021).

In this section we will briefly describe GAMs, limiting ourselves to log link GAMs, which are the ones most useful for pricing models. Log link GAMs without interacting variables can be written in this form:

$$\mathbb{E}(Y) = \exp\left[\beta_0 + \sum_j \Psi_j(X_j)\right]$$

The difference between each realisation of Y and its expected value is driven by noise similarly to generalised linear models. If interacting variables are present, we can add terms of the form $\Psi_{j,k}(X_j, X_k)$ or with more variables if necessary. However, these are difficult to calibrate.

The following normalisation condition ensures that $\exp(\beta_0)$ is the base premium: for all functions $\Psi_j(X_j)$ we demand that

$$\sum_r \Psi_j(x_{j,r}) = 0$$

where $x_{j,r}$ are the observations.

It is doubtful that GAM really brings a lot of improvements to the prediction error, even in the case of synthetic data. One advantage of using GAMs, however, is that, as mentioned in Wüthrich and Buser (2021): 'GAMs make it redundant to categorise non log-linear feature components in GLM applications. This has the advantage that the ordinal relationship is kept. The drawback of GAMs is that they are computationally more expensive than GLMs.'

12 However, the relativities between the levels will be consistent. For example, Section 27.4.7.7 shows that the estimated parameters Region A and Region D are 0 (Region A is the base) and 0.19451 respectively. The relevant ratio is therefore exp(0.19451)~1.21, to be compared with the true ratio 1.6/1.3~1.19. For the lasso, the same calculation gives ~1.20. These three ratios are the same within the margin of error due to parameter uncertainty.

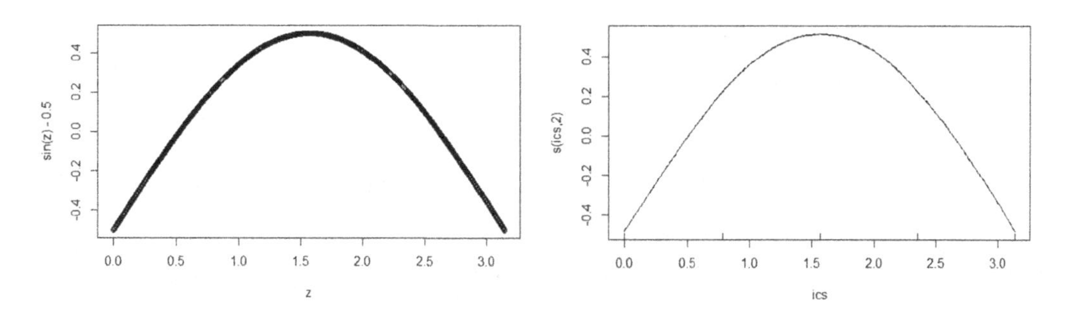

FIGURE 27.14
(Left) The function $\sin(x)$ between 0 and π, shifted downward by $(\sqrt{2}+1)/5$. (Right) The spline that goes through $\sin(x)$ calculated at points $0, \pi/4, \pi/2, 3\pi/4, \pi$.

27.5.1 What is the Difference Between GLMs and GAMs?

So far, we haven't shown much difference with standard GLM, at least in terms of the model structure. And indeed if all the functions $\Psi_j(X_j)$ are decided beforehand with no free parameters (no matter how complicated these functions are, e.g. $\Psi_1(X_1) = \sin(\log(X_1) + 5X_1^3)$), GAMs are basically be a special case of GLMs with some restriction on the type of functions to be used (typically cubic splines, as we'll see in a moment). The main difference, however, is that in the case of splines *some of the functions* $\Psi_j(X_j)$ have free internal parameters and these are automatically calibrated from the data using an optimisation procedure. The qualifier 'some of' is important – as we'll see in the practical example in Section 27.5.5, GAMs will typically mix standard categorical or continuous variables with functions with free parameters that need to be fitted.

This discussion about the functions $\Psi_j(X_j)$ may sound a bit abstract so let's look at the most common incarnation these functions take in practice – cubic splines.

27.5.2 Cubic Splines

Although GAMs could in principle be based on *any* functions, a typical implementation uses natural cubic splines. A *cubic spline* is a way of approximating a function by means of piecewise cubic polynomials that are joined at points called *knots* by imposing continuity up to the second derivative. A *natural cubic spline* is a cubic spline that extends to a straight line outside the range. The reason why this is often the chosen implementation is that the human brain is not good at perceiving discontinuities of higher order than the second derivative: in other terms, smooth up to the second derivative is *smooth enough*.

Because of the continuity conditions, fitting a spline through a number k of knots is strongly constrained, and it is easy to prove that only $k+2$ degrees of freedom are involved. In other terms, there exists only one natural cubic spline that goes through a given set of knots $(u_1, f_1^*), \ldots (u_m, f_m^*)$: $f(u_1) = f^{*1}, \ldots f(u_k) = f^{*k}$.

Also, you can simplify the description and write your spline as the following sum:

$$\Psi(X) = \alpha_0 + \alpha_1 X + \sum_{t=1}^{k} \gamma_t (X - u_t)_+^3$$

where $(X - u_t)_+$ is shorthand for $\max(X - u_t, 0)$.

27.5.3 A Simple Example of an Interpolating Spline

A simple example of a spline with three internal knots is that of fitting the function $f(x)=\sin(x)$ with x between 0 and π, with five knots at these values of x:. $0, \pi/4, \pi/2, 3\pi/4, \pi$. The spline goes through the points $(0,0), \left(\frac{\pi}{4}, \frac{\sqrt{2}}{2}\right), \left(\frac{\pi}{2}, 1\right), \left(\frac{3\pi}{4}, \frac{\sqrt{2}}{2}\right), (\pi, 0)$. The result is shown in Figure 27.14 (note that in the figure the chart is shifted downward by the quantity $(\sqrt{2}+1)/5$ as a result of imposing the condition $\sum_r \Psi_j(x_{j,r})=0$).

27.5.4 Smoothing (Regression) Splines

So far we have discussed *interpolating* splines – splines that provide a smooth way of connecting points that you know the function has to go through. However, if you have a large number of points, you don't necessarily want a function going through all those points, but a function that approximates the general behaviour of the data and is both as close as possible to the observed points and smooth enough – that is, having as few spikes in convexity/concavity as possible. This function is called a *smoother* and finding a good smoother is a generalisation of the problem of finding the regression line through a number of points, except that linearity is not a pre-requisite.

The problem of finding a good smoother is best described as an optimisation problem (Hastie and Tibshirani, 1990): specifically, we want to find the function (continuous up to the second derivative included) that minimises this regularised distance (technically, a functional):

$$d_{\text{REG}}(\mathcal{D}, f) = \sum_{i=1}^{n} (y_i - f(x_i))^2 + \lambda \int_a^b (f''(x))^2 \, dx \tag{27.8}$$

Where $\{(x_1, y_1), (x_2, y_2) \ldots (x_n, y_n)\}$ is a set of points that the function $f(x)$ needs to approximate, with the constraint $a \le x_1 \le x_2 \le \ldots \le x_n \le b$. The surprising result is that for a given value of λ there is a unique minimiser $f(x)$ and this minimiser is a natural cubic spline with knots at $x_1, x_2 \ldots x_n$.

The parameter λ controls the degree of smoothing: a high value of λ means that the function will have fewer and smaller 'ups and downs'. At the limit where λ goes to infinity, $f''(x)$ is forced to be zero everywhere and therefore the regularised spline is simply a straight line. When $\lambda = 0$, there is no smoothing and the spline will be an interpolating spline that goes through all points $(x_1, y_1), (x_2, y_2) \ldots (x_n, y_n)$.

This is a classic example of the bias-variance trade-off – you can either replicate the existing data with good accuracy (low λ) or increase the smoothness (high λ) but not both – and through cross-validation you can obtain an optimal value of λ which achieves a good trade-off between the two and minimises the prediction error.

An example of this in insurance is when modelling the dependency on the number of claims on age – we have a limited number of possible ages that we wish to differentiate but we may have plenty of data points.

Note that as part of the optimisation process, we can also decide the optimal number of knots – e.g. for ages we may find that having 70–80 different age levels is too much. A GAM algorithm will offer a number called 'effective degrees of freedom (edf)' – the determination of which is beyond the scope of this book – and if that is much less than the number of buckets into which we have split the data we may take that as a sign that we should reduce the number of buckets to a number similar to the edf.

Also note that while the regularised distance in Equation 27.8 is the one that produces cubic splines as the solution, depending on the application it may be useful to replace $\sum_{i=1}^{n}(y_i - f(x_i))^2$ with another metric – typically, the (negative) log-likelihood.

27.5.5 Practical Example

The following example uses the same artificial data set as for the GLM, which shows the number of losses from 10,000 policies, while recording for each policy the value of: Owner's Age, Location Type, Region, Vehicle's Age, Fuel Type, No-claim Years, Dummy1, Dummy2, Exposure.

For this data set, *we choose to model with splines only two variables*, Owner's Age ('age') and Vehicle's Age ('vehage'), using 15 knots for age and 30 knots for vehage. We'll comment later on whether this choice for the number of knots is suitable but for now let's just note that when using uniformly spaced knots we should be aiming to have the interval between successive knots reflect the scale at which changes happen, based on domain knowledge or simply on one-way analysis: in our case, Figure 27.7 shows that for variable 'age' this scale is around 5–6 years (the risk drops dramatically between 18 and 24 years of age, and then stabilises), whereas for vehage there is a sudden drop after around 1 year; the wobbliness towards the end is an artefact of data scarcity.

Using a basic selection method based on first fitting the complete model and then removing the non-significant variables, as written in R:

```
Family: poisson
Link function: log

y ~ s(Age, k = 15, bs = 'cr') + s(vehage, k = 30, bs = 'cr') +
    locurban + regiona + regionb + regionc + regiond + petrol +
    ncb_1 + ncb_2 + ncb_3 + ncb_4 + dummy1 + dummy2_0 + dummy2_1
```

where `s(z,k=k*,bs='cr')` denotes a cubic regression spline (bs means 'basis', and 'cr' stands for cubic regression) with k* knots approximating variable z. The fitted parameters are as follows:

```
Parametric coefficients:

              Estimate    Std. Error   z value   Pr(>|z|)
(Intercept)  -0.882636    0.024830     -35.547   < 2e-16 ***
locurban      0.332222    0.011822      28.103   < 2e-16 ***
regiona       0.267387    0.018414      14.521   < 2e-16 ***
regionb       0.385930    0.018950      20.366   < 2e-16 ***
regionc       0.193701    0.018578      10.427   < 2e-16 ***
regiond       0.458223    0.020923      21.901   < 2e-16 ***
petrol        0.076870    0.009269       8.293   < 2e-16 ***
ncb_1        -0.598431    0.016981     -35.242   < 2e-16 ***
ncb_2        -0.437763    0.016434     -26.637   < 2e-16 ***
ncb_3        -0.246805    0.014145     -17.448   < 2e-16 ***
```

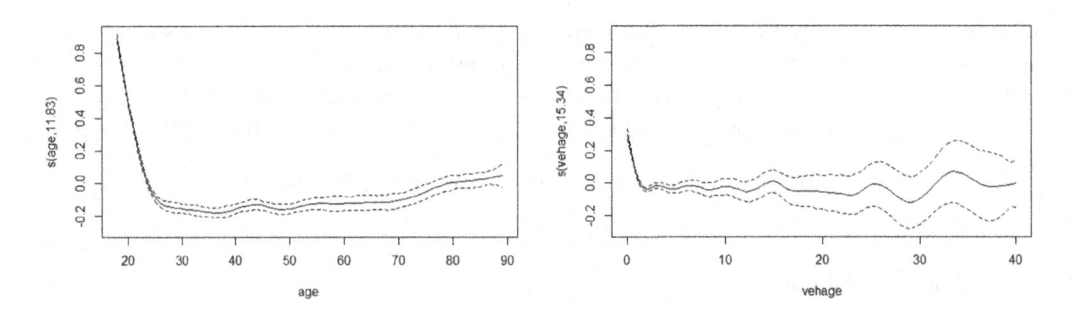

FIGURE 27.15
Smoothing splines for age (left) and vehage (right), based on 15 and 30 knots respectively. The wobbliness at the end of the chart for vehage is consistent with the extreme erraticity shown by the one-way analysis, which in turn is a side effect of limited sample size, especially for large vehicle ages.

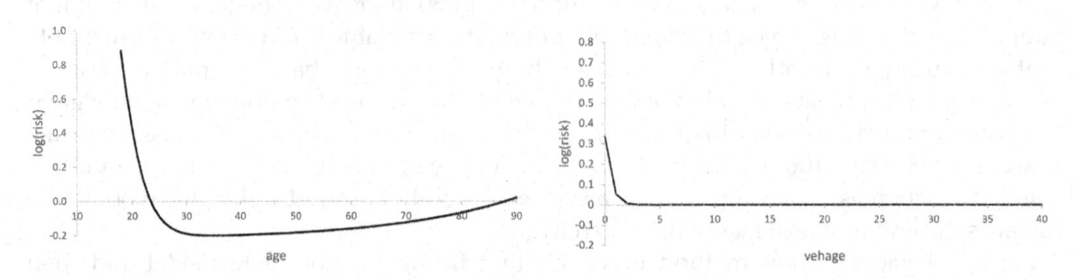

FIGURE 27.16
The *true* functional dependency (apart from a multiplicative constant) for age (left) and vehage (right), which we have used to generate the artificial data set.

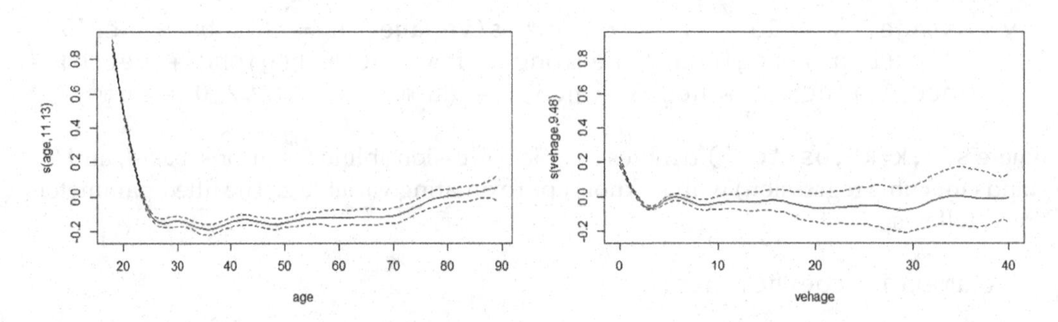

FIGURE 27.17
Smoothing splines for age (top) and vehage (bottom), based on 13 and 16 knots respectively.

```
ncb_4          -0.104522    0.017882        -5.845    5.06e-09 ***
dummy1          0.005896    0.009464         0.623    0.533
dummy2_0        0.001428    0.013019         0.110    0.913
dummy2_1        0.002103    0.010407         0.202    0.840

---
Signif. codes: 0 '***' 0.001 '**' 0.01 '*' 0.05 '.' 0.1 ' ' 1
```

```
Approximate significance of smooth terms:
                 edf Ref.df Chi.sq  p-value
s(age)         11.83     14  5897.7  <2e-16 ***
s(vehage)      15.33     29  343.9   <2e-16 ***
---
Signif. codes: 0 '***' 0.001 '**' 0.01 '*' 0.05 '.' 0.1 ' ' 1

R-sq.(adj) = 0.329 Deviance explained = 14.7%
UBRE = 0.12681 Scale est. = 1 n = 50000
```

Using the t-value criterion, we find that the variables Dummy1 and Dummy2 can be safely removed from the data set as being of little significance. The selected model along with its parameters is therefore the following:

```
Family: poisson
Link function: log

Formula:
y ~ s(age, k = k_age, bs = 'cr') + s(vehage, k = k_vehage,
    bs = 'cr') + locurban + regiona + regionb + regionc + regiond
    + petrol + ncb_1 + ncb_2 + ncb_3 + ncb_4

Parametric coefficients:

              Estimate    Std. Error   z value    Pr(>|z|)
(Intercept)   -0.877307   0.022911     -38.292    < 2e-16 ***
locurban       0.332184   0.011821      28.100    < 2e-16 ***
regiona        0.267345   0.018413      14.519    < 2e-16 ***
regionb        0.385913   0.018949      20.365    < 2e-16 ***
regionc        0.193632   0.018577      10.423    < 2e-16 ***
regiond        0.458228   0.020923      21.901    < 2e-16 ***
petrol         0.076899   0.009269       8.296    < 2e-16 ***
ncb_1         -0.598570   0.016979     -35.254    < 2e-16 ***
ncb_2         -0.437883   0.016433     -26.647    < 2e-16 ***
ncb_3         -0.246924   0.014142     -17.460    < 2e-16 ***
ncb_4         -0.104534   0.017881      -5.846    5.03e-09 ***

---
Signif. codes: 0 '***' 0.001 '**' 0.01 '*' 0.05 '.' 0.1 ' ' 1

Approximate significance of smooth terms:
               edf Ref.df Chi.sq  p-value
s(age)        11.83     14    5899  <2e-16 ***
s(vehage)     15.34     29     344  <2e-16 ***
---
Signif. codes: 0 '***' 0.001 '**' 0.01 '*' 0.05 '.' 0.1 ' ' 1

R-sq.(adj) = 0.329    Deviance explained = 14.7%
UBRE = 0.1267 Scale est. = 1       n = 50000
```

Figure 27.15 shows the spline fit for the variables age (left) and vehage (right). The results are broadly compatible with the functional shape used to generate the data (see

Figure 27.16) but with some wobbliness, which reflects the behaviour of the data set, in turn caused by data scarcity.

Note that the effective degrees of freedom edf for age and vehage is 11.83 and 15.34 respectively. This suggests that it might be worth trying a lower number of knots, for example $k = \lceil \text{edf} + 1 \rceil$: 13 for age and 16 for vehage. The results are shown in Figure 27.17.

As Figure 27.17 shows, the reduction in the degree of freedom hasn't brought much improvement for age. As for vehicleage, it is clear that while some wobbliness has now been removed, the results are less accurate once we compare them with the true chart. This is the standard bias vs variance trade-off. Note that we can play this game more than once since the new edfs are in general different from the original ones. If you do that with our example you'll see that the age smoothing spline stabilises at $k = 13$, while vehage spline keeps proposing lower and lower degrees of freedom which eventually produces a completely flat response.

A better solution to improve our smoothing splines would be to use cleverly placed knots: after all, for both age and vehage the change is dramatic at the beginning but then it is much slower. For example for vehage it would make sense to have denser knots at the beginning and fewer knots after that, because based on our domain knowledge (or sheer common sense) it would be strange for the risk to go up and down repeatedly between 20 and 40 years of vehicle age.[13]

27.6 Other Machine Learning Techniques for Rating

The use of machine learning techniques such as regularisation and neural networks has long been advocated as alternatives or complements to GLMs (among the earliest examples see Lowe and Pryor, 1996; and Dugas et al., 2003).

In this chapter we have seen how regularisation should not really be seen as an alternative to GLM: it can be fruitfully incorporated in GLM using regularised GLM to automate model selection.

As for neural networks, they have the doubtless advantage of working with very little need for model specification; however, their black-box nature (the model parameters are not immediately relatable to risk characteristics) has long been raised as an issue that will impair underwriters' acceptance. This might change as methods to increase interpretability of neural networks and improved visualisation techniques are being developed (see Richman (2012a, 2012b) for a discussion and more references).

More methodologies such as regression trees and ensemble boosting (along with GLMs, GAMs, and neural networks) are discussed in Wüthrich and Buser (2022). A comparison of the various techniques using a French Motor TPL case study is made in Noll et al. (2018).

Finally, it should also be noted that GLM and its relation, GAM, are unlikely to be made obsolete by other machine learning tools such as neural networks, regression trees, and ensemble boosting, at least outside Big Data territory, which is not the territory where general insurance pricing typically operates. And we should stop thinking that machine

13 Note that while the cubic regression that we have used does not use evenly spaced knots, but knots spaced at equal percentiles, for our artificial data set the two criteria are basically the same as we have sampled vehicle ages uniformly over the available range.

learning algorithms are an *alternative* to GLM – rather, GLM should be considered a full member of the machine learning toolkit.

27.7 Questions

1. Your company decided a few years ago to enter the lucrative business of selling alien abduction policies. This business has performed extremely well in the past, with an observed loss ratio consistently at 0% over the last five years.

   Building on this good fortune, the underwriting manager has decided to design a new policy that covers post-traumatic stress disorder (PTSD) arising from UFO sightings.

   The pricing factors of the policy (which is sold to individuals) are the following:
   - Distance of habitation from renowned sighting areas (≤100 km, between 100 km and 500 km, above 500 km)
   - Marriage status (married/not married)
   - Number of previous sightings by the policyholder (0, 1, more than 1)
   - Occupation of policyholder (pilot, other)

     The compensation is a flat FRD 10,000 (FRD = Freedonian dollars) for each episode (unlimited number of episodes per year).

   i. Find a suitable generalised linear model for the expected losses to the policy, which includes all the factors above. Assume that all factors are non-interacting.

   ii. Describe the process by which you could calibrate this model if you had sufficient data.

   iii. Repeat the exercise (a) by assuming that the number of previous sightings and the occupation are interacting.

   You soon find out that based on the relatively meagre loss data experience at your disposal, you will not be able to fully calibrate model (i), and you decide to use a simplified version of the model found in (i), i.e. a sub-model of (i) with fewer pricing factors.

   iv. Explain what strategy you would use to determine which sub-model of (i) is the most adequate to describe the risk of a policyholder.

2. A general insurance company that writes Fine Art insurance has decided to use its historical loss data to replace the current model (mostly based on underwriters' experience and judgment) with a generalised linear model calibrated using historical loss data, and you are in charge of this investigation.

   As a result of this investigation, you have produced a frequency/severity model where the severity distribution comes from an MBBEFD exposure curve. As for the frequency, you have modelled it as a Poisson distribution with Poisson rate depending on the following rating factors:
   - Territory X_1 (5 levels: US and Canada, Latin America, Europe, Asia, Middle East and Africa)
   - Security level X_2 (3 levels: low, medium, high)

     and proportional to the exposure X_3 (sum insured).

   i. Assuming the factors are non-interacting, write down an expression for the Poisson rate.

ii. Calculate the number of degrees of freedom of the model in (i).

According to the underwriters, the frequency of claims is not actually proportional to the exposure but sub-proportional. You then decide to explore a modified model in which the Poisson rate is proportional to X_3^α, with α to be determined.

iii. Modify the expression for the Poisson rate to accommodate this modified model.

iv. Explain how you would test whether this change is actually an improvement using (a) the Akaike Information Criterion (AIC), specifying what the relationship between the log-likelihood of the two models needs to be; (b) a hold-out sample (i.e. a test data set that has not been used for calibration of the model).

3. Explain (a) the difference between categorical and non-categorical factors; (b) the reason why, when the number of levels of a categorical factor becomes large, using GLM may become difficult; and (c) describe possible ways to address this problem. *(Note: The answer to part (c) of the question is the subject of the next chapter).*

28

Multilevel Factors and Smoothing

The clearest way into the Universe is through a forest wilderness.

John Muir

In our treatment of generalised linear models (GLMs) in Chapter 27, we have come across two different types of rating factors:

a. Continuous variables such as the annual mileage or the driver's age, which may or may not be related monotonically to the modelled variable (e.g. the claim frequency)
b. Categorical variables (i.e. variables with a limited number of possible values) that have in general no inherent order: examples are sex, occupation, car model, and geographical location

Categorical variables are straightforward to model when the number of levels is small, such as is the case with sex. However, they pose a particular challenge if the number of levels is very large and there is no obvious way of ordering them, such as is the case for car models[1] or postcodes. In this case, the calibration of a GLM becomes problematic because, for some of the values (in our example, some of the car models), loss data may be scanty. Also, the lack of a natural ordering means that there is no obvious way of merging categories together (as we can do with ages) to obtain categories with more data.

Following Ohlsson and Johansson (2010), we call these factors multilevel factors (MLF), and we describe them as nominal variables (i.e. with no particular ordering) that have too many levels to be analysed by standard GLM techniques.

Traditional approaches to MLFs are as follows:

1. Reduce the number of categories by merging together categories that are thought to be similar, in a heuristic fashion. For example, you could reduce the number of car models drastically by considering only the make of the car (e.g. BMW, Vauxhall, Fiat) rather than the full car model code, or use the Association of British Insurers' (ABI) grouping, which classifies all vehicles into one of 50 groups with similar characteristics (Institute and Faculty of Actuaries [IFoA] 2010). Analogously, postcodes at, say, town or borough level could be merged together. This is of course not ideal because a lot of useful information is discarded and is not the modern

1 Ohlsson and Johansson (2010) estimate 2500 car model codes in Sweden, and this estimate can probably be transferred to other countries as well.

DOI: 10.1201/9781003168881-32

approach to MLFs. However, it has traditionally been used when not enough computer power and adequate statistical techniques were available.
2. Find the underlying factors that explain the difference between the different categories captured by a MLF and use these as factors as inputs to the GLM instead of the original MLF. For example, in the case of car models, one can replace the car model code with factors such as the vehicle's cubic capacity, year of manufacturing, fuel type, body type and so on, and calibrate the GLM based on these underlying factors. The factors can be found by factor analysis (a statistical technique) or by using the factors already available in vehicle classification systems such as that of the ABI. The problem with this approach, of course, is that although these underlying factors might explain part of the variation across different car model codes, there will remain some residual variation that is specific to the car model because, for example, that car model may appeal to a more (or less) risky type of driver.

What these two traditional approaches have in common is that they attempt to work around the extreme granularity of the MLFs by reducing the number of categories. In the rest of this chapter, we want instead to consider more sophisticated approaches that endeavour to use the full granularity of the information available while ensuring that data credibility issues are taken into proper account: for example, that we do not draw hasty conclusion on the riskiness of a particular car model based on only a handful of observations. Both these approaches smooth the observations in the attempt of separating genuine effects from noise.

One such approach (Section 28.1) is to use credibility theory to smooth the values across the different levels, that is, to assign to each level a new value that depends partly on the observed value for that level and partly on the observed values of the other levels. The more data that is available for a particular level, the more credible the experience is.

In the case of geographical codes, smoothing can be achieved by exploiting the physical proximity of different postcodes, using the so-called spatial smoothing (Section 28.2).

In both cases, the idea is to deal with the MLF 'off-line', that is, outside the GLM machinery: in other terms, to deal with ordinary (i.e. non-multilevel) rating factors using GLM and MLFs using smoothing algorithms. The limitation of this approach is that it ignores the interaction between these two types of factors, and between their exposures. An iterative approach is however possible by which the results of the MLF analysis is influenced by the results of the GLM analysis and vice versa.

Finally, it will be mentioned (Section 28.3) that all these techniques to tackle MLFs can be understood in the framework of machine learning, which is the discipline that you need to refer to in order to access state-of-the-art methods for reducing the number of categories (clustering), for factor analysis, and for smoothing. A review of these more advanced techniques can be found in the review by Parodi (2012a).

28.1 Credibility Smoothing

The idea of credibility smoothing is to estimate the value of the modelled variable for a specific level of a MLF by a weighted average of the observed value for that level and the cross-level average. The weight given to the observation for a specific level is, as usual, based on the uncertainty of the observation and on the spread across the different levels. As an example, consider the case where the modelled variable is the frequency of claims and the MLF is the car model code.

28.1.1 An Example: Car Model Codes

Assume that:

- N_c is the random variable representing the number of claims for car model code c over a certain period, and n_c is the number of claims actually observed
- w_c is the exposure (vehicle-years) for car model code c
- The claim count can be modelled as a Poisson distribution

Using uncertainty-based credibility, the credibility-weighted estimate $\hat{\varphi}$ for the claim frequency (Equation 26.10) can be written as $\hat{\varphi} = \hat{Z} \times \hat{\varphi}_c + (1 - \hat{Z}) \times \varphi_m$, where $\hat{Z} = s_h^2 / (s_h^2 + \hat{se}_c^2)$ as in Equation 26.13 and

- $\hat{\varphi}_c = n_c / w_c$ is the estimated claim frequency for car model code c, based on the claim count for that code only
- φ_m is the claim frequency for the overall portfolio, which can be estimated as[2] $\hat{\varphi}_m = \sum_c n_c / \sum_c w_c$
- se_c is the standard error on the claim frequency for car model code c, which can be estimated as[3] $\hat{se}_c = \sqrt{n_c} / w_c$
- s_h is the estimated heterogeneity over the different car models, which is based on the standard deviation of the observed claim frequencies of the different car models, weighted by exposure

Based on the credibility-smoothed frequency, we can also calculate the ratio between that frequency and the overall frequency, $\hat{\rho} = \hat{\varphi} / \varphi_m$, which is a correction factor for the effect of the car model code. An actual calculation is shown in Figure 28.1. We have assumed for simplicity that we have only 10 car models, although if that were the case, we would probably not need a MLF approach!

28.1.2 Credibility Smoothing and the GLM Approach

How does this fit into the GLM approach? Assume, for example, that the dependent variable Y (claim frequency, in our case) can be written in the standard way with a multiplicative model that is a variant of Equation 26.8, using the log link function:

$$Y = \exp\left(\sum_j \beta_j \Psi_j (X_1, \ldots X_r)\right) \times \rho + \varepsilon \tag{28.1}$$

2 Consistently with what we said in Section 26.2.4, we ignore the error that we make by using $\hat{\varphi}_m$ as an estimate for φ_m, referring the interested reader to (Parodi and Bonche 2010) for a treatment that takes this error into account.

3 We have used the fact that the standard error for the claim count of a Poisson process N for which we have a single measurement n can be estimated as $se \sim \sqrt{n}$. The approximation is good when $n \geq 30$.

Credibility Smoothing – Calculation Example

Model	Exposure	No of claims	Observed frequency (×1000)	Standard error (×1000)	Credibility factor	Credibility-smoothed frequency (×1000)	Ratio to overall frequency
Comma 1.0	2640	172	65.2	5.0	0.70	64.7	1.02
Comma 1.6	788	39	49.5	7.9	0.48	56.8	0.89
Fuego 1100	6360	433	68.1	3.3	0.84	67.4	1.06
Fuego 700	8140	484	59.5	2.7	0.89	59.9	0.94
Ocelotto L	3199	162	50.6	4.0	0.79	53.4	0.84
Platypus Family	1569	94	59.9	6.2	0.60	61.3	0.97
Platypus Sport	8791	624	71.0	2.8	0.88	70.1	1.10
Roomy 7s	496	34	68.5	11.8	0.30	65.0	1.02
Swish Basic	654	43	65.7	10.0	0.37	64.3	1.01
Swish TDI	1178	63	53.5	6.7	0.56	57.9	0.91
Heterogeneity (weighted, ×1000)			7.6				
Overall frequency (×1000)			63.5				

FIGURE 28.1
Calculation of the credibility-smoothed frequency. Example for the Ocelotto L: the credibility factor can be calculated as $\hat{Z}$ = heterogeneity²/(heterogeneity² + (s.e.)²) = 7.6²/(7.6² + 4.0²) = 0.79, which leads to a credibility-smoothed frequency of 0.79 × 50.6 + 0.21 × 63.5 = 53.4, and a ratio to the overall frequency of 53.4/63.5 = 0.84.

where:

- $X_1, X_2 \ldots X_r$ are the ordinary rating factors (i.e. all the factors except for the MLF), which enter the model through the functions $\psi_1, \psi_2, \ldots$
- ρ is a MLF (the correction factor for the car model code)
- ε is Poisson noise

The simplest approach to combining the results of the GLM and credibility smoothing is as follows:

1. Find estimates $\hat{\beta}_j$ of the GLM by ignoring the car model code.
2. Find estimates $\hat{\rho}^c$ for each car model code c as explained in Section 28.1.1, ignoring the other rating factors.
3. Obtain the estimated claim frequency for a specific vehicle, i, as $\hat{Y}^{(i)} = \exp\left(\sum_j \hat{\beta}_j \Psi_j\left(X_1^{(i)}, \ldots X_r^i\right)\right) \times \hat{\rho}^{c(i)}$, where $\left(X_1^{(i)}, \ldots X_r^i\right)$ are the factors calculated for vehicle i and $\hat{\rho}^{c(i)}$ is the correction factor calculated for the car model code $c(i)$ of vehicle i.

As mentioned in the introduction to this chapter, this is not completely satisfactory because we are ignoring the interaction between the ordinary rating factors and the car model code, or between their underlying exposures. For example, we may find that a specific car model that is known to have a higher claim frequency than average is purchased mainly by young drivers, and that explains most of the higher claim frequency observed: disregarding this will lead to double-counting the effect of age.

A possible solution to this is using an iterative algorithm by which the correction factors for the MLF are updated based on the results of the GLM, and the parameters of the GLM are updated based on the results of the MLF calculation, until convergence is achieved. The details of a backfitting algorithm that does this are described in the book by Ohlsson and Johansson (2010), who also consider the case in which more than one MLF is present.

Finally, note that credibility smoothing can be combined with factor analysis by using, for example, the ABI vehicle classification factors as ordinary factors *and* using car model codes to analyse the residual variation. This should lead to a more accurate overall model, as shown in the book by Ohlsson and Johansson (2010).

28.2 Spatial Smoothing

Spatial smoothing is a form of smoothing that takes advantage of the spatial structure of geographical code information. In the following, we assume that the geographical code is simply the postcode. The use of a postcode for geographical coding is widespread in the United Kingdom as postcode information is quite granular, down to the street level (unlike, for example, in Italy, where a single postcode may cover a whole town with a population of 30,000 or more). It is, however, not a perfect coding system, as for example two locations at the top and the bottom of a hill might have the same postcode because they belong to the same street, but they may represent a very different flood risk.

Before looking at specific spatial smoothing techniques and explaining which one should be used depending on the circumstance, it is perhaps useful to look at a one-dimensional analogy.

28.2.1 One-Dimensional Analogy

Spatial smoothing is perhaps best understood with a one-dimensional analogy. Assume that we have a time series $f(t)$ (only discrete times t are considered), which could for example represent the claims frequency and which has an underlying increasing trend, as in Figure 28.2. The raw data series is rather ragged because of noise, hiding the underlying trend. However, if one replaces the time series with a moving average over $2r + 1$ elements:

$$f_r(t) = f(t-r)+\ldots+f(t-1)+f(t)+f(t+1)+\ldots+f(t+r)$$

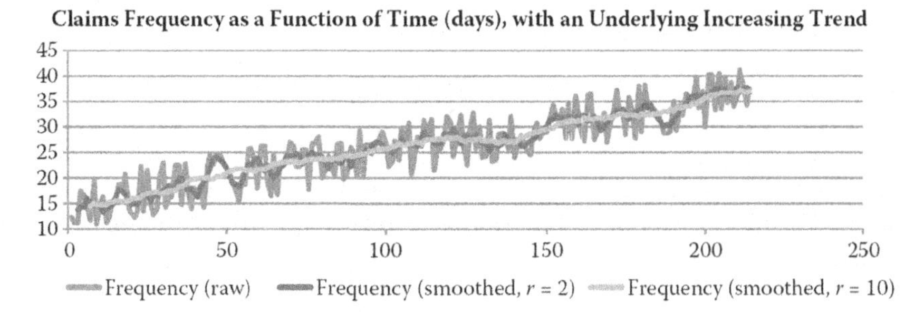

FIGURE 28.2
One-dimensional smoothing. The raw time series $f(t)$ is rather ragged, whereas the smoothed versions of it, obtained by averaging neighbouring data points, bring into relief the underlying trend of the series.

then the function is smoothed and the underlying trend is brought into relief, as shown in Figure 28.2, where the cases $r - 2$ and $r - 10$ are considered. The parameter r gives the degree of smoothing.

In the language of signal processing, what we have done is apply a 'filter' to the signal $f(t)$ and obtained a smoothed signal $f_r(t)$. The particular type of filter we have applied is a simple moving average, but many different filters are possible. For example, one might use a filter that gives less and less weight to the values of $f(t - \mathrm{j})$ as j becomes bigger. We can have triangular filters, exponential filters, Gaussian filters, and others, depending on the shape of the filter. The moving average is a rectangular filter in the sense that it gives the same weight to all neighbouring points.

It is possible to define the width of the filter and therefore the degree of smoothing. This must be chosen with care so as to bring into relief the features that we want to capture: if the filter size (e.g. the number r in our example above) is too small, a lot of noise will be captured and the underlying trend will not appear clearly; on the other hand, an excessively large filter size may drown those very trends that we are trying to capture.

Armed with these considerations, we can move into two dimensions, which is especially relevant in the case of postcodes. Following IFoA (2010), we will look at two types of spatial smoothing: distance-based smoothing and adjacency-based smoothing.

28.2.2 Distance-Based Smoothing

Distance-based smoothing averages the information of neighbouring postcodes, each postcode contributing in a measure that depends on its distance from the postcode under consideration. Note that the distance must be defined with some care. In the following, we assume that the distance between two postcodes is the great-circle distance between the central points of the two location codes.

28.2.2.1 How It Works

The value $P(x,y)$ of the observations (e.g. claim frequency) corresponding to the postcode at location (x,y) is replaced by a smoothed version $\tilde{P}(x,y)$, which is an average of the observations $P(x',y')$ at neighbouring postcodes, where the weight given to each postcode depends on the distance between the central points of the postcodes. A couple of examples:

i. All postcodes for locations whose central point (x',y') is at a distance within r from location (x,y) may be equally weighted (in other terms, the weight is 1 within a certain distance, and 0 otherwise):

$$\tilde{P}(x,y) = c \sum_{\text{all}(x',y')\text{s.t.}\,d((x',y'),(x,y)) \le r} P(x',y') \tag{28.2}$$

ii. All postcodes for locations at any distance from location (x,y) will contribute to the smoothed value $\tilde{P}(x,y)$ but with strength decreasing with their distance, with a so-called Gaussian filter:

$$\tilde{P}(x,y) = c \sum_{\text{all}(x',y')} P(x',y') \exp\left(-\frac{(x'-x)^2 + (y'-y)^2}{2\sigma^2} \right) \tag{28.3}$$

where σ is a parameter that affects how quickly the contribution of neighbouring postcodes decreases with the distance.

Note that $P(x,y)$ is not the postcode itself (such as EC3M 7DQ) but the value of the observations at the location (x,y). For instance, $P(x',y')$ could be the observed frequency of claims for the postcode at location (x',y').

28.2.2.2 Observations

Because the area of postcodes tends to be smaller in urban areas and larger in rural areas, distance-based methods may oversmooth urban areas and undersmooth rural areas (IFoA 2010). Also, distance-based methods are not ideal for sharply distinguishing locations that may be very different although geographically close; for example, two neighbourhoods separated by a river, very different in the profile of their population, or two locations at short distance but significant difference in altitude (for property risks). Note, however, that distance-based methods can be enhanced by amending the distance metric to include such dimensions as urban density or altitude. On the positive side, distance-based smoothing may be more effective than adjacency-based smoothing at dealing with weather risk, for which social factors are less relevant.

28.2.3 Adjacency-Based Smoothing

Adjacency-based smoothing averages the information of adjacent postcodes, regardless of the distance between the central points of the postcodes. Note that adjacency must also (as was the case with distance) be defined with some care, as explained below.

28.2.3.1 How It Works

The value $P(x,y)$ of the model corresponding to the postcode at location (x,y) is replaced by a smoothed version $\tilde{P}(x,y)$, which is an average of the adjacent postcodes $P(x',y')$. As an example:

$$\tilde{P}(x,y) = \frac{1}{\text{no. of postcodes adjacent to}(x,y)} \sum_{\text{all}(x',y')\text{adjacent to}(x,y)} P(x',y') \tag{28.4}$$

where two locations are adjacent if they share a boundary.

It may look like this is not just an example but the only possible form for $\tilde{P}(x,y)$ (after all, two postcodes either share a boundary or not!). However, adjacency can be defined in many different ways: for example, one might define (x,y) and (x',y') to be adjacent if and only if they either share a boundary or if there exists another location (x'',y'') that shares a boundary with both (x,y) and (x',y').

We may also define degrees of adjacency, as in Figure 28.3. For example, we can assign a degree 0 to the postcode P under consideration; a degree 1 to postcodes sharing a boundary with P; a degree 2 to postcodes sharing a boundary with postcodes with degree 1 and so on. We may then have each postcode contribute in a measure that depends on its degree of adjacency.

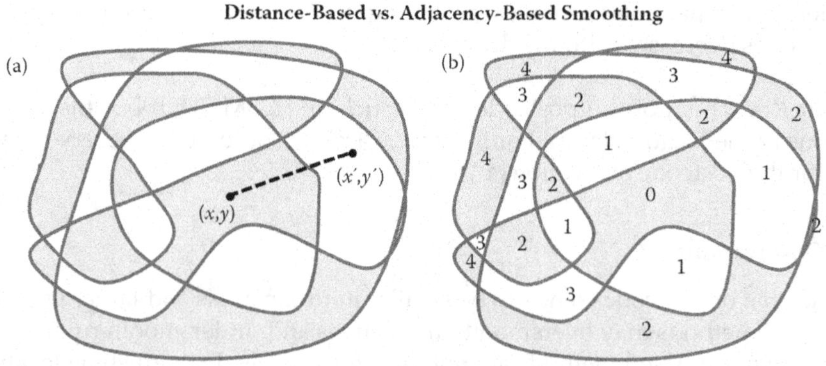

Distance-Based vs. Adjacency-Based Smoothing

FIGURE 28.3
In distance-based smoothing, the location codes contribute with a weight depending on the distance (in the picture, the length of the dashed segment), and which is possibly set to zero above a certain distance (a). In adjacency-based smoothing, all the location codes adjacent to a location contribute to the value for that location (b). Notice that 'adjacency' can be defined in different ways, of which the sharing of a boundary is only one. In the picture, two locations are adjacent if the location codes differ by 1. Also, note that one can define different degrees of adjacency to a given postcode, in this case, that marked as '0'.

28.2.3.2 Advantages and Disadvantages

Adjacency-based smoothing is less sensitive than distance-based smoothing to issues of urban density. However, it is perhaps less suitable when the underlying risk is weather-related because adjacent postcodes may be geographically distant, especially in rural areas.

Adjacency-based smoothing can be enhanced to take into account geographic features such as natural or administrative boundaries (IFoA 2010): for example, two postcodes separated by a river may be considered adjacent or non-adjacent depending on the peril modelled.

28.2.4 Degree of Smoothing

As mentioned in Section 28.2.1 about one-dimensional smoothing, it is important to calibrate the degree of smoothing so as to bring into relief the features that we are trying to capture. In two dimensions, the way in which the degree of smoothing is defined depends not only on the type of filter but also on the smoothing method.

Distance-Based Smoothing. In distance-based smoothing, the degree of smoothing is a measure of the distance within which two location codes affect each other. In Example i in Section 28.2.2.1, the degree of smoothing is related to (and can be defined as) r, whereas in Example ii (the Gaussian filter), the degree of smoothing can be defined as σ, the standard deviation of the Gaussian filter.

Adjacency-Based Smoothing. In adjacency-based smoothing, there seems to be less scope for fine-tuning the degree of smoothing. However, if adjacency to postcode P is defined as all postcodes having a degree of adjacency less than a given integer d, then d can be considered as a rough measure of the degree of smoothing.

28.2.5 Spatial Smoothing and the GLM Approach

The results of spatial smoothing can be combined with GLM in the same way that we discussed in Section 28.1.2 for credibility smoothing, by using an equation such as Equation

28.1. In this case, however, ρ represents the correction factor for spatial smoothing, which can be obtained as explained in Sections 28.2.2 through 28.2.4.

28.3 Machine Learning Techniques

As mentioned at the beginning, all these techniques to tackle MLFs can be understood in the framework of machine learning, and specifically can be seen as example of **unsupervised learning**.

Unsupervised learning – that is, learning without a teacher – is a form of exploratory analysis by which one finds patterns in the data and uses these patterns to simplify the representation of the data for further analysis. As explained in Hastie et al. (2001), one the main problems that unsupervised learning focuses on is clustering, which can be described informally as the attempt to 'segment the data points into sets such that points in the same set are as similar to each other and points in different sets are as dissimilar as possible' (Parodi, 2012a).

The problem of reducing the number of levels in multilevel factor is quite clearly a problem of clustering. Car models can be grouped into clusters of models with similar characteristics. Postcodes can be grouped into clusters with similar risk level. Once the clusters are obtained, these can be used as macro-levels for, e.g., a generalised linear model.

Examples of clustering techniques are (list lifted verbatim from Parodi, 2012a):

- Partitioning techniques (K-means, K-medoids, EM), which partition data based on a given dissimilarity measure.
- Hierarchical methods, which subdivide data by a successive number of splits.
- Spectral clustering, which works by graph partitioning techniques after a transformation into a suitable space.
- Dimensionality reduction techniques (principal component analysis, self-organising maps (SOMs), generative topographic mapping, etc), which are based on the observation that often the data points lie in (possibly non-linear) low-dimensional manifolds embedded in a data space with many more dimensions.
- Other: density-based methods, grid methods, kernel clustering.

Exploring these techniques further is beyond the scope of the book. The interested reader is referred to Parodi (2009, 2012a) for further details.

28.4 Questions

1. Consider the credibility-weighting example of car model codes discussed in Section 28.1.
   a. Derive the credibility-weighted claim frequencies for the car models Comma 1.0 and Comma 1.6 using uncertainty-based credibility and the numbers provided in Figure 28.1.

b. Repeat the exercise using the Bühlmann-Straub approach (Section 26.4.4), and assuming that the parameter Λ of the claim frequency is a random variable distributed as a Gamma distribution: $\Lambda \sim \text{Gamma}(\alpha, \beta)$. Comment on the difference between the results.

2. You want to incorporate the driver's postcode in your GLM pricing method for car insurance.
   a. Explain why it is difficult to deal with postcodes in GLM.
   b. State briefly how this difficulty can be overcome by employing a spatial smoothing technique.
   c. Briefly explain how two types of spatial smoothing (distance-based smoothing and adjacency-based smoothing) work.
   d. Describe the advantages and disadvantages of the two methods.
   e. State what the 'degree of smoothing' is and explain how it should be chosen.
   f. Can you think of a one-dimensional analogy to spatial smoothing? (Hint: think of time series instead of two-dimensional maps.)

29

Pricing Multiple Lines of Business and Risks

Until now we have generally assumed that policies to be priced could be described by a single risk model – that is, by the combination of a single frequency model and a single severity model. However, there are many common situations in which a single risk model is insufficient: this may happen if, for example, a particular policy covers different risks at the same time – say, property damage and business interruption, with different policy exclusions (such as deductibles) applying to the different risks. Section 29.1 will illustrate at some length situations in which the need to simultaneously analyse multiple risks arises.

The rest of the chapter will describe some of the modelling techniques that can be used to tackle this situation. Very often, it will be enough to assume that the losses from one risk are independent of the losses from the other risk(s). This is dealt with in Section 29.2.

In other cases, however, the different risks, or components of the risk, may be strongly interdependent, and this dependency needs to be modelled (Sections 29.3–29.6).

Section 29.9 looks at whether modelling dependency is really necessary in all situations in which some dependency is suspected – it will be argued that introducing complexities by modelling dependencies should not be done blindly but should be undertaken only if there is a business case for it.

29.1 Divide and Conquer?

Insurance policies will always have a number of different components (e.g. the sections, perils, or type of claims) and, to price a policy, it is tempting to 'divide and conquer', that is, to produce a different model for each of the sections of the policy and price the components separately. Furthermore, each risk can be further split by territory, by type of claim (slips and trips, serious accidents, industrial disease ...) and so on. The principle is not incorrect – but by doing this, we very quickly run into a credibility problem: there is not enough data to risk-cost each component separately.[1] Furthermore, if we do this, we need to investigate or make assumptions about the dependency structure among the different components.

The general recommendation to avoid this loss of credibility and the complications of dependency is therefore to avoid unnecessary splits – the number of components to model

1 We have analysed this problem in Chapter 26 when dealing with ratings factors, and we presented techniques such as GLM to make the most of our limited information: however, we were mainly concerned with estimating the expected losses for different risk categories, rather than producing an overall aggregate loss model.

DOI: 10.1201/9781003168881-33

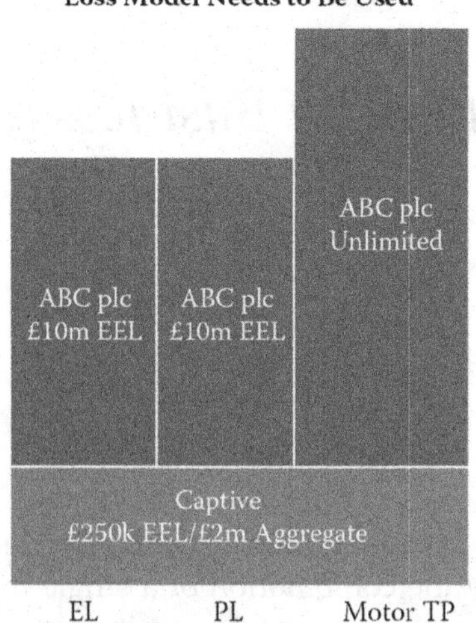

FIGURE 29.1
An example of a structure that requires a separate model for each risk (employers' liability [EL], public liability [PL], or motor third-party [TP] liability) because of different policy limits (and also because we may have different market information on the three risks), but also an aggregate model because there is a cross-class annual aggregate deductible of £2M. The first £250k of each employers' liability, public liability, or motor third-party liability loss are retained by the captive, whereas the rest is ceded to the market (in this case, to insurer ABC plc). Once the overall amount retained reaches £2M, everything is ceded to the market. The reason why a combined model is needed here is that simply summing the distribution of retained losses for the individual models do not give by themselves information on how likely it is that the £2M aggregate is breached and on how this should be priced.

should be kept at a minimum. However, there are situations in which the separate modelling of different components is inevitable. Reasons why this might be the case include different deductible/limits for the different components, a changing risk profile, and much more. We look at some of these reasons in a bit more detail below.

Examples of Reasons Why Different Components of a Risk May Need to Be Looked at Separately	
Different deductibles/limits	Different sections of a policy may be subject to different deductibles/limits. For example, we may have a combined liability policy covering employers' liability, public liability, and third-party motor liability, as in Figure 29.1, and each of these components may have a different deductible and limit, but may have a cross-class annual aggregate deductible that requires a combination of the models. Another example is a property damage and business interruption policy that may have a £1M deductible on property damage and a 15-day deductible on business interruption – hence, two separate gross loss models may be needed. Yet another example is when we need to analyse how much risk a company or its captive is retaining across a number of exposures (such as property, liability and operational risks), some of which may even be uninsurable. In this case, too, we need to be able to model the risks separately with their own separate insurance structures in place but calculate the combined aggregate loss distribution to determine the benefits of diversification.

Examples of Reasons Why Different Components of a Risk May Need to Be Looked at Separately	
Changing risk profile	The importance of different risk components may have changed over time and a single frequency/severity model based on loss experience may be inadequate to represent the current risk profile. For example, the exposure of a manufacturing industry to industrial disease may have improved dramatically by adopting safety procedures, for example, to prevent industrial deafness. Therefore, we may want to model industrial disease and accidents differently when costing employers' liability risk.
Scenario analysis	We might be required to analyse what happens if we exclude certain types of claims/ certain risks. For example, we may want to stop insuring own damage risk in a car fleet policy that includes third-party liability and own damage.
Availability of market information	As we have seen in past chapters, market information often needs to be used (where available) to complement a client's information to produce a fair assessment of a risk. An example is large loss information to model the tail of a severity distribution. However, this market information might be available separately for different components of a risk, and that forces us to produce different models in the first place.

We therefore need to develop techniques to deal with risks that need to be analysed separately but then priced together, which we are going to do in the rest of the chapter. A further complication is that the different components of the risk that we need to analyse will not, in general, be independent – that is, in a property damage and business interruption (PDBI) policy, a fire which causes a *large* property damage loss is likely to cause a *large* business interruption loss under typical circumstances. We will start with the relatively simple case of independent risks (Section 29.2), whereas the rest of the chapter will be devoted to the problem of dealing with dependency.

29.2 Independent Case

If two risks can be considered as independent, it is quite easy to obtain a combined aggregate loss model based on the aggregate loss models for each individual risks. Mathematically, we can express this as follows: if $S^{(1)} = X_1^{(1)} + \ldots + X_{N_1}^{(1)}$ is the random variable representing the total losses for risk 1 (e.g. an employers' liability risk) over a period, and $S^{(2)} = X_1^{(2)} + \ldots + X_{N_2}^{(2)}$ is the random variable representing the total losses for risk 2 (e.g. a public liability risk for the[2] same company), then the total losses for the combined risk will be simply represented by $S = S^{(1)} + S^{(2)}$. This is a completely general fact. If the two risks are independent, it is also possible to conclude that the cumulative distribution function of S, $F_S(s)$, is given by the convolution of the cumulative distribution function (CDF) of $S^{(1)}$ and the CDF of $S^{(2)}: F_S(s) = \left(F_{S^{(1)}} * F_{S^{(2)}}\right)(s)$.

The case where you need to combine $n > 2$ risks is no more complicated than the case where you only have two risks; if you have aggregate loss models for $S^{(1)}, S^{(2)} \ldots S^{(n)}$, the overall losses are $S = S^{(1)} + S^{(2)} + \ldots + S^{(n)}$ and the cumulative distribution of S is $F_S(s) = \left(F_{S^{(1)}} * F_{S^{(2)}} * \ldots * F_{S^{(n)}}\right)(s)$.

The practical consequence of all this is that you can estimate the CDF of S numerically quite easily by the Monte Carlo simulation, fast Fourier transform (FFT), or other methods.

29.2.1 Combining Aggregate Loss Models by the Monte Carlo Simulation

The Monte Carlo simulation can be used to calculate the combined aggregate loss model in the simple way illustrated in Box 29.1.

Although Box 29.1 shows the calculations for the gross aggregate loss model, it is straightforward to apply the same technique to the retained and ceded portion of the losses.

> *Question*: How would you rewrite the algorithm in Box 29.1 to calculate the *ceded* and *retained* aggregate loss models to cater to these policy features: each-and-every loss deductible D_j and aggregate loss deductible AAD, for risk j, each-and-every loss limit equal to L_j, then cross-class aggregate loss deductible (basically, an overall cap on the losses that can be retained) XAAD?

Figure 29.2 shows a simple illustration of this with just 10 simulations and two risks.

*29.2.2 Combining Aggregate Loss Models by FFT

This method exploits the useful fact that the Fourier transform of the convolution of two distributions is the product of the Fourier transforms of the two distribution; hence, $\tilde{F}_s(\omega) = \tilde{F}_{s^{(1)}}(\omega) \cdot \tilde{F}_{s^{(2)}}(\omega)$, where $\tilde{F}_s(\omega) = \mathcal{F}(F_s)(\omega)$ is the Fourier transform of $F_S(s)$. The method works as illustrated in Box 29.2.

The methodology outlined in Box 29.2 can be easily adapted to deal with the retained and ceded portion of the overall aggregate loss model, insofar as the Fourier transform of the retained and ceded distribution of the individual loss models of each risk can be calculated with the techniques discussed in Section 17.3; for example, the overall retained distribution (before applying any cross-class deductibles or other global features) is given by $F_S^{\text{ret}}(s) = \mathcal{F}^{-1}\left(\prod_{j=1}^{n} \tilde{F}_{S(j)}^{\text{ret}}\right)(s)$.

BOX 29.1 COMBINING AGGREGATE LOSS MODELS THROUGH MONTE CARLO SIMULATION

Input: Frequency and severity models for risks 1, ... n

1. For every scenario k, simulate a value of $S^{(1)}(k)$ of the total losses for risk 1 in the usual way, that is, by sampling a number of losses $N_1(k)$ from the frequency model, and the relevant loss amounts from the severity distributions: $X_1^{(1)}(k), X_2^{(1)}(k) \ldots X_{N_1(k)}^{(1)}(k)$, and summing them together: $S^{(1)}(k) = X_1^{(1)}(k) + X_2^{(1)}(k) + \ldots + X_{N_1(k)}^{(1)}(k)$.
2. Do the same for $S^{(2)}(k)$, ... $S^{(n)}(k)$.
3. Sum $S^{(1)}(k)$, $S^{(2)}(k)$ ... $S^{(n)}(k)$ together, obtaining the value of $S(k) = S^{(1)}(k) + S^{(2)}(k) + \ldots + S^{(n)}(k)$ for that scenario.
4. Repeat for a large number of scenarios, for example, 10,000 or 100,000.
5. Sort the values of $S(k)$ in ascending order, thus obtaining an empirical estimate of $F_S(s)$.

	Risk 1					Risk 2					Overall
Simulated scenario	No. of losses	$X(1,1)$	$X(1,2)$	$X(1,3)$	$S(1)$	No. of losses	$X(2,1)$	$X(2,2)$	$X(2,3)$	$S(2)$	$S = S(1) + S(2)$
1	1	78,070			78,070	1	77,211			77,211	155,281
2	3	28,659	17,473	19,152	65,284	0				0	65,284
3	0				0	3	60,321	32,686	47,398	140,405	140,405
4	1	18,292			18,292	0				0	18,292
5	1	101,130			101,130	1	98,083			98,083	199,213
6	0				0	3	25,538	34,270	24,509	84,316	84,316
7	1	34,048			34,048	2	88,997	46,196		135,192	169,240
8	2	11,782	33,860		45,642	0				0	45,642
9	2	73,911	20,151		94,062	1	69,740			69,740	163,802
10	0				0	0				0	0

FIGURE 29.2
An illustration of how the aggregate loss model of the combination of two risks can be simulated. The frequency has been chosen quite low so that all losses could be shown in the table. Only the first 10 simulations have been shown.

BOX 29.2 COMBINING AGGREGATE LOSS MODELS THROUGH FFT

Input: Frequency and severity models for risks 1, … *n*

1. Combine the frequency and severity model with FFT techniques as explained in Section 17.3.1, obtaining the (discretised) Fourier transforms of the aggregate loss models for each risk: $\tilde{F}_{s(1)}(\omega) = \mathcal{F}\left(F_{s(1)}\right)(\omega), \ldots \tilde{F}_{s(n)}(\omega) = \mathcal{F}\left(F_{s(n)}\right)(\omega)$.
2. Calculate the FFT of the combined aggregate model by multiplying the FFT of the *n* aggregate models: $\tilde{F}_s(\omega) = \prod_{j=1}^{n} \tilde{F}_{s(j)}(\omega)$.
3. Calculate the inverse of the FFT thus obtained, yielding the (discretised) CDF of the combined aggregate model: $F_S(s) = \mathcal{F}^{-1}\left(\tilde{F}_S\right)(s)$.

29.2.3 Limitations

The methodologies outlined in this section to combine the aggregate loss models of different risks are satisfactory if the risks are independent. This is often a good approximation, even when there *is* some dependence.

However, if dependency *does* need to be taken into account, the methods outlined above break down and we need to use different techniques. These may take a book of their own to write, but the core ideas are outlined in the rest of this chapter. Before illustrating how we can model dependency, let us consider how we can identify the presence of dependency in the data through some simple measures of dependency.

29.3 Measuring Dependency

When asked whether two random variables *X*, *Y* are dependent, the knee-jerk reaction for an actuary is typically to calculate the linear correlation (also called Pearson's correlation) between two vectors $x = (x_1, x_2 \ldots x_n)$ and $y = (y_1, y_2 \ldots y_n)$ of paired realisations of the two variables. This is given by

$$\hat{\rho}_P(x,y) = \frac{\sum_{i=1}^{n}(x_i - \bar{x})(y_i - \bar{y})}{\sqrt{\sum_{i=1}^{n}(x_i - \bar{x})^2}\sqrt{\sum_{i=1}^{n}(y_i - \bar{y})^2}} \tag{29.1}$$

(The reason for the 'hat' sign in $\hat{\rho}_P(x,y)$ is that $\hat{\rho}_P$ is an estimator based on a finite number of pairs of the theoretical correlation between the random variables *X* and *Y*, which is given by $\rho_P(X, Y) = \mathrm{Cov}(X, Y)/\sigma(X)\,\sigma(Y)$.) Despite this instinctive response, we all know from our statistical training that this is an insufficient way of looking at dependencies. For example, if you calculate the correlation between the vectors $x = (1, 2, \ldots 100)$ and $y = (\exp(1), \exp(2) \ldots \exp(100))$ this turns out to be approximately 25% (very poor), although it is obvious that y_k is known once x_k is known: $y_k = \exp(x_k)$, and therefore x_k and y_k are fully dependent.

The reason why linear correlation fails in this case is that linear correlation, as its name implies, estimates the dependency of two variables on the assumption that the relationship between them is linear. However, many dependencies in the real world are non-linear, and assuming that they are may result in missing these dependencies altogether.

Another problem with linear correlation is that although $\hat{\rho}_P(x,y)$ is a good estimator of $\rho_p(X, Y)$ when X and Y are known to be correlated, there is no general criterion to determine whether there *is* a linear correlation between two general random variables X, Y based on the value of $\hat{\rho}_P(x,y)$. The main exception to this is when X, Y can be approximated as normal variables, in which case the statistic

$$t = \hat{\rho}_P \sqrt{\frac{N-2}{1-\hat{\rho}_P^2}} \tag{29.2}$$

is distributed in the null hypothesis case (X, Y not linearly correlated, or formally $\rho_P(X, Y) = 0$) as a Student's t distribution with $\nu = N - 2df$, and the significance level of the calculated linear correlation can then be assessed. Further approximations can be obtained when N is large. The interested reader is referred to the book by Press et al. (2007), which has an in-depth discussion and formulae dealing with the different cases.

29.3.1 Rank Correlation

A much better way to look at the association between variables is through *rank* correlation. To calculate the rank correlation between two vectors, one simply sorts the elements of each vector in order and assigns a rank to each element depending on its position in the sorted vector (traditionally, the largest value has rank equal to 1).

There are different methods to calculate rank correlation. The most useful for our purposes is probably Spearman's ρ, which defines the rank correlation as the linear correlation of the ranks, after assigning an appropriate rank to tying values.[2] A common convention is to assign to all values that tie a rank equal to the average rank that these values would have if they were all different: for example, the ranks of values 9, 8, 6, 6 and 5 are 1, 2, 3.5, 3.5, and 5 (in this order). Other conventions are possible, but this particular one has the attractive property that the sum of the ranks of N values is always $N(N + 1)/2$, whether there are ties or not. The formula for Spearman's ρ is therefore

$$\hat{\rho}_P(x,y) = \frac{\sum_{i=1}^{n}\left(\text{rank}(x_i)-\bar{r}_x\right)\left(\text{rank}(y_i)-\bar{r}_y\right)}{\sqrt{\sum_{i=1}^{n}\left(\text{rank}(x_i)-\bar{r}_x\right)^2}\sqrt{\sum_{i=1}^{n}\left(\text{rank}(y_i)-\bar{r}_y\right)^2}} \tag{29.3}$$

where $\bar{r}_x = \sum_{i=1}^{n} \text{rank}(x_i)/n$, $\bar{r}_y = \sum_{i=1}^{n} \text{rank}(y_i)/n$. Again, $\hat{\rho}_S$ is an estimator – based on a finite number of pairs of points – of the theoretical Spearman's ρ of the random variables X and Y, which could be written as $\rho_S(X, Y) = \rho_P(F_X, F_Y)$.

Rank correlation does not use all the information in the vectors – specifically, it does not use the absolute size of the vector elements – but is more robust and is by definition invariant under transformations that leave the order of the values unchanged. So, for example:

2 A lazy and non-rigorous way to avoid the problem of ties is that of adding a very small random amount to each of the values so that the values are all distinct. The resulting rank correlation will depend on the random values sampled but will usually be close enough to the correct value.

- the rank correlation of $x = (x_1, \dots x_n)$ and $z = (z_1, \dots z_n)$ is the same as that of $x^2 = (x_1^2, \dots x_n^2)$ and z;
- the rank correlation of the vectors $x = (1, 2, \dots 100)$ and $y = (\exp(1), \exp(2) \dots \exp(100))$ is 1.

Another advantage of rank correlation is that the significance of a non-zero value of $\hat{\rho}_S(x,y)$ can be tested in a distribution-independent fashion. Similar to the case of linear correlation, the statistic

$$t = \hat{\rho}_S(x,y)\sqrt{\frac{N-2}{1-\hat{\rho}_S(x,y)^2}} \tag{29.4}$$

is approximately distributed according to a Student's t distribution with $N - 2$ *df*. However, this approximation works quite well regardless of the originating distributions of X and Y.

29.3.2 Tail Dependence

Tail dependence is a phenomenon by which the percentiles related to the tail (either one tail or both tails) of a distribution tend to be associated more strongly than the percentiles in the body of the distribution. The reason why this is interesting is that two risks may be reasonably independent except in the region of the very large losses.

Upper tail dependence can be defined rigorously based on $\tau(u)$, the conditional probability that Y is above a certain percentile u if X also is:

$$\tau(u) = \Pr\left(\mathbf{Y} > F^{-1}(u) | X > F^{-1}(u)\right) \tag{29.5}$$

(A similar definition holds for lower tail dependence.)

In practice, tail dependence can be estimated by counting the number of pairs that occur above a certain percentile: if we have n pairs (x_j, y_j) as usual, and we calculate the empirical severity distributions for $X\left(\hat{F}_X(x)\right)$ and $X\left(\hat{F}_Y(y)\right)$, we can estimate the tail dependence as

$$\hat{\tau}(u) = \frac{\text{no. of pairs}\left(x_j, y_j\right) \text{ for which } x_j > \hat{F}_X^{(-1)}(u) \text{ and } y_j > \hat{F}_Y^{(-1)}(u)}{\text{no. of pairs}\left(x_j, y_j\right) \text{ for which } x_j > \hat{F}_X^{(-1)}(u)} \tag{29.6}$$

We say that there is tail dependence if and only if $\tau^+ = \lim_{u \to 1^-} \tau(u) > 0$. Figure 29.3 shows a chart of $\tau(u)$ for two distributions with similar rank correlations, one with tail dependence and the other without tail dependence.

It should be noted that even if there is no linear correlation and no rank correlation, there may still be tail dependence between two variables, as it will be clear in the example of the t copula (see especially Equation 29.12).

Looking at rank correlation rather than at straightforward linear correlation may seem to some just another trick, or just another thing that one can be done with data. However, rank correlation actually captures something more profound about the dependency structure of two random variables, which is closer in spirit to the modern approach to modelling dependency.

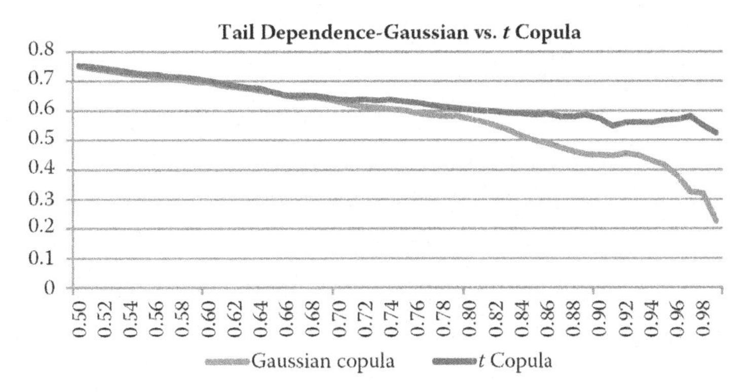

FIGURE 29.3
The chart shows $\tau(u)$ for two distributions with similar rank correlation, one of which (the black line, a Gaussian copula) has no tail dependence and for which therefore $\tau(u)$ decreases until becoming zero (except for some noise), and the other (the grey line, a *t* copula) has tail dependence and stabilises at a level of approximately 0.55.

BOX 29.3 PRACTICAL CALCULATION TIPS

In Excel, the linear correlation between two vectors x and y can be calculated with the function CORREL(x,y). The rank correlation between x and y can be calculated by first associating a rank to each element of x and y through the function RANK. AVG(element, vector), and then by calculating the correlation through the function CORREL(x,y), as in Figure 29.4.

In R, the linear correlation between two vectors x and y can be calculated with the function cor(x,y) and Spearman's rank correlation with the function cor

Rank Correlation −0.6075

`=CORREL(Range_Rank_X,Range_Rank_Y)`

ID	X	Y	rank(X)	rank(Y)
1	6	2	**4.5**	**9.5**
2	7	4	**2.5**	**7.0**
3	5	5	**6.0**	**5.0**
4	1	8	**10.0**	**3.5**
5	3	10	**8.5**	**1.0**
6	3	4	**8.5**	**7.0**
7	4	9	**7.0**	**2.0**
8	6	8	**4.5**	**3.5**
9	7	2	**2.5**	**9.5**
10	10	4	**1.0**	**7.0**

`Range_Rank_X`

`Range_Rank_Y`

FIGURE 29.4
An example of an Excel calculation of rank correlation between two vectors X and Y.

(x,y,method = 'spearman'): for example, for the vectors $x = (1, 2, \ldots 100)$ and $y = (\exp(1), \exp(2) \ldots \exp(100))$ mentioned above:

```
> x=c(1:100)
> y=exp(x)
> cor(x,y)
[1] 0.252032
> cor(x,y,method='spearman')
[1] 1
```

29.4 Modelling Dependency with Copulas

The modern way of looking at dependency is through copulas. Copulas go straight to the heart of things – because in statistics, we say that two variables are dependent when the joint distribution $F_{X,Y}(x, y)$ cannot be factored as $F_{X,Y}(x, y) = F_X(x)\, F_Y(y)$. In statistical jargon, this is equivalent to saying that you cannot get the joint distribution from the marginal distributions alone, that is, $F_X(x)$ and $F_Y(y)$.

From an abstract point of view, a copula is a function $C(u,v)$ that *does* allow one to write the joint distribution $F_{X,Y}(x, y)$ as a function of its marginal distributions:

$$F_{X,Y}(x, y) = C\left(F_X(x), F_Y(y)\right) \tag{29.7}$$

The simplest copula is possibly that which does not do anything – that is, it leaves the distributions $F_X(x)$, $F_Y(y)$ independent: $C(u,v) = uv$. This is normally indicated as $\Pi(u, v)$ and is called the product copula. Obviously, the resulting joint distribution is $F_{X,Y}(x, y) = \Pi(F_X(x), F_Y(y)) = F_X(x)F_Y(y)$.

For a more formal definition of a copula, and an explanation of its connection with the marginal distribution through Sklar's theorem (which is, essentially, Equation 29.7), refer to the book by Nelsen (2009). An influential introduction to modelling loss dependencies can be found in Wang (1998).

29.4.1 An Intuitive Approach to Copulas

In simple terms, copulas are – as the name suggests – 'mating systems', and they can be looked at as recipes for matching the percentiles of two (or more) distributions when sampling from the joint distribution. This is illustrated graphically in Figure 29.5 through three simple cases.

Figure 29.5a shows that if two random variables X and Y are completely dependent and concordant (this is the case if, for example, X and Y represent the two shares of a quota share arrangement), this means that the pth percentile of X will always be matched with the pth percentile of Y. Hence, the percentile values can be paired with straight lines that go from one variable to the other.

Figure 29.5b illustrates the case where the two distributions are completely independent, in which case the sample pth percentile of X can be matched with any percentile of Y with

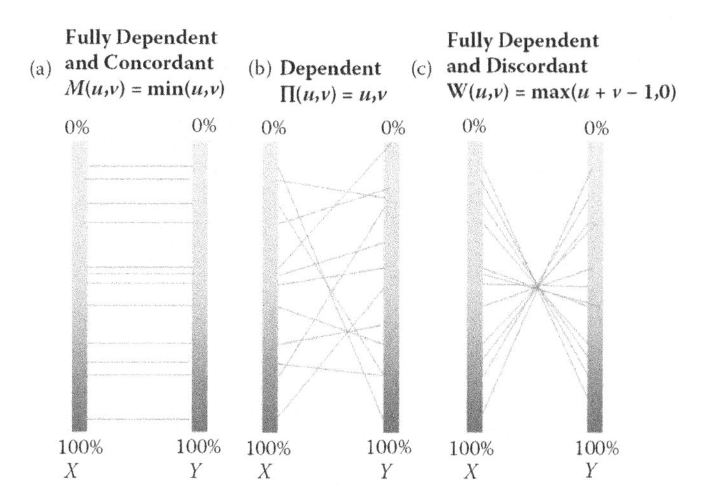

FIGURE 29.5
An illustration of copulas as rules for matching the percentiles of different distributions.

equal probability. As a result, the dots representing the percentiles of X and Y seem to be connected at random.

Figure 29.5c shows that if the two distributions are completely dependent and discordant (e.g. if X and Y' are two shares of a quota share arrangement, X and $Y = Y'$ will be fully dependent and discordant), this means that the pth percentile of X will always be matched with the $(1-p)$th percentile of Y.

Now the situation we have illustrated in Figure 29.5 actually corresponds to three 'trivial' copulas: the full dependence copula, $M(u, v) = \min(u, v)$, the full negative dependence copula $W(u, v) = \max(u + v - 1, 0)$, and the product copula, $\Pi(u, v) = uv$. Any other case will be somewhere in between the two extreme cases of $W(u,v)$ and $M(u,v)$. This is literally true because of the Frechet–Hoeffding bounds inequality, which states that $W(u, v) \leq \Pi(u, v) \leq M(u, v)$. Notice that $\Pi(u, v)$ can be considered to be 'right in the middle' between the full dependence and the full negative dependence cases.

If there is at least some degree of dependence, there will be some degree of regularity in the way that percentiles are associated, for example, for dependent and concordant distributions, the top percentiles of X will *tend* to be associated with the top percentiles of Y. Of course, there is always a degree of randomness, by which some of the top percentiles of X will associate with the bottom percentiles of Y, and the like; however, the association will not be completely chaotic.

This is illustrated in Figure 29.6a, which is based on a Gaussian copula with positive correlation. Note that both in Figure 29.5c and Figure 29.6a we do not see a deterministic association between two specific percentiles (as is the case in Figure 29.5a and b), but a probabilistic rule of association. If that helps, you can view the association rule as a stochastic process, and the particular graphs that we see in Figures 29.5 and 29.6 as *realisations of a stochastic process*, much in the same way as the throwing of a coin is a stochastic process but the particular sequence of heads and tails that you get in an experiment is a realisation of that process.

Finally, Figure 29.6b shows an example of *right-tail dependence*, by which the higher percentiles of two distributions tend to occur together. It should now be clear why we said, at the beginning of this section, that a copula is a mating system: it is a (probabilistic) rule for matching the percentiles of two marginal distributions (X and Y) – in graphical terms, a rule for producing graphs like those in Figures 29.5 and 29.6.

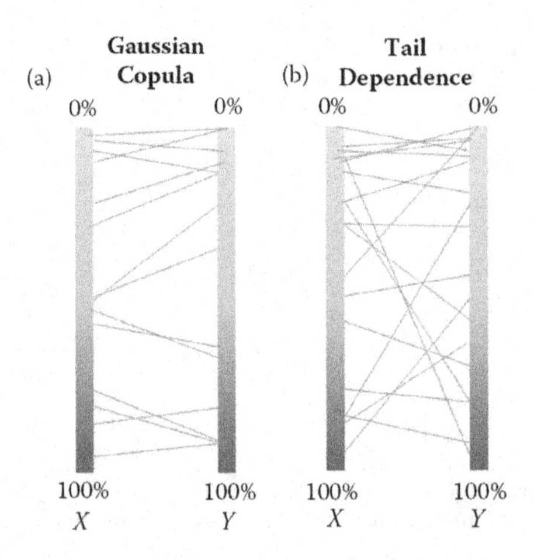

FIGURE 29.6
Gaussian copula with correlation 80% (a). Illustration of a copula with tail dependence (b). When two stochastic loss processes X and Y have strong (top) tail dependence, large losses of X tend to occur together with large losses of Y. Note that one can have right-tail dependence (large losses) or left-tail dependence (small losses), or both-tail dependence: the figure on the right is an illustration of right-tail dependence.

29.4.1.1 Relationship with Rank Correlation

As mentioned at the start, rank correlation is a concept which – being based on the percentiles of the distributions rather than on the absolute values of the random variables – is much closer in spirit to the modern copula approach. Indeed, it is straightforward to calculate Spearman's ρ based on the knowledge of the copula $C(u,v)$, which describes the dependency of two variables X, Y:

$$\rho_S(X, Y) = 12\int_0^1\int_0^1 (C(u,v) - uv)\,du\,dv = 12\int_0^1\int_0^1 C(u,v)\,du\,dv - 3 \tag{29.8}$$

29.4.2 Notable Examples of Copulas

Let us now look at some simple examples of copulas, which will also be the only types of copulas that we will need in this book: the Gaussian copula and the t copula.

29.4.2.1 Gaussian Copula

A Gaussian copula is a copula with the form:

$$C_\rho^\Phi(u,v) = \Phi_\rho\left(\Phi^{-1}(u), \Phi^{-1}(v)\right) \tag{29.9}$$

where $\Phi(x)$ is the CDF of the standardised normal distribution (mean = 0, standard deviation = 1) and $\Phi_\rho(x, y)$ is the CDF of the standardised binormal distribution with correlation ρ.

How do you create a dependency between two random variables X and Y (e.g. two risks) using a Gaussian copula? Assume X and Y are two random variables, representing, for example, the total losses over a year, or the severities of two different components of one loss, and you want to make sure that there is a rank correlation of value r_S.

Applying Equation 29.8 to the case of a Gaussian copula with linear correlation p yields the following value for the rank correlation:

$$\rho_S(X, Y) = \frac{6}{\pi}\arcsin\frac{\rho}{2} \tag{29.10}$$

Hence, to obtain a rank correlation of r_S, one must use a Gaussian copula with linear correlation $\rho = 2 \sin(r_S\pi/6)$. The algorithm for generating a pair of correlated values is outlined in Box 29.4.

Note that:

- By definition, x and y come from two distributions correlated in the sense of rank, and the rank correlation is r_S
- Note that F_X and F_Y can be either two models or two empirical distributions

Hopefully it is clear what the algorithm in Box 29.4 has managed to achieve: we have taken two random variables that are linearly correlated (with Pearson's correlation equal to ρ), and from them we have produced two numbers between 0 and 1 that are correlated in the sense of ranks (with Spearman's ρ correlation equal to r_S). We then use these correlated ranks to sample from two distributions that are not necessarily normal – they may simply be empirical distributions.

29.4.2.2 Student's t Copula

Although the Gaussian copula has the advantage of simplicity and of numerical tractability, it is of no use if we need to model tail dependency. A copula that is similar to the Gaussian copula and is still simple and tractable and is also able to model tail dependency is the *t* copula.

BOX 29.4 ALGORITHM FOR GENERATING TWO RANDOM VARIABLES WITH RANK CORRELATION R_S USING A GAUSSIAN COPULA

Input: The marginal distributions F_X and F_Y of two random variables; the desired rank correlation r_S.

1. Generate two random numbers z_1 and z_2 from a standard normal distribution: $Z_1 \sim N(0,1)$, $Z_2 \sim N(0,1)$
2. Choose $\rho = 2 \sin(r_S\pi/6)$
3. Produce a third number $z_2' = \rho z_1 + \sqrt{1-\rho^2}\, z_2$. This ensures that the two random variables Z_1 and Z_2' have a linear correlation of ρ
4. Calculate the value of the cumulative distribution of a normal distribution at the values z_1 and z_2': $u = \Phi(z_1), v = \Phi(z_2')$, where Φ is the symbol commonly used to represent the CDF of a standard normal distribution. By definition, u and v are numbers between 0 and 1
5. Choose two values from the distributions of X, Y as follows: $x = F_X^{-1}(u), y = F_X^{-1}(v)$. Note that F_X and F_Y are the CDFs for the two variables, which are either parametric distributions (such as gamma, or lognormal) or even just the empirical distributions

A t copula is formally defined as the function

$$C^{t}_{v,\rho}(u,v)=t_{v,\rho}\left(t_{v}^{-1}(u),t_{v}^{-1}(v)\right) \tag{29.11}$$

where t_{v} is the CDF of the one-dimensional Student's t distribution with v degrees of freedom, and $t_{v,\rho}$ is the CDF of the bivariate Student's t distribution with v degrees of freedom and linear correlation ρ. Note that although v is called the number of degrees of freedom with a good reason, it can actually be a non-integer, which adds extra flexibility to this copula. Also note that when the number of degrees of freedom tends to infinity, the Student's copula becomes a Gaussian copula.

The rank correlation is related to the linear correlation parameter p of the t copula in the same way as for the Gaussian copula: $\rho_S(X,Y)=\frac{6}{\pi}\arcsin(\rho/2)$.

As mentioned previously, the interesting thing about the t copula is that it exhibits tail dependence. The value of τ^{+} for a t copula with correlation ρ and ν degrees of freedom (remember that ν can actually be a non-integer) can be obtained through Equation 29.5, yielding:

$$\tau^{+}=2t_{\nu+1}\left(-\sqrt{\frac{(\nu+1)(1-\rho)}{1+\rho}}\right) \tag{29.12}$$

Note that τ^{-}, the limit for the lower tail dependence, is equal to τ^{+} for the t copula. Also note that the t copula exhibits tail dependence even in the case where $\rho = 0$ (and therefore the rank correlation is also 0).

To model the dependency of two risks using a t copula, you can proceed as follows: assume X and Y are two random variables, and you want to produce a rank correlation r_S and a tail dependence τ^{+}. One algorithm for generating a pair of correlated values is illustrated in Box 29.5.

As before, x and y come by definition from two correlated (in the sense of rank) distributions, and the rank correlation is r_S, and F_X and F_Y can either be two models or the empirical distributions.

29.4.2.3 Other Copulas

The Gaussian copula and the t copula are two very popular copulas amongst a large number of possible choices, each justified by their theoretical simplicity (such as the class of Archimedean copulas) or their appropriateness for a given context (such as extreme value copulas, obviously useful to model the dependency between large losses.

In practice, it will be quite difficult to be able to discern amongst many different types of copulas the one which is most suitable for a specific situation on the basis of data alone, and therefore the two basic types will normally suffice to capture the essential aspects of dependency. In this textbook, therefore, we will limit ourselves to the Gaussian and the t copula.

However, in some circumstances, there may be reasons to use different types of copula. The book by Nelsen (2009) is a good reference if you wish to become familiar with other types of copulas available.

BOX 29.5 ALGORITHM FOR GENERATING TWO RANDOM VARIABLES WITH RANK CORRELATION R_S AND TAIL DEPENDENCE T⁺ USING A *T* COPULA

Input: The marginal distributions F_X and F_Y of two random variables; the desired rank correlation r_S; the desired tail dependence τ^+.

1. Generate two random numbers z_1 and z_2 from a standard normal distribution: $Z_1 \sim N(0,1)$, $Z_2 \sim N(0,1)$
2. Choose $\rho = 2\sin(r_S\pi/6)$, and choose v with numerical methods so as to satisfy Equation 29.12 given the constraints on ρ and τ^+
3. Produce a third number $z_2' = \rho z_1 + \sqrt{1-\rho^2}\, z_2$. This makes sure that the two random variables Z_1 and Z_2' have a linear correlation of ρ
4. Generate a fourth random number s from a χ^2 distribution with v degrees of freedom: $s \sim Chi2(v)$
5. Calculate two new numbers: $z_1^v = \sqrt{v/s}\, z_1$, $z_2^v = \sqrt{v/s}\, z_2'$
6. Calculate the value of the cumulative distribution of a normal distribution at the values z_1 and z_2':

$$u = \begin{cases} 1 - t_v(z_1^v), & \text{if } z_1^v > 0 \\ t_v(-z_1^v), & \text{if } z_1^v > 0' \end{cases} \quad v = \begin{cases} 1 - t_v(z_2^v), & \text{if } z_2^v > 0 \\ t_v(-z_2^v), & \text{if } z_2^v > 0 \end{cases}$$

   where t_v represents the CDF of a Student *t* distribution with v degrees of freedom. By definition, u and v are numbers between 0 and 1
7. Choose two values from the distributions of X, Y as follows: $x = F_X^{-1}(u)$, $y = F_X^{-1}(v)$. Note that F_X and F_Y can either be two models or the empirical distributions

29.5 Aggregate Loss Distribution of Two Correlated Risks

Let us now see how it is possible in practice to produce an aggregate loss model for the combination of two correlated risks through the Monte Carlo simulation, which is a problem that we will have to deal with frequently in pricing.

There are several ways in which two risks can be correlated: the main ones are, as usual, by frequency or by severity, although subtler dependencies (e.g. between the reporting delays) are theoretically possible. Furthermore, in those cases in which we are unable or unwilling to disentangle the effects of frequency and severity dependency, one might consider directly the dependency between the aggregate losses.

29.5.1 Correlating the Severities of Two Risks

In many circumstances, it is the severities of two risks that are correlated. Some obvious examples include the following:

- The third-party liability and the own-damage components of a motor loss (a large liability claim seems likely to arise typically from a loss that also damages the driver's car)

- The third-party property damage and the third-party bodily injury components of a third-party liability motor loss
- The hull component and the liability component of an aviation loss arising from the same event (the more serious the damage to the hull is, the more likely it is that passengers or other members of the public suffer third-party damage and bodily injury)
- The property damage and the business interruption component of a commercial property loss (it is likely that more extensive property damage will cause longer periods of business interruption)

In all these cases, the issue of correlation arises because of the need to model the different components of a risk separately because, for example, they are subject to different retentions.

Special care must be taken in considering that in some cases, one of the two components of the loss might be zero, and that therefore the probability of a loss being zero must be modelled independently.

There are several ways of modelling the dependency of the severities of two risks A and B. We consider two of them.

29.5.1.1 Method 1: Split Aggregate Loss Model

In this method, we produce three different aggregate loss models:

(A,B) – One for losses for which there is non-zero element for both A and B
(A,0) – One for those losses that have a zero B component
(0,B) – One for those losses that have a zero A component

The overall method to create an aggregate loss model for two severity-correlated risks using the three-model approach and Monte Carlo simulation is described below.

Then, sort the values of $S^{(k)}$ in ascending order in the standard way to produce the empirical aggregate loss distribution. The retained and ceded distribution for a given insurance programme can also be easily produced during the simulation, by calculating the retained part of each $z_i^{(k)}$ and applying aggregate retentions/limits as necessary.

Of course, copulas other than the Gaussian and the t copula might have been used, but these have been used above for simplicity. As mentioned previously, it is very difficult to decide on the basis of empirical evidence alone which type of copula should be used.

29.5.1.2 Method 2: Two Severity Models

Another way is to produce two different severity distributions, each of which allows for the possibility of zero losses with finite probability, then model the dependency of those two severity distributions directly.

29.5.2 Correlating the Frequencies

Risks may also be correlated through the frequency. An example of this would be when some external factor (such as the state of the economy, or the weather) affects different risks (such as public liability and motor third-party liability), in the sense that in the periods where there are more public liability losses there tend to be a larger number of motor losses, and vice versa.

BOX 29.6 AGGREGATE LOSS MODEL OF TWO SEVERITY-CORRELATED RISKS VIA THE MONTE CARLO SIMULATION (METHOD 1)

1. Produce a frequency model for the number of non-zero losses (i.e. all losses for which at least one of the two components is non-zero)
2. Estimate from the data the probability $p_{(A,B)}, p_{(A,0)}, p_m$ ($p_{(A,B)} + p_{(A,0)} + p_{(0,B)} = 1$) that a particular non-zero loss will be of the type (A,B), (A,0) or (0,B): let $\hat{p}_{(A,B)}, \hat{p}_{(A,0)}, \hat{p}_{(0,B)}$ be such estimates
3. Produce a separate severity model for
   a. the A component of the losses of type (A,B), $\widehat{F}_X^{(A,B)}(x)$
   b. the A component of the losses of type (A,0), $\widehat{F}_X^{(A,0)}(x)$
   c. the B component of the losses of type (A,B), $\widehat{F}_Y^{(A,B)}(y)$
   d. the B component of the losses of type (0,B), $\widehat{F}_Y^{(0,B)}(y)$

   It might well be the case that the empirical distributions (a) and (b) (and [c] and [d]) are not significantly different, in which case a single model can be used for each component.
4. Estimate the *rank correlation* ρ of the A and B component of losses of type (A,B)
5. Produce an aggregate loss model by Monte Carlo simulation as follows: For each simulation $k = 1, \dots N$
   a. Sample a number n_k of losses from the overall frequency model produced in Step 1
   b. For each of the n_k losses, we need to select its loss amount $z_i^{(k)}$. To do this, allocate the loss randomly to one of the three types (A,B), (A,0), or (0,B) with probability $\hat{p}_{(A,B)}, \hat{p}_{(A,0)}, \hat{p}_{(0,B)}$
      - If the loss has been allocated the type (A,0), generate a random variable x from distribution $\hat{F}_X^{(A,0)}(x)$ and set $z_i^{(k)} = x$
      - If the loss has been allocated the type (0,B), generate a random variable y from distribution $\hat{F}_Y^{(0,B)}(y)$ and set $z_i^{(k)} = y$
      - If the loss has been allocated the type (A,B), generate two random variables x (from distribution $\hat{F}_X^{(A,B)}(x)$) and y (from distribution $\hat{F}_Y^{(A,B)}(y)$) with rank correlation p using either a Gaussian copula with correlation ρ or a t copula with correlation ρ and the desired tail dependence, as explained in Boxes 29.4 and 29.5, respectively, and set $z_i^{(k)} = x + y$
   c. Calculate the total losses $S^{(k)}$ for simulation k as $s^{(k)} = \sum_{i=1}^{n_k} z_i^{(k)}$

At an elementary level, correlating frequencies can be done exactly as for severities and aggregate losses. However, it is more difficult with frequencies to have a large data set on which to judge whether there is a dependency in the number of claims between two risks, let alone to judge whether there is tail dependence.

As a consequence, the calibration of the copulas that are used to correlate the frequency cannot normally be done based on data alone but will require a judgment call. Assuming we have decided to model paired frequencies with, for example, a Gaussian copula or a t copula, the algorithms in Boxes 29.4 and 29.5 can be used to produce correlated pairs of frequencies: the random variables X, Y are in this case replaced by N_1 and N_2, the

BOX 29.7 AGGREGATE LOSS MODEL OF TWO SEVERITY-CORRELATED RISKS VIA THE MONTE CARLO SIMULATION (METHOD 2)

1. Produce a frequency model for the number of non-zero losses (i.e. all losses for which at least one of the two components is non-zero)
2. Produce a separate severity model for
   a. the A component of the losses, regardless of whether the B component is zero, $\hat{F}_X^{(A)}(x)$
   b. the B component of the losses, regardless of whether the A component is zero, $\hat{F}_Y^{(B)}(y)$
3. Estimate the *rank correlation* ρ of the A and B component of losses, without excluding zero values
4. Produce an aggregate loss model by Monte Carlo simulation as follows:
   For each simulation $k = 1, \dots N$
   a. Sample a number n_k of losses from the overall frequency model produced in Step 1
   b. For each of the n_k losses,
      - If the loss has been allocated the type (A,B), generate two random variables x (from distribution $\hat{F}_X^{(A)}(x)$) and y (from distribution $\hat{F}_Y^{(B)}(y)$) with rank correlation ρ using either a Gaussian copula with correlation ρ or a t copula with correlation p and the desired tail dependence, as explained in Boxes 29.4 and 29.5, respectively, and set $z_i^{(k)} = x + y$
   c. Calculate the total losses $s^{(k)}$ for simulation k as $s^{(k)} = \sum_{i=1}^{n_k} z_i^{(k)}$

Then sort the values of $s^{(k)}$ in ascending order in the standard way to produce the empirical aggregate loss distribution. The retained and ceded distribution can be calculated as explained in Box 29.6.

discrete random variables describing the number of claims of the two risks. These pairs of correlated frequencies can then be used to produce frequency-correlated aggregate loss models, as illustrated in Box 29.8.

29.5.3 Correlating the Total Losses of Two Risks

In some cases, not enough information is known on the correlation of two risks by frequency or by severity, but we know (or we assume) that the two risks are correlated on the aggregate, that is, the total losses from one risk are not completely independent of the total losses from another risk.

As an example, we might have at one's disposal studies on the correlation between the loss ratios of different lines of business for different insurers or for the whole market, after correcting for effects such as the underwriting cycle.

Of the three cases (correlation by severity, by frequency, and by total losses) this is the most straightforward. However, attention must be paid to the way we correlate the ceded and retained components. Box 29.9 explains the procedure.

BOX 29.8 AGGREGATE LOSS MODEL OF TWO FREQUENCY-CORRELATED RISKS VIA MONTE CARLO SIMULATION

i. Choose the copula for correlating the frequency of risks A and B, and calibrate the copula based on a combination of data and judgment ii.
ii. For each MC simulation $k = 1, \ldots N$, produce a correlated pair of claim counts $n_k^{(A)}, n_k^{(B)}$ using the methods in Boxes 29.4 (Gaussian copula) and 29.5 (t copula)
iii. Choose $n_k^{(A)}$ severities from the severity distribution of A, and consider the sum of all these severities, $S_k^{(A)} = X_1^{(A)} + X_2^{(A)} + \ldots X_{n_k^{(A)}}^{(A)}$
iv. Choose $n_k^{(B)}$ severities from the severity distribution of A, and consider the sum of all these severities, $S_k^{(B)} = X_1^{(B)} + X_2^{(B)} + \ldots X_{n_k^{(B)}}^{(B)}$
v. Calculate $S_k = S_k^{(A)} + S_k^{(B)}$ and sort the results
vi. The outcome is a gross loss model for the correlated variables
vii. To obtain losses ceded/retained, simply amend the losses in steps (iii) and (iv) above as necessary, and follow the calculations through

*29.6 Aggregate Loss Distribution of an Arbitrary Number of Correlated Risks

In Section 29.5, we looked at how we could calculate the aggregate loss distribution of two correlated risks. This is perhaps the most typical in pricing situations, such as when dealing with the property and business interruption components of a PDBI risk, or the third party and the own-damage component of a motor risk, or the hull and liability components of an aviation risk. However, there are many meaningful cases in which we have to consider the interaction of a larger number of risks: for example, if we are looking at all the risks of a non-insurance organisation or of an insurer in a holistic way.

Conceptually, the generalisation from two risks to an arbitrary number of risks is straightforward. However, the added dimensionality of the problem requires the use of the formalism of linear algebra, and therefore of the appropriate computational tools to diagonalise matrices, to ascertain that they are positive-definite and so on.

To start with, the correlation of n risks is given in the form not of a single number but of a matrix. For example, if we use Spearman's ρ rank correlation, the matrix is $\left\{R_{i,j}^{s}\right\}_{i,j=1,\ldots n}$, where $R_{i,j}^{s}$ is the rank correlation between risk i and risk j.

The algorithm for generating an arbitrary number of correlated variables using a Gaussian and a t copula is outlined in Boxes 29.10 and 29.11. These algorithms can be found, for example, in the books by O'Kane (2010) and by Schmidt (2006). Given the additional technical difficulties with respect to the two-variable case, we have commented extensively (in italics) on each step of the algorithm, to explain why things are done in the way they are done.

The t copula case is only marginally different from the Gaussian case explained in Box 29.10 and the comments in Box 29.11 for the t copula case will therefore be much sparser, focusing on the aspects that are different.

BOX 29.9 AGGREGATE LOSS MODEL OF TWO RISKS CORRELATED THROUGH THE TOTAL LOSSES

i. Consider N simulations of the aggregate losses for risk $A: S_k^{(A)}$, $k = 1, \dots N$, and also calculate the retained $\left(S_{\text{ret},k}^{(A)}\right)$ and ceded $\left(S_{\text{ced},k}^{(A)}\right)$ components of $S_k^{(A)}$: $S_k^{(A)} = S_{\text{ret},k}^{(A)} + S_{\text{ced},k}^{(A)}$

ii. Sort the simulated losses in ascending order: $S_{(1)}^{(A)} \leq \dots \leq S_{(N)}^{(A)}$. This is the empirical distribution for $S^{(A)}$, which we can call $\hat{S}^{(A)}$. Note that the corresponding values of $S_{\text{ret},k}^{(A)}, S_{\text{ced},k}^{(A)}$ are not necessarily in ascending order!

iii. Do the same for risk B, calculating $S_k^{(B)}, S_{\text{ret},k}^{(B)}, S_{\text{ced},k}^{(B)}$, for $k = 1, \dots N$ and sorting them in ascending order:$S_{(1)}^{(B)} \leq \dots \leq S_{(N)}^{(B)}$, obtaining the empirical distribution for risk $B, \hat{S}^{(B)}$

iv. Simulate a large number N' of pairs of aggregate losses (N' is not necessarily equal to N), correlated with a Gaussian copula or a t copula, as required: $\left(\hat{S}_{(i_r)}^{(A)}, \hat{S}_{(j_r)}^{(B)}\right)$. The index r indicates here the rth simulation ($r = 1, \dots N'$) and i_r and j_r are the indices of the *sorted* aggregate loss distribution picked by the copula in that simulation. The retained and ceded values are also kept track of at this stage: $\left(\hat{S}_{\text{ret},(i_r)}^{(A)}, \hat{S}_{\text{ret},(j_r)}^{(B)}\right), \left(\hat{S}_{\text{ced},(i_r)}^{(A)}, \hat{S}_{\text{ced},(j_r)}^{(B)}\right)$

v. For each simulation, calculate the overall aggregate losses, on a gross basis $\left(\hat{S}_r^{(\text{overall})} = \hat{S}_{(i_r)}^{(A)}, + \hat{S}_{(j_r)}^{(B)}\right)$, on a retained basis $\left(\hat{S}_{\text{ret},r}^{(\text{overall})} = \hat{S}_{\text{ret},(i_r)}^{(A)}, + \hat{S}_{\text{ret},(j_r)}^{(B)}\right)$, and on a ceded basis $\left(\hat{S}_{\text{ced},r}^{(\text{overall})} = \hat{S}_{\text{ced},(i_r)}^{(A)}, + \hat{S}_{\text{ced},(j_r)}^{(B)}\right)$

vi. Sort the overall aggregate losses in ascending order in the usual way, on all bases (gross, retained and ceded), obtaining the empirical overall aggregate loss distribution

As can be imagined, implementing the algorithms above in a language without large mathematical and statistical libraries is complex and therefore providing code or pseudo-code here is not helpful. However, an R implementation – which has a library 'copula' dealing with the generation and the parameterisation of copulas – is relatively easy to illustrate and can be found in the appendix to this chapter.

The following R code shows a very simple case in which three lognormal distributions are correlated with a t copula (normally, the code will be more complicated because it has to allow for the severity distribution to be represented as a spliced distribution, or as a discrete vector).

29.7 Common Shock Modelling

One limitation of the copula approach to correlation is that it assumes a stationary situation. At any particular moment in time, there is a (stochastic) relationship between a number of risks, by which, for example, the frequency of one risk is more likely to be high when the frequency of the other is also high. However, it is often the case that there are

BOX 29.10 ALGORITHM FOR GENERATING A TUPLE OF *N* RANDOM VARIABLES WITH RANK CORRELATION MATRIX USING A GAUSSIAN COPULA

Input: The marginal distributions $F_{X(1)}, F_{X(2)} \ldots F_{X(n)}$ of n random variables; the desired rank correlation (Spearman's ᴀ ρ) matrix $R^s_{i,j}$.

1. Generate n random numbers $z_1, z_2 \ldots z_n$ from a standard normal distribution: $Z_j \sim N(0,1), j = 1, \ldots n$. For convenience, define the vector $z = (z_1, z_2 \ldots z_n)$
2. Choose $\rho_{i,j} = 2\sin\left(R^s_{i,j}\pi/6\right)$. This defines a new matrix $\Sigma = \left(\rho_{i,j}\right)_{i=1,\ldots n; j=1,\ldots n}$

   This is done because $\rho_{i,j}$ *is the correlation in the Gaussian copula (and the t copula as well) necessary to produce a Spearman's* ρ *equal to* $R^s_{i,j}$.
3. Put $\sum$ in diagonal form, obtaining $\mathbf{D} = \text{diag}(\lambda_1, \ldots \lambda_n)$ (that is, *a matrix for which all elements except those in the diagonal are zero, and the elements in the diagonal are* $\lambda_1, \ldots \lambda_n$).

   Since $\sum$ *is a symmetric matrix, this can always be done. The relationship between* $\sum$ *and* $\mathbf{D}$ *can be written as* $\mathbf{D} = \mathbf{V}^{-1}\Sigma\mathbf{V}$ *where* $\mathbf{V}$ *is the matrix, which has the eigenvectors of* $\sum$ *as its columns. This relationship can be inverted to yield* $\Sigma = \mathbf{VDV}^{-1}$.
4. If $\sum$ is not positive-semidefinite, repair $\sum$ by calculating the matrix $\Sigma' = \mathbf{VD'V}^{-1}$, where $\mathbf{D}' = \text{diag}(\max(\lambda_1, 0), \ldots \max(\lambda_n, 0))$ and V was defined in Step 3. $\sum'$ is in a way the 'closest' matrix to $\sum$, which is also positive-semidefinite.

   In order for $\sum$ *(which is a symmetric matrix) to be a valid correlation matrix, it must be positive-semidefinite,* that is, *its eigenvalues must all be non-negative:* $\lambda_j \geq 0$ *for all* $j = 1, \ldots n$.

   If Σ *is not positive-definite, technically, we cannot use that as a correlation matrix: however, what we can do is to assume that the matrix we have is a reasonable approximation of the true correlation matrix and calculate the 'closest' matrix to* Σ, *which is also positive-definite. This can be simply obtained by first setting to zero all eigenvalues that are less than zero:* $\mathbf{D}' = \text{diag}(\max[\lambda_1, 0], \ldots \max[\lambda_n, 0])$ *and then using the original eigenvectors to produce the corrected matrix:* $\Sigma' = \mathbf{VD'V}^{-1}$.

   As a result of doing this transformation, the matrix Σ' *does not have '1s' in the diagonal anymore if* Σ' *had negative eigenvalues, which makes it unsuitable to represent a correlation matrix. It is therefore necessary to normalise* Σ' *as follows.*
5. Normalise the matrix Σ', obtaining matrix $\Sigma'' = \sum'\mathbf{W}$ where $W_{ij} = 1/\sqrt{B_{ii}B_{jj}}$

   The matrix elements $\mathbf{W}$ *can be viewed as the external product of the* $\mathbf{w} = \left(1/\sqrt{B_{11}}, \ldots 1/\sqrt{B_{nn}}\right)$ *with itself.*
6. Calculate the Cholesky decomposition of Σ'', that is, the unique lower-triangular matrix (i.e. with all entries above the diagonal being zero) $\mathbf{B}$ such that $\Sigma'' = \mathbf{BB}^T$, and create the new vector $z' = Bz$

   The effect of applying the lower-triangular matrix B *to* z *is to create a vector* z', *which has covariance matrix* Σ'' *as desired. In the two-dimensional case in Section 29.4, the matrix* B *is*

$$\mathbf{B} = \begin{pmatrix} 1 & 0 \\ \rho & \sqrt{1-\rho^2} \end{pmatrix}$$

7. Calculate the value of the cumulative distribution of a standard normal distribution at the values $z'_1, z'_2 \dots z'_n : u_1 = \Phi(z'_1), \dots u_n = \Phi(z'_n)$.
8. Return n values from the distributions of X_1, X_2, … X_n as follows: $x_j = F_{X_j}^{-1}(u_j)\forall j = 1, \dots n$

   These x_j's are random variables correlated via a t copula with correlation matrix Σ'', as desired.

BOX 29.11 ALGORITHM FOR GENERATING A TUPLE OF N RANDOM VARIABLES WITH RANK CORRELATION MATRIX USING A T COPULA

Input: The marginal distributions $F_{X(1)}, F_{X(2)} \dots F_{X(n)}$ of n random variables; the desired rank correlation (Spearman's ρ) matrix $R^s_{i,j}$; the number of degrees of freedom ν > 0.

Ideally, the number of degrees of freedom should be chosen so as to produce the desired tail dependence, τ^+, between the variables. However, τ^+ depends on both the number of degrees of freedom and the correlation ρ (see Equation 29.12) and will therefore turn out to be different for every pair of variables. The problem of determining v is therefore overdetermined – a t copula simply cannot provide all the flexibility necessary to model a complex dependency structure. A good compromise is that of calculating the average correlation p between all nonidentical pairs of variables and use that to invert Equation 29.12.

1. Generate n random numbers z_1, z_2, … z_n from a standard normal distribution: $Z_j \sim N(0,1)$, $j = 1, \dots n$. For convenience, define the vector $z = (z_1, z_2, \dots z_n)$
2. Choose $\rho_{i,j} = 2\sin\left(R^s_{i,j}\pi/6\right)$. This defines a new matrix $\Sigma = \left(\rho_{i,j}\right)_{i=1,\dots n; j=1,\dots n}$
3. Put $\sum$ in diagonal form, obtaining $\mathbf{D} = \text{diag}(\lambda_1, \dots \lambda_n)$
4. If $\sum$ is not positive-semidefinite, repair $\sum$ by calculating the matrix $\Sigma' = \mathbf{VD'V^{-1}}$, where $\mathbf{D'} = \text{diag}(\max(\lambda_1, 0), \dots \max(\lambda_n, 0))$ and V was defined in Step 3 (Box 29.10), and normalise it so that its diagonal elements are 1, yielding $\sum''$ (Box 29.10, Step 5)
5. Calculate the Cholesky decomposition of $\sum''$, $\sum'' = \mathbf{BB}^T$, and create the new vector $z' = Bz$
6. Generate a random value ξ from a χ^2 distribution with $v \geq 0$ *df* (if v is an integer, this can be obtained as a sum of squares of normal random variables: $\xi = \sum_{i=1}^{v} X_j^2$)
7. Calculate n values $u_i = t_v\left(\frac{z_i}{\sqrt{\xi / v}}\right)$, thus producing n quantiles from the correlated t distribution
8. Return n values from the distributions of X_1, X_2, … X_n as follows: $x_j = F_{X_j}^{-1}(u_j)\forall j = 1, \dots n$

These x_j's are random variables correlated via a t copula with correlation matrix Σ', as desired.

common drivers at work that suddenly increase the frequency or the severity of claims across multiple risks.

As an example, an economic recession is likely to increase the general propensity to claim across many different lines of business, although there is no direct causal relationship between the number of claims in, say, household insurance and car insurance. What we have here is a so-called 'common shock' to the system (Meyers, 2007). There is therefore a causal relationship between the shock and each individual line of business, but the loss production process conditional to this common shock may well be independent.

Other examples of shocks are as follows:

- Economic deterioration causing simultaneous deterioration of the credit rating of different loans in a portfolio
- An epidemic causing a sudden increase in mortality in the livestock or bloodstock of an establishment
- A court decision affecting the severity distribution of all liability claims in a combined portfolio

29.7.1 Common Shock Model

The effect of a common shock is to produce correlation. As shown, for example, in the article by Meyers (2007), if two independent variables X and Y are multiplied by a common random variable β with mean 1 and variance $v\beta$, the resulting variables βX and βY are correlated with correlation:

$$\rho(X,Y)=\frac{\mathrm{Cov}(\beta X,\beta Y)}{\sqrt{\mathbb{V}(\beta X)\mathbb{V}(\beta Y)}}=\frac{\mu_X\mu_Y v_\beta}{\sqrt{\left((1+v_\beta)v_X+\mu_X^2 v_\beta\right)\left((1+v_\beta)v_Y+\mu_Y^2 v_\beta\right)}} \tag{29.13}$$

where $\mu_X=\mathbb{E}(X)$, $v_X=\mathbb{V}(X)$, $\mu_Y=\mathbb{E}(Y)$, $v_Y=\mathbb{V}(Y)$, $v_\beta=\mathbb{V}(\beta)$.

The effect of a common shock becomes particularly useful in the context of the collective risk model, when applied to either the claim count or the severity of multiple distributions. See the article by Meyers (2007) for specific analytical results after applying a common shock to either the frequency or the severity distribution or both.

29.7.2 Binary Common Shocks

In practice, it is quite difficult to calibrate common shocks in a meaningful way and, in some circumstances, it will be simpler and more appropriate to just apply a binary common shock, that is, a shock that increases the frequency or the severity as a consequence of the occurrence of a specified event/set of circumstances. For example, one might consider a common shock on all frequencies (e.g. an increase of 20% of the frequencies across all classes) as a consequence of the economy being in a recession, where 'recession' is a binary event. Another example is a common shock given to all probabilities of default for mortgage indemnity guarantee. Yet another example is the sudden increase of all liability claims because of a specific court/legislative decision, such as the decrease of the discount rate applied in the Ogden tables, used to determine long-term awards for injured parties in personal injury cases.

All one needs to specify in this framework is two numbers: (1) the probability that the binary even occurs and (2) the effect of the event. The framework can easily be extended to

n-ary shocks – for example, shocks leading into n possible different states (such as severe recession, mild recession, normal, or economic boom).

Because the probability of the binary/n-ary event might be difficult to gauge based on historical data, it can be argued that a truly stochastic model provides little further information than a straightforward scenario analysis, in which the 'normal' scenario is compared with a 'stressed' scenario (and possibly other scenarios) and for each scenario, a common shock is applied to the relevant random variable, but no attempt is made to run a single simulation with all the elements combined. Even when this combination is carried out, it is still useful – for communication purposes – to show the different scenarios separately.

An example of the application of a common shock in practice is shown in Box 29.12.

29.8 R Code for Generating an Arbitrary Number of Variables Correlated through a *t* Copula

Through the copula library and the related libraries, it is straightforward to produce correlated variables with R. Example code for the case of n statistical distributions (which could be severity, frequency, or aggregate loss distributions) correlated through a t copula is available through the online appendix to this chapter (Parodi, 2016).

29.9 Questions

1. Compare advantages and disadvantages of analysing a risk (such as motor) by splitting it into its main components (such as own damage, third-party liability for motor).
2. In the table below, X and Y represent the property damage and bodily injury components of 19 motor third-party liability losses.

    a. Calculate the linear correlation and the rank correlation (Spearman's ρ) between X and Y.
    b. For both linear correlation and rank correlation, determine (if possible) whether the correlation between X and Y is significant at a 5% significance level, stating all assumptions you are making.

X	Y
23,077	239,807
–	202,259
12,905	7467
37,618	82,353
1732	22,313
6219	18,975
41,585	346,581

BOX 29.12 EXAMPLE OF THE APPLICATION OF A COMMON SHOCK TO CLAIM COUNT ACROSS A NUMBER OF LINES OF BUSINESS WRITTEN BY AN INSURANCE COMPANY

An insurer ABC plc writes a combination of motor, employers' liability, public liability, property, and professional indemnity. It is assumed that upon the onset of an economic recession, the number of claims would increase as follows:

- Motor: +10%
- Employers' liability: +15%
- Public liability: +10%
- Property: +15%
- Professional indemnity: +25%

ABC plc's economic adviser estimates that there is an 8% likelihood that there will be a recession in the next year.

We therefore assume that there are two possible economic scenario: a 'normal' scenario and a 'stressed' scenario in which all the frequencies are increased as above. The losses from the different lines of business are assumed to be independent of one another in a given economic scenario.

Each of the above lines of business is modelled *in the normal scenario* with a collective risk model with these frequency models: $N_{motor} \sim Poi(\lambda_{motor})$, $N_{EL} \sim Poi(\lambda_{EL})$, $N_{PL} \sim Poi(\lambda_{PL})$, $N_{PDBI} \sim Poi(\lambda_{PDBI})$, $N_{PI} \sim Poi(\lambda_{PI})$. (A Poisson model is used for simplicity – but other models can be used with only a marginal increase in difficulty.) The severity models for each line of business are $X_{motor} \sim F_{motor}(x)$, $X_{EL} \sim F_{EL}(x)$, $X_{PL} \sim F_{PL}(x)$, $X_{PDBI} \sim F_{PDBI}(x)$ and $X_{PI} \sim F_{PI}(x)$. In the stressed economic scenario, the Poisson rates of the various lines of business are increased to $1.1 \times \lambda_{motor}$, $1.1 \times \lambda_{EL}$, $1.1 \times \lambda_{PL}$, $1.15 \times \lambda_{PDBI}$ and $1.25 \times \lambda_{PI}$, whereas the severity models remain the same.

In traditional scenario analysis, one runs a Monte Carlo simulation on two different bases, with normal and stressed Poisson rates, and presents two sets of results, possibly commenting on the relative likelihood of the two scenarios, like in Figure 29.7 below ('normal scenario' and 'stressed scenario' columns).

Common shocks can, however, also be modelled stochastically with the following procedure:

i. For each simulation $j = 1, \ldots N$, choose a stressed scenario with probability 8%, and a normal scenario with probability 92% (This can be done straightforwardly by generating a pseudo-random number between 0 and 1 and selecting a stressed scenario if that number is <0.08, a normal scenario otherwise.)
ii. If a normal scenario is selected, sample a number of losses independently for each line of business from the Poisson distributions with rates λ_{motor}, λ_{EL}, λ_{PL}, λ_{PDBI}, λ_{PI} and for each loss sample a loss amount from the relevant severity distribution. Calculate the total losses across all classes for this simulation, S_j
iii. If a stressed scenario is selected, sample a number of losses independently for each line of business from the Poisson distributions with rates $1.1 \times \lambda_{motor}$, 1.15

Percentile	Normal scenario: Number of losses	Normal scenario: Total loss amount	Stressed scenario: Number of losses	Stressed scenario: Total loss amount	Random systemic shock: Number of losses	Random systemic shock: Total loss amount
50%	95	9,293,838	108	10,574,203	95	9,334,894
75%	101	10,341,752	115	11,701,627	103	10,465,533
80%	103	10,617,382	117	12,030,864	104	10,765,303
90%	108	11,364,517	122	12,841,601	109	11,619,550
95%	111	12,021,990	126	13,550,152	113	12,337,546
98.0%	116	12,814,067	130	14,404,850	119	13,239,521
99.0%	118	13,489,902	133	15,126,906	122	13,814,775
99.5%	122	14,165,928	136	15,930,918	126	14,499,729
99.8%	125	15,042,238	140	17,029,757	130	15,298,086
99.9%	127	15,364,567	142	17,554,711	133	15,733,069
Mean	**94.9**	**9,372,627**	**108.0**	**10,673,560**	**95.9**	**9,454,347**
Std Dev	**9.8**	**1,547,566**	**10.5**	**1,675,898**	**10.4**	**1,636,484**

FIGURE 29.7
Cross-class aggregate loss distribution under a normal scenario ($N_{motor} \sim$ Poi(50), $N_{EL} \sim$ Poi(20), $N_{PL} \sim$ Poi(10), $N_{PDBI} \sim$ Poi(5), $N_{PI} \sim$ Poi(10)), a stressed scenario ($N_{motor} \sim$ Poi(55), $N_{EL} \sim$ Poi(23), $N_{PL} \sim$ Poi(11), $N_{PDBI} \sim$ Poi(6.5), $N_{PI} \sim$ Poi(12.5)), and the stochastic application of a common shock with a probability of 8%. The severity distribution is chosen for simplicity to be the same for all classes: a lognormal distribution with parameters $\mu = 11$, $\sigma = 1$. Note that the variance/mean ratio for the 'random common shock' scenario is significantly larger than 1, and the frequency could not therefore be modelled as a Poisson distribution.

$\times \lambda_{EL}$, $1.1 \times \lambda_{PL}$, $1.15 \times \lambda_{PDBI}$ and $1.25 \times \lambda_{PI}$, and for each loss sample a loss amount from the relevant severity distribution. Calculate the total losses across all classes for this simulation, S_j

iv. Repeat for all N simulations

v. Sort all total losses in ascending order: $S_{(1)} \leq \ldots \leq S_{(N)}$, obtaining the empirical aggregate loss distribution

The table in Figure 29.7 shows a few percentiles of the distribution thus obtained, and compares it with the distribution obtained under a normal and stressed scenario. Another example of the use of common shocks is for credit risk applications. An example of this was discussed in Chapter 23.

X	Y
25,135	34,940
18,365	34,563
–	28,474
24,355	35,531
5647	12,766
22,456	41,032
9520	34,327
29,236	17,499
10,023	124,720
19,893	9544
–	1555
7976	15,541

3. Create two vectors X, Y such that $Y = f(X)$ (and are therefore fully dependent) but have zero linear correlation.
   a. Can you produce an example for which the linear correlation is equal to 0?
   b. Can you produce an example for which the rank correlation is also equal to 0?
4. Using a spreadsheet, generate two random vectors of 1000 elements each, with each element coming from a lognormal distribution. Then combine them with a Gaussian copula, producing a *rank correlation* of around 0.6. Do the same using R code.
5. Do the same as in Question 3 but combining the two vectors with a t copula with a desired correlation of 0.6 and $v = 1.3$.
6. Calculate the theoretical tail correlation for the copula of Question 4.
7. *Monkey business.* Assume that you own an establishment with a troop of 120 macaques (on average) that you sell as domestic pets and that the probability of one of your monkeys dying in a normal year is approximately $p = 1.1\%$. However, this probability can increase to approximately $p = 15\%$ if there is an epidemic of tuberculosis. The value of each monkey is distributed roughly uniformly in your troop between £5000 and £8000 (depending on age, cuteness, and other factors). Your in-house veterinarian estimates that the probability of an epidemic of tuberculosis happening next year is around 4%. Using the Monte Carlo simulation or other methods, estimate the mean, standard deviation, 75th, 90th and 95th percentile of the aggregate distribution of losses next year under a normal scenario, an epidemic scenario, and a generic scenario, stating all assumptions you make.

30

Insurance Structure Optimisation

In Chapter 3 we looked at heuristic rules on how to choose the deductibles and limits of policies. These rules may suggest, for example, that the annual aggregate deductible (AAD) for a commercial liability policy should be hit with a probability of 5% to 20%, or that an each-and-every-loss deductible should filter out at least 90% of the claims. However, these heuristic rules are exactly what the name says – heuristics – and there is no rigorous justification for them, only common-sense reasoning. It is no surprise that these rules are often and cheerfully flouted.

Is it possible to improve on this and have a more scientific criterion for an insurance buyer to choose their insurance structure in an optimal way, or at least to prune out the 'inefficient' insurance structures and leave us with the 'efficient' ones? This chapter will illustrate some of the approaches that have been devised to identify *efficient* (Section 30.1) and *optimal* (Section 30.2) insurance structures. The limitations of these techniques – and the need to take their results with a pinch of salt – are also explained (Sections 30.1.4 and 30.2.6).

30.1 Efficient Frontier Approach

The idea of the efficient frontier approach is quite simple: use a cost-benefit analysis to compare different insurance structures.

30.1.1 Cost of Insurance

There is no doubt as to what the main cost of insurance is: the premium paid to the insurer (plus possibly some management costs for keeping track of the insurance purchase and monitoring its performance).

30.1.2 Benefits of Insurance

The calculation of the benefits of insurance is more complicated. General insurance (and reinsurance) policies are bought for many purposes:

DOI: 10.1201/9781003168881-34

- Reducing financial volatility
- Giving peace of mind
- Abiding by the law and regulatory constraints
- Satisfying shareholders
- Obtaining related risk management services from underwriters and brokers, and much more

For example, if you buy car insurance, you will see that it obviously satisfies the first three purposes: it avoids sudden financial outgoings to pay chunky claims, it gives you peace of mind and even if you didn't care about financial outgoings and your peace of mind is unaffected by the thought of the consequences of a car accident, you will have to buy it anyway because it is compulsory. If you own a pharmaceutical company, you may have other reasons to buy product liability insurance, for example, demonstrating to shareholders that you are taking care of that risk properly, and gain access (through a broker/consultant or directly through the insurer) to market information on product liability risk and different ways of managing it.

In practice, most of the benefits above are either not amenable to quantitative analysis (such as peace of mind) or do not help in differentiating between different insurance programmes (e.g. not abiding by laws and regulations is not really an option), and we can focus on volatility reduction.

30.1.3 Efficient Structures

Given a metric for financial volatility, some insurance programmes are obviously better than others:

- If A and B achieve the same financial volatility reduction, but A is cheaper than B, then A is better than B (A > B)
- If A and B have the same premium, but A achieves more volatility reduction than B, then A > B

Two programmes may, however, be 'not comparable':

- If A is cheaper than B but B achieves more volatility reduction than A, then A and B are not comparable.

In mathematical terms, one can say therefore that volatility reduction and premium impose a *partial ordering* on the set of insurance structures. This in turn is simply a consequence of the fact that we have two different criteria according to which we judge insurance structures, premium and volatility reduction, and no means is provided to weigh one against the other.

Having decided to focus on volatility reduction as the criterion for differentiating between different insurance programmes, we need a metric to measure volatility reduction (or, more simply, the volatility itself). Three measures of volatility reduction are commonly used:

- Reduction of the *standard deviation* (or *variance*) of the retained losses
- Reduction of the *value at risk* at the pth percentile, that is, of the maximum losses expected to be retained with probability p

- Reduction of the *tail value at risk* at the *p*th percentile, that is, of the mean losses expected to be retained above the *p*th percentile

Each of these measures has its own advantages and disadvantages. All of them, however, have one general disadvantage: they need a reference point – 'reduction' with respect to what? The obvious choice would be 'reduction with respect to the case where no insurance is purchased at all'. If we want to avoid this complication altogether, we can simply use the volatility itself instead of the *reduction* in volatility as a criterion to compare different structures.

As for the specific advantages and disadvantages of these three common measures of volatility (standard deviation/variance, value at risk, tail value at risk), see the table below.

Advantages and Disadvantages of Different Risk Measures for the Purpose of Comparing Insurance Structures

Measure	Advantages	Disadvantages
Standard deviation (variance)	Simple Does not have any parameters to agree on	Captures *all* volatility rather than only the volatility that hurts Two different solutions might have similar standard deviations but have very different tail behaviours One specific disadvantage of the variance (as opposed to the standard deviation) is that it is not a monetary amount (pounds squared are not pounds!), it is often uncomfortably large and has no immediately intuitive meaning
Value at risk at the *p*th percentile	Simple to understand Focuses on the risk that matters – the tail risk	Has one parameter (*p*) to agree on. Different values of the parameters might lead to different conclusions in comparing one structure to another Is not a good ('coherent') risk measure (see Section 19.2.2.1) Ignores whatever happens beyond the *p*th percentile
Tail value at risk at the *p*th percentile	Focuses on the risk that matters – the tail risk Takes into account the *whole* behaviour of the tail, not only the value at the *p*th percentile It is a *coherent* risk measure (see Section 19.2.2.1)	Not as easy to understand/communicate/calculate Has one parameter (*p*) to agree on. Different values of the parameters might lead to different conclusions in comparing one structure to another

Whatever measure we use to assess the benefit of insurance, we can place each conceivable insurance structure in a simple two-dimensional diagram, which has the cost of the insurance on one axis and its benefit on the other, as in Figure 30.1.

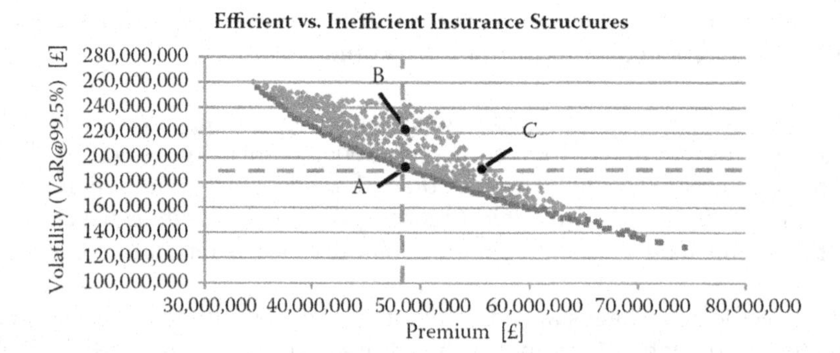

FIGURE 30.1
Each insurance structure can be represented as a light-grey dot in this cost-benefit chart, in which the cost is the insurance premium, and the benefit is in this case measured by the value at risk at the 99.5th percentile. The dark grey dots lie on the so-called efficient frontier. B and C are not efficient and they can both be improved by purchasing A instead.

One thing is immediately clear by looking at Figure 30.1: some insurance structures are obviously better than others. For example, if A and B have the same cost but the reduction in volatility is smaller for B, then B must be a worse structure. Analogously, if the reduction in volatility is the same for A and C but the cost for C is higher than the cost for A, C is immediately a worse choice. Actually, we only need one simple rule to get rid of the inefficient structures: given a structure X with cost C(X) and benefit B(X), all structures that lie both to the right of and above X are worse than X.

After plotting a sufficient number of structures, we find that all structures lie under a curved line, which is the set of structures for which it is not possible to find another structure that is simultaneously cheaper and less volatile. We call this line the efficient frontier. Constraints can be incorporated easily, for example, by imposing a limit to the premium spent or to the volatility accepted. Once you find the efficient frontier, the decision criterion for the choice of the insurance structure is clear: *choose an insurance structure that lies on the efficient frontier and satisfies all additional constraints.*

30.1.4 Limitations of the Efficient Frontier Approach

- The big problem with the efficient frontier approach is that you need to know the premium for a large number of structures. In practice, this is normally replaced with an estimated premium calculated with a formula. (For example, a loss ratio may be estimated based on a few actual quotes, and then the premium for the other structures may in turn be estimated by dividing the expected ceded losses by that loss ratio.) Therefore, once an actual quote is obtained for the chosen structure, it might turn out to be less (or more) convenient than initially thought.
- Using a different metric (or even the same metric with different parameters, such as VaR@95% and VaR@99%) will lead to a different efficient frontier.
- One issue that risks being concealed under the calculations and not communicated properly to clients is that of the uncertainty in the estimate of the expected retained and (especially) of the volatility. Because of data, parameter, and model uncertainty, the efficient frontier is actually a much fuzzier object than a line!
- The method does not offer us a 'best option' – just a range of noncomparable efficient options.

30.2 Minimising the Total Cost of (Insurable) Risk

One of the limitations that we have mentioned in Section 30.1.4 about the efficient frontier method is that it does not actually select a specific structure, it just yields a set of noncomparable efficient structure. This is a consequence of the fact that volatility and premium are two different criteria to judge insurance structures, and no way is provided to weigh one against the other. To have a criterion that allows us to compare *any* two insurance structures, we need to be able to assign a single real number to each insurance structure.

One commonly used method to achieve this is through the so-called 'total cost of risk'. First of all, note that we are interested here in the total cost of *insurable* risk: not because uninsurable risk is less significant, but because we are currently looking at how the selection of an appropriate level of retention for your insurance structure can be made optimal. Uninsurable risks are also important and are part of the problems that the risk management department of a company must face.

30.2.1 Risk Management Options

Before we define the total cost of risk, let us discuss what an individual or a company needs to do to address risks. After identifying and measuring one particular risk, one has several options for dealing with risks:

- *Ignore it* – it may sound silly to include this as an option, but this is a perfectly legitimate option in the face of risk, and arguably most risks are ignored by people and organisations if they are perceived as too remote.

  For example, a company may buy employers' liability cover up to £50M but ignore anything beyond that – judging the likelihood of a claim higher than that limit occurring too small to be concerned about. We do this all the time in our lives – we may be persuaded to follow a lower-fat diet to reduce the probability of heart disease, but we may decide to take no action to reduce the risk of rare diseases that affect one person in 100,000. One just accepts that awful things can happen without doing anything about it if the likelihood is small enough.
- *Retain it* – this might look very much like 'ignore it', because ultimately you are going to be the one who pays the consequences of either ignoring or retaining the risk. There is a difference, though: when you retain a risk, you manage it consciously, for example, by making sure you have money aside to pay for a large claim should such a claim happen.
- *Transfer it* – this normally means that you buy insurance to protect yourself against the risk. However, other ways of transferring are possible, for example, to the capital markets or to a parent company.
- *Mitigate/prevent it* – this means that you are setting up risk control mechanisms, such as (in the case of property risk) fire detection systems, firewalls, disaster recovery sites and the like, either to prevent the risk or to mitigate its effects after the event has occurred.
- *Adopt a mixture of the strategy above* – as we will see, the most common way of dealing with risk is by adopting a hybrid solution, that is, by actually doing some or all of the things above. Different risk strategies will simply shift the emphasis from one of the risk management options to another.

30.2.2 Total Cost of Risk: Definition

All of these options (except for 'ignore') have a cost, and all (including 'ignore') have consequences. In general, we can write down the *total cost of risk* (TCR) for a hybrid risk management strategy as the sum of all the costs for the options above:

$$\begin{aligned}\text{Total cost of risk} &= \text{Cost of retaining} + \text{Cost of transferring}\\ &\quad + \text{Cost of mitigating / preventing}\end{aligned} \tag{30.1}$$

Because we wish to focus on the differences between different insurance structures, we can assume that the cost of mitigating/preventing bit is roughly constant across different insurance structures (either because the company wills it or because the insurer requires a certain level of health and safety to issue insurance), and therefore ignore the term. Also, the cost of retaining a risk can be split into two terms, one for normal, day-to-day smaller losses (the so-called 'expected' losses) and the other for holding capital for a bad year. We can therefore rewrite the total cost of risk as

$$\begin{aligned}\text{Total cost of risk} &= \text{cost of retaining smaller losses}\\ &\quad + \text{cost of holding capital for a bad year}\\ &\quad + \text{cost of transferring}\end{aligned} \tag{30.2}$$

This is the general structure of the total cost of risk, which we can all relate to. After all, this is what we all do when dealing with risk as customers: insure part of it, keep the smaller losses that do not affect our ability to get to the end of the month, and possibly put some money aside just if an unexpected outgoing is called for.

For the sake of choosing between insurance structures, however, we need to be a bit more precise. What are smaller, normal losses? What is a bad year? What is the cost of holding capital?

There is no unique answer to these questions, and therefore no unique formula for the total cost of risk. However, a commonly adopted definition is this:

$$\begin{aligned}\text{Total cost of risk} &= \text{mean retained losses} + \text{CoC}\\ &\quad \times \left(\text{VaR}\,@\,p\% - \text{mean retained losses}\right)\\ &\quad + \text{insurance premium}\end{aligned} \tag{30.3}$$

where:

VaR@p% is the pth percentile of the retained loss distribution

CoC is the *cost of capital*, that is, the opportunity cost of freezing an amount equal to VaR@p% minus mean expected losses of capital to pay possible claims (rather than investing it into the business)

p is called the *safety level*, or risk tolerance level, or 'ignore' threshold, and represents the level of probability beyond which the company is happy to ignore the consequences of retaining risk.[1] For example, if $p = 99\%$, that means that the company is happy to be safe in 99% of the cases, and to ignore events that happen with less than 1% probability

Note that the 'frozen capital' is just the capital that you have to put aside *in excess* of the normal, expected losses.

One problem with this definition is that, especially in the case of excess layers with a large attachment point, some insurance structures may have a value at risk that is actually lower

1 This is true, at least, if the value at risk (which ignores whatever happens beyond a given percentile) is used as a risk measure. For other risk measures, the interpretation of the safety level is more complex.

than the mean retained. It would therefore make more sense to replace the mean retained losses with the *median* retained losses (which can also be written as VaR@50%), which are guaranteed to be less than or equal to the value at risk at any percentile higher than 50%.

Also, VaR@p% can be replaced with other measures of volatility, such as the tail value at risk. In this case, the tail value at risk for any p is *always* greater than the mean retained (which is, incidentally, equal to TVaR@0%), and therefore it is not necessary to replace the mean retained with the median retained. *Note that, in the case of the TVaR, the risk with probability less than 1–p is not really ignored!*

The list below contains some other legitimate definitions for the total cost of risk, each of which has its own advantages and disadvantages (Question 3).

$$\begin{aligned}
\text{a. } \mathrm{TCR} &= \text{median retained } + \text{premium} + \mathrm{CoC} \times (\mathrm{VaR}\ @\ p\% - \text{median losses}) \\
&= \mathrm{VaR}\ @\ 50\% + \text{premium} + \mathrm{CoC} \times (\mathrm{VaR}\ @\ p\% - \mathrm{VaR}\ @\ 50\%) \\
\text{b. } \mathrm{TCR} &= \text{median retained } + \text{premium} + \mathrm{CoC} \times (\mathrm{VaR}\ @\ p\% - \text{median losses}) \\
\text{c. } \mathrm{TCR} &= \text{median retained } + \text{premium} + \mathrm{CoC} \times (\mathrm{TVaR}\ @\ p\% - \text{median retained}) \\
&= \mathrm{TVaR}\ @\ 0\% + \text{premium} + \mathrm{CoC} \times (\mathrm{TVaR}\ @\ p\% - \mathrm{TVaR}\ @\ 0\%)
\end{aligned} \tag{30.4}$$

30.2.3 Choosing the Optimal Structure

Whatever the definition of TCR we use, TCR will enable us to associate a number to an insurance structure, no matter how complicated that is. As a result, this will make *any* two insurance structures comparable, and will give us a simple recipe for selecting the best insurance option.

Consider that, in doing this analysis, you are adopting the perspective of someone who is going to purchase insurance (or otherwise manage their risks) and you are advising them as to what their best option is. If you do this type of analysis, it will be because you are a consultant or a broker hired by a company, or because you are a member of the risk/insurance management team of that company (in any case, the client is the company itself). In the case of reinsurance, replace 'company' with 'insurer', and 'risk/insurance management team' with 'risk/reinsurance management team'.

30.2.3.1 Selecting the Best Insurance Option amongst K *Options*

a. Choose a safety level p, based on the client's risk tolerance level (this should normally be part of a discussion with the client). This level should be set as discussed in Section 30.2.1, by considering how unlikely the risk has to be for the client to be happy to ignore them.
b. Choose a CoC, again after discussion with the client, who is in the best position to advise what their CoC is.
c. Consider all the K alternative insurance structures and, for each of them, estimate the premium (if not available) and calculate the total cost of risk by Equation 30.3.
d. Impose any additional constraints as required by the client, such as maximum premium, minimum/maximum retention levels, or as dictated by the market, such as availability of certain limits of cover, and minimum premium.
e. The optimal insurance structure is that with the lowest value of the total costs of risk among those which satisfy the constraints.

Note that it is still possible to have more than one insurance structure with exactly the same total cost of risk. In this case, any of these structures would be equally viable.

30.2.4 Total Cost of Risk Calculation: A Simple Example

Assuming we use the standard definition of the total cost of risk given in Equation 30.3, let us see how the total cost of risk can be calculated in a very simple situation. First of all, consider a single insurance structure, a very simple one with no deductible, no policy limit, and an AAD of £10M. Assume that the distribution of gross, ceded, and retained losses is as in Figure 30.2 (the distribution has been deliberately chosen so that it has a very heavy tail, to make the effect of the AAD clear).

From a conversation with the client, it has emerged that their risk tolerance level is 97.5% (i.e. they wish to ignore anything that happens with a probability smaller than 2.5%), and that their CoC is 10%.

You do not have a quoted premium but you estimate that the underwriter will be seeking a loss ratio of 50%, which leads you to estimate a premium of £26.5M (premium = mean ceded losses/loss ratio).

Question: Before reading on, what is your estimate of the total cost of risk?

By using Equation 30.3, the total cost of risk can be calculated as

$$\begin{aligned} \text{TCR} &= \text{premium} + \text{mean retained} + \text{CoC} \times (\text{VaR}\,@97.5\% - \text{Mean Retained}) \\ &= £26.5\text{M} + £1.7\text{M} + 0.1 \times (£10.0\text{M} - £1.7\text{M}) = £29.0\text{M} \end{aligned}$$

Next, we wonder if this structure is optimal. We want to limit ourselves (for the sake of keeping the calculations simple) to structures that have no each-and-every-loss deductible and no policy limit, but different values of the AAD.

We then try out a large number of aggregate deductibles between £1M and £33M and, for each of them, we calculate the total cost of risk by recalculating the mean retained losses, the premium (by dividing the mean ceded by the loss ratio of 50%), and the value at risk at 97.5%. The results are shown in Figure 30.3.

	Gross	Retained	Ceded
10%	2853	2853	–
20%	13,022	13,022	–
30%	34,533	34,533	–
40%	87,656	87,656	–
50%	198,404	198,404	–
60%	438,636	438,636	–
70%	932,732	932,732	–
80%	2,354,382	2,354,382	–
90%	7,853,838	7,853,838	–
95%	20,327,166	10,000,000	10,327,166
97.50%	65,800,598	10,000,000	55,800,598
98%	79,013,718	10,000,000	69,013,718
99%	256,961,796	10,000,000	246,961,796
99.50%	923,918,924	10,000,000	913,918,924
99.90%	4,295,518,341	10,000,000	4,285,518,341
Mean	14,977,284	1,720,527	13,256,757
	Estimated premium (with LR=50%)		26,513,513

FIGURE 30.2
The aggregate distribution of gross, retained, and ceded losses for a very heavy-tailed risk.

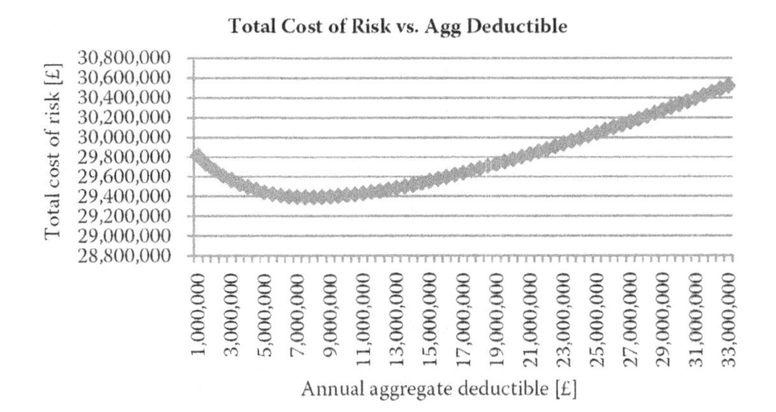

FIGURE 30.3
The estimated TCR for a large number of different structures, all of them identical except for the value of the AAD. Notice the minimum at approximately £8M.

It turns out that the 'optimal' structure has AAD ~ £8M (which corresponds to an annual cap that is going to be hit with a probability of roughly 10%). However, it should be noted that any value in the region of £5M to £10M will yield similar results in terms of TCR. Therefore, even though our current structure (AAD = £10M) is not optimal, it is near-optimal, and only a comparison of the actual (rather than estimated) premiums of the two structures will help us decide between the two.

30.2.5 Relationship between Total Cost of Risk and Efficient Frontier

How is the total cost of risk related to the efficient frontier? If you use the same measure of volatility and the same reference percentage (e.g. VaR@99%), then it seems sensible to expect that the structure with the optimal TCR will lie on the efficient frontier. Is that actually the case? First of all, let us rewrite the TCR as follows:

$$\begin{aligned}\text{TCR} &= P + \mathbb{E}(S_{\text{ret}}) + \text{CoC}\left(\text{VaR}@p - \mathbb{E}(S_{\text{ret}})\right)\\ &= P + (1-\text{CoC})\times\mathbb{E}(S_{\text{ret}}) + \text{CoC}\times\text{VaR}@p\\ &= P - (1-\text{CoC})\times\mathbb{E}(S_{\text{ced}}) + (1-\text{CoC})\times\mathbb{E}(S) + \text{CoC}\times\text{VaR}@p\end{aligned}$$

Assume that a structure $\sum$ has the lowest cost of risk but does not lie on the efficient frontier. Then there will be another structure $\sum'$ with the same value of VaR@p, which does lie on the efficient frontier and therefore has premium $P' < P$, and of course $\mathbb{E}(S'_{\text{ced}}) < \mathbb{E}(S_{\text{ced}})$ (unless it is possible to pay less by ceding more!).

Now if we can assume that $P - P' > (1-\text{CoC})\times\left(\mathbb{E}(S_{\text{ced}}) - \mathbb{E}(S'_{\text{ced}})\right)$, that is, that it is not possible to obtain £1 of reduction in the ceded losses by paying less than[2] £1, then TCR′ < TCR, which contradicts the assumption that $\sum$ was optimal. Therefore, under rather broad assumptions on the loss ratio, we see that the optimal structure from the TCR point of view also lies on the efficient frontier.

2 This is immediately true if we can assume that the same loss ratio will apply to both structures, and that such loss ratio is lower than $1/(1 - \text{CoC})$, which is a number larger than 1.

30.2.6 Limitations

The total cost of risk has several limitations, some of which are exactly the same as the limitations of the efficient frontier approach.

- As in the case of the efficient frontier methodology, it is only possible to obtain actual premiums or desk quotes for a few structures; most of the premiums will be obtained by a simple formula based on a few desk quotes/actual quotes, for example, assuming a loss ratio.
- The mean retained and the 'Retained@safety level' is also only an estimate based on a loss model, with its usual uncertainties (model, parameter, data.) – the total cost of risk is actually fuzzy.
- Using a different metric for the 'Retained@safety level' bit (or even the same metric with different parameters, for example, VaR@95% and VaR@99%) will yield a different TCR and a different optimal structure.
- Structures that are quite different will tend to yield very similar values of the TCR, which (added to the uncertainty around the TCR) makes one wonder about the discriminatory power of TCR minimisation as an optimality criterion.

Overall, these limitations suggest that while total-cost-of-risk optimisation may be useful to prune out absurd structures and get the analyst and the client thinking about a sensible insurance structure, the exact results should always be taken with a pinch of salt and balanced against more common-sense reasons for choosing one structure rather than another.

30.3 Questions

1. Consider the following list of insurance structures and highlight those that are surely inefficient using VaR@99% as a measure of volatility:
   a. Structure 1: premium = £1.2M, VaR@99% = £99M
   b. Structure 2: premium = £1.7M, VaR@99% = £78M
   c. Structure 3: premium = £1.3M, VaR@99% = £105M
   d. Structure 4: premium = £4M, VaR@99% = £50M
   e. Structure 5: premium = £4.5M, VaR@99% = £50M
2. Company X has traditionally purchased public liability insurance from the ground up, but to decrease the insurance premium (currently at £2,500,000) spend and in view of its financial robustness, it is now investigating the opportunity of retaining the portion of each loss, which is below a given amount ('each and every loss [EEL] deductible').

   The company has obtained these two quotes for insurance at these different levels of deductibles:
   a. EEL = £100k → Premium = £900,000
   b. EEL = £200k → Premium = £450,000

   The company asks a consultant what level of EEL they should choose, and the consultant proposes to choose the level of EEL that minimises the 'total cost of risk'.
   i. State the formula for the total cost of risk.

Percentile	Total loss	EEL = £100k		EEL = £200k		EEL = £500k	
		Retained	Ceded	Retained	Ceded	Retained	Ceded
50%	1,869,704	1,205,151	634,128	1,573,081	244,834	1,823,682	0
75%	2,380,186	1,424,166	983,601	1,887,698	516,466	2,268,864	37,675
80%	2,510,521	1,478,409	1,095,454	1,973,002	600,654	2,385,498	92,638
90%	2,903,459	1,633,268	1,434,902	2,196,113	885,675	2,682,736	278,422
95%	3,274,703	1,754,710	1,743,826	2,383,344	1,166,614	2,968,721	527,750
98.0%	3,779,592	1,892,509	2,164,576	2,608,209	1,547,230	3,316,971	880,550
99.0%	4,152,528	1,999,685	2,466,066	2,766,417	1,887,943	3,576,709	1,188,535
99.5%	4,495,618	2,103,179	2,842,877	2,897,132	2,210,172	3,740,428	1,476,792
99.8%	4,791,922	2,231,156	3,303,707	3,080,203	2,689,858	3,982,094	1,977,433
99.9%	4,991,777	2,319,460	3,468,427	3,204,756	2,918,832	4,121,569	2,236,319
Mean	1,969,444	1,223,767	745,676	1,602,097	367,346	1,882,367	87,077
Std Dev	718,000	308,145	525,032	448,452	412,922	610,099	243,299

The company gives the consultant this further information:

- They are not worried about overall level of retained losses that happen with a probability of less than 1% (on an annual basis).
- The company's CoC is 12%

ii. Explain what the CoC is.

An actuarial analysis gives the following estimates of the gross, retained and ceded distributions for different levels of EEL.

The company also asks the consultant to estimate what the premium will be if the EEL should be £500K. The consultant produces the following *desk quote:*

– EEL = £500k → Premium = £120,000

c. Explain what a desk quote is.
d. Based on the information provided, determine the best of the four options (no EEL, EEL = £100k, EEL = £200k, or EEL = £500k) from a total cost of risk's perspective, explaining your reasoning.
e. Explain what the main uncertainties around this result are.

3. Explain the rationale behind the four different possible formulations of the total cost of risk given in Equations 30.3 and 30.4, and their relative advantages and disadvantages.

31

An Introduction to Pricing Models

> [Pricing] tools have an unwelcome feature: sometimes they take over the body, preventing it from thinking, and turning the poor underwriter into a mindless tool at their mercy.
>
> **Ugo Serena**

> Que aqui começo a construir sempre buscando o belo e o Amaro
> […] but I've given up any attempts at perfection
>
> **Caetano Veloso, O estrangeiro**

In a way, this brief chapter is the point of at which most if not all the material we have looked at so far comes together: experience rating, exposure rating, credibility, generalised linear models, the premium formula, etc. – all these are possible ingredients of a pricing model.

A pricing model (also: rating model) is a tool that allows underwriters or actuaries to estimate the technical premium for a specific contract. (As we have seen before, the technical premium is the premium that covers losses, expenses and the desired level of profit, also taking into account any investment income that the insurance company can make. It is in general different from the actual premium, i.e. the premium that the company eventually charges to the client.) It may also do some extra jobs, such as calculating the rate change or allowing the underwriters to save non-pricing-related information, but at the very least a technical premium should be produced.

A pricing model can be anything ranging from a simple quoting system that produces rates that are deemed acceptable to the underwriters, to a sophisticated tool that produces an actuarially informed technical premium. Obviously, actuaries are specifically interested in the latter, although they may be involved in the building of the former.

The general architecture of a pricing model is described in Section 31.1. A pricing model can be based purely on exposure (Section 31.2), on experience (Section 31.3) or a combination of experience and exposure (Section 31.4), and can consider a number of rating factors. The calibration of the pricing model (especially an exposure-based model) may be carried out with mathematical methods (e.g. GLMs), by using underwriting judgment, or by a combination of the two (Section 31.7). A company will normally have some governance around the development and maintenance of models (Section 31.8).

Apart from the simplest cases, pricing models are normally developed via a collaboration of underwriters, pricing actuaries and IT professionals.

DOI: 10.1201/9781003168881-35

Spreadsheet implementations of some of the models described here are available in the Downloads section of the book's website (Parodi, 2016).

31.1 Generalities

A pricing model will typically include:

1. A **user interface** (UI) where the user enters (or uploads from another system) the contractual information (contract number, inception/expiry date of the contract, limit of liability, etc.) and the information of the client needed for pricing purposes (exposure, key risk factors, past losses). This would also include the results of the calculation engine, along with the premium actually charged (an input that might be filled in at a later stage or that can be read from the technical accounting system).
2. A **calculation engine** connected to the UI containing a pricing algorithm that calculates the technical premium based on information about the contract and the underlying risk. It could also calculate additional information for management such as rate change (discussed in Section 31.6) and other risk metrics (volatility, capital requirements, etc.) related to the policy.
3. A **contract/parameter database** to store information about contracts and parameters previously used in the model.
4. **Connections to other IT systems**. The UI and the database used to store information may also be connected to other components of the company's information system. The pricing model itself may be part of a general-purpose underwriting system, and it may be connected to the technical accounting system, one or more natural catastrophe commercial models, a document management system, and so on.

The typical architecture of a pricing model will look like that shown in Figure 31.1.

31.2 Exposure-Based Models

Exposure-based models are model that use exposure rating, i.e. they produce a technical premium based on the contract exposures, without regard to the past loss experience. Perhaps all this is best explained through a few simple examples ($\mathcal{M}_0$ to $\mathcal{M}_4$) of models of increasing complexity. These examples can also be found in spreadsheet format in the book's website (Parodi, 2016).

31.2.1 $\mathcal{M}_0$ – The Simplest Possible Model?

Let's assume we want to price employers' liability. Arguably the simplest model for this is one that has a simple measure of exposure (e.g. wage roll) as the only pricing input and produces a technical premium proportional to that exposure.

The parameters of this model are the (expected loss) base rate and the target loss ratio.

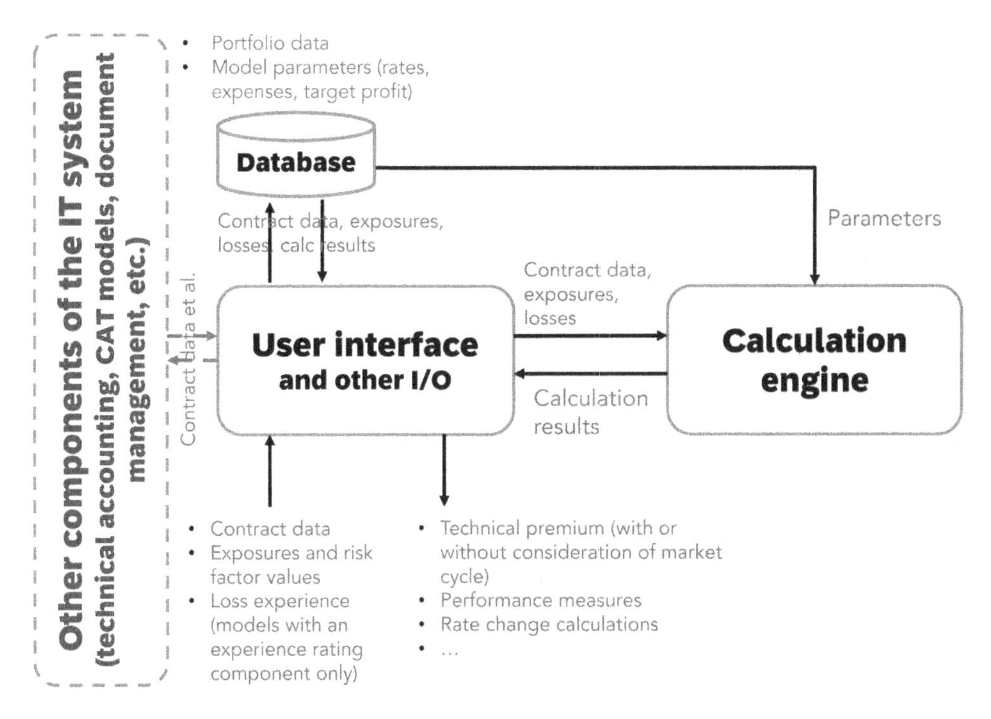

FIGURE 31.1
The conceptual architecture of a model. The actual IT architecture might of course be much more complex and detailed, but a model will typically need an engine, a UI (I/O facility), a place to store information, and connections to other systems: this will be sufficient for our purposes.

The calculation engine will consist of one formula giving the technical premium as a function of the two parameters above:

$$\text{Technical premium} = \frac{(\text{Expected loss})\,\text{base rate}}{\text{Target loss ratio}} \times \text{Wage roll} \tag{31.1}$$

The output is the technical premium calculated via Equation 31.1.

This model might or might not have a database – if it does have one, it may simply be a table (such as the one below) with a list of contracts giving basic contract information, exposure, parameters used and outputs. Note that the premium actually charged can be considered as part of the inputs, although it may be filled in later.

Contract ID	Client name	Inception date	Expiry date	Wage roll [CCY]	Exp loss base rate	Target loss ratio	Techn prem [CCY]	Written prem [CCY]
C0001	Paperoga Ltd	01/01/2021	31/12/2021	35,000,000	0.30%	60%	175,000	192,000
C0002	Dinamite BLA	19/02/2021	18/02/2022	110,808,920	0.30%	60%	554,045	495,000
C0003	Basettoni plc	18/05/2021	17/05/2022	82,330,858	0.30%	60%	411,654	526,500
...	...	...	...	...	...	...	...	...

Note that the model may be (as in this case) so simple that it may not exist as a tool or even a spreadsheet but simply as a formula in the underwriting manual, or in the underwriter's head. However, the underwriter or other personnel will certainly record the contract information and the written premium in an accounting system.

What is missing in this simple model? Almost everything actually, but two glaring limitations are these:

- all clients are treated in the same way, whether they are accountants or bomb disposal experts, and regardless of other risk factors (territory, type of industry, etc.);
- the model doesn't take into account deductibles/limits of liability or any other terms and conditions (Ts & Cs) of the policy.

Section 31.2.2 describes an example of a model for employers' liability that takes care of these shortcomings.

31.2.2 $\mathcal{M}_1$ – An Example of an Exposure-based Deterministic Model

Let us now consider a slightly more realistic (but drastically simplified) *deterministic* model for employers' liability. The model can be found (in spreadsheet format) in the book's website (Parodi, 2016). The reason for considering this model is that it is conceptually representative of many similar deterministic models of higher complexity.

The **user interface** (which is shown in Figure 31.2) allows the entry of the basic information about the contract (contract ID, client name, inception date, expiry date), the cover (deductible, limit), the exposure (wageroll, in this case) and even allows a couple of rating factors: territory and type of industry. It also records the premium actually charged, which can be entered manually or read from the technical account system.

The **pricing engine** is quite simple. The expected losses are calculated as:

$$\text{Exp losses} = f_{\text{terr}} \times f_{\text{ind}} \times ILF_{L_{\min}}(L) \times (\text{BR}_{\text{cler}} \times \mathcal{E}_{\text{cler}} + \text{BR}_{\text{man}} \times \mathcal{E}_{\text{man}})$$

INPUTS		
Contract information	Contract ID	C0010
	Client name	Dinamite BLA
	Inception date	01/01/2021
	Expiry date	31/12/2021
Exposure	Wage roll [FRD] (clerical)	20,000,000
	Wage roll [FRD] (manual)	18,000,000
Structure	Deductible	0
	Limit	10,000,000
Rating factors	Territory	Freedonian Isles
	Type of industry	Heavy
Brokerage/commission	Fee (% of premium)	11%

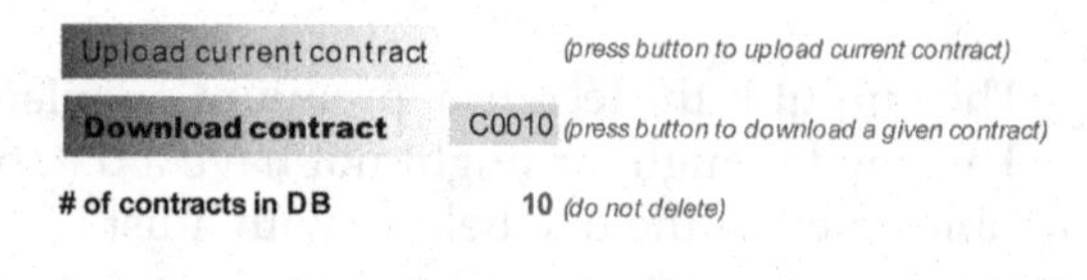

RESULTS	Currency	
Expected losses	FRD	274,752
Internal expenses	FRD	27,475
External expenses	FRD	41,658
Discount	FRD	-21,980
Profit	FRD	**56,807**
Tech prem (net)	FRD	337,054
Tech prem (gross)	FRD	**378,712**
Parameter version		v3

FIGURE 31.2
The user interface for model $\mathcal{M}_1$.

where:

- f_{terr} and f_{ind} are categorical factors that take on different values depending on the territory and the industry in which the client operates;
- $ILF_{L_{\min}}(L)$ is an increased limit factor which in this model uses the Riebesell curve (see § 22.6.3): $ILF_{L_{\min}}(L) = \max\left(1, (1+\rho)^{\log_2(L/L_{\min})}\right)$;
- $\mathcal{E}_{\text{cler}}, \mathcal{E}_{\text{man}}$ are the total wage roll for clerical and manual employees respectively;
- $\text{BR}_{\text{cler}}, \text{BR}_{\text{man}}$ are the base rates for clerical and manual employees respectively.

Note that there is no allowance for a deductible, as this model is assumed to be for use in jurisdictions such as the UK where employers' liability needs to be purchased from the ground up.

The internal expenses are assumed to be proportional to the expected losses:

$$\text{Int expenses} = \text{Expense ratio\%} \times \text{Exp losses}$$

The discount for investment income is given by:

$$\text{Discount} = -\text{Discount\%} \times \text{Exp losses}$$

The net technical premium is:

$$\text{Net technical premium} = \frac{(1 + \text{Expense ratio\%} - \text{Discount\%}) \times \text{Exp losses}}{1 - \text{Target Profit\%}} \tag{31.2}$$

And finally, the gross technical premium is:

$$\text{Gross technical premium} = \frac{\text{Net technical premium}}{1 - \text{Fee\%}} \tag{31.3}$$

As it will be obvious by looking at the formulae, the **general parameters** of this model are BR_{cler}, BR_{man}, f_{terr}, f_{ind}, ρ, $L_{\min}$, Expense ratio%, Discount% and Target Profit%. The **client-specific parameters** are $\mathcal{E}_{\text{cler}}$, $\mathcal{E}_{\text{man}}$, L and Fee%.

A simple spreadsheet implementation of this calculation engine can be found in the Downloads section of the book's website (Parodi, 2016).

The **database** is in this case simply one of the tabs of the spreadsheet. Contracts can be uploaded to the database and downloaded from it using the relevant buttons in the user interface.

31.2.2.1 Limitations

With respect to the trivial model $\mathcal{M}_0$, some features (ILF curve for limit impact, manual vs clerical, various risk factors) have been added. However, the model still suffers from several limitations, which are interesting to mention because they are common to many simple exposure rating models:

a. The model uses a simplistic base rate framework and is not linked explicitly to the loss generation process. For example, there is no distinction between frequency and severity of claims, and no information on what the underlying frequency and severity models are, from which the aggregate loss distribution could be calculated via Monte Carlo simulation or other techniques.

b. The type of policy structures one can model is limited – e.g. the model has no way of dealing with an aggregate limit. This is connected to (a), as Monte Carlo simulation (or other numerical techniques for the calculation of the aggregate loss distribution) is required for that.
c. The loading system is crude and does not reflect the fact that more volatile business, and especially business that contributes to the overall volatility of the portfolio more, should be written at lower loss ratios. This is also connected to (a), as Monte Carlo simulation is required for volatility calculations.
d. The model and the database are in this case a simple spreadsheet. A professional tool would be hosted centrally on a server and distributed to users via client or (more typically) web applications. The database would be hosted on a database management system (SQL Server, Oracle, etc.) and would contain both contract data and the parameters of the model.

Other limitations of the model may be related to the way it is calibrated – for example, if none of the parameters are data-driven that should certainly be considered a limitation.

Through the textbook, we have seen how more advanced models can be created once we master techniques such as frequency/severity modelling, GLM, and so on.

31.2.3 $\mathcal{M}_2$ – An Example of an Exposure-Based Stochastic Model

Section 31.2.2 showed an example of a traditional deterministic model. In this section we show a model with the same scope but a more modern conception – using the collective risk model (Chapter 6) to generate losses.

The **user interface** is a slight variation of that of Section 31.2.2, the main difference being that the exposure is here the number of employees and that it is possible to define not just a limit of liability L but an *aggregate limit* AL; also, we will assume that the policy structure allows for an each-and-every-loss deductible D. The Inputs section of the user interface is shown in Figure 31.3.

INPUTS

Contract information	Contract ID	C0010
	Client name	Dinamite BLA
	Inception date	01/01/2021
	Expiry date	31/12/2021
Exposure	No of employees (clerical)	1,000
	No of employees (manual)	1,800
Structure	Deductible	0
	Limit	5,000,000
	Agg limit	5,000,000
Rating factors	Territory	Freedonian Isles
	Type of industry	Heavy
Brokerage/commission	Fee (% of premium)	11%

FIGURE 31.3
The Inputs section of the user interface for model $\mathcal{M}_2$. Note how the way in which Exposure and Structure are defined differs from model $\mathcal{M}_1$.

The **pricing engine** is based on a simulation that draws samples from separate claim count (Poisson) and severity (spliced LogNormal/GPD) distributions for clerical and manual employees. We will assume that the risk factors f_{terr} and f_{ind} are (as in Section 31.2.2) categorical factors that depend on the territory and the industry in which the client operates, and that they only impact the frequency of losses, not the severity. We will also assume that the premium formula is as follows:

$$\text{Net technical premium} = (1 + \text{Expense ratio\%} - \text{Discount\%}) \times \text{Exp losses} + \text{Profit} \tag{31.4}$$

where the profit is proportional to the capital allocated to the policy, which in turn is chosen to be proportional to the volatility of the losses to policy on a stand-alone basis as measured by the excess expected shortfall at the 99th percentile, xES@99%. This is an example of the application of the *proportional spread* method (Section 19.2.3.1).

The gross technical premium is derived from the net technical premium as in Equation 31.3.

Box 31.1 shows the calculations in detail.

BOX 31.1 CALCULATION ENGINE FOR MODEL $\mathcal{M}_2$

Inputs

- Exposure (measured in employee-years) EY_{cler} and EY_{man} for clerical and manual employees respectively.
- Deductible (D), limit (L), aggregate limit (AL).

1. For each simulated scenario $k = 1, \dots K$:
   - Sample a number of claims separately for clerical employees: $N_{\text{cler}} \sim Poi(EY_{\text{cler}} \times \lambda_{\text{cler}})$ and manual employees: $N_{\text{man}} \sim Poi(EY_{\text{man}} \times \lambda_{\text{man}})$, where λ_{cler} and λ_{man} are the frequency per unit of exposure for clerical and manual employees respectively. Outputs: $n_{\text{cler}}^{(k)}, n_{\text{man}}^{(k)}$.
   - For each of the $n_{\text{cler}}^{(k)} \geq 0$ *clerical* claims, sample a loss amount from a spliced LogNormal/GPD distribution (see Chapter 16): $X_{\text{cler}} \sim \text{LogN}(\mu_{\text{cler}}, \sigma_{\text{cler}})/\text{GPD}(\theta_{\text{cler}}, \xi_{\text{cler}}, \nu_{\text{cler}})$, where θ_{cler} is the splicing threshold, ξ_{cler} is the shape parameter, and ν_{cler} is the scale parameter. Outputs: gross losses $x_{\text{cler}}^{(k,1)}, x_{\text{cler}}^{(k,2)} \dots x_{\text{cler}}^{(k,n_{\text{cler}}^{(k)})}$.
   - For each of the $n_{\text{man}}^{(k)}$ *manual* claims, sample a loss amount from a spliced LogNormal/GPD distribution (see Chapter 16): $X_{\text{man}} \sim \text{LogN}(\mu_{\text{man}}, \sigma_{\text{man}})/\text{GPD}(\theta_{\text{man}}, \xi_{\text{man}}, \nu_{\text{man}})$. Outputs: gross losses $x_{\text{man}}^{(k,1)}, x_{\text{man}}^{(k,2)} \dots x_{\text{man}}^{(k,n_{\text{man}}^{(k)})}$.
   - Calculate the total gross losses for each scenario as: $S^{(k)} = x_{\text{cler}}^{(k,1)} + \dots + x_{\text{cler}}^{(k,n_{\text{cler}}^{(k)})} + x_{\text{man}}^{(k,1)} + \dots + x_{\text{man}}^{(k,n_{\text{man}}^{(k)})}$ and (more importantly from the pricing perspective) the total insured (ceded) losses: $S_{\text{ced}}^{(k)}(D, L, AL) = \min\left(AL, \sum_{i=1}^{n_{\text{cler}}^{(k)}} x_{\text{cler}}^{(k,i)}[\text{ced}] + \sum_{j=1}^{n_{\text{man}}^{(k)}} x_{\text{man}}^{(k,j)}[\text{ced}]\right)$

where

$$x_{\text{cler}}^{(k,i)}[\text{ced}] = \min\left(\max\left(x_{\text{cler}}^{(k,i)} - D, 0\right), L\right)$$

$$x_{\text{man}}^{(k,j)}[\text{ced}] = \min\left(\max\left(x_{\text{man}}^{(k,j)} - D, 0\right), L\right)$$

2. Estimate the expected losses $\mathbb{E}\left(S_{\text{ced}}(D, L, AL)\right)$ to layer L xs D with aggregate limit AL by taking the average of the total losses across all simulations:

$$\mathbb{E}\left(S_{\text{ced}}(D, L, AL)\right)_{\text{est}} = \frac{\sum_{k=1}^{K} S_{\text{ced}}^{(k)}(D, L, AL)}{K}$$

3. Estimate the xES@99% by sorting the values of $S_{\text{ced}}^{(k)}(D, L, AL)$ in ascending order, selecting the empirical 99th percentile, and subtracting the estimated mean $\mathbb{E}\left(S_{\text{ced}}(D, L, AL)\right)_{\text{est}}$. The result is $(\text{xES}@99\%)_{\text{est}}$.
4. The net technical premium is calculated by using Equation 31.4 where:
   - Exp Losses $= \mathbb{E}\left(S_{\text{ced}}(D, L, AL)\right)_{\text{est}}$
   - Profit $= c \times (\text{xES}@99\%)_{\text{est}}$, where c is a constant that can be calibrated to ensure that at portfolio level the profit corresponds to the required rate on risk-adjusted capital (see Section 19.1.5).

The **database** is (as in the case of $\mathcal{M}_1$) simply one of the tabs of the spreadsheet.

Model $\mathcal{M}_2$ resolves most of the issues raised in Section 31.2.2.1 (link to the loss generation process, ability to price complex structures via simulation, accounting for volatility in the profit loading), but it is still in a way a toy model. Not only the model is overly simple in terms of information displayed and risk factors included (for illustration purposes), but the information technology solution adopted is not suitable for integration within the company information system and large-scale distribution to underwriters. Section 31.2.5 illustrates the features of a mature pricing model.

31.2.4 Exposure Models for different Types of Business

To illustrate exposure models we have so far used the example of an employers' liability model. More in general, the structure of exposure models will depend on the type of business written. Three common types of models are property-like models, liability-like models and credit-like models. This is by no means an exhaustive list, and many models will be hybrid – however these categories are so common that it is useful to highlight their main features.

For simplicity, we will limit ourselves to mature pricing models, which use suitable risk models and simulation.

31.2.4.1 Liability-Like Models

Models $\mathcal{M}_0, \mathcal{M}_1$ and $\mathcal{M}_2$ are examples of increasing complexity of a liability model. Mature liability models will typically have the following characteristics.

The **inputs** will be the exposures (turnover for public liability, wage roll or employee-years for employers' liability, vehicle-years for motor, turnover/fees/number of professionals for professional indemnity ...), the relevant risk factors, and the policy structure (deductibles, limits, round-the-clock reinstatements ...).

The **calculation engine** will be based on the collective risk model and will apply a system of frequencies per unit of exposure and of severity curves to derive the expected losses to the contract via Monte Carlo simulation. The frequencies and severities will be suitably modified by the relevant risk factors.

Where correlation between risks is high, such as in financial lines, copulas (Section 29.4) or mechanisms to deal with systemic shocks (Sections 6.1.3.1, 14.4.3.2.2, and 24.3.2.3) will be included.

31.2.4.2 $\mathcal{M}_3$ – An Example of a Property-Like Model

The main distinguishing feature of a property pricing model is that it deals with a portfolio of (normally physical) assets, with the loss to a specific asset being limited by its insured value. The concept of a standard severity curve for a portfolio doesn't make sense, since the severity curve will be unique for each account and will depend on the profile of the insured values and MPLs.

A simple example of a property model is the model $\mathcal{M}_3$ described below. An implementation of this model is provided in spreadsheet format in Parodi (2016). *The model only covers property damage risk for non-catastrophe perils.*

The **inputs** are the asset schedule (with type of occupancy (building, warehouse, plant), insured value, and MPL for each location), a single risk factor (quality of risk control at contract level – low, medium and high), plus the usual contract information, including the policy structure, which in this case is assumed to be a simple layer of insurance L xs D with an aggregate limit AL.

The **calculation engine** is still a frequency/severity model but the structure is different from a liability model.

The main ingredients of the model are a table of base rates by occupancy, giving the expected losses per unit of insured value, and a table of exposure curves by occupancy. The combination of base rates and exposure curves allows to derive the expected claim count for each location (alternatively, one could directly define a table of the frequency per unit of insured value for each location).

The model assumes a Poisson model for the number of claims and a Swiss Re 'c' exposure curve for the severity. It assumes the independence of losses from different locations (certainly a good approximation for non-cat losses). The model also assumes that no losses above the MPL are possible.

A description of the calculation engine for model $\mathcal{M}_3$ can be found in Box 31.2.

The **loadings** methodology is as for model $\mathcal{M}_2$.

31.2.4.3 $\mathcal{M}_4$ – An Example of a Credit-Like Model

Unlike liability and property models, credit risk models typically use the individual risk model, as the contract involves insuring a finite number of risks each of which can either fail during the insurance period or survive. We discussed this in Sections 6.1.1.2 and (more at length) in Section 24.3. A simpler application of the individual risk model (horse insurance) was also briefly discussed in Section 6.1.1.1.

A simple example of a credit risk model is the model $\mathcal{M}_4$ described below.

BOX 31.2 CALCULATION ENGINE FOR MODEL $\mathcal{M}_3$

Inputs

- List of m locations with insured values $IV_1, IV_2 \ldots IV_m$, maximum possible loss for each location $MPL_1, MPL_2 \ldots MPL_m$ (monetary amounts), and occupancy type
- Deductible (D), limit (L), aggregate limit (AL)

Model parameters

- Loss base rates giving the expected total losses per unit of insured value, by occupancy: $BR^{\text{building}}, BR^{\text{warehouse}}, BR^{\text{plant}}$
- Correction factor value for risk control: f^{RC} (=1 for 'medium', >1 for 'low', <1 for 'high'), which applies to the frequency of losses
- Parameter c for the MBBEFD exposure curves for each type of occupancy: $c^{\text{building}}, c^{\text{warehouse}}, c^{\text{plant}}$

Preliminary calculations (pre-simulation)

Calculate the Poisson rate for each location $j = 1, \ldots m$ as (Section 22.8): $\lambda_j = \dfrac{f_j^{\text{RC}} \times BR_j \times TIV_j \times G'(c_j;0)}{MPL_j}$ where $BR_j, TIV_j, MPL_j, c_j, f_j^{\text{RC}}$ are the values of the parameters assigned to each location, and $G(c;x)$ is the exposure curve. Calculate the overall Poisson rate for the contract as the sum of the individual Poisson rates: $\lambda_{\text{overall}} = \sum_{j=1}^{n} \lambda_j$.

Simulation algorithm

1. For each simulated scenario $k = 1, \ldots K$:
   a. Draw a random number n_k of losses from the Poisson distribution $N \sim \text{Poi}(\lambda_{\text{overall}})$.
   b. Allocate each of the n_k losses to one of the locations with probability proportional to the individual Poisson rates (note that each location can have more than one loss).
   c. For each of the n_k losses, draw a random amount between 0 and the MPL from the MBBEFD severity curve for that location: $x_1^{(k)}, x_2^{(k)} \ldots x_{n_k}^{(k)}$
   d. Calculate the overall loss for scenario k as $S^{(k)} = \sum_{i=1}^{n_k} x_i^{(k)}$ and the overall insured loss as

$$S_{\text{ced}}^{(k)}(D, L, AL) = \min\left(AL, \sum_{i=1}^{n_k} \min\left(\max\left(0, x_j^{(k)} - D\right), L\right)\right)$$

2. Estimate the expected losses $\mathbb{E}(S_{\text{ced}}(D, L, AL))$ to layer L xs D with aggregate limit AL by taking the average of the total losses across all simulations:

$$\mathbb{E}\left(S_{\text{ced}}\left(D,L,AL\right)\right)_{\text{est}}=\frac{\sum_{k=1}^{K}S_{\text{ced}}^{(k)}\left(D,L,AL\right)}{K}$$

3. The estimation of the profit loading is as described in Box 31.1.

Note that a real-life pricing tool for property will also need to include (see Section 22.8 for details):

- a method to simulate business interruption losses (normally done with base rates and exposure curves as for property damage);
- a method to correlate property damage and business interruption losses (business interruption losses will normally only be triggered by non-zero property damage losses);
- a method to produce losses beyond MPL when this is less than the insured value;
- an interface to a **cat modelling tool** to produce event-based catastrophe losses;
- a method for producing cat losses that are outside the scope of cat modelling tools ('non-modelled' cat losses).

The **inputs** to a credit risk model are a list of assets, each with their quality rating, plus amortisation schedules and other information.

The **calculation engine** applies a system of probabilities of failure (PD) and loss given default (LGD) curves (MBBEFD curves with a single parameter c) to derive the expected losses. Allowance is made for systemic shocks (e.g. economic recession) with a certain probability p_{rec}, which have the effect of driving upwards both the probabilities of failure and the mean loss given default.

The **loadings** methodology is as for model $\mathcal{M}_2$.

Box 31.3 gives a description of the calculation engine.

31.2.4.4 Hybrid Models

In general, pricing models will not be pure models such as $\mathcal{M}_0$ to $\mathcal{M}_4$, but will be a mixture of different components, each of which will *typically* be modelled like some of the models above: for example, a commercial motor pricing model for fleets might include a component for third party liability (TPL) – which will be modelled similarly to $\mathcal{M}_2$ – and a component for own damage (OD) – which will be modelled similarly to $\mathcal{M}_3$.

On top of this, however, there may be rules on how to deal with potential correlation between components and policy structures that aggregate across all components. For example, in the case mentioned above, we might be looking at generating K scenarios each of which produces separately a total loss for TPL and a total loss for OD for a fleet of vehicles, but there may be an annual aggregate limit across both TPL and OD for the fleet.

31.2.5 Characteristics of a Mature Exposure-based Pricing Model

Without going to the length of proposing a full maturity model for exposure pricing models (such as that in use by Lloyd's of London or other organisations), we list below what we view as the most important features of a mature pricing model.

BOX 31.3 CALCULATION ENGINE FOR MODEL $\mathcal{M}_4$

Inputs

- Asset values: $M_1, M_2 \ldots M_n$ (monetary amounts)
- Quality rating for each asset: $Q_1, Q_2 \ldots Q_n$ (low, medium, high)
- Deductible (D), limit (L) (we assume no aggregate limit for simplicity)

Model parameters

- Probability of being in a recession year: p_{rec}
- Probability of default for each quality rating and for both recession and non-recession years: $p_L^{\text{rec}}, p_M^{\text{rec}}, p_H^{\text{rec}}, p_L^{\text{norm}}, p_M^{\text{norm}}, p_H^{\text{norm}}$
- Parameter c for the MBBEFD curves for each recession status (we assume the value of c does not depend on the asset quality): $c^{\text{rec}}, c^{\text{norm}}$

Simulation algorithm

1. For each simulated scenario $k = 1, \ldots K$:
   a. Assign to scenario k a status 'recession' with probability p_{rec}.
   b. If the scenario has a 'recession' status:
   scan through all the assets $j = 1, \ldots n$;
   set them to default ($\delta_j = 1$) with probability $p_L^{\text{rec}}, p_M^{\text{rec}}$, or p_H^{rec} depending on whether the asset quality is low, medium or high respectively;
   sample a damage ratio r_j from an MBBEFD severity curve with parameter c^{rec};
   set the loss given default to $LGD_j = \delta_j \times r_j \times M_j$;
   apply the deductible and limit and sum over all assets to get the total insured loss $S_{\text{ced}}^{(k)}(D, L)$ for scenario k;
   c. Else, the scenario is normal, in which case:
   follow the same process as in Step b but using probabilities $p_L^{\text{norm}}, p_M^{\text{norm}}, p_H^{\text{norm}}$ and MBBEFD parameter c^{norm}.
2. Estimate the expected losses $\mathbb{E}(S_{\text{ced}}(D, L))$ to layer L xs D by taking the average of the total losses across all simulations:

$$\mathbb{E}(S_{\text{ced}}(D, L, AL))_{\text{est}} = \frac{\sum_{k=1}^{K} S_{\text{ced}}^{(k)}(D, L)}{K}$$

3. The estimation of the profit loading is as described in Box 31.1.

31.2.5.1 Model Structure

A mature model should have a structure that somehow reflects the loss generation process, using an adequate risk model such as the collective risk model or the individual risk model (Chapter 6). This normally means that the pricing model will have separate frequency and severity models, and where necessary a payment/settlement model. It will be clear whether the various risk adjustments apply to the frequency or to the severity of the losses. The model will have separate sub-models for the various components of the risk (e.g. third party liability vs property damage) where necessary, along with the necessary correlation structure (systemic shocks, copulas, etc.), with due consideration to the materiality of these features. The quest for the model to be 'realistic' (i.e. reflective of the loss generation principles) should be counterbalanced by the attempt to be as parsimonious as possible.

31.2.5.2 Ability to Price the Material Policy Features

A mature model should have the ability to price the policy structures (deductible, limits, aggregate limits/deductibles, index clauses, etc.) that are relevant to the line of business under examination, with due consideration to the materiality of these structures – e.g. it may not be possible to capture the effect of all sub-limits of a contract. This will normally translate into the need to incorporate Monte Carlo simulation or other numerical techniques in the pricing tool (Monte Carlo simulation being the most flexible), as only a few basic structures (normally not the ones with an aggregate element) can be priced analytically.

31.2.5.3 Loadings

The loadings for a contract should reflect the company policy. Ideally, they should include internal and external expenses (fixed and variable), discounts for investment income, and an adequate allowance for profit/cost of capital. The last item is the most complex as it will require an assessment of the volatility of the contract and, in the most mature model, an assessment of the contribution of the model to the volatility of the overall portfolio (Chapter 19). This is another reason to incorporate Monte Carlo simulation, since only very basic measures of volatility for basic policy features can be calculated analytically.

31.2.5.4 Information Technology Environment

In theory, a pricing model could be simply a set of calculation rules that lives in disembodied form in an underwriting manual or in the head of the underwriter. In practice, it will typically be embedded in an IT environment.

31.2.5.4.1 Technology

Pricing models often start as simple spreadsheets built by underwriters and actuaries. As the needs of the users widen, more features (such as macros written Visual Basic for Applications, sheet protection, database connections) are added. While this mode of operation works well with few users and limited processing needs, it becomes inadequate when the scope of the model expands and performance issues become crucial.

A mature model will therefore require the involvement of professional developers who can migrate the existing tool to (or create a new tool from scratch in) a professional environment – with the model hosted centrally on a server and distributed to users via client software or (more typically today) a web application. The tool will have adequate access control and security around the system.

Professional developers are then called in and the pricing tool is migrated to (or created from scratch in) a professional environment – with a centralised version of the tool on a server that is accessed via client software or a web application, a user interface, a database management system, and the necessary connections to the other software of the company, e.g. the technical accounting system and the cat modelling software. The role of actuaries and underwriters is *traditionally* limited to the specification and testing stages.

31.2.5.4.2 Integration

Basic pricing models work on a stand-alone basis, with no connection to the accounting system or to the other systems in use by the company.

A mature model will instead be integrated with the underwriting system, the technical accounting system and other systems that are relevant to the pricing of the segment, e.g. the catastrophe modelling software.

31.2.5.4.3 Database

Basic pricing tools are often stand-alone spreadsheets with no central repository for contract information, pricing results and model parameters – each pricing exercise will often have its own spreadsheet which might slightly differ from the other exercises. The contract information is still saved in the appropriate accounting system, but independently of the operations of the pricing tool.

In a mature model, however, the contract information to be used by the pricing tool is saved and retrieved from a database management system (SQL server, Oracle, etc.). This also contains the results of all pricing exercises and the parameters of the model.

31.2.5.4.4 Involvement of IT professionals

As already mentioned, a mature model is designed by actuaries and underwriters, but implemented and maintained by professional software developers and technologists.

31.2.5.5 Calibration

A basic model will normally use parameters based purely on underwriters' judgment and/or traditional market rates and curves.

A mature model will instead be calibrated in a data-driven fashion to the extent made possible by the availability of data (i.e. past loss experience derived from the company's portfolio or other market information).

Where the data is scant, it may only be possible to calibrate the general level of rates based on the observed loss ratio of a portfolio. At the other end of the spectrum, in data-intensive situations such as large personal lines insurers and some SME commercial lines insurers, the data may provide enough information to calibrate the loss curves via standard fitting techniques and the risk adjustments via GLM or other supervised learning algorithms. In the middle, an intermediate solution where the loss data is sufficient to calibrate some of the parameters but other parameters (e.g. the rate relativities between different categories of risk) will be based on judgment can be adopted.

Whatever the case, the calibration will need to be validated by the relevant stakeholders.

31.2.5.6 Model Development and Validation

A mature pricing model will be developed/updated according to agreed model development guidelines, which clarify what the structure of a model should be (rate-based, frequency/severity ...), the information to be displayed in the user interface, the loading methodology, the rules around how underwriters can adjust risk factors, the validation process, the calibration of the model and its documentation.

The model will be validated by all the relevant stakeholders (the underwriting function, the actuarial function, the IT function). Where appropriate, the risk management function will also review and validate the model.

31.2.5.7 Model Documentation and Governance

Basic pricing models are often spreadsheets whose workings are not documented but are handed over from one user to the other, and the actual logic might be lost.

A mature pricing model will instead be adequately documented in terms of (a) usage by underwriters; (b) the actuarial calculations; (c) the underlying programming code where applicable.

A mature model will also have some governance around it so that it is clear who is responsible for updating it and maintaining it. Governance will also include a process for defining and updating the model version.

31.3 Experience-Based Models

Underwriters will deal mostly with exposure-based pricing models, since experience rating models are inherently more difficult to use than exposure-based models as they require plenty of data exploration and manipulation and a fair extent of judgment calls during modelling – their operation, therefore, falling squarely into the skillset of actuaries. Indeed, a good chunk of this book (Chapters 6 to 17) is dedicated exactly to articulating the experience rating process.

Because of the need to explore and manipulate the data prior to modelling, and the need to adapt the modelling process to the data and the quirks of the accounts, experience-based models often do not come as a fully defined package, but as a process aided by a tool or a suite of tools. A solution often used is to separate the modelling part, which can be done flexibly by the actuary, and the pricing of various policy structures, which can be run by the underwriter.

Despite these caveats, we can still outline the main characteristics of experience rating models (Section 31.3.1), and present a couple of simple examples (Sections 31.3.2 and 31.3.3).

31.3.1 Structure of an Experience Rating Model

31.3.1.1 The User Interface

The inputs of an experience rating model will include contract/policy data and current exposures as for exposure rating, but will also include **historical loss experience** (a **'loss run'**) and **historical exposure information**. The user interface may be fully structured and constrained, but more likely than not it will need to allow for a fair degree of interaction with the user to accommodate for non-standard data formatting, pre-processing of loss and exposure data to ensure that the data is relevant to the exercise, and so on.

The user interface may need to play a role also in the modelling phase, to assist the user (typically, an actuary) making the right modelling choices, such as including/excluding specific years, choosing the threshold for large losses, trying out different severity distributions, etc.

The outputs will include the same outputs as for an exposure rating model (expected losses, technical premium, performance measures) but may also includes tables and charts to judge the quality of the fitting of the various models.

31.3.1.2 The Calculation Engine

The calculation engine will include mechanisms to choose the right statistical models for frequency, severity, payment, settlement, and the machinery to calibrate these models. For example, it may include mechanisms for:

- calculating the IBNR claim count, using development triangle or a triangle-free methodology (see Chapter 13);
- choosing a Poisson or a negative binomial distribution for frequency depending on the observed variance/mean ratio (Chapter 14);
- calculating the IBNER factors for the loss amounts (Chapter 15);
- choosing a severity distribution depending on some measure of fitness, and calibrate the parameters, including thresholds between attritional and large losses, and thresholds for market curves (Chapter 16);
- a Monte Carlo simulation to calculate the expected aggregate losses on a gross and insured basis (Chapter 17).

As mentioned in the previous section on the user interface, the operations of the calculation engine may be fully automated but will more often be interactive to some extent, allowing the analyst to make judgment calls during the process.

31.3.1.3 The Database

The simplest experience rating models will simply allow the user to enter the loss and exposure data relevant to the pricing exercise, and the contract information/pricing results may or may not be retrieved from and stored into a centralised system.

In the most mature models, the loss and exposure data of all contracts of a portfolio may be stored in a separate database, so that they can be retrieved for various types of analysis (year-on-year comparisons, derivation of portfolio curves …).

31.3.1.4 Connection to External Systems

The more mature models will also be connected to the relevant external systems, such as the technical accounting system or the cat modelling software.

31.3.2 $\mathcal{M}_5$ – A Simple Experience Rating Model: Burning Cost Model

The simplest example of an experience rating model ($\mathcal{M}_5$) is that of a model implementing a basic version of burning cost analysis (Chapter 11). This is available in spreadsheet format in Parodi (2016).

The **user interface** includes a tab where to enter:

- The list of past losses
- A list of exposures (both past and predicted for the renewal year)
- Past and projected inflationary adjustments for revaluing losses and on-levelling exposures
- IBNR factors from reserving
- A target loss ratio for the technical premium

It also shows the results of the exercise, i.e. the burning cost for each policy year and the technical premium.

The **calculation engine** sorts the losses automatically into policy years, adjusts past exposures for inflation, adjusts pass loss experience for inflation, IBNR, and differences in past exposures. It then calculates the expected losses for the renewal year based on a weighted average of the burning cost of each policy year and determines the technical premium that achieves the target loss ratio.

The model has no separate database or connection to an external system.

A summary description of model $\mathcal{M}_5$ can be found in Box 31.4. The algorithm is a simplified version of the burning cost algorithm described in Chapter 11 – a straightforward exposure-weighted average of past experience without adjustments for large losses, adjustments for changes in risk profile/terms and conditions, and the possibility of selecting a subset of past policy years. For simplicity of illustration, we have also made the following **assumptions**:

- the policy is written on an occurrence basis;
- every policy year (apart from the last one) considered in the calculation is complete, in the limited sense that the recording of losses has started before or at the beginning of the first policy year;
- the exposure period has already been aligned to the policy year;
- the loss development factors from reserving are already adjusted to take the number of development months into account between each policy year and the as-at date of the loss database;
- the past loss inflation and the past exposure inflation are constant and will remain the same in the renewal year;
- losses occur uniformly over the policy year, so that it makes sense to revalue to the mid-point of the policy.

31.3.3 $\mathcal{M}_6$ – A Simple Frequency/Severity Experience Rating Model

Next we consider $\mathcal{M}_6$, a stochastic frequency/severity model that takes the same inputs as the ones of $\mathcal{M}_5$, and based on this information estimates the parameters for the frequency and severity model, and estimates the aggregate loss distribution based on Monte Carlo simulation. The technical premium based on a target loss ratio, as for $\mathcal{M}_5$.

Again, the techniques used by this model have been the subject of Chapter 13 to 17. Depending on how complex the model is, a certain degree of interaction between the actuary and the algorithm is to be expected – e.g. in the choice of the splicing threshold between the attritional and large losses (Chapter 16), or the selection of the development factors if a triangle-based methodology is adopted.

BOX 31.4 CALCULATION ENGINE FOR MODEL $\mathcal{M}_5$

Inputs

- List of (unrevalued) losses over a certain time window: $X_1, X_2 \ldots X_n$, each with their time of occurrence $t(X_1), \ldots t(X_n)$
- Exposure (in monetary amount) for each policy year: $\mathcal{E}_1, \mathcal{E}_2 \ldots \mathcal{E}_{Y+1}$, where $1, 2 \ldots Y$ are the years used for pricing and $Y+1$ is the renewal year ($\mathcal{E}_{Y+1}$ is therefore only an estimated figure).
- Target loss ratio, TLR
- Loss inflation (past and future estimated): r_L
- Exposure inflation (past and future estimated): r_E
- Contract inception date t_{inc} (contract is assumed to be annual)
- Loss development factors LDF_i from reserving for each year
- Deductible (D), limit (L), aggregate limit (AL)

Pricing algorithm

1. On-level the exposure of each policy year i to the renewal year $Y+1$: $\mathcal{E}_{i,\text{REV}} = \mathcal{E}_i \times (1+r_E)^{Y+1-i}$
2. Revalue each loss X_j to the mid-point of the contract: $X_{j,\text{REV}} = X_j \times (1+r_L)^{t_{\text{inc}}+\frac{1}{2}-t(X_j)}$
3. Calculate the insured loss for every year after revaluation but *before the aggregate limit is imposed:* $S_{i,\text{CED}}(D,L) = \sum_{j=1}^{n_i} \min\left(L, \max\left(0, X_{j,REV} - D\right)\right)$, where n_i is the number of losses in year i.
4. Calculate the burning cost (per unit of exposure) for each year $1, \ldots Y$ *after the aggregate limit is imposed*:

$$\text{BC}_i(D,L,AL) = \frac{\min\left(AL, S_{i,\text{CED}}(D,L) \times \text{LDF}_i \times \dfrac{\mathcal{E}_{Y+1}}{\mathcal{E}_{i,\text{REV}}}\right)}{\mathcal{E}_{Y+1}}$$

5. Estimate the burning cost for the renewal year as the exposure-weighted average of the burning costs across all years:

$$\text{BC}_{Y+1}(D,L,AL) = \frac{\sum_{i=1}^{Y} \mathcal{E}_{i,\text{REV}} \times \text{BC}_i(D,L,AL)}{\sum_{i=1}^{Y} \mathcal{E}_{i,\text{REV}}}$$

6. Estimate the expected losses $\mathbb{E}\left(S_{\text{ced}}(D,L)\right)$ to layer L xs D(with aggregate limit AL) as:

$$\mathbb{E}\left(S_{\text{ced}}(D,L,AL)\right)_{\text{est}} = \mathcal{E}_{Y+1} \times \text{BC}_{Y+1}(D,L,AL)$$

7. Set the technical premium to

$$\text{Technical Premium} = \frac{\mathbb{E}\left(S_{\text{ced}}(D,L,AL)\right)_{\text{est}}}{\text{TLR}}$$

BOX 31.5 CALCULATION ENGINE FOR MODEL $\mathcal{M}_6$

Inputs

- List of (unrevalued) losses over a certain time window: $X_1, X_2 \ldots X_n$, each with their time of occurrence $t(X_1), t(X_2) \ldots t(X_n)$
- Exposure (in monetary amount) for each policy year: $\mathcal{E}_1, \mathcal{E}_2 \ldots \mathcal{E}_{Y+1}$, where $1,2 \ldots Y$ are the years used for pricing and $Y+1$ is the renewal year ($\mathcal{E}_{Y+1}$ is therefore only an estimated figure).
- Target loss ratio, TLR
- Loss inflation (past and future estimated): r_L
- Exposure inflation (past and future estimated): r_E
- Contract inception date t_{inc} (contract is assumed to be annual)
- Loss development factors LDF_i from reserving for each year
- Deductible (D), limit (L), aggregate limit (AL)

Pricing algorithm

1. On-level the exposure of each policy year i to the renewal year $Y+1$: $\mathcal{E}_{i,\text{REV}} = \mathcal{E}_i \times (1+r_E)^{Y+1-i}$
2. Revalue each loss X_j to the mid-point of the contract: $X_{j,\text{REV}} = X_j \times (1+r_L)^{t_{\text{inc}}+\frac{1}{2}-t(X_j)}$
3. Based on the number of reported losses $N_{\text{REP},i}$ *from the ground up* for year i, estimate the *ultimate* number of losses *per unit of exposure* for each year ($i = 1, \ldots Y$):

$$n_i = \frac{N_{\text{REP},i} \times \text{LDF}_i}{\mathcal{E}_{i,\text{REV}}}$$

4. Estimate the expected claim count $\hat{n}_{Y+1}$ and the expected variance $\hat{v}_{Y+1}$ for year $Y+1$ as the weighted average and variance of the n_i s calculated above. (See Box 11.1 for the calculation of the weighted variance.)
5. If $\hat{v}_{Y+1} / \hat{n}_{Y+1} \leq 1$, assume a Poisson frequency model with Poisson rate $\hat{n}_{Y+1}$. Else, assume a negative binomial model with estimated parameters $\hat{\beta}, \hat{r}$ derived using the method of moments (see Section 14.3.4.1).
6. Assume a lognormal severity model with parameters $\hat{\mu}, \hat{\sigma}$ calculated from the set of revalued losses $\{X_{1,\text{REV}}, X_{2,\text{REV}} \ldots X_{n,\text{REV}}\}$ using the standard MLE methodology (see, e.g., Section 16.4.1).
7. Simulate a large number K of scenarios. For each scenario $k = 1, \ldots K$:
   a. Sample a random number of claims $N(k)$ from the frequency model ($N(k) \sim \text{Poi}(\hat{n}_{Y+1})$ in the Poisson case, and $N(k) \sim NB(\hat{\beta}, \hat{r})$ in the negative binomial case).
   b. For each of the $N(k)$ losses, generate a loss amount $X(k,m)(m = 1, \ldots N(k))$ from a lognormal with parameters $\hat{\mu}, \hat{\sigma}$.
   c. For each of the $N(k)$ losses, calculate the amount ceded to the insurer as $X_{\text{ced}}(k,m) = \min(L, \max(0, X(k,m) - D))$.
   d. Calculate the total amount ceded for scenario k as $S_{\text{ced}}(D,L,AL;k) = \min\left(AL, \sum_{m=1}^{N(k)} X_{\text{ced}}(k,m)\right)$

8. Estimate the expected losses $\mathbb{E}\left(S_{\text{ced}}(D,L)\right)$ to layer L xs D (with aggregate limit AL) as:

$$\mathbb{E}\left(S_{\text{ced}}(D,L,AL)\right)_{\text{est}} = \frac{\sum_{k=1}^{K} S_{\text{ced}}(D,L,AL;k)}{K}$$

9. Set the technical premium to

$$\text{Technical Premium} = \frac{\mathbb{E}\left(S_{\text{ced}}(D,L,AL)\right)_{\text{est}}}{\text{TLR}}$$

For simplicity of illustration, we have made the same assumption as for model $\mathcal{M}_5$ (see Section 31.3.1). Furthermore, we have assumed that IBNER can be ignored and therefore the LDFs from reserving can be used to project to ultimate the total claim amount and not only the total number of claims.

31.3.4 User-Machine Interaction in Experience Rating Tools

In Sections 31.3.2 and 31.3.3 we have shown two experience rating models ($\mathcal{M}_5$ and $\mathcal{M}_6$) that run automatically once the correct inputs are entered. Achieving this automation has required some drastic simplifications, such as using for $\mathcal{M}_6$ a simple lognormal distribution for modelling severities (regardless of whether the fit is good and ignoring the dictates of extreme value theory on tail behaviour), or assuming that the correct LDFs are given as part of the inputs.

In general, however, as we mentioned earlier on, some level of user-machine interaction is necessary when using experience rating tools, and therefore these tools will need the flexibility to accommodate it. Specifically, in an experience rating tool (or, in general, in a pricing process) it is important to identify clearly what the decision points are where actuarial judgment may be needed. The tool should then allow the user to make these decisions interactively (for example by trial and error), or simply to confirm the recommendations made by the algorithm.

As an example, consider the case of our pricing process explained in Chapters 13 to 17. This process can be turned into an automated pricing model using triangle-free IBNR calculations by making some simple assumptions such as:

- using an exponential or a Type II Pareto for the reporting delay (Section 13.3);
- using a Poisson or NB model for the frequency, depending on the variance/mean ratio (Section 14.4);
- using an empirical/GPD severity model (Section 16.4);

with the following main decision points (details may vary depending on the case):

- choice of past/future loss inflation index;
- choice of past/future exposure on-levelling index;

- choice of the years to be included in the frequency and severity calculations;
- choice of the splicing point between the empirical distribution and GPD.

Note that if a triangle-based claim count IBNR calculation is used instead there are some additional decision points, such as selecting a tail factor, excluding some elements of the development triangle, and potentially smoothing some of the development factors.

31.4 Hybrid Exposure/Experience Models

It is possible for models to combine elements of exposure and experience rating. Without getting into the details of a specific hybrid model, let us mention a few characteristics of these models.

These models will have machinery for the calculation of both an exposure rate (base rates, exposure curves) and an experience rate (handling of loss runs, burning cost analysis or frequency/severity modelling).

In addition, they will have a machinery to combine the two rates and produce a selected rate. The most obvious mechanism by which one can do that is by using credibility theory (see Chapter 26). In simple terms, that means that the selected rate (or other selected variables such as frequency or severity) will be calculated as:

$$\text{Credibility rate} = Z \times \text{Experience rate} + (1 - Z) \times \text{Exposure rate}$$

where Z is the credibility factor, which in turn will be calculated by weighing the uncertainty around the experience rate against some measure of relevance (or lack of it) of the exposure rate to the situation at hand.

Although the credibility approach is the most popular approach to mixing experience and exposure in virtue of its simplicity, it is by no means the only one. Two alternatives are (a) selecting the experience rate for the lower layers of an insurance programme, and the exposure rate for the higher layers; (b) a full-blown Bayesian approach, as the one mentioned in Section 26.4.

31.5 Pricing Models by Type of Undertaking

31.5.1 Personal Lines Insurance

Models developed for personal lines insurance (e.g. private motor insurance or household insurance) will tend to have less focus on analysing the volatility of the outcomes for an individual client and will rather build on the insurer's ability to identify the factors that make one client a better or worse risk than the others. They will typically be exposure-based models as the loss experience of a single individual would be too scant by definition to produce a loss model – however, some models might include factors such as no-claims discounts that modify the premium based on the experience of the policyholder.

For example, a simple model for private motor might have a number of rating factors $X_1, \ldots X_n$ (age, owner's location, owner's occupation, type of car, car model … everything that you're normally asked when you're looking to buy a policy on an aggregator) as an input, and a simple GLM formula that gives the expected losses $\mathbb{E}(S)$:

$$\mathbb{E}(S) = \exp\left[\sum_j \beta_j \Psi_j (X_1, \ldots X_n)\right]$$

The way that such a model for the expected losses is selected and calibrated is discussed in Chapter 27. The technical premium can then be derived with the 'cost plus' methodology but in the case of personal lines insurance it will often be possible to move beyond that and optimise the price taking demand modelling into account (Section 19.3).

31.5.2 Commercial Lines Insurance

Models developed for commercial lines can either be exposure-based, experience-based or a combination of the two, as in the examples shown in Sections 31.2 to 31.4.

Conceptually, an exposure-based model for commercial lines will not be very different from a personal lines model, in that both will have an exposure measure (which for a personal line insurance may be equal to 1 or be proportional to the period of insurance) and risk factors (such as territory and industry in model $\mathcal{M}_1$, and owner's location, type of car, etc. in the private motor model in Section 31.5.1). However, for commercial lines model it will be more typical also to ask questions about the overall aggregate loss distribution and not just the expected losses.

Furthermore, experience rating will also be a possibility in commercial lines at least for larger clients with a substantial loss experience, where that is normally not possible for personal lines models.

31.5.3 Reinsurance

Reinsurance pricing models are largely overlapping with commercial lines models. However, there are some differences, which reflect the peculiarities of the risk and the policy structures commonly used in reinsurance, especially non-proportional reinsurance (see introduction to Chapter 21). To mention a few examples:

- reinsurance (exposure/experience) models may use original premium as a measure of exposure; hence, rate change information must also be used to on-level past exposures;
- a combination of loss ratios and (original/primary) premiums may replace base rates in (exposure) property models;
- we may need to cater for complex structures such as the indexation clause. As a result, the treatment of payment/settlement pattern may need to be incorporated in the Monte Carlo simulation.

31.6 Rate Change Calculations

Apart from the simplest models, pricing models will often need to include the ability to compare the results of pricing from one year to the next and calculate the rate change and the full walk from the expiring premium to the current premium. The model will therefore need access to the following information for both the expiring and the current year of account:

- Actual premium
- Technical premium
- Exposure
- Share
- Policy structure
- Model changes
- Other relevant information

and will need to have the machinery to do all the relevant calculations, as explained in Chapter 20.

31.7 Model Calibration

Wherever possible, the calibration of the model parameters (base rates, frequencies, severity parameters, risk factor adjustments) should be based at least partly on loss experience.

As a minimum, the **coarse structure of the model** (base rates, base frequency, base severity) should be consistent with observed data:

- the general level of base rates should be consistent with the observed loss ratio at the most granular level for which the data allow it;
- the frequency parameters can be derived from the observed number of losses over a number of years above a given threshold (Section 14.4);
- the severity parameters can be derived from the empirical distribution of historical losses after claims inflation adjustments using for example maximum likelihood estimation (Chapter 16).

Whenever there are sufficient data to allow it (as is typically the case in personal lines insurance), the **fine structure of the model** (e.g. the parameters of a generalised linear model that capture the necessary risk adjustments) should also be calibrated in a data-driven fashion. Note that in the case of GLMs and other models, the identification of the rating factors may go hand in hand with their calibration (see Chapter 27).

31.7.1 The Use of Expert Judgment in Calibration

Often, the model will have more parameters than can be reliably calibrated based on data. For the parameters that cannot be calibrated based on data, their calibration will need to be based to some extent on the underwriters' judgment.

The use of the term 'calibration' for this case is deliberate: the process of setting the value of parameters based on expert judgment is not as straightforward as simply asking the underwriters 'what should the value of this parameter be?' It often requires, among other things:

a. translating qualitative judgments ('builders are at more risk of falling from scaffolds than accountants') into quantitative relationships;
b. ensuring that different pieces of advice are incorporated in a coherent mathematical framework, which do not lead to contradictory results (e.g. the effect of doubling the deductible and then halving it needs to take us back to the starting point, at least from a technical point of view) and that have a sensible cumulative effect. E.g. underwriters may initially estimate that the effect of setting *each* of three independent adjustments to 'Yes' is to double the premium, but may find unacceptable that setting *all* the adjustments to 'Yes' will result in the premium going up by 8 times;
c. using an iterative process by which calibration proposals are first made based on early indications or no indications at all, and then these proposals are criticised by the experts and refined.

Plenty of ink has been spent by behavioural economists and psychologists on the validity and the incorporation of expert judgment. For an overview of the topic of expert intuition in general, see Chapters 20–22 in Kanheman (2011). For something more specific to non-life insurance, the main reference is the report of the GIRO 'Getting Better Judgment' working party (Tredger et al., 2015), which deals with the full process of eliciting, validating and documenting expert judgment in actuarial work, and includes a discussion on how to manage various sources of bias.

31.8 Model Governance

Since models are crucial to underwriting and to writing profitable business, there needs to be a model governance framework (Ordóñez-Sanz, 2014). This will include guidelines on how to set up a model inventory, assign ownership of the model, manage the model lifecycle, manage model risk, and provide adequate documentation.

31.8.1 The Model Landscape

The first thing we need to know about our models is what part of the business they cover – so the *first input is the inventory of business segments/types of policy that a company underwrites, along* with an inventory of the pricing models at our disposal and a mapping between the two. That provides the landscape for our models and is an input for our model strategy.

Drawing this model landscape will give information on what percentage of the business (in terms of number of contracts and especially of premium) is written using pricing models, and in which area it is necessary to concentrate to develop more models or refine the existing ones.

The model landscape may be captured by a table like this:

ID	*Type of business*	*Annual premium written [CCY]*	*No of contracts*	*Model name*	*% of business covered by model*	*Not covered*
1	Employers' liability	20,000,000	1,700	EL Pricer 3.5	100%	N/A
2	Commercial property	15,000,000	1,200	PDBI Rater 2.7	80%	CBI
3	Environmental liability	800,000	60	*No model*	*N/A*	*N/A*
4	Credit risk	6,000,000	500	CR Model 1.8	90%	Some territories
...	...	...	...	...		

Note how the table above shows that there are business segments (e.g. ID=3) for which no model is available.

While the model landscape table focuses on the business segments and how they are served by the models, the **model inventory** focuses on the currently existing models along with information relevant to them. It may look something like this.

Model ID	**1**	**2**	**3**	**4**	**...**
Model name	EL Pricer	BIT Wind	EnO	Bathtub	...
Current version	3.5	1.4	2.2	3.4	...
Business priced	Employers' Liability	Offshore Windfarms	Professional Indemnity	Extended Warranty	...
Platform	FirstRateR	Excel	FirstRateR	FirstRateR	
Owner	F Guaraño	F Bardee	J R Jahner	F Rossi	...
Status	In use	In use	In use	In use	...
Last major version rollout	11.7.2011	18.2.2019	1.10.2018	24.7.2017	...
Materiality	High	Low	Medium	Medium	...
Business written per annum					
No of users					
Documentation	(Location/link)	(Location/link)	(Location/link)	(Location/link)	...
Last calibrated	19.9.2022	17.1.2021	26.9.2022	10.10.2022	...
...	...	...	...	...	...
Notes	Needs revamp	Needs migration to professional environment (FirstRateR)			...

31.8.2 Model Development Guidelines

As pricing models are a crucial component of underwriting profitable business, it is important that a company has guidelines on how such models should be developed. The objective is to have reasonable consistency in both the development process and the pricing methodology across all models, in order to (a) reduce model risk and (b) be able to recycle methodologies for efficiency, documentation and maintenance purposes.

31.8.2.1 General Principles for Model Development

According to the Board of Actuarial Standards (BAS, 2008; TAS-M, 2010), there are several conditions that must be met so that the output of models can be used effectively for making decisions. Even though these standards are meant to apply to models used in preparing reports, they are useful for more general purposes as well. The principles below are a free rephrasing of material taken from TAS-M (2010) and BAS (2008).

1. **Representing the real world.** The models should reflect the aspects of the real world that affect the decisions that are to be made. All aspects of the world that have a *material* implication on those decisions should be reflected in the model.
2. **Model inputs and outputs.** It should be clear to the users what the inputs and outputs of the model represent, and the significance of the assumptions should be clear.
3. **Fitness for purpose in theory *and practice*.** The model should be implemented correctly to do what it is designed to do.
4. **Limitations.** The limitations of the model should be clear to the users of the model

These general principles are designed to be applicable across a variety of areas of practice including life insurance, non-life insurance and pensions and across different activities (pricing, reserving, capital modelling, etc.), and need to be translated into guidelines that apply specifically to pricing non-life insurance and to the context of a particular undertaking.

Principle 1 (*Representing the real world*) might be translated into the requirement that pricing models used by the company should use separate frequency and severity models (rather than a simple aggregate burning cost model) whenever the collective or individual risk mode is a good approximation of the real-world loss generating process. In the second part of Principle 1 (no aspects of the real world that have material implications should be left out) the key word is of course 'material' and there's a delicate balance to be struck between the need to incorporate essential aspects of reality and the need for parsimony (also addressed in TAS-M) – as we've seen before, a more 'realistic' model is not necessarily a better model: for example, to what extent is it necessary to capture the possible correlations between various sub-models?

Principle 2 (*Model inputs and outputs*) means, among other things, that

- it should be clear what the characteristics of the data used as inputs (both exposure and experience data) should be, and how the model is calibrated (what data was used for what, and where judgment was used);
- what outputs of a pricing model represent: for example, is the technical premium a purely technical expectation of the losses plus loadings, or does it attempt to incorporate the effects of the market cycle, as some underwriters expect?

Principle 3 (*Fitness in theory and practice*) may require – besides the obvious point of the formulae of the models to be correct – that Monte Carlo simulation (rather than analytical formulae) be used to calculate allocation of losses to the layers whenever aggregate features are included and these have a material impact on the results. It may also require that the model correctly identifies whether specific risk adjustment factors should apply to the base frequency or the severity parameters (it is not always obvious whether some requirements fall naturally under Principle 1 or Principle 3).

Finally, Principle 4 (*Limitations*) in the context of a pricing model means that users of the model need to be made aware of the scope of the model (what type of risk it covers, which policy structures it is able to price and which it isn't, and so on) and its technical limitations/domain of applicability (e.g. a property model might give funny results for very high attachment points, or a liability model might be unable to price an unlimited liability policy).

31.8.2.2 The Model Development Cycle

Section 31.8.2.1 discussed some general principles about models that need to be taken into account during the development of pricing models. A company will also need, however, a concrete process for dealing with model development and updates, with clear steps and responsibilities. Without being prescriptive, as different companies will have different processes and organisational structures, a model development cycle might look something like this:

	Task	*Responsibility*
1	Approve model development/update	Underwriting/actuarial/IT management
2	Agree model scope, business requirements and deadlines with stakeholders	Underwriters, actuaries, IT team
3	Design model (and possibly develop a prototype)	Actuaries
4	Calibrate model	Actuaries
5	Write technical specifications for use by software developers	Actuaries
6	Implement the model	IT team
7	Test and validate the model a. Test by software developers (unit testing, etc.) b. Test by actuaries c. User acceptance testing (underwriters) d. Validate model	As specified in the list
8	Complete user documentation	Actuaries/underwriters
9	Roll out model	IT team
10	Monitor model use and user feedback/requests for update	Actuaries
11	Go back to Step 1 as needed	

31.8.2.3 Model Risk Management

Once the model is operational, or sometimes during development, a second-line-of-defense model validation may be undertaken by the risk management function. This will include an assessment of whether:

- the methodology behind the model is sound;
- the calibration process (both data-driven and judgment-based) is adequate and well documented;
- the model has been implemented correctly;
- the maintenance process is adequate;
- the documentation is sufficient and up to date.

31.9 Questions

1. The underwriting manager in charge of Energy has asked you to build a model to cover operations all risks (OAR) for Offshore Windfarm. This model will cover property damage (PD) only.

   As a starting point, you decide to build a simple model in which each windfarm is modelled as a single property, based on these initial inputs from the underwriter, based on their market experience:

   - Base loss rate for PD is 0.5% of sum insured at the standard deductible of $100,000.
   - The main risk factors are the depth of the installation and its distance from shore.
     - Depth can be 'shallow' or 'deep'. 'Deep' attracts an extra 10% of premium.
     - Distance from the shore can be either 'close' or 'far'. 'Far' should attract an extra 15% of premium.
     - The base loss rate of 0.5% refers to the base case of shallow/close.
   - The deductible effect is based on Swiss Re exposure curve with c = 5.
   - Target loss ratio is 57%.

   The model is used to calculate the technical premium for any layer of insurance L xs AP where L is the limit and AP is the attachment point, measured in excess of the underlying deductible.

   The model should also calculate the performance excess, i.e. the (algebraic) difference between the commercial premium and the technical premium, both as a monetary amount and as a percentage.

   The model is on a spreadsheet and once the pricing is complete the spreadsheet is saved in a special folder that contains all completed pricings.

   a. Describe the architecture of the model, defining the inputs, outputs, parameters, and the calculation engine (i.e. the operations necessary to calculate the required outputs of the model).
   b. A client, Never2Windy Ltd, which owns a small windfarm of $20M of value far away from the shore and in deep waters, asks for two quotes for an OAR policy: one (i) for insuring the full value of its windfarm in excess of a $200k deductible, and one (ii) just for a layer of $15m xs $4.8m. Calculate the technical premium for cases (i) and (ii).
   c. Explain what the limitations of this simple model are.

References

Alaton, P., Djehiche, B., and Stillberger, D. (2007). *On Modelling and Pricing Weather Derivatives.* Available at www.treasury.nl/files/2007/10/treasury_301.pdf.

Anderson, D., Bolton, C., Callan, G., Cross, M., Howard, S., Mitchell, G., Murphy, K., Rakow, J., Stirling, P., and Welsh, G. (2007a). *General Insurance Premium Rating Issues Working Party (GRIP) Report.*

Anderson, D., Feldblum, S., Modlin, C., Schirmacher, D., Schirmacher, E., and Thandi, N. (2007b). *A Practitioner's Guide to Generalized Linear Models, A CAS Study Note.*

Antal, P. (2003). *Quantitative methods in reinsurance, Swiss Re.* Available at www.math.ethz.ch/finance/misc/MathMethodsReinsurance.pdf.

Antonio, K., and Plat, R. (2012). *Micro-Level Stochastic Loss Reserving.* KU Leuven. AFI-1270. Available at SSRN: http://ssrn.com/abstract=2111134 or http://dx.doi.org/10.2139/ssrn.2111134.

BAS (2008), Modelling: Consultation Paper. *Board of Actuarial Standards. Financial Reporting Council.* Available at www.frc.org.uk/document-library/bas/2008/tas-m-consultation-paper-(november-2008).

Beard, R.E., Pentikäinen, T., Pesonen, E. (1984). *Risk Theory: The Stochastic Basis of Insurance.* 3rd Edition. Chapman & Hall.

Bernegger, S. (1997). The Swiss Re exposure curves and the MBBEFD distribution class. *ASTIN Bulletin,* 27, 1, 99–111.

Bernstein, P.L. (1996). *Against the Gods: The Remarkable Story of Risk,* John Wiley & Sons, New York.

Blanchard, R.S. III (2005), Premium Accounting, Casualty Actuarial Society, Study Note, www.casact.org/library/studynotes/blanchard6a.pdf

Bodoff, N.M. (2008), Measuring the rate change of a non-static book of property and casualty insurance business, Society of Actuaries.

Bolancé C, Vernic R. (2020). Frequency and Severity Dependence in the Collective Risk Model: An Approach Based on Sarmanov Distribution. *Mathematics.* 2020; 8(9),1400. https://doi.org/10.3390/math8091400.

Boor, J. (1992). Credibility based on accuracy, *CAS Proceedings, LXXIX,* 2, 151.

Boor, J. (2000). A Macroeconomic View of the Insurance Marketplace, *CAS Exam 5 Study Kit.*

Bousquet, O., Boucheron, S., Lugosi, G. (2004). Introduction to Statistical Learning Theory. In: Bousquet, O., von Luxburg, U., Rätsch, G. (eds) *Advanced Lectures on Machine Learning. ML 2003. Lecture Notes in Computer Science,* vol 3176. Springer. https://doi.org/10.1007/978-3-540-28650-9_8.

Brix, A., Jewson, S., and Ziehmann, C. (2002). Weather derivative modelling and valuation: A statistical perspective, In Dischel, R.S. (ed.) *Climate Risk and the Weather Market, Financial Risk Management with Weather Hedges,* Risk Books, 127–150.

Briys, E., and De Varenne, F. (2001). *Insurance from Underwriting to Derivatives: Asset Liability Management in Insurance Companies,* John Wiley & Sons.

Brody, D.C., Syroka, J.J., and Zervos, M. (2002). Dynamical pricing of weather derivatives. *Quantitative Finance,* 2, 189–198.

Bühlmann, H. (1967). Experience rating and credibility. *ASTIN Bulletin,* 4, 199–207.

Bühlmann, H., and Straub, E. (1970). Glaubwürdigkeit für Schadensätze. *Bulletin of the Swiss Association of Actuaries,* 35–54.

Bühlmann, H., and Gisler, A. (2005). *A Course in Credibility Theory,* Springer-Verlag.

Burnham, K.P., and Anderson, D.R. (2002). *Model Selection and Multimodel Inference: A Practical Information-Theoretic Approach* (2nd ed.), Springer-Verlag.

Cairns, M., Allaire, F., Laird, A., Maurer, P., Sawhney, S., Skinner, J., Van Delm, R., Gilman, B., Lau, S., Rivers, T., Skelding, R., Trong, B., and Weston, R. (2008). *Variable Capital Loads in Pricing, GIRO, Working Party Report.*

Chhabra, A., and Parodi, P. (2010). Dealing with sparse data, *Proceedings of GIRO 2010.*

Clarks, D.R. (1996). Basics of reinsurance pricing, *CAS Library*. Available at www.casact.org/library/studynotes/clark6.pdf.

Cockcroft, M., Williams, N., Claughton, A., Gohil, A., Stocker, B., Yeates, G., Murphy, K., Yeates, P., Saunders, P., MacDonnell, S., Warsop, S., and Le Delliou-Viel, S. (2011). *Periodic Payment Orders Revisited, GIRO Working Party*.

Corry, L. (2009). *Writing the Ultimate Mathematical Textbook: Nicolas Bourbaki's Éléments de mathématique*. In Eleanor Robson et al. (eds.) *Handbook of the History of Mathematics*, Oxford University Press, 565–587.

Dai, B., Ding, S., & Wahba, G. (2013). Multivariate Bernoulli distribution. *Bernoulli*, 19(4), 1465–1483. www.jstor.org/stable/23525760.

Dean, C. (2014). Generalized Linear Models. In E. Frees, R. Derrig, & G. Meyers (Eds.), *Predictive Modeling Applications in Actuarial Science (International Series on Actuarial Science*, Cambridge University Press, 107–137, doi:10.1017/CBO9781139342674.005.

Derman, E. (2010). Metaphors, models and theories, *Edge*. Available at www.edge.org/3rd_culture/derman10.1/derman10.1_index.html.

Devroye, L. (1986). *Non-Uniform Random Variate Generators*, Springer-Verlag.

Dhesi, T. (2001). The freight demurrage and defence concept, *The Swedish Club Letter (3–2001)*, The Swedish Club Publications.

Duffie, D., and Singleton, K. (2003). *Credit Risk: Pricing, Measurement and Management*. Princeton University Press.

Dugas, C., Bengio, Y., Chapados, N., Vincent, P., Denoncourt, G., and Fournier, C. (2003). Statistical learning algorithms applied to automobile insurance ratemaking. *CAS Forum*, 1, 1, 179–214, Casualty Actuarial Society, Winter 2003.

Duguid, A. (2014). *On the Brink, Palgrave Macmillan Books, Palgrave Macmillan*, no. 978-1-137-29930-7.

Dutang, C., Goulet, V., and Pigeon, M. (2008). actuar: An R package for actuarial science. *Journal of Statistical Software*, 25, 7, 1–37. Available at www.jstatsoft.org/v25/i07.

Dvoretzky, A., Kiefer, J., and Wolfowitz, J. (1956). Asymptotic minimax character of the sample distribution function and of the classical multinomial estimator. *Annals of Mathematical Statistics*, 27, 3, 642–669.

Einstein, A. (1934). On the method of theoretical physics. *Philosophy of Science*, 1, 2, 163–169. The University of Chicago Press on behalf of the Philosophy of Science Association (stable URL: www.jstor.org/stable/184387).

Embrechts, P., and Frei, M. (2009). Panjer recursion vs FFT for compound distributions, *Mathematical Methods of Operations Research*, 69, 3, 497–508. Available at www.math.ethz.ch/~embrecht/papers.html.

Embrechts, P., Klüppelberg, C., and Mikosch, T. (1997). *Modelling Extremal Events: For Insurance and Finance*, Springer.

Fackler, M. (2009). Rating without data – How to estimate the loss frequency of loss-free risks, *ASTIN Colloquium, Helsinki*.

Fackler, M. (2011). Inflation and excess insurance, *ASTIN Colloquium, Madrid*.

Fackler, M. (2013). Reinventing Pareto: Fits for both small and large losses, *ASTIN Colloquium, The Hague*.

Farr, D., Subasinghe, H. et al. (2014), Marine and energy pricing, *GIRO Proceedings*, Institute and Faculty of Actuaries.

Fitzpatrick, S.M. (2004), Fear is the Key: A Behavioral Guide to Underwriting Cycles. *Connecticut Insurance Law Journal*, 10, 2, 255–275. Available at SSRN: https://ssrn.com/abstract=690316.

Flarend, A., and Hilborn, B. (2022). *Quantum Computing: From Alice to Bob*, Oxford University Press. ISBN: 978-0192857989.

Freund, J.E. (1999) *Mathematical Statistics*, 6th Edition, New Jersey, Prentice Hall International, Inc.

Friedman, J., Hastie, T., Tibshirani, R., Narasimhan, B., Tay, K., Simon, N., Qian, J., Yang, J. (2022). *Package 'glmnet'. Documentation*, available at https://cran.r-project.org/web/packages/glmnet/glmnet.pdf.

Frigg, R., and Hartmann, S. (2012). Models in science, *The Stanford Encyclopedia of Philosophy (Fall 2012 Edition)*, Zalta, E.N. (ed.). Available at http://plato.stanford.edu/archives/fall2012/entries/models-science/.

FSA General Insurance Handbook and Insurance Regulatory Reporting Instrument (2005). FSA 2005/3.

Geman, H. (ed.) (1999). *Insurance and Weather Derivatives: From Exotic Options to Exotic Underlyings*, Risk Publications, London.

Gnedenko, B.V. (1998). *Theory of Probability*, CRC Press, Boca Raton, FL.

Gönülal, S.O. (ed.) (2009). *Motor third-party liability insurance in developing countries – Raising awareness and improving safety*. The International Bank for Reconstruction and Development.

Grossi, P., and Kunreuther, H. (2005). Catastrophe Modeling: A New Approach to Managing Risk, *Huebner International Series on Risk, Insurance and Economic Security*, Springer, New York.

Grunwald, P.D., Myung, I.J., and Pitt, M.A. (eds.) (2005). *Advances in Minimum Description Length – Theory and Applications*, A Bradford Book, The MIT Press.

Hamisultane, H. (2008). *Which method for pricing weather derivatives?* Available at http://en.youscribe.com/catalogue/reports-and-theses/knowledge/humanities-and-social-sciences / which-method-for-pricing-weather-derivatives-1599572.

Hapgood, M. (2010). *Lloyd's 360° Risk Insight Space Weather: Its Impact on Earth and Implications for Business*, Lloyd's Publications.

Hastie, T. and Tibshirani, R. (1990). *Generalized Additive Models*, Chapman & Hall.

Hastie, T., Tibshirani, R., and Friedman, J. (2001). *The Elements of Statistical Learning, Springer Series in Statistics*, Springer New York Inc.

Homer, D., and Li, M. (2017). Notes on using property catastrophe model results, *Casualty Actuarial Society E-Forum, Spring 2017-Volume 2.*

Institute and Faculty of Actuaries (2010). *ST8-General Insurance Pricing*, Core Reading for the 2011 Examinations.

IIA (2013). IIA Position Paper: The Three Lines of Defense in *Effective Risk Management and Control.*

Isaacson, W. (2007). *Einstein: His Life and Universe, Pocket Books*, Source, www.theiia.org/3-lines-defense.

IUA/ABI. (2007), *Fourth UK Bodily Injury Awards Study*, International Underwriting Association of London.

Iwanik, J. (2011). Property and casualty insurance pricing with GLMs. In: Cizek, P., Härdle, W., Weron, R. (eds) *Statistical Tools for Finance and Insurance*, Springer. https://doi.org/10.1007/978-3-642-18062-0_11.

Jerome, J.K. (1900). *Three men on the bummel*. Available at www.gutenberg.org/files /2183/2183.txt.

Kaye, P. (2005). Risk measurement in insurance – A guide to risk measurement, capital allocation and related decision support issues. Casualty Actuarial Society, Discussion Paper Program. CAS Publications.

Kahneman, D. (2011). *Thinking, Fast and Slow*. Penguin Books Ltd.

Klugman, S.A., Panjer, H.H., and Willmot, G.E. (2008). *Loss Models. From Data to Decisions*, 3rd ed., Wiley.

Krikler, S., Dolberger, D., and Eckel, J. (2004). Method and tools for insurance price and revenue optimisation. *Journal of Financial Services Marketing*, 9, 1, 68–79.

Landsburg, S. E. (2002). *Price Theory and Applications (5th ed.)*, South-Western Pub.

Leigh, J. (2013). *An Introduction to Periodic Payment Orders, Staple Inn Presentation*, Staple Inn Actuarial Society.

Lee, G.Y., and Frees, E.W. (2016). General Insurance Deductible Ratemaking with Applications to the Local Government Property Insurance Fund. Available at: www.soa.org/49386b/globalassets/assets/files/research/projects/research-2016-gi-deductible-ratemaking.pdf.

Lewin, D. (2008). Indexation clauses in liability reinsurance treaties: A comparison across Europe. Available at www.gccapitalideas.com/2008/09/08/indexation-clauses-in-liability-reinsurance-treaties-a-comparison-across-europe/.

Lewis, M. (2007). In Nature's Casino. *The New York Times Magazine*, August 26, 2007. Available at: www.nytimes.com/2007/08/26/magazine/26neworleans-t.html.

Lloyd's (1998). *An Introduction to Lloyd's, The Training Center*, Lloyd's Publication.
Lloyd's (2018). *Performance Management Data Return, Instructions 2018 v1.1*, Lloyd's Publication.
Lloyd's (2017). *Performance Management Data Return, Renewal scenario examples, 2017 v1.0*, Lloyd's Publication.
Lowe, J., and Pryor, L. (1996). Neural networks v. GLMs in pricing general insurance. *General Insurance Convention 1996, Stratford-upon-Avon, UK*, 417–438.
Mack, T. (1993). Distribution-free calculation of the standard error of chain ladder reserves estimates. *ASTIN Bulletin*, 23/2, 213–225.
Mack, T., and Fackler, M. (2003). Exposure rating in liability reinsurance. *ASTIN Colloquium, Berlin*.
Marker, J.O., and Mohl, J.J. (1980). Rating claims-made insurance policies. *Casualty Actuarial Society Discussion Paper Program*.
McCullagh, P., and Nelder, J.A. (1990). *Generalized Linear Models*, 2nd ed., CRC Press
McLeod, A.I., and Xu, C. (2011). *bestglm: Best Subset GLM*. R package version 0.33. Available at http://CRAN.R-project.org/package=bestglm.
McNeil, A.J., Frey, R., and Embrechts, P. (2005). *Quantitative Risk Management*, Princeton Series in Finance, Princeton University Press.
Meyers, G. (2007). The common shock model for correlated insurance losses. *Variance*, 1, 1, 40–52.
Miccolis, R.S. (1977). On the theory of increased limits and excess of loss pricing. Available at www.casact.org/pubs/proceed/proceed77/77027.pdf.
Mildenhall, S. (1999). A systematic relationship between minimum bias and generalized linear models. *Proceedings of the Casualty Actuarial Society*, LXXXVI.
Mitchell-Wallace, K., Jones, M., Hillier, J., and Foote, M. (2017). *Natural Catastrophe Risk Management and Modelling: A Practitioner's Guide*, John Wiley & Sons.
Mowbray, A.H. (1914). How extensive a payroll exposure is necessary to give a dependable pure premium? *Proceedings of the Casualty Actuarial Society*, I, 24–30.
Murphy, K., and McLennan, A. (2006). A method for projecting individual large claims, *Casualty Actuarial Society Forum*, Fall 2006.
Nelsen, R.B. (2009). *An Introduction to Copulas*, Springer.
Noll, A., Salzmann, R. & Wüthrich, M. (2018). *Case Study: French Motor Third-Party Liability Claims*. Available online at the address https://ssrn.com/abstract=3164764 [accessed 29-Jul-2022].
NOAA/Space Weather Prediction Center. (2013). *Space Weather Prediction Center topic paper: Satellites and space weather*. Available at www.swpc.noaa.gov/info/Satellites.html.
Norberg, R. (1993). Prediction of outstanding liabilities in non-life insurance. *ASTIN Bulletin*, 23, 1, 95–115.
Norberg, R. (1999). Prediction of outstanding claims. II Model variations and extensions. *ASTIN Bulletin*, 29, 1, 5–25.
Ohlsson, E., and Johansson, B. (2010). Non-life insurance pricing with generalized linear models, *EAA Series*, Springer-Verlag.
Ordóñez-Sanz, G. (2014). Model governance. *Moody's Analytics: Risk Perspectives*, 4, 75–77.
O'Kane, D. (2010). Modelling single-name and multi-name credit derivatives, *The Wiley Finance Series*, Wiley, Chichester.
Palmer, J.M. (2006). *Increased limits ratemaking for liability insurance, CAS Proceedings*.
Panjer, H.H. (1981). Recursive evaluation of a family of compound distributions. *ASTIN Bulletin (International Actuarial Association)*, 12, 1, 22–26.
Parodi, P. (2009). *Computational intelligence techniques for general insurance*. SA0 dissertation for the Institute and Faculty of Actuaries. Available at www.actuaries.org.uk/system/files/documents/pdf/parodicomputationalintelligence.pdf.
Parodi, P., and Bonche, S. (2010). Uncertainty-based credibility and its applications. *Variance*, 4, 1, 18–29.
Parodi, P. (2011). Underwriters underwater, *The Actuary*, March.
Parodi, P. (2012a). Computational intelligence with applications to general insurance: A review. I – The role of statistical learning. *Annals of Actuarial Science*, September 2012.
Parodi, P. (2012b). Computational intelligence with applications to general insurance: A review. II – Dealing with uncertainty. *Annals of Actuarial Science*, September 2012.

Parodi, P. (2012c). Triangle-free reserving: A non-traditional protocol for estimating reserves and reserve uncertainty. *Proceedings of GIRO 2012.* Available at www.actuaries.org.uk/research-and-resources/documents/brian-hey-prizepaper-submission-triangle-free-reserving.

Parodi, P. (2014). Triangle-free reserving: A non-traditional framework for calculating reserves and reserve uncertainty, *British Actuarial Journal*, 19, 1, 168 – 218.

Parodi, P. (2016). https://pricingingeneralinsurance.net/.

Parodi, P. (2016b) Towards machine pricing. *The Actuary (Predictions supplement)* Sep 1, 2016.

Parodi, P. and Watson, P. (2019), Property graphs – A statistical model for fire and explosion losses based on graph theory, *ASTIN Bulletin*, 49, 2, 263–297.

Parodi, P. (2020), A generalised property exposure rating framework that incorporates scale-independent loss and maximum possible loss uncertainty, *ASTIN Bulletin,* 50, 2, 513–553.

Porter III, J.E., Coleman, J.W., and Moore, A.H. (1992). Modified KS, AD, and C-vM tests for the Pareto distribution with unknown location and scale parameters. *IEEE Transactions on Reliability*, 41, 1.

Press, W.H., Teukolsky, S.A., Vetterling, W.T., and Flannery, B.P. (2007). Numerical recipes, 3rd ed., *The Art of Scientific Computing*, Cambridge University Press.

Richman, R. (2021a). AI in actuarial science – a review of recent advances – part 1. *Annals of Actuarial Science*, 15, 2, 207–229. Doi:10.1017/S1748499520000238

Richman, R. (2021b). AI in actuarial science – a review of recent advances – part 2. *Annals of Actuarial Science*, 15, 2, 230–258. Doi:10.1017/S174849952000024X

Riegel, U. (2008). Generalizations of common ILF models, *Blätter der DGVFM*, 29, 1, 45–71.

Riegel, U. (2010). On fire exposure rating and the impact of the risk profile type. *ASTIN Bulletin*, 40, 2, 727–777.

Rivest, R. L.; Shamir, A.; Adleman, L. (1978). A Method for Obtaining Digital Signatures and Public-key Cryptosystems. *Communications of the ACM*. 21, 2, 120–126.

RMS. (2008). *A Guide to Catastrophe Modelling*. Available at www.rms.com/resources/publications/additional.

Ross, S.M. (2003), *Introduction to Probability Models*, 8th Ed., Academic Press.

Ruparelia, S., Pezzulli, S., and Looft, M. (2009). *Pricing optimisation issues and challenges (slides), GIRO Convention (6 October 2009).*

Schmidt, T. (2006). Coping with copulas, Department of Mathematics, University of Leipzig, Dec 2006, in *Copulas – From Theory to Application in Finance,* Risk Books, Incisive Financial Publishing Ltd.

Serena, U. (2018). *Underwriting: The Craft of Imagining*. Independently published.

Shevchenko, P.V. (2010). Calculation of aggregate loss distributions. *Journal of Operational Risk,* 5, 2, 3–40.

Smith, R.L. (1987). Estimating tails of probability distributions. *Annals of Statistics,* 15, 1174–1207.

Swiss Re. *Exposure Rating, Technical,* Publishing, Swiss Re Publications.

Tasche, D. (2008). Capital Allocation to Business Units and Sub-Portfolios: the Euler Principle, In: Resti, A. (ed.). *Pillar II in the New Basel Accord: The Challenge of Economic Capital* (Chapter 17), Risk Books.

TAS-M (2010), Technical Actuarial Standard M: Modelling. *Board of Actuarial Standards. Financial Reporting Council.* Available at www.frc.org.uk/getattachment/51f0f491-9ad4-4719-ad88-40e731b34632/TAS-M-Modelling-version-1-Apr-10.pdf.

Thoyts, R. (2010). *Insurance Theory and Practice*, Routledge.

Tredger, E.R.W., Lo, J.T.H., Haria, S., Lau, H.H.K., Bonello, N., Hlavka, B., and Scullion, C. (2016). Bias, guess and expert judgment in actuarial work. *British Actuarial Journal* 21, 03, 545–578.

Tsanakas, A. (2012). Modelling: The elephant in the room, *The Actuary*, September.

Turnbull, C. (2017). *A History of British Actuarial Thought*, Springer.

Wang, S. (1998). Aggregation of correlated risk portfolios: Models and algorithms, *Proceedings of the Casualty Actuarial Society*, 85, 163, 1–32.

Werner, G., Modlin, C. (2016). *Basic Ratemaking*, Fifth Edition. Casualty Actuarial Society. www.casact.org/library/studynotes/werner_modlin_ratemaking.pdf.

Wüthrich, M.V. and Buser, C. (2021). Data Analytics for Non-Life Insurance Pricing, *Swiss Finance Institute Research Paper No. 16–68,* Available at SSRN: https://ssrn.com/abstract=2870308 or http://dx.doi.org/10.2139/ssrn.2870308.

Wüthrich, M.V., and Merz, M. (2022). *Statistical Foundations of Actuarial Learning and its Applications.* https://ssrn.com/abstract=3822407.

Zemanian, A.H. (2003). Distribution Theory and Tranform Analysis: An Introduction to Generalized Functions, with Applications. Dover Books on Mathematics.

Zhou, F., Wright, D., and Owadally, I. (2012). Agent based modelling of insurance cycles: An update and results, *Risk and Investment Conference 2012.* Available at www.actuaries.org.uk/research-and-resources/documents/b02-agent-based-modelling-insurance-cycles-update-and-results.

Index

Note: Page numbers in *italics* indicate figures in the text.

D

H

I

K

L

M

T

U

V

W

Y

Z

Printed in the United States
by Baker & Taylor Publisher Services